Aircraft Structures

for engineering students

Aircraft Structures
for engineering students
Fourth Edition

T. H. G. Megson

AMSTERDAM • BOSTON • HEIDELBERG • LONDON • NEW YORK • OXFORD
PARIS • SAN DIEGO • SAN FRANCISCO • SINGAPORE • SYDNEY • TOKYO
Butterworth-Heinemann is an imprint of Elsevier

Butterworth-Heinemann is an imprint of Elsevier
Linacre House, Jordan Hill, Oxford OX2 8DP, UK
30 Corporate Drive, Suite 400, Burlington, MA 01803, USA

First edition 2007

British Library Cataloguing in Publication Data
A catalogue record for this book is available from the British Library

Library of Congress Cataloging-in-Publication Data
A catalog record for this book is availabe from the Library of Congress

ISBN-13: 978-0-75066-7395
ISBN-10: 0-750-667397

For information on all Butterworth-Heinemann publications visit our web site at books.elsevier.com

Typeset by Charon Tec Ltd (A Macmillan Company), Chennai, India
www.charontec.com
Printed and bound in Great Britain

07 08 09 10 10 9 8 7 6 5 4 3 2 1

Contents

Part B Analysis of Aircraft Structures 349

Preface

During my experience of teaching aircraft structures I have felt the need for a textbook written specifically for students of aeronautical engineering. Although there have been a number of excellent books written on the subject they are now either out of date or too specialist in content to fulfil the requirements of an undergraduate textbook. My aim, therefore, has been to fill this gap and provide a completely self-contained course in aircraft structures which contains not only the fundamentals of elasticity and aircraft structural analysis but also the associated topics of airworthiness and aeroelasticity.

The book in intended for students studying for degrees, Higher National Diplomas and Higher National Certificates in aeronautical engineering and will be found of value to those students in related courses who specialize in structures. The subject matter has been chosen to provide the student with a textbook which will take him from the beginning of the second year of his course, when specialization usually begins, up to and including his final examination. I have arranged the topics so that they may be studied to an appropriate level in, say, the second year and then resumed at a more advanced stage in the final year; for example, the instability of columns and beams may be studied as examples of structural instability at second year level while the instability of plates and stiffened panels could be studied in the final year. In addition, I have grouped some subjects under unifying headings to emphasize their interrelationship; thus, bending, shear and torsion of open and closed tubes are treated in a single chapter to underline the fact that they are just different loading cases of basic structural components rather than isolated topics. I realize however that the modern trend is to present methods of analysis in general terms and then consider specific applications. Nevertheless, I feel that in cases such as those described above it is beneficial for the student's understanding of the subject to see the close relationships and similarities amongst the different portions of theory.

Part I of the book, 'Fundamentals of Elasticity', Chapters 1–6, includes sufficient elasticity theory to provide the student with the basic tools of structural analysis. The work is standard but the presentation in some instances is original. In Chapter 4 I have endeavoured to clarify the use of energy methods of analysis and present a consistent, but general, approach to the various types of structural problem for which energy methods are employed. Thus, although a variety of methods are discussed, emphasis is placed on the methods of complementary and potential energy. Overall, my intention has been to given some indication of the role and limitations of each method of analysis.

Part II, 'Analysis of Aircraft Structures', Chapters 7–11, contains the analysis of the thin-walled, cellular type of structure peculiar to aircraft. In addition, Chapter 7 includes a discussion of structural materials, the fabrication and function of structural components and an introduction to structural idealization. Chapter 10 discusses the limitations of the theory presented in Chapters 8 and 9 and investigates modifications necessary to account for axial constraint effects. An introduction to computational methods of structural analysis is presented in Chapter 11 which also includes some elementary work on the relatively modern finite element method for continuum structures.

Finally, Part III, 'Airworthiness and Aeroelasticity', Chapters 12 and 13, are self explanatory.

Worked examples are used extensively in the text to illustrate the theory while numerous unworked problems with answers are listed at the end of each chapter; S.I. units are used throughout.

I am indebted to the Universities of London (L.U.) and Leeds for permission to include examples from their degree papers and also the Civil Engineering Department of the University of Leeds for allowing me any facilities I required during the preparation of the manuscript. I am also extremely indebted to my wife, Margaret, who willingly undertook the onerous task of typing the manuscript in addition to attending to the demands of a home and our three sons, Andrew, Richard and Antony.

T.H.G. Megson

Preface to Second Edition

The publication of a second edition has given me the opportunity to examine the contents of the book in detail and determine which parts required alteration and modernization. Aircraft structures, particularly in the field of materials, is a rapidly changing subject and, while the fundamentals of analysis remain essentially the same, clearly an attempt must be made to keep abreast of modern developments. At the same time I have examined the presentation making changes where I felt it necessary and including additional material which I believe will be useful for students of the subject.

The first five chapters remain essentially the same as in the first edition except for some minor changes in presentation.

In Chapter 6, Section 6.12 has been rewritten and extended to include flexural–torsional buckling of thin-walled columns; Section 6.13 has also been rewritten to present the theory of tension field beams in a more logical form.

The discussion of composite materials in Chapter 7 has been extended in the light of modern developments and the sections concerned with the function and fabrication of structural components now include illustrations of actual aircraft structures of different types. The topic of structural idealization has been removed to Chapter 8.

Chapter 8 has been retitled and the theory presented in a different manner. Matrix notation is used in the derivation of the expression for direct stress due to unsymmetrical bending and the 'bar' notation discarded. The theory of the torsion of closed sections has been extended to include a discussion of the mechanics of warping, and the theory for the secondary warping of open sections amended. Also included is the analysis of combined open and closed sections. Structural idealization has been removed from Chapter 7 and is introduced here so that the effects of structural idealization on the analysis follow on logically. An alternative method for the calculation of shear flow distributions is presented.

Chapter 9 has been retitled and extended to the analysis of actual structural components such as tapered spars and beams, fuselages and multicell wing sections. The method of successive approximations is included for the analysis of many celled wings and the effects of cut-outs in wings and fuselages are considered. In addition the calculation of loads on and the analysis of fuselage frames and wing ribs is presented. In addition to the analysis of structural components composite materials are considered with the determination of the elastic constants for a composite together with their use in the fabrication of plates.

Chapter 10 remains an investigation into structural constraint, although the presentation has been changed particularly in the case of the study of shear lag. The theory for the restrained warping of open section beams now includes general systems of loading and introduces the concept of a moment couple or bimoment.

Only minor changes have been made to Chapter 11 while Chapter 12 now includes a detailed study of fatigue, the fatigue strength of components, the prediction of fatigue life and crack propagation. Finally, Chapter 13 now includes a much more detailed investigation of flutter and the determination of critical flutter speed.

I am indebted to Professor D. J. Mead of the University of Southampton for many useful comments and suggestions. I am also grateful to Mr K. Broddle of British Aerospace for supplying photographs and drawings of aircraft structures.

T.H.G. Megson
1989

Preface to Third Edition

The publication of a third edition and its accompanying solutions manual has allowed me to take a close look at the contents of the book and also to test the accuracy of the answers to the examples in the text and the problems set at the end of each chapter.

I have reorganised the book into two parts as opposed, previously, to three. Part I, Elasticity, contains, as before, the first six chapters which are essentially the same except for the addition of two illustrative examples in Chapter 1 and one in Chapter 4.

Part II, Chapters 7 to 13, is retitled Aircraft structures, with Chapter 12, Airworthiness, now becoming Chapter 8, Airworthiness and airframe loads, since it is logical that loads on aircraft produced by different types of manoeuvre are considered before the stress distributions and displacements caused by these loads are calculated.

Chapter 7 has been updated to include a discussion of the latest materials used in aircraft construction with an emphasis on the different requirements of civil and military aircraft.

Chapter 8, as described above, now contains the calculation of airframe loads produced by different types of manoeuvre and has been extended to consider the inertia loads caused, for example, by ground manoeuvres such as landing.

Chapter 9 (previously Chapter 8) remains unchanged apart from minor corrections while Chapter 10 (9) is unchanged except for the inclusion of an example on the calculation of stresses and displacements in a laminated bar; an extra problem has been included at the end of the chapter.

Chapter 11 (10), Structural constraint, is unchanged while in Chapter 12 (11) the discussion of the finite element method has been extended to include the four node quadrilateral element together with illustrative examples on the calculation of element stiffnesses; a further problem has been added at the end of the chapter.

Chapter 13, Aeroelasticity, has not been changed from Chapter 13 in the second edition apart from minor corrections.

I am indebted to, formerly, David Ross and, latterly, Matthew Flynn of Arnold for their encouragement and support during this project.

T.H.G. Megson
1999

Preface to Fourth Edition

I have reviewed the three previous editions of the book and decided that a major overhaul would be beneficial, particularly in the light of developments in the aircraft industry and in the teaching of the subject. Present-day students prefer numerous worked examples and problems to solve so that I have included more worked examples in the text and more problems at the end of each chapter. I also felt that some of the chapters were too long. I have therefore broken down some of the longer chapters into shorter, more 'digestible' ones. For example, the previous Chapter 9 which covered bending, shear and torsion of open and closed section thin-walled beams plus the analysis of combined open and closed section beams, structural idealization and deflections now forms the contents of Chapters 16–20. Similarly, the Third Edition Chapter 10 'Stress Analysis of Aircraft Components' is now contained in Chapters 21–25 while 'Structural Instability', Chapter 6 in the Third Edition, is now covered by Chapters 8 and 9.

In addition to breaking down the longer chapters I have rearranged the material to emphasize the application of the fundamentals of structural analysis, contained in Part A, to the analysis of aircraft structures which forms Part B. For example, Matrix Methods, which were included in 'Part II, Aircraft Structures' in the Third Edition are now included in Part A since they are basic to general structural analysis; similarly for structural vibration.

Parts of the theory have been expanded. In Part A, virtual work now merits a chapter (Chapter 4) to itself since I believe this powerful and important method is worth an in-depth study. The work on tension field beams (Chapter 9) has become part of the chapter on thin plates and has been extended to include post-buckling behaviour. Materials, in Part B, now contains a section on material properties while, in response to readers' comments, the historical review has been discarded. The design of rivetted connections has been added to the consideration of structural components of aircraft in Chapter 12 while the work on crack propagation has been extended in Chapter 15. The method of successive approximations for multi-cellular wings has been dropped since, in these computer-driven times, it is of limited use and does not advance an understanding of the behaviour of structures. On the other hand the study of composite structures has been expanded as these form an increasing part of a modern aircraft's structure.

Finally, a Case Study, the design of part of the rear fuselage of a mythical trainer/semi-aerobatic aeroplane is presented in the Appendix to illustrate the application of some of the theory contained in this book.

I would like to thank Jonathan Simpson of Elsevier who initiated the project and who collated the very helpful readers' comments, Margaret, my wife, for suffering the long hours I sat at my word processor, and Jasmine, Lily, Tom and Bryony who are always an inspiration.

T.H.G. Megson

Supporting material accompanying this book

A full set of worked solutions for this book are available for teaching purposes.

Please visit http://www.textbooks.elsevier.com and follow the registration instructions to access this material, which is intended for use by lecturers and tutors.

Part A Fundamentals of Structural Analysis

Section A1 Elasticity

1

Basic elasticity

We shall consider, in this chapter, the basic ideas and relationships of the theory of elasticity. The treatment is divided into three broad sections: stress, strain and stress–strain relationships. The third section is deferred until the end of the chapter to emphasize the fact that the analysis of stress and strain, for example the equations of equilibrium and compatibility, does not assume a particular stress–strain law. In other words, the relationships derived in Sections 1.1–1.14 inclusive are applicable to non-linear as well as linearly elastic bodies.

1.1 Stress

Consider the arbitrarily shaped, three-dimensional body shown in Fig. 1.1. The body is in equilibrium under the action of externally applied forces $P_1, P_2, \ldots$ and is assumed to comprise a continuous and deformable material so that the forces are transmitted throughout its volume. It follows that at any internal point O there is a resultant force

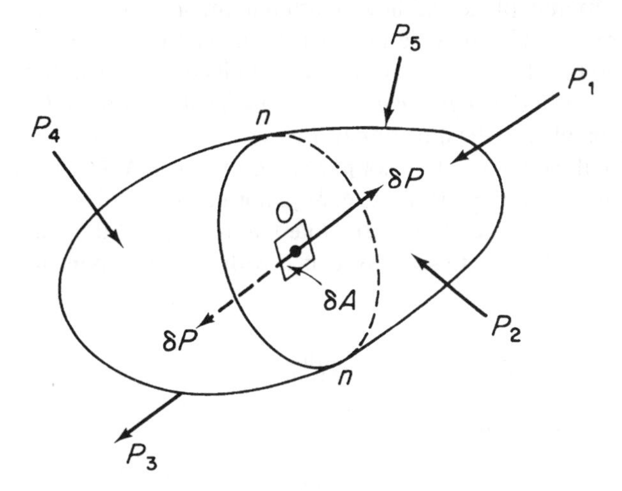

Fig. 1.1 Internal force at a point in an arbitrarily shaped body.

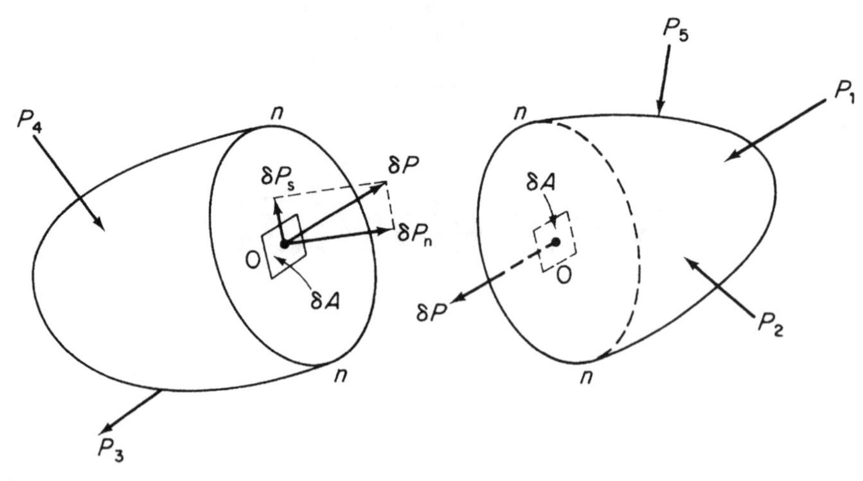

Fig. 1.2 Internal force components at the point O.

δP. The particle of material at O subjected to the force δP is in equilibrium so that there must be an equal but opposite force δP (shown dotted in Fig. 1.1) acting on the particle at the same time. If we now divide the body by any plane *nn* containing O then these two forces δP may be considered as being uniformly distributed over a small area δA of each face of the plane at the corresponding point O as in Fig. 1.2. The *stress* at O is then defined by the equation

$$\text{Stress} = \lim_{\delta A \to 0} \frac{\delta P}{\delta A} \tag{1.1}$$

The directions of the forces δP in Fig. 1.2 are such as to produce *tensile* stresses on the faces of the plane *nn*. It must be realized here that while the direction of δP is absolute the choice of plane is arbitrary, so that although the direction of the stress at O will always be in the direction of δP its magnitude depends upon the actual plane chosen since a different plane will have a different inclination and therefore a different value for the area δA. This may be more easily understood by reference to the bar in simple tension in Fig. 1.3. On the cross-sectional plane *mm* the uniform stress is given by P/A, *while on the inclined plane* $m'm'$ the stress is of magnitude P/A'. In both cases the stresses are parallel to the direction of P.

Generally, the direction of δP is not normal to the area δA, in which case it is usual to resolve δP into two components: one, δP_{n}, normal to the plane and the other, δP_{s}, acting in the plane itself (see Fig. 1.2). Note that in Fig. 1.2 the plane containing δP is perpendicular to δA. The stresses associated with these components are a *normal* or *direct stress* defined as

$$\sigma = \lim_{\delta A \to 0} \frac{\delta P_{\text{n}}}{\delta A} \tag{1.2}$$

and a *shear stress* defined as

$$\tau = \lim_{\delta A \to 0} \frac{\delta P_{\text{s}}}{\delta A} \tag{1.3}$$

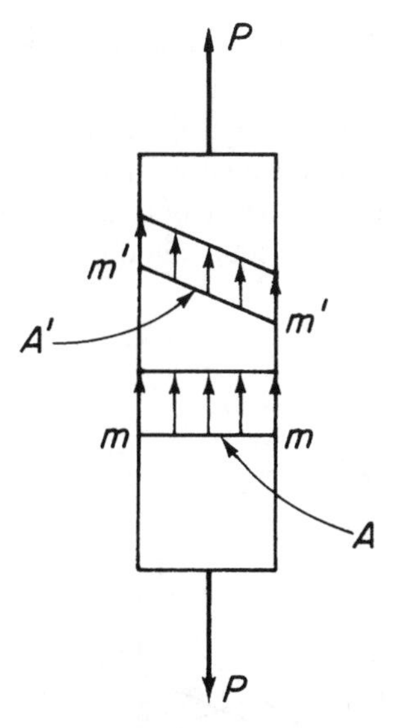

Fig. 1.3 Values of stress on different planes in a uniform bar.

The resultant stress is computed from its components by the normal rules of vector addition, namely

$$\text{Resultant stress} = \sqrt{\sigma^2 + \tau^2}$$

Generally, however, as indicated above, we are interested in the separate effects of σ and τ.

However, to be strictly accurate, stress is not a vector quantity for, in addition to magnitude and direction, we must specify the plane on which the stress acts. Stress is therefore a *tensor*, its complete description depending on the two vectors of force and surface of action.

1.2 Notation for forces and stresses

It is usually convenient to refer the state of stress at a point in a body to an orthogonal set of axes Oxyz. In this case we cut the body by planes parallel to the direction of the axes. The resultant force δP acting at the point O on one of these planes may then be resolved into a normal component and two in-plane components as shown in Fig. 1.4, thereby producing one component of direct stress and two components of shear stress.

The direct stress component is specified by reference to the plane on which it acts but the stress components require a specification of direction in addition to the plane. We therefore allocate a single subscript to direct stress to denote the plane on which it acts and two subscripts to shear stress, the first specifying the plane, the second direction. Therefore in Fig. 1.4, the shear stress components are τ_{zx} and τ_{zy} acting on the z plane and in the x and y directions, respectively, while the direct stress component is σ_z.

We may now completely describe the state of stress at a point O in a body by specifying components of shear and direct stress on the faces of an element of side δx, δy, δz, formed at O by the cutting planes as indicated in Fig. 1.5.

The sides of the element are infinitesimally small so that the stresses may be assumed to be uniformly distributed over the surface of each face. On each of the opposite faces there will be, to a first simplification, equal but opposite stresses.

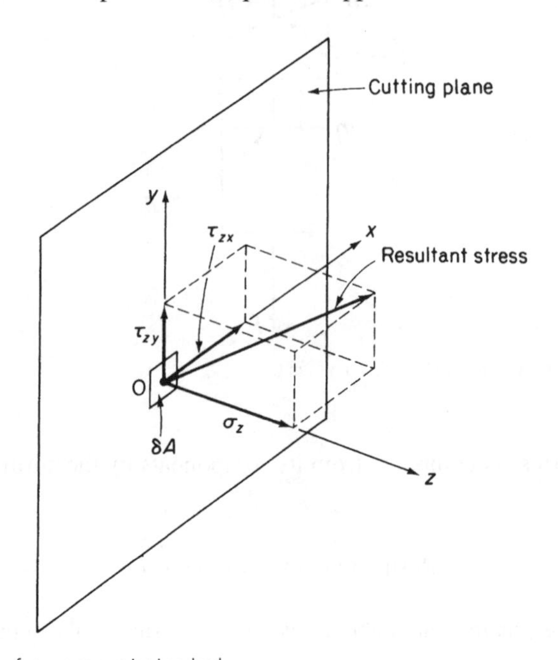

Fig. 1.4 Components of stress at a point in a body.

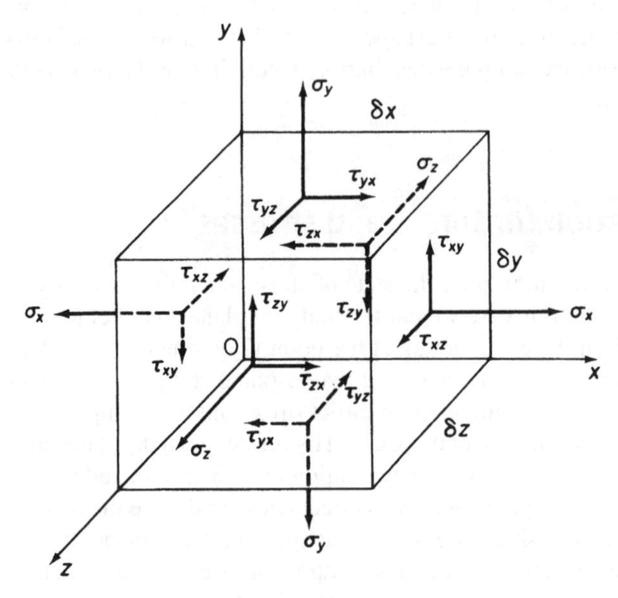

Fig. 1.5 Sign conventions and notation for stresses at a point in a body.

We shall now define the directions of the stresses in Fig. 1.5 as positive so that normal stresses directed away from their related surfaces are tensile and positive, opposite compressive stresses are negative. Shear stresses are positive when they act in the positive direction of the relevant axis in a plane on which the direct tensile stress is in the positive direction of the axis. If the tensile stress is in the opposite direction then positive shear stresses are in directions opposite to the positive directions of the appropriate axes.

Two types of external force may act on a body to produce the internal stress system we have already discussed. Of these, *surface forces* such as $P_1, P_2, \ldots,$ or hydrostatic pressure, are distributed over the surface area of the body. The surface force per unit area may be resolved into components parallel to our orthogonal system of axes and these are generally given the symbols $\overline{X}$, $\overline{Y}$ and $\overline{Z}$. The second force system derives from gravitational and inertia effects and the forces are known as *body forces*. These are distributed over the volume of the body and the components of body force per unit volume are designated X, Y and Z.

1.3 Equations of equilibrium

Generally, except in cases of uniform stress, the direct and shear stresses on opposite faces of an element are not equal as indicated in Fig. 1.5 but differ by small amounts. Therefore if, say, the direct stress acting on the z plane is σ_z then the direct stress acting on the $z + \delta z$ plane is, from the first two terms of a Taylor's series expansion, $\sigma_z + (\partial\sigma_z/\partial z)\delta z$.

We now investigate the equilibrium of an element at some internal point in an elastic body where the stress system is obtained by the method just described.

In Fig. 1.6 the element is in equilibrium under forces corresponding to the stresses shown and the components of body forces (not shown). Surface forces acting on the

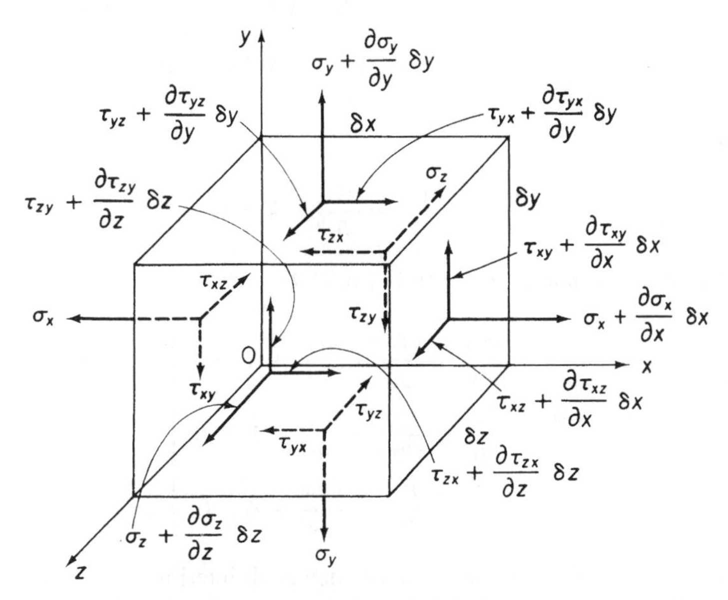

Fig. 1.6 Stresses on the faces of an element at a point in an elastic body.

boundary of the body, although contributing to the production of the internal stress system, do not directly feature in the equilibrium equations.

Taking moments about an axis through the centre of the element parallel to the z axis

$$\tau_{xy}\delta y\delta z\frac{\delta x}{2} + \left(\tau_{xy} + \frac{\partial \tau_{xy}}{\partial x}\delta x\right)\delta y\delta z\frac{\delta x}{2} - \tau_{yx}\delta x\delta z\frac{\delta y}{2} - \left(\tau_{yx} + \frac{\partial \tau_{yx}}{\partial y}\delta y\right)\delta x\delta z\frac{\delta y}{2} = 0$$

which simplifies to

$$\tau_{xy}\delta y\delta z\delta x + \frac{\partial \tau_{xy}}{\partial x}\delta y\delta z\frac{(\delta x)^2}{2} - \tau_{yx}\delta x\delta z\delta y - \frac{\partial \tau_{yx}}{\partial y}\delta x\,\delta z\frac{(\delta y)^2}{2} = 0$$

Dividing through by $\delta x\delta y\delta z$ and taking the limit as δx and δy approach zero

Similarly

$$\left.\begin{aligned}\tau_{xy} &= \tau_{yx}\\ \tau_{xz} &= \tau_{zx}\\ \tau_{yz} &= \tau_{zy}\end{aligned}\right\} \tag{1.4}$$

We see, therefore, that a shear stress acting on a given plane ($\tau_{xy}, \tau_{xz}, \tau_{yz}$) is always accompanied by an equal *complementary shear stress* ($\tau_{yx}, \tau_{zx}, \tau_{zy}$) acting on a plane perpendicular to the given plane and in the opposite sense.

Now considering the equilibrium of the element in the x direction

$$\left(\sigma_x + \frac{\partial \sigma_x}{\partial x}\delta x\right)\delta y\,\delta z - \sigma_x\delta y\delta z + \left(\tau_{yx} + \frac{\partial \tau_{yx}}{\partial y}\delta y\right)\delta x\delta z - \tau_{yx}\delta x\delta z + \left(\tau_{zx} + \frac{\partial \tau_{zx}}{\partial z}\delta z\right)\delta x\delta y - \tau_{zx}\delta x\delta y + X\delta x\delta y\delta z = 0$$

which gives

$$\frac{\partial \sigma_x}{\partial x} + \frac{\partial \tau_{yx}}{\partial y} + \frac{\partial \tau_{zx}}{\partial z} + X = 0$$

Or, writing $\tau_{xy} = \tau_{yx}$ and $\tau_{xz} = \tau_{zx}$ from Eq. (1.4)

Similarly

$$\left.\begin{aligned}\frac{\partial \sigma_x}{\partial x} + \frac{\partial \tau_{xy}}{\partial y} + \frac{\partial \tau_{xz}}{\partial z} + X &= 0\\ \frac{\partial \sigma_y}{\partial y} + \frac{\partial \tau_{yx}}{\partial x} + \frac{\partial \tau_{yz}}{\partial z} + Y &= 0\\ \frac{\partial \sigma_z}{\partial z} + \frac{\partial \tau_{zx}}{\partial x} + \frac{\partial \tau_{zy}}{\partial y} + Z &= 0\end{aligned}\right\} \tag{1.5}$$

The *equations of equilibrium* must be satisfied at all interior points in a deformable body under a three-dimensional force system.

1.4 Plane stress

Most aircraft structural components are fabricated from thin metal sheet so that stresses across the thickness of the sheet are usually negligible. Assuming, say, that the z axis is in the direction of the thickness then the three-dimensional case of Section 1.3 reduces to a two-dimensional case in which σ_z, τ_{xz} and τ_{yz} are all zero. This condition is known as *plane stress*; the equilibrium equations then simplify to

$$\left.\begin{aligned}\frac{\partial\sigma_x}{\partial x}+\frac{\partial\tau_{xy}}{\partial y}+X=0\\ \frac{\partial\sigma_y}{\partial y}+\frac{\partial\tau_{yx}}{\partial x}+Y=0\end{aligned}\right\} \tag{1.6}$$

1.5 Boundary conditions

The equations of equilibrium (1.5) (and also (1.6) for a two-dimensional system) satisfy the requirements of equilibrium at all internal points of the body. Equilibrium must also be satisfied at all positions on the boundary of the body where the components of the surface force per unit area are $\overline{X}$, $\overline{Y}$ and $\overline{Z}$. The triangular element of Fig. 1.7 at the boundary of a two-dimensional body of unit thickness is then in equilibrium under the action of surface forces on the elemental length AB of the boundary and internal forces on internal faces AC and CB.

Summation of forces in the x direction gives

$$\overline{X}\delta s-\sigma_x\delta y-\tau_{yx}\delta x+X\frac{1}{2}\delta x\delta y=0$$

which, by taking the limit as δx approaches zero, becomes

$$\overline{X}=\sigma_x\frac{\mathrm{d}y}{\mathrm{d}s}+\tau_{yx}\frac{\mathrm{d}x}{\mathrm{d}s}$$

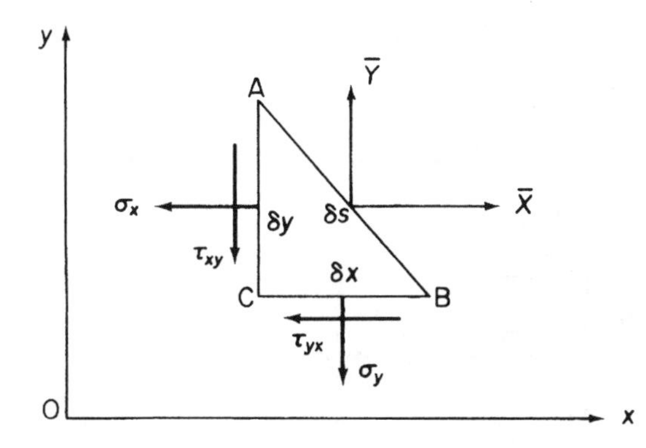

Fig. 1.7 Stresses on the faces of an element at the boundary of a two-dimensional body.

The derivatives dy/ds and dx/ds are the direction cosines l and m of the angles that a normal to AB makes with the x and y axes, respectively. It follows that

$$\bar{X} = \sigma_x l + \tau_{yx} m$$

and in a similar manner

$$\bar{Y} = \sigma_y m + \tau_{xy} l$$

A relatively simple extension of this analysis produces the boundary conditions for a three-dimensional body, namely

$$\left.\begin{aligned} \bar{X} &= \sigma_x l + \tau_{yx} m + \tau_{zx} n \\ \bar{Y} &= \sigma_y m + \tau_{xy} l + \tau_{zy} n \\ \bar{Z} &= \sigma_z n + \tau_{yz} m + \tau_{xz} l \end{aligned}\right\} \tag{1.7}$$

where l, m and n become the direction cosines of the angles that a normal to the surface of the body makes with the x, y and z axes, respectively.

1.6 Determination of stresses on inclined planes

The complex stress system of Fig. 1.6 is derived from a consideration of the actual loads applied to a body and is referred to a predetermined, though arbitrary, system of axes. The values of these stresses may not give a true picture of the severity of stress at that point so that it is necessary to investigate the state of stress on other planes on which the direct and shear stresses may be greater.

We shall restrict the analysis to the two-dimensional system of plane stress defined in Section 1.4.

Figure 1.8(a) shows a complex stress system at a point in a body referred to axes Ox, Oy. All stresses are positive as defined in Section 1.2. The shear stresses τ_{xy} and τ_{yx} were shown to be equal in Section 1.3. We now, therefore, designate them both τ_{xy}.

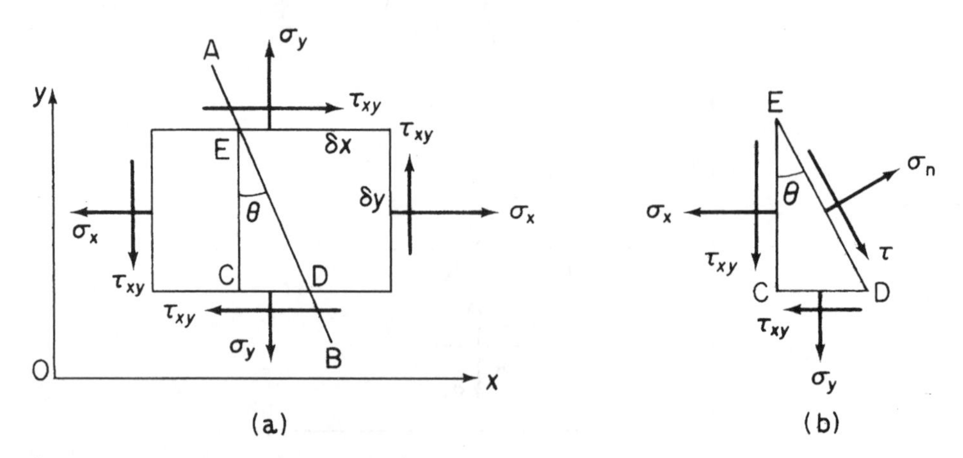

Fig. 1.8 (a) Stresses on a two-dimensional element; (b) stresses on an inclined plane at the point.

The element of side δx, δy and of unit thickness is small so that stress distributions over the sides of the element may be assumed to be uniform. Body forces are ignored since their contribution is a second-order term.

Suppose that we require to find the state of stress on a plane AB inclined at an angle θ to the vertical. The triangular element EDC formed by the plane and the vertical through E is in equilibrium under the action of the forces corresponding to the stresses shown in Fig. 1.8(b), where σ_n and τ are the direct and shear components of the resultant stress on AB. Then resolving forces in a direction perpendicular to ED we have

$$\sigma_{\rm n}{\rm ED} = \sigma_x{\rm EC}\cos\theta + \sigma_y{\rm CD}\sin\theta + \tau_{xy}{\rm EC}\sin\theta + \tau_{xy}{\rm CD}\cos\theta$$

Dividing through by ED and simplifying

$$\sigma_{\rm n} = \sigma_x\cos^2\theta + \sigma_y\sin^2\theta + \tau_{xy}\sin 2\theta \tag{1.8}$$

Now resolving forces parallel to ED

$$\tau{\rm ED} = \sigma_x{\rm EC}\sin\theta - \sigma_y{\rm CD}\cos\theta - \tau_{xy}{\rm EC}\cos\theta + \tau_{xy}{\rm CD}\sin\theta$$

Again dividing through by ED and simplifying

$$\tau = \frac{(\sigma_x - \sigma_y)}{2}\sin 2\theta - \tau_{xy}\cos 2\theta \tag{1.9}$$

Example 1.1

A cylindrical pressure vessel has an internal diameter of 2 m and is fabricated from plates 20 mm thick. If the pressure inside the vessel is 1.5 N/mm^2 and, in addition, the vessel is subjected to an axial tensile load of 2500 kN, calculate the direct and shear stresses on a plane inclined at an angle of 60° to the axis of the vessel. Calculate also the maximum shear stress.

The expressions for the longitudinal and circumferential stresses produced by the internal pressure may be found in any text on stress analysis[3] and are

$$\text{Longitudinal stress } (\sigma_x) = \frac{pd}{4t} = 1.5 \times 2 \times 10^3/4 \times 20 = 37.5\,\text{N/mm}^2$$

$$\text{Circumferential stress } (\sigma_y) = \frac{pd}{2t} = 1.5 \times 2 \times 10^3/2 \times 20 = 75\,\text{N/mm}^2$$

The direct stress due to the axial load will contribute to σ_x and is given by

$$\sigma_x\ (\text{axial load}) = 2500 \times 10^3/\pi \times 2 \times 10^3 \times 20 = 19.9\,\text{N/mm}^2$$

A rectangular element in the wall of the pressure vessel is then subjected to the stress system shown in Fig. 1.9. Note that there are no shear stresses acting on the x and y planes; in this case, σ_x and σ_y then form a *biaxial* stress system.

The direct stress, $\sigma_{\rm n}$, and shear stress, τ, on the plane AB which makes an angle of 60° with the axis of the vessel may be found from first principles by considering the

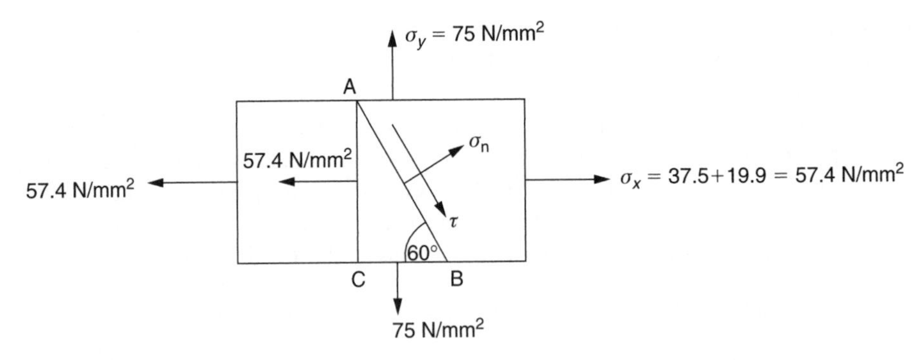

Fig. 1.9 Element of Example 1.1.

equilibrium of the triangular element ABC or by direct substitution in Eqs (1.8) and (1.9). Note that in the latter case $\theta = 30°$ and $\tau_{xy} = 0$. Then

$$\sigma_n = 57.4 \cos^2 30° + 75 \sin^2 30° = 61.8\,\text{N/mm}^2$$

$$\tau = (57.4 - 75)(\sin(2 \times 30°))/2 = -7.6\,\text{N/mm}^2$$

The negative sign for τ indicates that the shear stress is in the direction BA and not AB.

From Eq. (1.9) when $\tau_{xy} = 0$

$$\tau = (\sigma_x - \sigma_y)(\sin 2\theta)/2 \tag{i}$$

The maximum value of τ will therefore occur when $\sin 2\theta$ is a maximum, i.e. when $\sin 2\theta = 1$ and $\theta = 45°$. Then, substituting the values of σ_x and σ_y in Eq. (i)

$$\tau_{\max} = (57.4 - 75)/2 = -8.8\,\text{N/mm}^2$$

Example 1.2

A cantilever beam of solid, circular cross-section supports a compressive load of 50 kN applied to its free end at a point 1.5 mm below a horizontal diameter in the vertical plane of symmetry together with a torque of 1200 Nm (Fig. 1.10). Calculate the direct and shear stresses on a plane inclined at 60° to the axis of the cantilever at a point on the lower edge of the vertical plane of symmetry.

The direct loading system is equivalent to an axial load of 50 kN together with a bending moment of $50 \times 10^3 \times 1.5 = 75\,000$ Nmm in a vertical plane. Therefore, at any point on the lower edge of the vertical plane of symmetry there are compressive stresses due to the axial load and bending moment which act on planes perpendicular to the axis of the beam and are given, respectively, by Eqs (1.2) and (16.9), i.e.

$$\sigma_x\ (\text{axial load}) = 50 \times 10^3/\pi \times (60^2/4) = 17.7\,\text{N/mm}^2$$

$$\sigma_x\ (\text{bending moment}) = 75\,000 \times 30/\pi \times (60^4/64) = 3.5\,\text{N/mm}^2$$

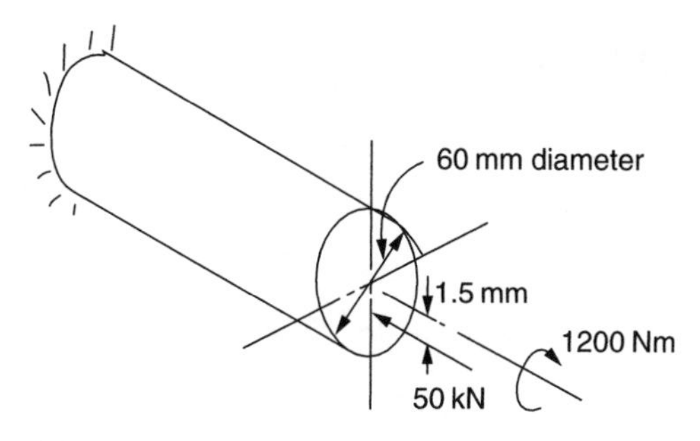

Fig. 1.10 Cantilever beam of Example 1.2.

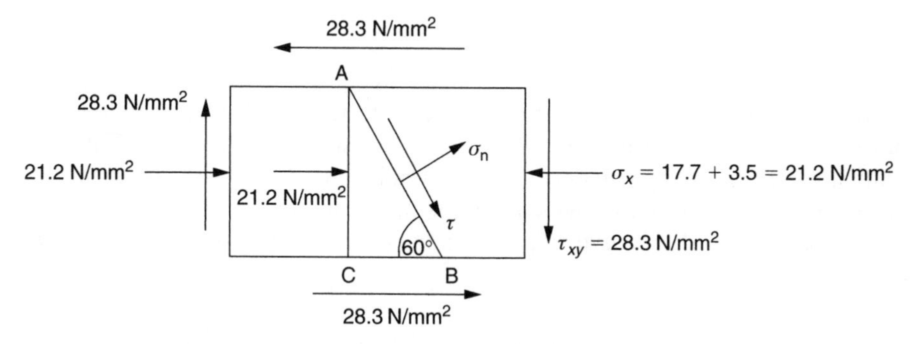

Fig. 1.11 Stress system on two-dimensional element of the beam of Example 1.2.

The shear stress, τ_{xy}, at the same point due to the torque is obtained from Eq. (iv) in Example 3.1, i.e.

$$\tau_{xy} = 1200 \times 10^3 \times 30/\pi \times (60^4/32) = 28.3\,\text{N/mm}^2$$

The stress system acting on a two-dimensional rectangular element at the point is shown in Fig. 1.11. Note that since the element is positioned at the bottom of the beam the shear stress due to the torque is in the direction shown and is negative (see Fig. 1.8).

Again σ_n and τ may be found from first principles or by direct substitution in Eqs (1.8) and (1.9). Note that $\theta = 30°$, $\sigma_y = 0$ and $\tau_{xy} = -28.3\,\text{N/mm}^2$ the negative sign arising from the fact that it is in the opposite direction to τ_{xy} in Fig. 1.8.

Then

$$\sigma_n = -21.2\cos^2 30° - 28.3\sin 60° = -40.4\,\text{N/mm}^2 \text{ (compression)}$$

$$\tau = (-21.2/2)\sin 60° + 28.3\cos 60° = 5.0\,\text{N/mm}^2 \text{ (acting in the direction AB)}$$

Different answers would have been obtained if the plane AB had been chosen on the opposite side of AC.

1.7 Principal stresses

For given values of σ_x, σ_y and τ_{xy}, in other words given loading conditions, σ_n varies with the angle θ and will attain a maximum or minimum value when $\mathrm{d}\sigma_{\mathrm{n}}/\mathrm{d}\theta = 0$. From Eq. (1.8)

$$\frac{\mathrm{d}\sigma_{\mathrm{n}}}{\mathrm{d}\theta} = -2\sigma_x \cos\theta \sin\theta + 2\sigma_y \sin\theta \cos\theta + 2\tau_{xy} \cos 2\theta = 0$$

Hence

$$-(\sigma_x - \sigma_y) \sin 2\theta + 2\tau_{xy} \cos 2\theta = 0$$

or

$$\tan 2\theta = \frac{2\tau_{xy}}{\sigma_x - \sigma_y} \tag{1.10}$$

Two solutions, θ and $\theta + \pi/2$, are obtained from Eq. (1.10) so that there are two mutually perpendicular planes on which the direct stress is either a maximum or a minimum. Further, by comparison of Eqs (1.9) and (1.10) it will be observed that these planes correspond to those on which there is no shear stress. The direct stresses on these planes are called *principal stresses* and the planes themselves, *principal planes*.

From Eq. (1.10)

$$\sin 2\theta = \frac{2\tau_{xy}}{\sqrt{(\sigma_x - \sigma_y)^2 + 4\tau_{xy}^2}} \qquad \cos 2\theta = \frac{\sigma_x - \sigma_y}{\sqrt{(\sigma_x - \sigma_y)^2 + 4\tau_{xy}^2}}$$

and

$$\sin 2(\theta + \pi/2) = \frac{-2\tau_{xy}}{\sqrt{(\sigma_x - \sigma_y)^2 + 4\tau_{xy}^2}} \qquad \cos 2(\theta + \pi/2) = \frac{-(\sigma_x - \sigma_y)}{\sqrt{(\sigma_x - \sigma_y)^2 + 4\tau_{xy}^2}}$$

Rewriting Eq. (1.8) as

$$\sigma_{\mathrm{n}} = \frac{\sigma_x}{2}(1 + \cos 2\theta) + \frac{\sigma_y}{2}(1 - \cos 2\theta) + \tau_{xy} \sin 2\theta$$

and substituting for $\{\sin 2\theta, \cos 2\theta\}$ and $\{\sin 2(\theta + \pi/2), \cos 2(\theta + \pi/2)\}$ in turn gives

$$\sigma_{\mathrm{I}} = \frac{\sigma_x + \sigma_y}{2} + \frac{1}{2}\sqrt{(\sigma_x - \sigma_y)^2 + 4\tau_{xy}^2} \tag{1.11}$$

and

$$\sigma_{\mathrm{II}} = \frac{\sigma_x + \sigma_y}{2} - \frac{1}{2}\sqrt{(\sigma_x - \sigma_y)^2 + 4\tau_{xy}^2} \tag{1.12}$$

where σ_{I} is the *maximum* or *major principal stress* and σ_{II} is the *minimum* or *minor principal stress*. Note that σ_{I} is algebraically the greatest direct stress at the point while σ_{II} is algebraically the least. Therefore, when σ_{II} is negative, i.e. compressive, it is possible for σ_{II} to be numerically greater than σ_{I}.

The maximum shear stress at this point in the body may be determined in an identical manner. From Eq. (1.9)

$$\frac{\mathrm{d}\tau}{\mathrm{d}\theta} = (\sigma_x - \sigma_y)\cos 2\theta + 2\tau_{xy}\sin 2\theta = 0$$

giving

$$\tan 2\theta = -\frac{(\sigma_x - \sigma_y)}{2\tau_{xy}} \tag{1.13}$$

It follows that

$$\sin 2\theta = \frac{-(\sigma_x - \sigma_y)}{\sqrt{(\sigma_x - \sigma_y)^2 + 4\tau_{xy}^2}} \qquad \cos 2\theta = \frac{2\tau_{xy}}{\sqrt{(\sigma_x - \sigma_y)^2 + 4\tau_{xy}^2}}$$

$$\sin 2(\theta + \pi/2) = \frac{(\sigma_x - \sigma_y)}{\sqrt{(\sigma_x - \sigma_y)^2 + 4\tau_{xy}^2}} \qquad \cos 2(\theta + \pi/2) = \frac{-2\tau_{xy}}{\sqrt{(\sigma_x - \sigma_y)^2 + 4\tau_{xy}^2}}$$

Substituting these values in Eq. (1.9) gives

$$\tau_{\max,\min} = \pm\frac{1}{2}\sqrt{(\sigma_x - \sigma_y)^2 + 4\tau_{xy}^2} \tag{1.14}$$

Here, as in the case of principal stresses, we take the maximum value as being the greater algebraic value.

Comparing Eq. (1.14) with Eqs (1.11) and (1.12) we see that

$$\tau_{\max} = \frac{\sigma_{\mathrm{I}} - \sigma_{\mathrm{II}}}{2} \tag{1.15}$$

Equations (1.14) and (1.15) give the maximum shear stress at the point in the body in *the plane of the given stresses*. For a three-dimensional body supporting a two-dimensional stress system this is not necessarily the maximum shear stress at the point.

Since Eq. (1.13) is the negative reciprocal of Eq. (1.10) then the angles 2θ given by these two equations differ by 90° or, alternatively, the planes of maximum shear stress are inclined at 45° to the principal planes.

1.8 Mohr's circle of stress

The state of stress at a point in a deformable body may be determined graphically by *Mohr's circle of stress.*

In Section 1.6 the direct and shear stresses on an inclined plane were shown to be given by

$$\sigma_{\mathrm{n}} = \sigma_x \cos^2\theta + \sigma_y \sin^2\theta + \tau_{xy}\sin 2\theta \tag{Eq. (1.8)}$$

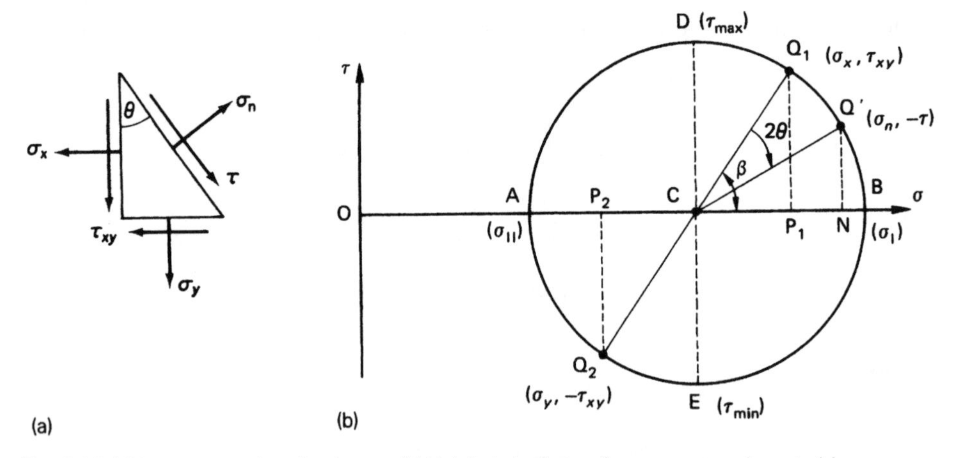

Fig. 1.12 (a) Stresses on a triangular element; (b) Mohr's circle of stress for stress system shown in (a).

and

$$\tau = \frac{(\sigma_x - \sigma_y)}{2} \sin 2\theta - \tau_{xy} \cos 2\theta \qquad \text{(Eq. (1.9))}$$

respectively. The positive directions of these stresses and the angle θ are defined in Fig. 1.12(a). Equation (1.8) may be rewritten in the form

$$\sigma_{\text{n}} = \frac{\sigma_x}{2}(1 + \cos 2\theta) + \frac{\sigma_y}{2}(1 - \cos 2\theta) + \tau_{xy} \sin 2\theta$$

or

$$\sigma_{\text{n}} - \frac{1}{2}(\sigma_x + \sigma_y) = \frac{1}{2}(\sigma_x - \sigma_y) \cos 2\theta + \tau_{xy} \sin 2\theta$$

Squaring and adding this equation to Eq. (1.9) we obtain

$$\left[\sigma_{\text{n}} - \frac{1}{2}(\sigma_x + \sigma_y)\right]^2 + \tau^2 = \left[\frac{1}{2}(\sigma_x - \sigma_y)\right]^2 + \tau_{xy}^2$$

which represents the equation of a circle of radius $\frac{1}{2}\sqrt{(\sigma_x - \sigma_y)^2 + 4\tau_{xy}^2}$ and having its centre at the point $((\sigma_x - \sigma_y)/2, 0)$.

The circle is constructed by locating the points $Q_1(\sigma_x, \tau_{xy})$ and $Q_2(\sigma_y, -\tau_{xy})$ referred to axes $O\sigma\tau$ as shown in Fig. 1.12(b). The centre of the circle then lies at C the intersection of Q_1Q_2 and the $O\sigma$ axis; clearly C is the point $((\sigma_x - \sigma_y)/2, 0)$ and the radius of the circle is $\frac{1}{2}\sqrt{(\sigma_x - \sigma_y)^2 + 4\tau_{xy}^2}$ as required. CQ′ is now set off at an angle 2θ (positive clockwise) to CQ_1, Q′ is then the point $(\sigma_n, -\tau)$ as demonstrated below. From Fig. 1.12(b) we see that

$$\text{ON} = \text{OC} + \text{CN}$$

or, since $\mathrm{OC} = (\sigma_x + \sigma_y)/2$, $\mathrm{CN} = \mathrm{CQ}' \cos(\beta - 2\theta)$ and $\mathrm{CQ}' = \mathrm{CQ}_1$ we have

$$\sigma_{\mathrm{n}} = \frac{\sigma_x + \sigma_y}{2} + \mathrm{CQ}_1(\cos\beta\cos 2\theta + \sin\beta\sin 2\theta)$$

But

$$\mathrm{CQ}_1 = \frac{\mathrm{CP}_1}{\cos\beta} \quad \text{and} \quad \mathrm{CP}_1 = \frac{(\sigma_x - \sigma_y)}{2}$$

Hence

$$\sigma_{\mathrm{n}} = \frac{\sigma_x + \sigma_y}{2} + \left(\frac{\sigma_x - \sigma_y}{2}\right)\cos 2\theta + \mathrm{CP}_1 \tan\beta\sin 2\theta$$

which, on rearranging, becomes

$$\sigma_{\mathrm{n}} = \sigma_x \cos^2\theta + \sigma_y \sin^2\theta + \tau_{xy}\sin 2\theta$$

as in Eq. (1.8). Similarly it may be shown that

$$\mathrm{Q'N} = \tau_{xy}\cos 2\theta - \left(\frac{\sigma_x - \sigma_y}{2}\right)\sin 2\theta = -\tau$$

as in Eq. (1.9). Note that the construction of Fig. 1.12(b) corresponds to the stress system of Fig. 1.12(a) so that any sign reversal must be allowed for. Also, the $\mathrm{O}\sigma$ and $\mathrm{O}\tau$ axes must be constructed to the same scale or the equation of the circle is not represented.

The maximum and minimum values of the direct stress, viz. the major and minor principal stresses σ_{I} and σ_{II}, occur when N (and Q′) coincide with B and A, respectively. Thus

$$\begin{aligned}\sigma_{\mathrm{I}} &= \mathrm{OC} + \text{radius of circle}\\ &= \frac{(\sigma_x + \sigma_y)}{2} + \sqrt{\mathrm{CP}_1^2 + \mathrm{P}_1\mathrm{Q}_1^2}\end{aligned}$$

or

$$\sigma_{\mathrm{I}} = \frac{(\sigma_x + \sigma_y)}{2} + \frac{1}{2}\sqrt{(\sigma_x - \sigma_y)^2 + 4\tau_{xy}^2}$$

and in the same fashion

$$\sigma_{\mathrm{II}} = \frac{(\sigma_x + \sigma_y)}{2} - \frac{1}{2}\sqrt{(\sigma_x - \sigma_y)^2 + 4\tau_{xy}^2}$$

The principal planes are then given by $2\theta = \beta(\sigma_{\mathrm{I}})$ and $2\theta = \beta + \pi(\sigma_{\mathrm{II}})$.

Also the maximum and minimum values of shear stress occur when Q′ coincides with D and E at the upper and lower extremities of the circle.

At these points Q′N is equal to the radius of the circle which is given by

$$\mathrm{CQ}_1 = \sqrt{\frac{(\sigma_x - \sigma_y)^2}{4} + \tau_{xy}^2}$$

Hence $\tau_{\max,\min} = \pm\frac{1}{2}\sqrt{(\sigma_x - \sigma_y)^2 + 4\tau_{xy}^2}$ as before. The planes of maximum and minimum shear stress are given by $2\theta = \beta + \pi/2$ and $2\theta = \beta + 3\pi/2$, these being inclined at 45° to the principal planes.

Example 1.3

Direct stresses of 160 N/mm^2 (tension) and 120 N/mm^2 (compression) are applied at a particular point in an elastic material on two mutually perpendicular planes. The principal stress in the material is limited to 200 N/mm^2 (tension). Calculate the allowable value of shear stress at the point on the given planes. Determine also the value of the other principal stress and the maximum value of shear stress at the point. Verify your answer using Mohr's circle.

The stress system at the point in the material may be represented as shown in Fig. 1.13 by considering the stresses to act uniformly over the sides of a triangular element ABC of unit thickness. Suppose that the direct stress on the principal plane AB is σ. For horizontal equilibrium of the element

$$\sigma \mathrm{AB}\cos\theta = \sigma_x \mathrm{BC} + \tau_{xy}\mathrm{AC}$$

which simplifies to

$$\tau_{xy}\tan\theta = \sigma - \sigma_x \qquad \text{(i)}$$

Considering vertical equilibrium gives

$$\sigma \mathrm{AB}\sin\theta = \sigma_y \mathrm{AC} + \tau_{xy}\mathrm{BC}$$

or

$$\tau_{xy}\cot\theta = \sigma - \sigma_y \qquad \text{(ii)}$$

Hence from the product of Eqs (i) and (ii)

$$\tau_{xy}^2 = (\sigma - \sigma_x)(\sigma - \sigma_y)$$

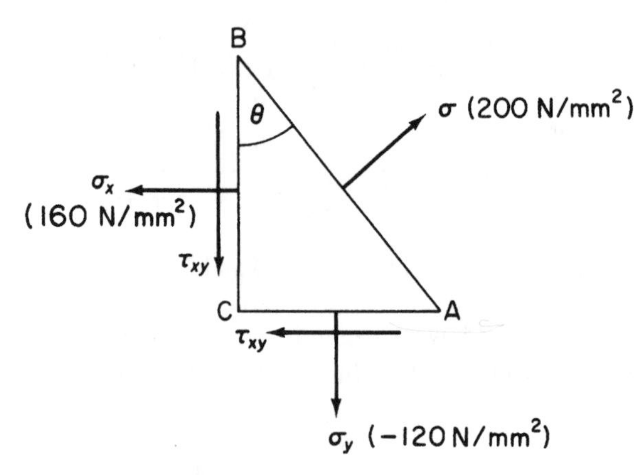

Fig. 1.13 Stress system for Example 1.3.

Now substituting the values $\sigma_x = 160\,\text{N/mm}^2$, $\sigma_y = -120\,\text{N/mm}^2$ and $\sigma = \sigma_\text{I} = 200\,\text{N/mm}^2$ we have

$$\tau_{xy} = \pm 113\,\text{N/mm}^2$$

Replacing $\cot\theta$ in Eq. (ii) by $1/\tan\theta$ from Eq. (i) yields a quadratic equation in σ

$$\sigma^2 - \sigma(\sigma_x - \sigma_y) + \sigma_x\sigma_y - \tau_{xy}^2 = 0 \qquad \text{(iii)}$$

The numerical solutions of Eq. (iii) corresponding to the given values of σ_x, σ_y and τ_{xy} are the principal stresses at the point, namely

$$\sigma_\text{I} = 200\,\text{N/mm}^2 \text{ (given)} \quad \sigma_\text{II} = -160\,\text{N/mm}^2$$

Having obtained the principal stresses we now use Eq. (1.15) to find the maximum shear stress, thus

$$\tau_\text{max} = \frac{200 + 160}{2} = 180\,\text{N/mm}^2$$

The solution is rapidly verified from Mohr's circle of stress (Fig. 1.14). From the arbitrary origin O, OP_1 and OP_2 are drawn to represent $\sigma_x = 160\,\text{N/mm}^2$ and $\sigma_y = -120\,\text{N/mm}^2$. The mid-point C of P_1P_2 is then located. $\text{OB} = \sigma_\text{I} = 200\,\text{N/mm}^2$ is marked out and the radius of the circle is then CB. OA is the required principal stress. Perpendiculars P_1Q_1 and P_2Q_2 to the circumference of the circle are equal to $\pm\tau_{xy}$ (to scale) and the radius of the circle is the maximum shear stress.

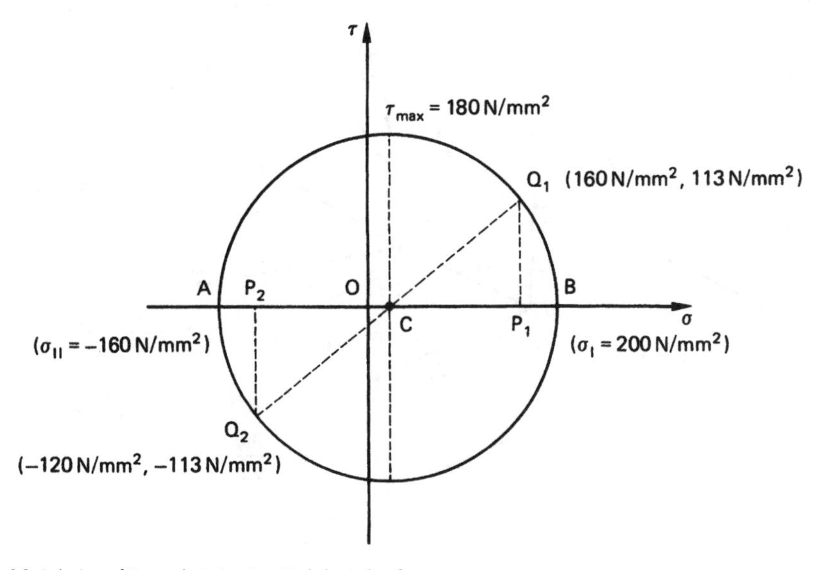

Fig. 1.14 Solution of Example 1.3 using Mohr's circle of stress.

1.9 Strain

The external and internal forces described in the previous sections cause linear and angular displacements in a deformable body. These displacements are generally defined in terms of *strain*. *Longitudinal* or *direct strains* are associated with direct stresses σ and relate to changes in length while *shear strains* define changes in angle produced by shear stresses. These strains are designated, with appropriate suffixes, by the symbols ε and γ, respectively, and have the same sign as the associated stresses.

Consider three mutually perpendicular line elements OA, OB and OC at a point O in a deformable body. Their original or unstrained lengths are δx, δy and δz, respectively. If, now, the body is subjected to forces which produce a complex system of direct and shear stresses at O, such as that in Fig. 1.6, then the line elements will deform to the positions O′A′, O′B′ and O′C′ shown in Fig. 1.15.

The coordinates of O in the unstrained body are (x, y, z) so that those of A, B and C are $(x+\delta x, y, z)$, $(x, y+\delta y, z)$ and $(x, y, z+\delta z)$. The components of the displacement of O to O′ parallel to the x, y and z axes are u, v and w. These symbols are used to designate these displacements throughout the book and are defined as positive in the positive directions of the axes. We again employ the first two terms of a Taylor's series expansion to determine the components of the displacements of A, B and C. Thus, the displacement of A in a direction parallel to the x axis is $u+(\partial u/\partial x)\delta x$. The remaining components are found in an identical manner and are shown in Fig. 1.15.

We now define direct strain in more quantitative terms. If a line element of length L at a point in a body suffers a change in length ΔL then the longitudinal strain at that

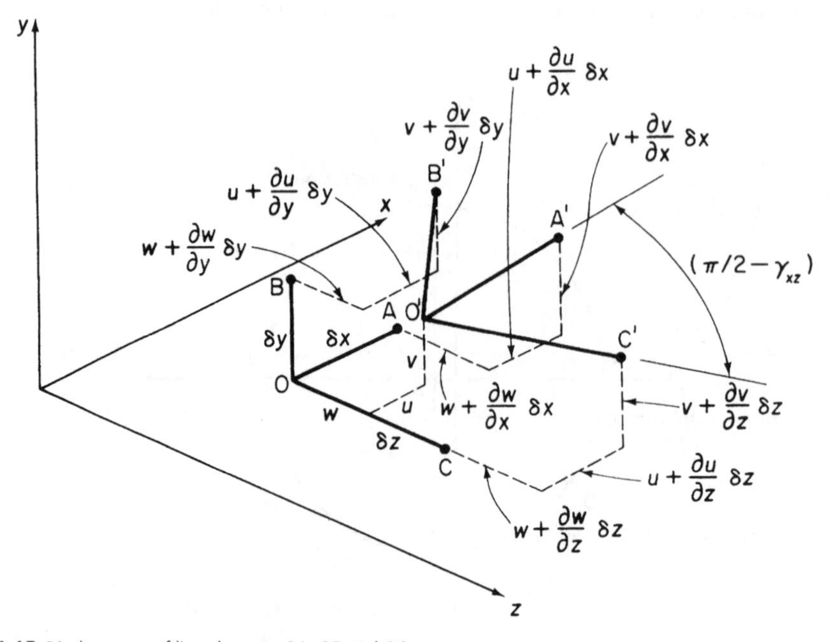

Fig. 1.15 Displacement of line elements OA, OB and OC.

point in the body in the direction of the line element is

$$\varepsilon = \lim_{L \to 0} \frac{\Delta L}{L}$$

The change in length of the element OA is (O′A′ − OA) so that the direct strain at O in the x direction is obtained from the equation

$$\varepsilon_x = \frac{\mathrm{O'A'} - \mathrm{OA}}{\mathrm{OA}} = \frac{\mathrm{O'A'} - \delta x}{\delta x} \tag{1.16}$$

Now

$$(\mathrm{O'A'})^2 = \left(\delta x + u + \frac{\partial u}{\partial x}\delta x - u\right)^2 + \left(v + \frac{\partial v}{\partial x}\delta x - v\right)^2 + \left(w + \frac{\partial w}{\partial x}\delta x - w\right)^2$$

or

$$\mathrm{O'A'} = \delta x \sqrt{\left(1 + \frac{\partial u}{\partial x}\right)^2 + \left(\frac{\partial v}{\partial x}\right)^2 + \left(\frac{\partial w}{\partial x}\right)^2}$$

which may be written when second-order terms are neglected

$$\mathrm{O'A'} = \delta x \left(1 + 2\frac{\partial u}{\partial x}\right)^{\frac{1}{2}}$$

Applying the binomial expansion to this expression we have

$$\mathrm{O'A'} = \delta x \left(1 + \frac{\partial u}{\partial x}\right) \tag{1.17}$$

in which squares and higher powers of $\partial u/\partial x$ are ignored. Substituting for O′A′ in Eq. (1.16) we have

It follows that

$$\left.\begin{aligned} \varepsilon_x &= \frac{\partial u}{\partial x} \\ \varepsilon_y &= \frac{\partial v}{\partial y} \\ \varepsilon_z &= \frac{\partial w}{\partial z} \end{aligned}\right\} \tag{1.18}$$

The shear strain at a point in a body is defined as the change in the angle between two mutually perpendicular lines at the point. Therefore, if the shear strain in the xz plane is γ_{xz} then the angle between the displaced line elements O′A′ and O′C′ in Fig. 1.15 is $\pi/2 - \gamma_{xz}$ radians.

Now $\cos \mathrm{A'O'C'} = \cos(\pi/2 - \gamma_{xz}) = \sin \gamma_{xz}$ and as γ_{xz} is small then $\cos \mathrm{A'O'C'} = \gamma_{xz}$. From the trigonometrical relationships for a triangle

$$\cos \mathrm{A'O'C'} = \frac{(\mathrm{O'A'})^2 + (\mathrm{O'C'})^2 - (\mathrm{A'C'})^2}{2(\mathrm{O'A'})(\mathrm{O'C'})} \tag{1.19}$$

We have previously shown, in Eq. (1.17), that

$$O'A' = \delta x\left(1 + \frac{\partial u}{\partial x}\right)$$

Similarly

$$O'C' = \delta z\left(1 + \frac{\partial w}{\partial z}\right)$$

But for small displacements the derivatives of u, v and w are small compared with l, so that, as we are concerned here with actual length rather than change in length, we may use the approximations

$$O'A' \approx \delta x \quad O'C' \approx \delta z$$

Again to a first approximation

$$(A'C')^2 = \left(\delta z - \frac{\partial w}{\partial x}\delta x\right)^2 + \left(\delta x - \frac{\partial u}{\partial z}\delta z\right)^2$$

Substituting for O′A′, O′C′ and A′C′ in Eq. (1.19) we have

$$\cos A'O'C' = \frac{(\delta x^2) + (\delta z)^2 - [\delta z - (\partial w/\partial x)\delta x]^2 - [\delta x - (\partial u/\partial z)\delta z]^2}{2\delta x \delta z}$$

Expanding and neglecting fourth-order powers gives

$$\cos A'O'C' = \frac{2(\partial w/\partial x)\delta x\delta z + 2(\partial u/\partial z)\delta x\delta z}{2\delta x\delta z}$$

or

Similarly

$$\left.\begin{aligned} \gamma_{xz} &= \frac{\partial w}{\partial x} + \frac{\partial u}{\partial z} \\ \gamma_{xy} &= \frac{\partial v}{\partial x} + \frac{\partial u}{\partial y} \\ \gamma_{yz} &= \frac{\partial w}{\partial y} + \frac{\partial v}{\partial z} \end{aligned}\right\} \tag{1.20}$$

It must be emphasized that Eqs (1.18) and (1.20) are derived on the assumption that the displacements involved are small. Normally these linearized equations are adequate for most types of structural problem but in cases where deflections are large, for example types of suspension cable, etc., the full, non-linear, large deflection equations, given in many books on elasticity, must be employed.

1.10 Compatibility equations

In Section 1.9 we expressed the six components of strain at a point in a deformable body in terms of the three components of displacement at that point, u, v and w. We have supposed that the body remains continuous during the deformation so that no voids are formed. It follows that each component, u, v and w, must be a continuous, single-valued function or, in quantitative terms

$$u = f_1(x, y, z) \quad v = f_2(x, y, z) \quad w = f_3(x, y, z)$$

If voids were formed then displacements in regions of the body separated by the voids would be expressed as different functions of x, y and z. The existence, therefore, of just three single-valued functions for displacement is an expression of the continuity or *compatibility* of displacement which we have presupposed.

Since the six strains are defined in terms of three displacement functions then they must bear some relationship to each other and cannot have arbitrary values. These relationships are found as follows. Differentiating γ_{xy} from Eq. (1.20) with respect to x and y gives

$$\frac{\partial^2 \gamma_{xy}}{\partial x\, \partial y} = \frac{\partial^2}{\partial x\, \partial y}\frac{\partial v}{\partial x} + \frac{\partial^2}{\partial x\, \partial y}\frac{\partial u}{\partial y}$$

or, since the functions of u and v are continuous

$$\frac{\partial^2 \gamma_{xy}}{\partial x\, \partial y} = \frac{\partial^2}{\partial x^2}\frac{\partial v}{\partial y} + \frac{\partial^2}{\partial y^2}\frac{\partial u}{\partial x}$$

which may be written, using Eq. (1.18)

$$\frac{\partial^2 \gamma_{xy}}{\partial x\, \partial y} = \frac{\partial^2 \varepsilon_y}{\partial x^2} + \frac{\partial^2 \varepsilon_x}{\partial y^2} \tag{1.21}$$

In a similar manner

$$\frac{\partial^2 \gamma_{yz}}{\partial y\, \partial z} = \frac{\partial^2 \varepsilon_y}{\partial z^2} + \frac{\partial^2 \varepsilon_z}{\partial y^2} \tag{1.22}$$

$$\frac{\partial^2 \gamma_{xz}}{\partial x\, \partial z} = \frac{\partial^2 \varepsilon_z}{\partial x^2} + \frac{\partial^2 \varepsilon_x}{\partial z^2} \tag{1.23}$$

If we now differentiate γ_{xy} with respect to x and z and add the result to γ_{zx}, differentiated with respect to y and x, we obtain

$$\frac{\partial^2 \gamma_{xy}}{\partial x\, \partial z} + \frac{\partial^2 \gamma_{xz}}{\partial y\, \partial x} = \frac{\partial^2}{\partial x\, \partial z}\left(\frac{\partial u}{\partial y} + \frac{\partial v}{\partial x}\right) + \frac{\partial^2}{\partial y\, \partial x}\left(\frac{\partial w}{\partial x} + \frac{\partial u}{\partial z}\right)$$

or

$$\frac{\partial}{\partial x}\left(\frac{\partial \gamma_{xy}}{\partial z} + \frac{\partial \gamma_{xz}}{\partial y}\right) = \frac{\partial^2}{\partial z\, \partial y}\frac{\partial u}{\partial x} + \frac{\partial^2}{\partial x^2}\left(\frac{\partial v}{\partial z} + \frac{\partial w}{\partial y}\right) + \frac{\partial^2}{\partial y\, \partial z}\frac{\partial u}{\partial x}$$

Substituting from Eqs (1.18) and (1.21) and rearranging

$$2\frac{\partial^2 \varepsilon_x}{\partial y\, \partial z} = \frac{\partial}{\partial x}\left(-\frac{\partial \gamma_{yz}}{\partial x} + \frac{\partial \gamma_{xz}}{\partial y} + \frac{\partial \gamma_{xy}}{\partial z}\right) \tag{1.24}$$

Similarly

$$2\frac{\partial^2 \varepsilon_y}{\partial x\,\partial z} = \frac{\partial}{\partial y}\left(\frac{\partial \gamma_{yz}}{\partial x} - \frac{\partial \gamma_{xz}}{\partial y} + \frac{\partial \gamma_{xy}}{\partial z}\right) \tag{1.25}$$

and

$$2\frac{\partial^2 \varepsilon_z}{\partial x\,\partial y} = \frac{\partial}{\partial z}\left(\frac{\partial \gamma_{yz}}{\partial x} + \frac{\partial \gamma_{xz}}{\partial y} - \frac{\partial \gamma_{xy}}{\partial z}\right) \tag{1.26}$$

Equations (1.21)–(1.26) are the six equations of *strain compatibility* which must be satisfied in the solution of three-dimensional problems in elasticity.

1.11 Plane strain

Although we have derived the compatibility equations and the expressions for strain for the general three-dimensional state of strain we shall be mainly concerned with the two-dimensional case described in Section 1.4. The corresponding state of strain, in which it is assumed that particles of the body suffer displacements in one plane only, is known as *plane strain*. We shall suppose that this plane is, as for plane stress, the xy plane. Then ε_z, γ_{xz} and γ_{yz} become zero and Eqs (1.18) and (1.20) reduce to

$$\varepsilon_x = \frac{\partial u}{\partial x} \qquad \varepsilon_y = \frac{\partial v}{\partial y} \tag{1.27}$$

and

$$\gamma_{xy} = \frac{\partial v}{\partial x} + \frac{\partial u}{\partial y} \tag{1.28}$$

Further, by substituting $\varepsilon_z = \gamma_{xz} = \gamma_{yz} = 0$ in the six equations of compatibility and noting that ε_x, ε_y and γ_{xy} are now purely functions of x and y, we are left with Eq. (1.21), namely

$$\frac{\partial^2 \gamma_{xy}}{\partial x\,\partial y} = \frac{\partial^2 \varepsilon_y}{\partial x^2} + \frac{\partial^2 \varepsilon_x}{\partial y^2}$$

as the only equation of compatibility in the two-dimensional or plane strain case.

1.12 Determination of strains on inclined planes

Having defined the strain at a point in a deformable body with reference to an arbitrary system of coordinate axes we may calculate direct strains in any given direction and the change in the angle (shear strain) between any two originally perpendicular directions at that point. We shall consider the two-dimensional case of plane strain described in Section 1.11.

An element in a two-dimensional body subjected to the complex stress system of Fig. 1.16(a) will distort into the shape shown in Fig. 1.16(b). In particular, the triangular element ECD will suffer distortion to the shape E′C′D′ with corresponding changes

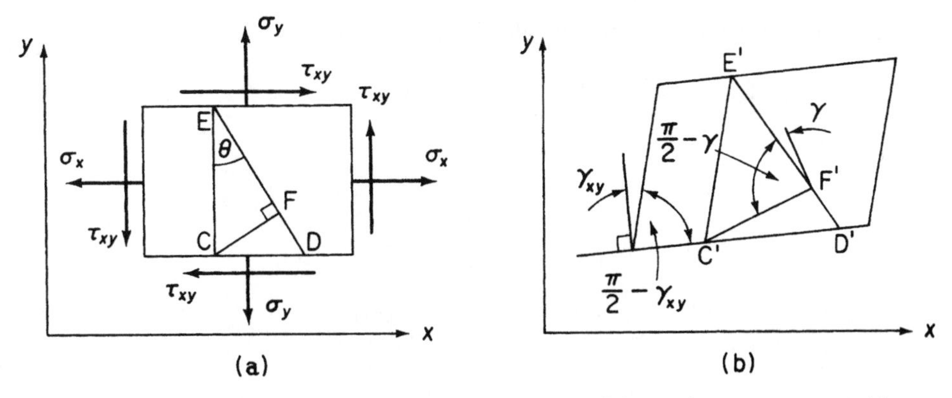

Fig. 1.16 (a) Stress system on rectangular element; (b) distorted shape of element due to stress system in (a).

in the length FC and angle EFC. Suppose that the known direct and shear strains associated with the given stress system are $\varepsilon_x, \varepsilon_y$ and γ_{xy} (the actual relationships will be investigated later) and that we require to find the direct strain ε_n in a direction normal to the plane ED and the shear strain γ produced by the shear stress acting on the plane ED.

To a first order of approximation

$$\left.\begin{aligned} C'D' &= CD(1+\varepsilon_x) \\ C'E' &= CE(1+\varepsilon_y) \\ E'D' &= ED(1+\varepsilon_{n+\pi/2}) \end{aligned}\right\} \tag{1.29}$$

where $\varepsilon_{n+\pi/2}$ is the direct strain in the direction ED. From the geometry of the triangle E′C′D′ in which angle $E'C'D' = \pi/2 - \gamma_{xy}$

$$(E'D')^2 = (C'D')^2 + (C'E')^2 - 2(C'D')(C'E')\cos(\pi/2 - \gamma_{xy})$$

or, substituting from Eqs (1.29)

$$(ED)^2(1+\varepsilon_{n+\pi/2})^2 = (CD)^2(1+\varepsilon_x)^2 + (CE)^2(1+\varepsilon_y)^2 - 2(CD)(CE)(1+\varepsilon_x)(1+\varepsilon_y)\sin\gamma_{xy}$$

Noting that $(ED)^2 = (CD)^2 + (CE)^2$ and neglecting squares and higher powers of small quantities this equation may be rewritten

$$2(ED)^2\varepsilon_{n+\pi/2} = 2(CD)^2\varepsilon_x + 2(CE)^2\varepsilon_y - 2(CE)(CD)\gamma_{xy}$$

Dividing through by $2(ED)^2$ gives

$$\varepsilon_{n+\pi/2} = \varepsilon_x \sin^2\theta + \varepsilon_y \cos^2\theta - \cos\theta\sin\theta\gamma_{xy} \tag{1.30}$$

The strain ε_n in the direction normal to the plane ED is found by replacing the angle θ in Eq. (1.30) by $\theta - \pi/2$. Hence

$$\varepsilon_n = \varepsilon_x \cos^2\theta + \varepsilon_y \sin^2\theta + \frac{\gamma_{xy}}{2}\sin 2\theta \tag{1.31}$$

Turning our attention now to the triangle C′F′E′ we have

$$(C'E')^2 = (C'F')^2 + (F'E')^2 - 2(C'F')(F'E')\cos(\pi/2 - \gamma) \tag{1.32}$$

in which

$$C'E' = CE(1 + \varepsilon_y)$$
$$C'F' = CF(1 + \varepsilon_n)$$
$$F'E' = FE(1 + \varepsilon_{n+\pi/2})$$

Substituting for C′E′, C′F′ and F′E′ in Eq. (1.32) and writing $\cos(\pi/2 - \gamma) = \sin\gamma$ we find

$$(CE)^2(1 + \varepsilon_y)^2 = (CF)^2(1 + \varepsilon_n)^2 + (FE)^2(1 + \varepsilon_{n+\pi/2})^2 - 2(CF)(FE)(1 + \varepsilon_n)(1 + \varepsilon_{n+\pi/2})\sin\gamma \tag{1.33}$$

All the strains are assumed to be small so that their squares and higher powers may be ignored. Further, $\sin\gamma \approx \gamma$ and Eq. (1.33) becomes

$$(CE)^2(1 + 2\varepsilon_y) = (CF)^2(1 + 2\varepsilon_n) + (FE)^2(1 + 2\varepsilon_{n+\pi/2}) - 2(CF)(FE)\gamma$$

From Fig. 1.16(a), $(CE)^2 = (CF)^2 + (FE)^2$ and the above equation simplifies to

$$2(CE)^2\varepsilon_y = 2(CF)^2\varepsilon_n + 2(FE)^2\varepsilon_{n+\pi/2} - 2(CF)(FE)\gamma$$

Dividing through by $2(CE)^2$ and transposing

$$\gamma = \frac{\varepsilon_n \sin^2\theta + \varepsilon_{n+\pi/2}\cos^2\theta - \varepsilon_y}{\sin\theta\cos\theta}$$

Substitution of ε_n and $\varepsilon_{n+\pi/2}$ from Eqs (1.31) and (1.30) yields

$$\frac{\gamma}{2} = \frac{(\varepsilon_x - \varepsilon_y)}{2}\sin 2\theta - \frac{\gamma_{xy}}{2}\cos 2\theta \tag{1.34}$$

1.13 Principal strains

If we compare Eqs (1.31) and (1.34) with Eqs (1.8) and (1.9) we observe that they may be obtained from Eqs (1.8) and (1.9) by replacing σ_n by ε_n, σ_x by ε_x, σ_y by ε_y, τ_{xy} by $\gamma_{xy}/2$ and τ by $\gamma/2$. Therefore, for each deduction made from Eqs (1.8) and (1.9) concerning σ_n and τ there is a corresponding deduction from Eqs (1.31) and (1.34) regarding ε_n and $\gamma/2$.

Therefore at a point in a deformable body, there are two mutually perpendicular planes on which the shear strain γ is zero and normal to which the direct strain is a maximum or minimum. These strains are the *principal strains* at that point and are given (from comparison with Eqs (1.11) and (1.12)) by

$$\varepsilon_{\mathrm{I}} = \frac{\varepsilon_x + \varepsilon_y}{2} + \frac{1}{2}\sqrt{(\varepsilon_x - \varepsilon_y)^2 + \gamma_{xy}^2} \tag{1.35}$$

and

$$\varepsilon_{\mathrm{II}} = \frac{\varepsilon_x + \varepsilon_y}{2} - \frac{1}{2}\sqrt{(\varepsilon_x - \varepsilon_y)^2 + \gamma_{xy}^2} \tag{1.36}$$

If the shear strain is zero on these planes it follows that the shear stress must also be zero and we deduce, from Section 1.7, that the directions of the principal strains and principal stresses coincide. The related planes are then determined from Eq. (1.10) or from

$$\tan 2\theta = \frac{\gamma_{xy}}{\varepsilon_x - \varepsilon_y} \tag{1.37}$$

In addition the maximum shear strain at the point is

$$\left(\frac{\gamma}{2}\right)_{\max} = \tfrac{1}{2}\sqrt{(\varepsilon_x - \varepsilon_y)^2 + \gamma_{xy}^2} \tag{1.38}$$

or

$$\left(\frac{\gamma}{2}\right)_{\max} = \frac{\varepsilon_{\mathrm{I}} - \varepsilon_{\mathrm{II}}}{2} \tag{1.39}$$

(*cf.* Eqs (1.14) and (1.15)).

1.14 Mohr's circle of strain

We now apply the arguments of Section 1.13 to the Mohr's circle of stress described in Section 1.8. A circle of strain, analogous to that shown in Fig. 1.12(b), may be drawn when σ_x, σ_y, etc. are replaced by $\varepsilon_x, \varepsilon_y$, etc. as specified in Section 1.13. The horizontal extremities of the circle represent the principal strains, the radius of the circle, half the maximum shear strain and so on.

1.15 Stress–strain relationships

In the preceding sections we have developed, for a three-dimensional deformable body, three equations of equilibrium (Eqs (1.5)) and six strain–displacement relationships (Eqs (1.18) and (1.20)). From the latter we eliminated displacements thereby deriving six auxiliary equations relating strains. These compatibility equations are an expression of the continuity of displacement which we have assumed as a prerequisite of the analysis. At this stage, therefore, we have obtained nine independent equations towards the solution of the three-dimensional stress problem. However, the number of unknowns totals 15, comprising six stresses, six strains and three displacements. An additional six equations are therefore necessary to obtain a solution.

So far we have made no assumptions regarding the force–displacement or stress–strain relationship in the body. This will, in fact, provide us with the required six equations but before these are derived it is worthwhile considering some general aspects of the analysis.

The derivation of the equilibrium, strain–displacement and compatibility equations does not involve any assumption as to the stress–strain behaviour of the material of the body. It follows that these basic equations are applicable to any type of continuous, deformable body no matter how complex its behaviour under stress. In fact we shall consider only the simple case of linearly elastic *isotropic* materials for which stress is directly proportional to strain and whose elastic properties are the same in all directions. A material possessing the same properties at all points is said to be *homogeneous.*

Particular cases arise where some of the stress components are known to be zero and the number of unknowns may then be no greater than the remaining equilibrium equations which have not identically vanished. The unknown stresses are then found from the conditions of equilibrium alone and the problem is said to be *statically determinate.* For example, the uniform stress in the member supporting a tensile load P in Fig. 1.3 is found by applying one equation of equilibrium and a boundary condition. This system is therefore statically determinate.

Statically indeterminate systems require the use of some, if not all, of the other equations involving strain–displacement and stress–strain relationships. However, whether the system be statically determinate or not, stress–strain relationships are necessary to determine deflections. The role of the six auxiliary compatibility equations will be discussed when actual elasticity problems are formulated in Chapter 2.

We now proceed to investigate the relationship of stress and strain in a three-dimensional, linearly elastic, isotropic body.

Experiments show that the application of a uniform direct stress, say σ_x, does not produce any shear distortion of the material and that the direct strain ε_x is given by the equation

$$\varepsilon_x = \frac{\sigma_x}{E} \tag{1.40}$$

where E is a constant known as the *modulus of elasticity* or *Young's modulus.* Equation (1.40) is an expression of *Hooke's law*. Further, ε_x is accompanied by lateral strains

$$\varepsilon_y = -\nu\frac{\sigma_x}{E} \qquad \varepsilon_z = -\nu\frac{\sigma_x}{E} \tag{1.41}$$

in which ν is a constant termed *Poisson's ratio.*

For a body subjected to direct stresses σ_x, σ_y and σ_z the direct strains are, from Eqs (1.40) and (1.41) and the *principle of superposition* (see Chapter 5, Section 5.9)

$$\left.\begin{aligned} \varepsilon_x &= \frac{1}{E}[\sigma_x - \nu(\sigma_y + \sigma_z)] \\ \varepsilon_y &= \frac{1}{E}[\sigma_y - \nu(\sigma_x + \sigma_z)] \\ \varepsilon_z &= \frac{1}{E}[\sigma_z - \nu(\sigma_x + \sigma_y)] \end{aligned}\right\} \tag{1.42}$$

Equations (1.42) may be transposed to obtain expressions for each stress in terms of the strains. The procedure adopted may be any of the standard mathematical approaches and gives

$$\sigma_x = \frac{\nu E}{(1+\nu)(1-2\nu)}e + \frac{E}{(1+\nu)}\varepsilon_x \tag{1.43}$$

$$\sigma_y = \frac{\nu E}{(1+\nu)(1-2\nu)}e + \frac{E}{(1+\nu)}\varepsilon_y \tag{1.44}$$

$$\sigma_z = \frac{\nu E}{(1+\nu)(1-2\nu)}e + \frac{E}{(1+\nu)}\varepsilon_z \tag{1.45}$$

in which

$$e = \varepsilon_x + \varepsilon_y + \varepsilon_z \quad \text{(see Eq. (1.53))}$$

For the case of plane stress in which $\sigma_z = 0$, Eqs (1.43) and (1.44) reduce to

$$\sigma_x = \frac{E}{1-\nu^2}(\varepsilon_x + \nu\varepsilon_y) \tag{1.46}$$

$$\sigma_y = \frac{E}{1-\nu^2}(\varepsilon_y + \nu\varepsilon_x) \tag{1.47}$$

Suppose now that, at some arbitrary point in a material, there are principal strains ε_{I} and ε_{II} corresponding to principal stresses σ_{I} and σ_{II}. If these stresses (and strains) are in the direction of the coordinate axes x and y, respectively, then $\tau_{xy} = \gamma_{xy} = 0$ and from Eq. (1.34) the shear strain on an arbitrary plane at the point inclined at an angle θ to the principal planes is

$$\gamma = (\varepsilon_{\text{I}} - \varepsilon_{\text{II}}) \sin 2\theta \tag{1.48}$$

Using the relationships of Eqs (1.42) and substituting in Eq. (1.48) we have

$$\gamma = \frac{1}{E}[(\sigma_{\text{I}} - \nu\sigma_{\text{II}}) - (\sigma_{\text{II}} - \nu\sigma_{\text{I}})] \sin 2\theta$$

or

$$\gamma = \frac{(1+\nu)}{E}(\sigma_{\text{I}} - \sigma_{\text{II}}) \sin 2\theta \tag{1.49}$$

Using Eq. (1.9) and noting that for this particular case $\tau_{xy} = 0$, $\sigma_x = \sigma_{\text{I}}$ and $\sigma_y = \sigma_{\text{II}}$

$$2\tau = (\sigma_{\text{I}} - \sigma_{\text{II}}) \sin 2\theta$$

from which we may rewrite Eq. (1.49) in terms of τ as

$$\gamma = \frac{2(1+\nu)}{E}\tau \tag{1.50}$$

The term $E/2(1+\nu)$ is a constant known as the *modulus of rigidity* G. Hence

$$\gamma = \tau/G$$

and the shear strains γ_{xy}, γ_{xz} and γ_{yz} are expressed in terms of their associated shear stresses as follows

$$\gamma_{xy} = \frac{\tau_{xy}}{G} \quad \gamma_{xz} = \frac{\tau_{xz}}{G} \quad \gamma_{yz} = \frac{\tau_{yz}}{G} \tag{1.51}$$

Equations (1.51), together with Eqs (1.42), provide the additional six equations required to determine the 15 unknowns in a general three-dimensional problem in elasticity. They are, however, limited in use to a linearly elastic isotropic body.

For the case of plane stress they simplify to

$$\left.\begin{aligned} \varepsilon_x &= \frac{1}{E}(\sigma_x - \nu\sigma_y) \\ \varepsilon_y &= \frac{1}{E}(\sigma_y - \nu\sigma_x) \\ \varepsilon_z &= \frac{-\nu}{E}(\sigma_x - \sigma_y) \\ \gamma_{xy} &= \frac{\tau_{xy}}{G} \end{aligned}\right\} \tag{1.52}$$

It may be seen from the third of Eqs (1.52) that the conditions of plane stress and plane strain do not necessarily describe identical situations.

Changes in the linear dimensions of a strained body may lead to a change in volume. Suppose that a small element of a body has dimensions δx, δy and δz. When subjected to a three-dimensional stress system the element will sustain a volumetric strain e (change in volume/unit volume) equal to

$$e = \frac{(1 + \varepsilon_x)\delta x(1 + \varepsilon_y)\delta y(1 + \varepsilon_z)\delta z - \delta x\delta y\delta z}{\delta x\delta y\delta z}$$

Neglecting products of small quantities in the expansion of the right-hand side of the above equation yields

$$e = \varepsilon_x + \varepsilon_y + \varepsilon_z \tag{1.53}$$

Substituting for $\varepsilon_x, \varepsilon_y$ and ε_z from Eqs (1.42) we find, for a linearly elastic, isotropic body

$$e = \frac{1}{E}[\sigma_x + \sigma_y + \sigma_z - 2\nu(\sigma_x + \sigma_y + \sigma_z)]$$

or

$$e = \frac{(1 - 2\nu)}{E}(\sigma_x + \sigma_y + \sigma_z)$$

In the case of a uniform hydrostatic pressure, $\sigma_x = \sigma_y = \sigma_z = -p$ and

$$e = -\frac{3(1 - 2\nu)}{E}p \tag{1.54}$$

The constant $E/3(1 - 2\nu)$ is known as the *bulk modulus* or *modulus of volume expansion* and is often given the symbol K.

An examination of Eq. (1.54) shows that $\nu \leq 0.5$ since a body cannot increase in volume under pressure. Also the lateral dimensions of a body subjected to uniaxial tension cannot increase so that $\nu > 0$. Therefore, for an isotropic material $0 \leq \nu \leq 0.5$ and for most isotropic materials ν is in the range 0.25–0.33 below the elastic limit. Above the limit of proportionality ν increases and approaches 0.5.

Example 1.4

A rectangular element in a linearly elastic isotropic material is subjected to tensile stresses of 83 and 65 N/mm^2 on mutually perpendicular planes. Determine the strain in the direction of each stress and in the direction perpendicular to both stresses. Find also the principal strains, the maximum shear stress, the maximum shear strain and their directions at the point. Take $E = 200\,000$ N/mm^2 and $\upsilon = 0.3$.

If we assume that $\sigma_x = 83$ N/mm^2 and $\sigma_y = 65$ N/mm^2 then from Eqs (1.52)

$$\varepsilon_x = \frac{1}{200\,000}(83 - 0.3 \times 65) = 3.175 \times 10^{-4}$$

$$\varepsilon_y = \frac{1}{200\,000}(65 - 0.3 \times 83) = 2.005 \times 10^{-4}$$

$$\varepsilon_z = \frac{-0.3}{200\,000}(83 + 65) = -2.220 \times 10^{-4}$$

In this case, since there are no shear stresses on the given planes, σ_x and σ_y are principal stresses so that ε_x and ε_y are the principal strains and are in the directions of σ_x and σ_y. It follows from Eq. (1.15) that the maximum shear stress (in the plane of the stresses) is

$$\tau_{\max} = \frac{83 - 65}{2} = 9\,\text{N/mm}^2$$

acting on planes at 45° to the principal planes.

Further, using Eq. (1.50), the maximum shear strain is

$$\gamma_{\max} = \frac{2 \times (1 + 0.3) \times 9}{200\,000}$$

so that $\gamma_{\max} = 1.17 \times 10^{-4}$ on the planes of maximum shear stress.

Example 1.5

At a particular point in a structural member a two-dimensional stress system exists where $\sigma_x = 60$ N/mm^2, $\sigma_y = -40$ N/mm^2 and $\tau_{xy} = 50$ N/mm^2. If Young's modulus $E = 200\,000$ N/mm^2 and Poisson's ratio $\nu = 0.3$ calculate the direct strain in the x and y directions and the shear strain at the point. Also calculate the principal strains at the point and their inclination to the plane on which σ_x acts; verify these answers using a graphical method.

From Eqs (1.52)

$$\varepsilon_x = \frac{1}{200\,000}(60 + 0.3 \times 40) = 360 \times 10^{-6}$$

$$\varepsilon_y = \frac{1}{200\,000}(-40 - 0.3 \times 60) = -290 \times 10^{-6}$$

From Eq. (1.50) the shear modulus, G, is given by

$$G = \frac{E}{2(1+\nu)} = \frac{200\,000}{2(1+0.3)} = 76\,923\,\text{N/mm}^2$$

Hence, from Eqs (1.52)

$$\gamma_{xy} = \frac{\tau_{xy}}{G} = \frac{50}{76\,923} = 650 \times 10^{-6}$$

Now substituting in Eq. (1.35) for $\varepsilon_x, \varepsilon_y$ and γ_{xy}

$$\varepsilon_{\text{I}} = 10^{-6}\left[\frac{360 - 290}{2} + \frac{1}{2}\sqrt{(360+290)^2 + 650^2}\right]$$

which gives

$$\varepsilon_{\text{I}} = 495 \times 10^{-6}$$

Similarly, from Eq. (1.36)

$$\varepsilon_{\text{II}} = -425 \times 10^{-6}$$

From Eq. (1.37)

$$\tan 2\theta = \frac{650 \times 10^{-6}}{360 \times 10^{-6} + 290 \times 10^{-6}} = 1$$

Therefore

$$2\theta = 45^\circ \text{ or } 225^\circ$$

so that

$$\theta = 22.5^\circ \text{ or } 112.5^\circ$$

The values of ε_{I}, ε_{II} and θ are verified using Mohr's circle of strain (Fig. 1.17). Axes Oε and Oγ are set up and the points Q_1 $(360 \times 10^{-6}, \frac{1}{2} \times 650 \times 10^{-6})$ and Q_2 $(-290 \times 10^{-6}, -\frac{1}{2} \times 650 \times 10^{-6})$ located. The centre C of the circle is the intersection of Q_1Q_2 and the Oε axis. The circle is then drawn with radius CQ_1 and the points B(ε_{I}) and A(ε_{II}) located. Finally angle $\text{Q}_1\text{CB} = 2\theta$ and angle $\text{Q}_1\text{CA} = 2\theta + \pi$.

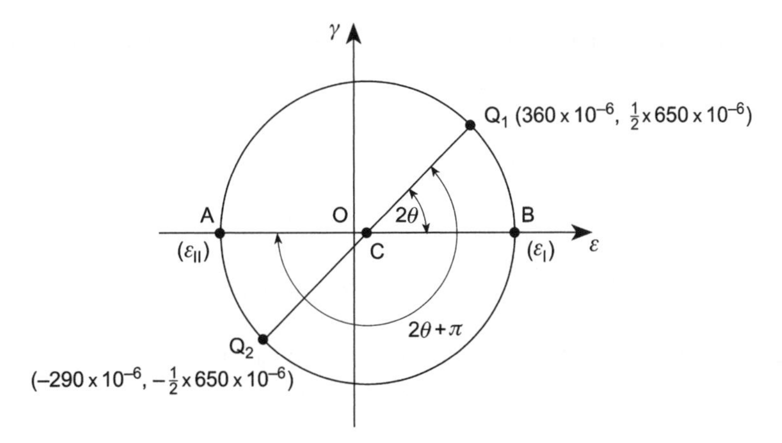

Fig. 1.17 Mohr's circle of strain for Example 1.5.

1.15.1 Temperature effects

The stress–strain relationships of Eqs (1.43)–(1.47) apply to a body or structural member at a constant uniform temperature. A temperature rise (or fall) generally results in an expansion (or contraction) of the body or structural member so that there is a change in size, i.e. a strain.

Consider a bar of uniform section, of original length L_o, and suppose that it is subjected to a temperature change ΔT along its length; ΔT can be a rise (+ve) or fall (−ve). If the coefficient of linear expansion of the material of the bar is α the final length of the bar is, from elementary physics

$$L = L_o(1 + \alpha\Delta T)$$

so that the strain, ε, is given by

$$\varepsilon = \frac{L - L_o}{L_o} = \alpha\Delta T \qquad (1.55)$$

Suppose now that a compressive axial force is applied to each end of the bar such that the bar returns to its original length. The *mechanical* strain produced by the axial force is therefore just large enough to offset the *thermal* strain due to the temperature change making the *total* strain zero. In general terms the total strain, ε, is the sum of the mechanical and thermal strains. Therefore, from Eqs (1.40) and (1.55)

$$\varepsilon = \frac{\sigma}{E} + \alpha\Delta T \qquad (1.56)$$

In the case where the bar is returned to its original length or if the bar had not been allowed to expand at all the total strain is zero and from Eq. (1.56)

$$\sigma = -E\alpha\Delta T \qquad (1.57)$$

Equations (1.42) may now be modified to include the contribution of thermal strain. Therefore, by comparison with Eq. (1.56)

$$\left.\begin{aligned}\varepsilon_x &= \frac{1}{E}[\sigma_x - \nu(\sigma_y + \sigma_z)] + \alpha\Delta T\\ \varepsilon_y &= \frac{1}{E}[\sigma_y - \nu(\sigma_x + \sigma_z)] + \alpha\Delta T\\ \varepsilon_z &= \frac{1}{E}[\sigma_z - \nu(\sigma_x + \sigma_y)] + \alpha\Delta T\end{aligned}\right\} \quad (1.58)$$

Equations (1.58) may be transposed in the same way as Eqs (1.42) to give stress–strain relationships rather than strain–stress relationships, i.e.

$$\left.\begin{aligned}\sigma_x &= \frac{\nu E}{(1+\nu)(1-2\nu)}e + \frac{E}{(1+\nu)}\varepsilon_x - \frac{E}{(1-2\nu)}\alpha\Delta T\\ \sigma_y &= \frac{\nu E}{(1+\nu)(1-2\nu)}e + \frac{E}{(1+\nu)}\varepsilon_y - \frac{E}{(1-2\nu)}\alpha\Delta T\\ \sigma_z &= \frac{\nu E}{(1+\nu)(1-2\nu)}e + \frac{E}{(1+\nu)}\varepsilon_z - \frac{E}{(1-2\nu)}\alpha\Delta T\end{aligned}\right\} \quad (1.59)$$

For the case of plane stress in which $\sigma_z = 0$ these equations reduce to

$$\left.\begin{aligned}\sigma_x &= \frac{E}{(1-\nu^2)}(\varepsilon_x + \nu\varepsilon_y) - \frac{E}{(1-\nu)}\alpha\Delta T\\ \sigma_y &= \frac{E}{(1-\nu^2)}(\varepsilon_y + \nu\varepsilon_x) - \frac{E}{(1-\nu)}\alpha\Delta T\end{aligned}\right\} \quad (1.60)$$

Example 1.6

A composite bar of length L has a central core of copper loosely inserted in a sleeve of steel; the ends of the steel and copper are attached to each other by rigid plates. If the bar is subjected to a temperature rise ΔT determine the stress in the steel and in the copper and the extension of the composite bar. The copper core has a Young's modulus E_c, a cross-sectional area A_c and a coefficient of linear expansion α_c; the corresponding values for the steel are E_s, A_s and α_s.

Assume that $\alpha_c > \alpha_s$.

If the copper core and steel sleeve were allowed to expand freely their final lengths would be different since they have different values of the coefficient of linear expansion. However, since they are rigidly attached at their ends one restrains the other and an axial stress is induced in each. Suppose that this stress is σ_x. Then in Eqs (1.58) $\sigma_x = \sigma_c$ or σ_s and $\sigma_y = \sigma_z = 0$; the total strain in the copper and steel is then, respectively

$$\varepsilon_c = \frac{\sigma_c}{E_c} + \alpha_c\Delta T \quad \text{(i)}$$

$$\varepsilon_s = \frac{\sigma_s}{E_s} + \alpha_s\Delta T \quad \text{(ii)}$$

The total strain in the copper and steel is the same since their ends are rigidly attached to each other. Therefore, from compatibility of displacement

$$\frac{\sigma_c}{E_c} + \alpha_c \Delta T = \frac{\sigma_s}{E_s} + \alpha_s \Delta T \tag{iii}$$

There is no external axial load applied to the bar so that

$$\sigma_c A_c + \sigma_s A_s = 0$$

i.e.

$$\sigma_s = -\frac{A_c}{A_s}\sigma_c \tag{iv}$$

Substituting for σ_s in Eq. (iii) gives

$$\sigma_c \left(\frac{1}{E_c} + \frac{A_c}{A_s E_s} \right) = \Delta T(\alpha_s - \alpha_c)$$

from which

$$\sigma_c = \frac{\Delta T(\alpha_s - \alpha_c) A_s E_s E_c}{A_s E_s + A_c E_c} \tag{v}$$

Also $\alpha_c > \sigma_s$ so that σ_c is negative and therefore compressive. Now substituting for σ_c in Eq. (iv)

$$\sigma_s = -\frac{\Delta T(\alpha_s - \alpha_c) A_c E_s E_c}{A_s E_s + A_c E_c} \tag{vi}$$

which is positive and therefore tensile as would be expected by a physical appreciation of the situation.

Finally the extension of the compound bar, δ, is found by substituting for σ_c in Eq. (i) or for σ_s in Eq. (ii). Then

$$\delta = \Delta T L \left(\frac{\alpha_c A_c E_c + \alpha_s A_s E_s}{A_s E_s + A_c E_c} \right) \tag{vii}$$

1.16 Experimental measurement of surface strains

Stresses at a point on the surface of a piece of material may be determined by measuring the strains at the point, usually by electrical resistance strain gauges arranged in the form of a rosette, as shown in Fig. 1.18. Suppose that ε_{I} and ε_{II} are the principal strains at the point, then if $\varepsilon_a, \varepsilon_b$ and ε_c are the measured strains in the directions θ, $(\theta + \alpha)$, $(\theta + \alpha + \beta)$ to ε_{I} we have, from the general direct strain relationship of Eq. (1.31)

$$\varepsilon_a = \varepsilon_{\text{I}} \cos^2 \theta + \varepsilon_{\text{II}} \sin^2 \theta \tag{1.61}$$

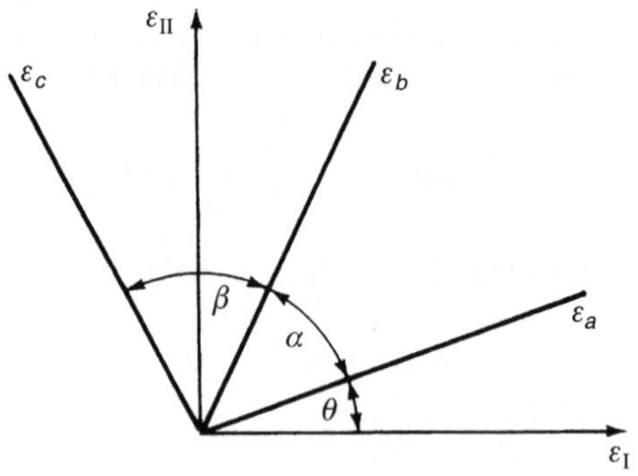

Fig. 1.18 Strain gauge rosette.

since ε_x becomes ε_I, ε_y becomes ε_{II} and γ_{xy} is zero since the x and y directions have become principal directions. Rewriting Eq. (1.61) we have

$$\varepsilon_a = \varepsilon_I\left(\frac{1+\cos 2\theta}{2}\right) + \varepsilon_{II}\left(\frac{1-\cos 2\theta}{2}\right)$$

or

$$\varepsilon_a = \tfrac{1}{2}(\varepsilon_I + \varepsilon_{II}) + \tfrac{1}{2}(\varepsilon_I - \varepsilon_{II})\cos 2\theta \tag{1.62}$$

Similarly

$$\varepsilon_b = \tfrac{1}{2}(\varepsilon_I + \varepsilon_{II}) + \tfrac{1}{2}(\varepsilon_I - \varepsilon_{II})\cos 2(\theta + \alpha) \tag{1.63}$$

and

$$\varepsilon_c = \tfrac{1}{2}(\varepsilon_I + \varepsilon_{II}) + \tfrac{1}{2}(\varepsilon_I - \varepsilon_{II})\cos 2(\theta + \alpha + \beta) \tag{1.64}$$

Therefore if ε_a, ε_b and ε_c are measured in given directions, i.e. given angles α and β, then ε_I, ε_{II} and θ are the only unknowns in Eqs (1.62)–(1.64).

The principal stresses are now obtained by substitution of ε_I and ε_{II} in Eqs (1.52). Thus

$$\varepsilon_I = \frac{1}{E}(\sigma_I - \nu\sigma_{II}) \tag{1.65}$$

and

$$\varepsilon_{II} = \frac{1}{E}(\sigma_{II} - \nu\sigma_I) \tag{1.66}$$

Solving Eqs (1.65) and (1.66) gives

$$\sigma_I = \frac{E}{1-\nu^2}(\varepsilon_I + \nu\varepsilon_{II}) \tag{1.67}$$

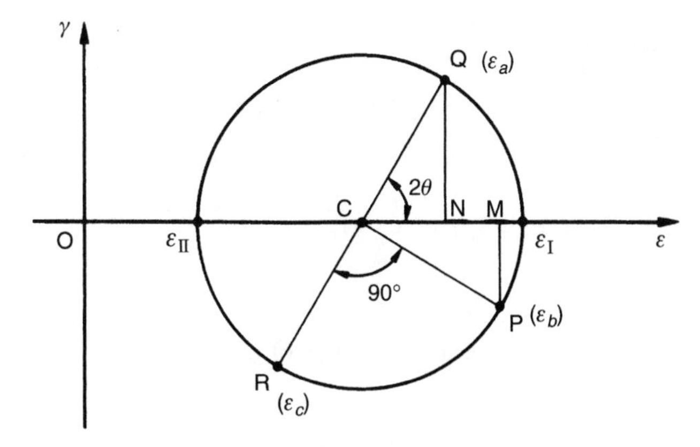

Fig. 1.19 Experimental values of principal strain using Mohr's circle.

and

$$\sigma_{II} = \frac{E}{1-\nu^2}(\varepsilon_{II} + \nu\varepsilon_I) \tag{1.68}$$

A typical rosette would have $\alpha = \beta = 45°$ in which case the principal strains are most conveniently found using the geometry of Mohr's circle of strain. Suppose that the arm a of the rosette is inclined at some unknown angle θ to the maximum principal strain as in Fig. 1.18. Then Mohr's circle of strain is as shown in Fig. 1.19; the shear strains γ_a, γ_b and γ_c do not feature in the analysis and are therefore ignored. From Fig. 1.19

$$\text{OC} = \tfrac{1}{2}(\varepsilon_a + \varepsilon_c)$$
$$\text{CN} = \varepsilon_a - \text{OC} = \tfrac{1}{2}(\varepsilon_a - \varepsilon_c)$$
$$\text{QN} = \text{CM} = \varepsilon_b - \text{OC} = \varepsilon_b - \tfrac{1}{2}(\varepsilon_a + \varepsilon_c)$$

The radius of the circle is CQ and

$$\text{CQ} = \sqrt{\text{CN}^2 + \text{QN}^2}$$

Hence

$$\text{CQ} = \sqrt{\left[\tfrac{1}{2}(\varepsilon_a - \varepsilon_c)\right]^2 + \left[\varepsilon_b - \tfrac{1}{2}(\varepsilon_a + \varepsilon_c)\right]^2}$$

which simplifies to

$$\text{CQ} = \frac{1}{\sqrt{2}}\sqrt{(\varepsilon_a - \varepsilon_b)^2 + (\varepsilon_c - \varepsilon_b)^2}$$

Therefore ε_{I}, which is given by

$$\varepsilon_{\text{I}} = \text{OC} + \text{radius of circle}$$

is

$$\varepsilon_{\text{I}} = \tfrac{1}{2}(\varepsilon_a + \varepsilon_c) + \frac{1}{\sqrt{2}}\sqrt{(\varepsilon_a - \varepsilon_b)^2 + (\varepsilon_c - \varepsilon_b)^2} \tag{1.69}$$

Also

$$\varepsilon_{\text{II}} = \text{OC} - \text{radius of circle}$$

i.e.

$$\varepsilon_{\text{II}} = \tfrac{1}{2}(\varepsilon_a + \varepsilon_c) - \frac{1}{\sqrt{2}}\sqrt{(\varepsilon_a - \varepsilon_b)^2 + (\varepsilon_c - \varepsilon_b)^2} \tag{1.70}$$

Finally the angle θ is given by

$$\tan 2\theta = \frac{\text{QN}}{\text{CN}} = \frac{\varepsilon_b - \frac{1}{2}(\varepsilon_a + \varepsilon_c)}{\frac{1}{2}(\varepsilon_a - \varepsilon_c)}$$

i.e.

$$\tan 2\theta = \frac{2\varepsilon_b - \varepsilon_a - \varepsilon_c}{\varepsilon_a - \varepsilon_c} \tag{1.71}$$

A similar approach may be adopted for a 60° rosette.

Example 1.7

A bar of solid circular cross-section has a diameter of 50 mm and carries a torque, T, together with an axial tensile load, P. A rectangular strain gauge rosette attached to the surface of the bar gave the following strain readings: $\varepsilon_a = 1000 \times 10^{-6}$, $\varepsilon_b = -200 \times 10^{-6}$ and $\varepsilon_c = -300 \times 10^{-6}$ where the gauges 'a' and 'c' are in line with, and perpendicular to, the axis of the bar, respectively. If Young's modulus, E, for the bar is 70 000 N/mm^2 and Poisson's ratio, ν, is 0.3, calculate the values of T and P.

Substituting the values of ε_a, ε_b and ε_c in Eq. (1.69)

$$\varepsilon_{\text{I}} = \frac{10^{-6}}{2}(1000 - 300) + \frac{10^{-6}}{\sqrt{2}}\sqrt{(1000 + 200)^2 + (-200 + 300)^2}$$

which gives

$$\varepsilon_{\text{I}} = 1202 \times 10^{-6}$$

Similarly, from Eq. (1.70)

$$\varepsilon_{\text{II}} = -502 \times 10^{-6}$$

Now substituting for $\varepsilon_{\rm I}$ and $\varepsilon_{\rm II}$ in Eq. (1.67)

$$\sigma_{\rm I} = \frac{70\,000 \times 10^{-6}}{1-(0.3)^2}(-502 + 0.3 \times 1202) = -80.9\,\text{N/mm}^2$$

Similarly, from Eq. (1.68)

$$\sigma_{\rm II} = -10.9\,\text{N/mm}^2$$

Since $\sigma_y = 0$, Eqs (1.11) and (1.12) reduce to

$$\sigma_{\rm I} = \frac{\sigma_x}{2} + \frac{1}{2}\sqrt{\sigma_x^2 + 4\tau_{xy}^2} \tag{i}$$

and

$$\sigma_{\rm II} = \frac{\sigma_x}{2} - \frac{1}{2}\sqrt{\sigma_x^2 + 4\tau_{xy}^2} \tag{ii}$$

respectively. Adding Eqs (i) and (ii) we obtain

$$\sigma_{\rm I} + \sigma_{\rm II} = \sigma_x$$

Thus

$$\sigma_x = 80.9 - 10.9 = 70\,\text{N/mm}^2$$

For an axial load P

$$\sigma_x = 70\,\text{N/mm}^2 = \frac{P}{A} = \frac{P}{\pi \times 50^2/4}$$

whence

$$P = 137.4\,\text{kN}$$

Substituting for σ_x in either of Eq. (i) or (ii) gives

$$\tau_{xy} = 29.7\,\text{N/mm}^2$$

From the theory of the torsion of circular section bars (see Eq. (iv) in Example 3.1)

$$\tau_{xy} = 29.7\,\text{N/mm}^2 = \frac{Tr}{J} = \frac{T \times 25}{\pi \times 50^4/32}$$

from which

$$T = 0.7\,\text{kN m}$$

Note that P could have been found directly in this particular case from the axial strain. Thus, from the first of Eqs (1.52)

$$\sigma_x = E\varepsilon_a = 70\,000 \times 1000 \times 10^{-6} = 70\,\text{N/mm}^2$$

as before.

References

1 Timoshenko, S. and Goodier, J. N., *Theory of Elasticity*, 2nd edition, McGraw-Hill Book Company, New York, 1951.
2 Wang, C. T., *Applied Elasticity*, McGraw-Hill Book Company, New York, 1953.
3 Megson, T. H. G., *Structural and Stress Analysis*, 2nd edition, Elsevier, 2005.

Problems

P.1.1 A structural member supports loads which produce, at a particular point, a direct tensile stress of 80 N/mm^2 and a shear stress of 45 N/mm^2 on the same plane. Calculate the values and directions of the principal stresses at the point and also the maximum shear stress, stating on which planes this will act.

Ans. $\sigma_{\text{I}} = 100.2\,\text{N/mm}^2 \quad \theta = 24°11'$
$\sigma_{\text{II}} = -20.2\,\text{N/mm}^2 \quad \theta = 114°11'$
$\tau_{\text{max}} = 60.2\,\text{N/mm}^2$ at 45° to principal planes.

P.1.2 At a point in an elastic material there are two mutually perpendicular planes, one of which carries a direct tensile stress at 50 N/mm^2 and a shear stress of 40 N/mm^2, while the other plane is subjected to a direct compressive stress of 35 N/mm^2 and a complementary shear stress of 40 N/mm^2. Determine the principal stresses at the point, the position of the planes on which they act and the position of the planes on which there is no normal stress.

Ans. $\sigma_{\text{I}} = 65.9\,\text{N/mm}^2 \quad \theta = 21°38'$
$\sigma_{\text{II}} = -50.9\,\text{N/mm}^2 \quad \theta = 111°38'$

No normal stress on planes at 70°21′ and −27°5′ to vertical.

P.1.3 Listed below are varying combinations of stresses acting at a point and referred to axes x and y in an elastic material. Using Mohr's circle of stress determine the principal stresses at the point and their directions for each combination.

	σ_x (N/mm^2)	σ_y (N/mm^2)	τ_{xy} (N/mm^2)
(i)	+54	+30	+5
(ii)	+30	+54	−5
(iii)	−60	−36	+5
(iv)	+30	−50	+30

Ans. (i) $\sigma_{\text{I}} = +55\,\text{N/mm}^2$ $\sigma_{\text{II}} = +29\,\text{N/mm}^2$ σ_{I} at 11.5° to x axis.
(ii) $\sigma_{\text{I}} = +55\,\text{N/mm}^2$ $\sigma_{\text{II}} = +29\,\text{N/mm}^2$ σ_{II} at 11.5° to x axis.
(iii) $\sigma_{\text{I}} = -34.5\,\text{N/mm}^2$ $\sigma_{\text{II}} = -61\,\text{N/mm}^2$ σ_{I} at 79.5° to x axis.
(iv) $\sigma_{\text{I}} = +40\,\text{N/mm}^2$ $\sigma_{\text{II}} = -60\,\text{N/mm}^2$ σ_{I} at 18.5° to x axis.

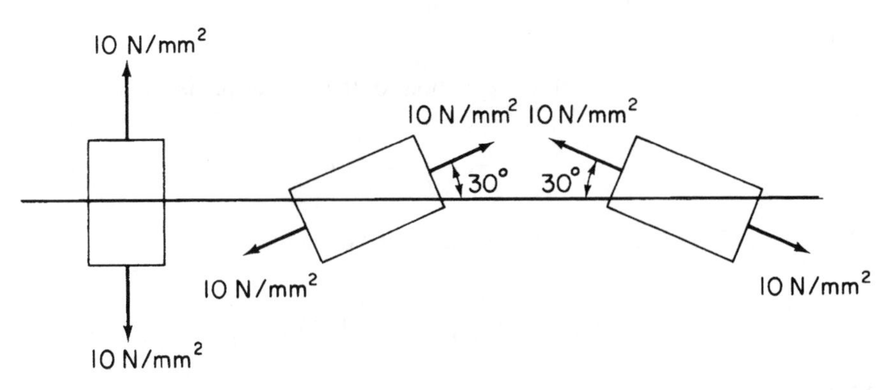

Fig. P.1.4

P.1.4 The state of stress at a point is caused by three separate actions, each of which produces a pure, unidirectional tension of $10\,\text{N/mm}^2$ individually but in three different directions as shown in Fig. P.1.4. By transforming the individual stresses to a common set of axes (x, y) determine the principal stresses at the point and their directions.

Ans. $\sigma_{\text{I}} = \sigma_{\text{II}} = 15\,\text{N/mm}^2$. All directions are principal directions.

P.1.5 A shear stress τ_{xy} acts in a two-dimensional field in which the maximum allowable shear stress is denoted by τ_{max} and the major principal stress by σ_{I}.

Derive, using the geometry of Mohr's circle of stress, expressions for the maximum values of direct stress which may be applied to the x and y planes in terms of the three parameters given above.

Ans.
$$\sigma_x = \sigma_{\text{I}} - \tau_{\text{max}} + \sqrt{\tau_{\text{max}}^2 - \tau_{xy}^2}$$
$$\sigma_y = \sigma_{\text{I}} - \tau_{\text{max}} - \sqrt{\tau_{\text{max}}^2 - \tau_{xy}^2}.$$

P.1.6 A solid shaft of circular cross-section supports a torque of 50 kNm and a bending moment of 25 kNm. If the diameter of the shaft is 150 mm calculate the values of the principal stresses and their directions at a point on the surface of the shaft.

Ans. $\sigma_{\text{I}} = 121.4\,\text{N/mm}^2$ $\theta = 31°43'$
$\sigma_{\text{II}} = -46.4\,\text{N/mm}^2$ $\theta = 121°43'$.

P.1.7 An element of an elastic body is subjected to a three-dimensional stress system σ_x, σ_y and σ_z. Show that if the direct strains in the directions x, y and z are $\varepsilon_x, \varepsilon_y$ and ε_z then

$$\sigma_x = \lambda e + 2G\varepsilon_x \quad \sigma_y = \lambda e + 2G\varepsilon_y \quad \sigma_z = \lambda e + 2G\varepsilon_z$$

where

$$\lambda = \frac{\nu E}{(1+\nu)(1-2\nu)} \quad \text{and} \quad e = \varepsilon_x + \varepsilon_y + \varepsilon_z$$

the volumetric strain.

P.1.8 Show that the compatibility equation for the case of plane strain, viz.

$$\frac{\partial^2 \gamma_{xy}}{\partial x\, \partial y} = \frac{\partial^2 \varepsilon_y}{\partial x^2} + \frac{\partial^2 \varepsilon_x}{\partial y^2}$$

may be expressed in terms of direct stresses σ_x and σ_y in the form

$$\left(\frac{\partial^2}{\partial x^2} + \frac{\partial^2}{\partial y^2}\right)(\sigma_x + \sigma_y) = 0$$

P.1.9 A bar of mild steel has a diameter of 75 mm and is placed inside a hollow aluminium cylinder of internal diameter 75 mm and external diameter 100 mm; both bar and cylinder are the same length. The resulting composite bar is subjected to an axial compressive load of 1000 kN. If the bar and cylinder contract by the same amount calculate the stress in each.

The temperature of the compressed composite bar is then reduced by 150°C but no change in length is permitted. Calculate the final stress in the bar and in the cylinder if E (steel) = 200 000 N/mm^2, E (aluminium) = 80 000 N/mm^2, α (steel) = 0.000012/°C and α (aluminium) = 0.000005/°C.

Ans. Due to load: σ (steel) = 172.6 N/mm^2 (compression)
σ (aluminium) = 69.1 N/mm^2 (compression).
Final stress: σ (steel) = 187.4 N/mm^2 (tension)
σ (aluminium) = 9.1 N/mm^2 (compression).

P.1.10 In Fig. P.1.10 the direct strains in the directions a, b, c are −0.002, −0.002 and +0.002, respectively. If I and II denote principal directions find ε_{I}, ε_{II} and θ.

Ans. $\varepsilon_{\text{I}} = +0.00283$ $\quad \varepsilon_{\text{II}} = -0.00283$ $\quad \theta = -22.5°$ or $+67.5°$.

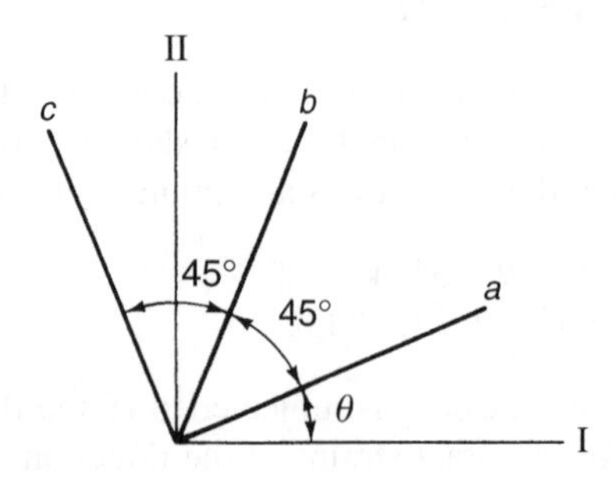

Fig. P.1.10

P.1.11 The simply supported rectangular beam shown in Fig. P.1.11 is subjected to two symmetrically placed transverse loads each of magnitude Q. A rectangular strain gauge rosette located at a point P on the centroidal axis on one vertical face of the beam gave strain readings as follows: $\varepsilon_a = -222 \times 10^{-6}$, $\varepsilon_b = -213 \times 10^{-6}$ and $\varepsilon_c = +45 \times 10^{-6}$. The longitudinal stress σ_x at the point P due to an external compressive force is 7 N/mm^2. Calculate the shear stress τ at the point P in the vertical plane and hence the transverse load Q:

$$(Q = 2bd\tau/3 \quad \text{where } b = \text{breadth}, d = \text{depth of beam})$$

$$E = 31\,000\,\text{N/mm}^2 \quad \nu = 0.2$$

Ans. $\tau = 3.17\,\text{N/mm}^2$ $Q = 95.1\,\text{kN}$.

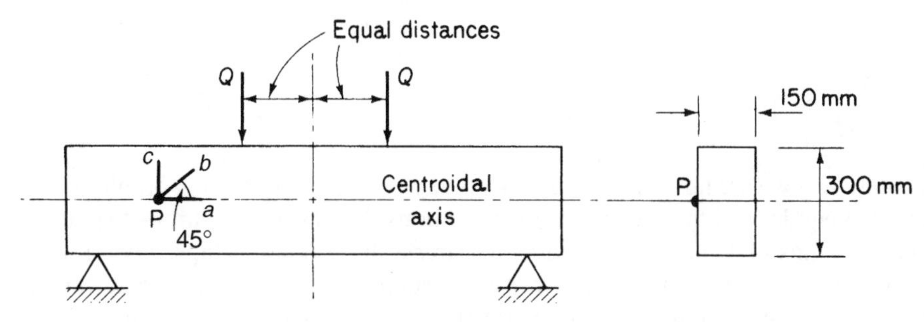

Fig. P.1.11

2

Two-dimensional problems in elasticity

Theoretically we are now in a position to solve any three-dimensional problem in elasticity having derived three equilibrium conditions, Eqs (1.5), six strain–displacement equations, Eqs (1.18) and (1.20), and six stress–strain relationships, Eqs (1.42) and (1.46). These equations are sufficient, when supplemented by appropriate boundary conditions, to obtain unique solutions for the six stress, six strain and three displacement functions. It is found, however, that exact solutions are obtainable only for some simple problems. For bodies of arbitrary shape and loading, approximate solutions may be found by numerical methods (e.g. finite differences) or by the Rayleigh–Ritz method based on energy principles (Chapter 7).

Two approaches are possible in the solution of elasticity problems. We may solve initially either for the three unknown displacements or for the six unknown stresses. In the former method the equilibrium equations are written in terms of strain by expressing the six stresses as functions of strain (see Problem P.1.7). The strain–displacement relationships are then used to form three equations involving the three displacements u, v and w. The boundary conditions for this method of solution must be specified as displacements. Determination of u, v and w enables the six strains to be computed from Eqs (1.18) and (1.20); the six unknown stresses follow from the equations expressing stress as functions of strain. It should be noted here that no use has been made of the compatibility equations. The fact that u, v and w are determined directly ensures that they are single-valued functions, thereby satisfying the requirement of compatibility.

In most structural problems the object is usually to find the distribution of stress in an elastic body produced by an external loading system. It is therefore more convenient in this case to determine the six stresses before calculating any required strains or displacements. This is accomplished by using Eqs (1.42) and (1.46) to rewrite the six equations of compatibility in terms of stress. The resulting equations, in turn, are simplified by making use of the stress relationships developed in the equations of equilibrium. The solution of these equations automatically satisfies the conditions of compatibility and equilibrium throughout the body.

2.1 Two-dimensional problems

For the reasons discussed in Chapter 1 we shall confine our actual analysis to the two-dimensional cases of plane stress and plane strain. The appropriate equilibrium conditions for plane stress are given by Eqs (1.6), viz.

$$\frac{\partial \sigma_x}{\partial x} + \frac{\partial \tau_{xy}}{\partial y} + X = 0$$

$$\frac{\partial \sigma_y}{\partial y} + \frac{\partial \tau_{yx}}{\partial y} + Y = 0$$

and the required stress–strain relationships obtained from Eqs (1.47), namely

$$\varepsilon_x = \frac{1}{E}(\sigma_x - \nu\sigma_y)$$

$$\varepsilon_y = \frac{1}{E}(\sigma_y - \nu\sigma_x)$$

$$\gamma_{xy} = \frac{2(1+\nu)}{E}\tau_{xy}$$

We find that although ε_z exists, Eqs (1.22)–(1.26) are identically satisfied leaving Eq. (1.21) as the required compatibility condition. Substitution in Eq. (1.21) of the above strains gives

$$2(1+\nu)\frac{\partial^2 \tau_{xy}}{\partial x\,\partial y} = \frac{\partial^2}{\partial x^2}(\sigma_y - \nu\sigma_x) + \frac{\partial^2}{\partial y^2}(\sigma_x - \nu\sigma_y) \tag{2.1}$$

From Eqs (1.6)

$$\frac{\partial^2 \tau_{xy}}{\partial y\,\partial x} = -\frac{\partial^2 \sigma_x}{\partial x^2} - \frac{\partial X}{\partial x} \tag{2.2}$$

and

$$\frac{\partial^2 \tau_{xy}}{\partial x\,\partial y} = -\frac{\partial^2 \sigma_y}{\partial y^2} - \frac{\partial Y}{\partial y} \quad (\tau_{yx} = \tau_{xy}) \tag{2.3}$$

Adding Eqs (2.2) and (2.3), then substituting in Eq. (2.1) for $2\partial^2\tau_{xy}/\partial x\partial y$, we have

$$-(1+\nu)\left(\frac{\partial X}{\partial x} + \frac{\partial Y}{\partial y}\right) = \frac{\partial^2 \sigma_x}{\partial x^2} + \frac{\partial^2 \sigma_y}{\partial y^2} + \frac{\partial^2 \sigma_y}{\partial x^2} + \frac{\partial^2 \sigma_x}{\partial y^2}$$

or

$$\left(\frac{\partial^2}{\partial x^2} + \frac{\partial^2}{\partial y^2}\right)(\sigma_x + \sigma_y) = -(1+\nu)\left(\frac{\partial X}{\partial x} + \frac{\partial Y}{\partial y}\right) \tag{2.4}$$

The alternative two-dimensional problem of plane strain may also be formulated in the same manner. We have seen in Section 1.11 that the six equations of compatibility reduce to the single equation (1.21) for the plane strain condition. Further, from the third of Eqs (1.42)

$$\sigma_z = \nu(\sigma_x + \sigma_y) \quad (\text{since } \varepsilon_z = 0 \text{ for plane strain})$$

so that

$$\varepsilon_x = \frac{1}{E}[(1 - \nu^2)\sigma_x - \nu(1 + \nu)\sigma_y]$$

and

$$\varepsilon_y = \frac{1}{E}[(1 - \nu^2)\sigma_y - \nu(1 + \nu)\sigma_x]$$

Also

$$\gamma_{xy} = \frac{2(1 + \nu)}{E}\tau_{xy}$$

Substituting as before in Eq. (1.21) and simplifying by use of the equations of equilibrium we have the compatibility equation for plane strain

$$\left(\frac{\partial^2}{\partial x^2} + \frac{\partial^2}{\partial y^2}\right)(\sigma_x + \sigma_y) = -\frac{1}{1 - \nu}\left(\frac{\partial X}{\partial x} + \frac{\partial Y}{\partial y}\right) \tag{2.5}$$

The two equations of equilibrium together with the boundary conditions, from Eqs (1.7), and one of the compatibility equations (2.4) or (2.5) are generally sufficient for the determination of the stress distribution in a two-dimensional problem.

2.2 Stress functions

The solution of problems in elasticity presents difficulties but the procedure may be simplified by the introduction of a *stress function*. For a particular two-dimensional case the stresses are related to a single function of x and y such that substitution for the stresses in terms of this function automatically satisfies the equations of equilibrium no matter what form the function may take. However, a large proportion of the infinite number of functions which fulfil this condition are eliminated by the requirement that the form of the stress function must also satisfy the two-dimensional equations of compatibility, (2.4) and (2.5), plus the appropriate boundary conditions.

For simplicity let us consider the two-dimensional case for which the body forces are zero. The problem is now to determine a stress–stress function relationship which satisfies the equilibrium conditions of

$$\left.\begin{aligned} \frac{\partial \sigma_x}{\partial x} + \frac{\partial \tau_{xy}}{\partial y} &= 0 \\ \frac{\partial \sigma_y}{\partial y} + \frac{\partial \tau_{yx}}{\partial x} &= 0 \end{aligned}\right\} \tag{2.6}$$

and a form for the stress function giving stresses which satisfy the compatibility equation

$$\left(\frac{\partial^2}{\partial x^2} + \frac{\partial^2}{\partial y^2}\right)(\sigma_x + \sigma_y) = 0 \tag{2.7}$$

The English mathematician Airy proposed a stress function ϕ defined by the equations

$$\sigma_x = \frac{\partial^2 \phi}{\partial y^2} \quad \sigma_y = \frac{\partial^2 \phi}{\partial x^2} \quad \tau_{xy} = -\frac{\partial^2 \phi}{\partial x \, \partial y} \tag{2.8}$$

Clearly, substitution of Eqs (2.8) into Eqs (2.6) verifies that the equations of equilibrium are satisfied by this particular stress–stress function relationship. Further substitution into Eq. (2.7) restricts the possible forms of the stress function to those satisfying the *biharmonic equation*

$$\frac{\partial^4 \phi}{\partial x^4} + 2\frac{\partial^4 \phi}{\partial x^2 \partial y^2} + \frac{\partial^4 \phi}{\partial y^4} = 0 \tag{2.9}$$

The final form of the stress function is then determined by the boundary conditions relating to the actual problem. Therefore, a two-dimensional problem in elasticity with zero body forces reduces to the determination of a function ϕ of x and y, which satisfies Eq. (2.9) at all points in the body and Eqs (1.7) reduced to two dimensions at all points on the boundary of the body.

2.3 Inverse and semi-inverse methods

The task of finding a stress function satisfying the above conditions is extremely difficult in the majority of elasticity problems although some important classical solutions have been obtained in this way. An alternative approach, known as the *inverse method*, is to specify a form of the function ϕ satisfying Eq. (2.9), assume an arbitrary boundary and then determine the loading conditions which fit the assumed stress function and chosen boundary. Obvious solutions arise in which ϕ is expressed as a polynomial. Timoshenko and Goodier[1] consider a variety of polynomials for ϕ and determine the associated loading conditions for a variety of rectangular sheets. Some of these cases are quoted here.

Example 2.1

Consider the stress function

$$\phi = Ax^2 + Bxy + Cy^2$$

where A, B and C are constants. Equation (2.9) is identically satisfied since each term becomes zero on substituting for ϕ. The stresses follow from

$$\sigma_x = \frac{\partial^2 \phi}{\partial y^2} = 2C$$

$$\sigma_y = \frac{\partial^2 \phi}{\partial x^2} = 2A$$

$$\tau_{xy} = -\frac{\partial^2 \phi}{\partial x \, \partial y} = -B$$

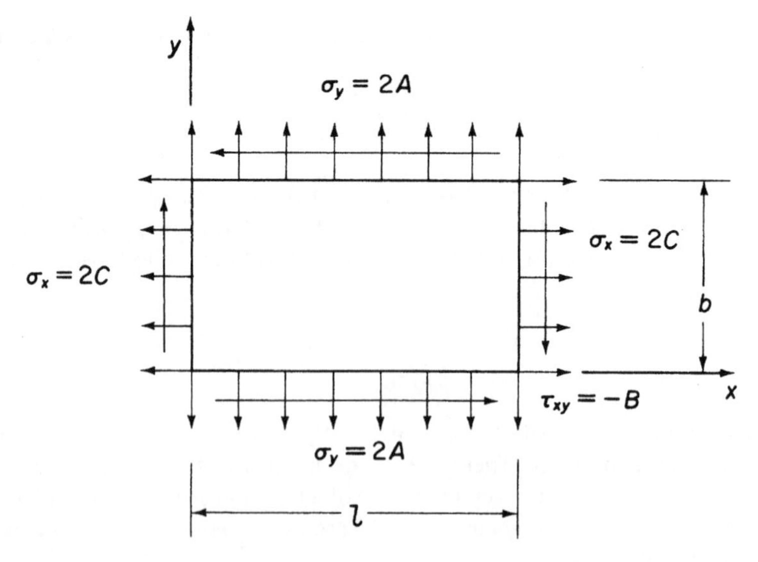

Fig. 2.1 Required loading conditions on rectangular sheet in Example 2.1.

To produce these stresses at any point in a rectangular sheet we require loading conditions providing the boundary stresses shown in Fig. 2.1.

Example 2.2

A more complex polynomial for the stress function is

$$\phi = \frac{Ax^3}{6} + \frac{Bx^2y}{2} + \frac{Cxy^2}{2} + \frac{Dy^3}{6}$$

As before

$$\frac{\partial^4\phi}{\partial x^4} = \frac{\partial^4\phi}{\partial x^2\,\partial y^2} = \frac{\partial^4\phi}{\partial y^4} = 0$$

so that the compatibility equation (2.9) is identically satisfied. The stresses are given by

$$\sigma_x = \frac{\partial^2\phi}{\partial y^2} = Cx + Dy$$

$$\sigma_y = \frac{\partial^2\phi}{\partial x^2} = Ax + By$$

$$\tau_{xy} = -\frac{\partial^2\phi}{\partial x\,\partial y} = -Bx - Cy$$

We may choose any number of values of the coefficients A, B, C and D to produce a variety of loading conditions on a rectangular plate. For example, if we assume $A = B = C = 0$ then $\sigma_x = Dy$, $\sigma_y = 0$ and $\tau_{xy} = 0$, so that for axes referred to an origin at

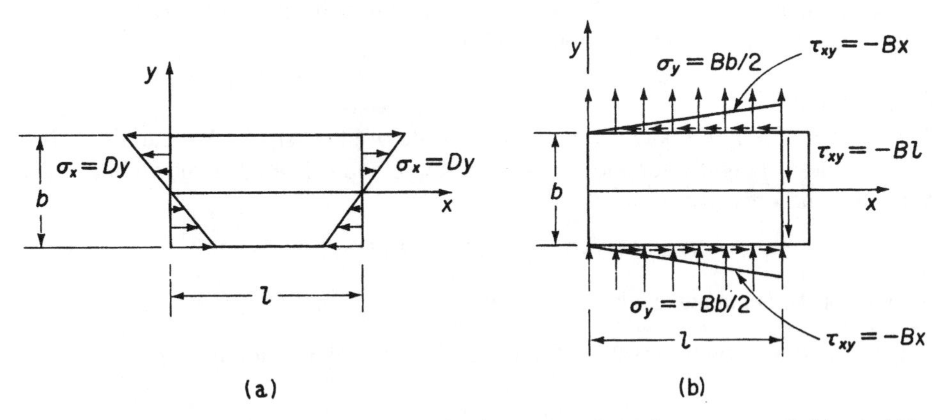

Fig. 2.2 (a) Required loading conditions on rectangular sheet in Example 2.2 for $A = B = C = 0$; (b) as in (a) but $A = C = D = 0$.

the mid-point of a vertical side of the plate we obtain the state of pure bending shown in Fig. 2.2(a). Alternatively, Fig. 2.2(b) shows the loading conditions corresponding to $A = C = D = 0$ in which $\sigma_x = 0$, $\sigma_y = By$ and $\tau_{xy} = -Bx$.

By assuming polynomials of the second or third degree for the stress function we ensure that the compatibility equation is identically satisfied whatever the values of the coefficients. For polynomials of higher degrees, compatibility is satisfied only if the coefficients are related in a certain way. For example, for a stress function in the form of a polynomial of the fourth degree

$$\phi = \frac{Ax^4}{12} + \frac{Bx^3y}{6} + \frac{Cx^2y^2}{2} + \frac{Dxy^3}{6} + \frac{Ey^4}{12}$$

and

$$\frac{\partial^4 \phi}{\partial x^4} = 2A \quad 2\frac{\partial^4 \phi}{\partial x^2 \partial y^2} = 4C \quad \frac{\partial^4 \phi}{\partial y^4} = 2E$$

Substituting these values in Eq. (2.9) we have

$$E = -(2C + A)$$

The stress components are then

$$\sigma_x = \frac{\partial^2 \phi}{\partial y^2} = Cx^2 + Dxy - (2C + A)y^2$$

$$\sigma_y = \frac{\partial^2 \phi}{\partial x^2} = Ax^2 + Bxy + Cy^2$$

$$\tau_{xy} = -\frac{\partial^2 \phi}{\partial x\, \partial y} = -\frac{Bx^2}{2} - 2Cxy - \frac{Dy^2}{2}$$

The coefficients A, B, C and D are arbitrary and may be chosen to produce various loading conditions as in the previous examples.

Example 2.3

A cantilever of length L and depth $2h$ is in a state of plane stress. The cantilever is of unit thickness, is rigidly supported at the end $x = L$ and is loaded as shown in Fig. 2.3. Show that the stress function

$$\phi = Ax^2 + Bx^2y + Cy^3 + D(5x^2y^3 - y^5)$$

is valid for the beam and evaluate the constants A, B, C and D.

The stress function must satisfy Eq. (2.9). From the expression for ϕ

$$\frac{\partial \phi}{\partial x} = 2Ax + 2Bxy + 10Dxy^3$$

$$\frac{\partial^2 \phi}{\partial x^2} = 2A + 2By + 10Dy^3 = \sigma_y \tag{i}$$

Also

$$\frac{\partial \phi}{\partial y} = Bx^2 + 3Cy^2 + 15Dx^2y^2 - 5Dy^4$$

$$\frac{\partial^2 \phi}{\partial y^2} = 6Cy + 30Dx^2y - 20Dy^3 = \sigma_x \tag{ii}$$

and

$$\frac{\partial^2 \phi}{\partial x\, \partial y} = 2Bx + 30Dxy^2 = -\tau_{xy} \tag{iii}$$

Further

$$\frac{\partial^4 \phi}{\partial x^4} = 0 \qquad \frac{\partial^4 \phi}{\partial y^4} = -120Dy \qquad \frac{\partial^4 \phi}{\partial x^2\, \partial y^2} = 60\,Dy$$

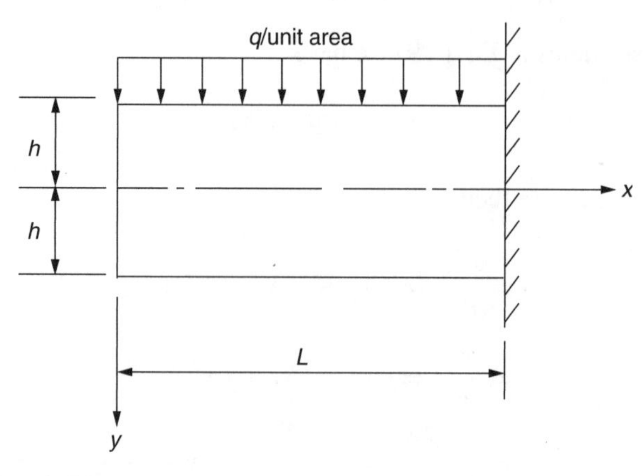

Fig. 2.3 Beam of Example 2.3.

Substituting in Eq. (2.9) gives

$$\frac{\partial^4\phi}{\partial x^4} + 2\frac{\partial^4\phi}{\partial x^2\partial y^2} + \frac{\partial^4\phi}{\partial y^4} = 2 \times 60Dy - 120Dy = 0$$

The biharmonic equation is therefore satisfied and the stress function is valid.

From Fig. 2.3, $\sigma_y = 0$ at $y = h$ so that, from Eq. (i)

$$2A + 2BH + 10Dh^3 = 0 \tag{iv}$$

Also from Fig. 2.3, $\sigma_y = -q$ at $y = -h$ so that, from Eq. (i)

$$2A - 2BH - 10Dh^3 = -q \tag{v}$$

Again from Fig. 2.3, $\tau_{xy} = 0$ at $y = \pm h$ giving, from Eq. (iii)

$$2Bx + 30Dxh^2 = 0$$

so that

$$2B + 30Dh^2 = 0 \tag{vi}$$

At $x = 0$ there is no resultant moment applied to the beam, i.e.

$$M_{x=0} = \int_{-h}^{h} \sigma_x y \, dy = \int_{-h}^{h} (6Cy^2 - 20Dy^4) \, dy = 0$$

i.e.

$$M_{x=0} = [2Cy^3 - 4Dy^5]_{-h}^{h} = 0$$

or

$$C - 2Dh^2 = 0 \tag{vii}$$

Subtracting Eq. (v) from (iv)

$$4Bh + 20Dh^3 = q$$

or

$$B + 5Dh^2 = \frac{q}{4h} \tag{viii}$$

From Eq. (vi)

$$B + 15Dh^2 = 0 \tag{ix}$$

so that, subtracting Eq. (viii) from Eq. (ix)

$$D = -\frac{q}{40h^3}$$

Then

$$B = \frac{3q}{8h} \quad A = -\frac{q}{4} \quad C = -\frac{q}{20h}$$

and

$$\phi = \frac{q}{40h^3}[-10h^3x^2 + 15h^2x^2y - 2h^2y^3 - (5x^2y^3 - y^5)]$$

The obvious disadvantage of the inverse method is that we are determining problems to fit assumed solutions, whereas in structural analysis the reverse is the case. However, in some problems the shape of the body and the applied loading allow simplifying assumptions to be made, thereby enabling a solution to be obtained. St. Venant suggested a *semi-inverse method* for the solution of this type of problem in which assumptions are made as to stress or displacement components. These assumptions may be based on experimental evidence or intuition. St. Venant first applied the method to the torsion of solid sections (Chapter 3) and to the problem of a beam supporting shear loads (Section 2.6).

2.4 St. Venant's principle

In the examples of Section 2.3 we have seen that a particular stress function form may be applicable to a variety of problems. Different problems are deduced from a given stress function by specifying, in the first instance, the shape of the body and then assigning a variety of values to the coefficients. The resulting stress functions give stresses which satisfy the equations of equilibrium and compatibility *at all points within and on the boundary of the body*. It follows that the applied loads must be distributed around the boundary of the body in the same manner as the internal stresses at the boundary. In the case of pure bending for example (Fig. 2.2(a)), the applied bending moment must be produced by tensile and compressive forces on the ends of the plate, their magnitudes being dependent on their distance from the neutral axis. If this condition is invalidated by the application of loads in an arbitrary fashion or by preventing the free distortion of any section of the body then the solution of the problem is no longer exact. As this is the

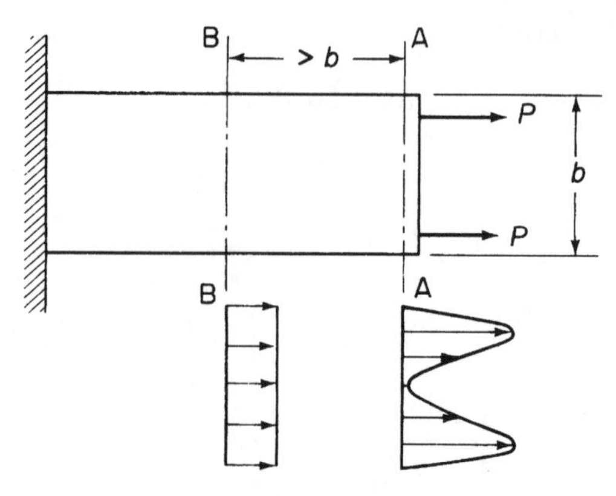

Fig. 2.4 Stress distributions illustrating St. Venant's principle.

case in practically every structural problem it would appear that the usefulness of the theory is strictly limited. To surmount this obstacle we turn to the important *principle of St. Venant* which may be summarized as stating:

> *that while statically equivalent systems of forces acting on a body produce substantially different local effects the stresses at sections distant from the surface of loading are essentially the same.*

Therefore at a section AA close to the end of a beam supporting two point loads P the stress distribution varies as shown in Fig. 2.4, whilst at the section BB, a distance usually taken to be greater than the dimension of the surface to which the load is applied, the stress distribution is uniform.

We may therefore apply the theory to sections of bodies away from points of applied loading or constraint. The determination of stresses in these regions requires, for some problems, separate calculation (see Chapters 26 and 27).

2.5 Displacements

Having found the components of stress, Eqs (1.47) (for the case of plane stress) are used to determine the components of strain. The displacements follow from Eqs (1.27) and (1.28). The integration of Eqs (1.27) yields solutions of the form

$$u = \varepsilon_x x + a - by \tag{2.10}$$

$$v = \varepsilon_y y + c + bx \tag{2.11}$$

in which a, b and c are constants representing movement of the body as a whole or *rigid body* displacements. Of these a and c represent pure translatory motions of the body while b is a small angular rotation of the body in the xy plane. If we assume that b is positive in an anticlockwise sense then in Fig. 2.5 the displacement v' due to the rotation is given by

$$\begin{aligned} v' &= \mathrm{P'Q'} - \mathrm{PQ} \\ &= \mathrm{OP}\sin(\theta + b) - \mathrm{OP}\sin\theta \end{aligned}$$

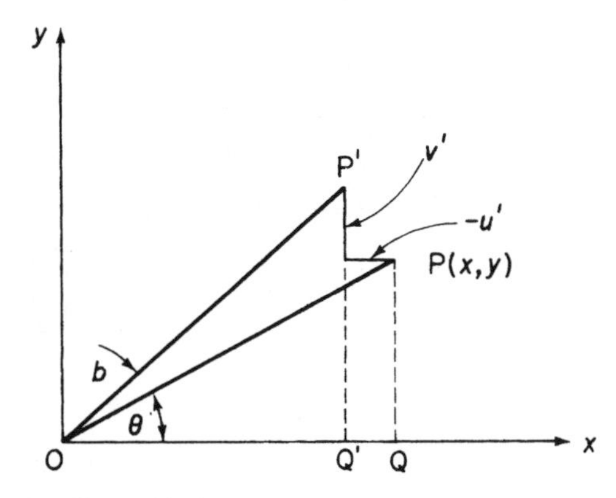

Fig. 2.5 Displacements produced by rigid body rotation.

which, since b is a small angle, reduces to

$$v' = bx$$

Similarly

$$u' = -by \text{ as stated}$$

2.6 Bending of an end-loaded cantilever

In his semi-inverse solution of this problem St. Venant based his choice of stress function on the reasonable assumptions that the direct stress is directly proportional to bending moment (and therefore distance from the free end) and height above the neutral axis. The portion of the stress function giving shear stress follows from the equilibrium condition relating σ_x and τ_{xy}. The appropriate stress function for the cantilever beam shown in Fig. 2.6 is then

$$\phi = Axy + \frac{Bxy^3}{6} \tag{i}$$

where A and B are unknown constants. Hence

$$\left.\begin{aligned} \sigma_x &= \frac{\partial^2\phi}{\partial y^2} = Bxy \\ \sigma_y &= \frac{\partial^2\phi}{\partial x^2} = 0 \\ \tau_{xy} &= -\frac{\partial^2\phi}{\partial x\,\partial y} = -A - \frac{By^2}{2} \end{aligned}\right\} \tag{ii}$$

Substitution for ϕ in the biharmonic equation shows that the form of the stress function satisfies compatibility for all values of the constants A and B. The actual values of A and B are chosen to satisfy the boundary condition, viz. $\tau_{xy} = 0$ along the upper and lower edges of the beam, and the resultant shear load over the free end is equal to P.

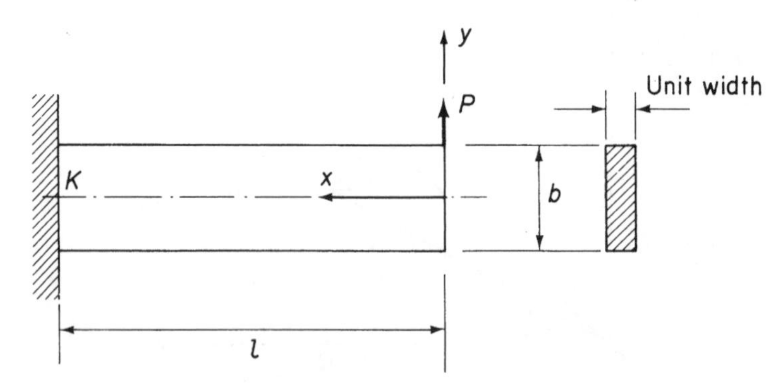

Fig. 2.6 Bending of an end-loaded cantilever.

From the first of these

$$\tau_{xy} = -A - \frac{By^2}{2} = 0 \quad \text{at } y = \pm\frac{b}{2}$$

giving

$$A = -\frac{Bb^2}{8}$$

From the second

$$-\int_{-b/2}^{b/2} \tau_{xy}\, \mathrm{d}y = P \quad (\text{see sign convention for } \tau_{xy})$$

or

$$-\int_{-b/2}^{b/2} \left(\frac{Bb^2}{8} - \frac{By^2}{2}\right) \mathrm{d}y = P$$

from which

$$B = -\frac{12P}{b^3}$$

The stresses follow from Eqs (ii)

$$\left.\begin{aligned} \sigma_x &= -\frac{12Pxy}{b^3} = -\frac{Px}{I}y \\ \sigma_y &= 0 \\ \tau_{xy} &= -\frac{12P}{8b^3}(b^2 - 4y^2) = -\frac{P}{8I}(b^2 - 4y^2) \end{aligned}\right\} \qquad \text{(iii)}$$

where $I = b^3/12$ the second moment of area of the beam cross-section.

We note from the discussion of Section 2.4 that Eq. (iii) represent an exact solution subject to the following conditions that:

(1) the shear force P is distributed over the free end in the same manner as the shear stress τ_{xy} given by Eqs (iii);
(2) the distribution of shear and direct stresses at the built-in end is the same as those given by Eqs (iii);
(3) all sections of the beam, including the built-in end, are free to distort.

In practical cases none of these conditions is satisfied, but by virtue of St. Venant's principle we may assume that the solution is exact for regions of the beam away from the built-in end and the applied load. For many solid sections the inaccuracies in these regions are small. However, for thin-walled structures, with which we are primarily concerned, significant changes occur and we shall consider the effects of structural and loading discontinuities on this type of structure in Chapters 26 and 27.

We now proceed to determine the displacements corresponding to the stress system of Eqs (iii). Applying the strain–displacement and stress–strain relationships, Eqs (1.27),

(1.28) and (1.47), we have

$$\varepsilon_x = \frac{\partial u}{\partial x} = \frac{\sigma_x}{E} = -\frac{Pxy}{EI} \tag{iv}$$

$$\varepsilon_y = \frac{\partial v}{\partial y} = -\frac{\nu\sigma_x}{E} = \frac{\nu Pxy}{EI} \tag{v}$$

$$\gamma_{xy} = \frac{\partial u}{\partial y} + \frac{\partial v}{\partial x} = \frac{\tau_{xy}}{G} = -\frac{P}{8IG}(b^2 - 4y^2) \tag{vi}$$

Integrating Eqs (iv) and (v) and noting that ε_x and ε_y are partial derivatives of the displacements, we find

$$u = -\frac{Px^2y}{2EI} + f_1(y) \quad v = \frac{\nu Pxy^2}{2EI} + f_2x \tag{vii}$$

where $f_1(y)$ and $f_2(x)$ are unknown functions of x and y. Substituting these values of u and v in Eq. (vi)

$$-\frac{Px^2}{2EI} + \frac{\partial f_1(y)}{\partial y} + \frac{\nu Py^2}{2EI} + \frac{\partial f_2(x)}{\partial x} = -\frac{P}{8IG}(b^2 - 4y^2)$$

Separating the terms containing x and y in this equation and writing

$$F_1(x) = -\frac{Px^2}{2EI} + \frac{\partial f_2(x)}{\partial x} \quad F_2(y) = \frac{\nu Py^2}{2EI} - \frac{Py^2}{2IG} + \frac{\partial f_1(y)}{\partial y}$$

we have

$$F_1(x) + F_2(y) = -\frac{Pb^2}{8IG}$$

The term on the right-hand side of this equation is a constant which means that $F_1(x)$ and $F_2(y)$ must be constants, otherwise a variation of either x or y would destroy the equality. Denoting $F_1(x)$ by C and $F_2(y)$ by D gives

$$C + D = -\frac{Pb^2}{8IG} \tag{viii}$$

and

$$\frac{\partial f_2(x)}{\partial x} = \frac{Px^2}{2EI} + C \quad \frac{\partial f_1(y)}{\partial y} = \frac{Py^2}{2IG} - \frac{\nu Py^2}{2EI} + D$$

so that

$$f_2(x) = \frac{Px^3}{6EI} + Cx + F$$

and

$$f_1(y) = \frac{Py^3}{6IG} - \frac{\nu Py^3}{6EI} + Dy + H$$

Therefore from Eqs (vii)

$$u = -\frac{Px^2y}{2EI} - \frac{\nu Py^3}{6EI} + \frac{Py^3}{6IG} + Dy + H \tag{ix}$$

$$v = \frac{\nu Pxy^2}{2EI} + \frac{Px^3}{6EI} + Cx + F \tag{x}$$

The constants C, D, F and H are now determined from Eq. (viii) and the displacement boundary conditions imposed by the support system. Assuming that the support prevents movement of the point K in the beam cross-section at the built-in end then $u = v = 0$ at $x = l$, $y = 0$ and from Eqs (ix) and (x)

$$H = 0 \quad F = -\frac{Pl^3}{6EI} - Cl$$

If we now assume that the slope of the neutral plane is zero at the built-in end then $\partial v/\partial x = 0$ at $x = l$, $y = 0$ and from Eq. (x)

$$C = -\frac{Pl^2}{2EI}$$

It follows immediately that

$$F = \frac{Pl^3}{2EI}$$

and, from Eq. (viii)

$$D = \frac{Pl^2}{2EI} - \frac{Pb^2}{8IG}$$

Substitution for the constants C, D, F and H in Eqs (ix) and (x) now produces the equations for the components of displacement at any point in the beam. Thus

$$u = -\frac{Px^2y}{2EI} - \frac{\nu Py^3}{6EI} + \frac{Py^3}{6IG} + \left(\frac{Pl^2}{2EI} - \frac{Pb^2}{8IG}\right)y \tag{xi}$$

$$v = \frac{\nu Pxy^2}{2EI} + \frac{Px^3}{6EI} - \frac{Pl^2x}{2EI} + \frac{Pl^3}{3EI} \tag{xii}$$

The deflection curve for the neutral plane is

$$(v)_{y=0} = \frac{Px^3}{6EI} - \frac{Pl^2x}{2EI} + \frac{Pl^3}{3EI} \tag{xiii}$$

from which the tip deflection ($x = 0$) is $Pl^3/3EI$. This value is that predicted by simple beam theory (Chapter 16) and does not include the contribution to deflection of the shear strain. This was eliminated when we assumed that the slope of the neutral plane

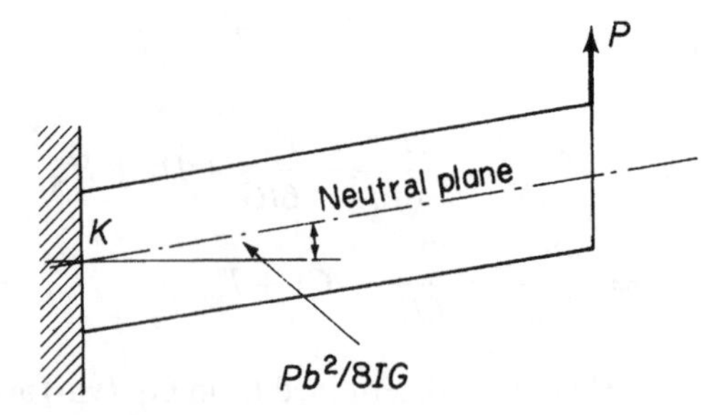

Fig. 2.7 Rotation of neutral plane due to shear in end-loaded cantilever.

at the built-in end was zero. A more detailed examination of this effect is instructive. The shear strain at any point in the beam is given by Eq. (vi)

$$\gamma_{xy} = -\frac{P}{8IG}(b^2 - 4y^2)$$

and is obviously independent of x. Therefore at all points on the neutral plane the shear strain is constant and equal to

$$\gamma_{xy} = -\frac{Pb^2}{8IG}$$

which amounts to a rotation of the neutral plane as shown in Fig. 2.7. The deflection of the neutral plane due to this shear strain at any section of the beam is therefore equal to

$$\frac{Pb^2}{8IG}(l - x)$$

and Eq. (xiii) may be rewritten to include the effect of shear as

$$(v)_{y=0} = \frac{Px^3}{6EI} - \frac{Pl^2x}{2EI} + \frac{Pl^3}{3EI} + \frac{Pb^2}{8IG}(l - x) \tag{xiv}$$

Let us now examine the distorted shape of the beam section which the analysis assumes is free to take place. At the built-in end when $x = l$ the displacement of any point is, from Eq. (xi)

$$u = \frac{\nu Py^3}{6EI} + \frac{Py^3}{6IG} - \frac{Pb^2y}{8IG} \tag{xv}$$

The cross-section would therefore, if allowed, take the shape of the shallow reversed S shown in Fig. 2.8(a). We have not included in Eq. (xv) the previously discussed effect of rotation of the neutral plane caused by shear. However, this merely rotates the beam section as indicated in Fig. 2.8(b).

The distortion of the cross-section is produced by the variation of shear stress over the depth of the beam. Thus the basic assumption of simple beam theory that plane sections remain plane is not valid when shear loads are present, although for long, slender beams bending stresses are much greater than shear stresses and the effect may be ignored.

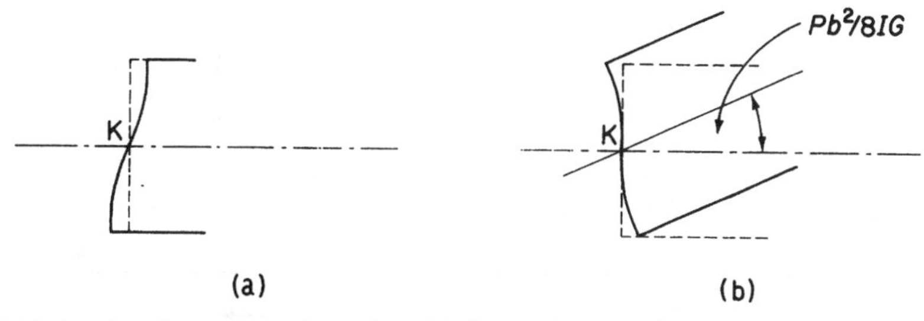

Fig. 2.8 (a) Distortion of cross-section due to shear; (b) effect on distortion of rotation due to shear.

It will be observed from Fig. 2.8 that an additional direct stress system will be imposed on the beam at the support where the section is constrained to remain plane. For most engineering structures this effect is small but, as mentioned previously, may be significant in thin-walled sections.

Reference

1 Timoshenko, S. and Goodier, J. N., *Theory of Elasticity*, 2nd edition, McGraw-Hill Book Company, New York, 1951.

Problems

P.2.1 A metal plate has rectangular axes Ox, Oy marked on its surface. The point O and the direction of Ox are fixed in space and the plate is subjected to the following uniform stresses:

compressive, $3p$, parallel to Ox,
tensile, $2p$, parallel to Oy,
shearing, $4p$, in planes parallel to Ox and Oy
in a sense tending to decrease the angle xOy.

Determine the direction in which a certain point on the plate will be displaced; the coordinates of the point are (2, 3) before straining. Poisson's ratio is 0.25.

Ans. 19.73° to Ox.

P.2.2 What do you understand by an Airy stress function in two dimensions? A beam of length l, with a thin rectangular cross-section, is built-in at the end $x = 0$ and loaded at the tip by a vertical force P (Fig. P.2.2). Show that the stress distribution, as calculated by simple beam theory, can be represented by the expression

$$\phi = Ay^3 + By^3x + Cyx$$

as an Airy stress function and determine the coefficients A, B and C.

Ans. $A = 2Pl/td^3, \quad B = -2P/td^3, \quad C = 3P/2td.$

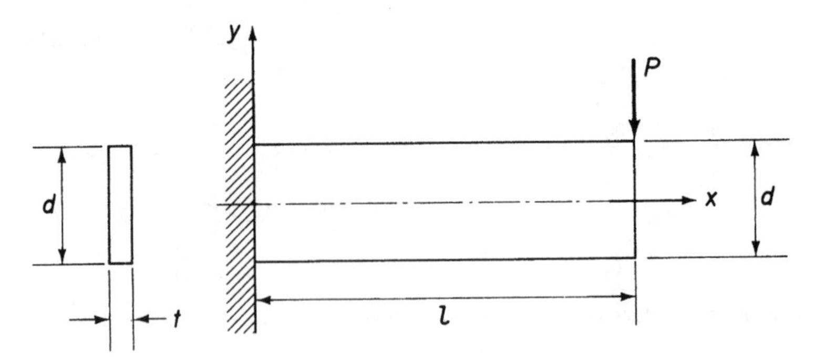

Fig. P.2.2

P.2.3 The cantilever beam shown in Fig. P.2.3 is in a state of plane strain and is rigidly supported at $x = L$. Examine the following stress function in relation to this problem:

$$\phi = \frac{w}{20h^3}(15h^2x^2y - 5x^2y^3 - 2h^2y^3 + y^5)$$

Show that the stresses acting on the boundaries satisfy the conditions except for a distributed direct stress at the free end of the beam which exerts no resultant force or bending moment.

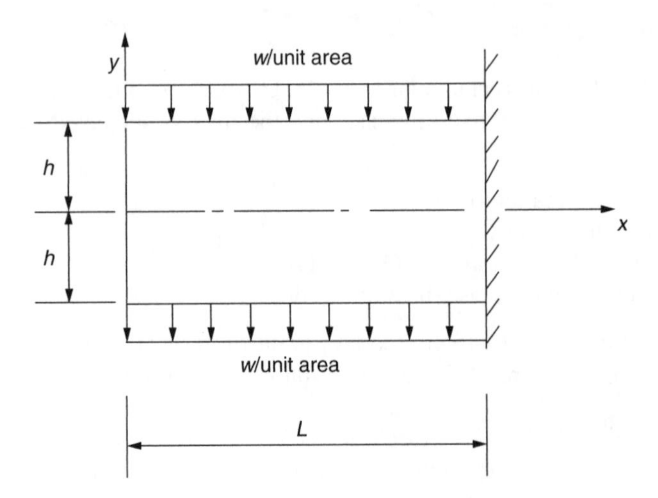

Fig. P.2.3

Ans. The stress function satisfies the biharmonic equation:

- At $y = h$, $\sigma_y = w$ and $\tau_{xy} = 0$, boundary conditions satisfied.
- At $y = -h$, $\sigma_y = -w$ and $\tau_{xy} = 0$, boundary conditions satisfied.

Direct stress at free end of beam is not zero, there is no resultant force or bending moment at the free end.

P.2.4 A thin rectangular plate of unit thickness (Fig. P.2.4) is loaded along the edge $y = +d$ by a linearly varying distributed load of intensity $w = px$ with corresponding equilibrating shears along the vertical edges at $x = 0$ and l. As a solution to the stress analysis problem an Airy stress function ϕ is proposed, where

$$\phi = \frac{p}{120d^3}[5(x^3 - l^2x)(y + d)^2(y - 2d) - 3yx(y^2 - d^2)^2]$$

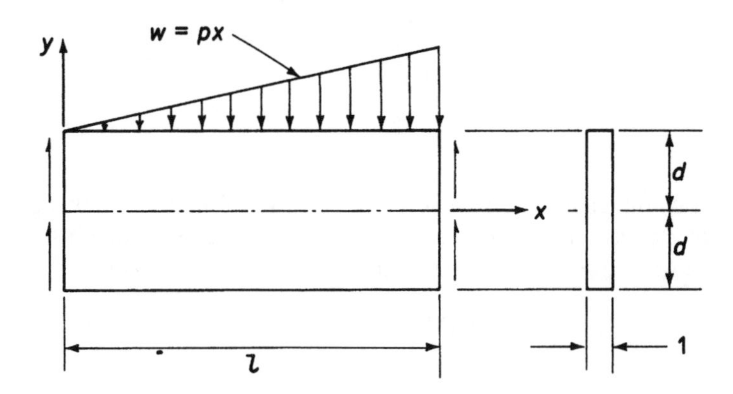

Fig. P.2.4

Show that ϕ satisfies the internal compatibility conditions and obtain the distribution of stresses within the plate. Determine also the extent to which the static boundary conditions are satisfied.

Ans.

$$\sigma_x = \frac{px}{20d^3}[5y(x^2 - l^2) - 10y^3 + 6d^2y]$$

$$\sigma_y = \frac{px}{4d^3}(y^3 - 3yd^2 - 2d^3)$$

$$\tau_{xy} = \frac{-p}{40d^3}[5(3x^2 - l^2)(y^2 - d^2) - 5y^4 + 6y^2d^2 - d^4].$$

The boundary stress function values of τ_{xy} do not agree with the assumed constant equilibrating shears at $x = 0$ and l.

P.2.5 The cantilever beam shown in Fig. P.2.5 is rigidly fixed at $x = L$ and carries loading such that the Airy stress function relating to the problem is

$$\phi = \frac{w}{40bc^3}(-10c^3x^2 - 15c^2x^2y + 2c^2y^3 + 5x^2y^3 - y^5)$$

Find the loading pattern corresponding to the function and check its validity with respect to the boundary conditions.

Ans. The stress function satisfies the biharmonic equation. The beam is a cantilever under a uniformly distributed load of intensity w/unit area with a self-equilibrating stress application given by $\sigma_x = w(12c^3y - 20y^3)/40bc^3$ at $x = 0$. There is zero shear stress at $y = \pm c$ and $x = 0$. At $y = +c$, $\sigma_y = -w/b$ and at $y = -c$, $\sigma_y = 0$.

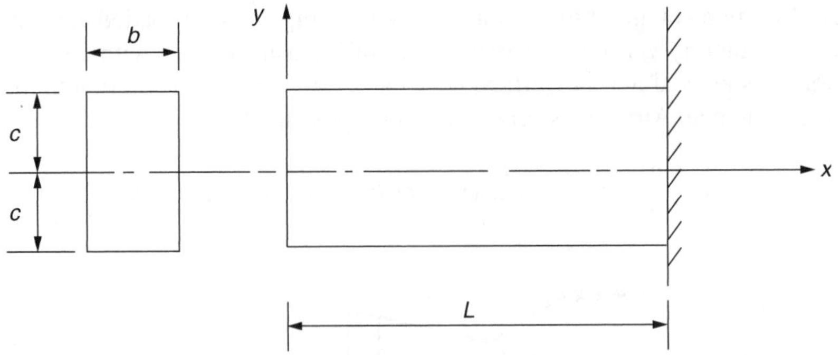

Fig. P.2.5

P.2.6 A two-dimensional isotropic sheet, having a Young's modulus E and linear coefficient of expansion α, is heated non-uniformly, the temperature being $T(x, y)$. Show that the Airy stress function ϕ satisfies the differential equation

$$\nabla^2(\nabla^2\phi + E\alpha T) = 0$$

where

$$\nabla^2 = \frac{\partial^2}{\partial x^2} + \frac{\partial^2}{\partial y^2}$$

is the Laplace operator.

P.2.7 Investigate the state of plane stress described by the following Airy stress function

$$\phi = \frac{3Qxy}{4a} - \frac{Qxy^3}{4a^3}$$

over the square region $x = -a$ to $x = +a$, $y = -a$ to $y = +a$. Calculate the stress resultants per unit thickness over each boundary of the region.

Ans. The stress function satisfies the biharmonic equation. Also,

when $x = a$,

$$\sigma_x = \frac{-3Qy}{2a^2}$$

when $x = -a$,

$$\sigma_x = \frac{3Qy}{2a^2}$$

and

$$\tau_{xy} = \frac{-3Q}{4a}\left(1 - \frac{y^2}{a^2}\right).$$

3

Torsion of solid sections

The elasticity solution of the torsion problem for bars of arbitrary but uniform cross-section is accomplished by the semi-inverse method (Section 2.3) in which assumptions are made regarding either stress or displacement components. The former method owes its derivation to Prandtl, the latter to St. Venant. Both methods are presented in this chapter, together with the useful membrane analogy introduced by Prandtl.

3.1 Prandtl stress function solution

Consider the straight bar of uniform cross-section shown in Fig. 3.1. It is subjected to equal but opposite torques T at each end, both of which are assumed to be free from restraint so that warping displacements w, that is displacements of cross-sections normal to and out of their original planes, are unrestrained. Further, we make the reasonable assumptions that since no direct loads are applied to the bar

$$\sigma_x = \sigma_y = \sigma_z = 0$$

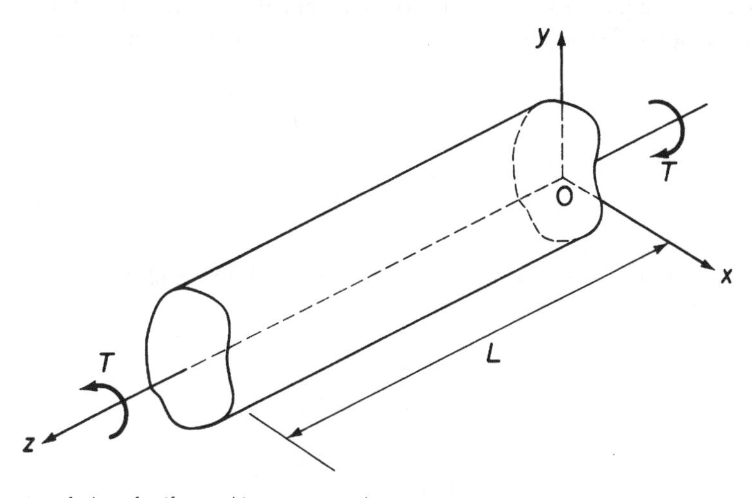

Fig. 3.1 Torsion of a bar of uniform, arbitrary cross-section.

and that the torque is resisted solely by shear stresses in the plane of the cross-section giving

$$\tau_{xy} = 0$$

To verify these assumptions we must show that the remaining stresses satisfy the conditions of equilibrium and compatibility at all points throughout the bar and, in addition, fulfil the equilibrium boundary conditions at all points on the surface of the bar.

If we ignore body forces the equations of equilibrium, (1.5), reduce, as a result of our assumptions, to

$$\frac{\partial \tau_{xz}}{\partial z} = 0 \quad \frac{\partial \tau_{yz}}{\partial z} = 0 \quad \frac{\partial \tau_{zx}}{\partial x} + \frac{\partial \tau_{yz}}{\partial y} = 0 \tag{3.1}$$

The first two equations of Eqs (3.1) show that the shear stresses τ_{xz} and τ_{yz} are functions of x and y only. They are therefore constant at all points along the length of the bar which have the same x and y coordinates. At this stage we turn to the stress function to simplify the process of solution. Prandtl introduced a stress function ϕ defined by

$$\frac{\partial \phi}{\partial x} = -\tau_{zy} \quad \frac{\partial \phi}{\partial y} = \tau_{zx} \tag{3.2}$$

which identically satisfies the third of the equilibrium equations (3.1) whatever form ϕ may take. We therefore have to find the possible forms of ϕ which satisfy the compatibility equations and the boundary conditions, the latter being, in fact, the requirement that distinguishes one torsion problem from another.

From the assumed state of stress in the bar we deduce that

$$\varepsilon_x = \varepsilon_y = \varepsilon_z = \gamma_{xy} = 0 \quad \text{(see Eqs (1.42) and (1.46))}$$

Further, since τ_{xz} and τ_{yz} and hence γ_{xz} and γ_{yz} are functions of x and y only then the compatibility equations (1.21)–(1.23) are identically satisfied as is Eq. (1.26). The remaining compatibility equations, (1.24) and (1.25), are then reduced to

$$\frac{\partial}{\partial x}\left(-\frac{\partial \gamma_{yz}}{\partial x} + \frac{\partial \gamma_{xz}}{\partial y}\right) = 0$$

$$\frac{\partial}{\partial y}\left(\frac{\partial \gamma_{yz}}{\partial x} - \frac{\partial \gamma_{xz}}{\partial y}\right) = 0$$

Substituting initially for γ_{yz} and γ_{xz} from Eqs (1.46) and then for $\tau_{zy}(=\tau_{yz})$ and $\tau_{zx}(=\tau_{xz})$ from Eqs (3.2) gives

$$\frac{\partial}{\partial x}\left(\frac{\partial^2 \phi}{\partial x^2} + \frac{\partial^2 \phi}{\partial y^2}\right) = 0$$

$$-\frac{\partial}{\partial y}\left(\frac{\partial^2 \phi}{\partial x^2} + \frac{\partial^2 \phi}{\partial y^2}\right) = 0$$

or

$$\frac{\partial}{\partial x}\nabla^2\phi = 0 \qquad -\frac{\partial}{\partial y}\nabla^2\phi = 0 \tag{3.3}$$

where ∇^2 is the two-dimensional Laplacian operator

$$\left(\frac{\partial^2}{\partial x^2} + \frac{\partial^2}{\partial y^2}\right)$$

The parameter $\nabla^2\phi$ is therefore constant at any section of the bar so that the function ϕ must satisfy the equation

$$\frac{\partial^2\phi}{\partial x^2} + \frac{\partial^2\phi}{\partial y^2} = \text{constant} = F \text{ (say)} \tag{3.4}$$

at all points within the bar.

Finally we must ensure that ϕ fulfils the boundary conditions specified by Eqs (1.7). On the cylindrical surface of the bar there are no externally applied forces so that $\bar{X} = \bar{Y} = \bar{Z} = 0$. The direction cosine n is also zero and therefore the first two equations of Eqs (1.7) are identically satisfied, leaving the third equation as the boundary condition, i.e.

$$\tau_{yz}m + \tau_{xz}l = 0 \tag{3.5}$$

The direction cosines l and m of the normal N to any point on the surface of the bar are, by reference to Fig. 3.2

$$l = \frac{\mathrm{d}y}{\mathrm{d}s} \qquad m = -\frac{\mathrm{d}x}{\mathrm{d}s} \tag{3.6}$$

Substituting Eqs (3.2) and (3.6) into (3.5) we have

$$\frac{\partial\phi}{\partial x}\frac{\mathrm{d}x}{\mathrm{d}s} + \frac{\partial\phi}{\partial y}\frac{\mathrm{d}y}{\mathrm{d}s} = 0$$

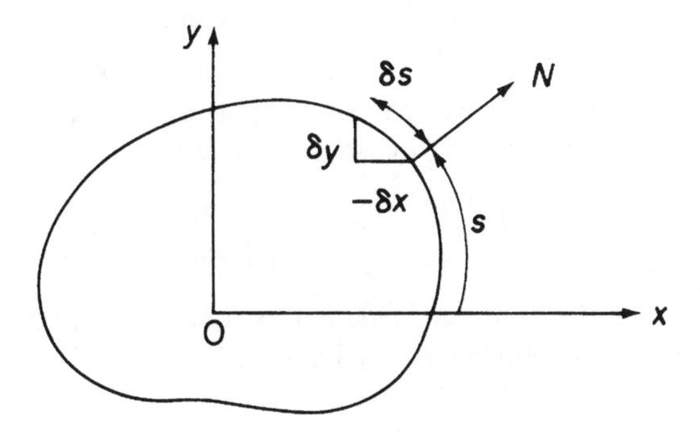

Fig. 3.2 Formation of the direction cosines l and m of the normal to the surface of the bar.

or

$$\frac{\partial \phi}{\mathrm{d}s} = 0$$

Thus ϕ is constant on the surface of the bar and since the actual value of this constant does not affect the stresses of Eq. (3.2) we may conveniently take the constant to be zero. Hence on the cylindrical surface of the bar we have the boundary condition

$$\phi = 0 \tag{3.7}$$

On the ends of the bar the direction cosines of the normal to the surface have the values $l = 0$, $m = 0$ and $n = 1$. The related boundary conditions, from Eqs (1.7), are then

$$\bar{X} = \tau_{zx}$$
$$\bar{Y} = \tau_{zy}$$
$$\bar{Z} = 0$$

We now observe that the forces on each end of the bar are shear forces which are distributed over the ends of the bar in the same manner as the shear stresses are distributed over the cross-section. The resultant shear force in the positive direction of the x axis, which we shall call S_x, is then

$$S_x = \iint \bar{X}\,\mathrm{d}x\,\mathrm{d}y = \iint \tau_{zx}\,\mathrm{d}x\,\mathrm{d}y$$

or, using the relationship of Eqs (3.2)

$$S_x = \iint \frac{\partial \phi}{\partial y}\,\mathrm{d}x\,\mathrm{d}y = \int \mathrm{d}x \int \frac{\partial \phi}{\partial y}\,\mathrm{d}y = 0$$

as $\phi = 0$ at the boundary. In a similar manner, S_y, the resultant shear force in the y direction, is

$$S_y = -\int \mathrm{d}y \int \frac{\partial \phi}{\partial x}\,\mathrm{d}x = 0$$

It follows that there is no resultant shear force on the ends of the bar and the forces represent a torque of magnitude, referring to Fig. 3.3

$$T = \iint (\tau_{zy}x - \tau_{zx}y)\,\mathrm{d}x\,\mathrm{d}y$$

in which we take the sign of T as being positive in the anticlockwise sense.

Rewriting this equation in terms of the stress function ϕ

$$T = -\iint \frac{\partial \phi}{\partial x}x\,\mathrm{d}x\,\mathrm{d}y - \iint \frac{\partial \phi}{\partial y}y\,\mathrm{d}x\,\mathrm{d}y$$

Integrating each term on the right-hand side of this equation by parts, and noting again that $\phi = 0$ at all points on the boundary, we have

$$T = 2\iint \phi\,\mathrm{d}x\,\mathrm{d}y \tag{3.8}$$

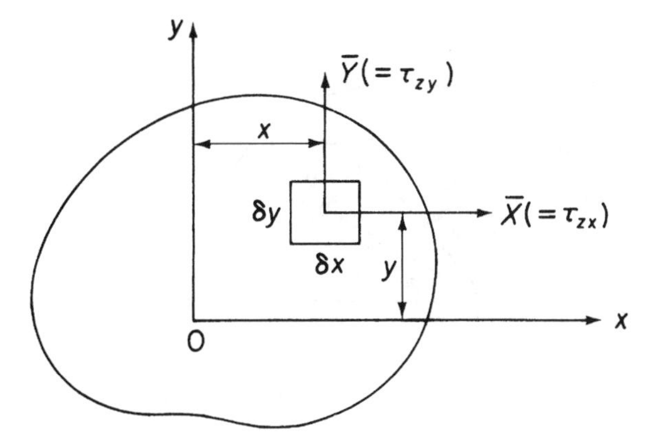

Fig. 3.3 Derivation of torque on cross-section of bar.

We are therefore in a position to obtain an exact solution to a torsion problem if a stress function $\phi(x, y)$ can be found which satisfies Eq. (3.4) at all points within the bar and vanishes on the surface of the bar, and providing that the external torques are distributed over the ends of the bar in an identical manner to the distribution of internal stress over the cross-section. Although the last proviso is generally impracticable we know from St. Venant's principle that only stresses in the end regions are affected; therefore, the solution is applicable to sections at distances from the ends usually taken to be greater than the largest cross-sectional dimension. We have now satisfied all the conditions of the problem without the use of stresses other than τ_{zy} and τ_{zx}, demonstrating that our original assumptions were justified.

Usually, in addition to the stress distribution in the bar, we require to know the angle of twist and the warping displacement of the cross-section. First, however, we shall investigate the mode of displacement of the cross-section. We have seen that as a result of our assumed values of stress

$$\varepsilon_x = \varepsilon_y = \varepsilon_z = \gamma_{xy} = 0$$

It follows, from Eqs (1.18) and the second of Eqs (1.20), that

$$\frac{\partial u}{\partial x} = \frac{\partial v}{\partial y} = \frac{\partial w}{\partial z} = \frac{\partial v}{\partial x} + \frac{\partial u}{\partial y} = 0$$

which result leads to the conclusions that each cross-section rotates as a rigid body in its own plane about a centre of rotation or twist, and that although cross-sections suffer warping displacements normal to their planes the values of this displacement at points having the same coordinates along the length of the bar are equal. Each longitudinal fibre of the bar therefore remains unstrained, as we have in fact assumed.

Let us suppose that a cross-section of the bar rotates through a small angle θ about its centre of twist assumed coincident with the origin of the axes Oxy (see Fig. 3.4). Some point P(r, α) will be displaced to P$'(r, \alpha + \theta)$, the components of its displacement being

$$u = -r\theta \sin \alpha \quad v = r\theta \cos \alpha$$

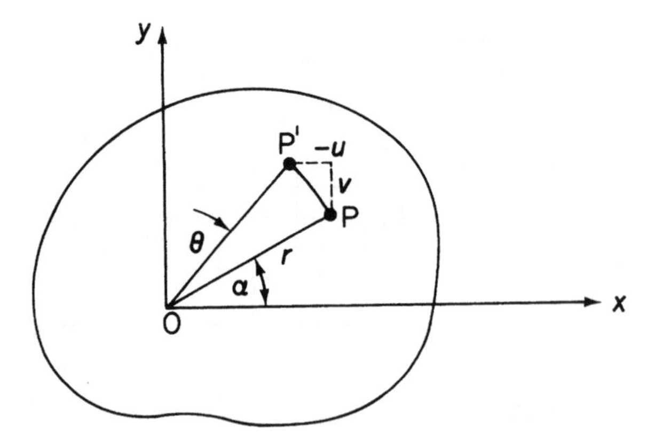

Fig. 3.4 Rigid body displacement in the cross-section of the bar.

or

$$u = -\theta y \quad v = \theta x \tag{3.9}$$

Referring to Eqs (1.20) and (1.46)

$$\gamma_{zx} = \frac{\partial u}{\partial z} + \frac{\partial w}{\partial x} = \frac{\tau_{zx}}{G} \quad \gamma_{zy} = \frac{\partial w}{\partial y} + \frac{\partial v}{\partial z} = \frac{\tau_{zy}}{G}$$

Rearranging and substituting for u and v from Eqs (3.9)

$$\frac{\partial w}{\partial x} = \frac{\tau_{zx}}{G} + \frac{\mathrm{d}\theta}{\mathrm{d}z}y \quad \frac{\partial w}{\partial y} = \frac{\tau_{zy}}{G} - \frac{\mathrm{d}\theta}{\mathrm{d}z}x \tag{3.10}$$

For a particular torsion problem Eqs (3.10) enable the warping displacement w of the originally plane cross-section to be determined. Note that since each cross-section rotates as a rigid body θ is a function of z only.

Differentiating the first of Eqs (3.10) with respect to y, the second with respect to x and subtracting we have

$$0 = \frac{1}{G}\left(\frac{\partial \tau_{zx}}{\partial y} - \frac{\partial \tau_{zy}}{\partial x}\right) + 2\frac{\mathrm{d}\theta}{\mathrm{d}z}$$

Expressing τ_{zx} and τ_{zy} in terms of ϕ gives

$$\frac{\partial^2 \phi}{\partial x^2} + \frac{\partial^2 \phi}{\partial y^2} = -2G\frac{\mathrm{d}\theta}{\mathrm{d}z}$$

or, from Eq. (3.4)

$$-2G\frac{\mathrm{d}\theta}{\mathrm{d}z} = \nabla^2 \phi = F \text{ (constant)} \tag{3.11}$$

It is convenient to introduce a *torsion constant* J defined by the general torsion equation

$$T = GJ\frac{\mathrm{d}\theta}{\mathrm{d}z} \tag{3.12}$$

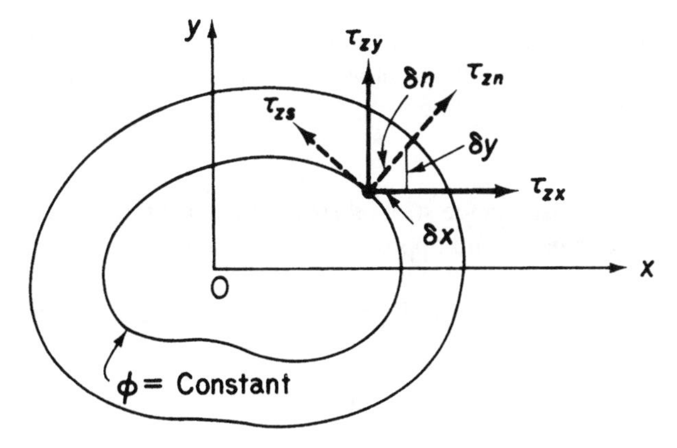

Fig. 3.5 Lines of shear stress.

The product GJ is known as the *torsional rigidity* of the bar and may be written, from Eqs (3.8) and (3.11)

$$GJ = -\frac{4G}{\nabla^2\phi}\iint \phi\,\mathrm{d}x\,\mathrm{d}y \tag{3.13}$$

Consider now the line of constant ϕ in Fig. 3.5. If s is the distance measured along this line from some arbitrary point then

$$\frac{\partial\phi}{\partial s} = 0 = \frac{\partial\phi}{\partial y}\frac{\mathrm{d}y}{\mathrm{d}s} + \frac{\partial\phi}{\partial x}\frac{\mathrm{d}x}{\mathrm{d}s}$$

Using Eqs (3.2) and (3.6) we may rewrite this equation as

$$\frac{\partial\phi}{\partial s} = \tau_{zx}l + \tau_{zy}m = 0 \tag{3.14}$$

From Fig. 3.5 the normal and tangential components of shear stress are

$$\tau_{zn} = \tau_{zx}l + \tau_{zy}m \quad \tau_{zs} = \tau_{zy}l - \tau_{zx}m \tag{3.15}$$

Comparing the first of Eqs (3.15) with Eq. (3.14) we see that the normal shear stress is zero so that the resultant shear stress at any point is tangential to a line of constant ϕ. These are known as *lines of shear stress* or *shear lines.*

Substituting ϕ in the second of Eqs (3.15) we have

$$\tau_{zs} = -\frac{\partial\phi}{\partial x}l - \frac{\partial\phi}{\partial y}m$$

which may be written, from Fig. 3.5, as

$$\tau_{zx} = -\frac{\partial\phi}{\partial x}\frac{\mathrm{d}x}{\mathrm{d}n} - \frac{\partial\phi}{\partial y}\frac{\mathrm{d}y}{\mathrm{d}n} = -\frac{\partial\phi}{\partial n} \tag{3.16}$$

where, in this case, the direction cosines l and m are defined in terms of an elemental normal of length δn.

We have therefore shown that the resultant shear stress at any point is tangential to the line of shear stress through the point and has a value equal to minus the derivative of ϕ in a direction normal to the line.

Example 3.1

Determine the rate of twist and the stress distribution in a circular section bar of radius R which is subjected to equal and opposite torques T at each of its free ends.

If we assume an origin of axes at the centre of the bar the equation of its surface is given by

$$x^2 + y^2 = R^2$$

If we now choose a stress function of the form

$$\phi = C(x^2 + y^2 - R^2) \tag{i}$$

the boundary condition $\phi = 0$ is satisfied at every point on the boundary of the bar and the constant C may be chosen to fulfil the remaining requirement of compatibility. Therefore from Eqs (3.11) and (i)

$$4C = -2G\frac{\mathrm{d}\theta}{\mathrm{d}z}$$

so that

$$C = -\frac{G}{2}\frac{\mathrm{d}\theta}{\mathrm{d}z}$$

and

$$\phi = -G\frac{\mathrm{d}\theta}{\mathrm{d}z}(x^2 + y^2 - R^2)|2 \tag{ii}$$

Substituting for ϕ in Eq. (3.8)

$$T = -G\frac{\mathrm{d}\theta}{\mathrm{d}z}\left(\iint x^2\,\mathrm{d}x\,\mathrm{d}y + \iint y^2\,\mathrm{d}x\,\mathrm{d}y - R^2\iint \mathrm{d}x\,\mathrm{d}y\right)$$

The first and second integrals in this equation both have the value $\pi R^4/4$ while the third integral is equal to πR^2, the area of cross-section of the bar. Then

$$T = -G\frac{\mathrm{d}\theta}{\mathrm{d}z}\left(\frac{\pi R^4}{4} + \frac{\pi R^4}{4} - \pi R^4\right)$$

which gives

$$T = \frac{\pi R^4}{2}G\frac{\mathrm{d}\theta}{\mathrm{d}z}$$

i.e.

$$T = GJ\frac{\mathrm{d}\theta}{\mathrm{d}z} \tag{iii}$$

in which $J = \pi R^4/2 = \pi D^4/32$ (D is the diameter), the *polar second moment of area* of the bar's cross-section.

Substituting for $G(\mathrm{d}\theta/\mathrm{d}z)$ in Eq. (ii) from (iii)

$$\phi = -\frac{T}{2J}(x^2 + y^2 - R^2)$$

and from Eqs (3.2)

$$\tau_{zy} = -\frac{\partial \phi}{\partial x} = \frac{Tx}{J} \qquad \tau_{zx} = \frac{\partial \phi}{\partial y} = -\frac{T}{J}y$$

The resultant shear stress at any point on the surface of the bar is then given by

$$\tau = \sqrt{\tau_{zy}^2 + \tau_{zx}^2}$$

i.e.

$$\tau = \frac{T}{J}\sqrt{x^2 + y^2}$$

i.e.

$$\tau = \frac{TR}{J} \tag{iv}$$

The above argument may be applied to any annulus of radius r within the cross-section of the bar so that the stress distribution is given by

$$\tau = \frac{Tr}{J}$$

and therefore increases linearly from zero at the centre of the bar to a maximum TR/J at the surface.

Example 3.2

A uniform bar has the elliptical cross-section shown in Fig. 3.6 and is subjected to equal and opposite torques T at each of its free ends. Derive expressions for the rate of twist in the bar, the shear stress distribution and the warping displacement of its cross-section.

The semi-major and semi-minor axes are a and b, respectively, so that the equation of its boundary is

$$\frac{x^2}{a^2} + \frac{y^2}{b^2} = 1$$

If we choose a stress function of the form

$$\phi = C\left(\frac{x^2}{a^2} + \frac{y^2}{b^2} - 1\right) \tag{i}$$

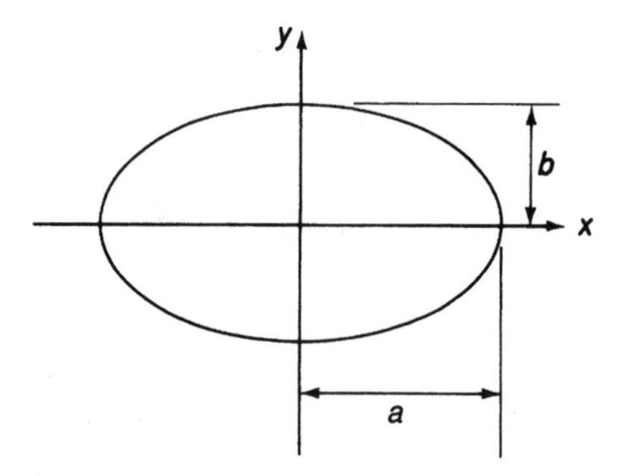

Fig. 3.6 Torsion of a bar of elliptical cross-section.

then the boundary condition $\phi = 0$ is satisfied at every point on the boundary and the constant C may be chosen to fulfil the remaining requirement of compatibility. Thus, from Eqs (3.11) and (i)

$$2C\left(\frac{1}{a^2}+\frac{1}{b^2}\right) = -2G\frac{\mathrm{d}\theta}{\mathrm{d}z}$$

or

$$C = -G\frac{\mathrm{d}\theta}{\mathrm{d}z}\frac{a^2b^2}{(a^2+b^2)} \tag{ii}$$

giving

$$\phi = -G\frac{\mathrm{d}\theta}{\mathrm{d}z}\frac{a^2b^2}{(a^2+b^2)}\left(\frac{x^2}{a^2}+\frac{y^2}{b^2}-1\right) \tag{iii}$$

Substituting this expression for ϕ in Eq. (3.8) establishes the relationship between the torque T and the rate of twist

$$T = -2G\frac{\mathrm{d}\theta}{\mathrm{d}z}\frac{a^2b^2}{(a^2+b^2)}\left(\frac{1}{a^2}\iint x^2\,\mathrm{d}x\,\mathrm{d}y + \frac{1}{b^2}\iint y^2\,\mathrm{d}x\,\mathrm{d}y - \iint \mathrm{d}x\,\mathrm{d}y\right)$$

The first and second integrals in this equation are the second moments of area $I_{yy} = \pi a^3b/4$ and $I_{xx} = \pi ab^3/4$, while the third integral is the area of the cross-section $A = \pi ab$. Replacing the integrals by these values gives

$$T = G\frac{\mathrm{d}\theta}{\mathrm{d}z}\frac{\pi a^3b^3}{(a^2+b^2)} \tag{iv}$$

from which (see Eq. (3.12))

$$J = \frac{\pi a^3b^3}{(a^2+b^2)} \tag{v}$$

The shear stress distribution is obtained in terms of the torque by substituting for the product $G\,(\mathrm{d}\theta/\mathrm{d}z)$ in Eq. (iii) from (iv) and then differentiating as indicated by the relationships of Eqs (3.2). Thus

$$\tau_{zx} = -\frac{2Ty}{\pi ab^3} \quad \tau_{zy} = \frac{2Tx}{\pi a^3 b} \tag{vi}$$

So far we have solved for the stress distribution, Eqs (vi), and the rate of twist, Eq. (iv). It remains to determine the warping distribution w over the cross-section. For this we return to Eqs (3.10) which become, on substituting from the above for τ_{zx}, τ_{zy} and $d\theta/dz$

$$\frac{\partial w}{\partial x} = -\frac{2Ty}{\pi ab^3 G} + \frac{T}{G}\frac{(a^2+b^2)}{\pi a^3 b^3}y \quad \frac{\partial w}{\partial y} = \frac{2Tx}{\pi a^3 bG} - \frac{T}{G}\frac{(a^2+b^2)}{\pi a^3 b^3}x$$

or

$$\frac{\partial w}{\partial x} = \frac{T}{\pi a^3 b^3 G}(b^2 - a^2)y \quad \frac{\partial w}{\partial y} = \frac{T}{\pi a^3 b^3 G}(b^2 - a^2)x \tag{vii}$$

Integrating both of Eqs (vii)

$$w = \frac{T(b^2 - a^2)}{\pi a^3 b^3 G}yx + f_1(y) \quad w = \frac{T(b^2 - a^2)}{\pi a^3 b^3 G}xy + f_2(x)$$

The warping displacement given by each of these equations must have the same value at identical points (x, y). It follows that $f_1(y) = f_2(x) = 0$. Hence

$$w = \frac{T(b^2 - a^2)}{\pi a^3 b^3 G}xy \tag{viii}$$

Lines of constant w therefore describe hyperbolas with the major and minor axes of the elliptical cross-section as asymptotes. Further, for a positive (anticlockwise) torque the warping is negative in the first and third quadrants ($a > b$) and positive in the second and fourth.

3.2 St. Venant warping function solution

In formulating his stress function solution Prandtl made assumptions concerned with the stress distribution in the bar. The alternative approach presented by St. Venant involves assumptions as to the mode of displacement of the bar; namely, that cross-sections of a bar subjected to torsion maintain their original unloaded shape although they may suffer warping displacements normal to their plane. The first of these assumptions leads to the conclusion that cross-sections rotate as rigid bodies about a centre of rotation or twist. This fact was also found to derive from the stress function approach of Section 3.1 so that, referring to Fig. 3.4 and Eq. (3.9), the components of displacement in the x and y directions of a point P in the cross-section are

$$u = -\theta y \quad v = \theta x$$

It is also reasonable to assume that the warping displacement w is proportional to the rate of twist and is therefore constant along the length of the bar. Hence we may define w by the equation

$$w = \frac{d\theta}{dz}\psi(x, y) \tag{3.17}$$

where $\psi(x, y)$ is the *warping function*.

The assumed form of the displacements u, v and w must satisfy the equilibrium and force boundary conditions of the bar. We note here that it is unnecessary to investigate compatibility as we are concerned with displacement forms which are single-valued functions and therefore automatically satisfy the compatibility requirement.

The components of strain corresponding to the assumed displacements are obtained from Eqs (1.18) and (1.20) and are

$$\left.\begin{aligned} &\varepsilon_x = \varepsilon_y = \varepsilon_z = \gamma_{xy} = 0 \\ &\gamma_{zx} = \frac{\partial w}{\partial x} + \frac{\partial u}{\partial z} = \frac{d\theta}{dz}\left(\frac{\partial \psi}{\partial x} - y\right) \\ &\gamma_{zy} = \frac{\partial w}{\partial y} + \frac{\partial v}{\partial z} = \frac{d\theta}{dz}\left(\frac{\partial \psi}{\partial y} + x\right) \end{aligned}\right\} \tag{3.18}$$

The corresponding components of stress are, from Eqs (1.42) and (1.46)

$$\left.\begin{aligned} &\sigma_x = \sigma_y = \sigma_z = \tau_{xy} = 0 \\ &\tau_{zx} = G\frac{d\theta}{dz}\left(\frac{\partial \psi}{\partial x} - y\right) \\ &\tau_{zy} = G\frac{d\theta}{dz}\left(\frac{\partial \psi}{\partial y} + x\right) \end{aligned}\right\} \tag{3.19}$$

Ignoring body forces we see that these equations identically satisfy the first two of the equilibrium equations (1.5) and also that the third is fulfilled if the warping function satisfies the equation

$$\frac{\partial^2 \psi}{\partial x^2} + \frac{\partial^2 \psi}{\partial y^2} = \nabla^2 \psi = 0 \tag{3.20}$$

The direction cosine n is zero on the cylindrical surface of the bar and so the first two of the boundary conditions (Eqs (1.7)) are identically satisfied by the stresses of Eqs (3.19). The third equation simplifies to

$$\left(\frac{\partial \psi}{\partial y} + x\right) m + \left(\frac{\partial \psi}{\partial x} - y\right) l = 0 \tag{3.21}$$

It may be shown, but not as easily as in the stress function solution, that the shear stresses defined in terms of the warping function in Eqs (3.19) produce zero resultant shear force over each end of the bar.[1] The torque is found in a similar manner to that in Section 3.1 where, by reference to Fig. 3.3, we have

$$T = \iint (\tau_{zy}x - \tau_{zx}y)\mathrm{d}x\,\mathrm{d}y$$

or

$$T = G\frac{\mathrm{d}\theta}{\mathrm{d}z}\iint \left[\left(\frac{\partial\psi}{\partial y} + x\right)x - \left(\frac{\partial\psi}{\partial x} - y\right)y\right]\mathrm{d}x\,\mathrm{d}y \tag{3.22}$$

By comparison with Eq. (3.12) the torsion constant J is now, in terms of ψ

$$J = \iint \left[\left(\frac{\partial\psi}{\partial y} + x\right)x - \left(\frac{\partial\psi}{\partial x} - y\right)y\right]\mathrm{d}x\,\mathrm{d}y \tag{3.23}$$

The warping function solution to the torsion problem reduces to the determination of the warping function ψ which satisfies Eqs (3.20) and (3.21). The torsion constant and the rate of twist follow from Eqs (3.23) and (3.22); the stresses and strains from Eqs (3.19) and (3.18) and, finally, the warping distribution from Eq. (3.17).

3.3 The membrane analogy

Prandtl suggested an extremely useful analogy relating the torsion of an arbitrarily shaped bar to the deflected shape of a membrane. The latter is a thin sheet of material which relies for its resistance to transverse loads on internal in-plane or membrane forces.

Suppose that a membrane has the same external shape as the cross-section of a torsion bar (Fig. 3.7(a)). It supports a transverse uniform pressure q and is restrained along its edges by a uniform tensile force N/unit length as shown in Fig. 3.7(a) and (b). It is assumed that the transverse displacements of the membrane are small so that N remains

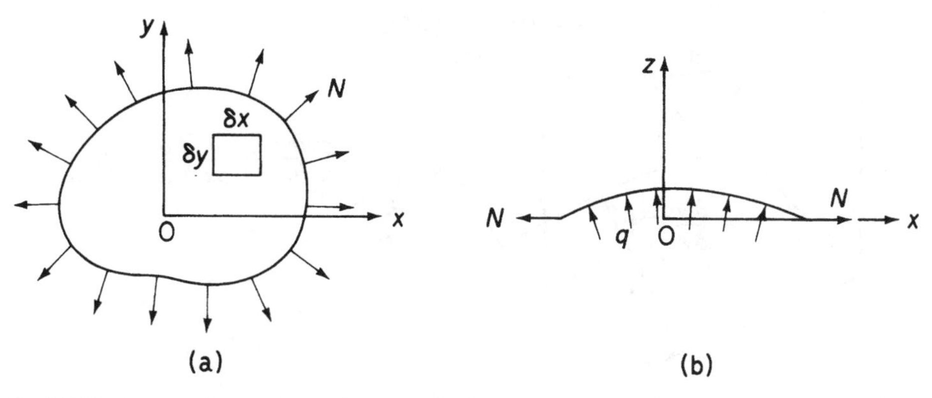

Fig. 3.7 Membrane analogy: in-plane and transverse loading.

unchanged as the membrane deflects. Consider the equilibrium of an element $\delta x\delta y$ of the membrane. Referring to Fig. 3.8 and summing forces in the z direction we have

$$-N\delta y\frac{\partial w}{\partial x} - N\delta y\left(-\frac{\partial w}{\partial x} - \frac{\partial^2 w}{\partial x^2}\delta x\right) - N\delta x\frac{\partial w}{\partial y} - N\delta x\left(-\frac{\partial w}{\partial y} - \frac{\partial^2 w}{\partial y^2}\delta x\right) + q\delta x\delta y = 0$$

or

$$\frac{\partial^2 w}{\partial x^2} + \frac{\partial^2 w}{\partial y^2} = \nabla^2 w = -\frac{q}{N} \tag{3.24}$$

Equation (3.24) must be satisfied at all points within the boundary of the membrane. Furthermore, at all points on the boundary

$$w = 0 \tag{3.25}$$

and we see that by comparing Eqs (3.24) and (3.25) with Eqs (3.11) and (3.7) w is analogous to ϕ when q is constant. Thus if the membrane has the same external shape as the cross-section of the bar then

$$w(x, y) = \phi(x, y)$$

and

$$\frac{q}{N} = -F = 2G\frac{\mathrm{d}\theta}{\mathrm{d}z}$$

The analogy now being established, we may make several useful deductions relating the deflected form of the membrane to the state of stress in the bar.

Contour lines or lines of constant w correspond to lines of constant ϕ or lines of shear stress in the bar. The resultant shear stress at any point is tangential to the membrane contour line and equal in value to the negative of the membrane slope, $\partial w/\partial n$, at that

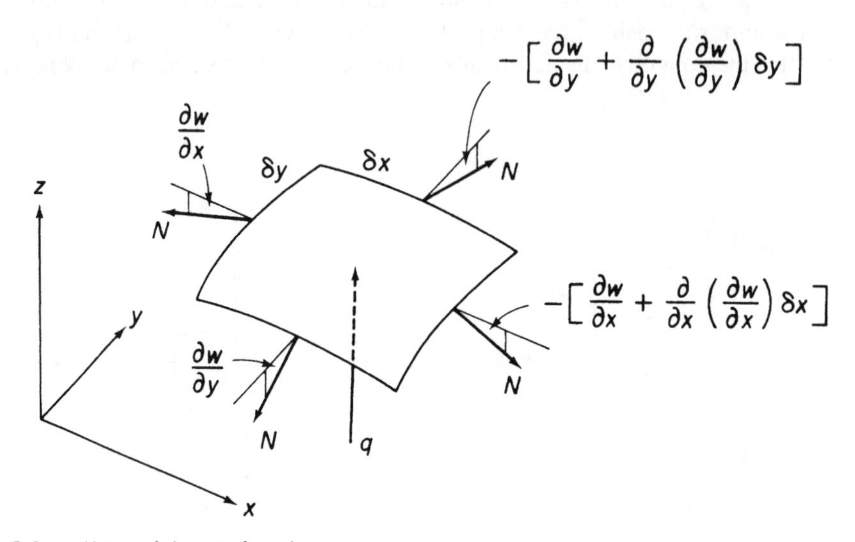

Fig. 3.8 Equilibrium of element of membrane.

point, the direction n being normal to the contour line (see Eq. (3.16)). The volume between the membrane and the xy plane is

$$\text{Vol} = \iint w \, \mathrm{d}x \, \mathrm{d}y$$

and we see that by comparison with Eq. (3.8)

$$T = 2\,\text{Vol}$$

The analogy therefore provides an extremely useful method of analysing torsion bars possessing irregular cross-sections for which stress function forms are not known. Hetényi[2] describes experimental techniques for this approach. In addition to the strictly experimental use of the analogy it is also helpful in the visual appreciation of a particular torsion problem. The contour lines often indicate a form for the stress function, enabling a solution to be obtained by the method of Section 3.1. Stress concentrations are made apparent by the closeness of contour lines where the slope of the membrane is large. These are in evidence at sharp internal corners, cut-outs, discontinuities, etc.

3.4 Torsion of a narrow rectangular strip

In Chapter 18 we shall investigate the torsion of thin-walled open section beams; the development of the theory being based on the analysis of a narrow rectangular strip subjected to torque. We now conveniently apply the membrane analogy to the torsion of such a strip shown in Fig. 3.9. The corresponding membrane surface has the same cross-sectional shape at all points along its length except for small regions near its ends where it flattens out. If we ignore these regions and assume that the shape of the

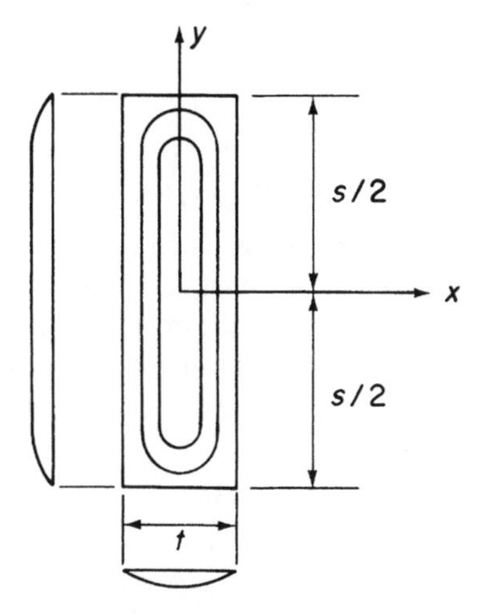

Fig. 3.9 Torsion of a narrow rectangular strip.

membrane is independent of y then Eq. (3.11) simplifies to

$$\frac{d^2\phi}{dx^2} = -2G\frac{d\theta}{dz}$$

Integrating twice

$$\phi = -G\frac{d\theta}{dz}x^2 + Bx + C$$

Substituting the boundary conditions $\phi = 0$ at $x = \pm t/2$ we have

$$\phi = -G\frac{d\theta}{dz}\left[x^2 - \left(\frac{t}{2}\right)^2\right] \tag{3.26}$$

Although ϕ does not disappear along the short edges of the strip and therefore does not give an exact solution, the actual volume of the membrane differs only slightly from the assumed volume so that the corresponding torque and shear stresses are reasonably accurate. Also, the maximum shear stress occurs along the long sides of the strip where the contours are closely spaced, indicating, in any case, that conditions in the end region of the strip are relatively unimportant.

The stress distribution is obtained by substituting Eq. (3.26) in Eqs (3.2), then

$$\tau_{zy} = 2Gx\frac{d\theta}{dz} \qquad \tau_{zx} = 0 \tag{3.27}$$

the shear stress varying linearly across the thickness and attaining a maximum

$$\tau_{zy,\max} = \pm Gt\frac{d\theta}{dz} \tag{3.28}$$

at the outside of the long edges as predicted. The torsion constant J follows from the substitution of Eq. (3.26) into (3.13), giving

$$J = \frac{st^3}{3} \tag{3.29}$$

and

$$\tau_{zy,\max} = \frac{3T}{st^3}$$

These equations represent exact solutions when the assumed shape of the deflected membrane is the actual shape. This condition arises only when the ratio s/t approaches infinity; however, for ratios in excess of 10 the error is of the order of only 6 per cent. Obviously the approximate nature of the solution increases as s/t decreases. Therefore, in order to retain the usefulness of the analysis, a factor μ is included in the torsion constant, i.e.

$$J = \frac{\mu st^3}{3}$$

Values of μ for different types of section are found experimentally and quoted in various references.[3,4] We observe that as s/t approaches infinity μ approaches unity.

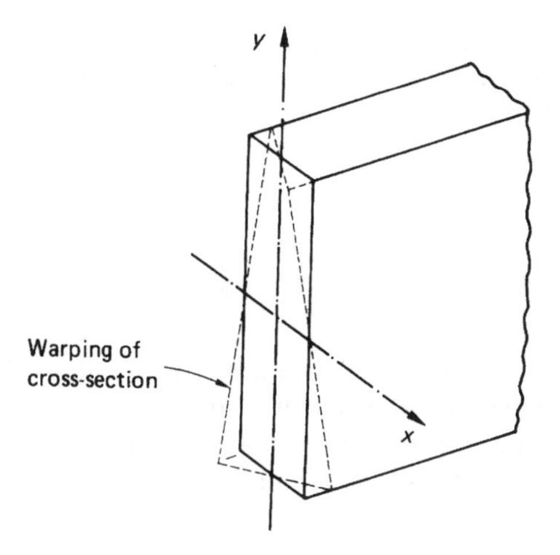

Fig. 3.10 Warping of a thin rectangular strip.

The cross-section of the narrow rectangular strip of Fig. 3.9 does not remain plane after loading but suffers warping displacements normal to its plane; this warping may be determined using either of Eqs (3.10). From the first of these equations

$$\frac{\partial w}{\partial x} = y\frac{\mathrm{d}\theta}{\mathrm{d}z} \tag{3.30}$$

since $\tau_{zx} = 0$ (see Eqs (3.27)). Integrating Eq. (3.30) we obtain

$$w = xy\frac{\mathrm{d}\theta}{\mathrm{d}z} + \text{constant} \tag{3.31}$$

Since the cross-section is doubly symmetrical $w = 0$ at $x = y = 0$ so that the constant in Eq. (3.31) is zero. Therefore

$$w = xy\frac{\mathrm{d}\theta}{\mathrm{d}z} \tag{3.32}$$

and the warping distribution at any cross-section is as shown in Fig. 3.10.

We should not close this chapter without mentioning alternative methods of solution of the torsion problem. These in fact provide approximate solutions for the wide range of problems for which exact solutions are not known. Examples of this approach are the numerical finite difference method and the Rayleigh–Ritz method based on energy principles.[5]

References

1 Wang, C. T., *Applied Elasticity,* McGraw-Hill Book Company, New York, 1953.
2 Hetényi, M., *Handbook of Experimental Stress Analysis,* John Wiley and Sons, Inc., New York, 1950.
3 Roark, R. J., *Formulas for Stress and Strain*, 4th edition, McGraw-Hill Book Company, New York, 1965.

4 *Handbook of Aeronautics, No. 1, Structural Principles and Data*, 4th edition. Published under the authority of the Royal Aeronautical Society, The New Era Publishing Co. Ltd., London, 1952.
5 Timoshenko, S. and Goodier, J. N., *Theory of Elasticity*, 2nd edition, McGraw-Hill Book Company, New York, 1951.

Problems

P.3.1 Show that the stress function $\phi = k(r^2 - a^2)$ is applicable to the solution of a solid circular section bar of radius a. Determine the stress distribution in the bar in terms of the applied torque, the rate of twist and the warping of the cross-section.

Is it possible to use this stress function in the solution for a circular bar of hollow section?

Ans. $\tau = Tr/I_p$ where $I_p = \pi a^4/2$,
$\mathrm{d}\theta/\mathrm{d}z = 2T/G\pi a^4$, $w = 0$ everywhere.

P.3.2 Deduce a suitable warping function for the circular section bar of P.3.1 and hence derive the expressions for stress distribution and rate of twist.

Ans. $\psi = 0, \quad \tau_{zx} = -\dfrac{Ty}{I_p}, \quad \tau_{zy} = \dfrac{Tx}{I_p}, \quad \tau_{zs} = \dfrac{Tr}{I_p}, \quad \dfrac{\mathrm{d}\theta}{\mathrm{d}z} = \dfrac{T}{GI_P}$

P.3.3 Show that the warping function $\psi = kxy$, in which k is an unknown constant, may be used to solve the torsion problem for the elliptical section of Example 3.2.

P.3.4 Show that the stress function

$$\phi = -G\frac{\mathrm{d}\theta}{\mathrm{d}z}\left[\frac{1}{2}(x^2 + y^2) - \frac{1}{2a}(x^3 - 3xy^2) - \frac{2}{27}a^2\right]$$

is the correct solution for a bar having a cross-section in the form of the equilateral triangle shown in Fig. P.3.4. Determine the shear stress distribution, the rate of twist and the warping of the cross-section. Find the position and magnitude of the maximum shear stress.

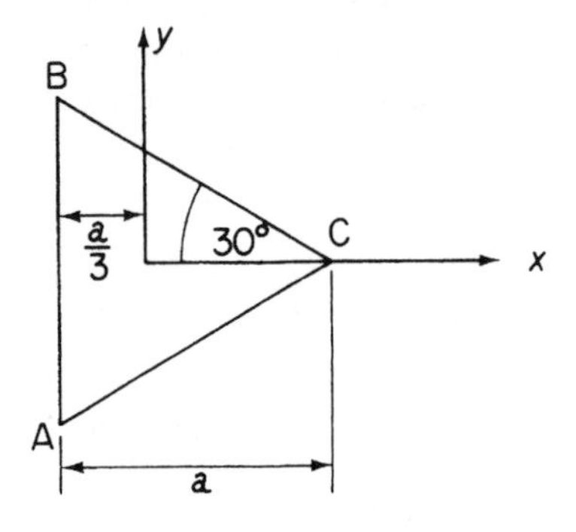

Fig. P.3.4

Ans.

$$\tau_{zy} = G\frac{d\theta}{dz}\left(x - \frac{3x^2}{2a} + \frac{3y^2}{2a}\right)$$

$$\tau_{zx} = -G\frac{d\theta}{dz}\left(y + \frac{3xy}{a}\right)$$

$$\tau_{max} \text{ (at centre of each side)} = -\frac{a}{2}G\frac{d\theta}{dz}$$

$$\frac{d\theta}{dz} = \frac{15\sqrt{3}T}{Ga^4}$$

$$w = \frac{1}{2a}\frac{d\theta}{dz}(y^3 - 3x^2y).$$

P.3.5 Determine the maximum shear stress and the rate of twist in terms of the applied torque T for the section comprising narrow rectangular strips shown in Fig. P.3.5.

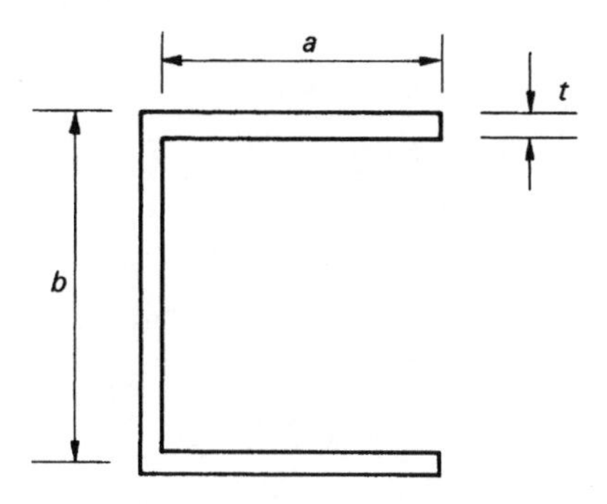

Fig. P.3.5

Ans. $\tau_{max} = 3T/(2a + b)t^2$, $d\theta/dz = 3T/G(2a + b)t^3$.

Section A2 Virtual Work, Energy and Matrix Methods

4

Virtual work and energy methods

Many structural problems are statically determinate, i.e., the support reactions and internal force systems may be found using simple statics where the number of unknowns is equal to the number of equations of equilibrium available. In cases where the number of unknowns exceeds the possible number of equations of equilibrium, for example, a propped cantilever beam, other methods of analysis are required.

The methods fall into two categories and are based on two important concepts; the first, which is presented in this chapter, is *the principle of virtual work*. This is the most fundamental and powerful tool available for the analysis of statically indeterminate structures and has the advantage of being able to deal with conditions other than those in the elastic range. The second, based on *strain energy*, can provide approximate solutions of complex problems for which exact solutions do not exist and is discussed in Chapter 5. In some cases the two methods are equivalent since, although the governing equations differ, the equations themselves are identical.

In modern structural analysis, computer-based techniques are widely used; these include the flexibility and stiffness methods (see Chapter 6). However, the formulation of, say, stiffness matrices for the elements of a complex structure is based on one of the above approaches so that a knowledge and understanding of their application is advantageous.

4.1 Work

Before we consider the principle of virtual work in detail, it is important to clarify exactly what is meant by *work*. The basic definition of work in elementary mechanics is that 'work is done when a force moves its point of application'. However, we shall require a more exact definition since we shall be concerned with work done by both forces and moments and with the work done by a force when the body on which it acts is given a displacement which is not coincident with the line of action of the force.

Consider the force, F, acting on a particle, A, in Fig. 4.1(a). If the particle is given a displacement, Δ, by some external agency so that it moves to A$'$ in a direction at an angle α to the line of action of F, the work, W_F, done by $F$is given by

$$W_F = F(\Delta \cos \alpha) \tag{4.1}$$

or

$$W_F = (F \cos \alpha)\Delta \tag{4.2}$$

We see therefore that the work done by the force, F, as the particle moves from A to A′ may be regarded as either the product of F and the component of Δ in the direction of F (Eq. (4.1)) or as the product of the component of F in the direction of Δ and Δ (Eq. (4.2)).

Now consider the couple (pure moment) in Fig. 4.1(b) and suppose that the couple is given a small rotation of θ radians. The work done by each force F is then $F(a/2)\theta$ so that the total work done, W_C, by the couple is

$$W_C = F\frac{a}{2}\theta + F\frac{a}{2}\theta = Fa\theta$$

It follows that the work done, W_M, by the pure moment, M, acting on the bar AB in Fig. 4.1(c) as it is given a small rotation, θ, is

$$W_M = M\theta \tag{4.3}$$

Note that in the above the force, F, and moment, M, are in position before the displacements take place and are not the cause of them. Also, in Fig. 4.1(a), the component of Δ parallel to the direction of F is in the same direction as F; if it had been in the opposite direction the work done would have been negative. The same argument applies to the work done by the moment, M, where we see in Fig. 4.1(c) that the rotation, θ, is in the same sense as M. Note also that if the displacement, Δ, had been perpendicular to the force, F, no work would have been done by F.

Finally it should be remembered that work is a scalar quantity since it is not associated with direction (in Fig. 4.1(a) the force F does work if the particle is moved in any direction). Thus the work done by a series of forces is the algebraic sum of the work done by each force.

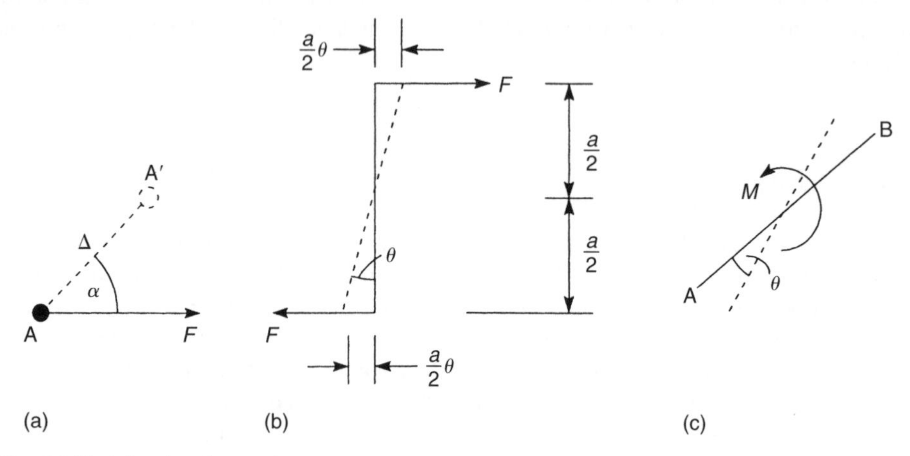

Fig. 4.1 Work done by a force and a moment.

4.2 Principle of virtual work

The establishment of the principle will be carried out in stages. First we shall consider a particle, then a rigid body and finally a deformable body, which is the practical application we require when analysing structures.

4.2.1 Principle of virtual work for a particle

In Fig. 4.2 a particle, A, is acted upon by a number of concurrent forces, $F_1, F_2, \ldots, F_k, \ldots, F_r$; the resultant of these forces is R. Suppose that the particle is given a small arbitrary displacement, Δ_v, to A′ in some specified direction; Δ_v is an imaginary or *virtual* displacement and is sufficiently small so that the directions of F_1, F_2, etc., are unchanged. Let θ_R be the angle that the resultant, R, of the forces makes with the direction of Δ_v and $\theta_1, \theta_2, \ldots, \theta_k, \ldots, \theta_r$ the angles that $F_1, F_2, \ldots, F_k, \ldots, F_r$ make with the direction of Δ_v, respectively. Then, from either of Eqs (4.1) or (4.2) the total virtual work, W_F, done by the forces Fas the particle moves through the virtual displacement, Δ_v, is given by

$$W_F = F_1\Delta_v \cos\theta_1 + F_2\Delta_v \cos\theta_2 + \cdots + F_k\Delta_v \cos\theta_k + \cdots + F_r\Delta_v \cos\theta_r$$

Thus

$$W_F = \sum_{k=1}^{r} F_k\Delta_v \cos\theta_k$$

or, since Δ_v is a fixed, although imaginary displacement

$$W_F = \Delta_v \sum_{k=1}^{r} F_k \cos\theta_k \tag{4.4}$$

In Eq. (4.4) $\sum_{k=1}^{r} F_k \cos\theta_k$ is the sum of all the components of the forces, F, in the direction of Δ_v and therefore must be equal to the component of the resultant, R, of the

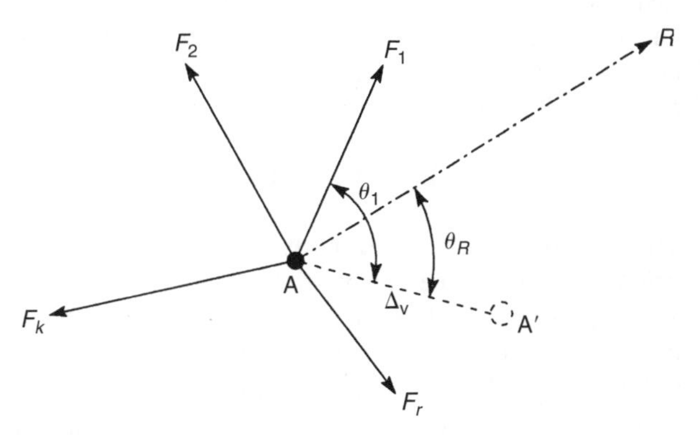

Fig. 4.2 Virtual work for a system of forces acting on a particle.

forces, F, in the direction of Δ_v, i.e.

$$W_F = \Delta_v \sum_{k=1}^{r} F_k \cos\theta_k = \Delta_v R \cos\theta_R \tag{4.5}$$

If the particle, A, is in equilibrium under the action of the forces, $F_1, F_2, \ldots, F_k, \ldots, Fr$, the resultant, R, of the forces is zero. It follows from Eq. (4.5) that the virtual work done by the forces, F, during the virtual displacement, Δ_v, is zero.

We can therefore state the *principle of virtual work* for a particle as follows:

If a particle is in equilibrium under the action of a number of forces the total work done by the forces for a small arbitrary displacement of the particle is zero.

It is possible for the total work done by the forces to be zero even though the particle is not in equilibrium if the virtual displacement is taken to be in a direction perpendicular to their resultant, R. We cannot, therefore, state the converse of the above principle unless we specify that the total work done must be zero for *any* arbitrary displacement. Thus:

A particle is in equilibrium under the action of a system of forces if the total work done by the forces is zero for any virtual displacement of the particle.

Note that in the above, Δ_v is a purely imaginary displacement and is not related in any way to the possible displacement of the particle under the action of the forces, F. Δ_v has been introduced purely as a device for setting up the work–equilibrium relationship of Eq. (4.5). The forces, F, therefore remain unchanged in magnitude and direction during this imaginary displacement; this would not be the case if the displacement were real.

4.2.2 Principle of virtual work for a rigid body

Consider the rigid body shown in Fig. 4.3, which is acted upon by a system of external forces, $F_1, F_2, \ldots, F_k, \ldots, F_r$. These external forces will induce internal forces in the body, which may be regarded as comprising an infinite number of particles; on adjacent particles, such as A_1 and A_2, these internal forces will be equal and opposite, in other words self-equilibrating. Suppose now that the rigid body is given a small, imaginary, that is virtual, displacement, Δ_v (or a rotation or a combination of both), in some specified direction. The external and internal forces then do virtual work and the total virtual work done, W_t, is the sum of the virtual work, W_e, done by the external forces and the virtual work, W_i, done by the internal forces. Thus

$$W_t = W_e + W_i \tag{4.6}$$

Since the body is rigid, all the particles in the body move through the same displacement, Δ_v, so that the virtual work done on all the particles is numerically the same. However, for a pair of adjacent particles, such as A_1 and A_2 in Fig. 4.3, the self-equilibrating forces are in opposite directions, which means that the work done on A_1 is opposite in sign to the work done on A_2. Therefore the sum of the virtual work done on A_1 and A_2 is zero. The argument can be extended to the infinite number of pairs of particles in the body from which we conclude that the internal virtual work produced by a virtual

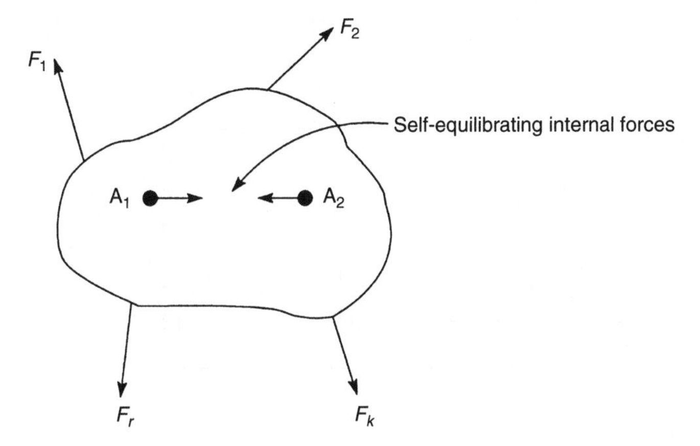

Fig. 4.3 Virtual work for a rigid body.

displacement in a rigid body is zero. Equation (4.6) then reduces to

$$W_t = W_e \tag{4.7}$$

Since the body is rigid and the internal virtual work is therefore zero, we may regard the body as a large particle. It follows that if the body is in equilibrium under the action of a set of forces, $F_1, F_2, \ldots, F_k, \ldots, F_r$, the total virtual work done by the external forces during an arbitrary virtual displacement of the body is zero.

Example 4.1

Calculate the support reactions in the simply supported beam shown in Fig. 4.4.

Only a vertical load is applied to the beam so that only vertical reactions, R_A and R_C, are produced.

Suppose that the beam at C is given a small imaginary, that is a virtual, displacement, $\Delta_{v,c}$, in the direction of R_C as shown in Fig. 4.4(b). Since we are concerned here solely with the *external* forces acting on the beam we may regard the beam as a rigid body. The beam therefore rotates about A so that C moves to C′ and B moves to B′. From similar triangles we see that

$$\Delta_{v,B} = \frac{a}{a+b}\Delta_{v,C} = \frac{a}{L}\Delta_{v,C} \tag{i}$$

The total virtual work, W_t, done by all the forces acting on the beam is then given by

$$W_t = R_C\Delta_{v,C} - W\Delta_{v,B} \tag{ii}$$

Note that the work done by the load, W, is negative since $\Delta_{v,B}$ is in the opposite direction to its line of action. Note also that the support reaction, R_A, does no work since the beam only rotates about A. Now substituting for $\Delta_{v,B}$ in Eq. (ii) from Eq. (i) we have

$$W_t = R_C\Delta_{v,C} - W\frac{a}{L}\Delta_{v,C} \tag{iii}$$

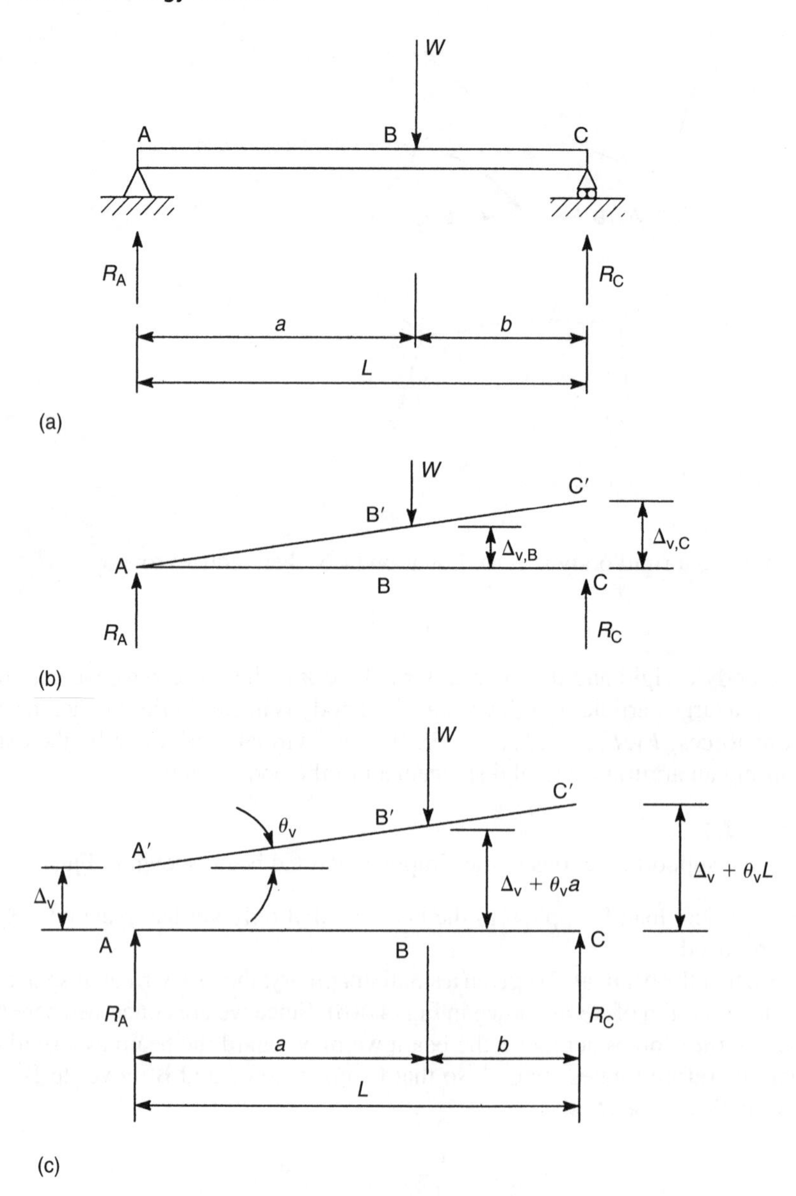

Fig. 4.4 Use of the principle of virtual work to calculate support reactions.

Since the beam is in equilibrium, W_t is zero from the principal of virtual work. Hence, from Eq. (iii)

$$R_C \Delta_{v,C} - W\frac{a}{L}\Delta_{v,C} = 0$$

which gives

$$R_C = W\frac{a}{L}$$

which is the result that would have been obtained from a consideration of the moment equilibrium of the beam about A. R_A follows in a similar manner. Suppose now that instead of the single displacement $\Delta_{v,C}$ the complete beam is given a vertical virtual displacement, Δ_v, together with a virtual rotation, θ_v, about A as shown in Fig. 4.4(c). The total virtual work, W_t, done by the forces acting on the beam is now given by

$$W_t = R_A\Delta_v - W(\Delta_v + a\theta_v) + R_C(\Delta_v + L\theta_v) = 0 \quad \text{(iv)}$$

since the beam is in equilibrium. Rearranging Eq. (iv)

$$(R_A + R_C - W)\Delta_v + (R_C L - Wa)\theta_v = 0 \quad \text{(v)}$$

Equation (v) is valid for all values of Δ_v and θ_v so that

$$R_A + R_C - W = 0 \quad R_C L - Wa = 0$$

which are the equations of equilibrium we would have obtained by resolving forces vertically and taking moments about A.

It is not being suggested here that the application of the principles of statics should be abandoned in favour of the principle of virtual work. The purpose of Example 4.1 is to illustrate the application of a virtual displacement and the manner in which the principle is used.

4.2.3 Virtual work in a deformable body

In structural analysis we are not generally concerned with forces acting on a rigid body. Structures and structural members deform under load, which means that if we assign a virtual displacement to a particular point in a structure, not all points in the structure will suffer the same virtual displacement as would be the case if the structure were rigid. This means that the virtual work produced by the internal forces is not zero as it is in the rigid body case, since the virtual work produced by the self-equilibrating forces on adjacent particles does not cancel out. The total virtual work produced by applying a virtual displacement to a deformable body acted upon by a system of external forces is therefore given by Eq. (4.6).

If the body is in equilibrium under the action of the external force system then every particle in the body is also in equilibrium. Therefore, from the principle of virtual work, the virtual work done by the forces acting on the particle is zero irrespective of whether the forces are external or internal. It follows that, since the virtual work is zero for all particles in the body, it is zero for the complete body and Eq. (4.6) becomes

$$W_e + W_i = 0 \quad (4.8)$$

Note that in the above argument only the conditions of equilibrium and the concept of work are employed. Equation (4.8) therefore does not require the deformable body to be linearly elastic (i.e. it need not obey Hooke's law) so that the principle of virtual work may be applied to any body or structure that is rigid, elastic or plastic. The principle does require that displacements, whether real or imaginary, must be small, so that we may assume that external and internal forces are unchanged in magnitude and direction during the displacements. In addition the virtual displacements must be compatible with the geometry of the structure and the constraints that are applied, such as those at a support. The exception is the situation we have in Example 4.1 where we apply a virtual displacement at a support. This approach is valid since we include the work done by the support reactions in the total virtual work equation.

4.2.4 Work done by internal force systems

The calculation of the work done by an external force is straightforward in that it is the product of the force and the displacement of its point of application in its own line of action (Eqs (4.1), (4.2) or (4.3)) whereas the calculation of the work done by an internal force system during a displacement is much more complicated. Generally no matter how complex a loading system is, it may be simplified to a combination of up to four load types: axial load, shear force, bending moment and torsion; these in turn produce corresponding internal force systems. We shall now consider the work done by these internal force systems during arbitrary virtual displacements.

Axial force

Consider the elemental length, δx, of a structural member as shown in Fig. 4.5 and suppose that it is subjected to a positive internal force system comprising a normal force (i.e. axial force), N, a shear force, S, a bending moment, M and a torque, T, produced by some external loading system acting on the structure of which the member is part. The stress distributions corresponding to these internal forces are related to an axis

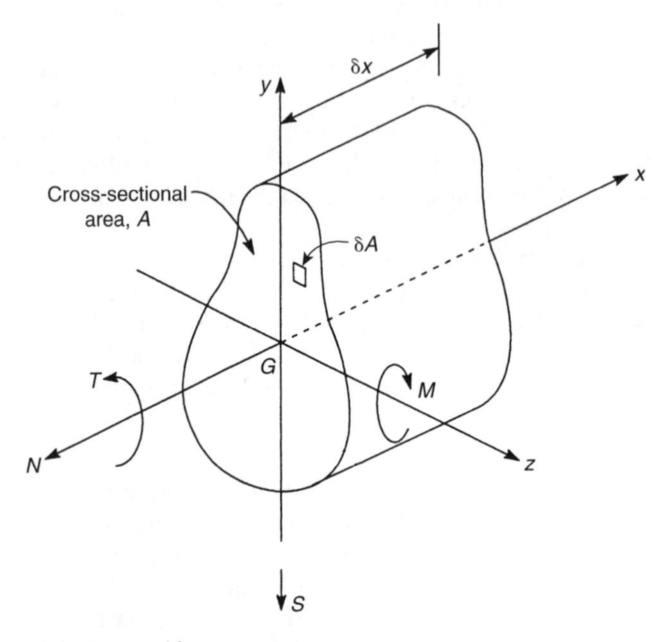

Fig. 4.5 Virtual work due to internal force system.

system whose origin coincides with the centroid of area of the cross-section. We shall, in fact, be using these stress distributions in the derivation of expressions for internal virtual work in linearly elastic structures so that it is logical to assume the same origin of axes here; we shall also assume that the y axis is an axis of symmetry. Initially we shall consider the normal force, N.

The direct stress, σ, at any point in the cross-section of the member is given by $\sigma = N/A$. Therefore the normal force on the element δA at the point (z, y) is

$$\delta N = \sigma \delta A = \frac{N}{A} \delta A$$

Suppose now that the structure is given an arbitrary virtual displacement which produces a virtual axial strain, ε_{v}, in the element. The internal virtual work, $\delta w_{\text{i},N}$, done by the axial force on the elemental length of the member is given by

$$\delta w_{\text{i},N} = \int_A \frac{N}{A} \mathrm{d}A \varepsilon_{\text{v}} \, \delta x$$

which, since $\int_A \mathrm{d}A = A$, reduces to

$$\delta w_{\text{i},N} = N \varepsilon_{\text{v}} \, \delta x \tag{4.9}$$

In other words, the virtual work done by N is the product of N and the virtual axial displacement of the element of the member. For a member of length L, the virtual work, $w_{\text{i},N}$, done during the arbitrary virtual strain is then

$$w_{\text{i},N} = \int_L N \varepsilon_{\text{v}} \, \mathrm{d}x \tag{4.10}$$

For a structure comprising a number of members, the total internal virtual work, $W_{\text{i},N}$, done by axial force is the sum of the virtual work of each of the members. Therefore

$$w_{\text{i},N} = \sum \int_L N \varepsilon_{\text{v}} \, \mathrm{d}x \tag{4.11}$$

Note that in the derivation of Eq. (4.11) we have made no assumption regarding the material properties of the structure so that the relationship holds for non-elastic as well as elastic materials. However, for a linearly elastic material, i.e. one that obeys Hooke's law, we can express the virtual strain in terms of an equivalent virtual normal force, i.e.

$$\varepsilon_{\text{v}} = \frac{\sigma_{\text{v}}}{E} = \frac{N_{\text{v}}}{EA}$$

Therefore, if we designate the *actual* normal force in a member by N_{A}, Eq. (4.11) may be expressed in the form

$$w_{\text{i},N} = \sum \int_L \frac{N_{\text{A}} N_{\text{v}}}{EA} \mathrm{d}x \tag{4.12}$$

Shear force

The shear force, S, acting on the member section in Fig. 4.5 produces a distribution of vertical shear stress which depends upon the geometry of the cross-section. However, since the element, δA, is infinitesimally small, we may regard the shear stress, τ, as constant over the element. The shear force, δS, on the element is then

$$\delta S = \tau\,\delta A \tag{4.13}$$

Suppose that the structure is given an arbitrary virtual displacement which produces a virtual shear strain, γ_{v}, at the element. This shear strain represents the angular rotation in a vertical plane of the element $\delta A \times \delta x$ relative to the longitudinal centroidal axis of the member. The vertical displacement at the section being considered is therefore $\gamma_{\text{v}}\,\delta x$. The internal virtual work, $\delta w_{\text{i},S}$, done by the shear force, S, on the elemental length of the member is given by

$$\delta w_{\text{i},S} = \int_A \tau\,\text{d}A\gamma_{\text{v}}\,\delta x$$

A uniform shear stress through the cross section of a beam may be assumed if we allow for the actual variation by including a form factor, β.[1] The expression for the internal virtual work in the member may then be written

$$\delta w_{\text{i},S} = \int_A \beta\left(\frac{S}{A}\right)\text{d}A\gamma_{\text{v}}\,\delta x$$

or

$$\delta w_{\text{i},S} = \beta S\gamma_{\text{v}}\,\delta x \tag{4.14}$$

Hence the virtual work done by the shear force during the arbitrary virtual strain in a member of length L is

$$w_{\text{i},S} = \beta\int_L S\gamma_{\text{v}}\,\text{d}x \tag{4.15}$$

For a linearly elastic member, as in the case of axial force, we may express the virtual shear strain, γ_{v}, in terms of an equivalent virtual shear force, S_{v}, i.e.

$$\gamma_{\text{v}} = \frac{\tau_{\text{v}}}{G} = \frac{S_{\text{v}}}{GA}$$

so that from Eq. (4.15)

$$w_{\text{i},S} = \beta\int_L \frac{S_{\text{A}}S_{\text{v}}}{GA}\,\text{d}x \tag{4.16}$$

For a structure comprising a number of linearly elastic members the total internal work, $W_{\text{i},S}$, done by the shear forces is

$$W_{\text{i},S} = \sum\beta\int_L \frac{S_{\text{A}}S_{\text{v}}}{GA}\,\text{d}x \tag{4.17}$$

Bending moment

The bending moment, M, acting on the member section in Fig. 4.5 produces a distribution of direct stress, σ, through the depth of the member cross-section. The normal force on the element, δA, corresponding to this stress is therefore $\sigma\,\delta A$. Again we shall suppose that the structure is given a small arbitrary virtual displacement which produces a virtual direct strain, ε_{v}, in the element $\delta A \times \delta x$. Thus the virtual work done by the normal force acting on the element δA is $\sigma\,\delta A\,\varepsilon_{\mathrm{v}}\,\delta x$. Hence, integrating over the complete cross-section of the member we obtain the internal virtual work, $\delta w_{\mathrm{i},M}$, done by the bending moment, M, on the elemental length of member, i.e.

$$\delta w_{\mathrm{i},M} = \int_A \sigma\,\mathrm{d}A\varepsilon_{\mathrm{v}}\,\delta x \tag{4.18}$$

The virtual strain, ε_{v}, in the element $\delta A \times \delta x$ is, from Eq. (16.2), given by

$$\varepsilon_{\mathrm{v}} = \frac{y}{R_{\mathrm{v}}}$$

where R_{v} is the radius of curvature of the member produced by the virtual displacement. Thus, substituting for ε_{v} in Eq. (4.18), we obtain

$$\delta w_{\mathrm{i},M} = \int_A \sigma \frac{y}{R_{\mathrm{v}}}\mathrm{d}A\,\delta x$$

or, since $\sigma y\,\delta A$ is the moment of the normal force on the element, δA, about the z axis

$$\delta w_{\mathrm{i},M} = \frac{M}{R_{\mathrm{v}}}\,\delta x$$

Therefore, for a member of length L, the internal virtual work done by an actual bending moment, M_{A}, is given by

$$w_{\mathrm{i},M} = \int_L \frac{M_{\mathrm{A}}}{R_{\mathrm{v}}}\,\mathrm{d}x \tag{4.19}$$

In the derivation of Eq. (4.19) no specific stress–strain relationship has been assumed, so that it is applicable to a non-linear system. For the particular case of a linearly elastic system, the virtual curvature $1/R_{\mathrm{v}}$ may be expressed in terms of an equivalent virtual bending moment, M_{v}, using the relationship of Eq. (16.20), i.e.

$$\frac{1}{R_{\mathrm{v}}} = \frac{M_{\mathrm{v}}}{EI}$$

Substituting for $1/R_{\mathrm{v}}$ in Eq. (4.19) we have

$$w_{\mathrm{i},M} = \int_L \frac{M_{\mathrm{A}}M_{\mathrm{v}}}{EI}\,\mathrm{d}x \tag{4.20}$$

so that for a structure comprising a number of members the total internal virtual work, $W_{\mathrm{i},M}$, produced by bending is

$$W_{\mathrm{i},M} = \sum \int_L \frac{M_{\mathrm{A}}M_{\mathrm{v}}}{EI}\,\mathrm{d}x \tag{4.21}$$

Torsion

The internal virtual work, $w_{i,T}$, due to torsion in the particular case of a linearly elastic circular section bar may be found in a similar manner and is given by

$$w_{i,T} = \int_L \frac{T_A T_v}{GI_o} \, dx \tag{4.22}$$

in which I_o is the polar second moment of area of the cross-section of the bar (see Example 3.1). For beams of non-circular cross-section, I_o is replaced by a torsion constant, J, which, for many practical beam sections is determined empirically.

Hinges

In some cases it is convenient to impose a virtual rotation, θ_v, at some point in a structural member where, say, the actual bending moment is M_A. The internal virtual work done by M_A is then $M_A\theta_v$ (see Eq. (4.3)); physically this situation is equivalent to inserting a hinge at the point.

Sign of internal virtual work

So far we have derived expressions for internal work without considering whether it is positive or negative in relation to external virtual work.

Suppose that the structural member, AB, in Fig. 4.6(a) is, say, a member of a truss and that it is in equilibrium under the action of two externally applied axial tensile loads, P; clearly the internal axial, that is normal, force at any section of the member is P. Suppose now that the member is given a virtual extension, δ_v, such that B moves to B′. Then the virtual work done by the applied load, P, is positive since the displacement, δ_v, is in the same direction as its line of action. However, the virtual work done by the internal force, N ($=P$), is negative since the displacement of B is in the opposite direction to its line of action; in other words work is done *on* the member. Thus, from Eq. (4.8), we see that in this case

$$W_e = W_i \tag{4.23}$$

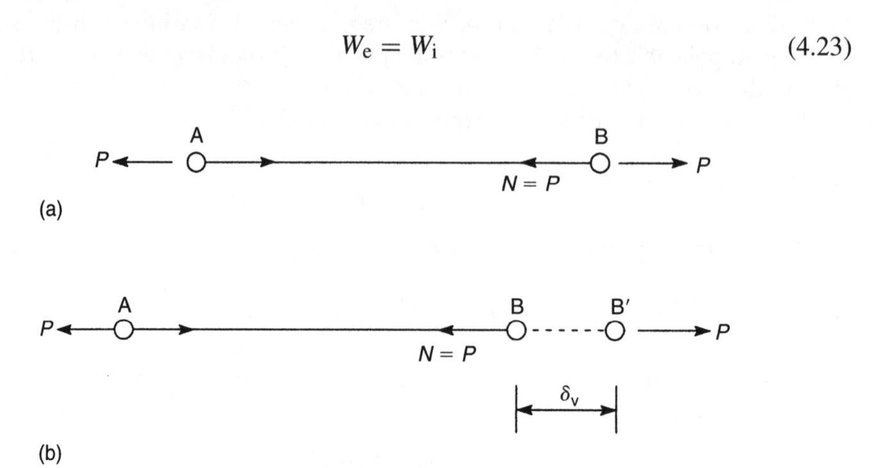

Fig. 4.6 Sign of the internal virtual work in an axially loaded member.

Equation (4.23) would apply if the virtual displacement had been a contraction and not an extension, in which case the signs of the external and internal virtual work in Eq. (4.8) would have been reversed. Clearly the above applies equally if P is a compressive load. The above arguments may be extended to structural members subjected to shear, bending and torsional loads, so that Eq. (4.23) is generally applicable.

4.2.5 Virtual work due to external force systems

So far in our discussion we have only considered the virtual work produced by externally applied concentrated loads. For completeness we must also consider the virtual work produced by moments, torques and distributed loads.

In Fig. 4.7 a structural member carries a distributed load, $w(x)$, and at a particular point a concentrated load, W, a moment, M and a torque, T. Suppose that at the point a virtual displacement is imposed that has translational components, $\Delta_{v,y}$ and $\Delta_{v,x}$, parallel to the y and x axes, respectively, and rotational components, θ_v and ϕ_v, in the yx and zy planes, respectively.

If we consider a small element, δx, of the member at the point, the distributed load may be regarded as constant over the length δx and acting, in effect, as a concentrated load $w(x)\delta x$. The virtual work, w_e, done by the complete external force system is therefore given by

$$w_e = W\Delta_{v,y} + P\Delta_{v,x} + M\theta_v + T\phi_v + \int_L w(x)\Delta_{v,y}\,dx$$

For a structure comprising a number of load positions, the total external virtual work done is then

$$W_e = \sum\left[W\Delta_{v,y} + P\Delta_{v,x} + M\theta_v + T\phi_v + \int_L w(x)\Delta_{v,y}\,dx\right] \tag{4.24}$$

In Eq. (4.24) there need not be a complete set of external loads applied at every loading point so, in fact, the summation is for the appropriate number of loads. Further, the virtual displacements in the above are related to forces and moments applied in a vertical plane. We could, of course, have forces and moments and components of the virtual

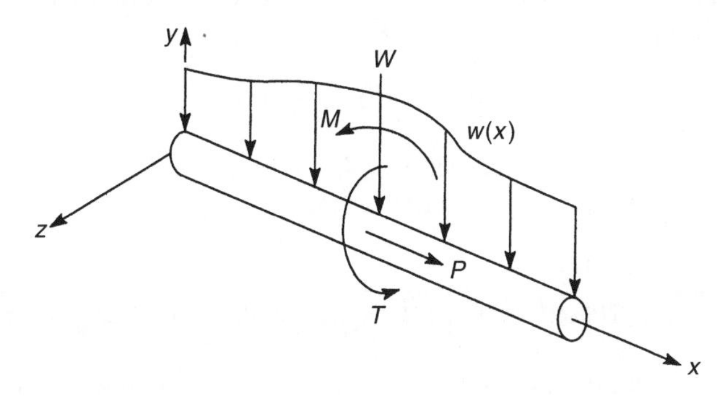

Fig. 4.7 Virtual work due to externally applied loads.

displacement in a horizontal plane, in which case Eq. (4.24) would be extended to include their contribution.

The internal virtual work equivalent of Eq. (4.24) for a linear system is, from Eqs (4.12), (4.17), (4.21) and (4.22)

$$W_\mathrm{i} = \sum \left[\int_L \frac{N_\mathrm{A} N_\mathrm{v}}{EA} \mathrm{d}x + \beta \int_L \frac{S_\mathrm{A} S_\mathrm{v}}{GA} \mathrm{d}x + \int_L \frac{M_\mathrm{A} M_\mathrm{v}}{EI} \mathrm{d}x + \int_L \frac{T_\mathrm{A} T_\mathrm{v}}{GJ} \mathrm{d}x + M_\mathrm{A}\theta_\mathrm{v} \right] \tag{4.25}$$

in which the last term on the right-hand side is the virtual work produced by an actual internal moment at a hinge (see above). Note that the summation in Eq. (4.25) is taken over all the *members* of the structure.

4.2.6 Use of virtual force systems

So far, in all the structural systems we have considered, virtual work has been produced by actual forces moving through imposed virtual displacements. However, the actual forces are not related to the virtual displacements in any way since, as we have seen, the magnitudes and directions of the actual forces are unchanged by the virtual displacements so long as the displacements are small. Thus the principle of virtual work applies for *any* set of forces in equilibrium and *any* set of displacements. Equally, therefore, we could specify that the forces are a set of virtual forces *in equilibrium* and that the displacements are actual displacements. Therefore, instead of relating actual external and internal force systems through virtual displacements, we can relate actual external and internal displacements through virtual forces.

If we apply a virtual force system to a deformable body it will induce an internal virtual force system which will move through the actual displacements; internal virtual work will therefore be produced. In this case, for example, Eq. (4.10) becomes

$$w_{\mathrm{i},N} = \int_L N_\mathrm{v} \varepsilon_\mathrm{A} \, \mathrm{d}x$$

in which N_v is the internal virtual normal force and ε_A is the actual strain. Then, for a linear system, in which the actual internal normal force is N_A, $\varepsilon_\mathrm{A} = N_\mathrm{A}/EA$, so that for a structure comprising a number of members the total internal virtual work due to a virtual normal force is

$$W_{\mathrm{i},N} = \sum \int_L \frac{N_\mathrm{v} N_\mathrm{A}}{EA} \mathrm{d}x$$

which is identical to Eq. (4.12). Equations (4.17), (4.21) and (4.22) may be shown to apply to virtual force systems in a similar manner.

4.3 Applications of the principle of virtual work

We have now seen that the principle of virtual work may be used either in the form of imposed virtual displacements or in the form of imposed virtual forces. Generally the former approach, as we saw in Example 4.1, is used to determine forces, while the latter is used to obtain displacements.

For statically determinate structures the use of virtual displacements to determine force systems is a relatively trivial use of the principle although problems of this type provide a useful illustration of the method. The real power of this approach lies in its application to the solution of statically indeterminate structures. However, the use of virtual forces is particularly useful in determining actual displacements of structures. We shall illustrate both approaches by examples.

Example 4.2

Determine the bending moment at the point B in the simply supported beam ABC shown in Fig. 4.8(a).

We determined the support reactions for this particular beam in Example 4.1. In this example, however, we are interested in the actual internal moment, M_B, at the point of application of the load. We must therefore impose a virtual displacement which will relate the internal moment at B to the applied load and which will exclude other unknown external forces such as the support reactions, and unknown internal force systems such as the bending moment distribution along the length of the beam. Therefore, if we imagine that the beam is hinged at B and that the lengths AB and BC are rigid, a virtual displacement, $\Delta_{v,B}$, at B will result in the displaced shape shown in Fig. 4.8(b).

Note that the support reactions at A and C do no work and that the internal moments in AB and BC do no work because AB and BC are rigid links. From Fig. 4.8(b)

$$\Delta_{v,B} = a\beta = b\alpha \tag{i}$$

Hence

$$\alpha = \frac{a}{b}\beta$$

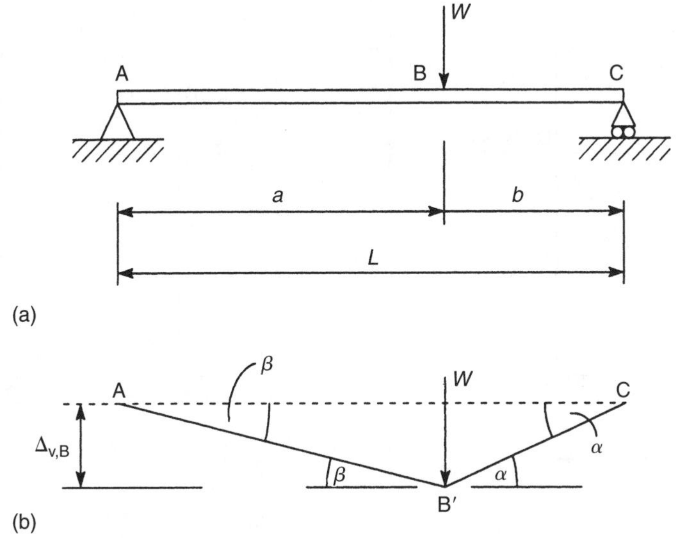

Fig. 4.8 Determination of bending moment at a point in the beam of Example 4.2 using virtual work.

and the angle of rotation of BC relative to AB is then

$$\theta_B = \beta + \alpha = \beta\left(1 + \frac{a}{b}\right) = \frac{L}{b}\beta \tag{ii}$$

Now equating the external virtual work done by W to the internal virtual work done by M_B (see Eq. (4.23)) we have

$$W\Delta_{v,B} = M_B\theta_B \tag{iii}$$

Substituting in Eq. (iii) for $\Delta_{v,B}$ from Eq. (i) and for θ_B from Eq. (ii) we have

$$Wa\beta = M_B\frac{L}{b}\beta$$

which gives

$$M_B = \frac{Wab}{L}$$

which is the result we would have obtained by calculating the moment of R_C (=Wa/L from Example 4.1) about B.

Example 4.3

Determine the force in the member AB in the truss shown in Fig. 4.9(a).

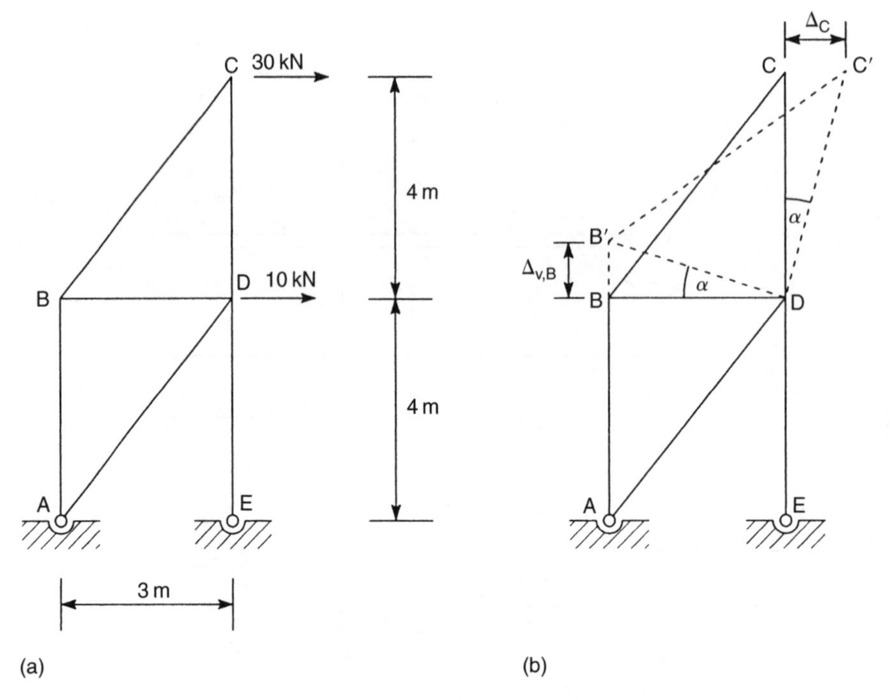

Fig. 4.9 Determination of the internal force in a member of a truss using virtual work.

We are required to calculate the force in the member AB, so that again we need to relate this internal force to the externally applied loads without involving the internal forces in the remaining members of the truss. We therefore impose a virtual extension, $\Delta_{v,B}$, at B in the member AB, such that B moves to B′. If we assume that the remaining members are rigid, the forces in them will do no work. Further, the triangle BCD will rotate as a rigid body about D to B′C′D as shown in Fig. 4.9(b). The horizontal displacement of C, Δ_C, is then given by

$$\Delta_C = 4\alpha$$

while

$$\Delta_{v,B} = 3\alpha$$

Hence

$$\Delta_C = \frac{4\Delta_{v,B}}{3} \tag{i}$$

Equating the external virtual work done by the 30 kN load to the internal virtual work done by the force, F_{BA}, in the member, AB, we have (see Eq. (4.23) and Fig. 4.6)

$$30\Delta_C = F_{BA}\Delta_{v,B} \tag{ii}$$

Substituting for ΔC from Eq. (i) in Eq. (ii),

$$30 \times \frac{4}{3}\Delta_{v,B} = F_{BA}\Delta_{v,B}$$

Whence

$$F_{BA} = +40\,\text{kN} \quad (\text{i.e. } F_{BA} \text{ is tensile})$$

In the above we are, in effect, assigning a positive (i.e. tensile) sign to F_{BA} by imposing a virtual extension on the member AB.

The actual sign of F_{BA} is then governed by the sign of the external virtual work. Thus, if the 30 kN load had been in the opposite direction to Δ_C the external work done would have been negative, so that F_{BA} would be negative and therefore compressive. This situation can be verified by inspection. Alternatively, for the loading as shown in Fig. 4.9(a), a contraction in AB would have implied that F_{BA} was compressive. In this case DC would have rotated in an anticlockwise sense, Δ_C would have been in the opposite direction to the 30 kN load so that the external virtual work done would be negative, resulting in a negative value for the compressive force F_{BA}; F_{BA} would therefore be tensile as before. Note also that the 10 kN load at D does no work since D remains undisplaced.

We shall now consider problems involving the use of virtual forces. Generally we shall require the displacement of a particular point in a structure, so that if we apply a virtual force to the structure at the point and in the direction of the required displacement the external virtual work done will be the product of the virtual force and the actual displacement, which may then be equated to the internal virtual work produced by the internal virtual force system moving through actual displacements. Since the choice of the virtual force is arbitrary, we may give it any convenient value; the simplest type of virtual force is therefore a unit load and the method then becomes the *unit load method* (see also Section 5.5).

Example 4.4

Determine the vertical deflection of the free end of the cantilever beam shown in Fig. 4.10(a).

Let us suppose that the actual deflection of the cantilever at B produced by the uniformly distributed load is υ_{B} and that a vertically downward virtual unit load was applied at B before the actual deflection took place. The external virtual work done by the unit load is, from Fig. 4.10(b), $1\upsilon_{\mathrm{B}}$. The deflection, υ_{B}, is assumed to be caused by bending only, i.e. we are ignoring any deflections due to shear. The internal virtual work is given by Eq. (4.21) which, since only one member is involved, becomes

$$W_{\mathrm{i},M} = \int_0^L \frac{M_{\mathrm{A}} M_{\mathrm{v}}}{EI}\,\mathrm{d}x \qquad \text{(i)}$$

The virtual moments, M_{v}, are produced by a unit load so that we shall replace M_{v} by M_1. Then

$$W_{\mathrm{i},M} = \int_0^L \frac{M_{\mathrm{A}} M_1}{EI}\,\mathrm{d}x \qquad \text{(ii)}$$

At any section of the beam a distance x from the built-in end

$$M_{\mathrm{A}} = -\frac{w}{2}(L-x)^2 \quad M_1 = -1(L-x)$$

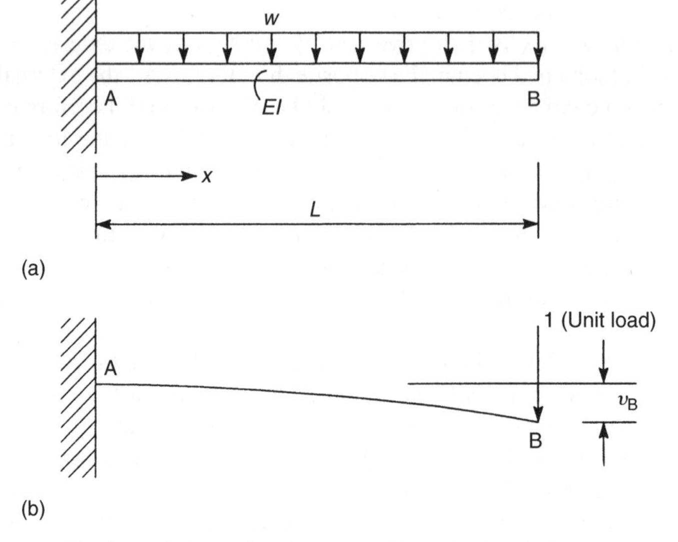

Fig. 4.10 Deflection of the free end of a cantilever beam using the unit load method.

Substituting for M_A and M_1 in Eq. (ii) and equating the external virtual work done by the unit load to the internal virtual work we have

$$1\upsilon_B = \int_0^L \frac{w}{2EI}(L-x)^3\,dx$$

which gives

$$\upsilon_B = -\frac{w}{2EI}\left[\frac{1}{4}(L-x)^4\right]_0^L$$

so that

$$\upsilon_B = \frac{wL^4}{8EI}$$

Note that υ_B is in fact negative but the positive sign here indicates that it is in the same direction as the unit load.

Example 4.5

Determine the rotation, i.e. the slope, of the beam ABC shown in Fig. 4.11(a) at A.

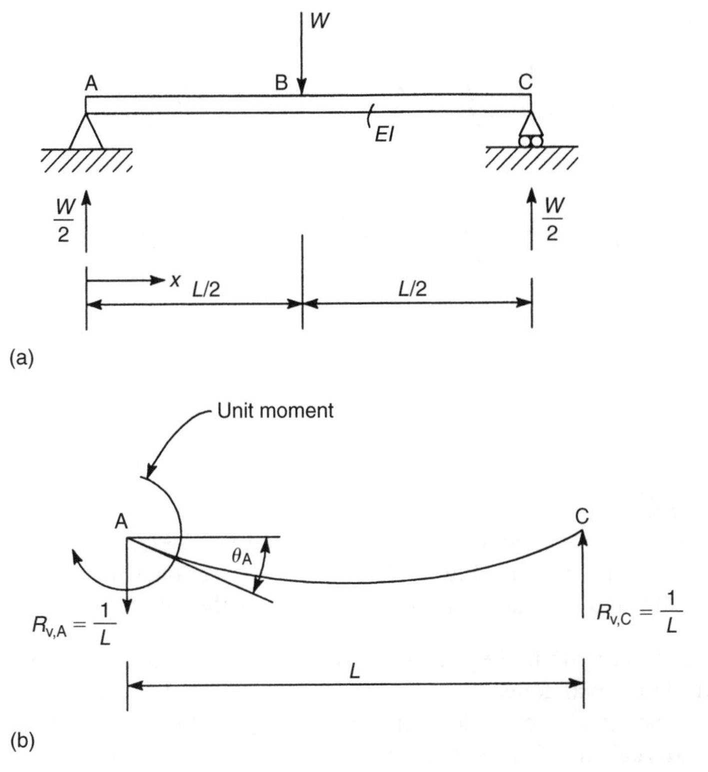

Fig. 4.11 Determination of the rotation of a simply supported beam at a support using the unit load method.

The actual rotation of the beam at A produced by the actual concentrated load, W, is θ_A. Let us suppose that a virtual unit moment is applied at A before the actual rotation takes place, as shown in Fig. 4.11(b). The virtual unit moment induces virtual support reactions of $R_{v,A}$ ($=1/L$) acting downwards and $R_{v,C}$ ($=1/L$) acting upwards. The actual internal bending moments are

$$M_A = +\frac{W}{2}x \quad 0 \leq x \leq L/2$$

$$M_A = +\frac{W}{2}(L-x) \quad L/2 \leq x \leq L$$

The internal virtual bending moment is

$$M_v = 1 - \frac{1}{L}x \quad 0 \leq x \leq L$$

The external virtual work done is $1\theta_A$ (the virtual support reactions do no work as there is no vertical displacement of the beam at the supports) and the internal virtual work done is given by Eq. (4.21). Hence

$$1\theta_A = \frac{1}{EI}\left[\int_0^{L/2} \frac{W}{2}x\left(1-\frac{x}{L}\right)\mathrm{d}x + \int_{L/2}^{L} \frac{W}{2}(L-x)\left(1-\frac{x}{L}\right)\mathrm{d}x\right] \qquad \text{(i)}$$

Simplifying Eq. (i) we have

$$\theta_A = \frac{W}{2EIL}\left[\int_0^{L/2} (Lx - x^2)\mathrm{d}x + \int_{L/2}^{L} (L-x)^2\mathrm{d}x\right] \qquad \text{(ii)}$$

Hence

$$\theta_A = \frac{W}{2EIL}\left\{\left[L\frac{x^2}{2} - \frac{x^3}{3}\right]_0^{L/2} - \frac{1}{3}\left[(L-x)^3\right]_{L/2}^{L}\right\}$$

from which

$$\theta_A = \frac{WL^2}{16EI}$$

Example 4.6

Calculate the vertical deflection of the joint B and the horizontal movement of the support D in the truss shown in Fig. 4.12(a). The cross-sectional area of each member is 1800 mm^2 and Young's modulus, E, for the material of the members is 200 000 N/mm^2.

The virtual force systems, i.e. unit loads, required to determine the vertical deflection of B and the horizontal deflection of D are shown in Fig. 4.12(b) and (c), respectively. Therefore, if the actual vertical deflection at B is $\delta_{B,v}$ and the horizontal deflection at D is $\delta_{D,h}$ the external virtual work done by the unit loads is $1\delta_{B,v}$ and $1\delta_{D,h}$, respectively. The internal actual and virtual force systems comprise axial forces in all the members.

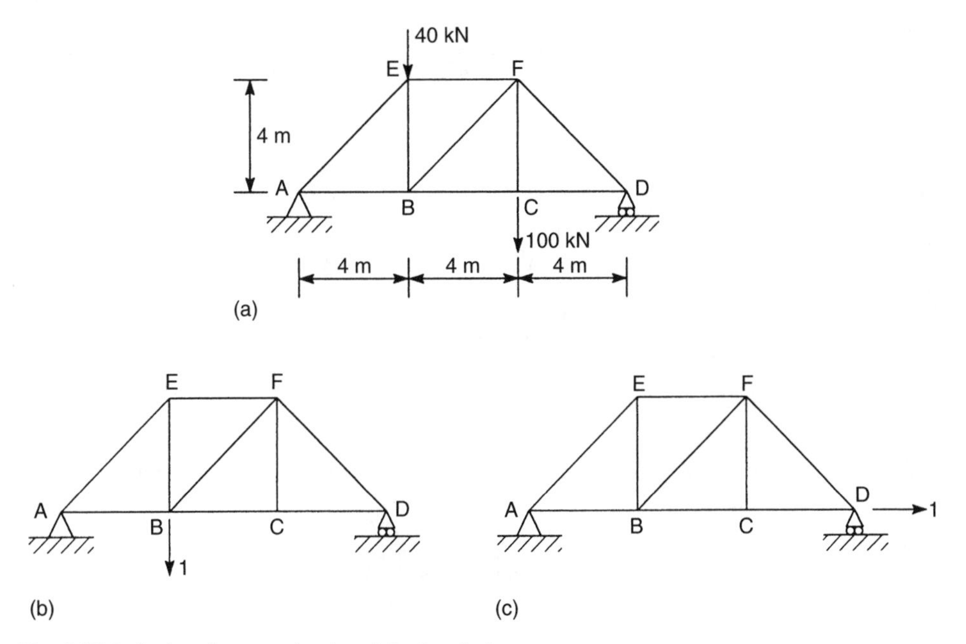

Fig. 4.12 Deflection of a truss using the unit load method.

These axial forces are constant along the length of each member so that for a truss comprising n members, Eq. (4.12) reduces to

$$W_{i,N} = \sum_{j=1}^{n} \frac{F_{A,j}F_{v,j}L_j}{E_jA_j} \tag{i}$$

in which $F_{A,j}$ and $F_{v,j}$ are the actual and virtual forces in the jth member which has a length L_j, an area of cross-section A_j and a Young's modulus E_j.

Since the forces $F_{v,j}$ are due to a unit load, we shall write Eq. (i) in the form

$$W_{i,N} = \sum_{j=1}^{n} \frac{F_{A,j}F_{1,j}L_j}{E_jA_j} \tag{ii}$$

Also, in this particular example, the area of cross-section, A, and Young's modulus, E, are the same for all members so that it is sufficient to calculate $\sum_{j=1}^{n} F_{A,j}F_{1,j}L_j$ and then divide by EA to obtain $W_{i,N}$.

The forces in the members, whether actual or virtual, may be calculated by the method of joints.[3] Note that the support reactions corresponding to the three sets of applied loads (one actual and two virtual) must be calculated before the internal force systems can be determined. However, in Fig. 4.12(c), it is clear from inspection that $F_{1,AB} = F_{1,BC} = F_{1,CD} = +1$ while the forces in all other members are zero. The calculations are presented in Table 4.1; note that positive signs indicate tension and negative signs compression.

Table 4.1

Member	L (m)	F_A (kN)	$F_{1,B}$	$F_{1,D}$	$F_A F_{1,B} L$ (kN m)	$F_A F_{1,D} L$ (kN m)
AE	5.7	−84.9	−0.94	0	+451.4	0
AB	4.0	+60.0	+0.67	+1.0	+160.8	+240.0
EF	4.0	−60.0	−0.67	0	+160.8	0
EB	4.0	+20.0	+0.67	0	+53.6	0
BF	5.7	−28.3	+0.47	0	−75.2	0
BC	4.0	+80.0	+0.33	+1.0	+105.6	+320.0
CD	4.0	+80.0	+0.33	+1.0	+105.6	+320.0
CF	4.0	+100.0	0	0	0	0
DF	5.7	−113.1	−0.47	0	+301.0	0
					$\sum = +1263.6$	$\sum = +880.0$

Thus equating internal and external virtual work done (Eq. (4.23)) we have

$$1\delta_{B,v} = \frac{1263.6 \times 10^6}{200\,000 \times 1800}$$

whence

$$\delta_{B,v} = 3.51 \text{ mm}$$

and

$$1\delta_{D,h} = \frac{880 \times 10^6}{200\,000 \times 1800}$$

which gives

$$\delta_{D,h} = 2.44 \text{ mm}$$

Both deflections are positive which indicates that the deflections are in the directions of the applied unit loads. Note that in the above it is unnecessary to specify units for the unit load since the unit load appears, in effect, on both sides of the virtual work equation (the internal F_1 forces are directly proportional to the unit load).

References

1 Megson, T. H. G., *Structural and Stress Analysis*, 2nd edition, Elsevier, Oxford, 2005.

Problems

P.4.1 Use the principle of virtual work to determine the support reactions in the beam ABCD shown in Fig. P.4.1.

Ans. $R_A = 1.25W \quad R_D = 1.75W$.

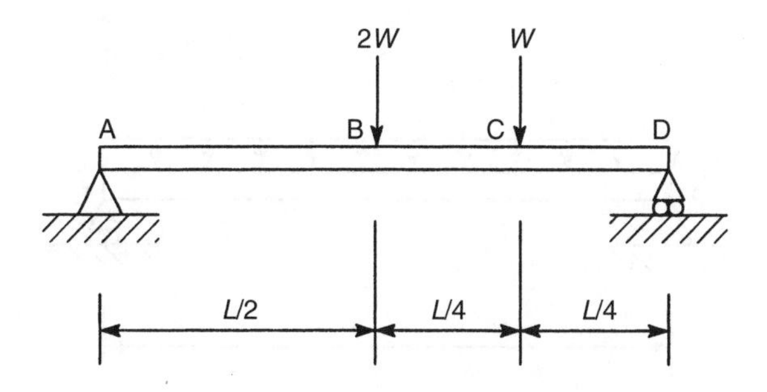

Fig. P.4.1

P.4.2 Find the support reactions in the beam ABC shown in Fig. P.4.2 using the principle of virtual work.

Ans. $R_{\mathrm{A}} = (W + 2wL)/4 \quad R_{\mathrm{c}} = (3w + 2wL)/4.$

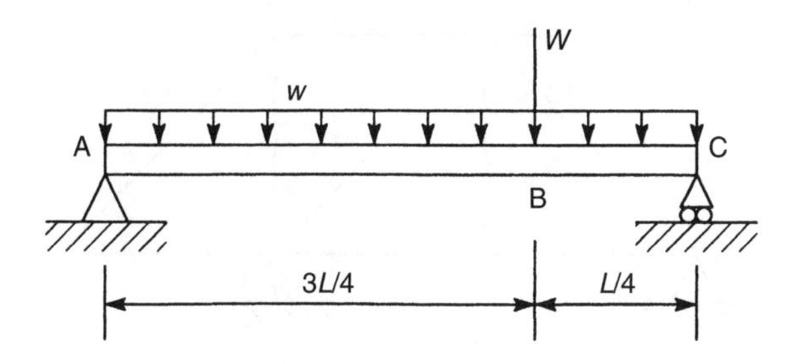

Fig. P.4.2

P.4.3 Determine the reactions at the built-in end of the cantilever beam ABC shown in Fig. P.4.3 using the principle of virtual work.

Ans. $R_{\mathrm{A}} = 3W \quad M_{\mathrm{A}} = 2.5WL.$

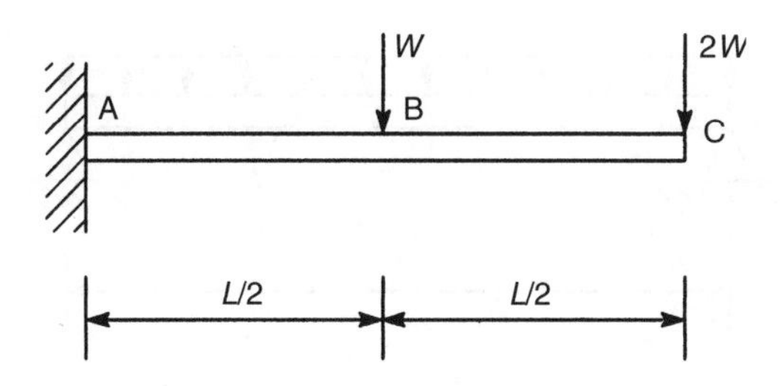

Fig. P.4.3

P.4.4 Find the bending moment at the three-quarter-span point in the beam shown in Fig. P.4.4. Use the principle of virtual work.

Ans. $3wL^2/32$.

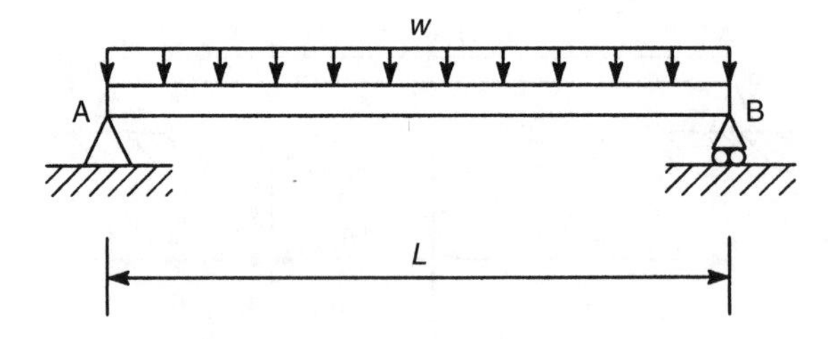

Fig. P.4.4

P.4.5 Calculate the forces in the members FG, GD and CD of the truss shown in Fig. P.4.5 using the principle of virtual work. All horizontal and vertical members are 1 m long.

Ans. FG = +20 kN GD = +28.3 kN CD = −20 kN.

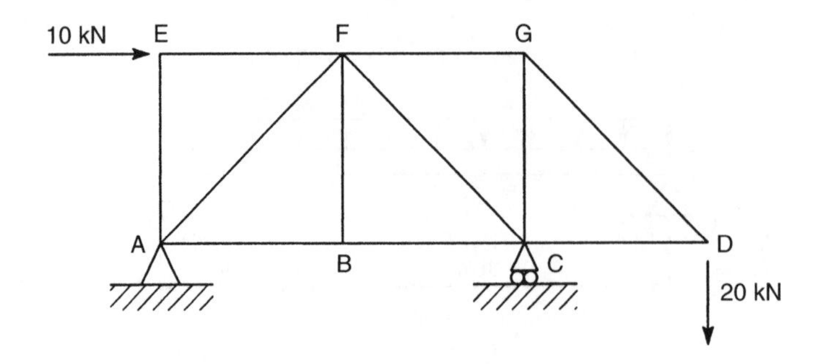

Fig. P.4.5

P.4.6 Use the principle of virtual work to calculate the vertical displacements at the quarter- and mid-span points in the beam shown in Fig. P.4.6.

Ans. $57wL^4/6144EI$ $5wL^4/384EI$ (both downwards).

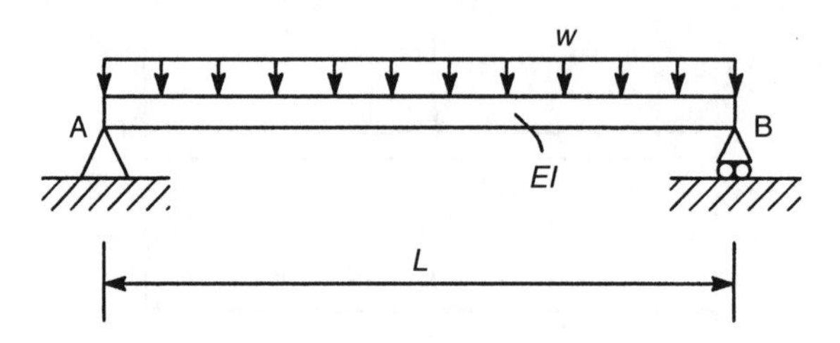

Fig. P.4.6

5

Energy methods

In Chapter 2 we have seen that the elasticity method of structural analysis embodies the determination of stresses and/or displacements by employing equations of equilibrium and compatibility in conjunction with the relevant force–displacement or stress–strain relationships. In addition, in Chapter 4, we investigated the use of virtual work in calculating forces, reactions and displacements in structural systems. A powerful alternative but equally fundamental approach is the use of energy methods. These, while providing exact solutions for many structural problems, find their greatest use in the rapid approximate solution of problems for which exact solutions do not exist. Also, many structures which are statically indeterminate, i.e. they cannot be analysed by the application of the equations of statical equilibrium alone, may be conveniently analysed using an energy approach. Further, energy methods provide comparatively simple solutions for deflection problems which are not readily solved by more elementary means.

Generally, as we shall see, modern analysis[1] uses the methods of *total complementary energy* and *total potential energy*. Either method may be employed to solve a particular problem, although as a general rule deflections are more easily found using complementary energy, and forces by potential energy.

Although energy methods are applicable to a wide range of structural problems and may even be used as indirect methods of forming equations of equilibrium or compatibility,[1,2] we shall be concerned in this chapter with the solution of deflection problems and the analysis of statically indeterminate structures. We shall also include some methods restricted to the solution of linear systems, i.e. the *unit load method*, the *principle of superposition* and the *reciprocal theorem*.

5.1 Strain energy and complementary energy

Figure 5.1(a) shows a structural member subjected to a steadily increasing load P. As the member extends, the load P does work and from the law of conservation of energy this work is stored in the member as *strain energy*. A typical load–deflection curve for a member possessing non-linear elastic characteristics is shown in Fig. 5.1(b). The strain energy U produced by a load P and corresponding extension y is then

$$U = \int_0^y P \, \mathrm{d}y \tag{5.1}$$

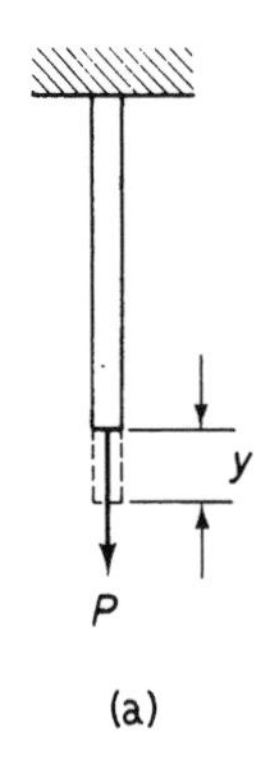

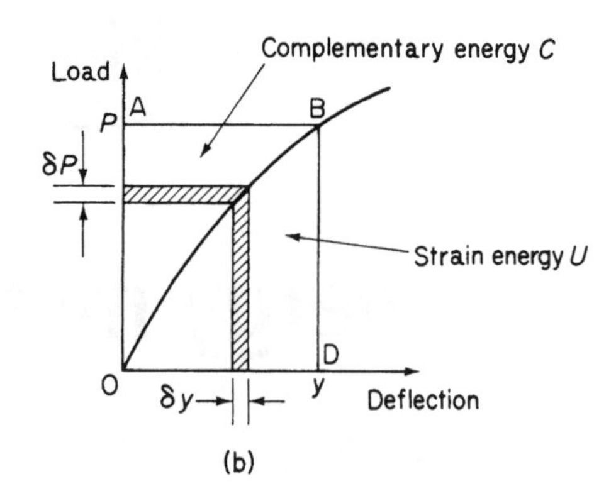

Fig. 5.1 (a) Strain energy of a member subjected to simple tension; (b) load–deflection curve for a nonlinearly elastic member.

and is clearly represented by the area OBD under the load–deflection curve. Engesser (1889) called the area OBA above the curve the *complementary energy* C, and from Fig. 5.1(b)

$$C = \int_0^P y \, \mathrm{d}P \tag{5.2}$$

Complementary energy, as opposed to strain energy, has no physical meaning, being purely a convenient mathematical quantity. However, it is possible to show that complementary energy obeys the law of conservation of energy in the type of situation usually arising in engineering structures, so that its use as an energy method is valid.

Differentiation of Eqs (5.1) and (5.2) with respect to y and P, respectively gives

$$\frac{\mathrm{d}U}{\mathrm{d}y} = P \quad \frac{\mathrm{d}C}{\mathrm{d}P} = y$$

Bearing these relationships in mind we can now consider the interchangeability of strain and complementary energy. Suppose that the curve of Fig. 5.1(b) is represented by the function

$$P = by^n$$

where the coefficient b and exponent n are constants. Then

$$U = \int_0^y P \, \mathrm{d}y = \frac{1}{n} \int_0^P \left(\frac{P}{b}\right)^{1/n} \mathrm{d}P$$

$$C = \int_0^P y \, \mathrm{d}P = n \int_0^y by^n \, \mathrm{d}y$$

Hence

$$\frac{\mathrm{d}U}{\mathrm{d}y} = P \quad \frac{\mathrm{d}U}{\mathrm{d}P} = \frac{1}{n}\left(\frac{P}{b}\right)^{1/n} = \frac{1}{n}y \tag{5.3}$$

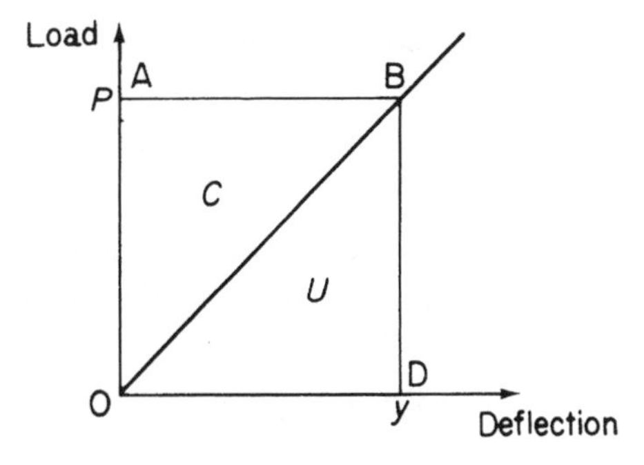

Fig. 5.2 Load–deflection curve for a linearly elastic member.

$$\frac{\mathrm{d}C}{\mathrm{d}P} = y \quad \frac{\mathrm{d}C}{\mathrm{d}y} = bny^n = nP \tag{5.4}$$

When $n = 1$

$$\left.\begin{aligned} \frac{\mathrm{d}U}{\mathrm{d}y} &= \frac{\mathrm{d}C}{\mathrm{d}y} = P \\ \frac{\mathrm{d}U}{\mathrm{d}P} &= \frac{\mathrm{d}C}{\mathrm{d}P} = y \end{aligned}\right\} \tag{5.5}$$

and the strain and complementary energies are completely interchangeable. Such a condition is found in a linearly elastic member; its related load–deflection curve being that shown in Fig. 5.2. Clearly, area OBD(U) is equal to area OBA(C).

It will be observed that the latter of Eqs (5.5) is in the form of what is commonly known as Castigliano's first theorem, in which the differential of the strain energy U of a structure with respect to a load is equated to the deflection of the load. To be mathematically correct, however, it is the differential of the complementary energy C which should be equated to deflection (compare Eqs (5.3) and (5.4)).

5.2 The principle of the stationary value of the total complementary energy

Consider an elastic system in equilibrium supporting forces $P_1, P_2, \ldots, P_n$ which produce real corresponding displacements $\Delta_1, \Delta_2, \ldots, \Delta_n$. If we impose virtual forces $\delta P_1, \delta P_2, \ldots, \delta P_n$ on the system acting through the real displacements then the total virtual work done by the system is (see Chapter 4)

$$-\int_{\text{vol}} y\,\mathrm{d}P + \sum_{r=1}^{n} \Delta_r \delta P_r$$

The first term in the above expression is the negative virtual work done by the particles in the elastic body, while the second term represents the virtual work of the externally applied virtual forces. From the principle of virtual work

$$-\int_{\text{vol}} y\,\mathrm{d}P + \sum_{r=1}^{n} \Delta_r \delta P_r = 0 \tag{5.6}$$

Comparing Eq. (5.6) with Eq. (5.2) we see that each term represents an increment in complementary energy; the first, of the internal forces, the second, of the external loads. Equation (5.6) may therefore be rewritten

$$\delta(C_\text{i} + C_\text{e}) = 0 \tag{5.7}$$

where

$$C_\text{i} = \int_{\text{vol}} \int_0^P y\,\mathrm{d}P \quad \text{and} \quad C_\text{e} = -\sum_{r=1}^{n} \Delta_r P_r \tag{5.8}$$

We shall now call the quantity $(C_\text{i} + C_\text{e})$ the total complementary energy C of the system.

The displacements specified in Eq. (5.6) are real displacements of a continuous elastic body; they therefore obey the condition of compatibility of displacement so that Eqs (5.6) and (5.7) are equations of geometrical compatibility. The *principle of the stationary value of the total complementary energy* may then be stated as:

> *For an elastic body in equilibrium under the action of applied forces the true internal forces (or stresses) and reactions are those for which the total complementary energy has a stationary value.*

In other words the true internal forces (or stresses) and reactions are those which satisfy the condition of compatibility of displacement. This property of the total complementary energy of an elastic system is particularly useful in the solution of statically indeterminate structures, in which an infinite number of stress distributions and reactive forces may be found to satisfy the requirements of equilibrium.

5.3 Application to deflection problems

Generally, deflection problems are most readily solved by the complementary energy approach, although for linearly elastic systems there is no difference between the methods of complementary and potential energy since, as we have seen, complementary and strain energy then become completely interchangeable. We shall illustrate the method by reference to the deflections of frames and beams which may or may not possess linear elasticity.

Let us suppose that we require to find the deflection Δ_2 of the load P_2 in the simple pin-jointed framework consisting, say, of k members and supporting loads P_1, $P_2, \ldots, P_n$, as shown in Fig. 5.3. From Eqs (5.8) the total complementary energy of the framework is given by

$$C = \sum_{i=1}^{k} \int_0^{F_i} \lambda_i\,\mathrm{d}F_i - \sum_{r=1}^{n} \Delta_r P_r \tag{5.9}$$

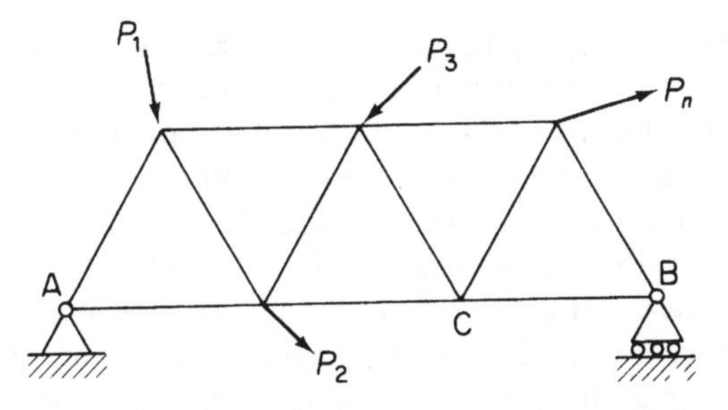

Fig. 5.3 Determination of the deflection of a point on a framework by the method of complementary energy.

where λ_i is the extension of the ith member, F_i the force in the ith member and Δ_r the corresponding displacement of the rth load P_r. From the principle of the stationary value of the total complementary energy

$$\frac{\partial C}{\partial P_2} = \sum_{i=1}^{k} \lambda_i \frac{\partial F_i}{\partial P_2} - \Delta_2 = 0 \tag{5.10}$$

from which

$$\Delta_2 = \sum_{i=1}^{k} \lambda_i \frac{\partial F_i}{\partial P_2} \tag{5.11}$$

Equation (5.10) is seen to be identical to the principle of virtual forces in which virtual forces δF and δP act through real displacements λ and Δ. Clearly the partial derivatives with respect to P_2 of the constant loads $P_1, P_2, \ldots, P_n$ vanish, leaving the required deflection Δ_2 as the unknown. At this stage, before Δ_2 can be evaluated, the load–displacement characteristics of the members must be known. For linear elasticity

$$\lambda_i = \frac{F_i L_i}{A_i E_i}$$

where L_i, A_i and E_i are the length, cross-sectional area and modulus of elasticity of the ith member. On the other hand, if the load–displacement relationship is of a non-linear form, say

$$F_i = b(\lambda_i)^c$$

in which b and c are known, then Eq. (5.11) becomes

$$\Delta_2 = \sum_{i=1}^{k} \left(\frac{F_i}{b}\right)^{1/c} \frac{\partial F_i}{\partial P_2}$$

The computation of Δ_2 is best accomplished in tabular form, but before the procedure is illustrated by an example some aspects of the solution merit discussion.

We note that the support reactions do not appear in Eq. (5.9). This convenient absence derives from the fact that the displacements $\Delta_1, \Delta_2, \ldots, \Delta_n$ are the real displacements of the frame and fulfil the conditions of geometrical compatibility and boundary restraint. The complementary energy of the reaction at A and the vertical reaction at B is therefore zero, since both of their corresponding displacements are zero. If we examine Eq. (5.11) we note that λ_i is the extension of the ith member of the framework due to the applied loads $P_1, P_2, \ldots, P_n$. Therefore, the loads F_i in the substitution for λ_i in Eq. (5.11) are those corresponding to the loads $P_1, P_2, \ldots, P_n$. The term $\partial F_i/\partial P_2$ in Eq. (5.11) represents the rate of change of F_i with P_2 and is calculated by applying the load P_2 to the *unloaded* frame and determining the corresponding member loads in terms of P_2. This procedure indicates a method for obtaining the displacement of either a point on the frame in a direction not coincident with the line of action of a load or, in fact, a point such as C which carries no load at all. We place at the point and in the required direction a *fictitious* or *dummy* load, say P_f, the original loads being removed. The loads in the members due to P_f are then calculated and $\partial F/\partial P_f$ obtained for each member. Substitution in Eq. (5.11) produces the required deflection.

It must be pointed out that it is not absolutely necessary to remove the actual loads during the application of P_f. The force in each member would then be calculated in terms of the actual loading and P_f. F_i follows by substituting $P_f = 0$ and $\partial F_i/\partial P_f$ is found by differentiation with respect to P_f. Obviously the two approaches yield the same expressions for F_i and $\partial F_i/\partial P_f$, although the latter is arithmetically clumsier.

Example 5.1

Calculate the vertical deflection of the point B and the horizontal movement of D in the pin-jointed framework shown in Fig. 5.4(a). All members of the framework are linearly

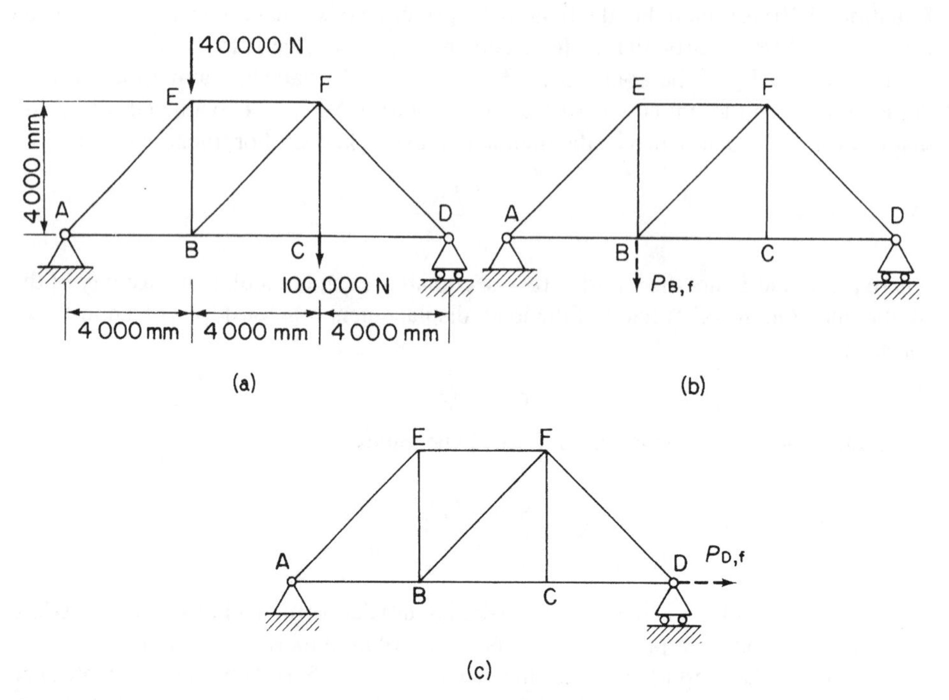

Fig. 5.4 (a) Actual loading of framework; (b) determination of vertical deflection of B; (c) determination of horizontal deflection of D.

elastic and have cross-sectional areas of 1800 mm^2. E for the material of the members is 200 000 N/mm^2.

The members of the framework are linearly elastic so that Eq. (5.11) may be written

$$\Delta = \sum_{i=1}^{k} \frac{F_i L_i}{A_i E_i} \frac{\partial F_i}{\partial P} \tag{i}$$

or, since each member has the same cross-sectional area and modulus of elasticity

$$\Delta = \frac{1}{AE} \sum_{i=1}^{k} F_i L_i \frac{\partial F_i}{\partial P} \tag{ii}$$

The solution is completed in Table 5.1, in which F are the member forces due to the actual loading of Fig. 5.4(a), $F_{\mathrm{B,f}}$ are the member forces due to the fictitious load $P_{\mathrm{B,f}}$ in Fig. 5.4(b) and $F_{\mathrm{D,f}}$ are the forces in the members produced by the fictitious load $P_{\mathrm{D,f}}$ in Fig. 5.4(c). We take tensile forces as positive and compressive forces as negative.

The vertical deflection of B is

$$\Delta_{\mathrm{B,v}} = \frac{1268 \times 10^6}{1800 \times 200\,000} = 3.52\,\mathrm{mm}$$

and the horizontal movement of D is

$$\Delta_{\mathrm{D,h}} = \frac{880 \times 10^6}{1800 \times 200\,000} = 2.44\,\mathrm{mm}$$

which agree with the virtual work solution (Example 4.6).

The positive values of $\Delta_{\mathrm{B,v}}$ and $\Delta_{\mathrm{D,h}}$ indicate that the deflections are in the directions of $P_{\mathrm{B,f}}$ and $P_{\mathrm{D,f}}$.

The analysis of beam deflection problems by complementary energy is similar to that of pin-jointed frameworks, except that we assume initially that displacements are caused primarily by bending action. Shear force effects are discussed later in the chapter. Figure 5.5 shows a tip loaded cantilever of uniform cross-section and length L. The tip load P produces a vertical deflection Δ_{v} which we require to find.

Table 5.1

① Member	② L (mm)	③ F(N)	④ $F_{\mathrm{B,f}}$ (N)	⑤ $\partial F_{\mathrm{B,f}}/\partial P_{\mathrm{B,f}}$	⑥ $F_{\mathrm{D,f}}$ (N)	⑦ $\partial F_{\mathrm{D,f}}/\partial P_{\mathrm{D,f}}$	⑧ $\times 10^6$ $FL\partial F_{\mathrm{B,f}}/\partial P_{\mathrm{B,f}}$	⑨ $\times 10^6$ $FL\partial F_{\mathrm{D,f}}/\partial P_{\mathrm{D,f}}$
AE	$4000\sqrt{2}$	$-60\,000\sqrt{2}$	$-2\sqrt{2}P_{\mathrm{B,f}}/3$	$-2\sqrt{2}/3$	0	0	$320\sqrt{2}$	0
EF	4000	$-60\,000$	$-2P_{\mathrm{B,f}}/3$	$-2/3$	0	0	160	0
FD	$4000\sqrt{2}$	$-80\,000\sqrt{2}$	$-\sqrt{2}P_{\mathrm{B,f}}/3$	$-\sqrt{2}/3$	0	0	$640\sqrt{2}/3$	0
DC	4000	80 000	$P_{\mathrm{B,f}}/3$	1/3	$P_{\mathrm{D,f}}$	1	320/3	320
CB	4000	80 000	$P_{\mathrm{B,f}}/3$	1/3	$P_{\mathrm{D,f}}$	1	320/3	320
BA	4000	60 000	$2P_{\mathrm{B,f}}/3$	2/3	$P_{\mathrm{D,f}}$	1	480/3	240
EB	4000	20 000	$2P_{\mathrm{B,f}}/3$	2/3	0	0	160/3	0
FB	$4000\sqrt{2}$	$-20\,000\sqrt{2}$	$\sqrt{2}P_{\mathrm{B,f}}/3$	$\sqrt{2}/3$	0	0	$-160\sqrt{2}/3$	0
FC	4000	100 000	0	0	0	0	0	0
							$\sum = 1268$	$\sum = 880$

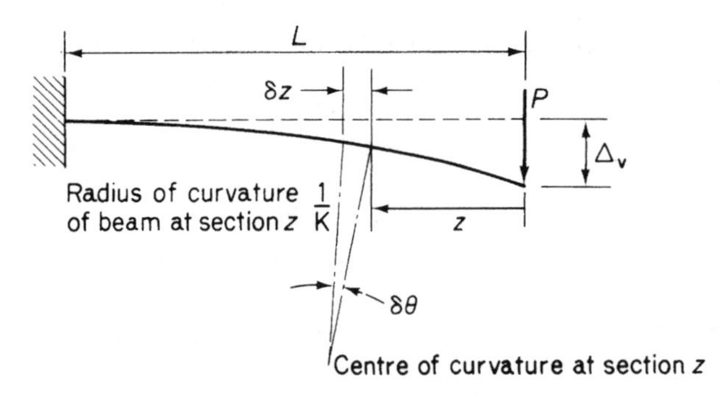

Fig. 5.5 Beam deflection by the method of complementary energy.

The total complementary energy C of the system is given by

$$C = \int_L \int_0^M \mathrm{d}\theta\,\mathrm{d}M - P\Delta_\mathrm{v} \tag{5.12}$$

in which $\int_0^M \mathrm{d}\theta\,\mathrm{d}M$ is the complementary energy of an element δz of the beam. This element subtends an angle $\delta\theta$ at its centre of curvature due to the application of the bending moment M. From the principle of the stationary value of the total complementary energy

$$\frac{\partial C}{\partial P} = \int_L \mathrm{d}\theta \frac{\mathrm{d}M}{\mathrm{d}P} - \Delta_\mathrm{v} = 0$$

or

$$\Delta_\mathrm{v} = \int_L \mathrm{d}\theta \frac{\mathrm{d}M}{\mathrm{d}P} \tag{5.13}$$

Equation (5.13) is applicable to either a non-linear or linearly elastic beam. To proceed further, therefore, we require the load–displacement (M–θ) and bending moment–load (M–P) relationships. It is immaterial for the purposes of this illustrative problem whether the system is linear or non-linear, since the mechanics of the solution are the same in either case. We choose therefore a linear M–θ relationship as this is the case in the majority of the problems we consider. Hence from Fig. 5.5

$$\delta\theta = K\delta z$$

or

$$\mathrm{d}\theta = \frac{M}{EI}\mathrm{d}z \quad \left(\frac{1}{K} = \frac{EI}{M} \text{ from simple beam theory}\right)$$

where the product *modulus of elasticity* × *second moment of area of the beam cross section* is known as the *bending* or *flexural rigidity* of the beam. Also

$$M = Pz$$

so that

$$\frac{\mathrm{d}M}{\mathrm{d}P} = z$$

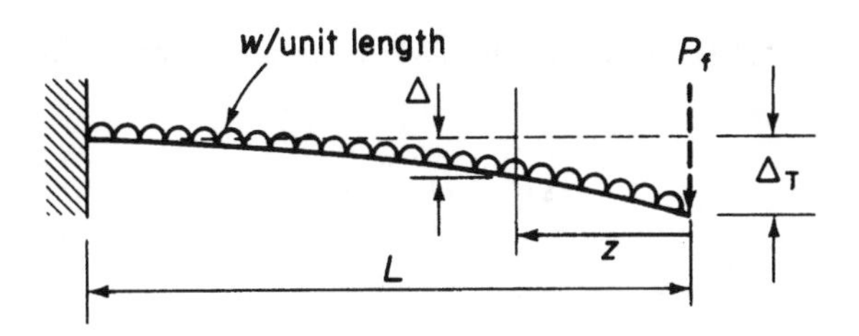

Fig. 5.6 Deflection of a uniformly loaded cantilever by the method of complementary energy.

Substitution for $\mathrm{d}\theta, M$ and $\mathrm{d}M/\mathrm{d}P$ in Eq. (5.13) gives

$$\Delta_{\mathrm{v}} = \int_0^L \frac{Pz^2}{EI}\,\mathrm{d}z$$

or

$$\Delta_{\mathrm{v}} = \frac{PL^3}{3EI}$$

The fictitious load method of the framework example may be employed in the solution of beam deflection problems where we require deflections at positions on the beam other than concentrated load points. Suppose that we are to find the tip deflection Δ_{T} of the cantilever of the previous example in which the concentrated load has been replaced by a uniformly distributed load of intensity w per unit length (see Fig. 5.6). First we apply a fictitious load P_{f} at the point where the deflection is required. The total complementary energy of the system is

$$C = \int_L \int_0^M \mathrm{d}\theta\,\mathrm{d}M - \Delta_{\mathrm{T}} P_{\mathrm{f}} - \int_0^L \Delta w\,\mathrm{d}z$$

where the symbols take their previous meanings and Δ is the vertical deflection of any point on the beam. Then

$$\frac{\partial C}{\partial P_{\mathrm{f}}} = \int_0^L \mathrm{d}\theta \frac{\partial M}{\partial P_{\mathrm{f}}} - \Delta_{\mathrm{T}} = 0 \tag{5.14}$$

As before

$$\mathrm{d}\theta = \frac{M}{EI}\mathrm{d}z$$

but

$$M = P_{\mathrm{f}} z + \frac{wz^2}{2} \quad (P_{\mathrm{f}} = 0)$$

Hence

$$\frac{\partial M}{\partial P_{\mathrm{f}}} = z$$

Substituting in Eq. (5.14) for $\mathrm{d}\theta, M$ and $\partial M/\partial P_{\mathrm{f}}$, and remembering that $P_{\mathrm{f}}=0$, we have

$$\Delta_{\mathrm{T}} = \int_0^L \frac{wz^3}{2EI}\,\mathrm{d}z$$

giving

$$\Delta_{\mathrm{T}} = \frac{wL^4}{8EI}$$

It will be noted that here, unlike the method for the solution of the pin-jointed framework, the fictitious load is applied to the loaded beam. There is, however, no arithmetical advantage to be gained by the former approach although the result would obviously be the same since M would equal $wz^2/2$ and $\partial M/\partial P_{\mathrm{f}}$ would have the value z.

Example 5.2

Calculate the vertical displacements of the quarter and mid-span points B and C of the simply supported beam of length L and flexural rigidity EI loaded, as shown in Fig. 5.7.

The total complementary energy C of the system including the fictitious loads $P_{\mathrm{B,f}}$ and $P_{\mathrm{C,f}}$ is

$$C = \int_L \int_0^M \mathrm{d}\theta\,\mathrm{d}M - P_{\mathrm{B,f}}\Delta_{\mathrm{B}} - P_{\mathrm{C,f}}\Delta_{\mathrm{C}} - \int_0^L \Delta w\,\mathrm{d}z \tag{i}$$

Hence

$$\frac{\partial C}{\partial P_{\mathrm{B,f}}} = \int_L \mathrm{d}\theta\frac{\partial M}{\partial P_{\mathrm{B,f}}} - \Delta_{\mathrm{B}} = 0 \tag{ii}$$

and

$$\frac{\partial C}{\partial P_{\mathrm{C,f}}} = \int_L \mathrm{d}\theta\frac{\partial M}{\partial P_{\mathrm{C,f}}} - \Delta_{\mathrm{C}} = 0 \tag{iii}$$

Assuming a linearly elastic beam, Eqs (ii) and (iii) become

$$\Delta_{\mathrm{B}} = \frac{1}{EI}\int_0^L M\frac{\partial M}{\partial P_{\mathrm{B,f}}}\,\mathrm{d}z \tag{iv}$$

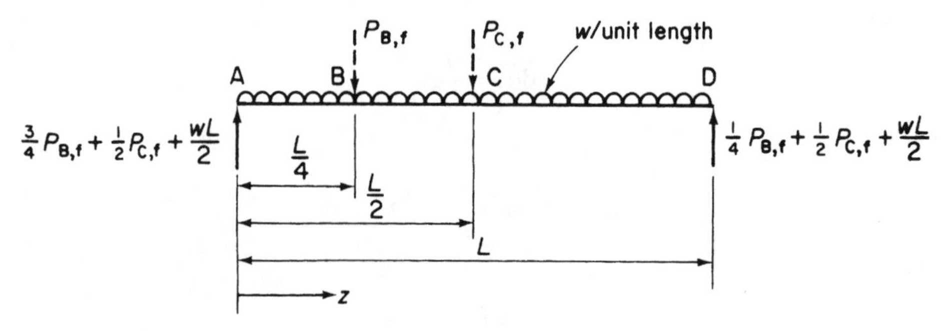

Fig. 5.7 Deflection of a simply supported beam by the method of complementary energy.

$$\Delta_C = \frac{1}{EI}\int_0^L M\frac{\partial M}{\partial P_{C,f}}\,dz \tag{v}$$

From A to B

$$M = \left(\tfrac{3}{4}P_{B,f} + \tfrac{1}{2}P_{C,f} + \frac{wL}{2}\right)z - \frac{wz^2}{2}$$

so that

$$\frac{\partial M}{\partial P_{B,f}} = \tfrac{3}{4}z, \quad \frac{\partial M}{\partial P_{C,f}} = \tfrac{1}{2}z$$

From B to C

$$M = \left(\tfrac{3}{4}P_{B,f} + \tfrac{1}{2}P_{C,f} + \frac{wL}{2}\right)z - \frac{wz^2}{2} - P_{B,f}\left(z - \frac{L}{4}\right)$$

giving

$$\frac{\partial M}{\partial P_{B,f}} = \frac{1}{4}(L-z), \quad \frac{\partial M}{\partial P_{C,f}} = \frac{1}{2}z$$

From C to D

$$M = \left(\frac{1}{4}P_{B,f} + \frac{1}{2}P_{C,f} + \frac{wL}{2}\right)(L-z) - \frac{w}{2}(L-z)^2$$

so that

$$\frac{\partial M}{\partial P_{B,f}} = \frac{1}{4}(L-z) \quad \frac{\partial M}{\partial P_{C,f}} = \frac{1}{2}(L-z)$$

Substituting these values in Eqs (iv) and (v) and remembering that $P_{B,f} = P_{C,f} = 0$ we have, from Eq. (iv)

$$\Delta_B = \frac{1}{EI}\left\{\int_0^{L/4}\left(\frac{wLz}{2} - \frac{wz^2}{2}\right)\tfrac{3}{4}z\,dz + \int_{L/4}^{L/2}\left(\frac{wLz}{2} - \frac{wz^2}{2}\right)\tfrac{1}{4}(L-z)dz \right.$$
$$\left. + \int_{L/2}^{L}\left(\frac{wLz}{2} - \frac{wz^2}{2}\right)\tfrac{1}{4}(L-z)dz\right\}$$

from which

$$\Delta_B = \frac{119wL^4}{24\,576EI}$$

Similarly

$$\Delta_C = \frac{5wL^4}{384EI}$$

The fictitious load method of determining deflections may be streamlined for linearly elastic systems and is then termed the *unit load method*; this we shall discuss later in the chapter.

5.4 Application to the solution of statically indeterminate systems

In a statically determinate structure the internal forces are determined uniquely by simple statical equilibrium considerations. This is not the case for a statically indeterminate system in which, as we have already noted, an infinite number of internal force or stress distributions may be found to satisfy the conditions of equilibrium. The true force system is, as we demonstrated in Section 5.2, the one satisfying the conditions of compatibility of displacement of the elastic structure or, alternatively, that for which the total complementary energy has a stationary value. We shall apply the principle to a variety of statically indeterminate structures, beginning with the relatively simple singly redundant pin-jointed frame shown in Fig. 5.8 in which each member has the same value of the product AE.

The first step is to choose the redundant member. In this example no advantage is gained by the choice of any particular member, although in some cases careful selection can result in a decrease in the amount of arithmetical labour. Taking BD as the redundant member we assume that it sustains a tensile force R due to the external loading. The total complementary energy of the framework is, with the notation of Eq. (5.9)

$$C = \sum_{i=1}^{k} \int_{0}^{F_i} \lambda_i \, \mathrm{d}F_i - P\Delta$$

Hence

$$\frac{\partial C}{\partial R} = \sum_{i=1}^{k} \lambda_i \frac{\partial F_i}{\partial R} = 0 \tag{5.15}$$

or, assuming linear elasticity

$$\frac{1}{AE} \sum_{i=1}^{k} F_i L_i \frac{\partial F_i}{\partial R} = 0 \tag{5.16}$$

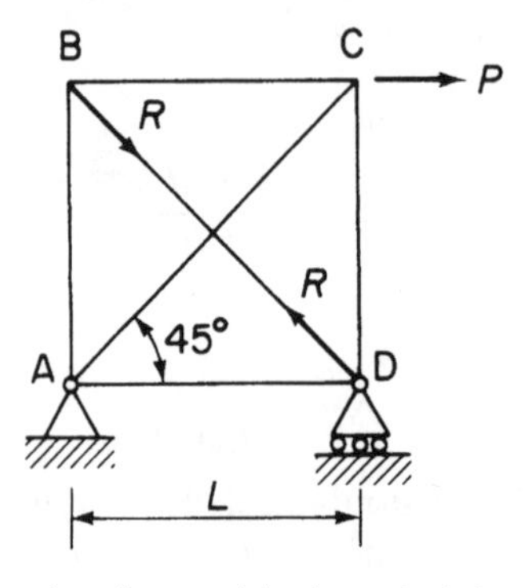

Fig. 5.8 Analysis of a statically indeterminate framework by the method of complementary energy.

The solution is now completed in Table 5.2 where, as in Table 5.1, positive signs indicate tension.

Hence from Eq. (5.16)

$$4.83RL + 2.707PL = 0$$

or

$$R = -0.56P$$

Substitution for R in column ③ of Table 5.2 gives the force in each member. Having determined the forces in the members then the deflection of any point on the framework may be found by the method described in Section 5.3.

Unlike the statically determinate type, statically indeterminate frameworks may be subjected to self-straining. Thus, internal forces are present before external loads are applied. Such a situation may be caused by a local temperature change or by an initial lack of fit of a member. Suppose that the member BD of the framework of Fig. 5.8 is short by a known amount Δ_R when the framework is assembled but is forced to fit. The load R in BD will then have suffered a displacement Δ_R in addition to that caused by the change in length of BD produced by the load P. The total complementary energy is then

$$C = \sum_{i=1}^{k} \int_0^{F_i} \lambda_i \, \mathrm{d}F_i - P\Delta - R\Delta_R$$

and

$$\frac{\partial C}{\partial R} = \sum_{i=1}^{k} \lambda_i \frac{\partial F_i}{\partial R} - \Delta_R = 0$$

or

$$\Delta_R = \frac{1}{AE} \sum_{i=1}^{k} F_i L_i \frac{\partial F_i}{\partial R} \tag{5.17}$$

Table 5.2

① Member	② Length	③ F	④ $\partial F/\partial R$	⑤ $FL\partial F/\partial R$
AB	L	$-R/\sqrt{2}$	$-1/\sqrt{2}$	$RL/2$
BC	L	$-R/\sqrt{2}$	$-1/\sqrt{2}$	$RL/2$
CD	L	$-(P+R/\sqrt{2})$	$-1/\sqrt{2}$	$L(P+R/\sqrt{2})/\sqrt{2}$
DA	L	$-R/\sqrt{2}$	$-1/\sqrt{2}$	$RL/2$
AC	$\sqrt{2}L$	$\sqrt{2}P+R$	1	$L(2P+\sqrt{2}R)$
BD	$\sqrt{2}L$	R	1	$\sqrt{2}RL$
				$\Sigma = 4.83RL + 2.707PL$

Obviously the summation term in Eq. (5.17) has the same value as in the previous case so that

$$R = -0.56P + \frac{AE}{4.83L}\Delta_R$$

Hence the forces in the members are due to both applied loads and an initial lack of fit.

Some care should be given to the sign of the lack of fit Δ_R. We note here that the member BD is short by an amount Δ_R so that the assumption of a positive sign for Δ_R is compatible with the tensile force R. If BD were initially too long then the total complementary energy of the system would be written

$$C = \sum_{i=1}^{k} \int_0^{F_i} \lambda_i \, \mathrm{d}F_i - P\Delta - R(-\Delta_R)$$

giving

$$-\Delta_R = \frac{1}{AE} \sum_{i=1}^{k} F_i L_i \frac{\partial F_i}{\partial R}$$

Example 5.3

Calculate the loads in the members of the singly redundant pin-jointed framework shown in Fig. 5.9. The members AC and BD are 30 mm^2 in cross-section, and all other members are 20 mm^2 in cross-section. The members AD, BC and DC are each 800 mm long. $E = 200\,000$ N/mm^2.

From the geometry of the framework $\widehat{ABD} = \widehat{CBD} = 30°$; therefore BD = AC = $800\sqrt{3}$ mm. Choosing CD as the redundant member and proceeding from Eq. (5.16) we have

$$\frac{1}{E} \sum_{i=1}^{k} \frac{F_i L_i}{A_i} \frac{\partial F_i}{\partial R} = 0 \qquad \text{(i)}$$

From Table 5.3 we have

$$\sum_{i=1}^{k} \frac{F_i L_i}{A_i} \frac{\partial F_i}{\partial R} = -268 + 129.2R = 0$$

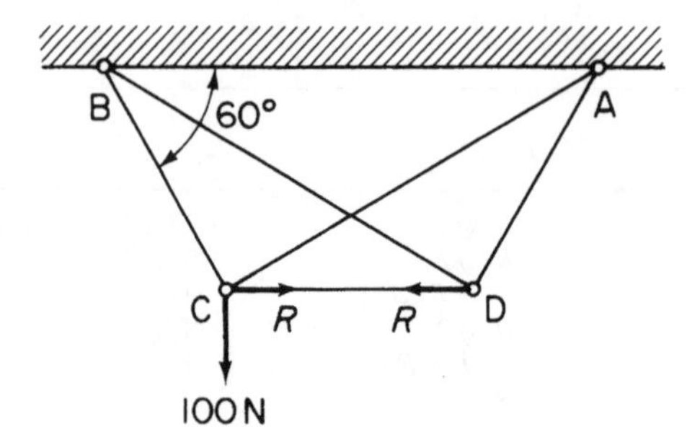

Fig. 5.9 Framework of Example 5.3.

Table 5.3 (Tension positive)

① Member	② L (mm)	③ A (mm^2)	④ F(N)	⑤ $\partial F/\partial R$	⑥ $(FL/A)\partial F/\partial R$	⑦ Force (N)
AC	$800\sqrt{3}$	30	$50-\sqrt{3}R/2$	$-\sqrt{3}/2$	$-2000+20\sqrt{3}R$	48.2
CB	800	20	$86.6+R/2$	1/2	$1732+10R$	87.6
BD	$800\sqrt{3}$	30	$-\sqrt{3}R/2$	$-\sqrt{3}/2$	$20\sqrt{3}R$	−1.8
CD	800	20	R	1	$40R$	2.1
AD	800	20	$R/2$	1/2	$10R$	1.0
					$\Sigma=-268+129.2R$	

Hence $R=2.1$ N and the forces in the members are tabulated in column ⑦ of Table 5.3.

Example 5.4

A plane, pin-jointed framework consists of six bars forming a rectangle ABCD 4000 mm by 3000 mm with two diagonals, as shown in Fig. 5.10. The cross-sectional area of each bar is 200 mm^2 and the frame is unstressed when the temperature of each member is the same. Due to local conditions the temperature of one of the 3000 mm members is raised by 30°C. Calculate the resulting forces in all the members if the coefficient of linear expansion α of the bars is 7×10^{-6}/°C. $E=200\,000$ N/mm^2.

Suppose that BC is the heated member, then the increase in length of BC $=3000\times30\times7\times10^{-6}=0.63$ mm. Therefore, from Eq. (5.17)

$$-0.63=\frac{1}{200\times200\,000}\sum_{i=1}^{k}F_iL_i\frac{\partial F_i}{\partial R} \tag{i}$$

Substitution from the summation of column ⑤ in Table 5.4 into Eq. (i) gives

$$R=\frac{-0.63\times200\times200\,000}{48\,000}=-525\text{ N}$$

Column ⑥ of Table 5.4 is now completed for the force in each member.

So far, our analysis has been limited to singly redundant frameworks, although the same procedure may be adopted to solve a multi-redundant framework of, say, m redundancies. Therefore, instead of a single equation of the type (5.15) we would have

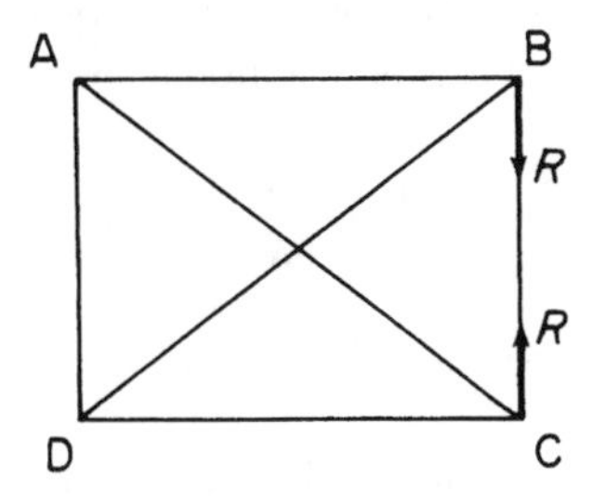

Fig. 5.10 Framework of Example 5.4.

Table 5.4 (Tension positive)

① Member	② L (mm)	③ F(N)	④ $\partial F/\partial R$	⑤ $FL\partial F/\partial R$	⑥ Force (N)
AB	4000	$4R/3$	4/3	$64\,000R/9$	−700
BC	3000	R	1	$3\,000R$	−525
CD	4000	$4R/3$	4/3	$64\,000R/9$	−700
DA	3000	R	1	$3\,000R$	−525
AC	5000	$-5R/3$	−5/3	$125\,000R/9$	875
DB	5000	$-5R/3$	−5/3	$125\,000R/9$	875
				$\Sigma = 48\,000R$	

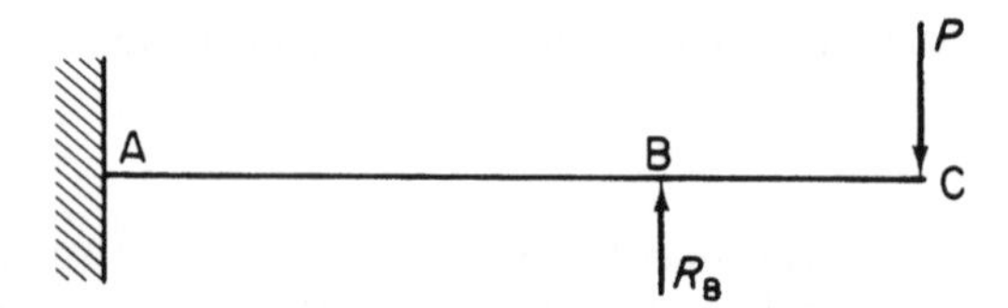

Fig. 5.11 Analysis of a propped cantilever by the method of complementary energy.

m simultaneous equations

$$\frac{\partial C}{\partial R_j} = \sum_{i=1}^{k} \lambda_i \frac{\partial F_i}{\partial R_j} = 0 \quad (j = 1, 2, \ldots, m)$$

from which the m unknowns $R_1, R_2, \ldots, R_m$ would be obtained. The forces F in the members follow, being expressed initially in terms of the applied loads and $R_1, R_2, \ldots, R_m$.

Other types of statically indeterminate structure are solved by the application of total complementary energy with equal facility. The propped cantilever of Fig. 5.11 is an example of a singly redundant beam structure for which total complementary energy readily yields a solution.

The total complementary energy of the system is, with the notation of Eq. (5.12)

$$C = \int_L \int_0^M \mathrm{d}\theta \, \mathrm{d}M - P\Delta_C - R_B\Delta_B$$

where Δ_C and Δ_B are the deflections at C and B, respectively. Usually, in problems of this type, Δ_B is either zero for a rigid support, or a known amount (sometimes in terms of R_B) for a sinking support. Hence, for a stationary value of C

$$\frac{\partial C}{\partial R_B} = \int_L \mathrm{d}\theta \frac{\partial M}{\partial R_B} - \Delta_B = 0$$

from which equation R_B may be found; R_B being contained in the expression for the bending moment M.

Obviously the same procedure is applicable to a beam having a multiredundant support system, e.g. a continuous beam supporting a series of loads $P_1, P_2, \ldots, P_n$.

The total complementary energy of such a beam would be given by

$$C = \int_L \int_0^M \mathrm{d}\theta \, \mathrm{d}M - \sum_{j=1}^{m} R_j \Delta_j - \sum_{r=1}^{n} P_r \Delta_r$$

where R_j and Δ_j are the reaction and known deflection (at least in terms of R_j) of the jth support point in a total of m supports. The stationary value of C gives

$$\frac{\partial C}{\partial R_j} = \int_L \mathrm{d}\theta \frac{\partial M}{\partial R_j} - \Delta_j = 0 \quad (j = 1, 2, \ldots, m)$$

producing m simultaneous equations for the m unknown reactions.

The intention here is not to suggest that continuous beams are best or most readily solved by the energy method; the moment distribution method produces a more rapid solution, especially for beams in which the degree of redundancy is large. Instead the purpose is to demonstrate the versatility and power of energy methods in their ready solution of a wide range of structural problems. A complete investigation of this versatility is impossible here due to restriction of space; in fact, whole books have been devoted to this topic. We therefore limit our analysis to problems peculiar to the field of aircraft structures with which we are primarily concerned. The remaining portion of this section is therefore concerned with the solution of frames and rings possessing varying degrees of redundancy.

The frameworks we considered in the earlier part of this section and in Section 5.3 comprised members capable of resisting direct forces only. Of a more general type are composite frameworks in which some or all of the members resist bending and shear loads in addition to direct loads. It is usual, however, except for the thin-walled structures in Part B of this book, to ignore deflections produced by shear forces. We only consider, therefore, bending and direct force contributions to the internal complementary energy of such structures. The method of analysis is illustrated in the following example.

Example 5.5

The simply supported beam ABC shown in Fig. 5.12 is stiffened by an arrangement of pin-jointed bars capable of sustaining axial loads only. If the cross-sectional area of the beam is A_{B} and that of the bars is A, calculate the forces in the members of the framework assuming that displacements are caused by bending and direct force action only.

We observe that if the beam were only capable of supporting direct loads then the structure would be a relatively simple statically determinate pin-jointed framework. Since the beam resists bending moments (we are ignoring shear effects) the system is statically indeterminate with a single redundancy, the bending moment at any section of the beam. The total complementary energy of the framework is given, with the notation previously developed, by

$$C = \int_{\mathrm{ABC}} \int_0^M \mathrm{d}\theta \, \mathrm{d}M + \sum_{i=1}^{k} \int_0^{F_i} \lambda_j \, \mathrm{d}F_i - P\Delta \qquad \text{(i)}$$

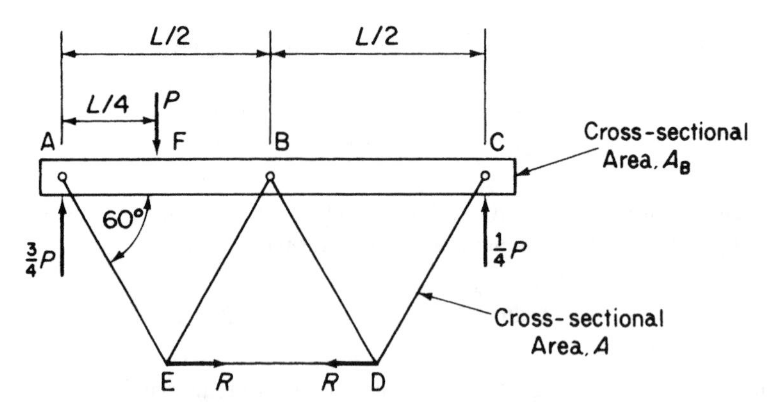

Fig. 5.12 Analysis of a trussed beam by the method of complementary energy.

Table 5.5 (Tension positive)

① Member	② Length	③ Area	④ F	⑤ $\partial F/\partial R$	⑥ $(F/A)\partial F/\partial R$
AB	$L/2$	A_B	$-R/2$	$-1/2$	$R/4A_B$
BC	$L/2$	A_B	$-R/2$	$-1/2$	$R/4A_B$
CD	$L/2$	A	R	1	R/A
DE	$L/2$	A	R	1	R/A
BD	$L/2$	A	$-R$	-1	R/A
EB	$L/2$	A	$-R$	-1	R/A
AE	$L/2$	A	R	1	R/A

If we suppose that the tensile load in the member ED is R then, for C to have a stationary value

$$\frac{\partial C}{\partial R} = \int_{\mathrm{ABC}} \mathrm{d}\theta \frac{\partial M}{\partial R} + \sum_{i=1}^{k} \lambda_i \frac{\partial F_i}{\partial R} = 0 \qquad \text{(ii)}$$

At this point we assume the appropriate load–displacement relationships; again we shall take the system to be linear so that Eq. (ii) becomes

$$\int_0^L \frac{M}{EI} \frac{\partial M}{\partial R} \mathrm{d}z + \sum_{i=1}^{k} \frac{F_i L_i}{A_i E} \frac{\partial F_i}{\partial R} = 0 \qquad \text{(iii)}$$

The two terms in Eq. (iii) may be evaluated separately, bearing in mind that only the beam ABC contributes to the first term while the complete structure contributes to the second. Evaluating the summation term by a tabular process we have Table 5.5.

Summation of column ⑥ in Table 5.5 gives

$$\sum_{i=1}^{k} \frac{F_i L_i}{A_i E} \frac{\partial F_i}{\partial R} = \frac{RL}{4E} \left(\frac{1}{A_B} + \frac{10}{A} \right) \qquad \text{(iv)}$$

The bending moment at any section of the beam between A and F is

$$M = \frac{3}{4}Pz - \frac{\sqrt{3}}{2}Rz \quad \text{hence} \quad \frac{\partial M}{\partial R} = -\frac{\sqrt{3}}{2}z$$

between F and B

$$M = \frac{P}{4}(L - z) - \frac{\sqrt{3}}{2}Rz \quad \text{hence} \quad \frac{\partial M}{\partial R} = -\frac{\sqrt{3}}{2}z$$

and between B and C

$$M = \frac{P}{4}(L - z) - \frac{\sqrt{3}}{2}R(L - z) \quad \text{hence} \quad \frac{\partial M}{\partial R} = -\frac{\sqrt{3}}{2}(L - z)$$

Thus

$$\int_0^L \frac{M}{EI}\frac{\partial M}{\partial R}\mathrm{d}z = \frac{1}{EI}\left\{\int_0^{L/4} -\left(\frac{3}{4}Pz - \frac{\sqrt{3}}{2}Rz\right)\frac{\sqrt{3}}{2}z\,\mathrm{d}z + \int_{L/4}^{L/2}\left[\frac{P}{4}(L - z) - \frac{\sqrt{3}}{2}Rz\right]\left(-\frac{\sqrt{3}}{2}z\right)\mathrm{d}z + \int_{L/2}^{L} -\left[\frac{P}{4}(L - z) - \frac{\sqrt{3}}{2}R(L - z)\right]\frac{\sqrt{3}}{2}(L - z)\mathrm{d}z\right\}$$

giving

$$\int_0^L \frac{M}{EI}\frac{\partial M}{\partial R}\mathrm{d}z = \frac{-11\sqrt{3}PL^3}{768EI} + \frac{RL^3}{16EI} \tag{v}$$

Substituting from Eqs (iv) and (v) into Eq. (iii)

$$-\frac{11\sqrt{3}PL^3}{768EI} + \frac{RL^3}{16EI} + \frac{RL}{4E}\left(\frac{A + 10A_{\mathrm{B}}}{A_{\mathrm{B}}A}\right) = 0$$

from which

$$R = \frac{11\sqrt{3}PL^2A_{\mathrm{B}}A}{48[L^2A_{\mathrm{B}}A + 4I(A + 10A_{\mathrm{B}})]}$$

Hence the forces in each member of the framework. The deflection Δ of the load P or any point on the framework may be obtained by the method of Section 5.3. For example, the stationary value of the total complementary energy of Eq. (i) gives Δ, i.e.

$$\frac{\partial C}{\partial P} = \int_{\mathrm{ABC}} \mathrm{d}\theta\frac{\partial M}{\partial R} + \sum_{i=1}^{k}\lambda_i\frac{\partial F_i}{\partial P} - \Delta = 0$$

Although braced beams are still found in modern light aircraft in the form of braced wing structures a much more common structural component is the ring frame. The role

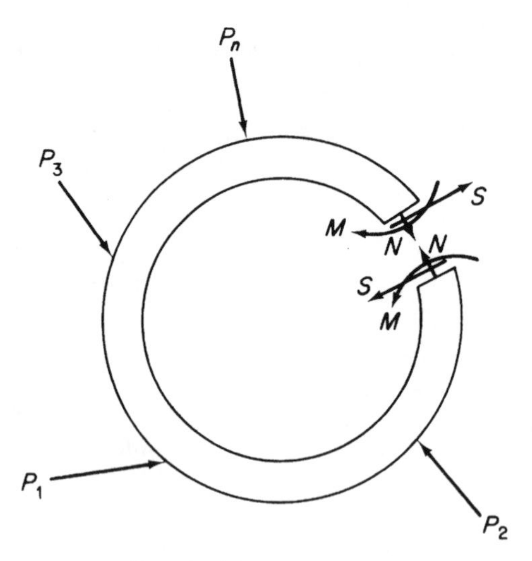

Fig. 5.13 Internal force system in a two-dimensional ring.

of this particular component is discussed in detail in Chapter 14; it is therefore sufficient for the moment to say that ring frames form the basic shape of semi-monocoque fuselages reacting shear loads from the fuselage skins, point loads from wing spar attachments and distributed loads from floor beams. Usually a ring is two-dimensional supporting loads applied in its own plane. Our analysis is limited to the two-dimensional case.

A two-dimensional ring has redundancies of direct load, bending moment and shear at any section, as shown in Fig. 5.13. However, in some special cases of loading the number of redundancies may be reduced. For example, on a plane of symmetry the shear loads and sometimes the normal or direct loads are zero, while on a plane of antisymmetry the direct loads and bending moments are zero. Let us consider the simple case of a doubly symmetrical ring shown in Fig. 5.14(a). At a section in the vertical plane of symmetry the internal shear and direct loads vanish, leaving one redundancy, the bending moment M_A (Fig. 5.14(b)). Note that in the horizontal plane of symmetry the internal shears are zero but the direct loads have a value $P/2$. The total complementary energy of the system is (again ignoring shear strains)

$$C = \int_{\text{ring}} \int_0^M \mathrm{d}\theta \, \mathrm{d}M - 2\left(\frac{P}{2}\Delta\right)$$

taking the bending moment as positive when it increases the curvature of the ring. In the above expression for C, Δ is the displacement of the top, A, of the ring relative to the bottom, B. Assigning a stationary value to C we have

$$\frac{\partial C}{\partial M_A} = \int_{\text{ring}} \mathrm{d}\theta \frac{\partial M}{\partial M_A} = 0$$

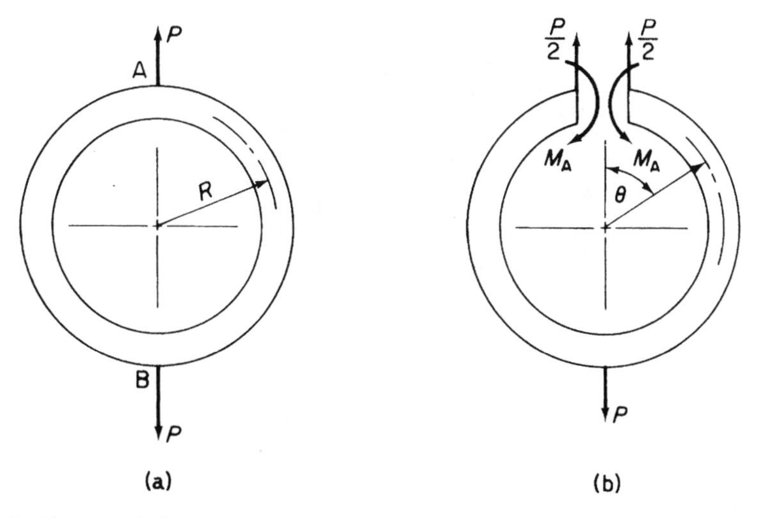

Fig. 5.14 Doubly symmetric ring.

or assuming linear elasticity and considering, from symmetry, half the ring

$$\int_0^{\pi R} \frac{M}{EI} \frac{\partial M}{\partial M_A} ds = 0$$

Thus since

$$M = M_A - \frac{P}{2} R \sin\theta \qquad \frac{\partial M}{\partial M_A} = 1$$

and we have

$$\int_0^{\pi} \left(M_A - \frac{P}{2} R \sin\theta \right) R \, d\theta = 0$$

or

$$\left[M_A \theta + \frac{P}{2} R \cos\theta \right]_0^{\pi} = 0$$

from which

$$M_A = \frac{PR}{\pi}$$

The bending moment distribution is then

$$M = PR \left(\frac{1}{\pi} - \frac{\sin\theta}{2} \right)$$

and is shown diagrammatically in Fig. 5.15.

Let us now consider a more representative aircraft structural problem. The circular fuselage frame of Fig. 5.16(a) supports a load P which is reacted by a shear flow q (i.e. a shear force per unit length: see Chapter 17), distributed around the circumference of the frame from the fuselage skin. The value and direction of this shear flow are quoted here

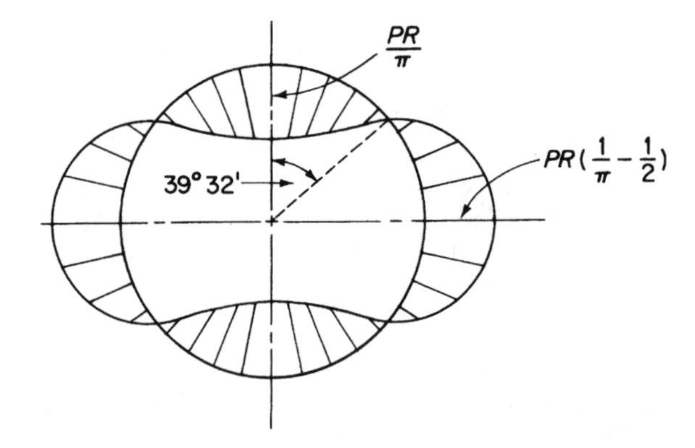

Fig. 5.15 Distribution of bending moment in a doubly symmetric ring.

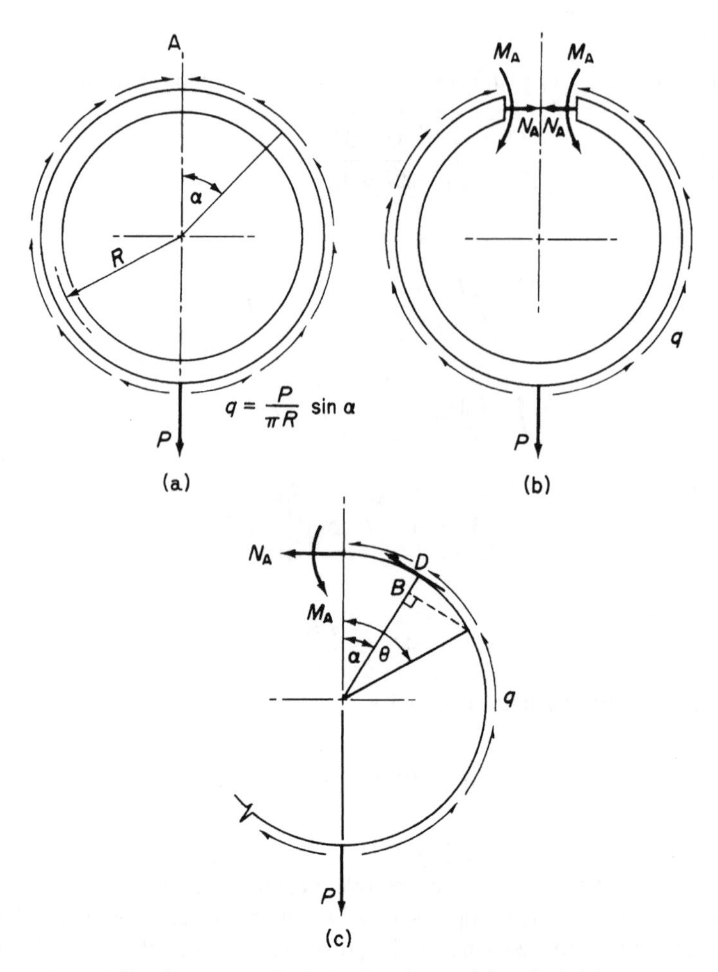

Fig. 5.16 Determination of bending moment distribution in a shear and direct loaded ring.

but are derived from theory established in Section 17.3. From our previous remarks on the effect of symmetry we observe that there is no shear force at the section A on the vertical plane of symmetry. The unknowns are therefore the bending moment M_A and normal force N_A. We proceed, as in the previous example, by writing down the total complementary energy C of the system. Then, neglecting shear strains

$$C = \int_{\text{ring}} \int_0^M \mathrm{d}\theta\, \mathrm{d}M - P\Delta \tag{i}$$

in which Δ is the deflection of the point of application of P relative to the top of the frame. Note that M_A and N_A do not contribute to the complement of the potential energy of the system since, by symmetry, the rotation and horizontal displacements at A are zero. From the principle of the stationary value of the total complementary energy

$$\frac{\partial C}{\partial M_A} = \int_{\text{ring}} \mathrm{d}\theta \frac{\partial M}{\partial M_A} = 0 \tag{ii}$$

and

$$\frac{\partial C}{\partial N_A} = \int_{\text{ring}} \mathrm{d}\theta \frac{\partial M}{\partial N_A} = 0 \tag{iii}$$

The bending moment at a radial section inclined at an angle θ to the vertical diameter is, from Fig. 5.16(c)

$$M = M_A + N_A R(1 - \cos\theta) + \int_0^\theta qBDR\, \mathrm{d}\alpha$$

or

$$M = M_A + N_A R(1 - \cos\theta) + \int_0^\theta \frac{P}{\pi R} \sin\alpha[R - R\cos(\theta - \alpha)]R\, \mathrm{d}\alpha$$

which gives

$$M = M_A + N_A R(1 - \cos\theta) + \frac{PR}{\pi}\left(1 - \cos\theta - \frac{1}{2}\theta\sin\theta\right) \tag{iv}$$

Hence

$$\frac{\partial M}{\partial M_A} = 1 \qquad \frac{\partial M}{\partial N_A} = R(1 - \cos\theta) \tag{v}$$

Assuming that the fuselage frame is linearly elastic we have, from Eqs (ii) and (iii)

$$2\int_0^\pi \frac{M}{EI}\frac{\partial M}{\partial M_A} R\, \mathrm{d}\theta = 2\int_0^\pi \frac{M}{EI}\frac{\partial M}{\partial N_A} R\, \mathrm{d}\theta = 0 \tag{vi}$$

Substituting from Eqs (iv) and (v) into Eq. (vi) gives two simultaneous equations

$$-\frac{PR}{2\pi} = M_A + N_A R \tag{vii}$$

$$-\frac{7PR}{8\pi} = M_A + \frac{3}{2}N_A R \qquad \text{(viii)}$$

These equations may be written in matrix form as follows

$$\frac{PR}{\pi}\begin{Bmatrix}-1/2\\-7/8\end{Bmatrix} = \begin{bmatrix}1 & R\\1 & 3R/2\end{bmatrix}\begin{Bmatrix}M_A\\N_A\end{Bmatrix} \qquad \text{(ix)}$$

so that

$$\begin{Bmatrix}M_A\\N_A\end{Bmatrix} = \frac{PR}{\pi}\begin{bmatrix}1 & R\\1 & 3R/2\end{bmatrix}^{-1}\begin{Bmatrix}-1/2\\-7/8\end{Bmatrix}$$

or

$$\begin{Bmatrix}M_A\\N_A\end{Bmatrix} = \frac{PR}{\pi}\begin{bmatrix}3 & -2\\-2/R & 2/R\end{bmatrix}\begin{Bmatrix}-1/2\\-7/8\end{Bmatrix}$$

which gives

$$M_A = \frac{PR}{4\pi} \quad N_A = \frac{-3P}{4\pi}$$

The bending moment distribution follows from Eq. (iv) and is

$$M = \frac{PR}{2\pi}(1 - \frac{1}{2}\cos\theta - \theta\sin\theta) \qquad \text{(x)}$$

The solution of Eq. (ix) involves the inversion of the matrix

$$\begin{bmatrix}1 & R\\1 & 3R/2\end{bmatrix}$$

which may be carried out using any of the standard methods detailed in texts on matrix analysis. In this example Eqs (vii) and (viii) are clearly most easily solved directly; however, the matrix approach illustrates the technique and serves as a useful introduction to the more detailed discussion in Chapter 6.

Example 5.6

A two-cell fuselage has circular frames with a rigidly attached straight member across the middle. The bending stiffness of the lower half of the frame is $2EI$, whilst that of the upper half and also the straight member is EI.

Calculate the distribution of the bending moment in each part of the frame for the loading system shown in Fig. 5.17(a). Illustrate your answer by means of a sketch and show clearly the bending moment carried by each part of the frame at the junction with the straight member. Deformations due only to bending strains need be taken into account.

The loading is antisymmetrical so that there are no bending moments or normal forces on the plane of antisymmetry; there remain three shear loads S_A, S_D and S_C,

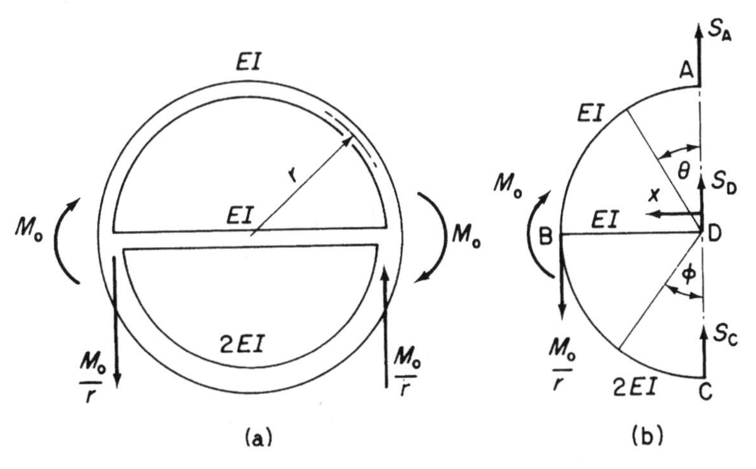

Fig. 5.17 Determination of bending moment distribution in an antisymmetrical fuselage frame.

as shown in Fig. 5.17(b). The total complementary energy of the half-frame is then (neglecting shear strains)

$$C = \int_{\text{half-frame}} \int_0^M \mathrm{d}\theta \, \mathrm{d}M - M_0\alpha_B - \frac{M_0}{r}\Delta_B \tag{i}$$

where α_B and Δ_B are the rotation and deflection of the frame at B caused by the applied moment M_0 and concentrated load M_0/r, respectively. From antisymmetry there is no deflection at A, D or C so that S_A, S_D and S_C make no contribution to the total complementary energy. In addition, overall equilibrium of the half-frame gives

$$S_A + S_D + S_C = \frac{M_0}{r} \tag{ii}$$

Assigning stationary values to the total complementary energy and considering the half-frame only, we have

$$\frac{\partial C}{\partial S_A} = \int_{\text{half-frame}} \mathrm{d}\theta \frac{\partial M}{\partial S_A} = 0$$

and

$$\frac{\partial C}{\partial S_D} = \int_{\text{half-frame}} \mathrm{d}\theta \frac{\partial M}{\partial S_D} = 0$$

or assuming linear elasticity

$$\int_{\text{half-frame}} \frac{M}{EI}\frac{\partial M}{\partial S_A}\mathrm{d}s = \int_{\text{half-frame}} \frac{M}{EI}\frac{\partial M}{\partial S_D}\mathrm{d}s = 0 \tag{iii}$$

In AB

$$M = -S_A r \sin\theta \quad \text{and} \quad \frac{\partial M}{\partial S_A} = -r\sin\theta, \quad \frac{\partial M}{\partial S_D} = 0$$

In DB

$$M = S_D x \quad \text{and} \quad \frac{\partial M}{\partial S_A} = 0, \quad \frac{\partial M}{\partial S_D} = x$$

In CB

$$M = S_C r \sin\phi = \left(\frac{M_0}{r} - S_A - S_D\right) r \sin\phi$$

Thus

$$\frac{\partial M}{\partial S_A} = -r\sin\phi \quad \text{and} \quad \frac{\partial M}{\partial S_D} = -r\sin\phi$$

Substituting these expressions in Eq. (iii) and integrating we have

$$3.365 S_A + S_C = M_0/r \tag{iv}$$

$$S_A + 2.178 S_C = M_0/r \tag{v}$$

which, with Eq. (ii), enable S_A, S_D and S_C to be found. In matrix form these equations are written

$$\begin{Bmatrix} M_0/r \\ M_0/r \\ M_0/r \end{Bmatrix} = \begin{bmatrix} 1 & 1 & 1 \\ 3.356 & 0 & 1 \\ 1 & 0 & 2.178 \end{bmatrix} \begin{Bmatrix} S_A \\ S_D \\ S_C \end{Bmatrix} \tag{vi}$$

from which we obtain

$$\begin{Bmatrix} S_A \\ S_D \\ S_C \end{Bmatrix} = \begin{bmatrix} 0 & 0.345 & -0.159 \\ 1 & -0.187 & -0.373 \\ 0 & -0.159 & 0.532 \end{bmatrix} \begin{Bmatrix} M_0/r \\ M_0/r \\ M_0/r \end{Bmatrix} \tag{vii}$$

which give

$$S_A = 0.187 M_0/r \quad S_D = 0.44 M_0/r \quad S_C = 0.373 M_0/r$$

Again the square matrix of Eq. (vi) has been inverted to produce Eq. (vii).

The bending moment distribution with directions of bending moment is shown in Fig. 5.18.

So far in this chapter we have considered the application of the principle of the stationary value of the total complementary energy of elastic systems in the analysis of various types of structure. Although the majority of the examples used to illustrate the method are of linearly elastic systems it was pointed out that generally they may be used with equal facility for the solution of non-linear systems.

In fact, the question of whether a structure possesses linear or non-linear characteristics arises only after the initial step of writing down expressions for the total potential or complementary energies. However, a great number of structures are linearly elastic and possess unique properties which enable solutions, in some cases, to be more easily obtained. The remainder of this chapter is devoted to these methods.

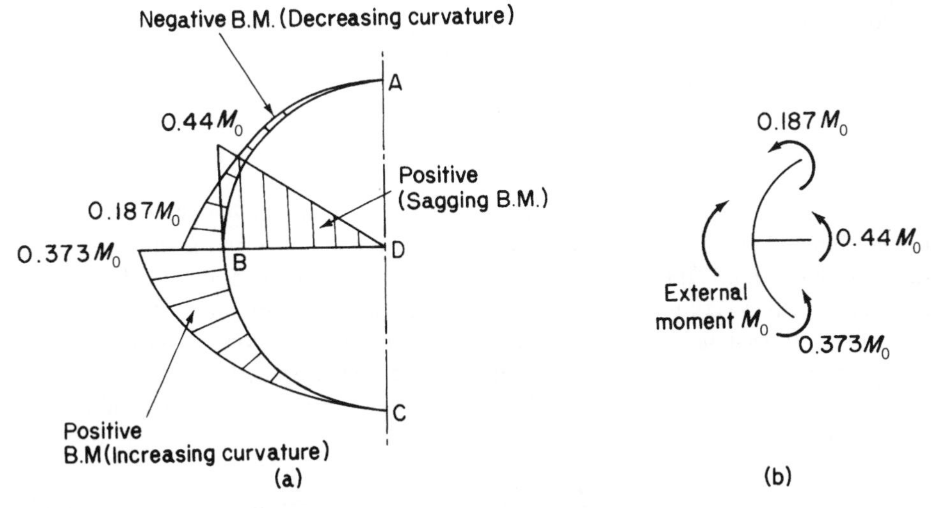

Fig. 5.18 Distribution of bending moment in frame of Example 5.6.

5.5 Unit load method

In Section 5.3 we discussed the *dummy* or *fictitious* load method of obtaining deflections of structures. For a linearly elastic structure the method may be stream-lined as follows.

Consider the framework of Fig. 5.3 in which we require, say, to find the vertical deflection of the point C. Following the procedure of Section 5.3 we would place a vertical dummy load P_{f} at C and write down the total complementary energy of the framework, i.e.

$$C = \sum_{i=1}^{k} \int_0^{F_i} \lambda_i \, \mathrm{d}F_i - \sum_{r=1}^{n} \Delta_r P_r \qquad \text{(see Eq. (5.9))}$$

For a stationary value of C

$$\frac{\partial C}{\partial P_{\text{f}}} = \sum_{i=1}^{k} \lambda_i \frac{\partial F_i}{\partial P_{\text{f}}} - \Delta_{\text{C}} = 0 \tag{5.18}$$

from which

$$\Delta_C = \sum_{i=1}^{k} \lambda_i \frac{\partial F_i}{\partial P_{\text{f}}} \quad \text{as before} \tag{5.19}$$

If instead of the arbitrary dummy load P_{f} we had placed a unit load at C, then the load in the ith linearly elastic member would be

$$F_i = \frac{\partial F_i}{\partial P_{\text{f}}} 1$$

Therefore, the term $\partial F_i/\partial P_f$ in Eq. (5.19) is equal to the load in the ith member due to a unit load at C, and Eq. (5.19) may be written

$$\Delta_C = \sum_{i=1}^{k} \frac{F_{i,0}F_{i,1}L_i}{A_i E_i} \tag{5.20}$$

where $F_{i,0}$ is the force in the ith member due to the actual loading and $F_{i,1}$ is the force in the ith member due to a unit load placed at the position and in the direction of the required deflection. Thus, in Example 5.1 columns ④ and ⑥ in Table 5.1 would be eliminated, leaving column ⑤ as $F_{B,1}$ and column ⑦ as $F_{D,1}$. Obviously column ③ is F_0.

Similar expressions for deflection due to bending and torsion of linear structures follow from the well-known relationships between bending and rotation and torsion and rotation. Hence, for a member of length L and flexural and torsional rigidities EI and GJ, respectively

$$\Delta_{B.M} = \int_L \frac{M_0 M_1}{EI} dz \quad \Delta_T = \int_L \frac{T_0 T_1}{GJ} dz \tag{5.21}$$

where M_0 is the bending moment at any section produced by the actual loading and M_1 is the bending moment at any section due to a unit load applied at the position and in the direction of the required deflection. Similarly for torsion.

Generally, shear deflections of slender beams are ignored but may be calculated when required for particular cases. Of greater interest in aircraft structures is the calculation of the deflections produced by the large shear stresses experienced by thin-walled sections. This problem is discussed in Chapter 17.

Example 5.7

A steel rod of uniform circular cross-section is bent as shown in Fig. 5.19, AB and BC being horizontal and CD vertical. The arms AB, BC and CD are of equal length. The rod is encastré at A and the other end D is free. A uniformly distributed load covers the length BC. Find the components of the displacement of the free end D in terms of EI and GJ.

Since the cross-sectional area A and modulus of elasticity E are not given we shall assume that displacements due to axial distortion are to be ignored. We place, in turn, unit loads in the assumed positive directions of the axes xyz.

First, consider the displacement in the direction parallel to the x axis. From Eqs (5.21)

$$\Delta_x = \int_L \frac{M_0 M_1}{EI} ds + \int_L \frac{T_0 T_1}{GJ} ds$$

Employing a tabular procedure

	M_0			M_1			T_0			T_1		
Plane	xy	xz	yz	xy	xz	yz	xy	xz	yz	xy	xz	yz
CD	0	0	0	y	0	0	0	0	0	0	0	0
CB	0	0	$-wz^2/2$	0	z	0	0	0	0	l	0	0
BA	$-wlx$	0	0	l	l	0	0	0	$wl^2/2$	0	0	0

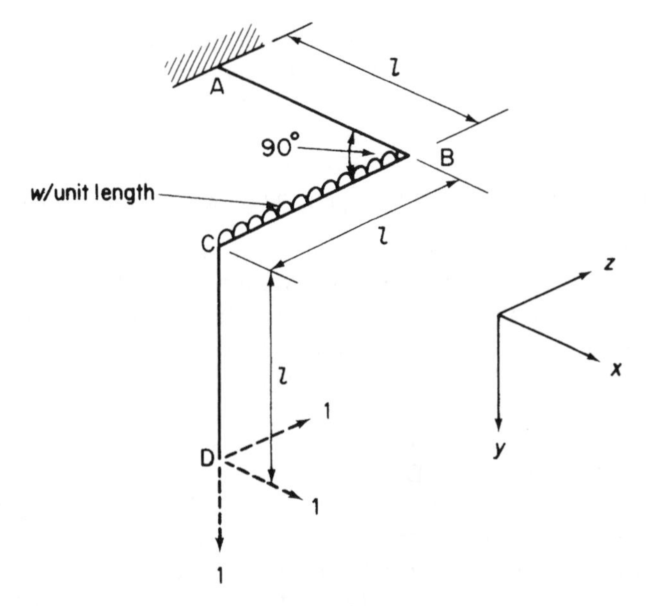

Fig. 5.19 Deflection of a bent rod.

Hence

$$\Delta_x = \int_0^l -\frac{wl^2x}{EI}\mathrm{d}x$$

or

$$\Delta_x = -\frac{wl^4}{2EI}$$

Similarly

$$\Delta_y = wl^4\left(\frac{11}{24EI} + \frac{1}{2GJ}\right)$$

$$\Delta_z = wl^4\left(\frac{1}{6EI} + \frac{1}{2GJ}\right)$$

5.6 Flexibility method

An alternative approach to the solution of statically indeterminate beams and frames is to release the structure, i.e. remove redundant members or supports, until the structure becomes statically determinate. The displacement of some point in the released structure is then determined by, say, the unit load method. The actual loads on the structure are removed and unknown forces applied to the points where the structure has been released; the displacement at the point produced by these unknown forces must, from compatibility, be the same as that in the released structure. The unknown forces are then obtained; this approach is known as *the flexibility method.*

Example 5.8

Determine the forces in the members of the truss shown in Fig. 5.20(a); the cross-sectional area A, and Young's modulus E, are the same for all members.

The truss in Fig. 5.20(a) is clearly externally statically determinate but has a degree of internal statical indeterminacy equal to 1. We therefore release the truss so that it becomes statically determinate by 'cutting' one of the members, say BD, as shown in Fig. 5.20(b). Due to the actual loads (P in this case) the cut ends of the member BD will separate or come together, depending on whether the force in the member (before it was cut) was tensile or compressive; we shall assume that it was tensile.

We are assuming that the truss is linearly elastic so that the relative displacement of the cut ends of the member BD (in effect the movement of B and D away from or towards each other along the diagonal BD) may be found using, say, the unit load method. Thus we determine the forces $F_{a,j}$, in the members produced by the actual loads. We then apply equal and opposite unit loads to the cut ends of the member BD as shown in Fig 5.20(c) and calculate the forces, $F_{1,j}$ in the members. The displacement of B relative to D, Δ_{BD}, is then given by

$$\Delta_{BD} = \sum_{j=1}^{n} \frac{F_{a,j}F_{1,j}L_j}{AE} \quad \text{(see Eq. (ii) in Example 4.6)}$$

The forces, $F_{a,j}$, are the forces in the members of the released truss due to the actual loads and are not, therefore, the actual forces in the members of the complete truss. We shall therefore redesignate the forces in the members of the released truss as $F_{0,j}$. The expression for Δ_{BD} then becomes

$$\Delta_{BD} = \sum_{j=1}^{n} \frac{F_{0,j}F_{1,j}L_j}{AE} \tag{i}$$

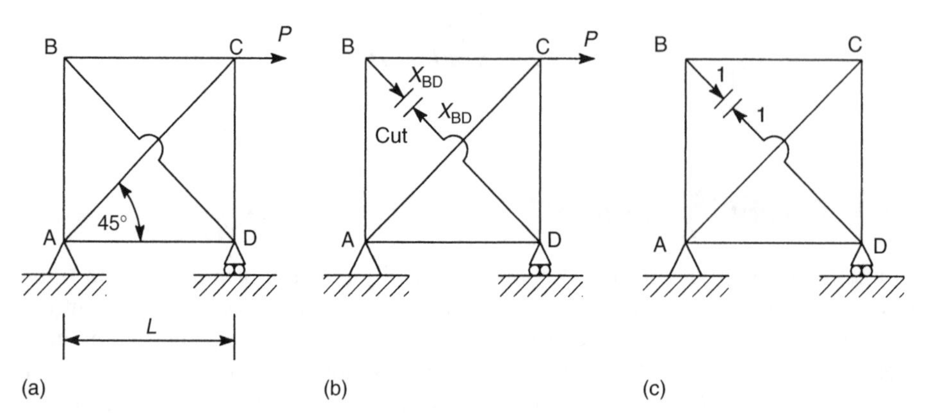

Fig. 5.20 Analysis of a statically indeterminate truss.

In the actual structure this displacement is prevented by the force, X_{BD}, in the redundant member BD. If, therefore, we calculate the displacement, a_{BD}, in the direction of BD produced by a unit value of X_{BD}, the displacement due to X_{BD} will be $X_{BD}a_{BD}$. Clearly, from compatibility

$$\Delta_{BD} + X_{BD}a_{BD} = 0 \tag{ii}$$

from which X_{BD} is found, a_{BD} is a *flexibility coefficient.* Having determined X_{BD}, the actual forces in the members of the complete truss may be calculated by, say, the method of joints or the method of sections.

In Eq. (ii), a_{BD} is the displacement of the released truss in the direction of BD produced by a unit load. Thus, in using the unit load method to calculate this displacement, the actual member forces ($F_{1,j}$) and the member forces produced by the unit load ($F_{l,j}$) are the same. Therefore, from Eq. (i)

$$a_{BD} = \sum_{j=1}^{n} \frac{F_{1,j}^2 L_j}{AE} \tag{iii}$$

The solution is completed in Table 5.6.

From Table 5.6

$$\Delta_{BD} = \frac{2.71PL}{AE} \qquad a_{BD} = \frac{4.82L}{AE}$$

Substituting these values in Eq. (i) we have

$$\frac{2.71PL}{AE} + X_{BD}\frac{4.82L}{AE} = 0$$

from which

$$X_{BD} = -0.56P \quad \text{(i.e. compression)}$$

The actual forces, $F_{a,j}$, in the members of the complete truss of Fig. 5.20(a) are now calculated using the method of joints and are listed in the final column of Table 5.6.

We note in the above that Δ_{BD} is positive, which means that Δ_{BD} is in the direction of the unit loads, i.e. B approaches D and the diagonal BD in the released structure decreases in length. Therefore in the complete structure the member BD, which prevents this shortening, must be in compression as shown; also a_{BD} will always be positive since

Table 5.6

Member	L_j (m)	$F_{0,j}$	$F_{1,j}$	$F_{0,j}F_{1,j}L_j$	$F_{1,j}^2 L_j$	$F_{a,j}$
AB	L	0	−0.71	0	$0.5L$	$+0.40P$
BC	L	0	−0.71	0	$0.5L$	$+0.40P$
CD	L	$-P$	−0.71	$0.71PL$	$0.5L$	$-0.60P$
BD	$1.41L$	–	1.0	–	$1.41L$	$-0.56P$
AC	$1.41L$	$1.41P$	1.0	$2.0PL$	$1.41L$	$+0.85P$
AD	L	0	−0.71	0	$0.5L$	$+0.40P$
				$\Sigma = 2.71\,PL$	$\Sigma = 4.82L$	

it contains the term $F_{1,j}^2$. Finally, we note that the cut member BD is included in the calculation of the displacements in the released structure since its deformation, under a unit load, contributes to a_{BD}.

Example 5.9

Calculate the forces in the members of the truss shown in Fig. 5.21(a). All members have the same cross-sectional area A, and Young's modulus E.

By inspection we see that the truss is both internally and externally statically indeterminate since it would remain stable and in equilibrium if one of the diagonals, AD or BD, and the support at C were removed; the degree of indeterminacy is therefore 2. Unlike the truss in Example 5.8, we could not remove *any* member since, if BC or CD were removed, the outer half of the truss would become a mechanism while the portion ABDE would remain statically indeterminate. Therefore we select AD and the support at C as the releases, giving the statically determinate truss shown in Fig. 5.21(b); we shall designate the force in the member AD as X_1 and the vertical reaction at C as R_2.

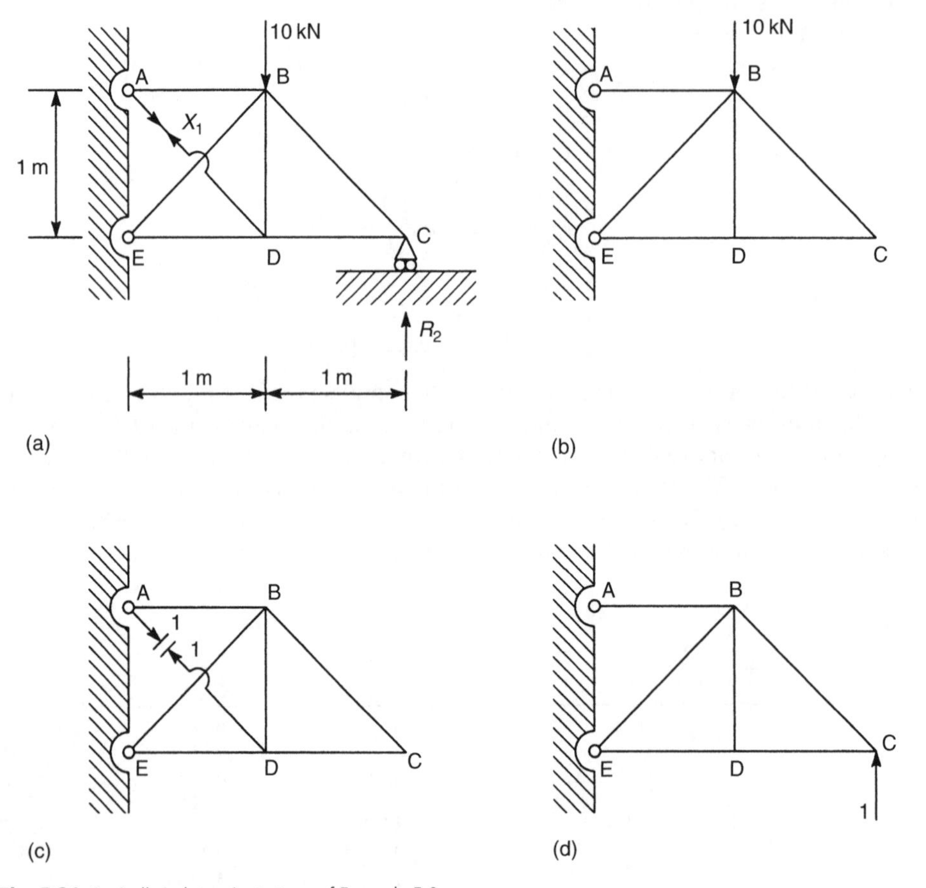

Fig. 5.21 Statically indeterminate truss of Example 5.9.

In this case we shall have two compatibility conditions, one for the diagonal AD and one for the support at C. We therefore need to investigate three loading cases: one in which the actual loads are applied to the released statically determinate truss in Fig. 5.21(b), a second in which unit loads are applied to the cut member AD (Fig. 5.21(c)) and a third in which a unit load is applied at C in the direction of R_2 (Fig. 5.21(d)). By comparison with the previous example, the compatibility conditions are

$$\Delta_{AD} + a_{11}X_1 + a_{12}R_2 = 0 \tag{i}$$

$$v_C + a_{21}X_1 + a_{22}R_2 = 0 \tag{ii}$$

in which Δ_{AD} and v_C are, respectively, the change in length of the diagonal AD and the vertical displacement of C due to the actual loads acting on the released truss, while a_{11}, a_{12}, etc., are flexibility coefficients, which we have previously defined. The calculations are similar to those carried out in Example 5.8 and are shown in Table 5.7.

From Table 5.7

$$\Delta_{AD} = \sum_{j=1}^{n} \frac{F_{0,j}F_{1,j}(X_1)L_j}{AE} = \frac{-27.1}{AE} \quad \text{(i.e. AD increases in length)}$$

$$v_C = \sum_{j=1}^{n} \frac{F_{0,j}F_{1,j}(R_2)L_j}{AE} = \frac{-48.11}{AE} \quad \text{(i.e. C displaced downwards)}$$

$$a_{11} = \sum_{j=1}^{n} \frac{F_{1,j}^2(X_1)L_j}{AE} = \frac{4.32}{AE}$$

$$a_{22} = \sum_{j=1}^{n} \frac{F_{1,j}^2(R_2)L_j}{AE} = \frac{11.62}{AE}$$

$$a_{12} = a_{21} \sum_{j=1}^{n} \frac{F_{1,j}(X_1)F_{1,j}(R_2)L_j}{AE} = \frac{2.7}{AE}$$

Table 5.7

Member	L_j	$F_{0,j}$	$F_{1,j}(X_1)$	$F_{1,j}(R_2)$	$F_{0,j}F_{1,j}(X_1)L_j$	$F_{0,j}F_{1,j}(R_2)L_j$	$F_{1,j}^2(X_1)L_j$	$F_{1,j}^2(R_2)L_j$	$F_{1,j}(X_1)F_{1,j}(R_2)L_j$	$F_{a,j}$
AB	1	10.0	−0.71	−2.0	−7.1	−20.0	0.5	4.0	1.41	0.67
BC	1.41	0	0	−1.41	0	0	0	2.81	0	−4.45
CD	1	0	0	1.0	0	0	0	1.0	0	3.15
DE	1	0	−0.71	1.0	0	0	0.5	1.0	−0.71	0.12
AD	1.41	0	1.0	0	0	0	1.41	0	0	4.28
BE	1.41	−14.14	1.0	1.41	−20.0	−28.11	1.41	2.81	2.0	−5.4
BD	1	0	−0.71	0	0	0	0.5	0	0	−3.03
					Σ = −27.1	Σ = −48.11	Σ = 4.32	Σ = 11.62	Σ = 2.7	

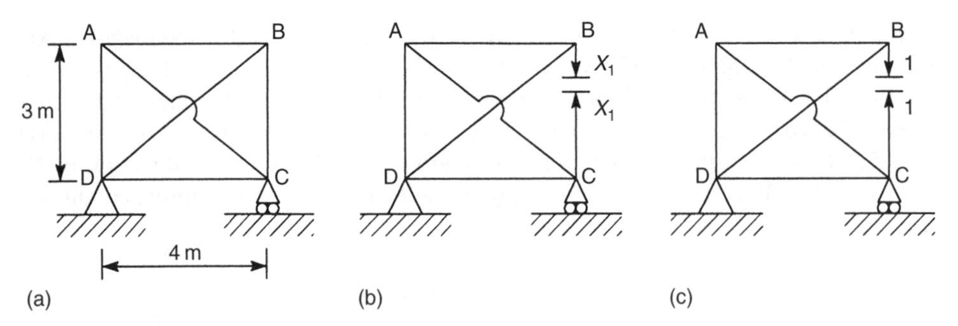

Fig. 5.22 Self-straining due to a temperature change.

Substituting in Eqs (i) and (ii) and multiplying through by AE we have

$$-27.1 + 4.32X_1 + 2.7R_2 = 0 \qquad \text{(iii)}$$

$$-48.11 + 2.7X_1 + 11.62R_2 = 0 \qquad \text{(iv)}$$

Solving Eqs (iii) and (iv) we obtain

$$X_1 = 4.28\,\text{kN} \quad R_2 = 3.15\,\text{kN}$$

The actual forces, $F_{a,j}$, in the members of the complete truss are now calculated by the method of joints and are listed in the final column of Table 5.7.

5.6.1 Self-straining trusses

Statically indeterminate trusses, unlike the statically determinate type, may be subjected to self-straining in which internal forces are present before external loads are applied. Such a situation may be caused by a local temperature change or by an initial lack of fit of a member. In cases such as these, the term on the right-hand side of the compatibility equations, Eq. (ii) in Example 5.8 and Eqs (i) and (ii) in Example 5.9, would not be zero.

Example 5.10

The truss shown in Fig. 5.22(a) is unstressed when the temperature of each member is the same, but due to local conditions the temperature in the member BC is increased by 30°C. If the cross-sectional area of each member is 200 mm^2 and the coefficient of linear expansion of the members is 7×10^{-6}/°C, calculate the resulting forces in the members; Young's modulus $E = 200\,000$ N/mm^2.

Due to the temperature rise, the increase in length of the member BC is $3 \times 10^3 \times 30 \times 7 \times 10^{-6} = 0.63$ mm. The truss has a degree of internal statical indeterminacy equal to 1 (by inspection). We therefore release the truss by cutting the member BC, which has experienced the temperature rise, as shown in Fig. 5.22(b); we shall suppose that the force in BC is X_1. Since there are no external loads on the truss, Δ_{BC} is zero

Table 5.8

Member	L_j (mm)	$F_{1,j}$	$F_{1,j}^2 L_j$	$F_{a,j}$ (N)
AB	4000	1.33	7111.1	−700
BC	3000	1.0	3000.0	−525
CD	4000	1.33	7111.1	−700
DA	3000	1.0	3000.0	−525
AC	5000	−1.67	13 888.9	875
DB	5000	−1.67	13 888.9	875
			$\Sigma = 48\,000.0$	

and the compatibility condition becomes

$$a_{11}X_1 = -0.63\,\text{mm} \tag{i}$$

in which, as before

$$a_{11} = \sum_{j=1}^{n} \frac{F_{1,j}^2 L_j}{AE}$$

Note that the extension of BC is negative since it is opposite in direction to X_1. The solution is now completed in Table 5.8. Hence

$$a_{11} = \frac{48\,000}{200 \times 200\,000} = 1.2 \times 10^{-3}$$

Then, from Eq. (i)

$$X_1 = -525\,\text{N}$$

The forces, $F_{a,j}$, in the members of the complete truss are given in the final column of Table 5.8. Compare the above with the solution of Ex. 5.4.

5.7 Total potential energy

In the spring–mass system shown in its unstrained position in Fig. 5.23(a) we normally define the *potential energy* of the mass as the product of its weight, Mg, and its height, h, above some arbitrarily fixed datum. In other words it possesses energy by virtue of its position. After deflection to an equilibrium state (Fig. 5.23(b)), the mass has lost an amount of potential energy equal to Mgy. Thus we may associate deflection with a loss of potential energy. Alternatively, we may argue that the gravitational force acting on the mass does work during its displacement, resulting in a loss of energy. Applying this reasoning to the elastic system of Fig. 5.1(a) and assuming that the potential energy of the system is zero in the unloaded state, then the *loss* of potential energy of the load P as it produces a deflection y is Py. Thus, the potential energy V of P in the deflected equilibrium state is given by

$$V = -Py$$

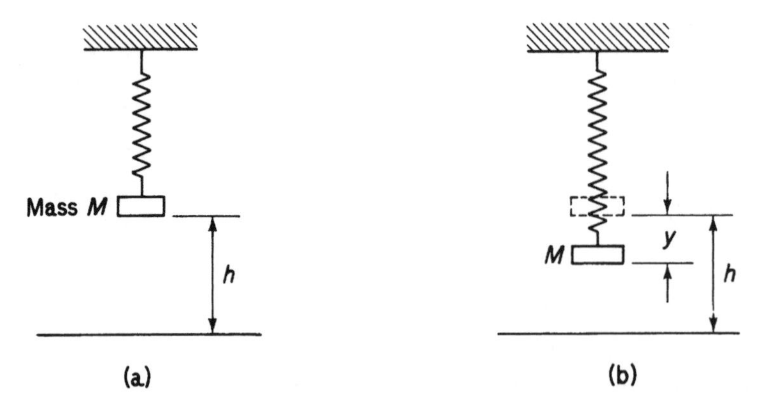

Fig. 5.23 (a) Potential energy of a spring–mass system; (b) loss in potential energy due to change in position.

We now define the *total potential energy* (TPE) of a system in its deflected equilibrium state as the sum of its internal or strain energy and the potential energy of the applied external forces. Hence, for the single member–force configuration of Fig. 5.1(a)

$$\text{TPE} = U + V = \int_0^y P\,\mathrm{d}y - Py$$

For a general system consisting of loads $P_1, P_2, \ldots, P_n$ producing *corresponding displacements* (i.e. displacements in the directions of the loads: see Section 5.10) $\Delta_1, \Delta_2, \ldots, \Delta_n$ the potential energy of all the loads is

$$V = \sum_{r=1}^{n} V_r = \sum_{r=1}^{n} (-P_r \Delta_r)$$

and the total potential energy of the system is given by

$$\text{TPE} = U + V = U + \sum_{r=1}^{n} (-P_r \Delta_r) \tag{5.22}$$

5.8 The principle of the stationary value of the total potential energy

Let us now consider an elastic body in equilibrium under a series of external loads, $P_1, P_2, \ldots, P_n$, and suppose that we impose small virtual displacements $\delta\Delta_1, \delta\Delta_2, \ldots, \delta\Delta_n$ in the directions of the loads. The virtual work done by the loads is then

$$\sum_{r=1}^{n} P_r \delta\Delta_r$$

This work will be accompanied by an increment of strain energy δU in the elastic body since by specifying virtual displacements of the loads we automatically impose virtual displacements on the particles of the body itself, as the body is continuous and is assumed to remain so. This increment in strain energy may be regarded as negative virtual work done by the particles so that the total work done during the virtual displacement is

$$-\delta U + \sum_{r=1}^{n} P_r \delta \Delta_r$$

The body is in equilibrium under the applied loads so that by the principle of virtual work the above expression must be equal to zero. Hence

$$\delta U - \sum_{r=1}^{n} P_r \delta \Delta_r = 0 \tag{5.23}$$

The loads P_r remain constant during the virtual displacement; therefore, Eq. (5.23) may be written

$$\delta U - \delta \sum_{r=1}^{n} P_r \Delta_r = 0$$

or, from Eq. (5.22)

$$\delta(U + V) = 0 \tag{5.24}$$

Thus, the total potential energy of an elastic system has a stationary value for all small displacements if the system is in equilibrium. It may also be shown that if the stationary value is a minimum the equilibrium is stable. A qualitative demonstration of this fact is sufficient for our purposes, although mathematical proofs exist.[1] In Fig. 5.24 the positions A, B and C of a particle correspond to different equilibrium states. The total potential energy of the particle in each of its three positions is proportional to its height h above some arbitrary datum, since we are considering a single particle for which the strain energy is zero. Clearly at each position the first order variation, $\partial(U + V)/\partial u$, is zero (indicating equilibrium), but only at B where the total potential energy is a minimum is the equilibrium stable. At A and C we have unstable and neutral equilibrium, respectively.

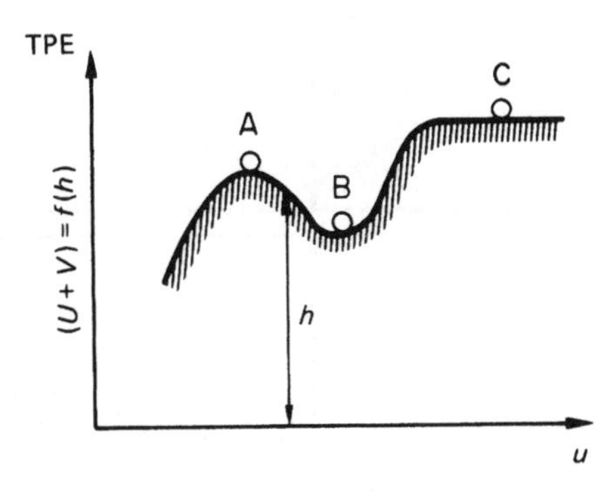

Fig. 5.24 States of equilibrium of a particle.

To summarize, *the principle of the stationary value of the total potential energy* may be stated as:

The total potential energy of an elastic system has a stationary value for all small displacements when the system is in equilibrium; further, the equilibrium is stable if the stationary value is a minimum.

This principle may often be used in the approximate analysis of structures where an exact analysis does not exist. We shall illustrate the application of the principle in Example 5.11 below, where we shall suppose that the displaced form of the beam is unknown and must be assumed; this approach is called the *Rayleigh–Ritz* method.

Example 5.11

Determine the deflection of the mid-span point of the linearly elastic, simply supported beam shown in Fig. 5.25; the flexural rigidity of the beam is EI.

The assumed displaced shape of the beam must satisfy the boundary conditions for the beam. Generally, trigonometric or polynomial functions have been found to be the most convenient where, however, the simpler the function the less accurate the solution. Let us suppose that the displaced shape of the beam is given by

$$v = v_{\mathrm{B}} \sin \frac{\pi z}{L} \tag{i}$$

in which v_{B} is the displacement at the mid-span point. From Eq. (i) we see that $v = 0$ when $z = 0$ and $z = L$ and that $v = v_{\mathrm{B}}$ when $z = L/2$. Also $\mathrm{d}v/\mathrm{d}z = 0$ when $z = L/2$ so that the displacement function satisfies the boundary conditions of the beam.

The strain energy, U, due to bending of the beam, is given by (see Ref. [3])

$$U = \int_L \frac{M^2}{2EI} \mathrm{d}z \tag{ii}$$

Also

$$M = -EI \frac{\mathrm{d}^2 v}{\mathrm{d}z^2} \quad \text{(see Chapter 16)} \tag{iii}$$

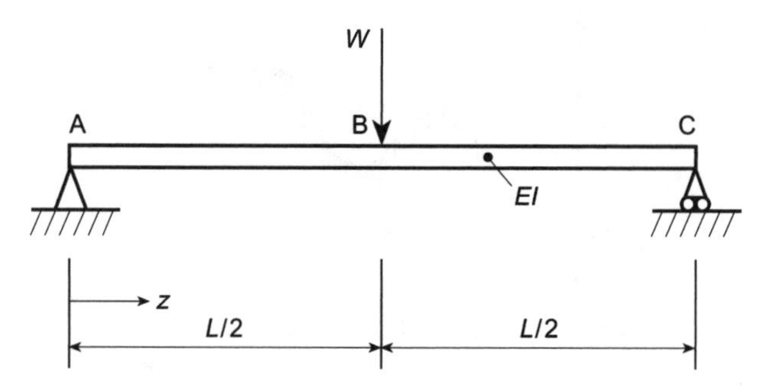

Fig. 5.25 Approximate determination of beam deflection using total potential energy.

Substituting in Eq. (iii) for v from Eq. (i) and for M in Eq. (ii) from (iii)

$$U = \frac{EI}{2}\int_0^L \frac{v_B^2\pi^4}{L^4}\sin^2\frac{\pi z}{L}\,dz$$

which gives

$$U = \frac{\pi^4 EI v_B^2}{4L^3}$$

The total potential energy of the beam is then given by

$$\text{TPE} = U + V = \frac{\pi^4 EI v_B^2}{4L^3} - W v_B$$

Then, from the principle of the stationary value of the total potential energy

$$\frac{\partial(U+V)}{\partial v_B} = \frac{\pi^4 EI v_B}{2L^3} - W = 0$$

whence

$$v_B = \frac{2WL^3}{\pi^4 EI} = 0.02053\frac{WL^3}{EI} \quad \text{(iv)}$$

The exact expression for the mid-span displacement is (Ref. [3])

$$v_B = \frac{WL^3}{48EI} = 0.02083\frac{WL^3}{EI} \quad \text{(v)}$$

Comparing the exact (Eq. (v)) and approximate results (Eq. (iv)) we see that the difference is less than 2 per cent. Further, the approximate displacement is less than the exact displacement since, by assuming a displaced shape, we have, in effect, forced the beam into taking that shape by imposing restraint; the beam is therefore stiffer.

5.9 Principle of superposition

An extremely useful principle employed in the analysis of linearly elastic structures is that of superposition. The principle states that if the displacements at all points in an elastic body are proportional to the forces producing them, i.e. the body is linearly elastic, the effect on such a body of a number of forces is the sum of the effects of the forces applied separately. We shall make immediate use of the principle in the derivation of the reciprocal theorem in the following section.

5.10 The reciprocal theorem

The reciprocal theorem is an exceptionally powerful method of analysis of linearly elastic structures and is accredited in turn to Maxwell, Betti and Rayleigh. However,

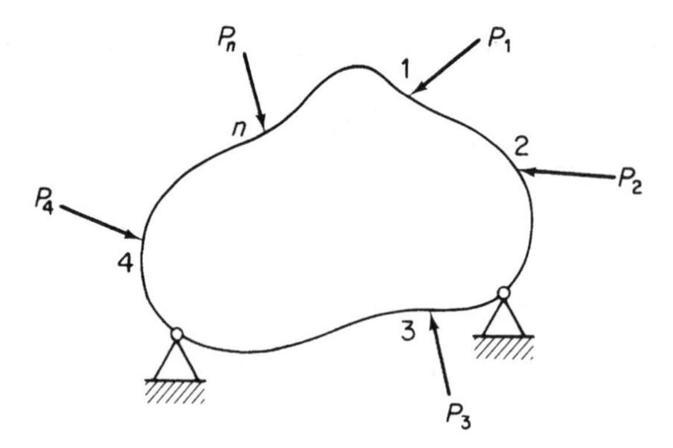

Fig. 5.26 Linearly elastic body subjected to loads $P_1, P_2, P_3, \ldots, P_n$.

before we establish the theorem we first consider a useful property of linearly elastic systems resulting from the principle of superposition. The principle enables us to express the deflection of any point in a structure in terms of a constant coefficient and the applied loads. For example, a load P_1 applied at a point 1 in a linearly elastic body will produce a deflection Δ_1 at the point given by

$$\Delta_1 = a_{11} P_1$$

in which the *influence* or *flexibility* coeffcient a_{11} is defined as the deflection at the point 1 in the direction of P_1, produced by a unit load at the point 1 applied in the direction of P_1. Clearly, if the body supports a system of loads such as those shown in Fig. 5.26, each of the loads $P_1, P_2, \ldots, P_n$ will contribute to the deflection at the point 1. Thus, the *corresponding deflection* Δ_1 at the point 1 (i.e. the total deflection in the direction of P_1 produced by all the loads) is then

$$\Delta_1 = a_{11} P_1 + a_{12} P_2 + \cdots + a_{1n} P_n$$

where a_{12} is the deflection at the point 1 in the direction of P_1, produced by a unit load at the point 2 in the direction of the load P_2 and so on. The corresponding deflections at the points of application of the complete system of loads are then

$$\left.\begin{aligned} \Delta_1 &= a_{11}P_1 + a_{12}P_2 + a_{13}P_3 + \cdots + a_{1n}P_n \\ \Delta_2 &= a_{21}P_1 + a_{22}P_2 + a_{23}P_3 + \cdots + a_{2n}P_n \\ \Delta_3 &= a_{31}P_1 + a_{32}P_2 + a_{33}P_3 + \cdots + a_{3n}P_n \\ &\vdots \\ \Delta_n &= a_{n1}P_1 + a_{n2}P_2 + a_{n3}P_3 + \cdots + a_{nn}P_n \end{aligned}\right\} \tag{5.25}$$

or, in matrix form

$$\begin{Bmatrix} \Delta_1 \\ \Delta_2 \\ \Delta_3 \\ \vdots \\ \Delta_n \end{Bmatrix} = \begin{bmatrix} a_{11} & a_{12} & a_{13} & \ldots & a_{1n} \\ a_{21} & a_{22} & a_{23} & \ldots & a_{2n} \\ a_{31} & a_{32} & a_{33} & \ldots & a_{3n} \\ \vdots & \vdots & \vdots & & \vdots \\ a_{n1} & a_{n2} & a_{n3} & \ldots & a_{nn} \end{bmatrix} \begin{Bmatrix} P_1 \\ P_2 \\ P_3 \\ \vdots \\ P_n \end{Bmatrix}$$

which may be written in shorthand matrix notation as

$$\{\Delta\} = [A]\{P\}$$

Suppose now that an elastic body is subjected to a gradually applied force P_1 at a point 1 and then, while P_1 remains in position, a force P_2 is gradually applied at another point 2. The total strain energy U of the body is given by

$$U_1 = \frac{P_1}{2}(a_{11}P_1) + \frac{P_2}{2}(a_{22}P_2) + P_1(a_{12}P_2) \qquad (5.26)$$

The third term on the right-hand side of Eq. (5.26) results from the additional work done by P_1 as it is displaced through a further distance $a_{12}P_2$ by the action of P_2. If we now remove the loads and apply P_2 followed by P_1 we have

$$U_2 = \frac{P_2}{2}(a_{22}P_2) + \frac{P_1}{2}(a_{11}P_1) + P_2(a_{21}P_1) \qquad (5.27)$$

By the principle of superposition the strain energy stored is independent of the order in which the loads are applied. Hence

$$U_1 = U_2$$

and it follows that

$$a_{12} = a_{21} \qquad (5.28)$$

Thus in its simplest form the reciprocal theorem states that:

The deflection at a point 1 in a given direction due to a unit load at a point 2 in a second direction is equal to the deflection at the point 2 in the second direction due to a unit load at the point 1 in the first direction.

In a similar manner, we derive the relationship between moments and rotations, thus:

The rotation at a point 1 due to a unit moment at a point 2 is equal to the rotation at the point 2 produced by a unit moment at the point 1.

Finally, we have:

The rotation at a point 1 due to a unit load at a point 2 is numerically equal to the deflection at the point 2 in the direction of the unit load due to a unit moment at the point 1.

Example 5.12

A cantilever 800 mm long with a prop 500 mm from the wall deflects in accordance with the following observations when a point load of 40 N is applied to its end.

Distance (mm)	0	100	200	300	400	500	600	700	800
Deflection (mm)	0	−0.3	−1.4	−2.5	−1.9	0	2.3	4.8	10.6

What will be the angular rotation of the beam at the prop due to a 30 N load applied 200 mm from the wall, together with a 10 N load applied 350 mm from the wall?

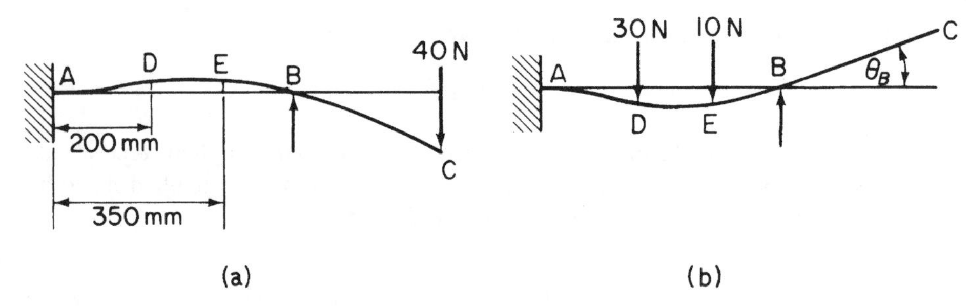

Fig. 5.27 (a) Given deflected shape of propped cantilever; (b) determination of the deflection of C.

The initial deflected shape of the cantilever is plotted as shown in Fig. 5.27(a) and the deflections at D and E produced by the 40 N load determined. The solution then proceeds as follows.

Deflection at D due to 40 N load at C = −1.4 mm.

Hence from the reciprocal theorem the deflection at C due to a 40 N load at D = −1.4 mm.

It follows that the deflection at C due to a 30 N load at D $= -\frac{3}{4} \times 1.4 = -1.05$ mm.

Similarly the deflection at C due to a 10 N load at E $= -\frac{1}{4} \times 2.4 = -0.6$ mm.

Therefore, the total deflection at C, produced by the 30 and 10 N loads acting simultaneously (Fig. 5.27(b)), is $-1.05 - 0.6 = -1.65$ mm from which the angular rotation of the beam at B, θ_B, is given by

$$\theta_B = \tan^{-1} \frac{1.65}{300} = \tan^{-1} 0.0055$$

or

$$\theta_B = 0°19'$$

Example 5.13

An elastic member is pinned to a drawing board at its ends A and B. When a moment M is applied at A, A rotates θ_A, B rotates θ_B and the centre deflects δ_1. The same moment M applied to B rotates B, θ_C and deflects the centre through δ_2. Find the moment induced at A when a load W is applied to the centre in the direction of the measured deflections, both A and B being restrained against rotation.

The three load conditions and the relevant displacements are shown in Fig. 5.28. Thus from Fig. 5.28(a) and (b) the rotation at A due to M at B is, from the reciprocal theorem, equal to the rotation at B due to M at A. Hence

$$\theta_{A(b)} = \theta_B$$

It follows that the rotation at A due to M_B at B is

$$\theta_{A(c),1} = \frac{M_B}{M}\theta_B \tag{i}$$

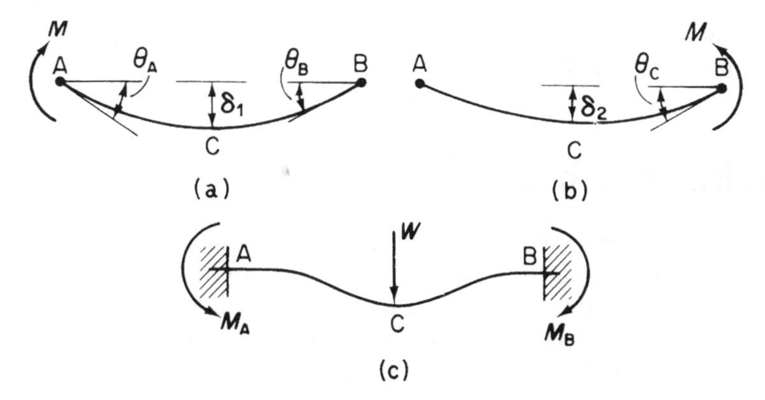

Fig. 5.28 Model analysis of a fixed beam.

Also the rotation at A due to unit load at C is equal to the deflection at C due to unit moment at A. Therefore

$$\frac{\theta_{A(c),2}}{W} = \frac{\delta_1}{M}$$

or

$$\theta_{A(c),2} = \frac{W}{M}\delta_1 \tag{ii}$$

where $\theta_{A(c),2}$ is the rotation at A due to W at C. Finally, the rotation at A due to M_A at A is, from Fig. 5.28(a) and (c)

$$\theta_{A(c),3} = \frac{M_A}{M}\theta_A \tag{iii}$$

The total rotation at A produced by M_A at A, W at C and M_B at B is, from Eqs (i), (ii) and (iii)

$$\theta_{A(c),1} + \theta_{A(c),2} + \theta_{A(c),3} = \frac{M_B}{M}\theta_B + \frac{W}{M}\delta_1 + \frac{M_A}{M}\theta_A = 0 \tag{iv}$$

since the end A is restrained from rotation. Similarly the rotation at B is given by

$$\frac{M_B}{M}\theta_C + \frac{W}{M}\delta_2 + \frac{M_A}{M}\theta_B = 0 \tag{v}$$

Solving Eqs (iv) and (v) for M_A gives

$$M_A = W\left(\frac{\delta_2\theta_B - \delta_1\theta_C}{\theta_A\theta_C - \theta_B^2}\right)$$

The fact that the arbitrary moment M does not appear in the expression for the restraining moment at A (similarly it does not appear in M_B), produced by the load W, indicates an extremely useful application of the reciprocal theorem, namely the model analysis of statically indeterminate structures. For example, the fixed beam of Fig. 5.28(c) could possibly be a full-scale bridge girder. It is then only necessary to construct a model, say of Perspex, having the same flexural rigidity EI as the full-scale beam and measure rotations and displacements produced by an arbitrary moment M to obtain fixing moments in the full-scale beam supporting a full-scale load.

5.11 Temperature effects

A uniform temperature applied across a beam section produces an expansion of the beam, as shown in Fig. 5.29, provided there are no constraints. However, a linear temperature gradient across the beam section causes the upper fibres of the beam to expand more than the lower ones, producing a bending strain as shown in Fig. 5.30 without the associated bending stresses, again provided no constraints are present.

Consider an element of the beam of depth h and length δz subjected to a linear temperature gradient over its depth, as shown in Fig. 5.31(a). The upper surface

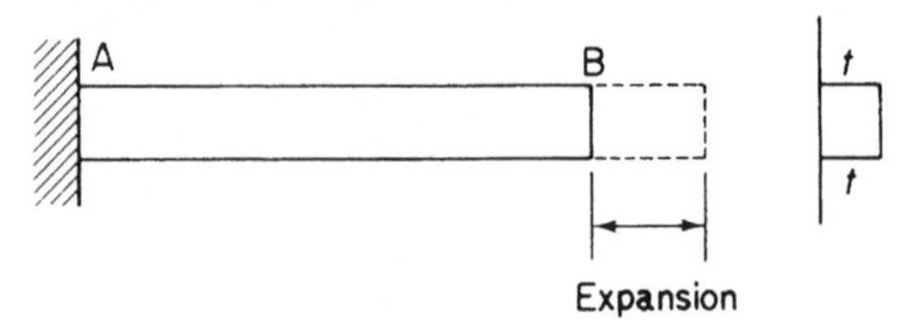

Fig. 5.29 Expansion of beam due to uniform temperature.

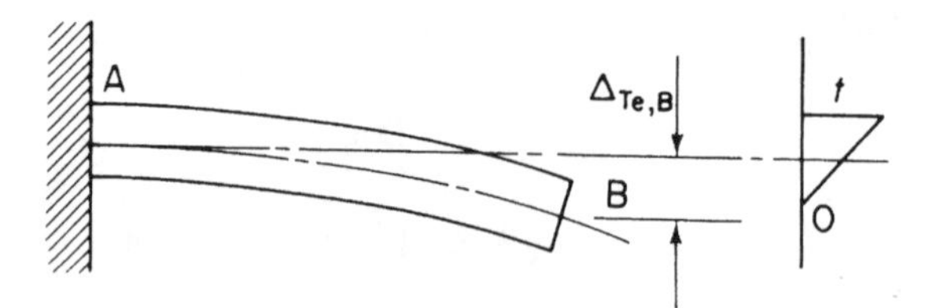

Fig. 5.30 Bending of beam due to linear temperature gradient.

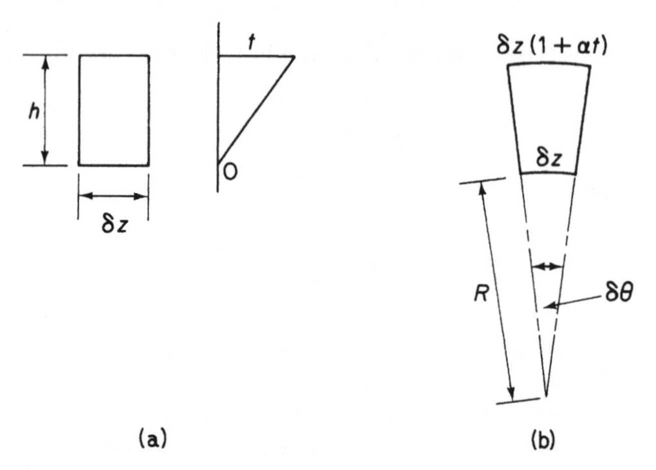

Fig. 5.31 (a) Linear temperature gradient applied to beam element; (b) bending of beam element due to temperature gradient.

of the element will increase in length to $\delta z(1+\alpha t)$ (see Section 1.15.1) where α is the coefficient of linear expansion of the material of the beam. Thus from Fig. 5.31(b)

$$\frac{R}{\delta z} = \frac{R+h}{\delta z(1+\alpha t)}$$

giving

$$R = h/\alpha t \tag{5.29}$$

Also

$$\delta\theta = \delta z/R$$

so that, from Eq. (5.29)

$$\delta\theta = \frac{\delta z \alpha t}{h} \tag{5.30}$$

We may now apply the principle of the stationary value of the total complementary energy in conjunction with the unit load method to determine the deflection Δ_{Te}, due to the temperature of any point of the beam shown in Fig. 5.30. We have seen that the above principle is equivalent to the application of the principle of virtual work where virtual forces act through real displacements. Therefore, we may specify that the displacements are those produced by the temperature gradient while the virtual force system is the unit load. Thus, the deflection $\Delta_{\mathrm{Te,B}}$ of the tip of the beam is found by writing down the increment in total complementary energy caused by the application of a virtual unit load at B and equating the resulting expression to zero (see Eqs (5.7) and (5.12)). Thus

$$\delta C = \int_L M_1 \mathrm{d}\theta - 1\Delta_{\mathrm{Te,B}} = 0$$

or

$$\Delta_{\mathrm{Te,B}} = \int_L M_1 \, \mathrm{d}\theta \tag{5.31}$$

where M_1 is the bending moment at any section due to the unit load. Substituting for $\mathrm{d}\theta$ from Eq. (5.30) we have

$$\Delta_{\mathrm{Te,B}} = \int_L M_1 \frac{\alpha t}{h} \mathrm{d}z \tag{5.32}$$

where t can vary arbitrarily along the span of the beam, but only linearly with depth. For a beam supporting some form of external loading the total deflection is given by the superposition of the temperature deflection from Eq. (5.32) and the bending deflection from Eq. (5.21); thus

$$\Delta = \int_L M_1 \left(\frac{M_0}{EI} + \frac{\alpha t}{h} \right) \mathrm{d}z \tag{5.33}$$

Example 5.14

Determine the deflection of the tip of the cantilever in Fig. 5.32 with the temperature gradient shown.

Applying a unit load vertically downwards at B, $M_1 = 1 \times z$. Also the temperature t at a section z is $t_0(l - z)/l$. Substituting in Eq. (5.32) gives

$$\Delta_{\mathrm{Te,B}} = \int_0^l z \frac{\alpha}{h} \frac{t_0}{l}(l - z)\mathrm{d}z \tag{i}$$

Integrating Eq. (i) gives

$$\Delta_{\mathrm{Te,B}} = \frac{\alpha t_0 l^2}{6h} \quad \text{(i.e. downwards)}$$

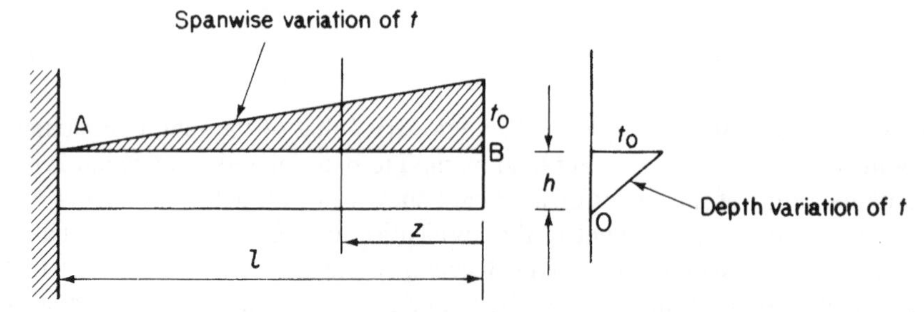

Fig. 5.32 Beam of Example 5.14.

References

1 Charlton, T. M., *Energy Principles in Applied Statics*, Blackie, London, 1959.
2 Gregory, M. S., *Introduction to Extremum Principles*, Butterworths, London, 1969.
3 Megson, T. H. G., *Structural and Stress Analysis*, 2nd edition, Elsevier, Oxford, 2005.

Further reading

Argyris, J. H. and Kelsey, S., *Energy Theorems and Structural Analysis*, Butterworths, London, 1960.
Hoff, N. J., *The Analysis of Structures*, John Wiley and Sons, Inc., New York, 1956.
Timoshenko, S. P. and Gere, J. M., *Theory of Elastic Stability*, McGraw-Hill Book Company, New York, 1961.

Problems

P.5.1 Find the magnitude and the direction of the movement of the joint C of the plane pin-jointed frame loaded as shown in Fig. P.5.1. The value of L/AE for each member is 1/20 mm/N.

Ans. 5.24 mm at 14.7° to left of vertical.

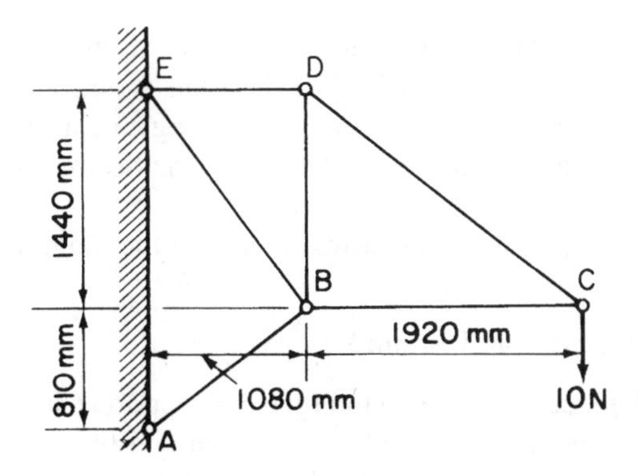

Fig. P.5.1

P.5.2 A rigid triangular plate is suspended from a horizontal plane by three vertical wires attached to its corners. The wires are each 1 mm diameter, 1440 mm long, with a modulus of elasticity of 196 000 N/mm^2. The ratio of the lengths of the sides of the plate is 3:4:5. Calculate the deflection at the point of application due to a 100 N load placed at a point equidistant from the three sides of the plate.

Ans. 0.33 mm.

P.5.3 The pin-jointed space frame shown in Fig. P.5.3 is attached to rigid supports at points 0, 4, 5 and 9, and is loaded by a force P in the x direction and a force $3P$ in the negative y direction at the point 7. Find the rotation of member 27 about the z axis due to this loading. Note that the plane frames 01234 and 56789 are identical. All members have the same cross-sectional area A and Young's modulus E.

Ans. $382P/9AE$.

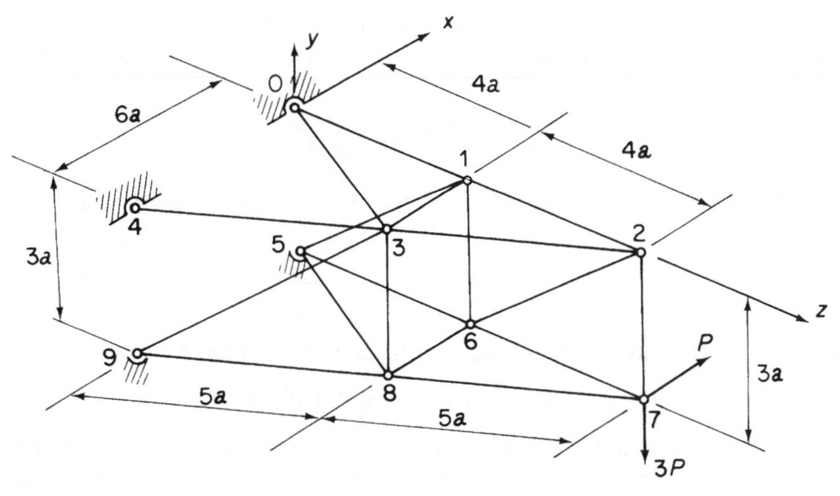

Fig. P.5.3

P.5.4 A horizontal beam is of uniform material throughout, but has a second moment of area of I for the central half of the span L and $I/2$ for each section in both outer quarters of the span. The beam carries a single central concentrated load P.

(a) Derive a formula for the central deflection of the beam, due to P, when simply supported at each end of the span.

(b) If both ends of the span are encastré determine the magnitude of the fixed end moments.

Ans. $3PL^3/128EI$, $5PL/48$ (hogging).

P.5.5 The tubular steel post shown in Fig. P.5.5 supports a load of 250 N at the free end C. The outside diameter of the tube is 100 mm and the wall thickness is 3 mm. Neglecting the weight of the tube find the horizontal deflection at C. The modulus of elasticity is 206 000 N/mm^2.

Ans. 53.3 mm.

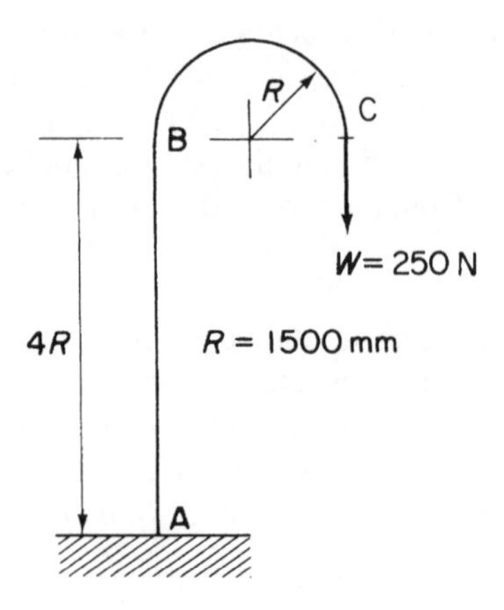

Fig. P.5.5

P.5.6 A simply supported beam AB of span L and uniform section carries a distributed load of intensity varying from zero at A to w_0/unit length at B according to the law

$$w = \frac{2w_0 z}{L}\left(1 - \frac{z}{2L}\right)$$

per unit length. If the deflected shape of the beam is given approximately by the expression

$$v = a_1 \sin\frac{\pi z}{L} + a_2 \sin\frac{2\pi z}{L}$$

evaluate the coefficients a_1 and a_2 and find the deflection of the beam at mid-span.

Ans. $a_1 = 2w_0L^4(\pi^2 + 4)/EI\pi^7$, $a_2 = -w_0L^4/16EI\pi^5$, $0.00918\, w_0L^4/EI$.

P.5.7 A uniform simply supported beam, span L, carries a distributed loading which varies according to a parabolic law across the span. The load intensity is zero at both ends of the beam and w_0 at its mid-point. The loading is normal to a principal axis of the beam cross-section and the relevant flexural rigidity is EI. Assuming that the deflected shape of the beam can be represented by the series

$$v = \sum_{i=1}^{\infty} a_i \sin \frac{i\pi z}{L}$$

find the coefficients a_i and the deflection at the mid-span of the beam using the first term only in the above series.

Ans. $a_i = 32w_0L^4/EI\pi^7 i^7$ (i odd), $w_0L^4/94.4EI$.

P.5.8 Figure P.5.8 shows a plane pin-jointed framework pinned to a rigid foundation. All its members are made of the same material and have equal cross-sectional area A, except member 12 which has area $A\sqrt{2}$.

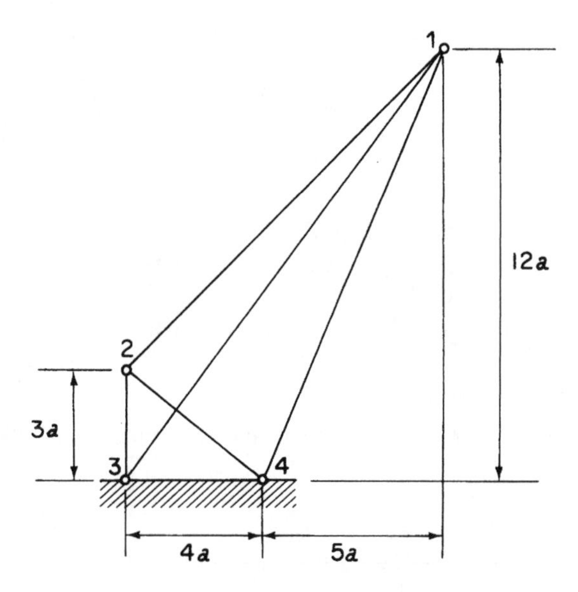

Fig. P.5.8

Under some system of loading, member 14 carries a tensile stress of 0.7 N/mm^2. Calculate the change in temperature which, if applied to member 14 only, would reduce the stress in that member to zero. Take the coefficient of linear expansion as $\alpha = 24 \times 10^{-6}/°C$ and Young's modulus $E = 70\,000$ N/mm^2.

Ans. 5.6°C.

P.5.9 The plane, pin-jointed rectangular framework shown in Fig. P.5.9(a) has one member (24) which is loosely attached at joint 2, so that relative movement between the end of the member and the joint may occur when the framework is loaded. This movement is a maximum of 0.25 mm and takes place only in the direction 24. Figure P.5.9(b) shows joint 2 in detail when the framework is unloaded. Find the value of the load P at which member 24 just becomes an effective part of the structure and also the loads in all the members when P is 10 000 N. All bars are of the same material (E = 70 000 N/mm^2) and have a cross-sectional area of 300 mm^2.

Ans. $P = 294$ N, $F_{12} = 2481.6$ N(T), $F_{23} = 1861.2$ N(T), $F_{34} = 2481.6$ N(T), $F_{41} = 5638.9$ N(C), $F_{13} = 9398.1$ N(T), $F_{24} = 3102.0$ N(C).

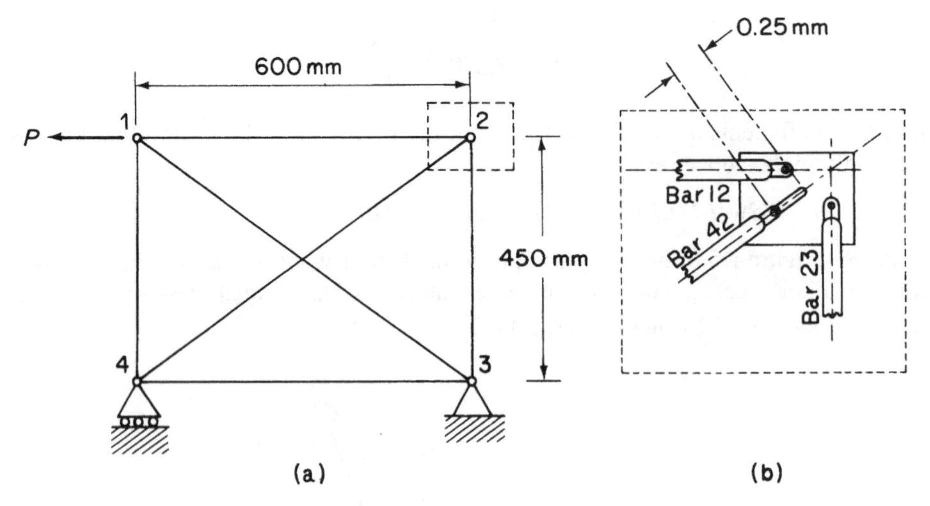

Fig. P.5.9

P.5.10 The plane frame ABCD of Fig. P.5.10 consists of three straight members with rigid joints at B and C, freely hinged to rigid supports at A and D. The flexural rigidity of AB and CD is twice that of BC. A distributed load is applied to AB, varying linearly in intensity from zero at A to w per unit length at B.

Determine the distribution of bending moment in the frame, illustrating your results with a sketch showing the principal values.

Ans. $M_B = 7\,wl^2/45$, $M_C = 8\,wl^2/45$, Cubic distribution on AB, linear on BC and CD.

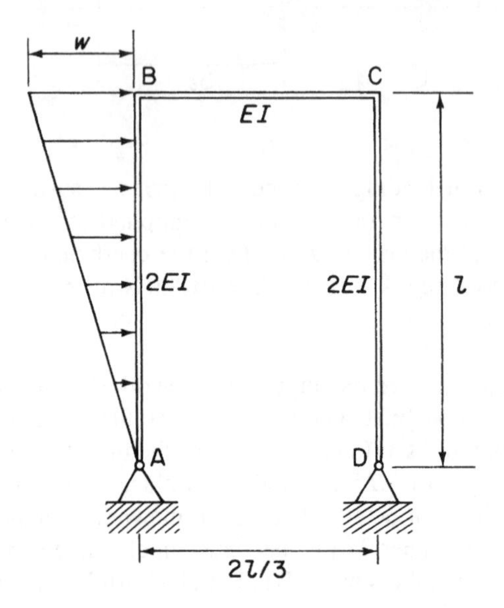

Fig. P.5.10

P.5.11 A bracket BAC is composed of a circular tube AB, whose second moment of area is $1.5I$, and a beam AC, whose second moment of area is I and which has negligible resistance to torsion. The two members are rigidly connected together at A and built into a rigid abutment at B and C as shown in Fig. P.5.11. A load P is applied at A in a direction normal to the plane of the figure.

Determine the fraction of the load which is supported at C. Both members are of the same material for which $G = 0.38E$.

Ans. $0.72P$.

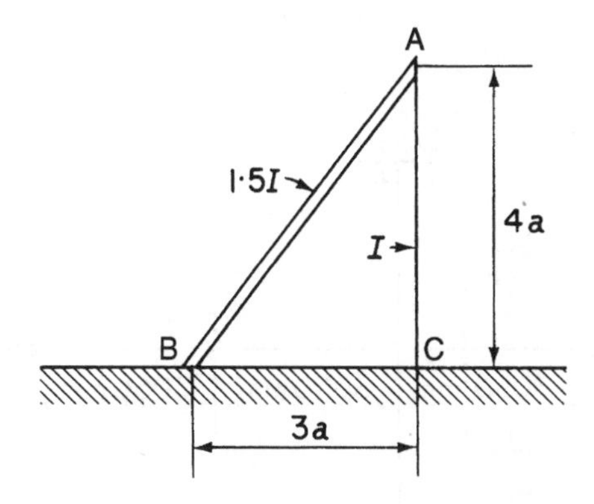

Fig. P.5.11

P.5.12 In the plane pin-jointed framework shown in Fig. P.5.12, bars 25, 35, 15 and 45 are linearly elastic with modulus of elasticity E. The remaining three bars obey a non-linear elastic stress–strain law given by

$$\varepsilon = \frac{\tau}{E}\left[1 + \left(\frac{\tau}{\tau_0}\right)^n\right]$$

where τ is the stress corresponding to strain ε. Bars 15, 45 and 23 each have a cross-sectional area A, and each of the remainder has an area of $A/\sqrt{3}$. The length of member 12 is equal to the length of member $34 = 2L$.

If a vertical load P_0 is applied at joint 5 as shown, show that the force in the member 23, i.e. F_{23}, is given by the equation

$$\alpha^n x^{n+1} + 3.5x + 0.8 = 0$$

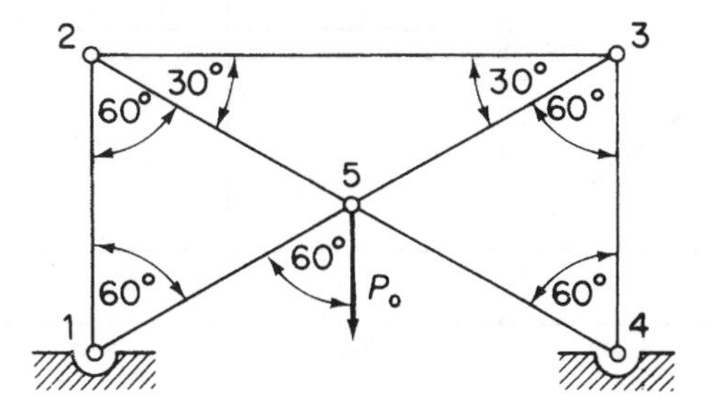

Fig. P.5.12

where

$$x = F_{23}/P_0 \quad \text{and} \quad \alpha = P_0/A\tau_0$$

P.5.13 Figure P.5.13 shows a plan view of two beams, AB 9150 mm long and DE 6100 mm long. The simply supported beam AB carries a vertical load of 100 000 N applied at F, a distance one-third of the span from B. This beam is supported at C on the encastré beam DE. The beams are of uniform cross-section and have the same second moment of area 83.5×10^6 mm^4. $E = 200\,000$ N/mm^2. Calculate the deflection of C.

Ans. 5.6 mm

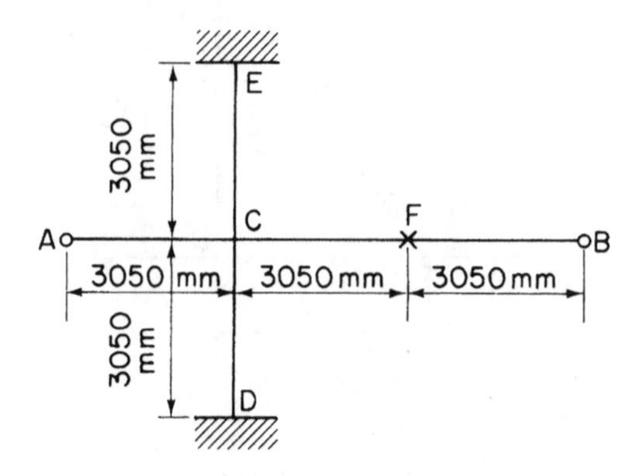

Fig. P.5.13

P.5.14 The plane structure shown in Fig. P.5.14 consists of a uniform continuous beam ABC pinned to a fixture at A and supported by a framework of pin-jointed members. All members other than ABC have the same cross-sectional area A. For ABC, the area is $4A$ and the second moment of area for bending is $Aa^2/16$. The material is the same throughout. Find (in terms of w, A, a and Young's modulus E) the vertical displacement of point D under the vertical loading shown. Ignore shearing strains in the beam ABC.

Ans. $30\,232\,wa^2/3AE$.

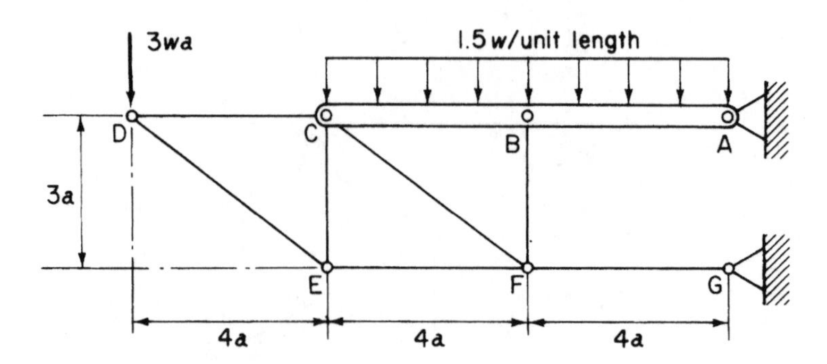

Fig. P.5.14

P.5.15 The fuselage frame shown in Fig. P.5.15 consists of two parts, ACB and ADB, with frictionless pin joints at A and B. The bending stiffness is constant in each part, with value *EI* for ACB and *xEI* for ADB. Find *x* so that the maximum bending moment in ADB will be one half of that in ACB. Assume that the deflections are due to bending strains only.

Ans. 0.092.

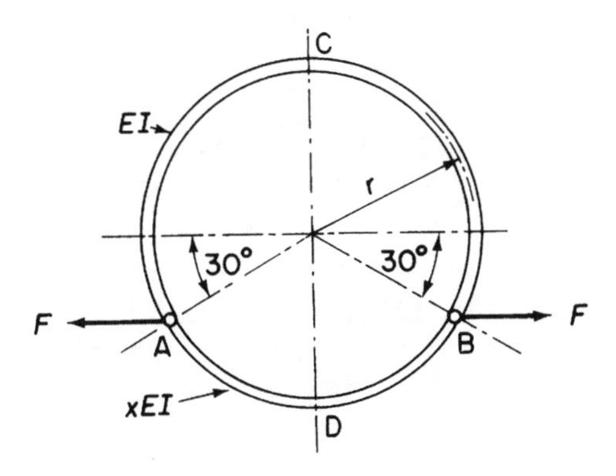

Fig. P.5.15

P.5.16 A transverse frame in a circular section fuel tank is of radius *r* and constant bending stiffness *EI*. The loading on the frame consists of the hydrostatic pressure due to the fuel and the vertical support reaction *P*, which is equal to the weight of fuel carried by the frame, shown in Fig. P.5.16.

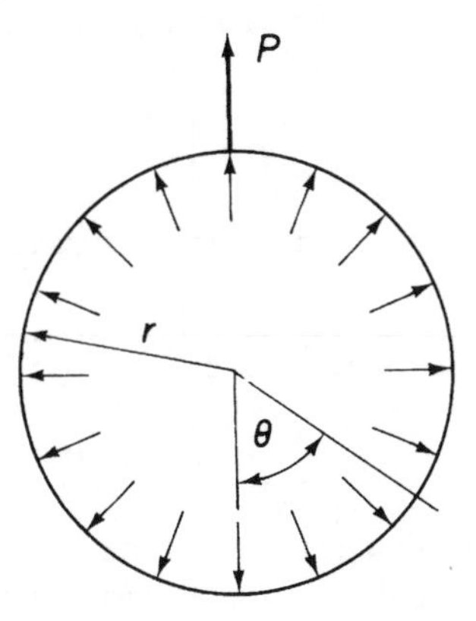

Fig. P.5.16

Taking into account only strains due to bending, calculate the distribution of bending moment around the frame in terms of the force *P*, the frame radius *r* and the angle θ.

Ans. $M = Pr(0.160 - 0.080\cos\theta - 0.159\theta\sin\theta)$

P.5.17 The frame shown in Fig. P.5.17 consists of a semi-circular arc, centre B, radius *a*, of constant flexural rigidity *EI* jointed rigidly to a beam of constant flexural rigidity $2EI$. The frame is subjected to an outward loading as shown arising from an internal pressure p_0.

Find the bending moment at points A, B and C and locate any points of contraflexure. A is the mid point of the arc. Neglect deformations of the frame due to shear and normal forces.

Ans. $M_A = -0.057p_0a^2$, $M_B = -0.292p_0a^2$, $M_C = 0.208p_0a^2$.

Points of contraflexure: in AC, at 51.7° from horizontal; in BC, 0.764a from B.

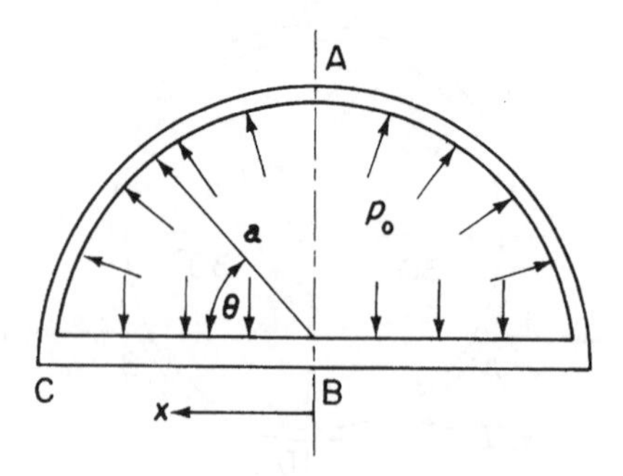

Fig. P.5.17

P.5.18 The rectangular frame shown in Fig. P.5.18 consists of two horizontal members 123 and 456 rigidly joined to three vertical members 16, 25 and 34. All five members have the same bending stiffness EI.

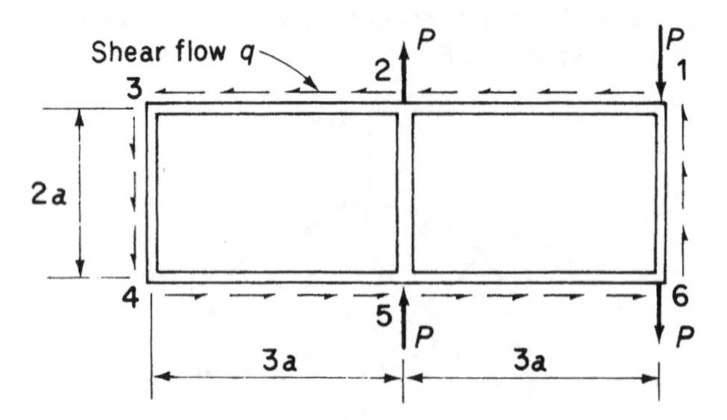

Fig. P.5.18

The frame is loaded in its own plane by a system of point loads P which are balanced by a constant shear flow q around the outside. Determine the distribution of the bending moment in the frame and sketch the bending moment diagram. In the analysis take bending deformations only into account.

Ans. Shears only at mid-points of vertical members. On the lower half of the frame $S_{43} = 0.27P$ to right, $S_{52} = 0.69P$ to left, $S_{61} = 1.08P$ to left; the bending moment diagram follows.

P.5.19 A circular fuselage frame shown in Fig. P.5.19, of radius r and constant bending stiffness EI, has a straight floor beam of length $r\sqrt{2}$, bending stiffness EI,

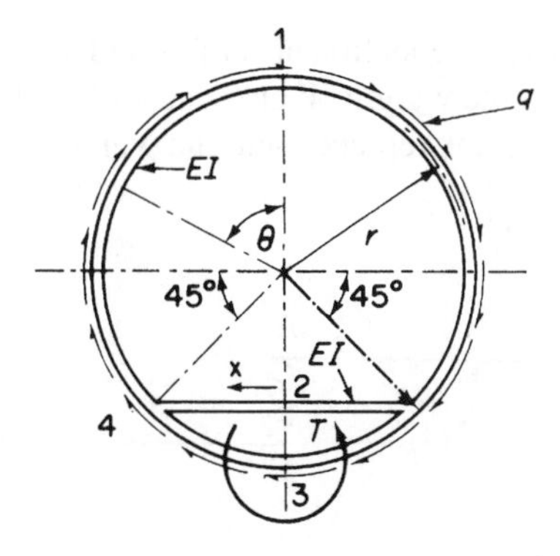

Fig. P.5.19

rigidly fixed to the frame at either end. The frame is loaded by a couple T applied at its lowest point and a constant equilibrating shear flow q around its periphery. Determine the distribution of the bending moment in the frame, illustrating your answer by means of a sketch.

In the analysis, deformations due to shear and end load may be considered negligible. The depth of the frame cross-section in comparison with the radius r may also be neglected.

Ans. $M_{14} = T(0.29 \sin\theta - 0.16\theta)$, $M_{24} = 0.30Tx/r$, $M_{43} = T(0.59 \sin\theta - 0.16\theta)$.

P.5.20 A thin-walled member BCD is rigidly built-in at D and simply supported at the same level at C, as shown in Fig. P.5.20.

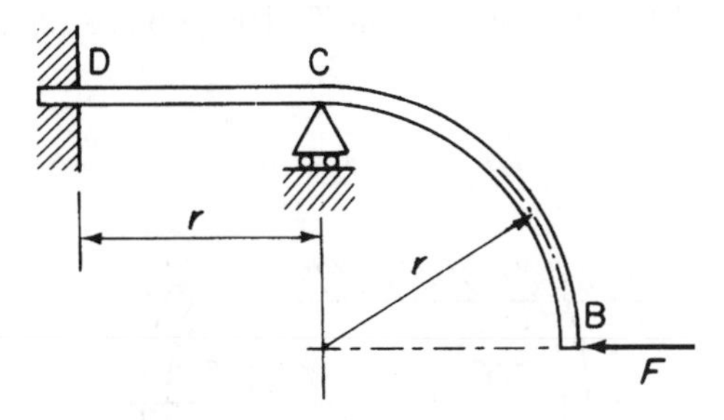

Fig. P.5.20

Find the horizontal deflection at B due to the horizontal force F. Full account must be taken of deformations due to shear and direct strains, as well as to bending.

The member is of uniform cross-section, of area A, relevant second moment of area in bending $I = Ar^2/400$ and 'reduced' effective area in shearing $A' = A/4$. Poisson's ratio for the material is $\nu = 1/3$.

Give the answer in terms of F, r, A and Young's modulus E.

Ans. $448\,Fr/EA$.

P.5.21 Figure P.5.21 shows two cantilevers, the end of one being vertically above the other and connected to it by a spring AB. Initially the system is unstrained. A weight

W placed at A causes a vertical deflection at A of δ_1 and a vertical deflection at B of δ_2. When the spring is removed the weight W at A causes a deflection at A of δ_3. Find the extension of the spring when it is replaced and the weight W is transferred to B.

Ans. $\delta_2(\delta_1 - \delta_2)/(\delta_3 - \delta_1)$.

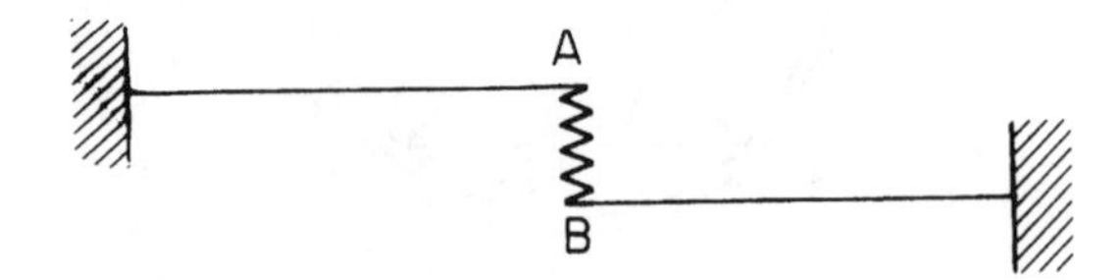

Fig. P.5.21

P.5.22 A beam 2400 mm long is supported at two points A and B which are 1440 mm apart; point A is 360 mm from the left-hand end of the beam and point B is 600 mm from the right-hand end; the value of EI for the beam is 240×10^8 N mm^2. Find the slope at the supports due to a load of 2000 N applied at the mid-point of AB.

Use the reciprocal theorem in conjunction with the above result, to find the deflection at the mid-point of AB due to loads of 3000 N applied at each of the extreme ends of the beam.

Ans. 0.011, 15.8 mm.

P.5.23 Figure P.5.23 shows a frame pinned to its support at A and B. The frame centre-line is a circular arc and the section is uniform, of bending stiffness EI and depth d. Find an expression for the maximum stress produced by a uniform temperature gradient through the depth, the temperatures on the outer and inner surfaces being respectively raised and lowered by amount T. The points A and B are unaltered in position.

Ans. $1.30ET\alpha$.

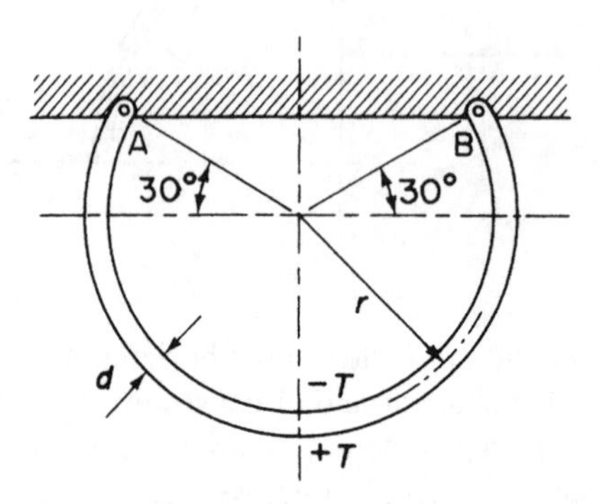

Fig. P.5.23

P.5.24 A uniform, semi-circular fuselage frame is pin-jointed to a rigid portion of the structure and is subjected to a given temperature distribution on the inside as shown in Fig. P.5.24. The temperature falls linearly across the section of the frame to zero on the outer surface. Find the values of the reactions at the pin-joints and show that the distribution of the bending moment in the frame is

$$M = \frac{0.59\, EI\alpha\theta_0 \cos\psi}{h}$$

given that:

(a) the temperature distribution is

$$\theta = \theta_0 \cos 2\psi \quad \text{for } -\pi/4 < \psi < \pi/4$$
$$\theta = 0 \quad \text{for } -\pi/4 > \psi > \pi/4$$

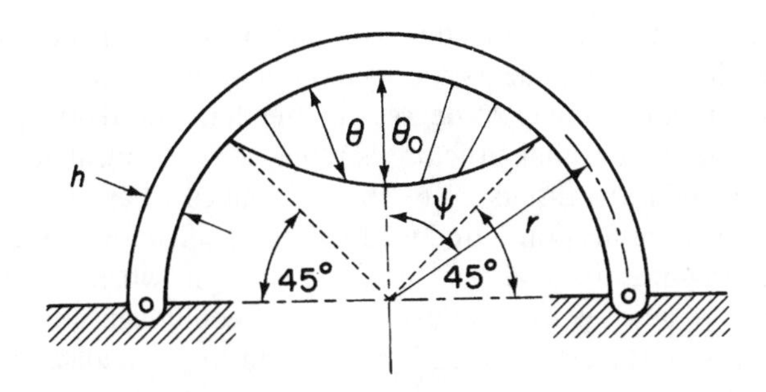

Fig. P.5.24

(b) bending deformations only are to be taken into account:

α = coefficient of linear expansion of frame material
EI = bending rigidity of frame
h = depth of cross-section
r = mean radius of frame.

6

Matrix methods

Actual aircraft structures consist of numerous components generally arranged in an irregular manner. These components are usually continuous and therefore, theoretically, possess an infinite number of degrees of freedom and redundancies. Analysis is then only possible if the actual structure is replaced by an idealized approximation or model. This procedure is discussed to some extent in Chapter 20 where we note that the greater the simplification introduced by the idealization the less complex but more inaccurate becomes the analysis. In aircraft design, where structural weight is of paramount importance, an accurate knowledge of component loads and stresses is essential so that at some stage in the design these must be calculated as accurately as possible. This accuracy may only be achieved by considering an idealized structure which closely represents the actual structure. Standard methods of structural analysis are inadequate for coping with the necessary degree of complexity in such idealized structures. It was this situation which led, in the late 1940s and early 1950s, to the development of *matrix methods* of analysis and at the same time to the emergence of high-speed, electronic, digital computers. Conveniently, matrix methods are ideally suited for expressing structural theory and for expressing the theory in a form suitable for numerical solution by computer.

A structural problem may be formulated in either of two different ways. One approach proceeds with the displacements of the structure as the unknowns, the internal forces then follow from the determination of these displacements, while in the alternative approach forces are treated as being initially unknown. In the language of matrix methods these two approaches are known as the *stiffness* (or *displacement*) method and the *flexibility* (or *force*) method, respectively. The most widely used of these two methods is the stiffness method and for this reason, we shall concentrate on this particular approach. Argyris and Kelsey,[1] however, showed that complete duality exists between the two methods in that the form of the governing equations is the same whether they are expressed in terms of displacements or forces.

Generally, actual structures must be idealized to some extent before they become amenable to analysis. Examples of some simple idealizations and their effect on structural analysis are presented in Chapter 20 for aircraft structures. Outside the realms of aeronautical engineering the representation of a truss girder by a pin-jointed framework is a well-known example of the idealization of what are known as 'skeletal' structures. Such structures are assumed to consist of a number of elements joined at points called *nodes*. The behaviour of each element may be determined by basic methods of structural analysis and hence the behaviour of the complete structure is obtained by superposition. Operations such as this are easily carried out by matrix methods as we shall see later in this chapter.

A more difficult type of structure to idealize is the continuum structure; in this category are dams, plates, shells and, obviously, aircraft fuselage and wing skins. A method, extending the matrix technique for skeletal structures, of representing continua by any desired number of elements connected at their nodes was developed by Clough *et al.*[2] at the Boeing Aircraft Company and the University of Berkeley in California. The elements may be of any desired shape but the simplest, used in plane stress problems, are the triangular and quadrilateral elements. We shall discuss the *finite element method*, as it is known, in greater detail later.

Initially, we shall develop the matrix stiffness method of solution for simple skeletal and beam structures. The fundamentals of matrix algebra are assumed.

6.1 Notation

Generally we shall consider structures subjected to forces, $F_{x,1}, F_{y,1}, F_{z,1}, F_{x,2}, F_{y,2}, F_{z,2}, \ldots, F_{x,n}, F_{y,n}, F_{z,n}$, at nodes $1, 2, \ldots, n$ at which the displacements are u_1, v_1, w_1, u_2, v_2, $w_2, \ldots,$ u_n, v_n, w_n. The numerical suffixes specify nodes while the algebraic suffixes relate the direction of the forces to an arbitrary set of axes, x, y, z. Nodal displacements u, v, w represent displacements in the positive directions of the x, y and z axes, respectively. The forces and nodal displacements are written as column matrices (alternatively known as column vectors)

$$\begin{Bmatrix} F_{x,1} \\ F_{y,1} \\ F_{z,1} \\ F_{x,2} \\ F_{y,2} \\ F_{z,2} \\ \vdots \\ F_{x,n} \\ F_{y,n} \\ F_{z,n} \end{Bmatrix} \quad \begin{Bmatrix} u_1 \\ v_1 \\ w_1 \\ u_2 \\ v_2 \\ w_2 \\ \vdots \\ u_n \\ v_n \\ w_n \end{Bmatrix}$$

which, when once established for a particular problem, may be abbreviated to

$$\{F\} \quad \{\delta\}$$

The generalized force system $\{F\}$ can contain moments M and torques T in addition to direct forces in which case $\{\delta\}$ will include rotations θ. Therefore, in referring simply to a nodal force system, we imply the possible presence of direct forces, moments and torques, while the corresponding nodal displacements can be translations and rotations.

For a complete structure the nodal forces and nodal displacements are related through a *stiffness matrix* $[K]$. We shall see that, in general

$$\{F\} = [K]\{\delta\} \tag{6.1}$$

where $[K]$ is a symmetric matrix of the form

$$[K] = \begin{bmatrix} k_{11} & k_{12} & \cdots & k_{1n} \\ k_{21} & k_{22} & \cdots & k_{2n} \\ \cdots & \cdots & \cdots & \cdots \\ k_{n1} & k_{n2} & \cdots & k_{nn} \end{bmatrix} \tag{6.2}$$

The element k_{ij} (that is the element located on row i and in column j) is known as the *stiffness influence coefficient* (note $k_{ij} = k_{ji}$). Once the stiffness matrix $[K]$ has been formed the complete solution to a problem follows from routine numerical calculations that are carried out, in most practical cases, by computer.

6.2 Stiffness matrix for an elastic spring

The formation of the stiffness matrix $[K]$ is the most crucial step in the matrix solution of any structural problem. We shall show in the subsequent work how the stiffness matrix for a complete structure may be built up from a consideration of the stiffness of its individual elements. First, however, we shall investigate the formation of $[K]$ for a simple spring element which exhibits many of the characteristics of an actual structural member.

The spring of stiffness k shown in Fig. 6.1 is aligned with the x axis and supports forces $F_{x,1}$ and $F_{x,2}$ at its nodes 1 and 2 where the displacements are u_1 and u_2. We build up the stiffness matrix for this simple case by examining different states of nodal displacement. First we assume that node 2 is prevented from moving such that $u_1 = u_1$ and $u_2 = 0$. Hence

$$F_{x,1} = ku_1$$

and from equilibrium we see that

$$F_{x,2} = -F_{x,1} = -ku_1 \tag{6.3}$$

which indicates that $F_{x,2}$ has become a reactive force in the opposite direction to $F_{x,1}$. Secondly, we take the reverse case where $u_1 = 0$ and $u_2 = u_2$ and obtain

$$F_{x,2} = ku_2 = -F_{x,1} \tag{6.4}$$

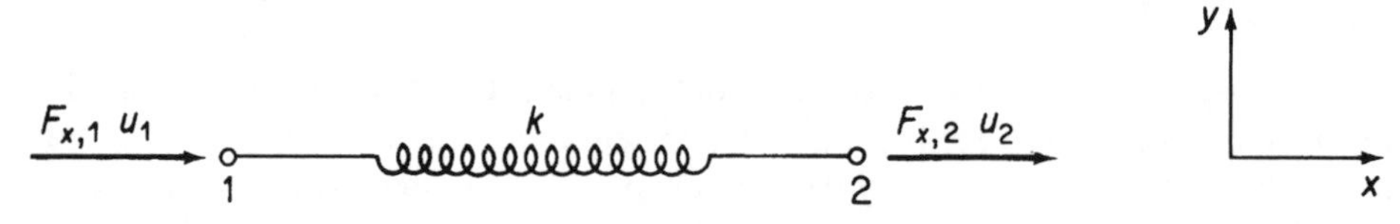

Fig. 6.1 Determination of stiffness matrix for a single spring.

By superposition of these two conditions we obtain relationships between the applied forces and the nodal displacements for the state when $u_1 = u_1$ and $u_2 = u_2$. Thus

$$\left.\begin{aligned} F_{x,1} &= ku_1 - ku_2 \\ F_{x,2} &= -ku_1 + ku_2 \end{aligned}\right\} \tag{6.5}$$

Writing Eq. (6.5) in matrix form we have

$$\begin{Bmatrix} F_{x,1} \\ F_{x,2} \end{Bmatrix} = \begin{bmatrix} k & -k \\ -k & k \end{bmatrix} \begin{Bmatrix} u_1 \\ u_2 \end{Bmatrix} \tag{6.6}$$

and by comparison with Eq. (6.1) we see that the stiffness matrix for this spring element is

$$[K] = \begin{bmatrix} k & -k \\ -k & k \end{bmatrix} \tag{6.7}$$

which is a symmetric matrix of order 2×2.

6.3 Stiffness matrix for two elastic springs in line

Bearing in mind the results of the previous section we shall now proceed, initially by a similar process, to obtain the stiffness matrix of the composite two-spring system shown in Fig. 6.2. The notation and sign convention for the forces and nodal displacements are identical to those specified in Section 6.1.

First let us suppose that $u_1 = u_1$ and $u_2 = u_3 = 0$. By comparison with the single spring case we have

$$F_{x,1} = k_a u_1 = -F_{x,2} \tag{6.8}$$

but, in addition, $F_{x,3} = 0$ since $u_2 = u_3 = 0$.

Secondly, we put $u_1 = u_3 = 0$ and $u_2 = u_2$. Clearly, in this case, the movement of node 2 takes place against the combined spring stiffnesses k_a and k_b. Hence

$$\left.\begin{aligned} F_{x,2} &= (k_a + k_b)u_2 \\ F_{x,1} &= -k_a u_2, \quad F_{x,3} = -k_b u_2 \end{aligned}\right\} \tag{6.9}$$

Hence the reactive force $F_{x,1} (= -k_a u_2)$ is not directly affected by the fact that node 2 is connected to node 3, but is determined solely by the displacement of node 2. Similar conclusions are drawn for the reactive force $F_{x,3}$.

Finally, we set $u_1 = u_2 = 0$, $u_3 = u_3$ and obtain

$$\left.\begin{aligned} F_{x,3} &= k_b u_3 = -F_{x,2} \\ F_{x,1} &= 0 \end{aligned}\right\} \tag{6.10}$$

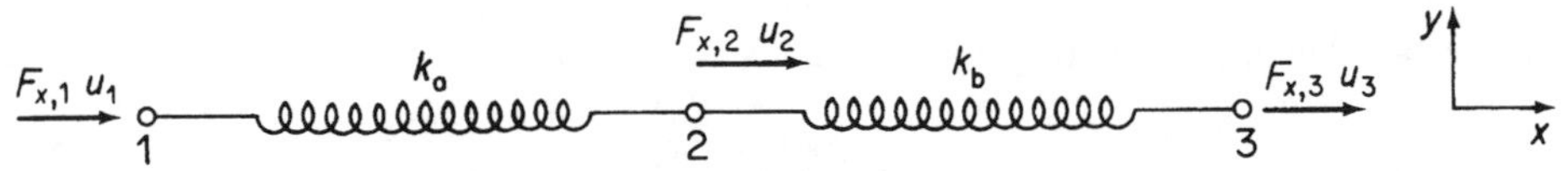

Fig. 6.2 Stiffness matrix for a two-spring system.

Superimposing these three displacement states we have, for the condition $u_1 = u_1$, $u_2 = u_2$, $u_3 = u_3$

$$\left.\begin{aligned} F_{x,1} &= k_a u_1 - k_a u_2 \\ F_{x,2} &= -k_a u_1 + (k_a + k_b)u_2 - k_b u_3 \\ F_{x,3} &= -k_b u_2 + k_b u_3 \end{aligned}\right\} \tag{6.11}$$

Writing Eqs (6.11) in matrix form gives

$$\begin{Bmatrix} F_{x,1} \\ F_{x,2} \\ F_{x,3} \end{Bmatrix} = \begin{bmatrix} k_a & -k_a & 0 \\ -k_a & k_a + k_b & -k_b \\ 0 & -k_b & k_b \end{bmatrix} \begin{Bmatrix} u_1 \\ u_2 \\ u_3 \end{Bmatrix} \tag{6.12}$$

Comparison of Eqs (6.12) with Eq. (6.1) shows that the stiffness matrix $[K]$ of this two-spring system is

$$[K] = \begin{bmatrix} k_a & -k_a & 0 \\ -k_a & k_a + k_b & -k_b \\ 0 & -k_b & k_b \end{bmatrix} \tag{6.13}$$

Equation (6.13) is a symmetric matrix of order 3×3.

It is important to note that the order of a stiffness matrix may be predicted from a knowledge of the number of nodal forces and displacements. For example, Eq. (6.7) is a 2×2 matrix connecting *two* nodal forces with *two* nodal displacements; Eq. (6.13) is a 3×3 matrix relating *three* nodal forces to *three* nodal displacements. We deduce that a stiffness matrix for a structure in which n nodal forces relate to n nodal displacements will be of order $n \times n$. The order of the stiffness matrix does not, however, bear a direct relation to the number of nodes in a structure since it is possible for more than one force to be acting at any one node.

So far we have built up the stiffness matrices for the single- and two-spring assemblies by considering various states of displacement in each case. Such a process would clearly become tedious for more complex assemblies involving a large number of springs so that a shorter, alternative, procedure is desirable. From our remarks in the preceding paragraph and by reference to Eq. (6.2) we could have deduced at the outset of the analysis that the stiffness matrix for the two-spring assembly would be of the form

$$[K] = \begin{bmatrix} k_{11} & k_{12} & k_{13} \\ k_{21} & k_{22} & k_{23} \\ k_{31} & k_{32} & k_{33} \end{bmatrix} \tag{6.14}$$

The element k_{11} of this matrix relates the force at node 1 to the displacement at node 1 and so on. Hence, remembering the stiffness matrix for the single spring (Eq. (6.7)) we may write down the stiffness matrix for an elastic element connecting nodes 1 and 2 in a structure as

$$[K_{12}] = \begin{bmatrix} k_{11} & k_{12} \\ k_{21} & k_{22} \end{bmatrix} \tag{6.15}$$

and for the element connecting nodes 2 and 3 as

$$[K_{23}] = \begin{bmatrix} k_{22} & k_{23} \\ k_{32} & k_{33} \end{bmatrix} \tag{6.16}$$

In our two-spring system the stiffness of the spring joining nodes 1 and 2 is k_a and that of the spring joining nodes 2 and 3 is k_b. Therefore, by comparison with Eq. (6.7), we may rewrite Eqs (6.15) and (6.16) as

$$[K_{12}] = \begin{bmatrix} k_a & -k_a \\ -k_a & k_a \end{bmatrix} \quad [K_{23}] = \begin{bmatrix} k_b & -k_b \\ -k_b & k_b \end{bmatrix} \tag{6.17}$$

Substituting in Eq. (6.14) gives

$$[K] = \begin{bmatrix} k_a & -k_a & 0 \\ -k_a & k_a + k_b & -k_b \\ 0 & -k_b & k_b \end{bmatrix}$$

which is identical to Eq. (6.13). We see that only the k_{22} term (linking the force at node 2 to the displacement at node 2) receives contributions from both springs. This results from the fact that node 2 is directly connected to both nodes 1 and 3 while nodes 1 and 3 are each joined directly only to node 2. Also, the elements k_{13} and k_{31} of $[K]$ are zero since nodes 1 and 3 are not directly connected and are therefore not affected by each other's displacement.

The formation of a stiffness matrix for a complete structure thus becomes a relatively simple matter of the superposition of individual or element stiffness matrices. The procedure may be summarized as follows: terms of the form k_{ii} on the main diagonal consist of the sum of the stiffnesses of all the structural elements meeting at node i while off-diagonal terms of the form k_{ij} consist of the sum of the stiffnesses of all the elements connecting node i to node j.

An examination of the stiffness matrix reveals that it possesses certain properties. For example, the sum of the elements in any column is zero, indicating that the conditions of equilibrium are satisfied. Also, the non-zero terms are concentrated near the leading diagonal while all the terms in the leading diagonal are positive; the latter property derives from the physical behaviour of any actual structure in which positive nodal forces produce positive nodal displacements.

Further inspection of Eq. (6.13) shows that its determinant vanishes. As a result the stiffness matrix $[K]$ is singular and its inverse does not exist. We shall see that this means that the associated set of simultaneous equations for the unknown nodal displacements cannot be solved for the simple reason that we have placed no limitation on any of the displacements u_1, u_2 or u_3. Thus the application of external loads results in the system moving as a rigid body. Sufficient boundary conditions must therefore be specified to enable the system to remain stable under load. In this particular problem we shall demonstrate the solution procedure by assuming that node 1 is fixed, i.e. $u_1 = 0$.

The first step is to rewrite Eq. (6.13) in partitioned form as

$$\begin{Bmatrix} F_{x,1} \\ F_{x,2} \\ F_{x,3} \end{Bmatrix} = \left[\begin{array}{c:cc} k_a & -k_a & 0 \\ \hdashline -k_a & k_a + k_b & -k_b \\ 0 & -k_b & k_b \end{array}\right] \begin{Bmatrix} u_1 = 0 \\ u_2 \\ u_3 \end{Bmatrix} \tag{6.18}$$

In Eq. (6.18) $F_{x,1}$ is the unknown reaction at node 1, u_1 and u_2 are unknown nodal displacements, while $F_{x,2}$ and $F_{x,3}$ are known applied loads. Expanding Eq. (6.18) by matrix multiplication we obtain

$$\{F_{x,1}\} = [-k_a \quad 0]\begin{Bmatrix} u_2 \\ u_3 \end{Bmatrix} \quad \begin{Bmatrix} F_{x,2} \\ F_{x,3} \end{Bmatrix} = \begin{bmatrix} k_a + k_b & -k_b \\ -k_b & k_b \end{bmatrix}\begin{Bmatrix} u_2 \\ u_3 \end{Bmatrix} \tag{6.19}$$

Inversion of the second of Eqs (6.19) gives u_2 and u_3 in terms of $F_{x,2}$ and $F_{x,3}$. Substitution of these values in the first equation then yields $F_{x,1}$.

Thus

$$\begin{Bmatrix} u_2 \\ u_3 \end{Bmatrix} = \begin{bmatrix} k_a + k_b & -k_b \\ -k_b & k_b \end{bmatrix}^{-1}\begin{Bmatrix} F_{x,2} \\ F_{x,3} \end{Bmatrix}$$

or

$$\begin{Bmatrix} u_2 \\ u_3 \end{Bmatrix} = \begin{bmatrix} 1/k_a & 1/k_a \\ 1/k_a & 1/k_b + 1/k_a \end{bmatrix}\begin{Bmatrix} F_{x,2} \\ F_{x,3} \end{Bmatrix}$$

Hence

$$\{F_{x,1}\} = [-k_a \quad 0]\begin{bmatrix} 1/k_a & 1/k_a \\ 1/k_a & 1/k_b + 1/k_a \end{bmatrix}\begin{Bmatrix} F_{x,2} \\ F_{x,3} \end{Bmatrix}$$

which gives

$$F_{x,1} = -F_{x,2} - F_{x,3}$$

as would be expected from equilibrium considerations. In problems where reactions are not required, equations relating known applied forces to unknown nodal displacements may be obtained by deleting the rows and columns of $[K]$ corresponding to zero displacements. This procedure eliminates the necessity of rearranging rows and columns in the original stiffness matrix when the fixed nodes are not conveniently grouped together.

Finally, the internal forces in the springs may be determined from the force–displacement relationship of each spring. Thus, if S_a is the force in the spring joining nodes 1 and 2 then

$$S_a = k_a(u_2 - u_1)$$

Similarly for the spring between nodes 2 and 3

$$S_b = k_b(u_3 - u_2)$$

6.4 Matrix analysis of pin-jointed frameworks

The formation of stiffness matrices for pin-jointed frameworks and the subsequent determination of nodal displacements follow a similar pattern to that described for a spring assembly. A member in such a framework is assumed to be capable of carrying axial forces only and obeys a unique force–deformation relationship given by

$$F = \frac{AE}{L}\delta$$

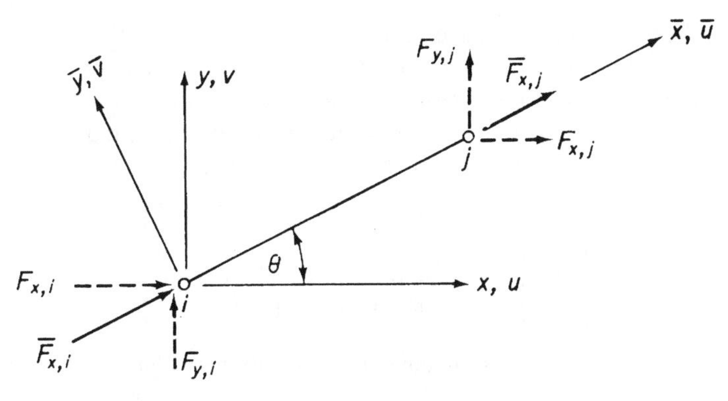

Fig. 6.3 Local and global coordinate systems for a member of a plane pin-jointed framework.

where F is the force in the member, δ its change in length, A its cross-sectional area, L its unstrained length and E its modulus of elasticity. This expression is seen to be equivalent to the spring–displacement relationships of Eqs (6.3) and (6.4) so that we may immediately write down the stiffness matrix for a member by replacing k by AE/L in Eq. (6.7). Thus

$$[K] = \begin{bmatrix} AE/L & -AE/L \\ -AE/L & AE/L \end{bmatrix}$$

or

$$[K] = \frac{AE}{L}\begin{bmatrix} 1 & -1 \\ -1 & 1 \end{bmatrix} \tag{6.20}$$

so that for a member aligned with the x axis, joining nodes i and j subjected to nodal forces $F_{x,i}$ and $F_{x,j}$, we have

$$\begin{Bmatrix} F_{x,i} \\ F_{x,j} \end{Bmatrix} = \frac{AE}{L}\begin{bmatrix} 1 & -1 \\ -1 & 1 \end{bmatrix}\begin{Bmatrix} u_i \\ u_j \end{Bmatrix} \tag{6.21}$$

The solution proceeds in a similar manner to that given in the previous section for a spring or spring assembly. However, some modification is necessary since frameworks consist of members set at various angles to one another. Figure 6.3 shows a member of a framework inclined at an angle θ to a set of arbitrary reference axes x, y. We shall refer every member of the framework to this *global coordinate* system, as it is known, when we are considering the complete structure but we shall use a *member* or *local* coordinate system $\bar{x}$, $\bar{y}$ when considering individual members. Nodal forces and displacements referred to local coordinates are written as $\bar{F}$, $\bar{u}$ etc. so that Eq. (6.21) becomes, in terms of local coordinates

$$\begin{Bmatrix} \overline{F_{x,i}} \\ \overline{F_{x,j}} \end{Bmatrix} = \frac{AE}{L}\begin{bmatrix} 1 & -1 \\ -1 & 1 \end{bmatrix}\begin{Bmatrix} \overline{u_i} \\ \overline{u_j} \end{Bmatrix} \tag{6.22}$$

where the element stiffness matrix is written $[\overline{K_{ij}}]$.

In Fig. 6.3 external forces $\overline{F_{x,i}}$ and $\overline{F_{x,j}}$ are applied to nodes i and j. It should be noted that $\overline{F_{y,i}}$, and $\overline{F_{y,j}}$, do not exist since the member can only support axial forces. However, $\overline{F_{x,i}}$ and $\overline{F_{x,j}}$ have components $F_{x,i}, F_{y,i}$ and $F_{x,j}, F_{y,j}$ respectively, so that, whereas only two force components appear for the member in terms of local coordinates, four components are present when global coordinates are used. Therefore, if we are to transfer from local to global coordinates, Eq. (6.22) must be expanded to an order consistent with the use of global coordinates, i.e.

$$\begin{Bmatrix} \overline{F_{x,i}} \\ \overline{F_{y,i}} \\ \overline{F_{x,j}} \\ \overline{F_{y,j}} \end{Bmatrix} = \frac{AE}{L} \begin{bmatrix} 1 & 0 & -1 & 0 \\ 0 & 0 & 0 & 0 \\ -1 & 0 & 1 & 0 \\ 0 & 0 & 0 & 0 \end{bmatrix} \begin{Bmatrix} \overline{u_i} \\ \overline{v_i} \\ \overline{u_j} \\ \overline{v_j} \end{Bmatrix} \tag{6.23}$$

Equation (6.23) does not change the basic relationship between $\overline{F_{x,i}}$, $\overline{F_{x,j}}$ and $\overline{u_i}$, $\overline{u_j}$ as defined in Eq. (6.22).

From Fig. 6.3 we see that

$$\overline{F_{x,i}} = F_{x,i}\cos\theta + F_{y,i}\sin\theta$$
$$\overline{F_{y,i}} = -F_{x,i}\sin\theta + F_{y,i}\cos\theta$$

and

$$\overline{F_{x,j}} = F_{x,j}\cos\theta + F_{y,j}\sin\theta$$
$$\overline{F_{y,j}} = -F_{x,j}\sin\theta + F_{y,j}\cos\theta$$

Writing λ for $\cos\theta$ and μ for $\sin\theta$ we express the above equations in matrix form as

$$\begin{Bmatrix} \overline{F_{x,i}} \\ \overline{F_{y,i}} \\ \overline{F_{x,j}} \\ \overline{F_{y,j}} \end{Bmatrix} = \begin{bmatrix} \lambda & \mu & 0 & 0 \\ -\mu & \lambda & 0 & 0 \\ 0 & 0 & \lambda & \mu \\ 0 & 0 & -\mu & \lambda \end{bmatrix} \begin{Bmatrix} F_{x,i} \\ F_{y,i} \\ F_{x,j} \\ F_{y,j} \end{Bmatrix} \tag{6.24}$$

or, in abbreviated form

$$\{\overline{F}\} = [T]\{F\} \tag{6.25}$$

where $[T]$ is known as the *transformation matrix*. A similar relationship exists between the sets of nodal displacements. Thus, again using our shorthand notation

$$\{\bar{\delta}\} = [T]\{\delta\} \tag{6.26}$$

Substituting now for $\{\bar{F}\}$ and $\{\bar{\delta}\}$ in Eq. (6.23) from Eqs (6.25) and (6.26), we have

$$[T]\{F\} = [\overline{K_{ij}}][T]\{\delta\}$$

Hence

$$\{F\} = [T^{-1}][\overline{K_{ij}}][T]\{\delta\} \tag{6.27}$$

It may be shown that the inverse of the transformation matrix is its transpose, i.e.

$$[T^{-1}] = [T]^{\mathrm{T}}$$

Thus we rewrite Eq. (6.27) as

$$\{F\} = [T]^{\mathrm{T}}[\overline{K_{ij}}][T]\{\delta\} \tag{6.28}$$

The nodal force system referred to global coordinates, $\{F\}$ is related to the corresponding nodal displacements by

$$\{F\} = [K_{ij}]\{\delta\} \tag{6.29}$$

where $[K_{ij}]$ is the member stiffness matrix referred to global coordinates. Comparison of Eqs (6.28) and (6.29) shows that

$$[K_{ij}] = [T]^{\mathrm{T}}[\overline{K_{ij}}][T]$$

Substituting for $[T]$ from Eq. (6.24) and $[\overline{K_{ij}}]$ from Eq. (6.23), we obtain

$$[K_{ij}] = \frac{AE}{L}\begin{bmatrix} \lambda^2 & \lambda\mu & -\lambda^2 & -\lambda\mu \\ \lambda\mu & \mu^2 & -\lambda\mu & -\mu^2 \\ -\lambda^2 & -\lambda\mu & \lambda^2 & \lambda\mu \\ -\lambda\mu & -\mu^2 & \lambda\mu & \mu^2 \end{bmatrix} \tag{6.30}$$

By evaluating $\lambda(=\cos\theta)$ and $\mu(=\sin\theta)$ for each member and substituting in Eq. (6.30) we obtain the stiffness matrix, referred to global coordinates, for each member of the framework.

In Section 6.3 we determined the internal force in a spring from the nodal displacements. Applying similar reasoning to the framework member we may write down an expression for the internal force S_{ij} in terms of the local coordinates. Thus

$$S_{ij} = \frac{AE}{L}(\overline{u_j} - \overline{u_i}) \tag{6.31}$$

Now

$$\overline{u_j} = \lambda u_j + \mu v_j$$

$$\overline{u_i} = \lambda u_i + \mu v_i$$

Hence

$$\overline{u_j} - \overline{u_i} = \lambda(u_j - u_i) + \mu(v_j - v_i)$$

Substituting in Eq. (6.31) and rewriting in matrix form, we have

$$S_{ij} = \frac{AE}{L}\begin{bmatrix} \lambda & \mu \end{bmatrix}_{ij}\begin{Bmatrix} u_j - u_i \\ v_j - v_i \end{Bmatrix} \tag{6.32}$$

Example 6.1

Determine the horizontal and vertical components of the deflection of node 2 and the forces in the members of the pin-jointed framework shown in Fig. 6.4. The product AE is constant for all members.

We see in this problem that nodes 1 and 3 are pinned to a fixed foundation and are therefore not displaced. Hence, with the global coordinate system shown

$$u_1 = v_1 = u_3 = v_3 = 0$$

The external forces are applied at node 2 such that $F_{x,2} = 0$, $F_{y,2} = -W$; the nodal forces at 1 and 3 are then unknown reactions.

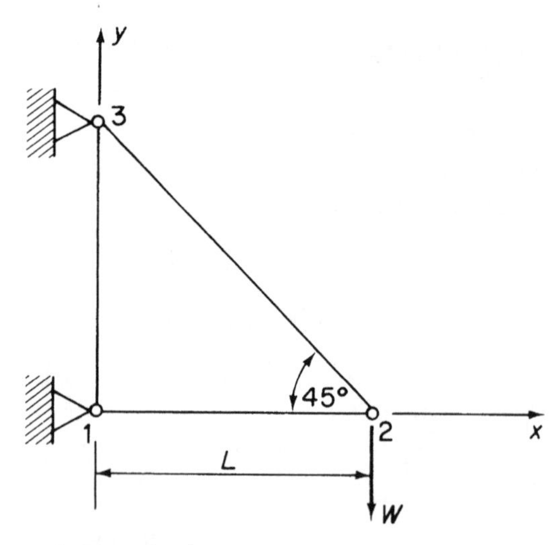

Fig. 6.4 Pin-jointed framework of Example 6.1.

The first step in the solution is to assemble the stiffness matrix for the complete framework by writing down the member stiffness matrices referred to the global coordinate system using Eq. (6.30). The direction cosines λ and μ take different values for each of the three members, therefore remembering that the angle θ is measured anticlockwise from the positive direction of the x axis we have the following:

Member	θ	λ	μ
1–2	0	1	0
1–3	90	0	1
2–3	135	$-1/\sqrt{2}$	$1/\sqrt{2}$

The member stiffness matrices are therefore

$$[K_{12}] = \frac{AE}{L}\begin{bmatrix} 1 & 0 & -1 & 0 \\ 0 & 0 & 0 & 0 \\ -1 & 0 & 1 & 0 \\ 0 & 0 & 0 & 0 \end{bmatrix} \quad [K_{13}] = \frac{AE}{L}\begin{bmatrix} 0 & 0 & 0 & 0 \\ 0 & 1 & 0 & -1 \\ 0 & 0 & 0 & 0 \\ 0 & -1 & 0 & 1 \end{bmatrix}$$

$$[K_{23}] = \frac{AE}{\sqrt{2}L}\begin{bmatrix} \frac{1}{2} & -\frac{1}{2} & -\frac{1}{2} & \frac{1}{2} \\ -\frac{1}{2} & \frac{1}{2} & \frac{1}{2} & -\frac{1}{2} \\ -\frac{1}{2} & \frac{1}{2} & \frac{1}{2} & -\frac{1}{2} \\ \frac{1}{2} & -\frac{1}{2} & -\frac{1}{2} & \frac{1}{2} \end{bmatrix} \qquad \text{(i)}$$

The next stage is to add the member stiffness matrices to obtain the stiffness matrix for the complete framework. Since there are six possible nodal forces producing six possible nodal displacements the complete stiffness matrix is of the order 6×6. Although the addition is not difficult in this simple problem care must be taken, when solving more complex structures, to ensure that the matrix elements are placed in the correct position in the complete stiffness matrix. This may be achieved by expanding each member stiffness matrix to the order of the complete stiffness matrix by inserting appropriate rows and columns of zeros. Such a method is, however, time and space consuming. An alternative procedure is suggested here. The complete stiffness matrix is of the form shown in Eq. (ii)

$$\begin{Bmatrix} F_{x,1} \\ F_{y,1} \\ F_{x,2} \\ F_{y,2} \\ F_{x,3} \\ F_{y,3} \end{Bmatrix} = \begin{bmatrix} [k_{11}] & [k_{12}] & [k_{13}] \\ [k_{21}] & [k_{22}] & [k_{23}] \\ [k_{31}] & [k_{32}] & [k_{33}] \end{bmatrix} \begin{Bmatrix} u_1 \\ v_1 \\ u_2 \\ v_2 \\ u_3 \\ v_3 \end{Bmatrix} \tag{ii}$$

The complete stiffness matrix has been divided into a number of submatrices in which $[k_{11}]$ is a 2×2 matrix relating the nodal forces $F_{x,1}$, $F_{y,1}$ to the nodal displacements u_1, v_1 and so on. It is a simple matter to divide each member stiffness matrix into submatrices of the form $[k_{11}]$, as shown in Eqs (iii). All that remains is to insert each submatrix into its correct position in Eq. (ii), adding the matrix elements where they overlap; for example, the $[k_{11}]$ submatrix in Eq. (ii) receives contributions from $[K_{12}]$ and $[K_{13}]$. The complete stiffness matrix is then of the form shown in Eq. (iv). It is sometimes helpful, when considering the stiffness matrix separately, to write the nodal displacement above the appropriate column (see Eq. (iv)). We note that $[K]$ is symmetrical, that all the diagonal terms are positive and that the sum of each row and column is zero

$$[K_{12}] = \frac{AE}{L} \begin{bmatrix} \underbrace{\begin{bmatrix} 1 & 0 \\ 0 & 0 \end{bmatrix}}_{k_{11}} & \underbrace{\begin{bmatrix} -1 & 0 \\ 0 & 0 \end{bmatrix}}_{k_{12}} \\ \underbrace{\begin{bmatrix} -1 & 0 \\ 0 & 0 \end{bmatrix}}_{k_{21}} & \underbrace{\begin{bmatrix} 1 & 0 \\ 0 & 0 \end{bmatrix}}_{k_{22}} \end{bmatrix}$$

$$[K_{13}] = \frac{AE}{L} \begin{bmatrix} \underbrace{\begin{bmatrix} 0 & 0 \\ 0 & 1 \end{bmatrix}}_{k_{11}} & \underbrace{\begin{bmatrix} 0 & 0 \\ 0 & -1 \end{bmatrix}}_{k_{13}} \\ \underbrace{\begin{bmatrix} 0 & 0 \\ 0 & -1 \end{bmatrix}}_{k_{31}} & \underbrace{\begin{bmatrix} 0 & 0 \\ 0 & 1 \end{bmatrix}}_{k_{33}} \end{bmatrix} \tag{iii}$$

$$[K_{23}] = \frac{AE}{\sqrt{2}L}\begin{bmatrix} \begin{bmatrix} \frac{1}{2} & k_{22} & -\frac{1}{2} \\ -\frac{1}{2} & & \frac{1}{2} \end{bmatrix} & \begin{bmatrix} -\frac{1}{2} & k_{23} & \frac{1}{2} \\ \frac{1}{2} & & -\frac{1}{2} \end{bmatrix} \\ \begin{bmatrix} -\frac{1}{2} & k_{32} & \frac{1}{2} \\ \frac{1}{2} & & -\frac{1}{2} \end{bmatrix} & \begin{bmatrix} \frac{1}{2} & k_{33} & -\frac{1}{2} \\ -\frac{1}{2} & & \frac{1}{2} \end{bmatrix} \end{bmatrix} \tag{iii}$$

$$\begin{Bmatrix} F_{x,1} \\ F_{y,1} \\ F_{x,2} \\ F_{y,2} \\ F_{x,3} \\ F_{y,3} \end{Bmatrix} = \frac{AE}{L}\begin{bmatrix} 1 & 0 & -1 & 0 & 0 & 0 \\ 0 & 1 & 0 & 0 & 0 & -1 \\ -1 & 0 & 1+\frac{1}{2\sqrt{2}} & -\frac{1}{2\sqrt{2}} & -\frac{1}{2\sqrt{2}} & \frac{1}{2\sqrt{2}} \\ 0 & 0 & -\frac{1}{2\sqrt{2}} & \frac{1}{2\sqrt{2}} & \frac{1}{2\sqrt{2}} & -\frac{1}{2\sqrt{2}} \\ 0 & 0 & -\frac{1}{2\sqrt{2}} & \frac{1}{2\sqrt{2}} & \frac{1}{2\sqrt{2}} & -\frac{1}{2\sqrt{2}} \\ 0 & -1 & \frac{1}{2\sqrt{2}} & -\frac{1}{2\sqrt{2}} & -\frac{1}{2\sqrt{2}} & 1+\frac{1}{2\sqrt{2}} \end{bmatrix}\begin{Bmatrix} u_1 = 0 \\ v_1 = 0 \\ u_2 \\ v_2 \\ u_3 = 0 \\ v_3 = 0 \end{Bmatrix} \tag{iv}$$

(columns labelled u_1, v_1, u_2, v_2, u_3, v_3)

If we now delete rows and columns in the stiffness matrix corresponding to zero displacements, we obtain the unknown nodal displacements u_2 and v_2 in terms of the applied loads $F_{x,2}$ $(=0)$ and $F_{y,2}$ $(=-W)$. Thus

$$\begin{Bmatrix} F_{x,2} \\ F_{y,2} \end{Bmatrix} = \frac{AE}{L}\begin{bmatrix} 1+\frac{1}{2\sqrt{2}} & -\frac{1}{2\sqrt{2}} \\ -\frac{1}{2\sqrt{2}} & \frac{1}{2\sqrt{2}} \end{bmatrix}\begin{Bmatrix} u_2 \\ v_2 \end{Bmatrix} \tag{v}$$

Inverting Eq. (v) gives

$$\begin{Bmatrix} u_2 \\ v_2 \end{Bmatrix} = \frac{L}{AE}\begin{bmatrix} 1 & 1 \\ 1 & 1+2\sqrt{2} \end{bmatrix}\begin{Bmatrix} F_{x,2} \\ F_{y,2} \end{Bmatrix} \tag{vi}$$

from which

$$u_2 = \frac{L}{AE}(F_{x,2} + F_{y,2}) = -\frac{WL}{AE} \tag{vii}$$

$$v_2 = \frac{L}{AE}[F_{x,2} + (1+2\sqrt{2})F_{y,2}] = -\frac{WL}{AE}(1+2\sqrt{2}) \tag{viii}$$

The reactions at nodes 1 and 3 are now obtained by substituting for u_2 and v_2 from Eq. (vi) into Eq. (iv). Thus

$$\begin{Bmatrix} F_{x,1} \\ F_{y,1} \\ F_{x,3} \\ F_{y,3} \end{Bmatrix} = \begin{bmatrix} -1 & 0 \\ 0 & 0 \\ -\dfrac{1}{2\sqrt{2}} & \dfrac{1}{2\sqrt{2}} \\ \dfrac{1}{2\sqrt{2}} & -\dfrac{1}{2\sqrt{2}} \end{bmatrix} \begin{bmatrix} 1 & 1 \\ 1 & 1+2\sqrt{2} \end{bmatrix} \begin{Bmatrix} F_{x,2} \\ F_{y,2} \end{Bmatrix}$$

$$= \begin{bmatrix} -1 & -1 \\ 0 & 0 \\ 0 & 1 \\ 0 & -1 \end{bmatrix} \begin{Bmatrix} F_{x,2} \\ F_{y,2} \end{Bmatrix}$$

giving

$$F_{x,1} = -F_{x,2} - F_{y,2} = W$$
$$F_{y,1} = 0$$
$$F_{x,3} = F_{y,2} = -W$$
$$F_{y,3} = W$$

Finally, the forces in the members are found from Eqs (6.32), (vii) and (viii)

$$S_{12} = \frac{AE}{L}[1 \quad 0] \begin{Bmatrix} u_2 - u_1 \\ v_2 - v_1 \end{Bmatrix} = -W \text{ (compression)}$$

$$S_{13} = \frac{AE}{L}[0 \quad 1] \begin{Bmatrix} u_3 - u_1 \\ v_3 - v_1 \end{Bmatrix} = 0 \text{ (as expected)}$$

$$S_{23} = \frac{AE}{\sqrt{2}L}\left[-\frac{1}{\sqrt{2}} \quad \frac{1}{\sqrt{2}}\right] \begin{Bmatrix} u_3 - u_2 \\ v_3 - v_2 \end{Bmatrix} = \sqrt{2}W \text{ (tension)}$$

6.5 Application to statically indeterminate frameworks

The matrix method of solution described in the previous sections for spring and pin-jointed framework assemblies is completely general and is therefore applicable to any structural problem. We observe that at no stage in Example 6.1 did the question of the degree of indeterminacy of the framework arise. It follows that problems involving statically indeterminate frameworks (and other structures) are solved in an identical manner to that presented in Example 6.1, the stiffness matrices for the redundant members being included in the complete stiffness matrix as before.

6.6 Matrix analysis of space frames

The procedure for the matrix analysis of space frames is similar to that for plane pin-jointed frameworks. The main difference lies in the transformation of the member stiffness matrices from local to global coordinates since, as we see from Fig. 6.5, axial nodal forces $\overline{F_{x,i}}$ and $\overline{F_{x,j}}$ have each now three global components $F_{x,i}, F_{y,i}, F_{z,i}$ and $F_{x,j}, F_{y,j}, F_{z,j}$, respectively. The member stiffness matrix referred to global coordinates is therefore of the order 6×6 so that $[K_{ij}]$ of Eq. (6.22) must be expanded to the same order to allow for this. Hence

$$[\overline{K_{ij}}] = \frac{AE}{L} \begin{array}{c} \begin{matrix} \bar{u}_i & \bar{v}_i & \bar{w}_i & \bar{u}_j & \bar{v}_j & \bar{w}_j \end{matrix} \\ \begin{bmatrix} 1 & 0 & 0 & -1 & 0 & 0 \\ 0 & 0 & 0 & 0 & 0 & 0 \\ 0 & 0 & 0 & 0 & 0 & 0 \\ -1 & 0 & 0 & 1 & 0 & 0 \\ 0 & 0 & 0 & 0 & 0 & 0 \\ 0 & 0 & 0 & 0 & 0 & 0 \end{bmatrix} \end{array} \tag{6.33}$$

In Fig. 6.5 the member ij is of length L, cross-sectional area A and modulus of elasticity E. Global and local coordinate systems are designated as for the two-dimensional case. Further, we suppose that

$$\theta_{x\bar{x}} = \text{angle between } x \text{ and } \bar{x}$$
$$\theta_{x\bar{y}} = \text{angle between } x \text{ and } \bar{y}$$
$$\vdots$$
$$\theta_{z\bar{y}} = \text{angle between } z \text{ and } \bar{y}$$
$$\vdots$$

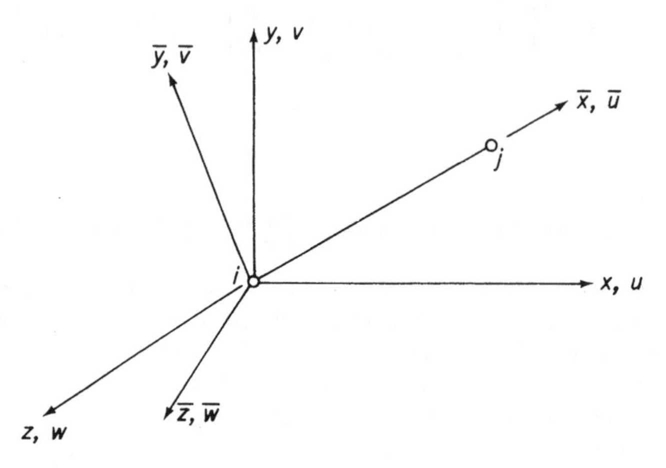

Fig. 6.5 Local and global coordinate systems for a member in a pin-jointed space frame.

Therefore, nodal forces referred to the two systems of axes are related as follows

$$\left.\begin{aligned}\overline{F_x} &= F_x\cos\theta_{x\bar{x}} + F_y\cos\theta_{x\bar{y}} + F_z\cos\theta_{x\bar{z}}\\ \overline{F_y} &= F_x\cos\theta_{y\bar{x}} + F_y\cos\theta_{y\bar{y}} + F_z\cos\theta_{y\bar{z}}\\ \overline{F_z} &= F_x\cos\theta_{z\bar{x}} + F_y\cos\theta_{z\bar{y}} + F_z\cos\theta_{z\bar{z}}\end{aligned}\right\} \tag{6.34}$$

Writing

$$\left.\begin{aligned}\lambda_{\bar{x}} &= \cos\theta_{x\bar{x}}, & \lambda_{\bar{y}} &= \cos\theta_{x\bar{y}}, & \lambda_{\bar{z}} &= \cos\theta_{x\bar{z}}\\ \mu_{\bar{x}} &= \cos\theta_{y\bar{x}}, & \mu_{\bar{y}} &= \cos\theta_{y\bar{y}}, & \mu_{\bar{z}} &= \cos\theta_{y\bar{z}}\\ \nu_{\bar{x}} &= \cos\theta_{z\bar{x}}, & \nu_{\bar{y}} &= \cos\theta_{z\bar{y}}, & \nu_{\bar{z}} &= \cos\theta_{z\bar{z}}\end{aligned}\right\} \tag{6.35}$$

we may express Eq. (6.34) for nodes i and j in matrix form as

$$\begin{Bmatrix}\overline{F_{x,i}}\\ \overline{F_{y,i}}\\ \overline{F_{z,i}}\\ \overline{F_{x,j}}\\ \overline{F_{y,j}}\\ \overline{F_{z,j}}\end{Bmatrix} = \begin{bmatrix}\lambda_{\bar{x}} & \mu_{\bar{x}} & \nu_{\bar{x}} & 0 & 0 & 0\\ \lambda_{\bar{y}} & \mu_{\bar{y}} & \nu_{\bar{y}} & 0 & 0 & 0\\ \lambda_{\bar{z}} & \mu_{\bar{z}} & \nu_{\bar{z}} & 0 & 0 & 0\\ 0 & 0 & 0 & \lambda_{\bar{x}} & \mu_{\bar{x}} & \nu_{\bar{x}}\\ 0 & 0 & 0 & \lambda_{\bar{y}} & \mu_{\bar{y}} & \nu_{\bar{y}}\\ 0 & 0 & 0 & \lambda_{\bar{z}} & \mu_{\bar{z}} & \nu_{\bar{z}}\end{bmatrix}\begin{Bmatrix}F_{x,i}\\ F_{y,i}\\ F_{z,i}\\ F_{x,j}\\ F_{y,j}\\ F_{z,j}\end{Bmatrix} \tag{6.36}$$

or in abbreviated form

$$\{\overline{F}\} = [T]\{F\}$$

The derivation of $[K_{ij}]$ for a member of a space frame proceeds on identical lines to that for the plane frame member. Thus, as before

$$[K_{ij}] = [T]^{\mathrm{T}}[\overline{K_{ij}}][T]$$

Substituting for $[T]$ and $[\overline{Kij}]$ from Eqs (6.36) and (6.33) gives

$$[K_{ij}] = \frac{AE}{L}\begin{bmatrix}\lambda_{\bar{x}}^2 & \lambda_{\bar{x}}\mu_{\bar{x}} & \lambda_{\bar{x}}\nu_{\bar{x}} & -\lambda_{\bar{x}}^2 & -\lambda_{\bar{x}}\mu_{\bar{x}} & -\lambda_{\bar{x}}\nu_{\bar{x}}\\ \lambda_{\bar{x}}\mu_{\bar{x}} & \mu_{\bar{x}}^2 & \mu_{\bar{x}}\nu_{\bar{x}} & -\lambda_{\bar{x}}\mu_{\bar{x}} & -\mu_{\bar{x}}^2 & -\mu_{\bar{x}}\nu_{\bar{x}}\\ \lambda_{\bar{x}}\nu_{\bar{x}} & \mu_{\bar{x}}\nu_{\bar{x}} & \nu_{\bar{x}}^2 & -\lambda_{\bar{x}}\nu_{\bar{x}} & -\mu_{\bar{x}}\nu_{\bar{x}} & -\nu_{\bar{x}}^2\\ -\lambda_{\bar{x}}^2 & -\lambda_{\bar{x}}\mu_{\bar{x}} & -\lambda_{\bar{x}}\nu_{\bar{x}} & \lambda_{\bar{x}}^2 & \lambda_{\bar{x}}\mu_{\bar{x}} & \lambda_{\bar{x}}\nu_{\bar{x}}\\ -\lambda_{\bar{x}}\mu_{\bar{x}} & -\mu_{\bar{x}}^2 & -\mu_{\bar{x}}\nu_{\bar{x}} & \lambda_{\bar{x}}\mu_{\bar{x}} & \mu_{\bar{x}}^2 & \mu_{\bar{x}}\nu_{\bar{x}}\\ -\lambda_{\bar{x}}\nu_{\bar{x}} & -\mu_{\bar{x}}\nu_{\bar{x}} & -\nu_{\bar{x}}^2 & \lambda_{\bar{x}}\nu_{\bar{x}} & \mu_{\bar{x}}\nu_{\bar{x}} & \nu_{\bar{x}}^2\end{bmatrix} \tag{6.37}$$

All the suffixes in Eq. (6.37) are $\bar{x}$ so that we may rewrite the equation in simpler form, namely

$$[K_{ij}] = \frac{AE}{L}\left[\begin{array}{ccc:ccc} \lambda^2 & & & & & \text{SYM} \\ \lambda\mu & \mu^2 & & & & \\ \lambda\nu & \mu\nu & \nu^2 & & & \\ \hdashline -\lambda^2 & -\lambda\mu & -\lambda\nu & \lambda^2 & & \\ -\lambda\mu & -\mu^2 & -\mu\nu & \lambda\mu & \mu^2 & \\ -\lambda\nu & -\mu\nu & -\nu^2 & \lambda\nu & \mu\nu & \nu^2 \end{array}\right] \tag{6.38}$$

where λ, μ and ν are the direction cosines between the x, y, z and $\bar{x}$ axes, respectively.

The complete stiffness matrix for a space frame is assembled from the member stiffness matrices in a similar manner to that for the plane frame and the solution completed as before.

6.7 Stiffness matrix for a uniform beam

Our discussion so far has been restricted to structures comprising members capable of resisting axial loads only. Many structures, however, consist of beam assemblies in which the individual members resist shear and bending forces, in addition to axial loads. We shall now derive the stiffness matrix for a uniform beam and consider the solution of rigid jointed frameworks formed by an assembly of beams, or beam elements as they are sometimes called.

Figure 6.6 shows a uniform beam ij of flexural rigidity EI and length L subjected to nodal forces $F_{y,i}$, $F_{y,j}$ and nodal moments M_i, M_j in the xy plane. The beam suffers nodal displacements and rotations v_i, v_j and θ_i, θ_j. We do not include axial forces here since their effects have already been determined in our investigation of pin-jointed frameworks.

The stiffness matrix $[K_{ij}]$ may be built up by considering various deflected states for the beam and superimposing the results, as we did initially for the spring assemblies

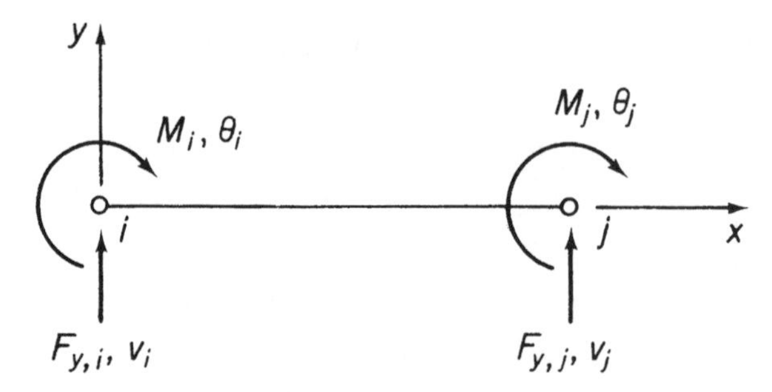

Fig. 6.6 Forces and moments on a beam element.

of Figs 6.1 and 6.2 or, alternatively, it may be written down directly from the well-known beam slope–deflection equations.[3] We shall adopt the latter procedure. From slope–deflection theory we have

$$M_i = -\frac{6EI}{L^2}v_i + \frac{4EI}{L}\theta_i + \frac{6EI}{L^2}v_j + \frac{2EI}{L}\theta_j \tag{6.39}$$

and

$$M_j = -\frac{6EI}{L^2}v_i + \frac{2EI}{L}\theta_i + \frac{6EI}{L^2}v_j + \frac{4EI}{L}\theta_j \tag{6.40}$$

Also, considering vertical equilibrium we obtain

$$F_{y,i} + F_{y,j} = 0 \tag{6.41}$$

and from moment equilibrium about node j we have

$$F_{y,i}L + M_i + M_j = 0 \tag{6.42}$$

Hence the solution of Eqs (6.39)–(6.42) gives

$$-F_{y,i} = F_{y,j} = -\frac{12EI}{L^3}v_i + \frac{6EI}{L^2}\theta_i + \frac{12EI}{L^3}v_j + \frac{6EI}{L^2}\theta_j \tag{6.43}$$

Expressing Eqs (6.39), (6.40) and (6.43) in matrix form yields

$$\begin{Bmatrix} F_{y,i} \\ M_i \\ F_{y,j} \\ M_j \end{Bmatrix} = EI \begin{bmatrix} 12/L^3 & -6/L^2 & -12/L^3 & -6/L^2 \\ -6/L^2 & 4/L & 6/L^2 & 2/L \\ -12/L^3 & 6/L^2 & 12/L^3 & 6/L^2 \\ -6/L^2 & 2/L & 6/L^2 & 4/L \end{bmatrix} \begin{Bmatrix} v_i \\ \theta_i \\ v_j \\ \theta_j \end{Bmatrix} \tag{6.44}$$

which is of the form

$$\{F\} = [K_{ij}]\{\delta\}$$

where $[K_{ij}]$ is the stiffness matrix for the beam.

It is possible to write Eq. (6.44) in an alternative form such that the elements of $[K_{ij}]$ are pure numbers. Thus

$$\begin{Bmatrix} F_{y,i} \\ M_i/L \\ F_{y,j} \\ M_j/L \end{Bmatrix} = \frac{EI}{L^3} \begin{bmatrix} 12 & -6 & -12 & -6 \\ -6 & 4 & 6 & 2 \\ -12 & 6 & 12 & 6 \\ -6 & 2 & 6 & 4 \end{bmatrix} \begin{Bmatrix} v_i \\ \theta_i L \\ v_j \\ \theta_j L \end{Bmatrix}$$

This form of Eq. (6.44) is particularly useful in numerical calculations for an assemblage of beams in which EI/L^3 is constant.

Equation (6.44) is derived for a beam whose axis is aligned with the x axis so that the stiffness matrix defined by Eq. (6.44) is actually $[\overline{K_{ij}}]$ the stiffness matrix referred to a local coordinate system. If the beam is positioned in the xy plane with its axis arbitrarily inclined to the x axis then the x and y axes form a global coordinate system and it becomes necessary to transform Eq. (6.44) to allow for this. The procedure is similar to that for the pin-jointed framework member of Section 6.4 in that $[\overline{K_{ij}}]$ must be expanded to allow for the fact that nodal displacements $\bar{u}_i$ and $\bar{u}_j$, which are irrelevant for the beam in local coordinates, have components u_i, v_i and u_j, v_j in global coordinates. Thus

$$[\overline{K_{ij}}] = EI \begin{array}{c} \begin{matrix} u_i & v_i & \theta_i & u_j & v_j & \theta_j \end{matrix} \\ \begin{bmatrix} 0 & 0 & 0 & 0 & 0 & 0 \\ 0 & 12/L^3 & -6/L^2 & 0 & -12/L^3 & -6/L^2 \\ 0 & -6/L^2 & 4/L & 0 & 6/L^2 & 2/L \\ 0 & 0 & 0 & 0 & 0 & 0 \\ 0 & -12/L^3 & 6/L^2 & 0 & 12/L^3 & 6/L^2 \\ 0 & -6/L^2 & 2/L & 0 & 6/L^2 & 4/L \end{bmatrix} \end{array} \quad (6.45)$$

We may deduce the transformation matrix $[T]$ from Eq. (6.24) if we remember that although u and v transform in exactly the same way as in the case of a pin-jointed member the rotations θ remain the same in either local or global coordinates.

Hence

$$[T] = \begin{bmatrix} \lambda & \mu & 0 & 0 & 0 & 0 \\ -\mu & \lambda & 0 & 0 & 0 & 0 \\ 0 & 0 & 1 & 0 & 0 & 0 \\ 0 & 0 & 0 & \lambda & \mu & 0 \\ 0 & 0 & 0 & -\mu & \lambda & 0 \\ 0 & 0 & 0 & 0 & 0 & 1 \end{bmatrix} \quad (6.46)$$

where λ and μ have previously been defined. Thus since

$$[K_{ij}] = [T]^{\mathrm{T}}[\overline{K_{ij}}][T] \quad \text{(see Section 6.4)}$$

we have, from Eqs (6.45) and (6.46)

$$[K_{ij}] = EI \begin{bmatrix} 12\mu^2/L^3 & & & & & \text{SYM} \\ -12\lambda\mu/L^3 & 12\lambda^2/L^3 & & & & \\ 6\mu/L^2 & -6\lambda/L^2 & 4/L & & & \\ -12\mu^2/L^3 & 12\lambda\mu/L^3 & -6\mu/L^2 & 12\mu^2/L^3 & & \\ 12\lambda\mu/L^3 & -12\lambda^2/L^3 & 6\lambda/L^2 & -12\lambda\mu/L^3 & 12\lambda^2/L^3 & \\ 6\mu/L^2 & -6\lambda/L^2 & 2/L & 6\mu/L^2 & 6\lambda/L^2 & 4\lambda/L \end{bmatrix} \quad (6.47)$$

Again the stiffness matrix for the complete structure is assembled from the member stiffness matrices, the boundary conditions are applied and the resulting set of equations solved for the unknown nodal displacements and forces.

The internal shear forces and bending moments in a beam may be obtained in terms of the calculated nodal displacements. Thus, for a beam joining nodes i and j we shall have obtained the unknown values of v_i, θ_i and v_j, θ_j. The nodal forces $F_{y,i}$ and M_i are

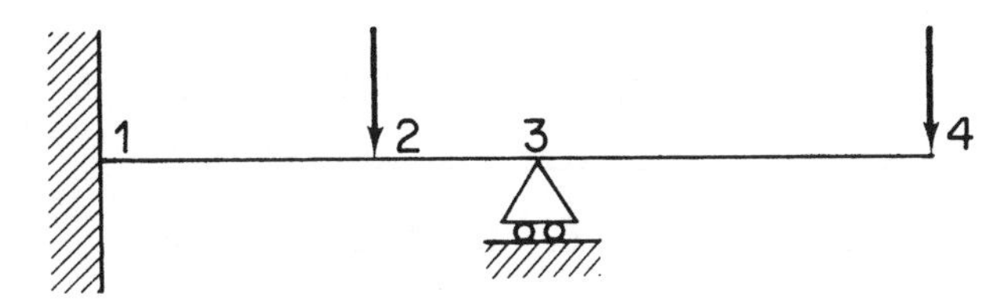

Fig. 6.7 Idealization of a beam into beam–elements.

then obtained from Eq. (6.44) if the beam is aligned with the x axis. Hence

$$\left.\begin{aligned} F_{y,i} &= EI\left(\frac{12}{L^3}v_i - \frac{6}{L^2}\theta_i - \frac{12}{L^3}v_j - \frac{6}{L^2}\theta_j\right) \\ M_i &= EI\left(-\frac{6}{L^2}v_i + \frac{4}{L}\theta_i + \frac{6}{L^2}v_j + \frac{2}{L}\theta_j\right) \end{aligned}\right\} \tag{6.48}$$

Similar expressions are obtained for the forces at node j. From Fig. 6.6 we see that the shear force S_y and bending moment M in the beam are given by

$$\left.\begin{aligned} S_y &= F_{y,i} \\ M &= F_{y,i}x + M_i \end{aligned}\right\} \tag{6.49}$$

Substituting Eq. (6.48) into Eq. (6.49) and expressing in matrix form yields

$$\begin{Bmatrix} S_y \\ M \end{Bmatrix} = EI\begin{bmatrix} \dfrac{12}{L^3} & -\dfrac{6}{L^2} & -\dfrac{12}{L^3} & -\dfrac{6}{L^2} \\ \dfrac{12}{L^3}x - \dfrac{6}{L^2} & -\dfrac{6}{L^2}x + \dfrac{4}{L} & -\dfrac{12}{L^3}x + \dfrac{6}{L^2} & -\dfrac{6}{L^2}x + \dfrac{2}{L} \end{bmatrix}\begin{Bmatrix} v_i \\ \theta_i \\ v_j \\ \theta_j \end{Bmatrix} \tag{6.50}$$

The matrix analysis of the beam in Fig. 6.6 is based on the condition that no external forces are applied between the nodes. Obviously in a practical case a beam supports a variety of loads along its length and therefore such beams must be idealized into a number of *beam–elements* for which the above condition holds. The idealization is accomplished by merely specifying nodes at points along the beam such that any element lying between adjacent nodes carries, at the most, a uniform shear and a linearly varying bending moment. For example, the beam of Fig. 6.7 would be idealized into beam–elements 1–2, 2–3 and 3–4 for which the unknown nodal displacements are $v_2, \theta_2, \theta_3, v_4$ and θ_4 ($v_1 = \theta_1 = v_3 = 0$).

Beams supporting distributed loads require special treatment in that the distributed load is replaced by a series of statically equivalent point loads at a selected number of nodes. Clearly the greater the number of nodes chosen, the more accurate but more complicated and therefore time consuming will be the analysis. Figure 6.8 shows a typical idealization of a beam supporting a uniformly distributed load. Details of the analysis of such beams may be found in Martin.[4]

Many simple beam problems may be idealized into a combination of two beam–elements and three nodes. A few examples of such beams are shown in Fig. 6.9. If we therefore assemble a stiffness matrix for the general case of a two beam–element system we may use it to solve a variety of problems simply by inserting the appropriate

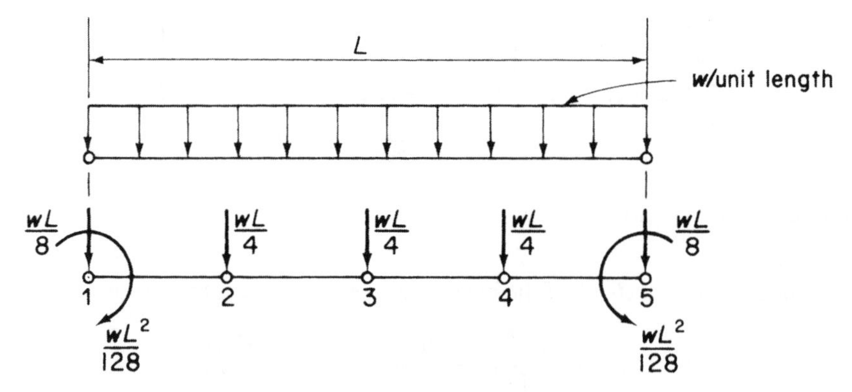

Fig. 6.8 Idealization of a beam supporting a uniformly distributed load.

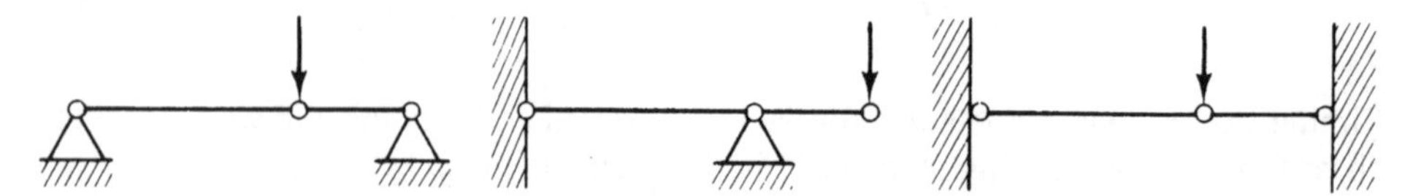

Fig. 6.9 Idealization of beams into beam–elements.

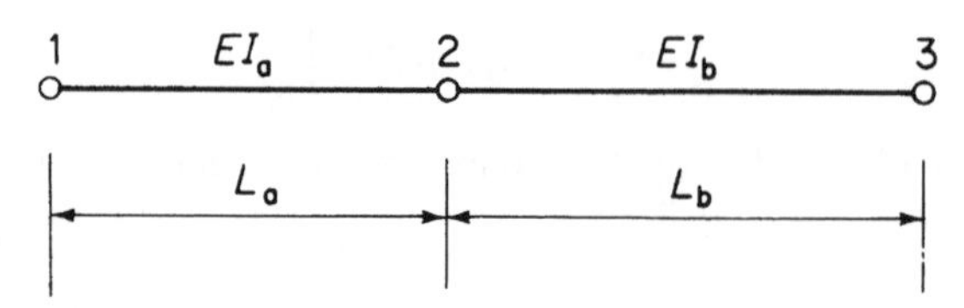

Fig. 6.10 Assemblage of two beam–elements.

loading and support conditions. Consider the assemblage of two beam–elements shown in Fig. 6.10. The stiffness matrices for the beam–elements 1–2 and 2–3 are obtained from Eq. (6.44); thus

$$[K_{12}] = EI_{\mathrm{a}} \begin{array}{c} \begin{array}{cccc} v_1 & \theta_1 & v_2 & \theta_2 \end{array} \\ \left[\begin{array}{cc|cc} 12/L_{\mathrm{a}}^3 & -6/L_{\mathrm{a}}^2 & -12/L_{\mathrm{a}}^3 & -6/L_{\mathrm{a}}^2 \\ -6/L_{\mathrm{a}}^2 & 4/L_{\mathrm{a}} & 6/L_{\mathrm{a}}^2 & 2/L_{\mathrm{a}} \\ \hline -12/L_{\mathrm{a}}^3 & 6/L_{\mathrm{a}}^2 & 12/L_{\mathrm{a}}^3 & 6/L_{\mathrm{a}}^2 \\ -6/L_{\mathrm{a}}^2 & 2/L_{\mathrm{a}} & 6/L_{\mathrm{a}}^2 & 4/L_{\mathrm{a}} \end{array}\right] \end{array} \quad \begin{array}{cc} k_{11} & k_{12} \\ k_{21} & k_{22} \end{array} \tag{6.51}$$

$$[K_{23}] = EI_{\mathrm{b}} \begin{array}{c} \begin{array}{cccc} v_2 & \theta_2 & v_3 & \theta_3 \end{array} \\ \left[\begin{array}{cc|cc} 12/L_{\mathrm{b}}^3 & -6/L_{\mathrm{b}}^2 & -12/L_{\mathrm{b}}^3 & -6/L_{\mathrm{b}}^2 \\ -6/L_{\mathrm{b}}^2 & 4/L_{\mathrm{b}} & 6/L_{\mathrm{b}}^2 & 2/L_{\mathrm{b}} \\ \hline -12/L_{\mathrm{b}}^3 & 6/L_{\mathrm{b}}^2 & 12/L_{\mathrm{b}}^3 & 6/L_{\mathrm{b}}^2 \\ -6/L_{\mathrm{b}}^2 & 2/L_{\mathrm{b}} & 6/L_{\mathrm{b}}^2 & 4/L_{\mathrm{b}} \end{array}\right] \end{array} \quad \begin{array}{cc} k_{22} & k_{23} \\ k_{32} & k_{33} \end{array} \tag{6.52}$$

The complete stiffness matrix is formed by superimposing $[K_{12}]$ and $[K_{23}]$ as described in Example 6.1. Hence

$$[K] = E\begin{bmatrix} \frac{12I_a}{L_a^3} & -\frac{6I_a}{L_a^2} & -\frac{12I_a}{L_a^3} & -\frac{6I_a}{L_a^2} & 0 & 0 \\ -\frac{6I_a}{L_a^2} & \frac{4I_a}{L_a} & \frac{6I_a}{L_a^2} & \frac{2I_a}{L_a} & 0 & 0 \\ -\frac{12I_a}{L_a^3} & \frac{6I_a}{L_a^2} & 12\left(\frac{I_a}{L_a^3}+\frac{I_b}{L_b^3}\right) & 6\left(\frac{I_a}{L_a^2}-\frac{I_b}{L_b^2}\right) & -\frac{12I_b}{L_b^3} & -\frac{6I_b}{L_b^2} \\ -\frac{6I_a}{L_a^2} & \frac{2I_a}{L_a} & 6\left(\frac{I_a}{L_a^2}-\frac{I_b}{L_b^2}\right) & 4\left(\frac{I_a}{L_a}+\frac{I_b}{L_b}\right) & \frac{6I_b}{L_b^2} & \frac{2I_b}{L_b} \\ 0 & 0 & -\frac{12I_b}{L_b^3} & \frac{6I_b}{L_b^2} & \frac{12I_b}{L_b^3} & \frac{6I_b}{L_b^2} \\ 0 & 0 & -\frac{6I_b}{L_b^2} & \frac{2I_b}{L_b} & \frac{6I_b}{L_b^2} & \frac{4I_b}{L_b} \end{bmatrix} \tag{6.53}$$

Example 6.2

Determine the unknown nodal displacements and forces in the beam shown in Fig. 6.11. The beam is of uniform section throughout.

The beam may be idealized into two beam–elements, 1–2 and 2–3. From Fig. 6.11 we see that $v_1 = v_3 = 0, F_{y,2} = -W, M_2 = +M$. Therefore, eliminating rows and columns corresponding to zero displacements from Eq. (6.53), we obtain

$$\begin{Bmatrix} F_{y,2} = -W \\ M_2 = M \\ M_1 = 0 \\ M_3 = 0 \end{Bmatrix} = EI\begin{bmatrix} 27/2L^3 & 9/2L^2 & 6/L^2 & -3/2L^2 \\ 9/2L^2 & 6/L & 2/L & 1/L \\ 6/L^2 & 2/L & 4/L & 0 \\ -3/2L^2 & 1/L & 0 & 2/L \end{bmatrix}\begin{Bmatrix} v_2 \\ \theta_2 \\ \theta_1 \\ \theta_3 \end{Bmatrix} \tag{i}$$

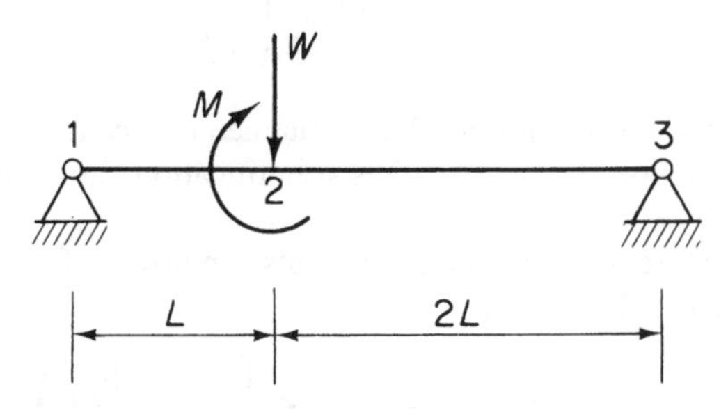

Fig. 6.11 Beam of Example 6.2.

Equation (i) may be written such that the elements of [K] are pure numbers

$$\begin{Bmatrix} F_{y,2} = -W \\ M_2/L = M/L \\ M_1/L = 0 \\ M_3/L = 0 \end{Bmatrix} = \frac{EI}{2L^3} \begin{bmatrix} 27 & 9 & 12 & -3 \\ 9 & 12 & 4 & 2 \\ 12 & 4 & 8 & 0 \\ -3 & 2 & 0 & 4 \end{bmatrix} \begin{Bmatrix} v_2 \\ \theta_2 L \\ \theta_1 L \\ \theta_3 L \end{Bmatrix} \tag{ii}$$

Expanding Eq. (ii) by matrix multiplication we have

$$\begin{Bmatrix} -W \\ M/L \end{Bmatrix} = \frac{EI}{2L^3} \left(\begin{bmatrix} 27 & 9 \\ 9 & 12 \end{bmatrix} \begin{Bmatrix} v_2 \\ \theta_2 L \end{Bmatrix} + \begin{bmatrix} 12 & -3 \\ 4 & 2 \end{bmatrix} \begin{Bmatrix} \theta_1 L \\ \theta_3 L \end{Bmatrix} \right) \tag{iii}$$

and

$$\begin{Bmatrix} 0 \\ 0 \end{Bmatrix} = \frac{EI}{2L^3} \left(\begin{bmatrix} 12 & 4 \\ -3 & 2 \end{bmatrix} \begin{Bmatrix} v_2 \\ \theta_2 L \end{Bmatrix} + \begin{bmatrix} 8 & 0 \\ 0 & 4 \end{bmatrix} \begin{Bmatrix} \theta_1 L \\ \theta_3 L \end{Bmatrix} \right) \tag{iv}$$

Equation (iv) gives

$$\begin{Bmatrix} \theta_1 L \\ \theta_3 L \end{Bmatrix} = \begin{bmatrix} -\frac{3}{2} & -\frac{1}{2} \\ -\frac{3}{4} & -\frac{1}{2} \end{bmatrix} \begin{Bmatrix} v_2 \\ \theta_2 L \end{Bmatrix} \tag{v}$$

Substituting Eq. (v) in Eq. (iii) we obtain

$$\begin{Bmatrix} v_2 \\ \theta_2 L \end{Bmatrix} = \frac{L^3}{9EI} \begin{bmatrix} -4 & -2 \\ -2 & 3 \end{bmatrix} \begin{Bmatrix} -W \\ M/L \end{Bmatrix} \tag{vi}$$

from which the unknown displacements at node 2 are

$$v_2 = -\frac{4}{9}\frac{WL^3}{EI} - \frac{2}{9}\frac{ML^2}{EI}$$

$$\theta_2 = \frac{2}{9}\frac{WL^2}{EI} + \frac{1}{3}\frac{ML}{EI}$$

In addition, from Eq. (v) we find that

$$\theta_1 = \frac{5}{9}\frac{WL^2}{EI} + \frac{1}{6}\frac{ML}{EI}$$

$$\theta_3 = -\frac{4}{9}\frac{WL^2}{EI} - \frac{1}{3}\frac{ML}{EI}$$

It should be noted that the solution has been obtained by inverting two 2×2 matrices rather than the 4×4 matrix of Eq. (ii). This simplification has been brought about by the fact that $M_1 = M_3 = 0$.

The internal shear forces and bending moments can now be found using Eq. (6.50). For the beam–element 1–2 we have

$$S_{y,12} = EI\left(\frac{12}{L^3}v_1 - \frac{6}{L^2}\theta_1 - \frac{12}{L^3}v_2 - \frac{6}{L^2}\theta_2\right)$$

or

$$S_{y,12} = \frac{2}{3}W - \frac{1}{3}\frac{M}{L}$$

and

$$M_{12} = EI\left[\left(\frac{12}{L^3}x - \frac{6}{L^2}\right)v_1 + \left(-\frac{6}{L^2}x + \frac{4}{L}\right)\theta_1 + \left(-\frac{12}{L^3}x + \frac{6}{L^2}\right)v_2 + \left(-\frac{6}{L^2}x + \frac{2}{L}\right)\theta_2\right]$$

which reduces to

$$M_{12} = \left(\frac{2}{3}W - \frac{1}{3}\frac{M}{L}\right)x$$

6.8 Finite element method for continuum structures

In the previous sections we have discussed the matrix method of solution of structures composed of elements connected only at nodal points. For skeletal structures consisting of arrangements of beams these nodal points fall naturally at joints and at positions of concentrated loading. Continuum structures, such as flat plates, aircraft skins, shells etc., do not possess such natural subdivisions and must therefore be artificially idealized into a number of elements before matrix methods can be used. These *finite elements*, as they are known, may be two- or three-dimensional but the most commonly used are two-dimensional triangular and quadrilateral shaped elements. The idealization may be carried out in any number of different ways depending on such factors as the type of problem, the accuracy of the solution required and the time and money available. For example, a *coarse* idealization involving a small number of large elements would provide a comparatively rapid but very approximate solution while a *fine* idealization of small elements would produce more accurate results but would take longer and consequently cost more. Frequently, *graded meshes* are used in which small elements are placed in regions where high stress concentrations are expected, for example around cut-outs and loading points. The principle is illustrated in Fig. 6.12 where a graded system of triangular elements is used to examine the stress concentration around a circular hole in a flat plate.

Although the elements are connected at an infinite number of points around their boundaries it is assumed that they are only interconnected at their corners or nodes. Thus, compatibility of displacement is only ensured at the nodal points. However, in the finite element method a displacement pattern is chosen for each element which may satisfy some, if not all, of the compatibility requirements along the sides of adjacent elements.

Since we are employing matrix methods of solution we are concerned initially with the determination of nodal forces and displacements. Thus, the system of loads on the structure must be replaced by an equivalent system of nodal forces. Where these

Fig. 6.12 Finite element idealization of a flat plate with a central hole.

loads are concentrated the elements are chosen such that a node occurs at the point of application of the load. In the case of distributed loads, equivalent nodal concentrated loads must be calculated.[4]

The solution procedure is identical in outline to that described in the previous sections for skeletal structures; the differences lie in the idealization of the structure into finite elements and the calculation of the stiffness matrix for each element. The latter procedure, which in general terms is applicable to all finite elements, may be specified in a number of distinct steps. We shall illustrate the method by establishing the stiffness matrix for the simple one-dimensional beam–element of Fig. 6.6 for which we have already derived the stiffness matrix using slope–deflection.

6.8.1 Stiffness matrix for a beam–element

The first step is to choose a suitable coordinate and node numbering system for the element and define its nodal displacement vector $\{\delta^e\}$ and nodal load vector $\{F^e\}$. Use is made here of the superscript e to denote element vectors since, in general, a finite element possesses more than two nodes. Again we are not concerned with axial or shear displacements so that for the beam–element of Fig. 6.6 we have

$$\{\delta^e\} = \begin{Bmatrix} v_i \\ \theta_i \\ v_j \\ \theta_j \end{Bmatrix} \qquad \{F^e\} = \begin{Bmatrix} F_{y,i} \\ M_i \\ F_{y,j} \\ M_j \end{Bmatrix}$$

Since each of these vectors contains four terms the element stiffness matrix $[K^e]$ will be of order 4×4.

In the second step we select a displacement function which uniquely defines the displacement of all points in the beam–element in terms of the nodal displacements. This displacement function may be taken as a polynomial which must include four arbitrary constants corresponding to the four nodal degrees of freedom of the element.

Thus

$$v(x) = \alpha_1 + \alpha_2 x + \alpha_3 x^2 + \alpha_4 x^3 \tag{6.54}$$

Equation (6.54) is of the same form as that derived from elementary bending theory for a beam subjected to concentrated loads and moments and may be written in matrix form as

$$\{v(x)\} = [1 \ x \ x^2 \ x^3] \begin{Bmatrix} \alpha_1 \\ \alpha_2 \\ \alpha_3 \\ \alpha_4 \end{Bmatrix}$$

or in abbreviated form as

$$\{v(x)\} = [f(x)]\{\alpha\} \tag{6.55}$$

The rotation θ at any section of the beam–element is given by $\partial v/\partial x$; therefore

$$\theta = \alpha_2 + 2\alpha_3 x + 3\alpha_4 x^2 \tag{6.56}$$

From Eqs (6.54) and (6.56) we can write down expressions for the nodal displacements v_i, θ_i and v_j, θ_j at $x = 0$ and $x = L$, respectively. Hence

$$\left.\begin{aligned} v_i &= \alpha_1 \\ \theta_i &= \alpha_2 \\ v_j &= \alpha_1 + \alpha_2 L + \alpha_3 L^2 + \alpha_4 L^3 \\ \theta_j &= \alpha_2 + 2\alpha_3 L + 3\alpha_4 L^2 \end{aligned}\right\} \tag{6.57}$$

Writing Eqs (6.57) in matrix form gives

$$\begin{Bmatrix} v_i \\ \theta_i \\ v_j \\ \theta_j \end{Bmatrix} = \begin{bmatrix} 1 & 0 & 0 & 0 \\ 0 & 1 & 0 & 0 \\ 1 & L & L^2 & L^3 \\ 0 & 1 & 2L & 3L^2 \end{bmatrix} \begin{Bmatrix} \alpha_1 \\ \alpha_2 \\ \alpha_3 \\ \alpha_4 \end{Bmatrix} \tag{6.58}$$

or

$$\{\delta^e\} = [A]\{\alpha\} \tag{6.59}$$

The third step follows directly from Eqs (6.58) and (6.55) in that we express the displacement at any point in the beam–element in terms of the nodal displacements. Using Eq. (6.59) we obtain

$$\{\alpha\} = [A^{-1}]\{\delta^e\} \tag{6.60}$$

Substituting in Eq. (6.55) gives

$$\{v(x)\} = [f(x)][A^{-1}]\{\delta^e\} \tag{6.61}$$

where $[A^{-1}]$ is obtained by inverting $[A]$ in Eq. (6.58) and may be shown to be given by

$$[A^{-1}] = \begin{bmatrix} 1 & 0 & 0 & 0 \\ 0 & 1 & 0 & 0 \\ -3/L^2 & -2/L & 3/L^2 & -1/L \\ 2/L^3 & 1/L^2 & -2/L^3 & 1/L^2 \end{bmatrix} \tag{6.62}$$

In step four we relate the strain $\{\varepsilon(x)\}$ at any point x in the element to the displacement $\{v(x)\}$ and hence to the nodal displacements $\{\delta^e\}$. Since we are concerned here with bending deformations only we may represent the strain by the curvature $\partial^2 v/\partial x^2$. Hence from Eq. (6.54)

$$\frac{\partial^2 v}{\partial x^2} = 2\alpha_3 + 6\alpha_4 x \tag{6.63}$$

or in matrix form

$$\{\varepsilon\} = [0 \;\; 0 \;\; 2 \;\; 6x] \begin{Bmatrix} \alpha_1 \\ \alpha_2 \\ \alpha_3 \\ \alpha_4 \end{Bmatrix} \tag{6.64}$$

which we write as

$$\{\varepsilon\} = [C]\{\alpha\} \tag{6.65}$$

Substituting for $\{\alpha\}$ in Eq. (6.65) from Eq. (6.60) we have

$$\{\varepsilon\} = [C][A^{-1}]\{\delta^e\} \tag{6.66}$$

Step five relates the internal stresses in the element to the strain $\{\varepsilon\}$ and hence, using Eq. (6.66), to the nodal displacements $\{\delta^e\}$. In our beam–element the stress distribution at any section depends entirely on the value of the bending moment M at that section. Thus we may represent a 'state of stress' $\{\sigma\}$ at any section by the bending moment M, which, from simple beam theory, is given by

$$M = EI\frac{\partial^2 v}{\partial x^2}$$

or

$$\{\sigma\} = [EI]\{\varepsilon\} \tag{6.67}$$

which we write as

$$\{\sigma\} = [D]\{\varepsilon\} \tag{6.68}$$

The matrix $[D]$ in Eq. (6.68) is the 'elasticity' matrix relating 'stress' and 'strain'. In this case $[D]$ consists of a single term, the flexural rigidity EI of the beam. Generally, however, $[D]$ is of a higher order. If we now substitute for $\{\varepsilon\}$ in Eq. (6.68) from Eq. (6.66) we obtain the 'stress' in terms of the nodal displacements, i.e.

$$\{\sigma\} = [D][C][A^{-1}]\{\delta^e\} \tag{6.69}$$

The element stiffness matrix is finally obtained in step six in which we replace the internal 'stresses' $\{\sigma\}$ by a statically equivalent nodal load system $\{F^e\}$, thereby relating nodal loads to nodal displacements (from Eq. (6.69)) and defining the element stiffness matrix $[K^e]$. This is achieved by employing the principle of the stationary value of the total potential energy of the beam (see Section 5.8) which comprises the internal strain energy U and the potential energy V of the nodal loads. Thus

$$U + V = \frac{1}{2}\int_{\text{vol}} \{\varepsilon\}^{\text{T}}\{\sigma\}\text{d(vol)} - \{\delta^e\}^{\text{T}}\{F^e\} \tag{6.70}$$

Substituting in Eq. (6.70) for $\{\varepsilon\}$ from Eq. (6.66) and $\{\sigma\}$ from Eq. (6.69) we have

$$U + V = \frac{1}{2}\int_{\text{vol}} \{\delta^e\}^{\text{T}}[A^{-1}]^{\text{T}}[C]^{\text{T}}[D][C][A^{-1}]\{\delta^e\}\text{d(vol)} - \{\delta^e\}^{\text{T}}\{F^e\} \tag{6.71}$$

The total potential energy of the beam has a stationary value with respect to the nodal displacements $\{\delta^e\}^T$; hence, from Eq. (6.71)

$$\frac{\partial(U+V)}{\partial\{\delta^e\}^{\text{T}}} = \int_{\text{vol}} [A^{-1}]^{\text{T}}[C]^{\text{T}}[D][C][A^{-1}]\{\delta^e\}\text{d(vol)} - \{F^e\} = 0 \tag{6.72}$$

whence

$$\{F^e\} = \left[\int_{\text{vol}} [C]^{\text{T}}[A^{-1}]^{\text{T}}[D][C][A^{-1}]\text{d(vol)}\right]\{\delta^e\} \tag{6.73}$$

or writing $[C][A^{-1}]$ as $[B]$ we obtain

$$\{F^e\} = \left[\int_{\text{vol}} [B]^{\text{T}}[D][B]\text{d(vol)}\right]\{\delta^e\} \tag{6.74}$$

from which the element stiffness matrix is clearly

$$[K^e] = \left[\int_{\text{vol}} [B]^{\text{T}}[D][B]\text{d(vol)}\right] \tag{6.75}$$

From Eqs (6.62) and (6.64) we have

$$[B] = [C][A^{-1}] = [0 \quad 0 \quad 2 \quad 6x]\begin{bmatrix} 1 & 0 & 0 & 0 \\ 0 & 1 & 0 & 0 \\ -3/L^2 & -2/L & 3/L^2 & -1/L \\ 2/L^3 & 1/L^2 & -2/L^3 & 1/L^2 \end{bmatrix}$$

or

$$[B]^T = \begin{bmatrix} -\dfrac{6}{L^2} + \dfrac{12x}{L^3} \\ -\dfrac{4}{L} + \dfrac{6x}{L^2} \\ \dfrac{6}{L^2} - \dfrac{12x}{L^3} \\ -\dfrac{2}{L} + \dfrac{6x}{L^2} \end{bmatrix} \tag{6.76}$$

Hence

$$[K^{\mathrm{e}}] = \int_0^L \begin{bmatrix} -\dfrac{6}{L^2} + \dfrac{12x}{L^3} \\ -\dfrac{4}{L} + \dfrac{6x}{L^2} \\ \dfrac{6}{L^2} - \dfrac{12x}{L^3} \\ -\dfrac{2}{L} + \dfrac{6x}{L^2} \end{bmatrix} [EI] \left[-\frac{6}{L^2} + \frac{12x}{L^3} \quad -\frac{4}{L} + \frac{6x}{L^2} \frac{6}{L^2} - \frac{12x}{L^3} \quad -\frac{2}{L} + \frac{6x}{L^2} \right] \mathrm{d}x$$

which gives

$$[K^{\mathrm{e}}] = \frac{EI}{L^3} \begin{bmatrix} 12 & -6L & -12 & -6L \\ -6L & 4L^2 & 6L & 2L^2 \\ -12 & 6L & 12 & 6L \\ -6L & 2L^2 & 6L & 4L^2 \end{bmatrix} \tag{6.77}$$

Equation (6.77) is identical to the stiffness matrix (see Eq. (6.44)) for the uniform beam of Fig. 6.6.

Finally, in step seven, we relate the internal 'stresses', $\{\sigma\}$, in the element to the nodal displacements $\{\delta^{\mathrm{e}}\}$. This has in fact been achieved to some extent in Eq. (6.69), namely

$$\{\sigma\} = [D][C][A^{-1}]\{\delta^{\mathrm{e}}\}$$

or, from the above

$$\{\sigma\} = [D][B]\{\delta^{\mathrm{e}}\} \tag{6.78}$$

Equation (6.78) is usually written

$$\{\sigma\} = [H]\{\delta^{\mathrm{e}}\} \tag{6.79}$$

in which $[H]=[D][B]$ is the stress–displacement matrix. For this particular beam–element $[D]=EI$ and $[B]$ is defined in Eq. (6.76). Thus

$$[H] = EI \left[-\frac{6}{L^2} + \frac{12x}{L^3} \quad -\frac{4}{L} + \frac{6x}{L^2} \frac{6}{L^2} - \frac{12x}{L^3} \quad -\frac{2}{L} + \frac{6x}{L^2} \right] \tag{6.80}$$

6.8.2 Stiffness matrix for a triangular finite element

Triangular finite elements are used in the solution of plane stress and plane strain problems. Their advantage over other shaped elements lies in their ability to represent irregular shapes and boundaries with relative simplicity.

In the derivation of the stiffness matrix we shall adopt the step by step procedure of the previous example. Initially, therefore, we choose a suitable coordinate and node numbering system for the element and define its nodal displacement and nodal force

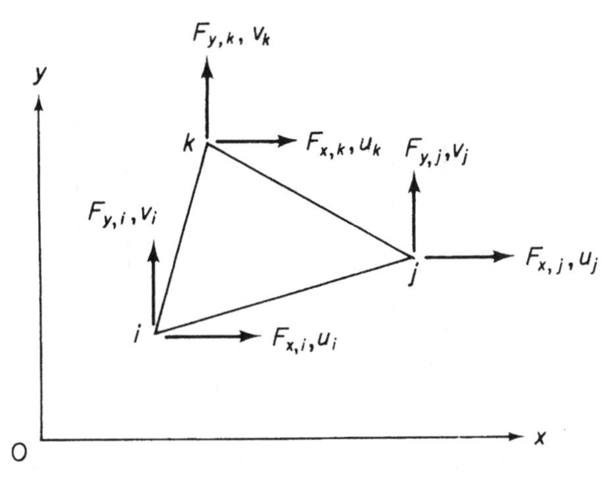

Fig. 6.13 Triangular element for plane elasticity problems.

vectors. Figure 6.13 shows a triangular element referred to axes Oxy and having nodes i, j and k lettered anticlockwise. It may be shown that the inverse of the $[A]$ matrix for a triangular element contains terms giving the actual area of the element; this area is positive if the above node lettering or numbering system is adopted. The element is to be used for plane elasticity problems and has therefore two degrees of freedom per node, giving a total of six degrees of freedom for the element, which will result in a 6×6 element stiffness matrix $[K^e]$. The nodal forces and displacements are shown and the complete displacement and force vectors are

$$\{\delta^e\} = \begin{Bmatrix} u_i \\ v_i \\ u_j \\ v_j \\ u_k \\ v_k \end{Bmatrix} \qquad \{F^e\} = \begin{Bmatrix} F_{x,i} \\ F_{y,i} \\ F_{x,j} \\ F_{y,j} \\ F_{x,k} \\ F_{y,k} \end{Bmatrix} \tag{6.81}$$

We now select a displacement function which must satisfy the boundary conditions of the element, i.e. the condition that each node possesses two degrees of freedom. Generally, for computational purposes, a polynomial is preferable to, say, a trigonometric series since the terms in a polynomial can be calculated much more rapidly by a digital computer. Furthermore, the total number of degrees of freedom is six, so that only six coefficients in the polynomial can be obtained. Suppose that the displacement function is

$$\left.\begin{aligned} u(x,y) &= \alpha_1 + \alpha_2 x + \alpha_3 y \\ v(x,y) &= \alpha_4 + \alpha_5 x + \alpha_6 y \end{aligned}\right\} \tag{6.82}$$

The constant terms, α_1 and α_4, are required to represent any in-plane rigid body motion, i.e. motion without strain, while the linear terms enable states of constant strain to be specified; Eqs (6.82) ensure compatibility of displacement along the edges of adjacent elements. Writing Eqs (6.82) in matrix form gives

$$\begin{Bmatrix} u(x,y) \\ v(x,y) \end{Bmatrix} = \begin{bmatrix} 1 & x & y & 0 & 0 & 0 \\ 0 & 0 & 0 & 1 & x & y \end{bmatrix} \begin{Bmatrix} \alpha_1 \\ \alpha_2 \\ \alpha_3 \\ \alpha_4 \\ \alpha_5 \\ \alpha_6 \end{Bmatrix} \tag{6.83}$$

Comparing Eq. (6.83) with Eq. (6.55) we see that it is of the form

$$\begin{Bmatrix} u(x,y) \\ v(x,y) \end{Bmatrix} = [f(x,y)]\{\alpha\} \tag{6.84}$$

Substituting values of displacement and coordinates at each node in Eq. (6.84) we have, for node i

$$\begin{Bmatrix} u_i \\ v_i \end{Bmatrix} = \begin{bmatrix} 1 & x_i & y_i & 0 & 0 & 0 \\ 0 & 0 & 0 & 1 & x_i & y_i \end{bmatrix} \{\alpha\}$$

Similar expressions are obtained for nodes j and k so that for the complete element we obtain

$$\begin{Bmatrix} u_i \\ v_i \\ u_j \\ v_j \\ u_k \\ v_k \end{Bmatrix} = \begin{bmatrix} 1 & x_i & y_i & 0 & 0 & 0 \\ 0 & 0 & 0 & 1 & x_i & y_i \\ 1 & x_j & y_j & 0 & 0 & 0 \\ 0 & 0 & 0 & 1 & x_j & y_j \\ 1 & x_k & y_k & 0 & 0 & 0 \\ 0 & 0 & 0 & 1 & x_k & y_k \end{bmatrix} \begin{Bmatrix} \alpha_1 \\ \alpha_2 \\ \alpha_3 \\ \alpha_4 \\ \alpha_5 \\ \alpha_6 \end{Bmatrix} \tag{6.85}$$

From Eq. (6.81) and by comparison with Eqs (6.58) and (6.59) we see that Eq. (6.85) takes the form

$$\{\delta^e\} = [A]\{\alpha\}$$

Hence (step 3) we obtain

$$\{\alpha\} = [A^{-1}]\{\delta^e\} \quad \text{(compare with Eq. (6.60))}$$

The inversion of $[A]$, defined in Eq. (6.85), may be achieved algebraically as illustrated in Example 6.3. Alternatively, the inversion may be carried out numerically for a particular element by computer. Substituting for $\{\alpha\}$ from the above into Eq. (6.84) gives

$$\begin{Bmatrix} u(x,y) \\ v(x,y) \end{Bmatrix} = [f(x,y)][A^{-1}]\{\delta^e\} \tag{6.86}$$

(compare with Eq. (6.61)).

The strains in the element are

$$\{\varepsilon\} = \begin{Bmatrix} \varepsilon_x \\ \varepsilon_y \\ \gamma_{xy} \end{Bmatrix} \tag{6.87}$$

From Eqs (1.18) and (1.20) we see that

$$\varepsilon_x = \frac{\partial u}{\partial x} \quad \varepsilon_y = \frac{\partial v}{\partial y} \quad \gamma_{xy} = \frac{\partial u}{\partial y} + \frac{\partial v}{\partial x} \tag{6.88}$$

Substituting for u and v in Eqs (6.88) from Eqs (6.82) gives

$$\begin{aligned} \varepsilon_x &= \alpha_2 \\ \varepsilon_y &= \alpha_6 \\ \gamma_{xy} &= \alpha_3 + \alpha_5 \end{aligned}$$

or in matrix form

$$\{\varepsilon\} = \begin{bmatrix} 0 & 1 & 0 & 0 & 0 & 0 \\ 0 & 0 & 0 & 0 & 0 & 1 \\ 0 & 0 & 1 & 0 & 1 & 0 \end{bmatrix} \begin{Bmatrix} \alpha_1 \\ \alpha_2 \\ \alpha_3 \\ \alpha_4 \\ \alpha_5 \\ \alpha_6 \end{Bmatrix} \tag{6.89}$$

which is of the form

$$\{\varepsilon\} = [C]\{\alpha\} \quad \text{(see Eqs (6.64) and (6.65))}$$

Substituting for $\{\alpha\}(=[A^{-1}]\{\delta^{e}\})$ we obtain

$$\{\varepsilon\} = [C][A^{-1}]\{\delta^{e}\} \quad \text{(compare with Eq. (6.66))}$$

or

$$\{\varepsilon\} = [B]\{\delta^{e}\} \quad \text{(see Eq. (6.76))}$$

where $[C]$ is defined in Eq. (6.89).

In step five we relate the internal stresses $\{\sigma\}$ to the strain $\{\varepsilon\}$ and hence, using step four, to the nodal displacements $\{\delta^{e}\}$. For plane stress problems

$$\{\sigma\} = \begin{Bmatrix} \sigma_x \\ \sigma_y \\ \tau_{xy} \end{Bmatrix} \tag{6.90}$$

and

$$\left.\begin{aligned} \varepsilon_x &= \frac{\sigma_x}{E} - \frac{\nu\sigma_y}{E} \\ \varepsilon_y &= \frac{\sigma_y}{E} - \frac{\nu\sigma_x}{E} \\ \gamma_{xy} &= \frac{\tau_{xy}}{G} = \frac{2(1+\nu)}{E}\tau_{xy} \end{aligned}\right\} \text{(see Chapter 1)}$$

Thus, in matrix form,

$$\{\varepsilon\} = \begin{Bmatrix} \varepsilon_x \\ \varepsilon_y \\ \gamma_{xy} \end{Bmatrix} = \frac{1}{E} \begin{bmatrix} 1 & -\nu & 0 \\ -\nu & 1 & 0 \\ 0 & 0 & 2(1+\nu) \end{bmatrix} \begin{Bmatrix} \sigma_x \\ \sigma_y \\ \tau_{xy} \end{Bmatrix} \tag{6.91}$$

It may be shown that (see Chapter 1)

$$\{\sigma\} = \begin{Bmatrix} \sigma_x \\ \sigma_y \\ \tau_{xy} \end{Bmatrix} = \frac{E}{1-\nu^2} \begin{bmatrix} 1 & \nu & 0 \\ \nu & 1 & 0 \\ 0 & 0 & \frac{1}{2}(1-\nu) \end{bmatrix} \begin{Bmatrix} \varepsilon_x \\ \varepsilon_y \\ \gamma_{xy} \end{Bmatrix} \tag{6.92}$$

which has the form of Eq. (6.68), i.e.

$$\{\sigma\} = [D]\{\varepsilon\}$$

Substituting for $\{\varepsilon\}$ in terms of the nodal displacements $\{\delta^{\mathrm{e}}\}$ we obtain

$$\{\sigma\} = [D][B]\{\delta^{\mathrm{e}}\} \quad \text{(see Eq. (6.69))}$$

In the case of plane strain the elasticity matrix $[D]$ takes a different form to that defined in Eq. (6.92). For this type of problem

$$\varepsilon_x = \frac{\sigma_x}{E} - \frac{\nu\sigma_y}{E} - \frac{\nu\sigma_z}{E}$$

$$\varepsilon_y = \frac{\sigma_y}{E} - \frac{\nu\sigma_x}{E} - \frac{\nu\sigma_z}{E}$$

$$\varepsilon_z = \frac{\sigma_z}{E} - \frac{\nu\sigma_x}{E} - \frac{\nu\sigma_y}{E} = 0$$

$$\gamma_{xy} = \frac{\tau_{xy}}{G} = \frac{2(1+\nu)}{E}\tau_{xy}$$

Eliminating σ_z and solving for σ_x, σ_y and τ_{xy} gives

$$\{\sigma\} = \begin{Bmatrix} \sigma_x \\ \sigma_y \\ \tau_{xy} \end{Bmatrix} = \frac{E(1-\nu)}{(1+\nu)(1-2\nu)} \begin{bmatrix} 1 & \dfrac{\nu}{1-\nu} & 0 \\ \dfrac{\nu}{1-\nu} & 1 & 0 \\ 0 & 0 & \dfrac{(1-2\nu)}{2(1-\nu)} \end{bmatrix} \begin{Bmatrix} \varepsilon_x \\ \varepsilon_y \\ \gamma_{xy} \end{Bmatrix} \tag{6.93}$$

which again takes the form

$$\{\sigma\} = [D]\{\varepsilon\}$$

Step six, in which the internal stresses $\{\sigma\}$ are replaced by the statically equivalent nodal forces $\{F^{\mathrm{e}}\}$ proceeds, in an identical manner to that described for the beam-element. Thus

$$\{F^{\mathrm{e}}\} = \left[\int_{\mathrm{vol}} [B]^{\mathrm{T}}[D][B]\mathrm{d(vol)}\right]\{\delta^{\mathrm{e}}\}$$

as in Eq. (6.74), whence

$$[K^e] = \left[\int_{\text{vol}} [B]^T[D][B]\text{d(vol)}\right]$$

In this expression $[B] = [C][A^{-1}]$ where $[A]$ is defined in Eq. (6.85) and $[C]$ in Eq. (6.89). The elasticity matrix $[D]$ is defined in Eq. (6.92) for plane stress problems or in Eq. (6.93) for plane strain problems. We note that the $[C]$, $[A]$ (therefore $[B]$) and $[D]$ matrices contain only constant terms and may therefore be taken outside the integration in the expression for $[K^e]$, leaving only $\int \text{d(vol)}$ which is simply the area A, of the triangle times its thickness t. Thus

$$[K^e] = [[B]^T[D][B]At] \tag{6.94}$$

Finally the element stresses follow from Eq. (6.79), i.e.

$$\{\sigma\} = [H]\{\delta^e\}$$

where $[H] = [D][B]$ and $[D]$ and $[B]$ have previously been defined. It is usually found convenient to plot the stresses at the centroid of the element.

Of all the finite elements in use the triangular element is probably the most versatile. It may be used to solve a variety of problems ranging from two-dimensional flat plate structures to three-dimensional folded plates and shells. For three-dimensional applications the element stiffness matrix $[K^e]$ is transformed from an in-plane xy coordinate system to a three-dimensional system of global coordinates by the use of a transformation matrix similar to those developed for the matrix analysis of skeletal structures. In addition to the above, triangular elements may be adapted for use in plate flexure problems and for the analysis of bodies of revolution.

Example 6.3

A constant strain triangular element has corners 1(0, 0), 2(4, 0) and 3(2, 2) referred to a Cartesian Oxy axes system and is 1 unit thick. If the elasticity matrix $[D]$ has elements $D_{11} = D_{22} = a$, $D_{12} = D_{21} = b$, $D_{13} = D_{23} = D_{31} = D_{32} = 0$ and $D_{33} = c$, derive the stiffness matrix for the element.

From Eq. (6.82)

$$u_1 = \alpha_1 + \alpha_2(0) + \alpha_3(0)$$

i.e.

$$u_1 = \alpha_1 \tag{i}$$

$$u_2 = \alpha_1 + \alpha_2(4) + \alpha_3(0)$$

i.e.

$$u_2 = \alpha_1 + 4\alpha_2 \tag{ii}$$

$$u_3 = \alpha_1 + \alpha_2(2) + \alpha_3(2)$$

i.e.

$$u_3 = \alpha_1 + 2\alpha_2 + 2\alpha_3 \tag{iii}$$

From Eq. (i)

$$\alpha_1 = u_1 \tag{iv}$$

and from Eqs (ii) and (iv)

$$\alpha_2 = \frac{u_2 - u_1}{4} \tag{v}$$

Then, from Eqs (iii) to (v)

$$\alpha_3 = \frac{2u_3 - u_1 - u_2}{4} \tag{vi}$$

Substituting for α_1, α_2 and α_3 in the first of Eqs (6.82) gives

$$u = u_1 + \left(\frac{u_2 - u_1}{4}\right)x + \left(\frac{2u_3 - u_1 - u_2}{4}\right)y$$

or

$$u = \left(1 - \frac{x}{4} - \frac{y}{4}\right)u_1 + \left(\frac{x}{4} - \frac{y}{4}\right)u_2 + \frac{y}{2}u_3 \tag{vii}$$

Similarly

$$v = \left(1 - \frac{x}{4} - \frac{y}{4}\right)v_1 + \left(\frac{x}{4} - \frac{y}{4}\right)v_2 + \frac{y}{2}v_3 \tag{viii}$$

Now from Eq. (6.88)

$$\varepsilon_x = \frac{\partial u}{\partial x} = -\frac{u_1}{4} + \frac{u_2}{4}$$

$$\varepsilon_y = \frac{\partial v}{\partial y} = -\frac{v_1}{4} - \frac{v_2}{4} + \frac{v_3}{2}$$

and

$$\gamma_{xy} = \frac{\partial u}{\partial y} + \frac{\partial v}{\partial x} = -\frac{u_1}{4} - \frac{u_2}{4} - \frac{v_1}{4} + \frac{v_2}{4}$$

Hence

$$[B]\{\delta^e\} = \begin{bmatrix} \dfrac{\partial u}{\partial x} \\ \dfrac{\partial v}{\partial y} \\ \dfrac{\partial u}{\partial y} + \dfrac{\partial v}{\partial x} \end{bmatrix} = \frac{1}{4}\begin{bmatrix} -1 & 0 & 1 & 0 & 0 & 0 \\ 0 & -1 & 0 & -1 & 0 & 2 \\ -1 & -1 & -1 & 1 & 2 & 0 \end{bmatrix} \begin{Bmatrix} u_1 \\ v_1 \\ u_2 \\ v_2 \\ u_3 \\ v_3 \end{Bmatrix} \tag{ix}$$

Also

$$[D] = \begin{bmatrix} a & b & 0 \\ b & a & 0 \\ 0 & 0 & c \end{bmatrix}$$

Hence

$$[D][B] = \frac{1}{4}\begin{bmatrix} -a & -b & a & -b & 0 & 2b \\ -b & -a & b & -a & 0 & 2a \\ -c & -c & -c & c & 2c & 0 \end{bmatrix}$$

and

$$[B]^{\mathrm{T}}[D][B] = \frac{1}{16}\begin{bmatrix} a+c & b+c & -a+c & b-c & -2c & -2b \\ b+c & a+c & -b+c & a-c & -2c & -2a \\ -a+c & -b+c & a+c & -b-c & -2c & 2b \\ b-c & a-c & -b-c & a+c & 2c & -2a \\ -2c & -2c & -2c & 2c & 4c & 0 \\ -2b & -2a & 2b & -2a & 0 & 4a \end{bmatrix}$$

Then, from Eq. (6.94)

$$[K^{\mathrm{e}}] = \frac{1}{4}\begin{bmatrix} a+c & b+c & -a+c & b-c & -2c & -2b \\ b+c & a+c & -b+c & a-c & -2c & -2a \\ -a+c & -b+c & a+c & -b-c & -2c & 2b \\ b-c & a-c & -b-c & a+c & 2c & -2a \\ -2c & -2c & -2c & 2c & 4c & 0 \\ -2b & -2a & 2b & -2a & 0 & 4a \end{bmatrix}$$

6.8.3 Stiffness matrix for a quadrilateral element

Quadrilateral elements are frequently used in combination with triangular elements to build up particular geometrical shapes.

Figure 6.14 shows a quadrilateral element referred to axes Oxy and having corner nodes, i, j, k and l; the nodal forces and displacements are also shown and the displacement and force vectors are

$$\{\delta^{\mathrm{e}}\} = \begin{Bmatrix} u_i \\ v_i \\ u_j \\ v_j \\ u_k \\ v_k \\ u_l \\ v_l \end{Bmatrix} \qquad \{F^{\mathrm{e}}\} = \begin{Bmatrix} F_{x,i} \\ F_{y,i} \\ F_{x,j} \\ F_{y,j} \\ F_{x,k} \\ F_{y,k} \\ F_{x,l} \\ F_{y,l} \end{Bmatrix} \tag{6.95}$$

As in the case of the triangular element we select a displacement function which satisfies the total of eight degrees of freedom of the nodes of the element; again this displacement function will be in the form of a polynomial with a maximum of eight coefficients. Thus

$$\left.\begin{aligned} u(x, y) &= \alpha_1 + \alpha_2 x + \alpha_3 y + \alpha_4 xy \\ v(x, y) &= \alpha_5 + \alpha_6 x + \alpha_7 y + \alpha_8 xy \end{aligned}\right\} \tag{6.96}$$

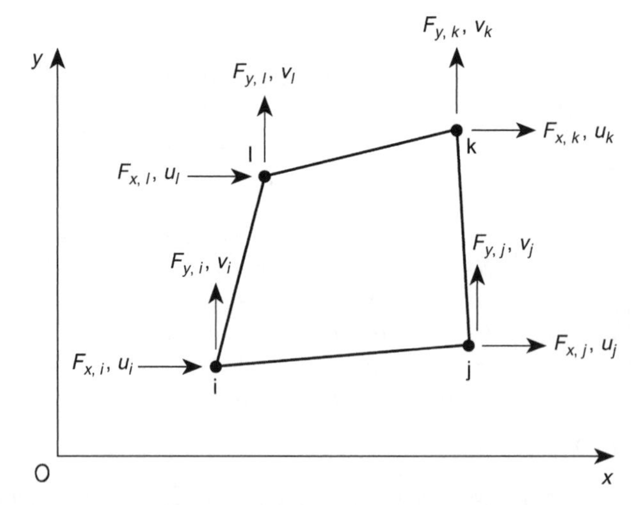

Fig. 6.14 Quadrilateral element subjected to nodal in-plane forces and displacements.

The constant terms, α_1 and α_5, are required, as before, to represent the in-plane rigid body motion of the element while the two pairs of linear terms enable states of constant strain to be represented throughout the element. Further, the inclusion of the xy terms results in both the $u(x, y)$ and $v(x, y)$ displacements having the same algebraic form so that the element behaves in exactly the same way in the x direction as it does in the y direction.

Writing Eqs (6.96) in matrix form gives

$$\begin{Bmatrix} u(x,y) \\ v(x,y) \end{Bmatrix} = \begin{bmatrix} 1 & x & y & xy & 0 & 0 & 0 & 0 \\ 0 & 0 & 0 & 0 & 1 & x & y & xy \end{bmatrix} \begin{Bmatrix} \alpha_1 \\ \alpha_2 \\ \alpha_3 \\ \alpha_4 \\ \alpha_5 \\ \alpha_6 \\ \alpha_7 \\ \alpha_8 \end{Bmatrix} \tag{6.97}$$

or

$$\begin{Bmatrix} u(x,y) \\ v(x,y) \end{Bmatrix} = [f(x,y)]\{\alpha\} \tag{6.98}$$

Now substituting the coordinates and values of displacement at each node we obtain

$$\begin{Bmatrix} u_i \\ v_i \\ u_j \\ v_j \\ u_k \\ v_k \\ u_l \\ v_l \end{Bmatrix} = \begin{bmatrix} 1 & x_i & y_i & x_iy_i & 0 & 0 & 0 & 0 \\ 0 & 0 & 0 & 0 & 1 & x_i & y_i & x_iy_i \\ 1 & x_j & y_j & x_jy_j & 0 & 0 & 0 & 0 \\ 0 & 0 & 0 & 0 & 1 & x_j & y_j & x_jy_j \\ 1 & x_k & y_k & x_ky_k & 0 & 0 & 0 & 0 \\ 0 & 0 & 0 & 0 & 1 & x_k & y_k & x_ky_k \\ 1 & x_l & y_l & x_ly_l & 0 & 0 & 0 & 0 \\ 0 & 0 & 0 & 0 & 1 & x_l & y_l & x_ly_l \end{bmatrix} \begin{Bmatrix} \alpha_1 \\ \alpha_2 \\ \alpha_3 \\ \alpha_4 \\ \alpha_5 \\ \alpha_6 \\ \alpha_7 \\ \alpha_8 \end{Bmatrix} \tag{6.99}$$

which is of the form

$$\{\delta^{e}\} = [A]\{\alpha\}$$

Then

$$\{\alpha\} = [A^{-1}]\{\delta^{e}\} \qquad (6.100)$$

The inversion of $[A]$ is illustrated in Example 6.4 but, as in the case of the triangular element, is most easily carried out by means of a computer. The remaining analysis is identical to that for the triangular element except that the $\{\varepsilon\}$–$\{\alpha\}$ relationship (see Eq. (6.89)) becomes

$$\{\varepsilon\} = \begin{bmatrix} 0 & 1 & 0 & y & 0 & 0 & 0 & 0 \\ 0 & 0 & 0 & 0 & 0 & 0 & 1 & x \\ 0 & 0 & 1 & x & 0 & 1 & 0 & y \end{bmatrix} \begin{Bmatrix} \alpha_1 \\ \alpha_2 \\ \alpha_3 \\ \alpha_4 \\ \alpha_5 \\ \alpha_6 \\ \alpha_7 \\ \alpha_8 \end{Bmatrix} \qquad (6.101)$$

Example 6.4

A rectangular element used in a plane stress analysis has corners whose coordinates (in metres), referred to an Oxy axes system, are 1(−2, −1), 2(2, −1), 3(2, 1) and 4(−2, 1); the displacements (also in metres) of the corners were

$$\begin{aligned} &u_1 = 0.001, \quad &&u_2 = 0.003, \quad &&u_3 = -0.003, \quad &&u_4 = 0 \\ &v_1 = -0.004, \quad &&v_2 = -0.002, \quad &&v_3 = 0.001, \quad &&v_4 = 0.001 \end{aligned}$$

If Young's modulus $E = 200\,000\ \text{N/mm}^2$ and Poisson's ratio $\nu = 0.3$; calculate the stresses at the centre of the element.

From the first of Eqs (6.96)

$$u_1 = \alpha_1 - 2\alpha_2 - \alpha_3 + 2\alpha_4 = 0.001 \qquad \text{(i)}$$

$$u_2 = \alpha_1 + 2\alpha_2 - \alpha_3 - 2\alpha_4 = 0.003 \qquad \text{(ii)}$$

$$u_3 = \alpha_1 + 2\alpha_2 + \alpha_3 + 2\alpha_4 = -0.003 \qquad \text{(iii)}$$

$$u_4 = \alpha_1 - 2\alpha_2 + \alpha_3 - 2\alpha_4 = 0 \qquad \text{(iv)}$$

Subtracting Eq. (ii) from Eq. (i)

$$\alpha_2 - \alpha_4 = 0.0005 \qquad \text{(v)}$$

Now subtracting Eq. (iv) from Eq. (iii)

$$\alpha_2 + \alpha_4 = -0.00075 \qquad \text{(vi)}$$

Then subtracting Eq. (vi) from Eq. (v)

$$\alpha_4 = -0.000625 \tag{vii}$$

whence, from either of Eqs (v) or (vi)

$$\alpha_2 = -0.000125 \tag{viii}$$

Adding Eqs (i) and (ii)

$$\alpha_1 - \alpha_3 = 0.002 \tag{ix}$$

Adding Eqs (iii) and (iv)

$$\alpha_1 + \alpha_3 = -0.0015 \tag{x}$$

Then adding Eqs (ix) and (x)

$$\alpha_1 = 0.00025 \tag{xi}$$

and, from either of Eqs (ix) or (x)

$$\alpha_3 = -0.00175 \tag{xii}$$

The second of Eqs (6.96) is used to determine $\alpha_5, \alpha_6, \alpha_7, \alpha_8$ in an identical manner to the above. Thus

$$\alpha_5 = -0.001$$
$$\alpha_6 = 0.00025$$
$$\alpha_7 = 0.002$$
$$\alpha_8 = -0.00025$$

Now substituting for $\alpha_1, \alpha_2, \ldots, \alpha_8$ in Eqs (6.96)

$$u_i = 0.00025 - 0.000125x - 0.00175y - 0.000625xy$$

and

$$v_i = -0.001 + 0.00025x + 0.002y - 0.00025xy$$

Then, from Eqs (6.88)

$$\varepsilon_x = \frac{\partial u}{\partial x} = -0.000125 - 0.000625y$$

$$\varepsilon_y = \frac{\partial v}{\partial y} = 0.002 - 0.00025x$$

$$\gamma_{xy} = \frac{\partial u}{\partial y} + \frac{\partial v}{\partial x} = -0.0015 - 0.000625x - 0.00025y$$

Therefore, at the centre of the element ($x = 0$, $y = 0$)

$$\varepsilon_x = -0.000125$$

$$\varepsilon_y = 0.002$$

$$\gamma_{xy} = -0.0015$$

so that, from Eqs (6.92)

$$\sigma_x = \frac{E}{1-\nu^2}(\varepsilon_x + \nu\varepsilon_y) = \frac{200\,000}{1-0.3^2}(-0.000125 + (0.3 \times 0.002))$$

i.e.

$$\sigma_x = 104.4\,\text{N/mm}^2$$

$$\sigma_y = \frac{E}{1-\nu^2}(\varepsilon_y + \nu\varepsilon_x) = \frac{200\,000}{1-0.3^2}(0.002 + (0.3 \times 0.000125))$$

i.e.

$$\sigma_y = 431.3\,\text{N/mm}^2$$

and

$$\tau_{xy} = \frac{E}{1-\nu^2} \times \frac{1}{2}(1-\nu)\gamma_{xy} = \frac{E}{2(1+\nu)}\gamma_{xy}$$

Thus

$$\tau_{xy} = \frac{200\,000}{2(1+0.3)} \times (-0.0015)$$

i.e.

$$\tau_{xy} = -115.4\,\text{N/mm}^2$$

The application of the finite element method to three-dimensional solid bodies is a straightforward extension of the analysis of two-dimensional structures. The basic

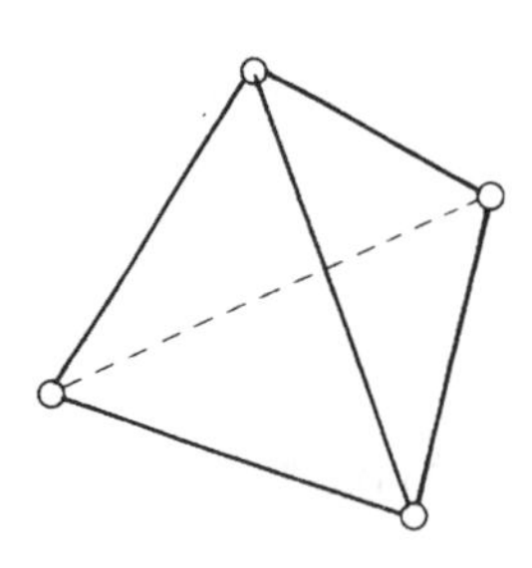
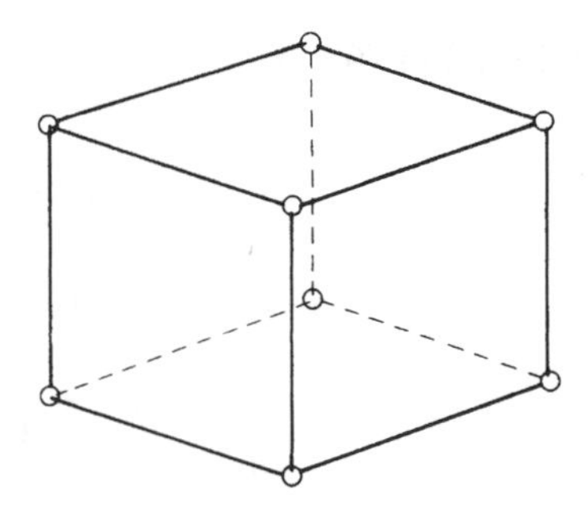

Fig. 6.15 Tetrahedron and rectangular prism finite elements for three-dimensional problems.

three-dimensional elements are the tetrahedron and the rectangular prism, both shown in Fig. 6.15. The tetrahedron has four nodes each possessing three degrees of freedom, a total of 12 for the element, while the prism has 8 nodes and therefore a total of 24 degrees of freedom. Displacement functions for each element require polynomials in x, y and z; for the tetrahedron the displacement function is of the first degree with 12 constant coefficients, while that for the prism may be of a higher order to accommodate the 24 degrees of freedom. A development in the solution of three-dimensional problems has been the introduction of curvilinear coordinates. This enables the tetrahedron and prism to be distorted into arbitrary shapes that are better suited for fitting actual boundaries. For more detailed discussions of the finite element method reference should be made to the work of Jenkins,[5] Zienkiewicz[6] and to the many research papers published on the method.

New elements and new applications of the finite element method are still being developed, some of which lie outside the field of structural analysis. These fields include soil mechanics, heat transfer, fluid and seepage flow, magnetism and electricity.

References

1 Argyris, J. H. and Kelsey, S., *Energy Theorems and Structural Analysis*, Butterworth Scientific Publications, London, 1960.

2 Clough, R. W., Turner, M. J., Martin, H. C. and Topp, L. J., Stiffness and deflection analysis of complex structures, *J. Aero. Sciences*, **23**(9), 1956.

3 Megson, T. H. G., *Structural and Stress Analysis*, 2nd edition, Elsevier, Oxford, 2005.

4 Martin, H. C., *Introduction to Matrix Methods of Structural Analysis*, McGraw-Hill Book Company, New York, 1966.

5 Jenkins, W. M., *Matrix and Digital Computer Methods in Structural Analysis*, McGraw-Hill Publishing Co. Ltd., London, 1969.

6 Zienkiewicz, O. C. and Cheung, Y. K., *The Finite Element Method in Structural and Continuum Mechanics*, McGraw-Hill Publishing Co. Ltd., London, 1967.

Further reading

Zienkiewicz, O. C. and Holister, G. S., *Stress Analysis*, John Wiley and Sons Ltd., London, 1965.

Problems

P.6.1 Figure P.6.1 shows a square symmetrical pin-jointed truss 1234, pinned to rigid supports at 2 and 4 and loaded with a vertical load at 1. The axial rigidity *EA* is the same for all members.

Use the stiffness method to find the displacements at nodes 1 and 3 and hence solve for all the internal member forces and support reactions.

Ans. $v_1 = -PL/\sqrt{2}AE,\ v_3 = -0.293PL/AE,\ S_{12} = P/2 = S_{14},$

$S_{23} = -0.207P = S_{43},\ S_{13} = 0.293P$

$F_{x,2} = -F_{x,4} = 0.207P,\ F_{y,2} = F_{y,4} = P/2.$

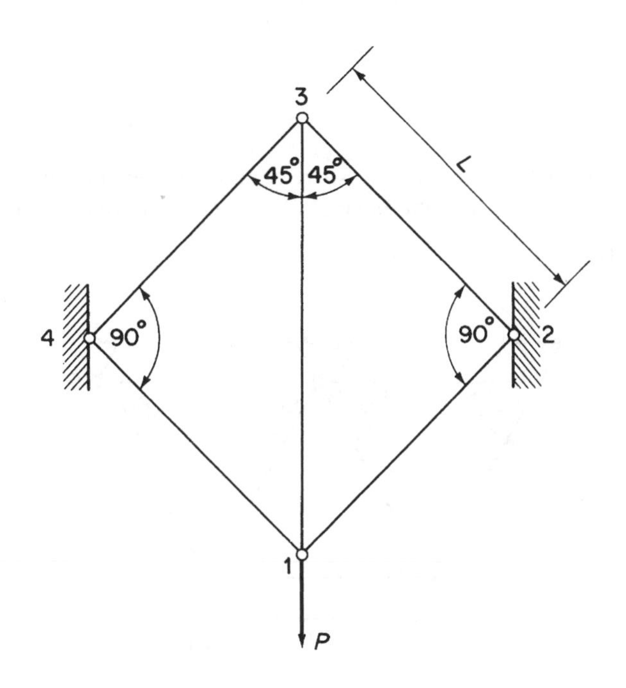

Fig. P.6.1

P.6.2 Use the stiffness method to find the ratio *H*/*P* for which the displacement of node 4 of the plane pin-jointed frame shown loaded in Fig. P.6.2 is zero, and for that case give the displacements of nodes 2 and 3.

All members have equal axial rigidity *EA*.

Ans. $H/P = 0.449,\ v_2 = -4Pl/(9 + 2\sqrt{3})AE,$

$v_3 = -6PL/(9 + 2\sqrt{3})AE.$

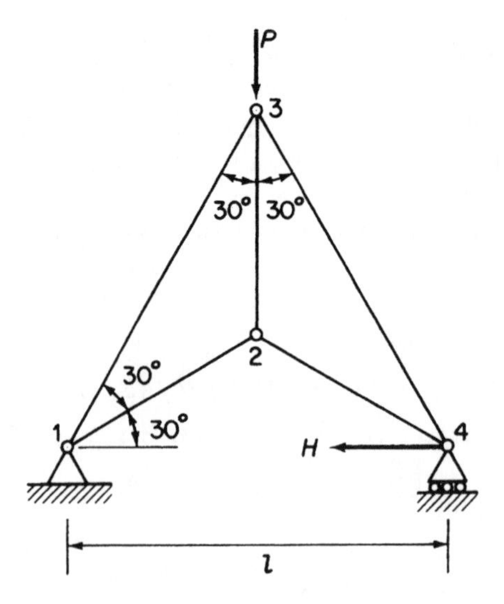

Fig. P.6.2

P.6.3 Form the matrices required to solve completely the plane truss shown in Fig. P.6.3 and determine the force in member 24. All members have equal axial rigidity.

Ans. $S_{24} = 0$.

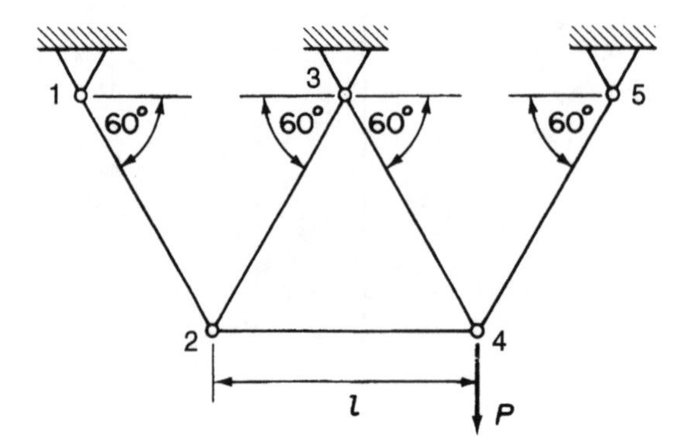

Fig. P.6.3

P.6.4 The symmetrical plane rigid jointed frame 1234567, shown in Fig. P.6.4, is fixed to rigid supports at 1 and 5 and supported by rollers inclined at 45° to the horizontal at nodes 3 and 7. It carries a vertical point load P at node 4 and a uniformly distributed load w per unit length on the span 26. Assuming the same flexural rigidity EI for all members, set up the stiffness equations which, when solved, give the nodal displacements of the frame.

Explain how the member forces can be obtained.

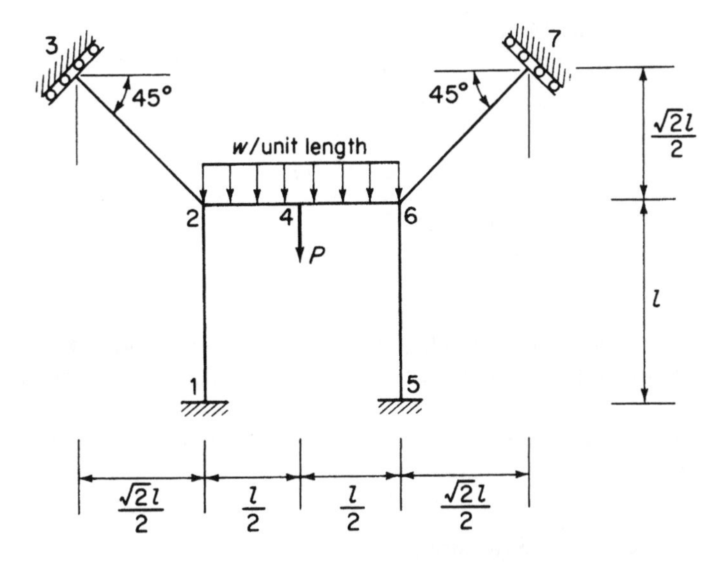

Fig. P.6.4

P.6.5 The frame shown in Fig. P.6.5 has the planes xz and yz as planes of symmetry. The nodal coordinates of one quarter of the frame are given in Table P.6.5(i).

In this structure the deformation of each member is due to a single effect, this being axial, bending or torsional. The mode of deformation of each member is given in Table P.6.5(ii), together with the relevant rigidity.

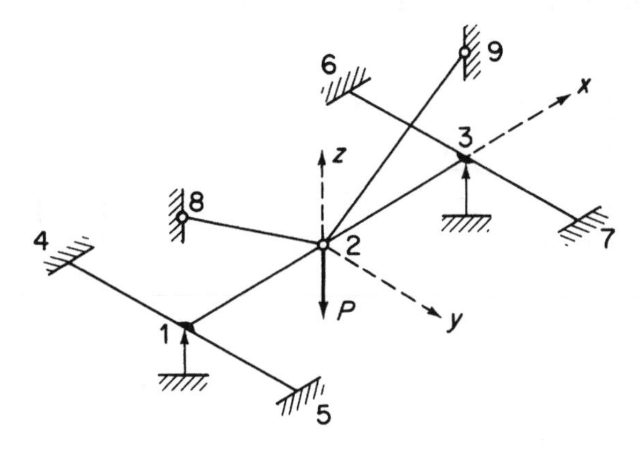

Fig. P.6.5

Table P.6.5(i)

Node	x	y	z
2	0	0	0
3	L	0	0
7	L	$0.8L$	0
9	L	0	L

Table P.6.5(ii)

	Effect		
Member	Axial	Bending	Torsional
23	–	EI	–
37	–	–	$GJ=0.8\,EI$
29	$EA=6\sqrt{2}\frac{EI}{L^2}$	–	–

Use the *direct stiffness* method to find all the displacements and hence calculate the forces in all the members. For member 123 plot the shear force and bending moment diagrams.

Briefly outline the sequence of operations in a typical computer program suitable for linear frame analysis.

Ans. $S_{29}=S_{28}=\sqrt{2}P/6$ (tension)

$M_3=-M_1=PL/9$ (hogging), $M_2=2PL/9$(sagging)

$SF_{12}=-SF_{23}=P/3$

Twisting moment in 37, $PL/18$ (anticlockwise).

P.6.6 Given that the force–displacement (stiffness) relationship for the beam element shown in Fig. P.6.6(a) may be expressed in the following form:

$$\begin{Bmatrix} F_{y,1} \\ M_1/L \\ F_{y,2} \\ M_2/L \end{Bmatrix} = \frac{EI}{L^3}\begin{bmatrix} 12 & -6 & -12 & -6 \\ -6 & 4 & 6 & 2 \\ -12 & 6 & 12 & 6 \\ -6 & 2 & 6 & 4 \end{bmatrix}\begin{Bmatrix} v_1 \\ \theta_1 L \\ v_2 \\ \theta_2 L \end{Bmatrix}$$

Obtain the force–displacement (stiffness) relationship for the variable section beam (Fig. P.6.6(b)), composed of elements 12, 23 and 34.

Such a beam is loaded and supported symmetrically as shown in Fig. P.6.6(c). Both ends are rigidly fixed and the ties FB, CH have a cross-section area a_1 and the ties EB, CG a cross-section area a_2. Calculate the deflections under the loads, the forces in the ties and all other information necessary for sketching the bending moment and shear force diagrams for the beam.

Neglect axial effects in the beam. The ties are made from the same material as the beam.

Ans. $v_B=v_C=-5PL^3/144EI,\ \ \theta_B=-\theta_C=PL^2/24EI,$

$S_1=2P/3,\ \ S_2=\sqrt{2}P/3,$

$F_{y,A}=P/3,\ \ M_A=-PL/4.$

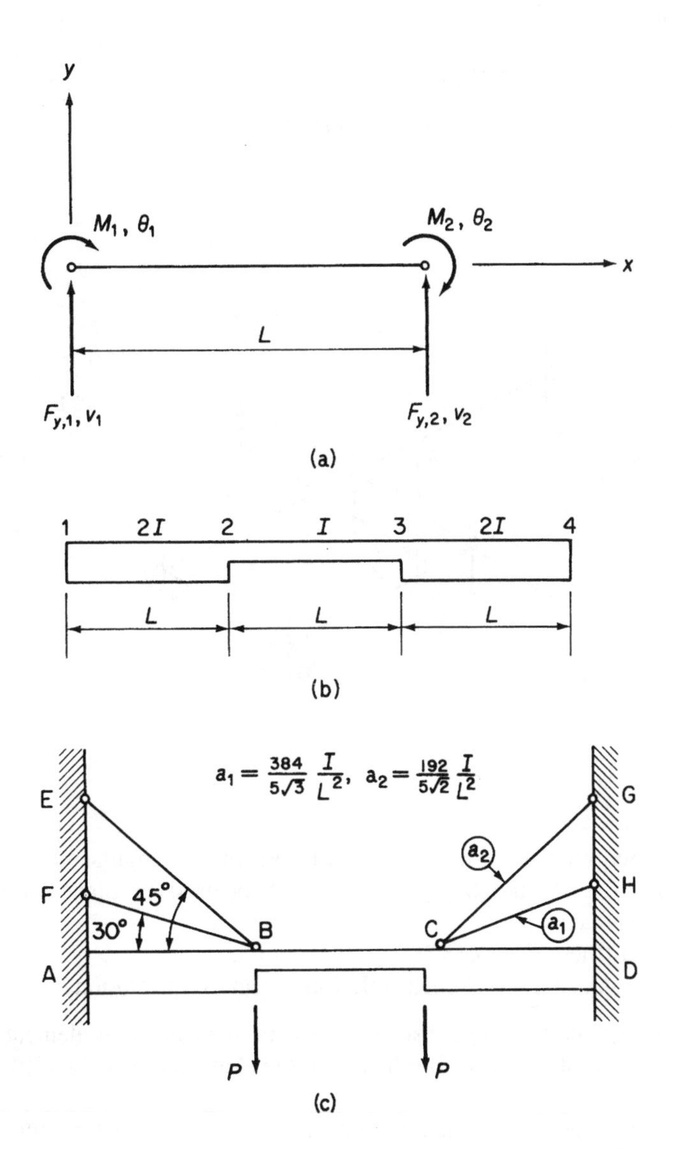

Fig. P.6.6

P.6.7 The symmetrical rigid jointed grillage shown in Fig. P.6.7 is encastré at 6, 7, 8 and 9 and rests on simple supports at 1, 2, 4 and 5. It is loaded with a vertical point load P at 3.

Use the stiffness method to find the displacements of the structure and hence calculate the support reactions and the forces in all the members. Plot the bending moment diagram for 123. All members have the same section properties and $GJ = 0.8EI$.

Ans. $F_{y,1} = F_{y,5} = -P/16$

$F_{y,2} = F_{y,4} = 9P/16$

$M_{21} = M_{45} = -Pl/16$ (hogging)

$$M_{23} = M_{43} = -Pl/12 \text{ (hogging)}$$

Twisting moment in 62, 82, 74 and 94 is $Pl/96$.

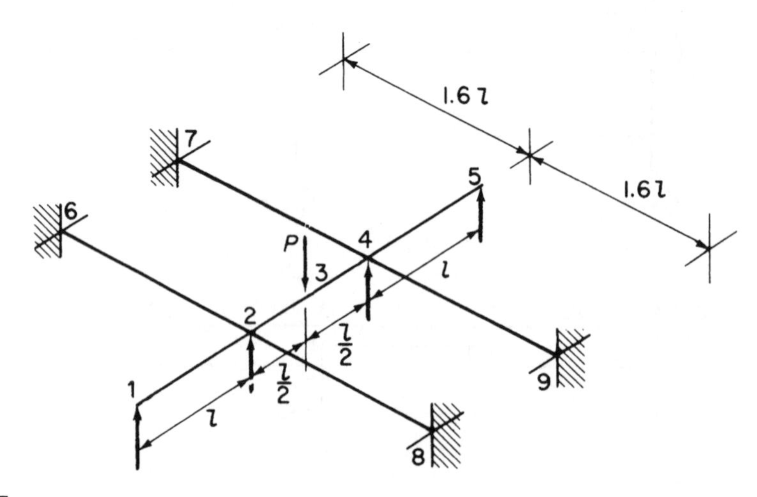

Fig. P.6.7

P.6.8 It is required to formulate the stiffness of a triangular element 123 with coordinates (0, 0), $(a, 0)$, and $(0, a)$, respectively, to be used for 'plane stress' problems.
(a) Form the $[B]$ matrix.
(b) Obtain the stiffness matrix $[K^{\mathrm{e}}]$.
Why, in general, is a finite element solution not an exact solution?

P.6.9 It is required to form the stiffness matrix of a triangular element 123 for use in stress analysis problems. The coordinates of the element are (1, 1), (2, 1), and (2, 2), respectively.
(a) Assume a suitable displacement field explaining the reasons for your choice.
(b) Form the $[B]$ matrix.
(c) Form the matrix which gives, when multiplied by the element nodal displacements, the stresses in the element. Assume a general $[D]$ matrix.

P.6.10 It is required to form the stiffness matrix for a rectangular element of side $2a \times 2b$ and thickness t for use in 'plane stress' problems.
(a) Assume a suitable displacement field.
(b) Form the $[C]$ matrix.
(c) Obtain $\int_{\mathrm{vol}} [C]^{\mathrm{T}}[D][C]\,\mathrm{d}V$.
Note that the stiffness matrix may be expressed as

$$[K^{\mathrm{e}}] = [A^{-1}]^{\mathrm{T}} \left[\int_{\mathrm{vol}} [C]^{\mathrm{T}}[D][C]\,\mathrm{d}V \right] [A^{-1}]$$

P.6.11 A square element 1234, whose corners have coordinates x, y (in metres) of $(-1, -1)$, $(1, -1)$, $(1, 1)$, and $(-1, 1)$, respectively, was used in a plane stress finite element analysis. The following nodal displacements (mm) were obtained:

$$\begin{array}{llll} u_1 = 0.1 & u_2 = 0.3 & u_3 = 0.6 & u_4 = 0.1 \\ v_1 = 0.1 & v_2 = 0.3 & v_3 = 0.7 & v_4 = 0.5 \end{array}$$

If Young's modulus $E = 200\ 000\ \text{N/mm}^2$ and Poisson's ratio $\nu = 0.3$, calculate the stresses at the centre of the element.

Ans. $\sigma_x = 51.65\,\text{N/mm}^2$, $\sigma_y = 55.49\,\text{N/mm}^2$, $\tau_{xy} = 13.46\,\text{N/mm}^2$.

P.6.12 A rectangular element used in plane stress analysis has corners whose coordinates in metres referred to an Oxy axes system are $1(-2, -1)$, $2(2, -1)$, $3(2, 1)$, $4(-2, 1)$. The displacements of the corners (in metres) are

$$\begin{array}{llll} u_1 = 0.001 & u_2 = 0.003 & u_3 = -0.003 & u_4 = 0 \\ v_1 = -0.004 & v_2 = -0.002 & v_3 = 0.001 & v_4 = 0.001 \end{array}$$

If Young's modulus is $200\,000\ \text{N/mm}^2$ and Poisson's ratio is 0.3 calculate the strains at the centre of the element.

Ans. $\varepsilon_x = -0.000125$, $\varepsilon_y = 0.002$, $\gamma_{xy} = -0.0015$.

P.6.13 A constant strain triangular element has corners 1(0,0), 2(4,0) and 3(2,2) and is 1 unit thick. If the elasticity matrix $[D]$ has elements $D_{11} = D_{22} = a$, $D_{12} = D_1 = b$, $D_{13} = D_{23} = D_{31} = D_{32} = 0$ and $D_{33} = c$ derive the stiffness matrix for the element.

Ans.

$$[K^e] = \frac{1}{4}\begin{bmatrix} a+c & & & & & \\ b+c & a+c & & & & \\ -a+c & -b+c & a+c & & & \\ b-c & a-c & -b-c & a+c & & \\ -2c & -2c & -2c & 2c & 4c & \\ -2b & -2a & 2b & -2a & 0 & 4a \end{bmatrix}$$

P.6.14 The following interpolation formula is suggested as a displacement function for deriving the stiffness of a plane stress rectangular element of uniform thickness t shown in Fig. P.6.14.

$$u = \frac{1}{4ab}[(a-x)(b-y)u_1 + (a+x)(b-y)u_2 + (a+x)(b+y)u_3 + (a-x)(b+y)u_1]$$

Form the strain matrix and obtain the stiffness coefficients K_{11} and K_{12} in terms of the material constants c, d and e defined below.

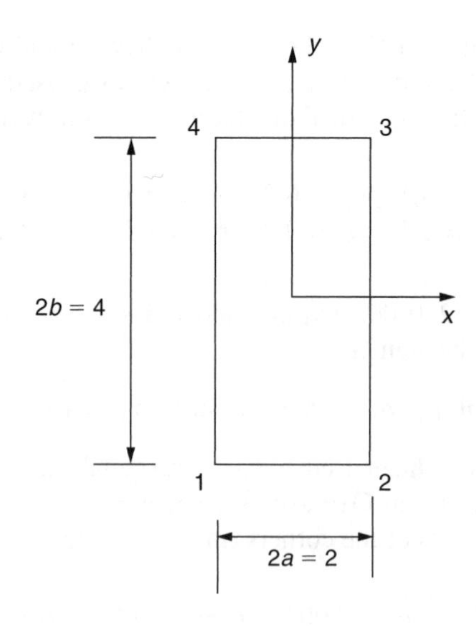

Fig. P.6.14

In the elasticity matrix $[D]$

$$D_{11} = D_{22} = c \quad D_{12} = d \quad D_{33} = e \quad \text{and} \quad D_{13} = D_{23} = 0$$

Ans. $K_{11} = t(4c + e)/6, \; K_{12} = t(d + e)/4.$

SECTION A3 THIN PLATE THEORY

7

Bending of thin plates

Generally, we define a thin plate as a sheet of material whose thickness is small compared with its other dimensions but which is capable of resisting bending in addition to membrane forces. Such a plate forms a basic part of an aircraft structure, being, for example, the area of stressed skin bounded by adjacent stringers and ribs in a wing structure or by adjacent stringers and frames in a fuselage.

In this chapter we shall investigate the effect of a variety of loading and support conditions on the small deflection of rectangular plates. Two approaches are presented: an 'exact' theory based on the solution of a differential equation and an energy method relying on the principle of the stationary value of the total potential energy of the plate and its applied loading. The latter theory will subsequently be used in Chapter 9 to determine buckling loads for unstiffened and stiffened panels.

7.1 Pure bending of thin plates

The thin rectangular plate of Fig. 7.1 is subjected to pure bending moments of intensity M_x and M_y per unit length uniformly distributed along its edges. The former bending moment is applied along the edges parallel to the y axis, the latter along the edges

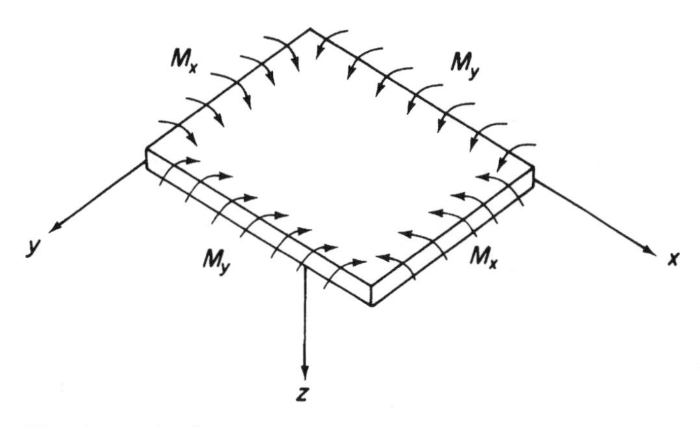

Fig. 7.1 Plate subjected to pure bending.

parallel to the x axis. We shall assume that these bending moments are positive when they produce compression at the upper surface of the plate and tension at the lower.

If we further assume that the displacement of the plate in a direction parallel to the z axis is small compared with its thickness t and that sections which are plane before bending remain plane after bending, then, as in the case of simple beam theory, the middle plane of the plate does not deform during the bending and is therefore a *neutral plane*. We take the neutral plane as the reference plane for our system of axes.

Let us consider an element of the plate of side $\delta x \delta y$ and having a depth equal to the thickness t of the plate as shown in Fig. 7.2(a). Suppose that the radii of curvature of the neutral plane n are ρ_x and ρ_y in the xz and yz planes respectively (Fig. 7.2(b)). Positive curvature of the plate corresponds to the positive bending moments which produce displacements in the positive direction of the z or downward axis. Again, as in simple beam theory, the direct strains ε_x and ε_y corresponding to direct stresses σ_x and σ_y of an elemental lamina of thickness δz a distance z below the neutral plane are given by

$$\varepsilon_x = \frac{z}{\rho_x} \qquad \varepsilon_y = \frac{z}{\rho_y} \tag{7.1}$$

Referring to Eqs (1.52) we have

$$\varepsilon_x = \frac{1}{E}(\sigma_x - \nu\sigma_y) \qquad \varepsilon_y = \frac{1}{E}(\sigma_y - \nu\sigma_x) \tag{7.2}$$

Substituting for ε_x and ε_y from Eqs (7.1) into (7.2) and rearranging gives

$$\left.\begin{aligned} \sigma_x &= \frac{Ez}{1-\nu^2}\left(\frac{1}{\rho_x} + \frac{\nu}{\rho_y}\right) \\ \sigma_y &= \frac{Ez}{1-\nu^2}\left(\frac{1}{\rho_y} + \frac{\nu}{\rho_x}\right) \end{aligned}\right\} \tag{7.3}$$

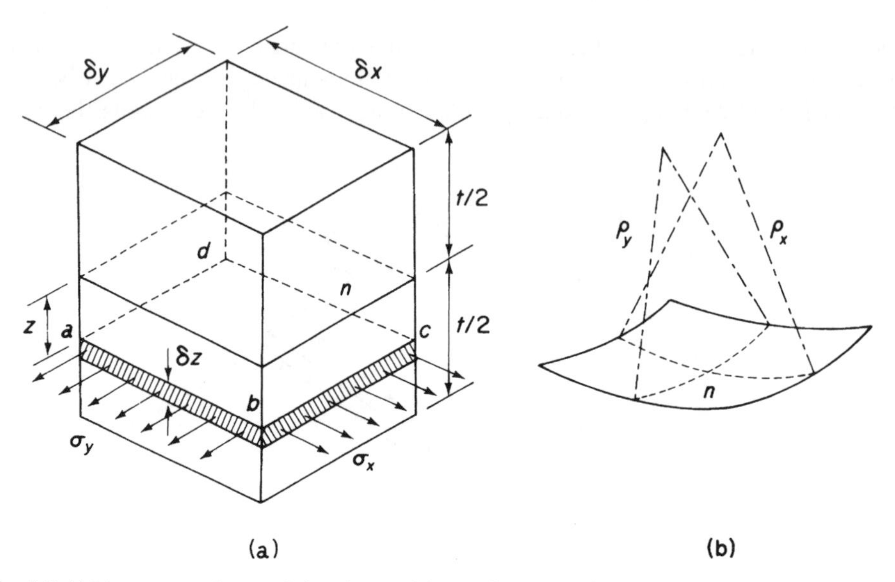

Fig. 7.2 (a) Direct stress on lamina of plate element; (b) radii of curvature of neutral plane.

As would be expected from our assumption of plane sections remaining plane the direct stresses vary linearly across the thickness of the plate, their magnitudes depending on the curvatures (i.e. bending moments) of the plate. The internal direct stress distribution on each vertical surface of the element must be in equilibrium with the applied bending moments. Thus

$$M_x \delta y = \int_{-t/2}^{t/2} \sigma_x z \delta y \, dz$$

and

$$M_y \delta x = \int_{-t/2}^{t/2} \sigma_y z \delta x \, dz$$

Substituting for σ_x and σ_y from Eqs (7.3) gives

$$M_x = \int_{-t/2}^{t/2} \frac{Ez^2}{1-\nu^2} \left(\frac{1}{\rho_x} + \frac{\nu}{\rho_y} \right) dz$$

$$M_y = \int_{-t/2}^{t/2} \frac{Ez^2}{1-\nu^2} \left(\frac{1}{\rho_y} + \frac{\nu}{\rho_x} \right) dz$$

Let

$$D = \int_{-t/2}^{t/2} \frac{Ez^2}{1-\nu^2} dz = \frac{Et^3}{12(1-\nu^2)} \tag{7.4}$$

Then

$$M_x = D \left(\frac{1}{\rho_x} + \frac{\nu}{\rho_y} \right) \tag{7.5}$$

$$M_y = D \left(\frac{1}{\rho_y} + \frac{\nu}{\rho_x} \right) \tag{7.6}$$

in which D is known as the *flexural rigidity* of the plate.

If w is the deflection of any point on the plate in the z direction, then we may relate w to the curvature of the plate in the same manner as the well-known expression for beam curvature. Hence

$$\frac{1}{\rho_x} = -\frac{\partial^2 w}{\partial x^2} \quad \frac{1}{\rho_y} = -\frac{\partial^2 w}{\partial y^2}$$

the negative signs resulting from the fact that the centres of curvature occur above the plate in which region z is negative. Equations (7.5) and (7.6) then become

$$M_x = -D \left(\frac{\partial^2 w}{\partial x^2} + \nu \frac{\partial^2 w}{\partial y^2} \right) \tag{7.7}$$

$$M_y = -D \left(\frac{\partial^2 w}{\partial y^2} + \nu \frac{\partial^2 w}{\partial x^2} \right) \tag{7.8}$$

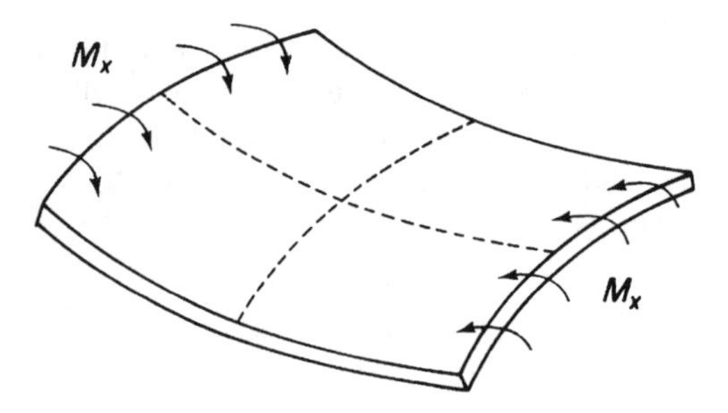

Fig. 7.3 Anticlastic bending.

Equations (7.7) and (7.8) define the deflected shape of the plate provided that M_x and M_y are known. If either M_x or M_y is zero then

$$\frac{\partial^2 w}{\partial x^2} = -\nu \frac{\partial^2 w}{\partial y^2} \quad \text{or} \quad \frac{\partial^2 w}{\partial y^2} = -\nu \frac{\partial^2 w}{\partial x^2}$$

and the plate has curvatures of opposite signs. The case of $M_y = 0$ is illustrated in Fig. 7.3. A surface possessing two curvatures of opposite sign is known as an *anticlastic surface*, as opposed to a *synclastic surface* which has curvatures of the same sign. Further, if $M_x = M_y = M$ then from Eqs (7.5) and (7.6)

$$\frac{1}{\rho_x} = \frac{1}{\rho_y} = \frac{1}{\rho}$$

Therefore, the deformed shape of the plate is spherical and of curvature

$$\frac{1}{\rho} = \frac{M}{D(1+\nu)} \tag{7.9}$$

7.2 Plates subjected to bending and twisting

In general, the bending moments applied to the plate will not be in planes perpendicular to its edges. Such bending moments, however, may be resolved in the normal manner into tangential and perpendicular components, as shown in Fig. 7.4. The perpendicular components are seen to be M_x and M_y as before, while the tangential components M_{xy} and M_{yx} (again these are moments per unit length) produce twisting of the plate about axes parallel to the x and y axes. The system of suffixes and the sign convention for these twisting moments must be clearly understood to avoid confusion. M_{xy} is a twisting moment intensity in a vertical x plane parallel to the y axis, while M_{yx} is a twisting moment intensity in a vertical y plane parallel to the x axis. Note that the first suffix gives the direction of the axis of the twisting moment. We also define positive twisting moments as being clockwise when viewed along their axes in directions parallel to the positive directions of the corresponding x or y axis. In Fig. 7.4, therefore, all moment intensities are positive.

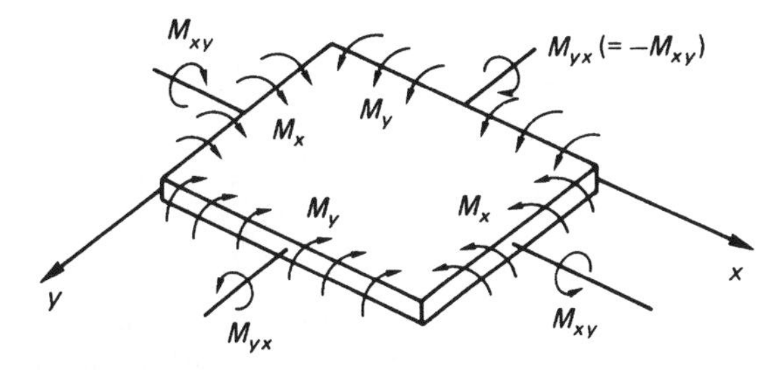

Fig. 7.4 Plate subjected to bending and twisting.

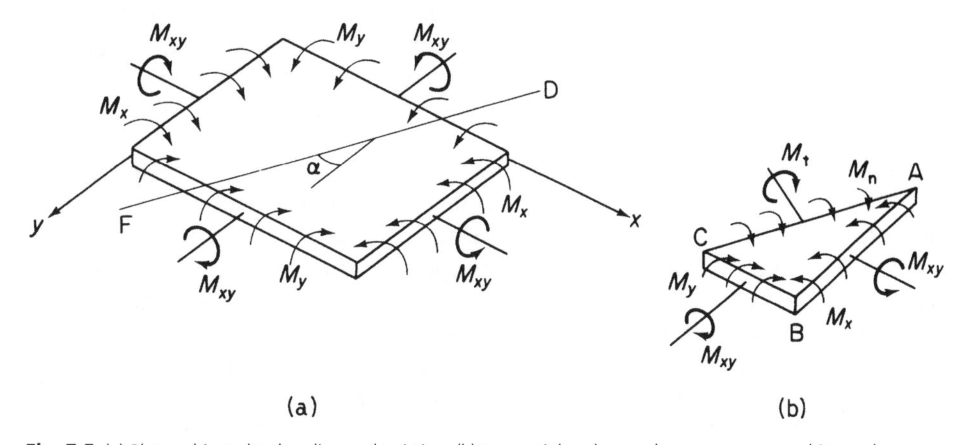

Fig. 7.5 (a) Plate subjected to bending and twisting; (b) tangential and normal moments on an arbitrary plane.

Since the twisting moments are tangential moments or torques they are resisted by a system of horizontal shear stresses τ_{xy}, as shown in Fig. 7.6. From a consideration of complementary shear stresses (see Fig. 7.6) $M_{xy} = -M_{yx}$, so that we may represent a general moment application to the plate in terms of M_x, M_y and M_{xy} as shown in Fig. 7.5(a). These moments produce tangential and normal moments, M_t and M_n, on an arbitrarily chosen diagonal plane FD. We may express these moment intensities (in an analogous fashion to the complex stress systems of Section 1.6) in terms of M_x, M_y and M_{xy}. Thus, for equilibrium of the triangular element ABC of Fig. 7.5(b) in a plane perpendicular to AC

$$M_\mathrm{n}\mathrm{AC} = M_x\mathrm{AB}\cos\alpha + M_y\mathrm{BC}\sin\alpha - M_{xy}\mathrm{AB}\sin\alpha - M_{xy}\mathrm{BC}\cos\alpha$$

giving

$$M_\mathrm{n} = M_x\cos^2\alpha + M_y\sin^2\alpha - M_{xy}\sin 2\alpha \tag{7.10}$$

Similarly for equilibrium in a plane parallel to CA

$$M_\mathrm{t}\mathrm{AC} = M_x\mathrm{AB}\sin\alpha - M_y\mathrm{BC}\cos\alpha + M_{xy}\mathrm{AB}\cos\alpha - M_{xy}\mathrm{BC}\sin\alpha$$

or

$$M_\mathrm{t} = \frac{(M_x - M_y)}{2}\sin 2\alpha + M_{xy}\cos 2\alpha \tag{7.11}$$

(compare Eqs (7.10) and (7.11) with Eqs (1.8) and (1.9)). We observe from Eq. (7.11) that there are two values of α, differing by 90° and given by

$$\tan 2\alpha = -\frac{2M_{xy}}{M_x - M_y}$$

for which $M_t = 0$, leaving normal moments of intensity M_n on two mutually perpendicular planes. These moments are termed *principal moments* and their corresponding curvatures *principal curvatures*. For a plate subjected to pure bending and twisting in which M_x, M_y and M_{xy} are invariable throughout the plate, the principal moments are the algebraically greatest and least moments in the plate. It follows that there are no shear stresses on these planes and that the corresponding direct stresses, for a given value of z and moment intensity, are the algebraically greatest and least values of direct stress in the plate.

Let us now return to the loaded plate of Fig. 7.5(a). We have established, in Eqs (7.7) and (7.8), the relationships between the bending moment intensities M_x and M_y and the deflection w of the plate. The next step is to relate the twisting moment M_{xy} to w. From the principle of superposition we may consider M_{xy} acting separately from M_x and M_y. As stated previously M_{xy} is resisted by a system of horizontal complementary shear stresses on the vertical faces of sections taken throughout the thickness of the plate parallel to the x and y axes. Consider an element of the plate formed by such sections, as shown in Fig. 7.6. The complementary shear stresses on a lamina of the element a distance z below the neutral plane are, in accordance with the sign convention of Section 1.2, τ_{xy}. Therefore, on the face ABCD

$$M_{xy}\delta y = -\int_{-t/2}^{t/2} \tau_{xy}\delta y z \, dz$$

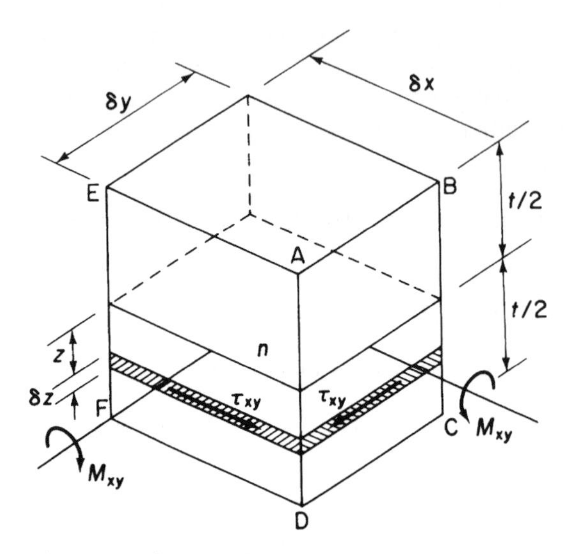

Fig. 7.6 Complementary shear stresses due to twisting moments M_{xy}.

and on the face ADFE

$$M_{xy}\delta x = -\int_{-t/2}^{t/2} \tau_{xy}\delta x z \, dz$$

giving

$$M_{xy} = -\int_{-t/2}^{t/2} \tau_{xy} z \, dz$$

or in terms of the shear strain γ_{xy} and modulus of rigidity G

$$M_{xy} = -G\int_{-t/2}^{t/2} \gamma_{xy} z \, dz \tag{7.12}$$

Referring to Eqs (1.20), the shear strain γ_{xy} is given by

$$\gamma_{xy} = \frac{\partial v}{\partial x} + \frac{\partial u}{\partial y}$$

We require, of course, to express γ_{xy} in terms of the deflection w of the plate; this may be accomplished as follows. An element taken through the thickness of the plate will suffer rotations equal to $\partial w/\partial x$ and $\partial w/\partial y$ in the xz and yz planes respectively. Considering the rotation of such an element in the xz plane, as shown in Fig. 7.7, we see that the displacement u in the x direction of a point a distance z below the neutral plane is

$$u = -\frac{\partial w}{\partial x} z$$

Similarly, the displacement v in the y direction is

$$v = -\frac{\partial w}{\partial y} z$$

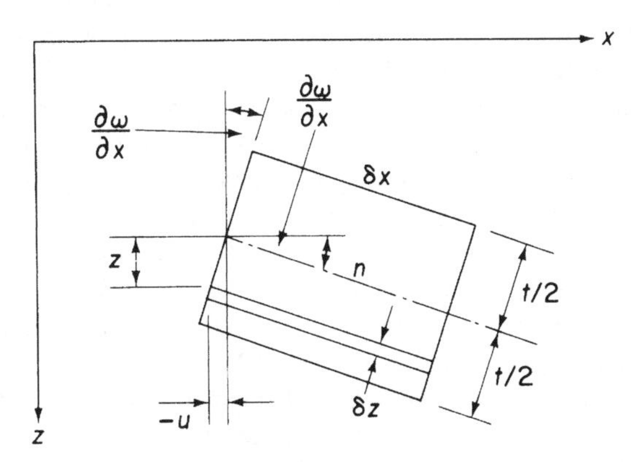

Fig. 7.7 Determination of shear strain γ_{xy}.

Hence, substituting for u and v in the expression for γ_{xy} we have

$$\gamma_{xy} = -2z\frac{\partial^2 w}{\partial x \partial y} \tag{7.13}$$

whence from Eq. (7.12)

$$M_{xy} = G\int_{-t/2}^{t/2} 2z^2 \frac{\partial^2 w}{\partial x \partial y} dz$$

or

$$M_{xy} = \frac{Gt^3}{6}\frac{\partial^2 w}{\partial x \partial y}$$

Replacing G by the expression $E/2(1+\nu)$ established in Eq. (1.50) gives

$$M_{xy} = \frac{Et^3}{12(1+\nu)}\frac{\partial^2 w}{\partial x \partial y}$$

Multiplying the numerator and denominator of this equation by the factor $(1-\nu)$ yields

$$M_{xy} = D(1-\nu)\frac{\partial^2 w}{\partial x \partial y} \tag{7.14}$$

Equations (7.7), (7.8) and (7.14) relate the bending and twisting moments to the plate deflection and are analogous to the bending moment-curvature relationship for a simple beam.

7.3 Plates subjected to a distributed transverse load

The relationships between bending and twisting moments and plate deflection are now employed in establishing the general differential equation for the solution of a thin rectangular plate, supporting a distributed transverse load of intensity q per unit area (see Fig. 7.8). The distributed load may, in general, vary over the surface of the plate and is therefore a function of x and y. We assume, as in the preceding analysis, that the middle plane of the plate is the neutral plane and that the plate deforms such that plane sections remain plane after bending. This latter assumption introduces an apparent inconsistency in the theory. For plane sections to remain plane the shear strains γ_{xz} and γ_{yz} must be zero. However, the transverse load produces transverse shear forces (and therefore stresses) as shown in Fig. 7.9. We therefore assume that although $\gamma_{xz} = \tau_{xz}/G$ and $\gamma_{yz} = \tau_{yz}/G$ are negligible the corresponding shear forces are of the same order of magnitude as the applied load q and the moments M_x, M_y and M_{xy}. This assumption is analogous to that made in a slender beam theory in which shear strains are ignored.

The element of plate shown in Fig. 7.9 supports bending and twisting moments as previously described and, in addition, vertical shear forces Q_x and Q_y per unit length on faces perpendicular to the x and y axes, respectively. The variation of shear stresses τ_{xz} and τ_{yz} along the small edges δx, δy of the element is neglected and the resultant

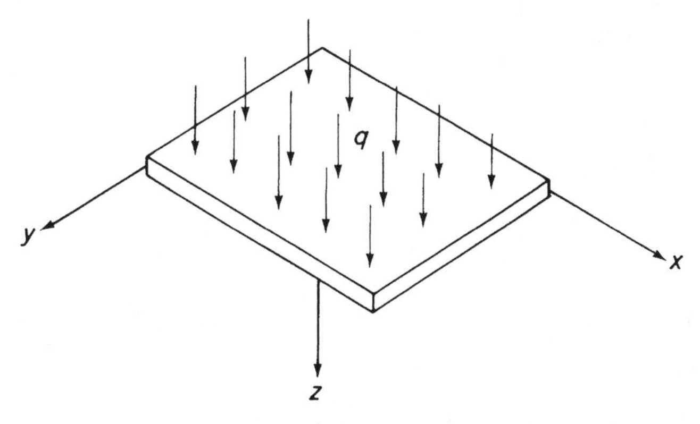

Fig. 7.8 Plate supporting a distributed transverse load.

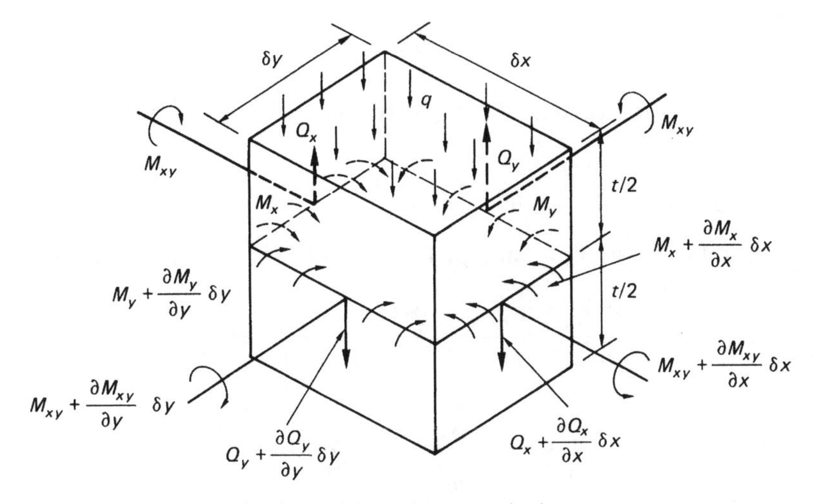

Fig. 7.9 Plate element subjected to bending, twisting and transverse loads.

shear forces $Q_x\delta y$ and $Q_y\delta x$ are assumed to act through the centroid of the faces of the element. From the previous sections

$$M_x = \int_{-t/2}^{t/2} \sigma_x z\,\mathrm{d}z \quad M_y = \int_{-t/2}^{t/2} \sigma_y z\,\mathrm{d}z \quad M_{xy} = (-M_{yx}) = -\int_{-t/2}^{t/2} \tau_{xy} z\,\mathrm{d}z$$

In a similar fashion

$$Q_x = \int_{-t/2}^{t/2} \tau_{xz}\,\mathrm{d}z \quad Q_y = \int_{-t/2}^{t/2} \tau_{yz}\,\mathrm{d}z \tag{7.15}$$

For equilibrium of the element parallel to Oz and assuming that the weight of the plate is included in q

$$\left(Q_x + \frac{\partial Q_x}{\partial x}\delta x\right)\delta y - Q_x\delta y + \left(Q_y + \frac{\partial Q_y}{\partial y}\delta y\right)\delta x - Q_y\delta x + q\delta x\delta y = 0$$

or, after simplification

$$\frac{\partial Q_x}{\partial x} + \frac{\partial Q_y}{\partial y} + q = 0 \tag{7.16}$$

Taking moments about the x axis

$$M_{xy}\delta y - \left(M_{xy} + \frac{\partial M_{xy}}{\partial x}\delta x\right)\delta y - M_y\delta x + \left(M_y + \frac{\partial M_y}{\partial y}\delta y\right)\delta x$$
$$- \left(Q_y + \frac{\partial Q_y}{\partial y}\delta y\right)\delta x\delta y + Q_x\frac{\delta y^2}{2} - \left(Q_x + \frac{\partial Q_x}{\partial x}\delta x\right)\frac{\delta y^2}{2} - q\delta x\frac{\delta y^2}{2} = 0$$

Simplifying this equation and neglecting small quantities of a higher order than those retained gives

$$\frac{\partial M_{xy}}{\partial x} - \frac{\partial M_y}{\partial y} + Q_y = 0 \tag{7.17}$$

Similarly taking moments about the y axis we have

$$\frac{\partial M_{xy}}{\partial y} - \frac{\partial M_x}{\partial x} + Q_x = 0 \tag{7.18}$$

Substituting in Eq. (7.16) for Q_x and Q_y from Eqs (7.18) and (7.17) we obtain

$$\frac{\partial^2 M_x}{\partial x^2} - \frac{\partial^2 M_{xy}}{\partial x \partial y} + \frac{\partial^2 M_y}{\partial y^2} - \frac{\partial^2 M_{xy}}{\partial x \partial y} = -q$$

or

$$\frac{\partial^2 M_x}{\partial x^2} - 2\frac{\partial^2 M_{xy}}{\partial x \partial y} + \frac{\partial^2 M_y}{\partial y^2} = -q \tag{7.19}$$

Replacing M_x, M_{xy} and M_y in Eq. (7.19) from Eqs (7.7), (7.14) and (7.8) gives

$$\frac{\partial^4 w}{\partial x^4} + 2\frac{\partial^4 w}{\partial x^2 \partial y^2} + \frac{\partial^4 w}{\partial y^4} = \frac{q}{D} \tag{7.20}$$

This equation may also be written

$$\left(\frac{\partial^2}{\partial x^2} + \frac{\partial^2}{\partial y^2}\right)\left(\frac{\partial^2 w}{\partial x^2} + \frac{\partial^2 w}{\partial y^2}\right) = \frac{q}{D}$$

or

$$\left(\frac{\partial^2}{\partial x^2} + \frac{\partial^2}{\partial y^2}\right)^2 w = \frac{q}{D}$$

The operator $(\partial^2/\partial x^2 + \partial^2/\partial y^2)$ is the well-known Laplace operator in two dimensions and is sometimes written as ∇^2. Thus

$$(\nabla^2)^2 w = \frac{q}{D}$$

Generally, the transverse distributed load q is a function of x and y so that the determination of the deflected form of the plate reduces to obtaining a solution of Eq. (7.20), which satisfies the known boundary conditions of the problem. The bending and twisting moments follow from Eqs (7.7), (7.8) and (7.14), and the shear forces per unit length Q_x and Q_y are found from Eqs (7.17) and (7.18) by substitution for M_x, M_y and M_{xy} in terms of the deflection w of the plate; thus

$$Q_x = \frac{\partial M_x}{\partial x} - \frac{\partial M_{xy}}{\partial y} = -D\frac{\partial}{\partial x}\left(\frac{\partial^2 w}{\partial x^2} + \frac{\partial^2 w}{\partial y^2}\right) \tag{7.21}$$

$$Q_y = \frac{\partial M_y}{\partial y} - \frac{\partial M_{xy}}{\partial x} = -D\frac{\partial}{\partial y}\left(\frac{\partial^2 w}{\partial x^2} + \frac{\partial^2 w}{\partial y^2}\right) \tag{7.22}$$

Direct and shear stresses are then calculated from the relevant expressions relating them to M_x, M_y, M_{xy}, Q_x and Q_y.

Before discussing the solution of Eq. (7.20) for particular cases we shall establish boundary conditions for various types of edge support.

7.3.1 The simply supported edge

Let us suppose that the edge $x = 0$ of the thin plate shown in Fig. 7.10 is free to rotate but not to deflect. The edge is then said to be simply supported. The bending moment along this edge must be zero and also the deflection $w = 0$. Thus

$$(w)_{x=0} = 0 \quad \text{and} \quad (M_x)_{x=0} = -D\left(\frac{\partial^2 w}{\partial x^2} + \nu\frac{\partial^2 w}{\partial y^2}\right)_{x=0} = 0$$

The condition that $w = 0$ along the edge $x = 0$ also means that

$$\frac{\partial w}{\partial y} = \frac{\partial^2 w}{\partial y^2} = 0$$

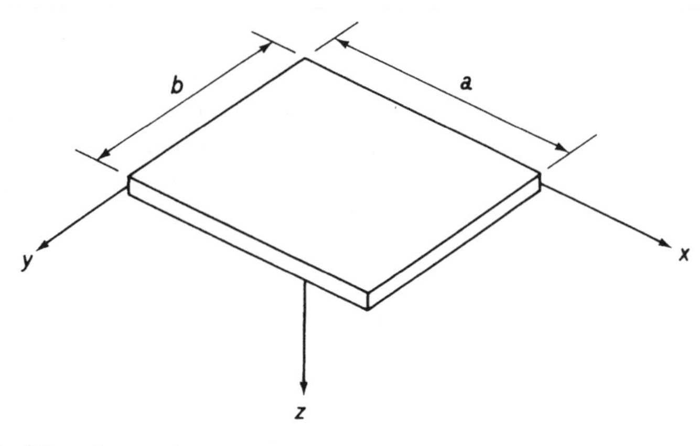

Fig. 7.10 Plate of dimensions $a \times b$.

along this edge. The above boundary conditions therefore reduce to

$$(w)_{x=0} = 0 \quad \left(\frac{\partial^2 w}{\partial x^2}\right)_{x=0} = 0 \tag{7.23}$$

7.3.2 The built-in edge

If the edge $x = 0$ is built-in or firmly clamped so that it can neither rotate nor deflect, then, in addition to w, the slope of the middle plane of the plate normal to this edge must be zero. That is

$$(w)_{x=0} = 0 \quad \left(\frac{\partial w}{\partial x}\right)_{x=0} = 0 \tag{7.24}$$

7.3.3 The free edge

Along a free edge there are no bending moments, twisting moments or vertical shearing forces, so that if $x = 0$ is the free edge then

$$(M_x)_{x=0} = 0 \quad (M_{xy})_{x=0} = 0 \quad (Q_x)_{x=0} = 0$$

giving, in this instance, three boundary conditions. However, Kirchhoff (1850) showed that only two boundary conditions are necessary to obtain a solution of Eq. (7.20), and that the reduction is obtained by replacing the two requirements of zero twisting moment and zero shear force by a single equivalent condition. Thomson and Tait (1883) gave a physical explanation of how this reduction may be effected. They pointed out that the horizontal force system equilibrating the twisting moment M_{xy} may be replaced along the edge of the plate by a vertical force system.

Consider two adjacent elements δy_1 and δy_2 along the edge of the thin plate of Fig. 7.11. The twisting moment $M_{xy}\delta y_1$ on the element δy_1 may be replaced by *forces* M_{xy} a distance δy_1 apart. Note that M_{xy}, being a twisting moment per unit length, has the dimensions of force. The twisting moment on the adjacent element δy_2 is $[M_{xy} + (\partial M_{xy}/\partial y)\delta y]\delta y_2$. Again this may be replaced by forces $M_{xy} + (\partial M_{xy}/\partial y)\delta y$. At the common surface of the two adjacent elements there is now a resultant force $(\partial M_{xy}/\partial y)\delta y$ or a vertical force per unit length of $\partial M_{xy}/\partial y$. For the sign convention for Q_x shown in Fig. 7.9 we have a statically equivalent vertical force per unit length of $(Q_x - \partial M_{xy}/\partial y)$. The separate conditions for a free edge of $(M_{xy})_{x=0} = 0$ and $(Q_x)_{x=0} = 0$ are therefore replaced by the equivalent condition

$$\left(Q_x - \frac{\partial M_{xy}}{\partial y}\right)_{x=0} = 0$$

or in terms of deflection

$$\left[\frac{\partial^3 w}{\partial x^3} + (2 - \nu)\frac{\partial^3 w}{\partial x \partial y^2}\right]_{x=0} = 0 \tag{7.25}$$

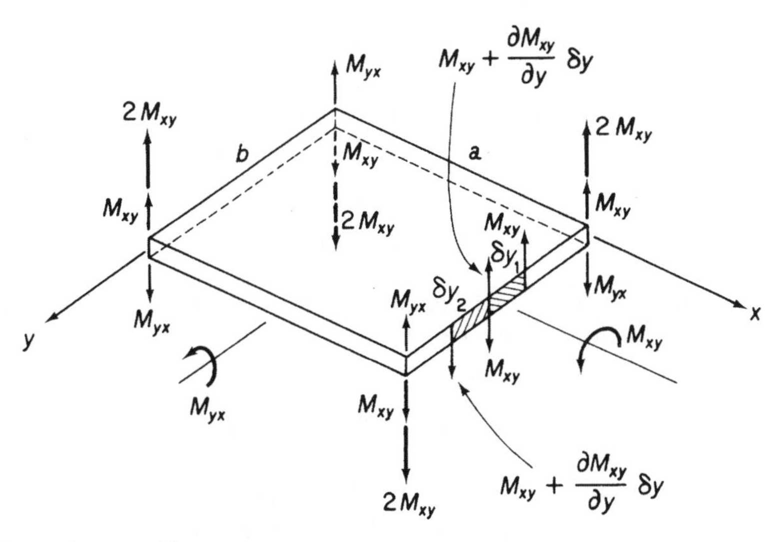

Fig. 7.11 Equivalent vertical force system.

Also, for the bending moment along the free edge to be zero

$$(M_x)_{x=0} = \left(\frac{\partial^2 w}{\partial x^2} + \nu \frac{\partial^2 w}{\partial y^2} \right)_{x=0} = 0 \tag{7.26}$$

The replacement of the twisting moment M_{xy} along the edges $x = 0$ and $x = a$ of a thin plate by a vertical force distribution results in leftover concentrated forces at the corners of M_{xy} as shown in Fig. 7.11. By the same argument there are concentrated forces M_{yx} produced by the replacement of the twisting moment M_{yx}. Since $M_{xy} = -M_{yx}$, then resultant forces $2M_{xy}$ act at each corner as shown and must be provided by external supports if the corners of the plate are not to move. The directions of these forces are easily obtained if the deflected shape of the plate is known. For example, a thin plate simply supported along all four edges and uniformly loaded has $\partial w/\partial x$ positive and numerically increasing, with increasing y near the corner $x = 0$, $y = 0$. Hence $\partial^2 w/\partial x \partial y$ is positive at this point and from Eq. (7.14) we see that M_{xy} is positive and M_{yx} negative; the resultant force $2M_{xy}$ is therefore downwards. From symmetry the force at each remaining corner is also $2M_{xy}$ downwards so that the tendency is for the corners of the plate to rise.

Having discussed various types of boundary conditions we shall proceed to obtain the solution for the relatively simple case of a thin rectangular plate of dimensions $a \times b$, simply supported along each of its four edges and carrying a distributed load $q(x, y)$.We have shown that the deflected form of the plate must satisfy the differential equation

$$\frac{\partial^4 w}{\partial x^4} + 2 \frac{\partial^4 w}{\partial x^2 \partial y^2} + \frac{\partial^4 w}{\partial y^4} = \frac{q(x, y)}{D}$$

with the boundary conditions

$$(w)_{x=0,a} = 0 \quad \left(\frac{\partial^2 w}{\partial x^2}\right)_{x=0,a} = 0$$

$$(w)_{y=0,b} = 0 \quad \left(\frac{\partial^2 w}{\partial y^2}\right)_{x=0,b} = 0$$

Navier (1820) showed that these conditions are satisfied by representing the deflection w as an infinite trigonometrical or Fourier series

$$w = \sum_{m=1}^{\infty}\sum_{n=1}^{\infty} A_{mn} \sin\frac{m\pi x}{a}\sin\frac{n\pi y}{b} \tag{7.27}$$

in which m represents the number of half waves in the x direction and n the corresponding number in the y direction. Further, A_{mn} are unknown coefficients which must satisfy the above differential equation and may be determined as follows.

We may also represent the load $q(x, y)$ by a Fourier series, thus

$$q(x,y) = \sum_{m=1}^{\infty}\sum_{n=1}^{\infty} a_{mn} \sin\frac{m\pi x}{a}\sin\frac{n\pi y}{b} \tag{7.28}$$

A particular coefficient $a_{m'n'}$ is calculated by first multiplying both sides of Eq. (7.28) by $\sin(m'\pi x/a)\sin(n'\pi y/b)$ and integrating with respect to x from 0 to a and with respect to y from 0 to b. Thus

$$\begin{aligned}
&\int_0^a\int_0^b q(x,y)\sin\frac{m'\pi x}{a}\sin\frac{n'\pi y}{b}\,\mathrm{d}x\,\mathrm{d}y \\
&= \sum_{m=1}^{\infty}\sum_{n=1}^{\infty}\int_0^a\int_0^b a_{mn}\sin\frac{m\pi x}{a}\sin\frac{m'\pi x}{a}\sin\frac{n\pi y}{b}\sin\frac{n'\pi y}{b}\,\mathrm{d}x\,\mathrm{d}y \\
&= \frac{ab}{4}a_{m'n'}
\end{aligned}$$

since

$$\begin{aligned}
\int_0^a \sin\frac{m\pi x}{a}\sin\frac{m'\pi x}{a}\,\mathrm{d}x &= 0 \quad \text{when} \quad m \neq m' \\
&= \frac{a}{2} \quad \text{when} \quad m = m'
\end{aligned}$$

and

$$\begin{aligned}
\int_0^b \sin\frac{n\pi y}{b}\sin\frac{n'\pi y}{b}\,\mathrm{d}y &= 0 \quad \text{when} \quad n \neq n' \\
&= \frac{b}{2} \quad \text{when} \quad n = n'
\end{aligned}$$

It follows that

$$a_{m'n'} = \frac{4}{ab}\int_0^a \int_0^b q(x,y)\sin\frac{m'\pi x}{a}\sin\frac{n'\pi y}{b}\,\mathrm{d}x\,\mathrm{d}y \tag{7.29}$$

Substituting now for w and $q(x, y)$ from Eqs (7.27) and (7.28) into the differential equation for w we have

$$\sum_{m=1}^{\infty}\sum_{n=1}^{\infty}\left\{A_{mn}\left[\left(\frac{m\pi}{a}\right)^4 + 2\left(\frac{m\pi}{a}\right)^2\left(\frac{n\pi}{b}\right)^2 + \left(\frac{n\pi}{b}\right)^4\right] - \frac{a_{mn}}{D}\right\}\sin\frac{m\pi x}{a}\sin\frac{n\pi y}{b} = 0$$

This equation is valid for all values of x and y so that

$$A_{mn}\left[\left(\frac{m\pi}{a}\right)^4 + 2\left(\frac{m\pi}{a}\right)^2\left(\frac{n\pi}{b}\right)^2 + \left(\frac{n\pi}{b}\right)^4\right] - \frac{a_{mn}}{D} = 0$$

or in alternative form

$$A_{mn}\pi^4\left(\frac{m^2}{a^2} + \frac{n^2}{b^2}\right)^2 - \frac{a_{mn}}{D} = 0$$

giving

$$A_{mn} = \frac{1}{\pi^4 D}\frac{a_{mn}}{[(m^2/a^2) + (n^2/b^2)]^2}$$

Hence

$$w = \frac{1}{\pi^4 D}\sum_{m=1}^{\infty}\sum_{n=1}^{\infty}\frac{a_{mn}}{[(m^2/a^2) + (n^2/b^2)]^2}\sin\frac{m\pi x}{a}\sin\frac{n\pi y}{b} \tag{7.30}$$

in which a_{mn} is obtained from Eq. (7.29). Equation (7.30) is the general solution for a thin rectangular plate under a transverse load $q(x, y)$.

Example 7.1

A thin rectangular plate $a \times b$ is simply supported along its edges and carries a uniformly distributed load of intensity q_0. Determine the deflected form of the plate and the distribution of bending moment.

Since $q(x, y) = q_0$ we find from Eq. (7.29) that

$$a_{mn} = \frac{4q_0}{ab}\int_0^a \int_0^b \sin\frac{m\pi x}{a}\sin\frac{n\pi y}{b}\,\mathrm{d}x\,\mathrm{d}y = \frac{16q_0}{\pi^2 mn}$$

where m and n are odd integers. For m or n even, $a_{mn} = 0$. Hence from Eq. (7.30)

$$w = \frac{16q_0}{\pi^6 D}\sum_{m=1,3,5}^{\infty}\sum_{n=1,3,5}^{\infty}\frac{\sin(m\pi x/a)\sin(n\pi y/b)}{mn[(m^2/a^2) + (n^2/b^2)]^2} \tag{i}$$

The maximum deflection occurs at the centre of the plate where $x = a/2$, $y = b/2$. Thus

$$w_{\max} = \frac{16q_0}{\pi^6 D} \sum_{m=1,3,5}^{\infty} \sum_{n=1,3,5}^{\infty} \frac{\sin(m\pi/2)\sin(n\pi/2)}{mn[(m^2/a^2) + (n^2/b^2)]^2} \tag{ii}$$

This series is found to converge rapidly, the first few terms giving a satisfactory answer. For a square plate, taking $\nu = 0.3$, summation of the first four terms of the series gives

$$w_{\max} = 0.0443 q_0 \frac{a^4}{Et^3}$$

Substitution for w from Eq. (i) into the expressions for bending moment, Eqs (7.7) and (7.8), yields

$$M_x = \frac{16q_0}{\pi^4} \sum_{m=1,3,5}^{\infty} \sum_{n=1,3,5}^{\infty} \frac{[(m^2/a^2) + \nu(n^2/b^2)]}{mn[(m^2/a^2) + (n^2/b^2)]^2} \sin\frac{m\pi x}{a} \sin\frac{n\pi y}{b} \tag{iii}$$

$$M_y = \frac{16q_0}{\pi^4} \sum_{m=1,3,5}^{\infty} \sum_{n=1,3,5}^{\infty} \frac{[\nu(m^2/a^2) + (n^2/b^2)]}{mn[(m^2/a^2) + (n^2/b^2)]^2} \sin\frac{m\pi x}{a} \sin\frac{n\pi y}{b} \tag{iv}$$

Maximum values occur at the centre of the plate. For a square plate $a = b$ and the first five terms give

$$M_{x,\max} = M_{y,\max} = 0.0479 q_0 a^2$$

Comparing Eqs (7.3) with Eqs (7.5) and (7.6) we observe that

$$\sigma_x = \frac{12M_x z}{t^3} \qquad \sigma_y = \frac{12M_y z}{t^3}$$

Again the maximum values of these stresses occur at the centre of the plate at $z = \pm t/2$ so that

$$\sigma_{x,\max} = \frac{6M_x}{t^2} \qquad \sigma_{y,\max} = \frac{6M_y}{t^2}$$

For the square plate

$$\sigma_{x,\max} = \sigma_{y,\max} = 0.287 q_0 \frac{a^2}{t^2}$$

The twisting moment and shear stress distributions follow in a similar manner.

The infinite series (Eq. (7.27)) assumed for the deflected shape of a plate gives an exact solution for displacements and stresses. However, a more rapid, but approximate, solution may be obtained by assuming a displacement function in the form of a polynomial. The polynomial must, of course, satisfy the governing differential equation (Eq. (7.20)) and the boundary conditions of the specific problem. The "guessed" form of the deflected shape of a plate is the basis for the energy method of solution described in Section 7.6.

Example 7.2

Show that the deflection function

$$w = A(x^2y^2 - bx^2y - axy^2 + abxy)$$

is valid for a rectangular plate of sides a and b, built in on all four edges and subjected to a uniformly distributed load of intensity q. If the material of the plate has a Young's modulus E and is of thickness t determine the distributions of bending moment along the edges of the plate.

Differentiating the deflection function gives

$$\frac{\partial^4 w}{\partial x^4} = 0 \quad \frac{\partial^4 w}{\partial y^4} = 0 \quad \frac{\partial^4 w}{\partial x^2 \partial y^2} = 4A$$

Substituting in Eq. (7.20) we have

$$0 + 2 \times 4A + 0 = \text{constant} = \frac{q}{D}$$

The deflection function is therefore valid and

$$A = \frac{q}{8D}$$

The bending moment distributions are given by Eqs (7.7) and (7.8), i.e.

$$M_x = -\frac{q}{4}[y^2 - by + \nu(x^2 - ax)] \tag{i}$$

$$M_y = -\frac{q}{4}[x^2 - ax + \nu(y^2 - by)] \tag{ii}$$

For the edges $x = 0$ and $x = a$

$$M_x = -\frac{q}{4}(y^2 - by) \quad M_y = -\frac{\nu q}{4}(y^2 - by)$$

For the edges $y = 0$ and $y = b$

$$M_x = -\frac{\nu q}{4}(x^2 - ax) \quad M_y = -\frac{q}{4}(x^2 - ax)$$

7.4 Combined bending and in-plane loading of a thin rectangular plate

So far our discussion has been limited to small deflections of thin plates produced by different forms of transverse loading. In these cases we assumed that the middle or neutral plane of the plate remained unstressed. Additional in-plane tensile, compressive or shear loads will produce stresses in the middle plane, and these, if of sufficient magnitude, will affect the bending of the plate. Where the in-plane stresses are small

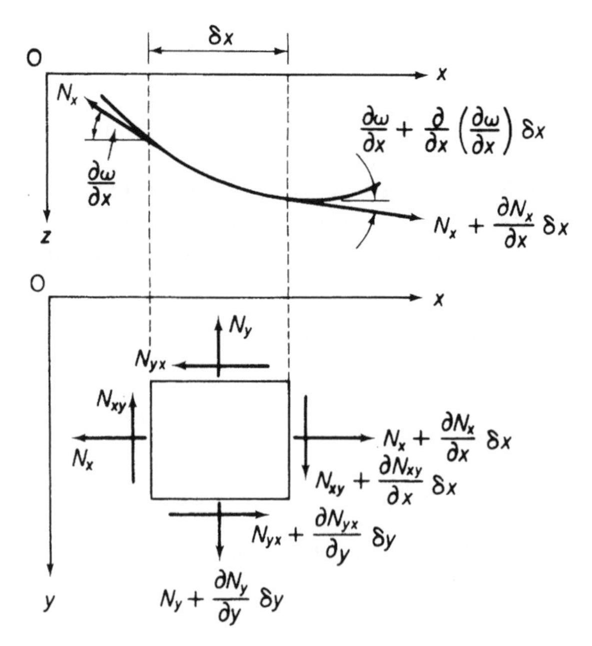

Fig. 7.12 In-plane forces on plate element.

compared with the critical buckling stresses it is sufficient to consider the two systems separately; the total stresses are then obtained by superposition. On the other hand, if the in-plane stresses are not small then their effect on the bending of the plate must be considered.

The elevation and plan of a small element $\delta x\delta y$ of the middle plane of a thin deflected plate are shown in Fig. 7.12. Direct and shear forces per unit length produced by the in-plane loads are given the notation N_x, N_y and N_{xy} and are assumed to be acting in positive senses in the directions shown. Since there are no resultant forces in the x or y directions from the transverse loads (see Fig. 7.9) we need only include the in-plane loads shown in Fig. 7.12 when considering the equilibrium of the element in these directions. For equilibrium parallel to Ox

$$\left(N_x + \frac{\partial N_x}{\partial x}\delta x\right)\delta y \cos\left(\frac{\partial w}{\partial x} + \frac{\partial^2 w}{\partial x^2}\delta x\right) - N_x \delta y \cos\frac{\partial w}{\partial x}$$
$$+\left(N_{yx} + \frac{\partial N_{yx}}{\partial y}\delta y\right)\delta x - N_{yx}\delta x = 0$$

For small deflections $\partial w/\partial x$ and $(\partial w/\partial x) + (\partial^2 w/\partial x^2)\delta x$ are small and the cosines of these angles are therefore approximately equal to one. The equilibrium equation thus simplifies to

$$\frac{\partial N_x}{\partial x} + \frac{\partial N_{yx}}{\partial y} = 0 \tag{7.31}$$

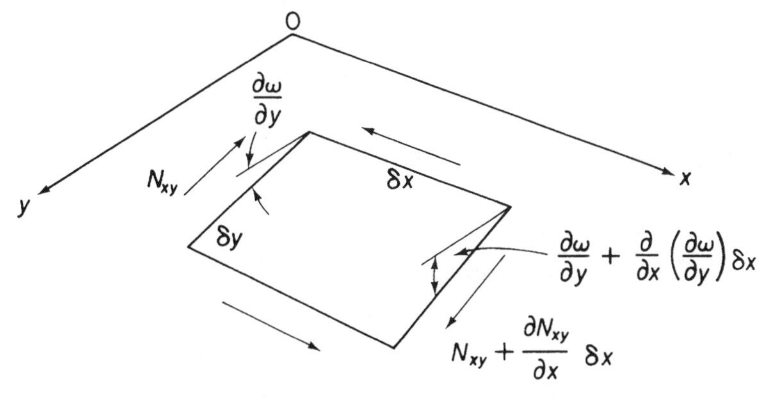

Fig. 7.13 Component of shear loads in the *z* direction.

Similarly for equilibrium in the y direction we have

$$\frac{\partial N_y}{\partial y} + \frac{\partial N_{xy}}{\partial x} = 0 \qquad (7.32)$$

Note that the components of the in-plane shear loads per unit length are, to a first order of approximation, the value of the shear load multiplied by the projection of the element on the relevant axis.

The determination of the contribution of the shear loads to the equilibrium of the element in the z direction is complicated by the fact that the element possesses curvature in both xz and yz planes. Therefore, from Fig. 7.13 the component in the z direction due to the N_{xy} shear loads only is

$$\left(N_{xy} + \frac{\partial N_{xy}}{\partial x}\delta x\right)\delta y\left(\frac{\partial w}{\partial y} + \frac{\partial^2 w}{\partial x\,\partial y}\delta x\right) - N_{xy}\delta y\frac{\partial w}{\partial y}$$

or

$$N_{xy}\frac{\partial^2 w}{\partial x\,\partial y}\delta x\,\delta y + \frac{\partial N_{xy}}{\partial x}\frac{\partial w}{\partial y}\delta x\,\delta y$$

neglecting terms of a lower order. Similarly, the contribution of N_{yx} is

$$N_{yx}\frac{\partial^2 w}{\partial x\,\partial y}\delta x\,\delta y + \frac{\partial N_{yx}}{\partial y}\frac{\partial w}{\partial x}\delta x\,\delta y$$

The components arising from the direct forces per unit length are readily obtained from Fig. 7.12, namely

$$\left(N_x + \frac{\partial N_x}{\partial x}\delta x\right)\delta y\left(\frac{\partial w}{\partial x} + \frac{\partial^2 w}{\partial x^2}\delta x\right) - N_x\delta y\frac{\partial w}{\partial x}$$

or

$$N_x\frac{\partial^2 w}{\partial x^2}\delta x\,\delta y + \frac{\partial N_x}{\partial x}\frac{\partial w}{\partial x}\delta x\,\delta y$$

and similarly

$$N_y\frac{\partial^2 w}{\partial y^2}\delta x\,\delta y + \frac{\partial N_y}{\partial y}\frac{\partial w}{\partial y}\delta x\,\delta y$$

The total force in the z direction is found from the summation of these expressions and is

$$N_x\frac{\partial^2 w}{\partial x^2}\delta x\,\delta y + \frac{\partial N_x}{\partial x}\frac{\partial w}{\partial x}\delta x\,\delta y + N_y\frac{\partial^2 w}{\partial y^2}\delta x\,\delta y + \frac{\partial N_y}{\partial y}\frac{\partial w}{\partial y}\delta x\,\delta y$$

$$+\frac{\partial N_{xy}}{\partial x}\frac{\partial w}{\partial y}\delta x\,\delta y + 2N_{xy}\frac{\partial^2 w}{\partial x\,\partial y}\delta x\,\delta y + \frac{\partial N_{xy}}{\partial y}\frac{\partial w}{\partial x}\delta x\,\delta y$$

in which N_{yx} is equal to and is replaced by N_{xy}. Using Eqs (7.31) and (7.32) we reduce this expression to

$$\left(N_x\frac{\partial^2 w}{\partial x^2} + N_y\frac{\partial^2 w}{\partial y^2} + 2N_{xy}\frac{\partial^2 w}{\partial x\,\partial y}\right)\delta x\,\delta y$$

Since the in-plane forces do not produce moments along the edges of the element then Eqs (7.17) and (7.18) remain unaffected. Further, Eq. (7.16) may be modified simply by the addition of the above vertical component of the in-plane loads to $q\delta x\delta y$. Therefore, the governing differential equation for a thin plate supporting transverse and in-plane loads is, from Eq. (7.20)

$$\frac{\partial^4 w}{\partial x^4} + 2\frac{\partial^4 w}{\partial x^2\,\partial y^2} + \frac{\partial^4 w}{\partial y^4} = \frac{1}{D}\left(q + N_x\frac{\partial^2 w}{\partial x^2} + N_y\frac{\partial^2 w}{\partial y^2} + 2N_{xy}\frac{\partial^2 w}{\partial x\,\partial y}\right) \tag{7.33}$$

Example 7.3

Determine the deflected form of the thin rectangular plate of Example 7.1 if, in addition to a uniformly distributed transverse load of intensity q_0, it supports an in-plane tensile force N_x per unit length.

The uniform transverse load may be expressed as a Fourier series (see Eq. (7.28) and Example 7.1), i.e.

$$q = \frac{16q_0}{\pi^2}\sum_{m=1,3,5}^{\infty}\sum_{n=1,3,5}^{\infty}\frac{1}{mn}\sin\frac{m\pi x}{a}\sin\frac{n\pi y}{b}$$

Equation (7.33) then becomes, on substituting for q

$$\frac{\partial^4 w}{\partial x^4} + 2\frac{\partial^4 w}{\partial x^2\,\partial y^2} + \frac{\partial^4 w}{\partial y^4} - \frac{N_x}{D}\frac{\partial^2 w}{\partial x^2} = \frac{16q_0}{\pi^2 D}\sum_{m=1,3,5}^{\infty}\sum_{n=1,3,5}^{\infty}\frac{1}{mn}\sin\frac{m\pi x}{a}\sin\frac{n\pi y}{b} \tag{i}$$

The appropriate boundary conditions are

$$w = \frac{\partial^2 w}{\partial x^2} = 0 \quad \text{at} \quad x = 0 \quad \text{and} \quad a$$

$$w = \frac{\partial^2 w}{\partial y^2} = 0 \quad \text{at} \quad y = 0 \quad \text{and} \quad b$$

These conditions may be satisfied by the assumption of a deflected form of the plate given by

$$w = \sum_{m=1}^{\infty}\sum_{n=1}^{\infty} A_{mn} \sin\frac{m\pi x}{a}\sin\frac{n\pi y}{b}$$

Substituting this expression into Eq. (i) gives

$$A_{mn} = \frac{16q_0}{\pi^6 Dmn\left[\left(\frac{m^2}{a^2}+\frac{n^2}{b^2}\right)^2 + \frac{N_x m^2}{\pi^2 D a^2}\right]} \quad \text{for odd } m \text{ and } n$$

$$A_{mn} = 0 \quad \text{for even } m \text{ and } n$$

Therefore

$$w = \frac{16q_0}{\pi^6 D}\sum_{m=1,3,5}^{\infty}\sum_{n=1,3,5}^{\infty}\frac{1}{mn\left[\left(\frac{m^2}{a^2}+\frac{n^2}{b^2}\right)^2 + \frac{N_x m^2}{\pi^2 D a^2}\right]}\sin\frac{m\pi x}{a}\sin\frac{n\pi y}{b} \tag{ii}$$

Comparing Eq. (ii) with Eq. (i) of Example 7.1 we see that, as a physical inspection would indicate, the presence of a tensile in-plane force decreases deflection. Conversely a compressive in-plane force would increase the deflection.

7.5 Bending of thin plates having a small initial curvature

Suppose that a thin plate has an initial curvature so that the deflection of any point in its middle plane is w_0. We assume that w_0 is small compared with the thickness of the plate. The application of transverse and in-plane loads will cause the plate to deflect a further amount w_1 so that the total deflection is then $w = w_0 + w_1$. However, in the derivation of Eq. (7.33) we note that the left-hand side was obtained from expressions for bending moments which themselves depend on the change of curvature. We therefore use the deflection w_1 on the left-hand side, not w. The effect on bending of the in-plane forces depends on the total deflection w so that we write Eq. (7.33)

$$\begin{aligned}\frac{\partial^4 w_1}{\partial x^4} + 2\frac{\partial^4 w_1}{\partial x^2 \partial y^2} + \frac{\partial^4 w_1}{\partial y^4}\\ = \frac{1}{D}\left[q + N_x\frac{\partial^2(w_0+w_1)}{\partial x^2} + N_y\frac{\partial^2(w_0+w_1)}{\partial y^2} + 2N_{xy}\frac{\partial^2(w_0+w_1)}{\partial x\,\partial y}\right]\end{aligned} \tag{7.34}$$

The effect of an initial curvature on deflection is therefore equivalent to the application of a transverse load of intensity

$$N_x\frac{\partial^2 w_0}{\partial x^2} + N_y\frac{\partial^2 w_0}{\partial y^2} + 2N_{xy}\frac{\partial^2 w_0}{\partial x\,\partial y}$$

Thus, in-plane loads alone produce bending provided there is an initial curvature.

Assuming that the initial form of the deflected plate is

$$w_0 = \sum_{m=1}^{\infty}\sum_{n=1}^{\infty} A_{mn} \sin\frac{m\pi x}{a} \sin\frac{n\pi y}{b} \tag{7.35}$$

then by substitution in Eq. (7.34) we find that if N_x is compressive and $N_y = N_{xy} = 0$

$$w_1 = \sum_{m=1}^{\infty}\sum_{n=1}^{\infty} B_{mn} \sin\frac{m\pi x}{a} \sin\frac{n\pi y}{b} \tag{7.36}$$

where

$$B_{mn} = \frac{A_{mn}N_x}{(\pi^2 D/a^2)[m + (n^2a^2/mb^2)]^2 - N_x}$$

We shall return to the consideration of initially curved plates in the discussion of the experimental determination of buckling loads of flat plates in Chapter 9.

7.6 Energy method for the bending of thin plates

Two types of solution are obtainable for thin plate bending problems by the application of the principle of the stationary value of the total potential energy of the plate and its external loading. The first, in which the form of the deflected shape of the plate is known, produces an exact solution; the second, the *Rayleigh–Ritz* method, assumes an approximate deflected shape in the form of a series having a finite number of terms chosen to satisfy the boundary conditions of the problem and also to give the kind of deflection pattern expected.

In Chapter 5 we saw that the total potential energy of a structural system comprised the internal or strain energy of the structural member, plus the potential energy of the applied loading. We now proceed to derive expressions for these quantities for the loading cases considered in the preceding sections.

7.6.1 Strain energy produced by bending and twisting

In thin plate analysis we are concerned with deflections normal to the loaded surface of the plate. These, as in the case of slender beams, are assumed to be primarily due to bending action so that the effects of shear strain and shortening or stretching of the middle plane of the plate are ignored. Therefore, it is sufficient for us to calculate the strain energy produced by bending and twisting only as this will be applicable, for the reason of the above assumption, to all loading cases. It must be remembered that we are only neglecting the contributions of shear and direct *strains* on the deflection of the plate; the stresses producing them must not be ignored.

Consider the element $\delta x \times \delta y$ of a thin plate $a \times b$ shown in elevation in the xz plane in Fig. 7.14(a). Bending moments M_x per unit length applied to its δy edge produce

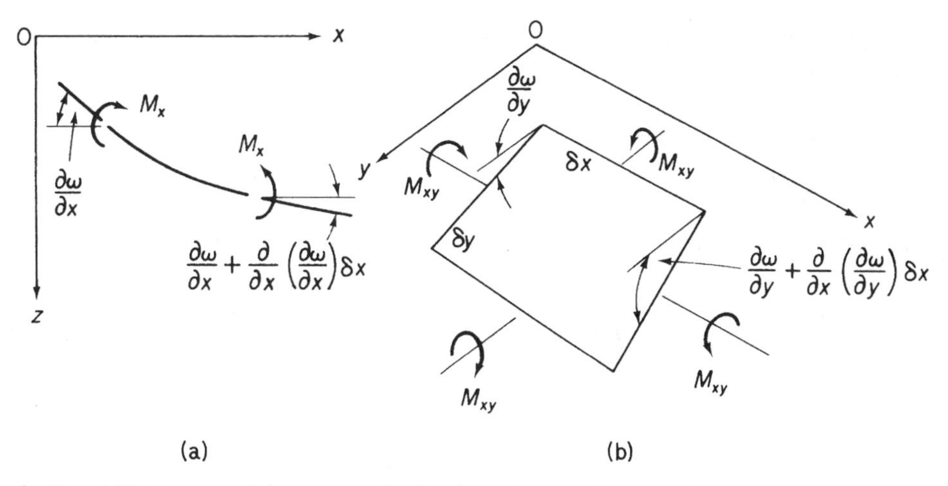

Fig. 7.14 (a) Strain energy of element due to bending; (b) strain energy due to twisting.

a change in slope between its ends equal to $(\partial^2 w/\partial x^2)\delta x$. However, since we regard the moments M_x as positive in the sense shown, then this change in slope, or relative rotation, of the ends of the element is negative as the slope decreases with increasing x. The bending strain energy due to M_x is then

$$\frac{1}{2}M_x\delta y\left(-\frac{\partial^2 w}{\partial x^2}\delta x\right)$$

Similarly, in the yz plane the contribution of M_y to the bending strain energy is

$$\frac{1}{2}M_y\delta x\left(-\frac{\partial^2 w}{\partial y^2}\delta y\right)$$

The strain energy due to the twisting moment per unit length, M_{xy}, applied to the δy edges of the element, is obtained from Fig. 7.14(b). The relative rotation of the δy edges is $(\partial^2 w/\partial x\partial y)\delta x$ so that the corresponding strain energy is

$$\frac{1}{2}M_{xy}\delta y\frac{\partial^2 w}{\partial x\,\partial y}\delta x$$

Finally, the contribution of the twisting moment M_{xy} on the δx edges is, in a similar fashion

$$\frac{1}{2}M_{xy}\delta x\frac{\partial^2 w}{\partial x\,\partial y}\delta y$$

The total strain energy of the element from bending and twisting is thus

$$\frac{1}{2}\left(-M_x\frac{\partial^2 w}{\partial x^2}-M_y\frac{\partial^2 w}{\partial y^2}+2M_{xy}\frac{\partial^2 w}{\partial x\,\partial y}\right)\delta x\delta y$$

Substitution for M_x, M_y and M_{xy} from Eqs (7.7), (7.8) and (7.14) gives the total strain energy of the element as

$$\frac{D}{2}\left[\left(\frac{\partial^2 w}{\partial x^2}\right)^2+\left(\frac{\partial^2 w}{\partial y^2}\right)^2+2\nu\frac{\partial^2 w}{\partial x^2}\frac{\partial^2 w}{\partial y^2}+2(1-\nu)\left(\frac{\partial^2 w}{\partial x\,\partial y}\right)^2\right]\delta x\,\delta y$$

which on rearranging becomes

$$\frac{D}{2}\left\{\left(\frac{\partial^2 w}{\partial x^2}+\frac{\partial^2 w}{\partial y^2}\right)^2-2(1-\nu)\left[\frac{\partial^2 w}{\partial x^2}\frac{\partial^2 w}{\partial y^2}-\left(\frac{\partial^2 w}{\partial x\,\partial y}\right)^2\right]\right\}\delta x\,\delta y$$

Hence the total strain energy U of the rectangular plate $a \times b$ is

$$U=\frac{D}{2}\int_0^a\int_0^b\left\{\left(\frac{\partial^2 w}{\partial x^2}+\frac{\partial^2 w}{\partial y^2}\right)^2-2(1-\nu)\left[\frac{\partial^2 w}{\partial x^2}\frac{\partial^2 w}{\partial y^2}-\left(\frac{\partial^2 w}{\partial x\,\partial y}\right)^2\right]\right\}\mathrm{d}x\,\mathrm{d}y \tag{7.37}$$

Note that if the plate is subject to pure bending only, then $M_{xy}=0$ and from Eq. (7.14) $\partial^2 w/\partial x\partial y=0$, so that Eq. (7.37) simplifies to

$$U=\frac{D}{2}\int_0^a\int_0^b\left[\left(\frac{\partial^2 w}{\partial x^2}\right)^2+\left(\frac{\partial^2 w}{\partial y^2}\right)^2+2\nu\frac{\partial^2 w}{\partial x^2}\frac{\partial^2 w}{\partial y^2}\right]\mathrm{d}x\,\mathrm{d}y \tag{7.38}$$

7.6.2 Potential energy of a transverse load

An element $\delta x \times \delta y$ of the transversely loaded plate of Fig. 7.8 supports a load $q\delta x\delta y$. If the displacement of the element normal to the plate is w then the potential energy δV of the load on the element referred to the undeflected plate position is

$$\delta V=-wq\delta x\,\delta y \qquad \text{(See Section 5.7)}$$

Therefore, the potential energy V of the total load on the plate is given by

$$V=-\int_0^a\int_0^b wq\,\mathrm{d}x\,\mathrm{d}y \tag{7.39}$$

7.6.3 Potential energy of in-plane loads

We may consider each load N_x, N_y and N_{xy} in turn, then use the principle of superposition to determine the potential energy of the loading system when they act simultaneously. Consider an elemental strip of width δy along the length a of the plate in Fig. 7.15(a). The compressive load on this strip is $N_x\delta y$ and due to the bending of the plate the horizontal length of the strip decreases by an amount λ, as shown in

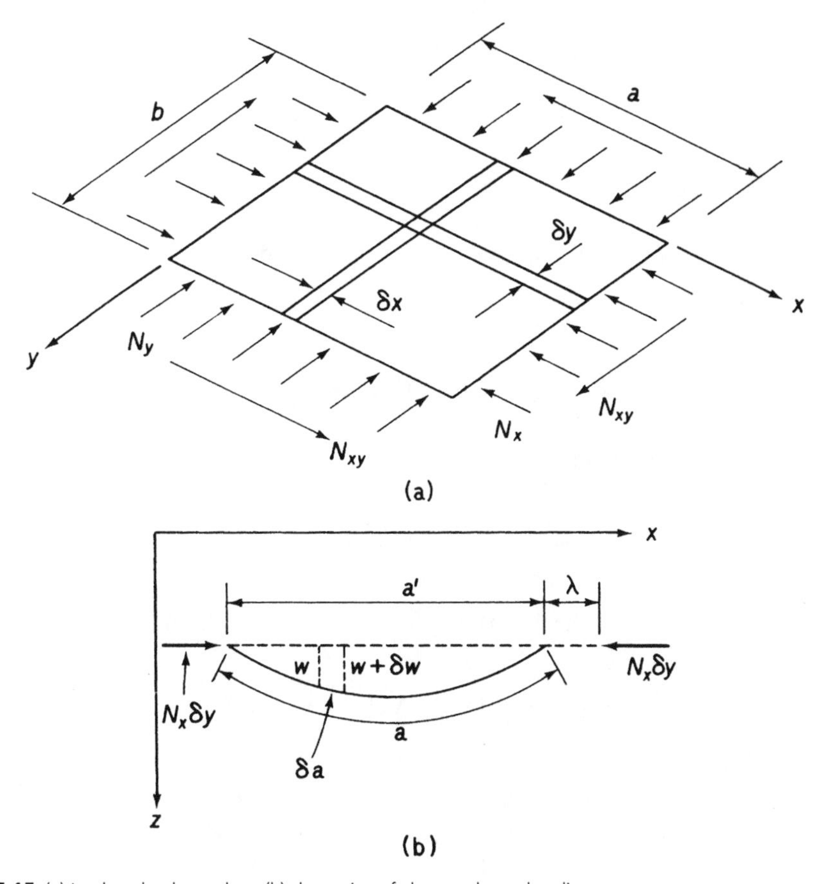

Fig. 7.15 (a) In-plane loads on plate; (b) shortening of element due to bending.

Fig. 7.15(b). The potential energy δV_x of the load $N_x\delta y$, referred to the undeflected position of the plate as the datum, is then

$$\delta V_x = -N_x\lambda\delta y \tag{7.40}$$

From Fig. 7.15(b) the length of a small element δa of the strip is

$$\delta a = (\delta x^2 + \delta w^2)^{\frac{1}{2}}$$

and since $\partial w/\partial x$ is small then

$$\delta a \approx \delta x\left[1 + \frac{1}{2}\left(\frac{\partial w}{\partial x}\right)^2\right]$$

Hence

$$a = \int_0^{a'}\left[1 + \frac{1}{2}\left(\frac{\partial w}{\partial x}\right)^2\right]\mathrm{d}x$$

giving

$$a = a' + \int_0^{a'} \frac{1}{2}\left(\frac{\partial w}{\partial x}\right)^2 \mathrm{d}x$$

and

$$\lambda = a - a' = \int_0^{a'} \frac{1}{2}\left(\frac{\partial w}{\partial x}\right)^2 \mathrm{d}x$$

Since

$$\int_0^{a'} \frac{1}{2}\left(\frac{\partial w}{\partial x}\right)^2 \mathrm{d}x \quad \text{only differs from} \quad \int_0^{a} \frac{1}{2}\left(\frac{\partial w}{\partial x}\right)^2 \mathrm{d}x$$

by a term of negligible order we write

$$\lambda = \int_0^{a} \frac{1}{2}\left(\frac{\partial w}{\partial x}\right)^2 \mathrm{d}x \tag{7.41}$$

The potential energy V_x of the N_x loading follows from Eqs (7.40) and (7.41), thus

$$V_x = -\frac{1}{2}\int_0^a \int_0^b N_x \left(\frac{\partial w}{\partial x}\right)^2 \mathrm{d}x\,\mathrm{d}y \tag{7.42}$$

Similarly

$$V_y = -\frac{1}{2}\int_0^a \int_0^b N_y \left(\frac{\partial w}{\partial y}\right)^2 \mathrm{d}x\,\mathrm{d}y \tag{7.43}$$

The potential energy of the in-plane shear load N_{xy} may be found by considering the work done by N_{xy} during the shear distortion corresponding to the deflection w of an element. This shear strain is the reduction in the right angle C_2AB_1 to the angle C_1AB_1 of the element in Fig. 7.16 or, rotating C_2A with respect to AB_1 to AD in the plane C_1AB_1, the angle DAC_1. The displacement C_2D is equal to $(\partial w/\partial y)\delta y$ and the angle DC_2C_1 is $\partial w/\partial x$. Thus C_1D is equal to

$$\frac{\partial w}{\partial x}\frac{\partial w}{\partial y}\delta y$$

and the angle DAC_1 representing the shear strain corresponding to the bending displacement w is

$$\frac{\partial w}{\partial x}\frac{\partial w}{\partial y}$$

so that the work done on the element by the shear force $N_{xy}\delta x$ is

$$\frac{1}{2}N_{xy}\delta x\frac{\partial w}{\partial x}\frac{\partial w}{\partial y}$$

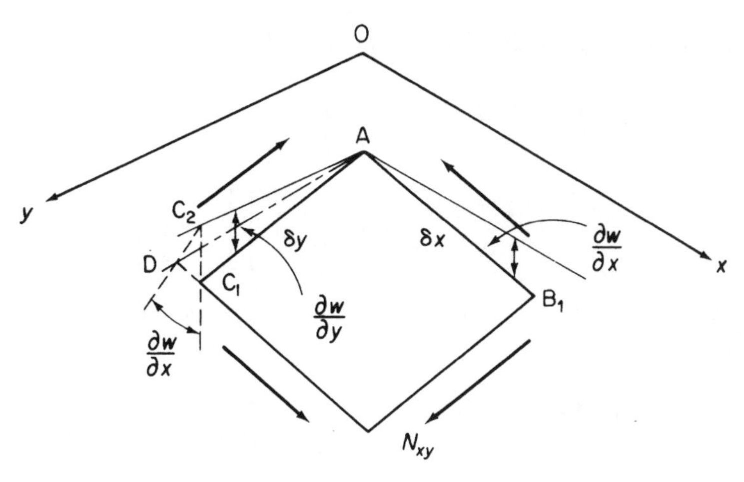

Fig. 7.16 Calculation of shear strain corresponding to bending deflection.

Similarly, the work done by the shear force $N_{xy}\delta y$ is

$$\frac{1}{2}N_{xy}\delta y\frac{\partial w}{\partial x}\frac{\partial w}{\partial y}$$

and the total work done taken over the complete plate is

$$\frac{1}{2}\int_0^a\int_0^b 2N_{xy}\frac{\partial w}{\partial x}\frac{\partial w}{\partial y}\mathrm{d}x\,\mathrm{d}y$$

It follows immediately that the potential energy of the N_{xy} loads is

$$V_{xy} = -\frac{1}{2}\int_0^a\int_0^b 2N_{xy}\frac{\partial w}{\partial x}\frac{\partial w}{\partial y}\mathrm{d}x\,\mathrm{d}y \tag{7.44}$$

and for the complete in-plane loading system we have, from Eqs (7.42), (7.43) and (7.44), a potential energy of

$$V = -\frac{1}{2}\int_0^a\int_0^b\left[N_x\left(\frac{\partial w}{\partial x}\right)^2 + N_y\left(\frac{\partial w}{\partial y}\right)^2 + 2N_{xy}\frac{\partial w}{\partial x}\frac{\partial w}{\partial y}\right]\mathrm{d}x\,\mathrm{d}y \tag{7.45}$$

We are now in a position to solve a wide range of thin plate problems provided that the deflections are small, obtaining exact solutions if the deflected form is known or approximate solutions if the deflected shape has to be 'guessed'.

Considering the rectangular plate of Section 7.3, simply supported along all four edges and subjected to a uniformly distributed transverse load of intensity q_0, we know that its deflected shape is given by Eq. (7.27), namely

$$w = \sum_{m=1}^{\infty}\sum_{n=1}^{\infty}A_{mn}\sin\frac{m\pi x}{a}\sin\frac{n\pi y}{b}$$

The total potential energy of the plate is, from Eqs (7.37) and (7.39)

$$U + V = \int_0^a \int_0^b \left\{ \frac{D}{2} \left[\left(\frac{\partial^2 w}{\partial x^2} + \frac{\partial^2 w}{\partial y^2} \right)^2 \right.\right.$$
$$\left.\left. -2(1-\nu) \left\{ \frac{\partial^2 w}{\partial x^2} \frac{\partial^2 w}{\partial y^2} - \left(\frac{\partial^2 w}{\partial x \, \partial y} \right)^2 \right\} \right] - wq_0 \right\} \mathrm{d}x \, \mathrm{d}y \tag{7.46}$$

Substituting in Eq. (7.46) for w and realizing that 'cross-product' terms integrate to zero, we have

$$U + V = \int_0^a \int_0^b \left\{ \frac{D}{2} \sum_{m=1}^{\infty} \sum_{n=1}^{\infty} A_{mn}^2 \left[\pi^4 \left(\frac{m^2}{a^2} + \frac{n^2}{b^2} \right)^2 \sin^2 \frac{m\pi x}{a} \sin^2 \frac{n\pi y}{b} \right.\right.$$
$$\left. - 2(1-\nu) \frac{m^2 n^2 \pi^4}{a^2 b^2} \left(\sin^2 \frac{m\pi x}{a} \sin^2 \frac{n\pi y}{b} - \cos^2 \frac{m\pi x}{a} \cos^2 \frac{n\pi y}{b} \right) \right]$$
$$\left. - q_0 \sum_{m=1}^{\infty} \sum_{n=1}^{\infty} A_{mn} \sin \frac{m\pi x}{a} \sin \frac{n\pi y}{b} \right\} \mathrm{d}x \, \mathrm{d}y$$

The term multiplied by $2(1-\nu)$ integrates to zero and the mean value of $\sin^2$ or $\cos^2$ over a complete number of half waves is $\frac{1}{2}$, thus integration of the above expression yields

$$U + V = \frac{D}{2} \sum_{m=1,3,5}^{\infty} \sum_{n=1,3,5}^{\infty} A_{mn}^2 \frac{\pi^4 ab}{4} \left(\frac{m^2}{a^2} + \frac{n^2}{b^2} \right)^2 - q_0 \sum_{m=1,3,5}^{\infty} \sum_{n=1,3,5}^{\infty} A_{mn} \frac{4ab}{\pi^2 mn} \tag{7.47}$$

From the principle of the stationary value of the total potential energy we have

$$\frac{\partial (U+V)}{\partial A_{mn}} = \frac{D}{2} 2A_{mn} \frac{\pi^4 ab}{4} \left(\frac{m^2}{a^2} + \frac{n^2}{b^2} \right)^2 - q_0 \frac{4ab}{\pi^2 mn} = 0$$

so that

$$A_{mn} = \frac{16 q_0}{\pi^6 D mn[(m^2/a^2) + (n^2/b^2)]^2}$$

giving a deflected form

$$w = \frac{16 q_0}{\pi^6 D} \sum_{m=1,3,5}^{\infty} \sum_{n=1,3,5}^{\infty} \frac{\sin(m\pi x/a) \sin(n\pi y/b)}{mn[(m^2/a^2) + (n^2/b^2)]^2}$$

which is the result obtained in Eq. (i) of Example 7.1.

The above solution is exact since we know the true deflected shape of the plate in the form of an infinite series for w. Frequently, the appropriate infinite series is not known so that only an approximate solution may be obtained. The method of solution, known as the *Rayleigh–Ritz* method, involves the selection of a series for w containing a finite number of functions of x and y. These functions are chosen to satisfy the boundary conditions of the problem as far as possible and also to give the type of deflection pattern expected. Naturally, the more representative the 'guessed' functions are the more accurate the solution becomes.

Suppose that the 'guessed' series for w in a particular problem contains three different functions of x and y. Thus

$$w = A_1 f_1(x, y) + A_2 f_2(x, y) + A_3 f_3(x, y)$$

where A_1, A_2 and A_3 are unknown coefficients. We now substitute for w in the appropriate expression for the total potential energy of the system and assign stationary values with respect to A_1, A_2 and A_3 in turn. Thus

$$\frac{\partial(U+V)}{\partial A_1} = 0 \qquad \frac{\partial(U+V)}{\partial A_2} = 0 \qquad \frac{\partial(U+V)}{\partial A_3} = 0$$

giving three equations which are solved for A_1, A_2 and A_3.

Example 7.4

A rectangular plate $a \times b$, is simply supported along each edge and carries a uniformly distributed load of intensity q_0. Assuming a deflected shape given by

$$w = A_{11} \sin\frac{\pi x}{a} \sin\frac{\pi y}{b}$$

determine the value of the coefficient A_{11} and hence find the maximum value of deflection.

The expression satisfies the boundary conditions of zero deflection and zero curvature (i.e. zero bending moment) along each edge of the plate. Substituting for w in Eq. (7.46) we have

$$U + V = \int_0^a \int_0^b \left[\frac{DA_{11}^2}{2} \left\{ \frac{\pi^4}{(a^2b^2)^2}(a^2+b^2)^2 \sin^2\frac{\pi x}{a} \sin^2\frac{\pi y}{b} - 2(1-\nu) \right.\right.$$
$$\left.\left. \times \left[\frac{\pi^4}{a^2b^2} \sin^2\frac{\pi x}{a} \sin^2\frac{\pi y}{b} - \frac{\pi^4}{a^2b^2} \cos^2\frac{\pi x}{a} \cos^2\frac{\pi y}{b} \right] \right\} \right.$$
$$\left. - q_0 A_{11} \sin\frac{\pi x}{a} \sin\frac{\pi y}{b} \right] \mathrm{d}x\,\mathrm{d}y$$

whence

$$U + V = \frac{DA_{11}^2}{2} \frac{\pi^4}{4a^3b^3}(a^2+b^2)^2 - q_0 A_{11} \frac{4ab}{\pi^2}$$

so that

$$\frac{\partial(U+V)}{\partial A_{11}} = \frac{DA_{11}\pi^4}{4a^3b^3}(a^2+b^2)^2 - q_0 \frac{4ab}{\pi^2} = 0$$

and

$$A_{11} = \frac{16q_0a^4b^4}{\pi^6D(a^2+b^2)^2}$$

giving

$$w = \frac{16q_0a^4b^4}{\pi^6D(a^2+b^2)^2} \sin\frac{\pi x}{a} \sin\frac{\pi y}{b}$$

At the centre of the plate w is a maximum and

$$w_{\text{max}} = \frac{16q_0a^4b^4}{\pi^6D(a^2+b^2)^2}$$

For a square plate and assuming $\nu = 0.3$

$$w_{\text{max}} = 0.0455q_0\frac{a^4}{Et^3}$$

which compares favourably with the result of Example 7.1.

In this chapter we have dealt exclusively with small deflections of thin plates. For a plate subjected to large deflections the middle plane will be stretched due to *bending* so that Eq. (7.33) requires modification. The relevant theory is outside the scope of this book but may be found in a variety of references.

References

1 Jaeger, J. C., *Elementary Theory of Elastic Plates*, Pergamon Press, New York, 1964.
2 Timoshenko, S. P. and Woinowsky-Krieger, S., *Theory of Plates and Shells*, 2nd edition, McGraw-Hill Book Company, New York, 1959.
3 Timoshenko, S. P. and Gere, J. M., *Theory of Elastic Stability*, 2nd edition, McGraw-Hill Book Company, New York, 1961.
4 Wang, Chi-Teh, *Applied Elasticity*, McGraw-Hill Book Company, New York, 1953.

Problems

P.7.1 A plate 10 mm thick is subjected to bending moments M_x equal to 10 Nm/mm and M_y equal to 5 Nm/mm. Calculate the maximum direct stresses in the plate.

Ans. $\sigma_{x,\text{max}} = \pm 600\,\text{N/mm}^2$, $\sigma_{y,\text{max}} = \pm 300\,\text{N/mm}^2$.

P.7.2 For the plate and loading of problem P.7.1 find the maximum twisting moment per unit length in the plate and the direction of the planes on which this occurs.

Ans. 2.5 N m/mm at 45° to the x and y axes.

P.7.3 The plate of the previous two problems is subjected to a twisting moment of 5 Nm/mm along each edge, in addition to the bending moments of $M_x = 10$ N m/mm and $M_y = 5$ N m/mm. Determine the principal moments in the plate, the planes on which they act and the corresponding principal stresses.

Ans. 13.1 N m/mm, 1.9 N m/mm, $\alpha = -31.7°$, $\alpha = +58.3°$, ± 786 N/mm^2, ± 114 N/mm^2.

P.7.4 A thin rectangular plate of length a and width $2a$ is simply supported along the edges $x = 0$, $x = a$, $y = -a$ and $y = +a$. The plate has a flexural rigidity D, a Poisson's ratio of 0.3 and carries a load distribution given by $q(x, y) = q_0 \sin(\pi x/a)$. If the deflection of the plate may be represented by the expression

$$w = \frac{qa^4}{D\pi^4}\left(1 + A\cosh\frac{\pi y}{a} + B\frac{\pi y}{a}\sinh\frac{\pi y}{a}\right)\sin\frac{\pi x}{a}$$

determine the values of the constants A and B.

Ans. $A = -0.2213$, $B = 0.0431$.

P.7.5 A thin, elastic square plate of side a is simply supported on all four sides and supports a uniformly distributed load q. If the origin of axes coincides with the centre of the plate show that the deflection of the plate can be represented by the expression

$$w = \frac{q}{96(1-\nu)D}[2(x^4 + y^4) - 3a^2(1-\nu)(x^2 + y^2) - 12\nu x^2 y^2 + A]$$

where D is the flexural rigidity, ν is Poisson's ratio and A is a constant. Calculate the value of A and hence the central deflection of the plate.

Ans. $A = a^4(5 - 3\nu)/4$, Cen. def. $= qa^4(5 - 3\nu)/384D(1-\nu)$

P.7.6 The deflection of a square plate of side a which supports a lateral load represented by the function $q(x, y)$ is given by

$$w(x, y) = w_0 \cos\frac{\pi x}{a}\cos\frac{3\pi y}{a}$$

where x and y are referred to axes whose origin coincides with the centre of the plate and w_0 is the deflection at the centre.

If the flexural rigidity of the plate is D and Poisson's ratio is ν determine the loading function q, the support conditions of the plate, the reactions at the plate corners and the bending moments at the centre of the plate.

Ans. $q(x, y) = w_0 D 100\dfrac{\pi^4}{a^4}\cos\dfrac{\pi x}{a}\cos\dfrac{3\pi y}{a}$

The plate is simply supported on all edges.

Reactions: $-6w_0 D\left(\dfrac{\pi}{a}\right)^2(1-\nu)$

$M_x = w_0 D\left(\dfrac{\pi}{a}\right)^2(1 + 9\nu)$, $M_y = w_0 D\left(\dfrac{\pi}{a}\right)^2(9 + \nu)$.

P.7.7 A simply supported square plate $a \times a$ carries a distributed load according to the formula

$$q(x, y) = q_0\frac{x}{a}$$

where q_0 is its intensity at the edge $x = a$. Determine the deflected shape of the plate.

Ans. $$w = \frac{8q_0a^4}{\pi^6 D} \sum_{m=1,2,3}^{\infty} \sum_{n=1,3,5}^{\infty} \frac{(-1)^{m+1}}{mn(m^2+n^2)^2} \sin\frac{m\pi x}{a} \sin\frac{n\pi y}{a}$$

P.7.8 An elliptic plate of major and minor axes $2a$ and $2b$ and of small thickness t is clamped along its boundary and is subjected to a uniform pressure difference p between the two faces. Show that the usual differential equation for normal displacements of a thin flat plate subject to lateral loading is satisfied by the solution

$$w = w_0\left(1 - \frac{x^2}{a^2} - \frac{y^2}{b^2}\right)^2$$

where w_0 is the deflection at the centre which is taken as the origin.

Determine w_0 in terms of p and the relevant material properties of the plate and hence expressions for the greatest stresses due to bending at the centre and at the ends of the minor axis.

Ans. $$w_0 = \frac{3p(1-\nu^2)}{2Et^3\left(\dfrac{3}{a^4} + \dfrac{2}{a^2b^2} + \dfrac{3}{b^4}\right)}$$

Centre, $$\sigma_{x,\max} = \frac{\pm 3pa^2b^2(b^2 + \nu a^2)}{t^2(3b^4 + 2a^2b^2 + 3a^4)}, \quad \sigma_{y,\max} = \frac{\pm 3pa^2b^2(a^2 + \nu b^2)}{t^2(3b^4 + 2a^2b^2 + 3a^4)}$$

Ends of minor axis

$$\sigma_{x,\max} = \frac{\pm 6pa^4b^2}{t^2(3b^4 + 2a^2b^2 + 3a^4)}, \quad \sigma_{y,\max} = \frac{\pm 6pb^4a^2}{t^2(3b^4 + 2a^2b^2 + 3a^4)}$$

P.7.9 Use the energy method to determine the deflected shape of a rectangular plate $a \times b$, simply supported along each edge and carrying a concentrated load W at a position (ξ, η) referred to axes through a corner of the plate. The deflected shape of the plate can be represented by the series

$$w = \sum_{m=1}^{\infty} \sum_{n=1}^{\infty} A_{mn} \sin\frac{m\pi x}{a} \sin\frac{n\pi y}{b}$$

Ans. $$A_{mn} = \frac{4W \sin\dfrac{m\pi\xi}{a} \sin\dfrac{n\pi\eta}{b}}{\pi^4 Dab[(m^2/a^2) + (n^2/b^2)]^2}$$

P.7.10 If, in addition to the point load W, the plate of problem P.7.9 supports an in-plane compressive load of N_x per unit length on the edges $x=0$ and $x=a$, calculate the resulting deflected shape.

$$Ans. \quad A_{mn} = \frac{4W \sin \dfrac{m\pi\xi}{a} \sin \dfrac{n\pi\eta}{b}}{abD\pi^4 \left[\left(\dfrac{m^2}{a^2} + \dfrac{n^2}{b^2} \right)^2 - \dfrac{m^2 N_x}{\pi^2 a^2 D} \right]}$$

P.7.11 A square plate of side a is simply supported along all four sides and is subjected to a transverse uniformly distributed load of intensity q_0. It is proposed to determine the deflected shape of the plate by the Rayleigh–Ritz method employing a 'guessed' form for the deflection of

$$w = A_{11} \left(1 - \frac{4x^2}{a^2} \right) \left(1 - \frac{4y^2}{a^2} \right)$$

in which the origin is taken at the centre of the plate.

Comment on the degree to which the boundary conditions are satisfied and find the central deflection assuming $\nu = 0.3$.

$$Ans. \quad \frac{0.0389 q_0 a^4}{Et^3}$$

P.7.12 A rectangular plate $a \times b$, simply supported along each edge, possesses a small initial curvature in its unloaded state given by

$$w_0 = A_{11} \sin \frac{\pi x}{a} \sin \frac{\pi y}{b}$$

Determine, using the energy method, its final deflected shape when it is subjected to a compressive load N_x per unit length along the edges $x=0$, $x=a$.

$$Ans. \quad w = \frac{A_{11}}{\left[1 - \dfrac{N_x a^2}{\pi^2 D} \Big/ \left(1 + \dfrac{a^2}{b^2} \right)^2 \right]} \sin \frac{\pi x}{a} \sin \frac{\pi y}{b}$$

SECTION A4 STRUCTURAL INSTABILITY

8

Columns

A large proportion of an aircraft's structure comprises thin webs stiffened by slender longerons or stringers. Both are susceptible to failure by buckling at a buckling stress or critical stress, which is frequently below the limit of proportionality and seldom appreciably above the yield stress of the material. Clearly, for this type of structure, buckling is the most critical mode of failure so that the prediction of buckling loads of columns, thin plates and stiffened panels is extremely important in aircraft design. In this chapter we consider the buckling failure of all these structural elements and also the flexural–torsional failure of thin-walled open tubes of low torsional rigidity.

Two types of structural instability arise: *primary* and *secondary*. The former involves the complete element, there being no change in cross-sectional area while the wavelength of the buckle is of the same order as the length of the element. Generally, solid and thick-walled columns experience this type of failure. In the latter mode, changes in cross-sectional area occur and the wavelength of the buckle is of the order of the cross-sectional dimensions of the element. Thin-walled columns and stiffened plates may fail in this manner.

8.1 Euler buckling of columns

The first significant contribution to the theory of the buckling of columns was made as early as 1744 by Euler. His classical approach is still valid, and likely to remain so, for slender columns possessing a variety of end restraints. Our initial discussion is therefore a presentation of the Euler theory for the small elastic deflection of perfect columns. However, we investigate first the nature of buckling and the difference between theory and practice.

It is common experience that if an increasing axial compressive load is applied to a slender column there is a value of the load at which the column will suddenly bow or buckle in some unpredetermined direction. This load is patently the buckling load of the column or something very close to the buckling load. Clearly this displacement implies a degree of asymmetry in the plane of the buckle caused by geometrical and/or material imperfections of the column and its load. However, in our theoretical stipulation of a perfect column in which the load is applied precisely along the perfectly straight centroidal axis, there is perfect symmetry so that, theoretically, there can be no sudden bowing or buckling. We therefore require a precise definition of buckling load which may be used in our analysis of the perfect column.

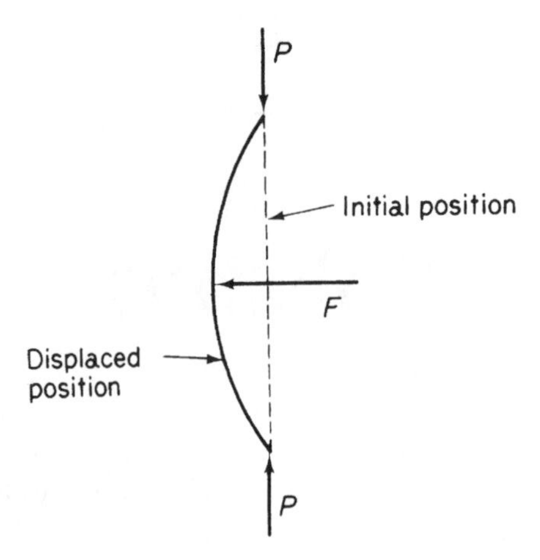

Fig. 8.1 Definition of buckling load for a perfect column.

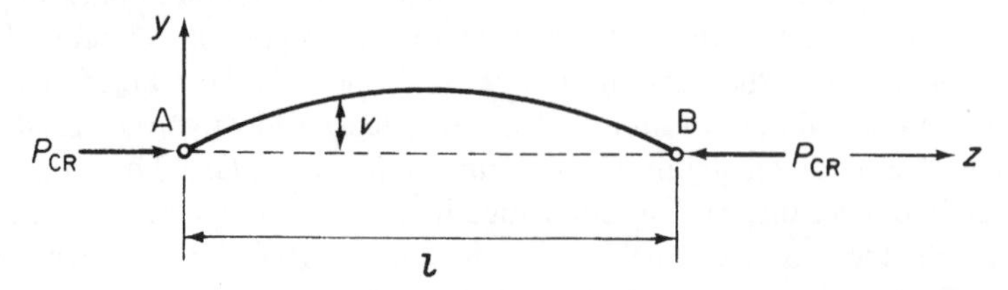

Fig. 8.2 Determination of buckling load for a pin-ended column.

If the perfect column of Fig. 8.1 is subjected to a compressive load P, only shortening of the column occurs no matter what the value of P. However, if the column is displaced a small amount by a lateral load F then, at values of P below the critical or buckling load, P_{CR}, removal of F results in a return of the column to its undisturbed position, indicating a state of stable equilibrium. At the critical load the displacement does not disappear and, in fact, the column will remain in *any* displaced position as long as the displacement is small. Thus, the buckling load P_{CR} is associated with a state of *neutral equilibrium.* For $P > P_{CR}$ enforced lateral displacements increase and the column is unstable.

Consider the pin-ended column AB of Fig. 8.2. We assume that it is in the displaced state of neutral equilibrium associated with buckling so that the compressive load P has attained the critical value P_{CR}. Simple bending theory (see Chapter 16) gives

$$EI\frac{d^2v}{dz^2} = -M$$

or

$$EI\frac{d^2v}{dz^2} = -P_{CR}v \tag{8.1}$$

so that the differential equation of bending of the column is

$$\frac{\mathrm{d}^2 v}{\mathrm{d}z^2} + \frac{P_{\mathrm{CR}}}{EI} v = 0 \tag{8.2}$$

The well-known solution of Eq. (8.2) is

$$v = A\cos\mu z + B\sin\mu z \tag{8.3}$$

where $\mu^2 = P_{\mathrm{CR}}/EI$ and A and B are unknown constants. The boundary conditions for this particular case are $v=0$ at $z=0$ and l. Thus $A=0$ and

$$B\sin\mu l = 0$$

For a non-trivial solution (i.e. $v \neq 0$) then

$$\sin\mu l = 0 \quad \text{or} \quad \mu l = n\pi \quad \text{where } n = 1, 2, 3, \ldots$$

giving

$$\frac{P_{\mathrm{CR}} l^2}{EI} = n^2\pi^2$$

or

$$P_{\mathrm{CR}} = \frac{n^2\pi^2 EI}{l^2} \tag{8.4}$$

Note that Eq. (8.3) cannot be solved for v no matter how many of the available boundary conditions are inserted. This is to be expected since the neutral state of equilibrium means that v is indeterminate.

The smallest value of buckling load, in other words the smallest value of P which can maintain the column in a neutral equilibrium state, is obtained by substituting $n=1$ in Eq. (8.4). Hence

$$P_{\mathrm{CR}} = \frac{\pi^2 EI}{l^2} \tag{8.5}$$

Other values of P_{CR} corresponding to $n = 2, 3, \ldots,$ are

$$P_{\mathrm{CR}} = \frac{4\pi^2 EI}{l^2}, \frac{9\pi^2 EI}{l^2}, \ldots$$

These higher values of buckling load cause more complex modes of buckling such as those shown in Fig. 8.3. The different shapes may be produced by applying external restraints to a very slender column at the points of contraflexure to prevent lateral movement. If no restraints are provided then these forms of buckling are unstable and have little practical meaning.

The critical stress, σ_{CR}, corresponding to P_{CR}, is, from Eq. (8.5)

$$\sigma_{\mathrm{CR}} = \frac{\pi^2 E}{(l/r)^2} \tag{8.6}$$

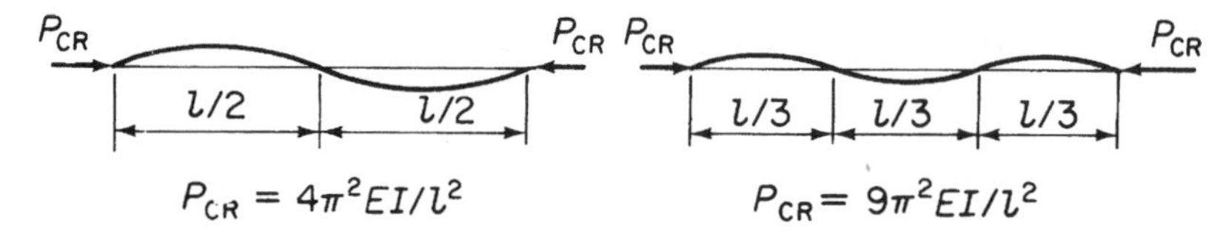

Fig. 8.3 Buckling loads for different buckling modes of a pin-ended column.

Table 8.1

Ends	l_e/l	Boundary conditions
Both pinned	1.0	$v=0$ at $z=0$ and l
Both fixed	0.5	$v=0$ at $z=0$ and $z=l$, $dv/dz=0$ at $z=l$
One fixed, the other free	2.0	$v=0$ and $dv/dz=0$ at $z=0$
One fixed, the other pinned	0.6998	$dv/dz=0$ at $z=0$, $v=0$ at $z=l$ and $z=0$

where r is the radius of gyration of the cross-sectional area of the column. The term l/r is known as the *slenderness ratio* of the column. For a column that is not doubly symmetrical, r is the least radius of gyration of the cross-section since the column will bend about an axis about which the flexural rigidity EI is least. Alternatively, if buckling is prevented in all but one plane then EI is the flexural rigidity in that plane.

Equations (8.5) and (8.6) may be written in the form

$$P_{CR} = \frac{\pi^2 EI}{l_e^2} \tag{8.7}$$

and

$$\sigma_{CR} = \frac{\pi^2 E}{(l_e/r)^2} \tag{8.8}$$

where l_e is the *effective length* of the column. This is the length of a *pin-ended column* that would have the same critical load as that of a column of length l, but with different end conditions. The determination of critical load and stress is carried out in an identical manner to that for the pin-ended column except that the boundary conditions are different in each case. Table 8.1 gives the solution in terms of effective length for columns having a variety of end conditions. In addition, the boundary conditions referred to the coordinate axes of Fig. 8.2 are quoted. The last case in Table 8.1 involves the solution of a transcendental equation; this is most readily accomplished by a graphical method.

Let us now examine the buckling of the perfect pin-ended column of Fig. 8.2 in greater detail. We have shown, in Eq. (8.4), that the column will buckle at *discrete* values of axial load and that associated with each value of buckling load there is a particular buckling mode (Fig. 8.3). These discrete values of buckling load are called *eigenvalues*, their associated functions (in this case $v = B \sin n\pi z/l$) are called *eigenfunctions* and the problem itself is called an *eigenvalue problem.*

Further, suppose that the lateral load F in Fig. 8.1 is removed. Since the column is perfectly straight, homogeneous and loaded exactly along its axis, it will suffer only axial compression as P is increased. This situation, theoretically, would continue until yielding of the material of the column occurred. However, as we have seen,

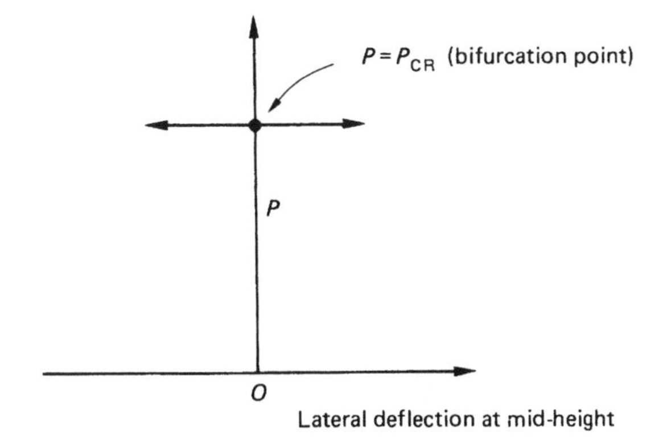

Fig. 8.4 Behaviour of a perfect pin-ended column.

for values of P below P_{CR} the column is in stable equilibrium whereas for $P > P_{CR}$ the column is unstable. A plot of load against lateral deflection at mid-height would therefore have the form shown in Fig. 8.4 where, at the point $P = P_{CR}$, it is theoretically possible for the column to take one of three deflection paths. Thus, if the column remains undisturbed the deflection at mid-height would continue to be zero but unstable (i.e. the trivial solution of Eq. (8.3), $v = 0$) or, if disturbed, the column would buckle in either of two lateral directions; the point at which this possible branching occurs is called a *bifurcation point*; further bifurcation points occur at the higher values of $P_{CR}(4\pi^2 EI/l^2, 9\pi^2 EI/l^2, \ldots)$.

Example 8.1

A uniform column of length L and flexural stiffness EI is simply supported at its ends and has an additional elastic support at midspan. This support is such that if a lateral displacement v_c occurs at this point a restoring force kv_c is generated at the point. Derive an equation giving the buckling load of the column. If the buckling load is $4\pi^2 EI/L^2$ find the value of k. Also if the elastic support is infinitely stiff show that the buckling load is given by the equation $\tan \lambda L/2 = \lambda L/2$ where $\lambda = \sqrt{P/EI}$.

The column is shown in its displaced position in Fig. 8.5.The bending moment at any section of the column is given by

$$M = Pv - \frac{kv_c}{2}z$$

so that, by comparison with Eq. (8.1)

$$EI\frac{d^2v}{dz^2} = -Pv + \frac{kv_c}{2}z$$

giving

$$\frac{d^2v}{dz^2} + \lambda^2 v = \frac{kv_c}{2EI}z \tag{i}$$

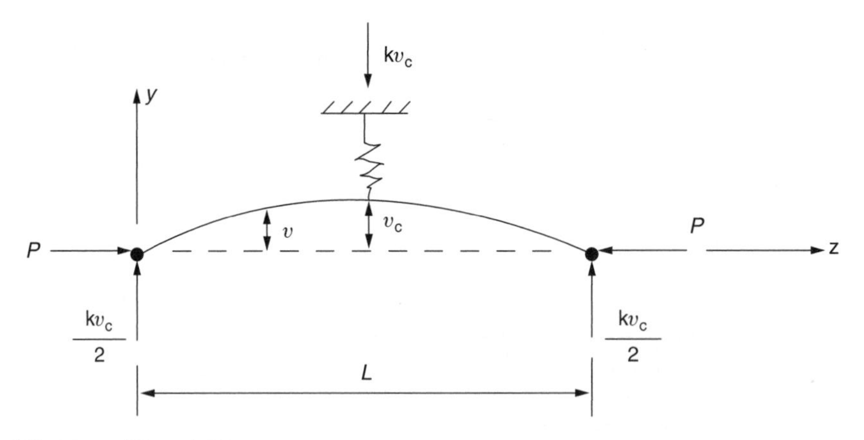

Fig. 8.5 Column of Example 8.1.

The solution of Eq. (i) is of standard form and is

$$v = A\cos\lambda z + B\sin\lambda z + \frac{kv_c}{2P}z$$

The constants A and B are found using the boundary conditions of the column which are: $v=0$ when $z=0$, $v=v_c$, when $z=L/2$ and $(\mathrm{d}v/\mathrm{d}z)=0$ when $z=L/2$.

From the first of these, $A=0$ while from the second

$$B = \frac{v_c}{\sin(\lambda L/2)}\left(1 - \frac{k\lambda}{4P}\right)$$

The third boundary condition gives, since $v_c \neq 0$, the required equation, i.e.

$$\left(1 - \frac{kL}{4P}\right)\cos\frac{\lambda L}{2} + \frac{k}{2P\lambda}\sin\frac{\lambda L}{2} = 0$$

Rearranging

$$P = \frac{kL}{4}\left(1 - \frac{\tan(\lambda L/2)}{\lambda L/2}\right)$$

If P (buckling load) $= 4\pi^2 EI/L^2$ then $\lambda L/2 = \pi$ so that $k = 4P/L$.
Finally, if $k \to \infty$

$$\tan\frac{\lambda L}{2} = \frac{\lambda L}{2} \tag{ii}$$

Note that Eq. (ii) is the transcendental equation which would be derived when determining the buckling load of a column of length $L/2$, built in at one end and pinned at the other.

8.2 Inelastic buckling

We have shown that the critical stress, Eq. (8.8), depends only on the elastic modulus of the material of the column and the slenderness ratio l/r. For a given material the critical stress increases as the slenderness ratio decreases; i.e. as the column becomes shorter and thicker. A point is then reached when the critical stress is greater than the yield stress of the material so that Eq. (8.8) is no longer applicable. For mild steel this point occurs at a slenderness ratio of approximately 100, as shown in Fig. 8.6. We therefore require some alternative means of predicting column behaviour at low values of slenderness ratio.

It was assumed in the derivation of Eq. (8.8) that the stresses in the column remained within the elastic range of the material so that the modulus of elasticity $E(=\mathrm{d}\sigma/\mathrm{d}\varepsilon)$ was constant. Above the elastic limit $\mathrm{d}\sigma/\mathrm{d}\varepsilon$ depends upon the value of stress and whether the stress is increasing or decreasing. Thus, in Fig. 8.7 the elastic modulus at the point A is the *tangent modulus* E_t if the stress is increasing but E if the stress is decreasing.

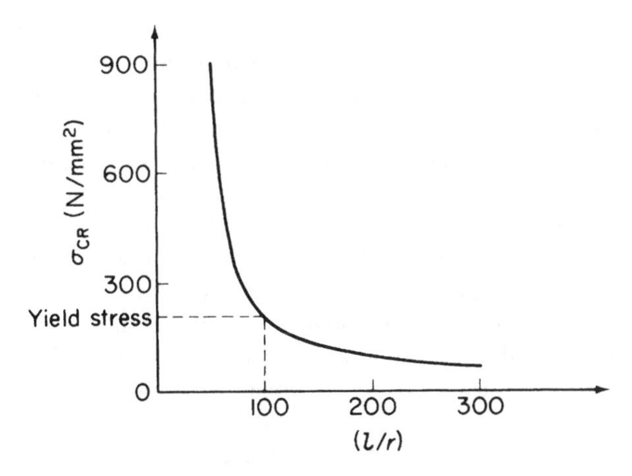

Fig. 8.6 Critical stress–slenderness ratio for a column.

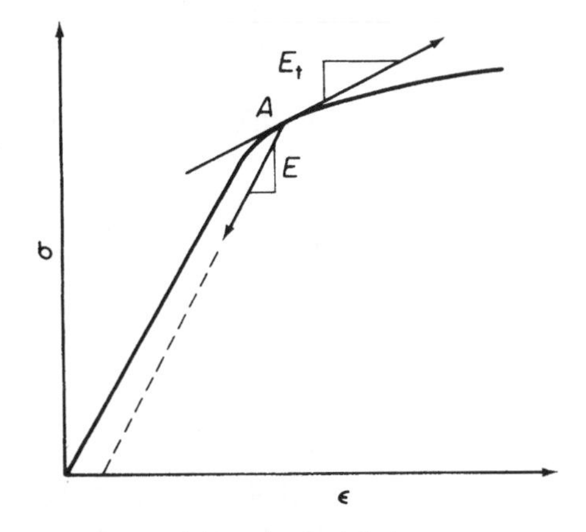

Fig. 8.7 Elastic moduli for a material stressed above the elastic limit.

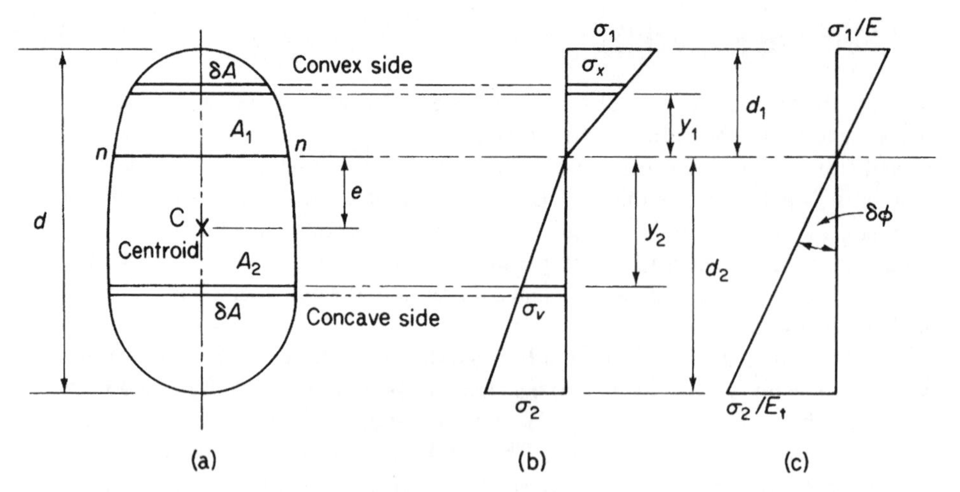

Fig. 8.8 Determination of reduced elastic modulus.

Consider a column having a plane of symmetry and subjected to a compressive load P such that the direct stress in the column P/A is above the elastic limit. If the column is given a small deflection, v, in its plane of symmetry, then the stress on the concave side increases while the stress on the convex side decreases. Thus, in the cross-section of the column shown in Fig. 8.8(a) the compressive stress decreases in the area A_1 and increases in the area A_2, while the stress on the line nn is unchanged. Since these changes take place outside the elastic limit of the material, we see, from our remarks in the previous paragraph, that the modulus of elasticity of the material in the area A_1 is E while that in A_2 is E_t. The homogeneous column now behaves as if it were non-homogeneous, with the result that the stress distribution is changed to the form shown in Fig. 8.8(b); the linearity of the distribution follows from an assumption that plane sections remain plane.

As the axial load is unchanged by the disturbance

$$\int_0^{d_1} \sigma_x \, \mathrm{d}A = \int_0^{d_2} \sigma_v \, \mathrm{d}A \tag{8.9}$$

Also, P is applied through the centroid of each end section a distance e from nn so that

$$\int_0^{d_1} \sigma_x(y_1 + e) \, \mathrm{d}A + \int_0^{d_2} \sigma_v(y_2 - e) \, \mathrm{d}A = -Pv \tag{8.10}$$

From Fig. 8.8(b)

$$\sigma_x = \frac{\sigma_1}{\mathrm{d}_1} y_1 \quad \sigma_v = \frac{\sigma_2}{\mathrm{d}_2} y_2 \tag{8.11}$$

The angle between two close, initially parallel, sections of the column is equal to the change in slope $\mathrm{d}^2v/\mathrm{d}z^2$ of the column between the two sections. This, in turn, must be equal to the angle $\delta\phi$ in the strain diagram of Fig. 8.8(c). Hence

$$\frac{d^2v}{dz^2} = \frac{\sigma_1}{Ed_1} = \frac{\sigma_2}{E_t d_2} \tag{8.12}$$

and Eq. (8.9) becomes, from Eqs (8.11) and (8.12)

$$E\frac{d^2v}{dz^2}\int_0^{d_1} y_1 dA - E_t\frac{d^2v}{dz^2}\int_0^{d_2} y_2\, dA = 0 \tag{8.13}$$

Further, in a similar manner, from Eq. (8.10)

$$\frac{d^2v}{dz^2}\left(E\int_0^{d_1} y_1^2\, dA + E_t\int_0^{d_2} y_2^2\, dA\right) + e\frac{d^2v}{dz^2}\left(E\int_0^{d_1} y_1\, dA - E_t\int_0^{d_2} y_2\, dA\right) = -Pv \tag{8.14}$$

The second term on the left-hand side of Eq. (8.14) is zero from Eq. (8.13). Therefore we have

$$\frac{d^2v}{dz^2}(EI_1 + E_t I_2) = -Pv \tag{8.15}$$

in which

$$I_1 = \int_0^{d_1} y_1^2\, dA \quad \text{and} \quad I_2 = \int_0^{d_2} y_2^2\, dA$$

the second moments of area about *nn* of the convex and concave sides of the column respectively. Putting

$$E_r I = EI_1 + E_t I_2$$

or

$$E_r = E\frac{I_1}{I} + E_t\frac{I_2}{I} \tag{8.16}$$

where E_r is known as the *reduced modulus*, gives

$$E_r I\frac{d^2v}{dz^2} + Pv = 0$$

Comparing this with Eq. (8.2) we see that if P is the critical load P_{CR} then

$$P_{CR} = \frac{\pi^2 E_r I}{l_e^2} \tag{8.17}$$

and

$$\sigma_{CR} = \frac{\pi^2 E_r}{(l_e/r)^2} \tag{8.18}$$

The above method for predicting critical loads and stresses outside the elastic range is known as the *reduced modulus theory*. From Eq. (8.13) we have

$$E \int_0^{d_1} y_1 \, \mathrm{d}A - E_\mathrm{t} \int_0^{d_2} y_2 \, \mathrm{d}A = 0 \tag{8.19}$$

which, together with the relationship $d = d_1 + d_2$, enables the position of nn to be found.

It is possible that the axial load P is increased at the time of the lateral disturbance of the column such that there is no strain reversal on its convex side. The compressive stress therefore increases over the complete section so that the tangent modulus applies over the whole cross-section. The analysis is then the same as that for column buckling within the elastic limit except that E_t is substituted for E. Hence the *tangent modulus theory* gives

$$P_\mathrm{CR} = \frac{\pi^2 E_\mathrm{t} I}{l_\mathrm{e}^2} \tag{8.20}$$

and

$$\sigma_\mathrm{CR} = \frac{\pi^2 E_\mathrm{t}}{(l_\mathrm{e}/r^2)} \tag{8.21}$$

By a similar argument, a reduction in P could result in a decrease in stress over the whole cross-section. The elastic modulus applies in this case and the critical load and stress are given by the standard Euler theory; namely, Eqs (8.7) and (8.8).

In Eq. (8.16), I_1 and I_2 are together greater than I while E is greater than E_t. It follows that the reduced modulus E_r is greater than the tangent modulus E_t. Consequently, buckling loads predicted by the reduced modulus theory are greater than buckling loads derived from the tangent modulus theory, so that although we have specified theoretical loading situations where the different theories would apply there still remains the difficulty of deciding which should be used for design purposes.

Extensive experiments carried out on aluminium alloy columns by the aircraft industry in the 1940s showed that the actual buckling load was approximately equal to the tangent modulus load. Shanley (1947) explained that for columns with small imperfections, an increase of axial load and bending occur simultaneously. He then showed analytically that after the tangent modulus load is reached, the strain on the concave side of the column increases rapidly while that on the convex side decreases slowly. The large deflection corresponding to the rapid strain increase on the concave side, which occurs soon after the tangent modulus load is passed, means that it is only possible to exceed the tangent modulus load by a small amount. It follows that the buckling load of columns is given most accurately for practical purposes by the tangent modulus theory.

Empirical formulae have been used extensively to predict buckling loads, although in view of the close agreement between experiment and the tangent modulus theory they would appear unnecessary. Several formulae are in use; for example, the *Rankine, Straight-line* and *Johnson's parabolic* formulae are given in many books on elastic stability.[1]

8.3 Effect of initial imperfections

Obviously it is impossible in practice to obtain a perfectly straight homogeneous column and to ensure that it is exactly axially loaded. An actual column may be bent with some eccentricity of load. Such imperfections influence to a large degree the behaviour of the column which, unlike the perfect column, begins to bend immediately the axial load is applied.

Let us suppose that a column, initially bent, is subjected to an increasing axial load P as shown in Fig. 8.9. In this case the bending moment at any point is proportional to the change in curvature of the column from its initial bent position. Thus

$$EI\frac{d^2v}{dz^2} - EI\frac{d^2v_0}{dz^2} - Pv \tag{8.22}$$

which, on rearranging, becomes

$$\frac{d^2v}{dz^2} + \lambda^2 v = \frac{d^2v_0}{dz^2} \tag{8.23}$$

where $\lambda^2 = P/EI$. The final deflected shape, v, of the column depends upon the form of its unloaded shape, v_0. Assuming that

$$v_0 = \sum_{n=1}^{\infty} A_n \sin\frac{n\pi z}{l} \tag{8.24}$$

and substituting in Eq. (8.23) we have

$$\frac{d^2v}{dz^2} + \lambda^2 v = -\frac{\pi^2}{l^2}\sum_{n=1}^{\infty} n^2 A_n \sin\frac{n\pi z}{l}$$

The general solution of this equation is

$$v = B\cos\lambda z + D\sin\lambda z + \sum_{n=1}^{\infty}\frac{n^2 A_n}{n^2-\alpha}\sin\frac{n\pi z}{l}$$

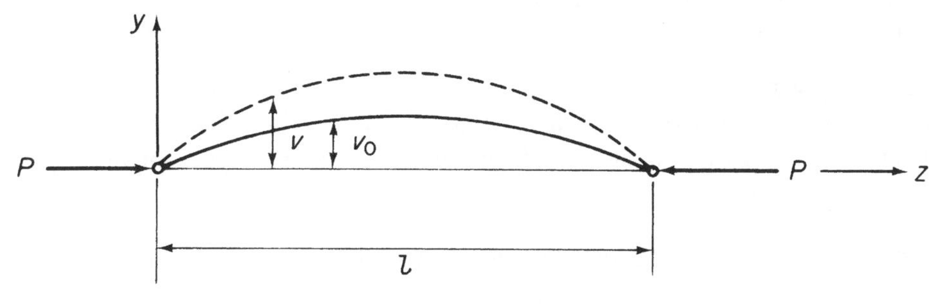

Fig. 8.9 Initially bent column.

where B and D are constants of integration and $\alpha = \lambda^2 l^2/\pi^2$. The boundary conditions are $v = 0$ at $z = 0$ and l, giving $B = D = 0$ whence

$$v = \sum_{n=1}^{\infty} \frac{n^2 A_n}{n^2 - \alpha} \sin \frac{n\pi z}{l} \tag{8.25}$$

Note that in contrast to the perfect column we are able to obtain a non-trivial solution for deflection. This is to be expected since the column is in stable equilibrium in its bent position at all values of P.

An alternative form for α is

$$\alpha = \frac{Pl^2}{\pi^2 EI} = \frac{P}{P_{CR}} \qquad \text{(see Eq. (8.5))}$$

Thus α is always less than one and approaches unity when P approaches P_{CR} so that the first term in Eq. (8.25) usually dominates the series. A good approximation, therefore, for deflection when the axial load is in the region of the critical load is

$$v = \frac{A_1}{1 - \alpha} \sin \frac{\pi z}{l} \tag{8.26}$$

or at the centre of the column where $z = l/2$

$$v = \frac{A_1}{1 - P/P_{CR}} \tag{8.27}$$

in which A_1 is seen to be the initial central deflection. If central deflections $\delta(= v - A_1)$ are measured from the initially bowed position of the column then from Eq. (8.27) we obtain

$$\frac{A_1}{1 - P/P_{CR}} - A_1 = \delta$$

which gives on rearranging

$$\delta = P_{CR} \frac{\delta}{P} - A_1 \tag{8.28}$$

and we see that a graph of δ plotted against δ/P has a slope, in the region of the critical load, equal to P_{CR} and an intercept equal to the initial central deflection. This is the well known *Southwell plot* for the experimental determination of the elastic buckling load of an imperfect column.

Timoshenko[1] also showed that Eq. (8.27) may be used for a perfectly straight column with small eccentricities of column load.

Example 8.2

The pin-jointed column shown in Fig. 8.10 carries a compressive load P applied eccentrically at a distance e from the axis of the column. Determine the maximum bending moment in the column.

The bending moment at any section of the column is given by

$$M = P(e + v)$$

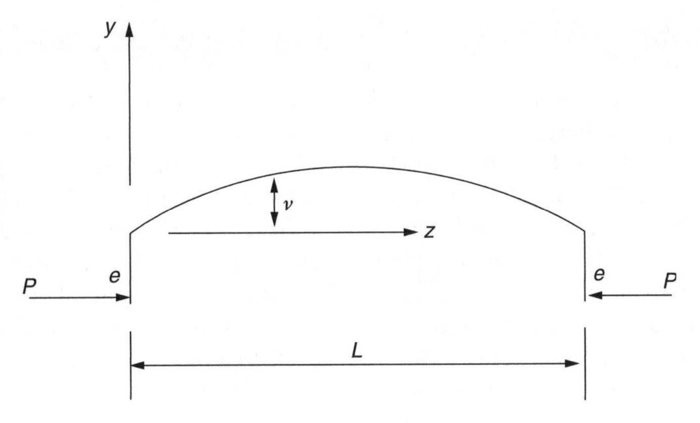

Fig. 8.10 Eccentrically loaded column of Example 8.2

Then, by comparison with Eq. (8.1)

$$EI\frac{\mathrm{d}^2 v}{\mathrm{d}z^2} = -P(e + v)$$

giving

$$\frac{\mathrm{d}^2 v}{\mathrm{d}z^2} + \mu^2 v = -\frac{Pe}{EI} \quad (\mu^2 = P/EI) \tag{i}$$

The solution of Eq. (i) is of standard form and is

$$v = A\cos\mu z + B\sin\mu z - e$$

The boundary conditions are: $v = 0$ when $z = 0$ and $(\mathrm{d}v/\mathrm{d}z) = 0$ when $z = L/2$. From the first of these $A = e$ while from the second

$$B = e\tan\frac{\mu L}{2}$$

The equation for the deflected shape of the column is then

$$v = e\left[\frac{\cos\mu(z - L/2)}{\cos\mu L/2} - 1\right]$$

The maximum value of v occurs at midspan where $z = L/2$, i.e.

$$v_{\max} = e\left(\sec\frac{\mu L}{2} - 1\right)$$

The maximum bending moment is given by

$$M(\max) = Pe + Pv_{\max}$$

so that

$$M(\max) = Pe\sec\frac{\mu L}{2}$$

8.4 Stability of beams under transverse and axial loads

Stresses and deflections in a linearly elastic beam subjected to transverse loads as predicted by simple beam theory, are directly proportional to the applied loads. This relationship is valid if the deflections are small such that the slight change in geometry produced in the loaded beam has an insignificant effect on the loads themselves. This situation changes drastically when axial loads act simultaneously with the transverse loads. The internal moments, shear forces, stresses and deflections then become dependent upon the magnitude of the deflections as well as the magnitude of the external loads. They are also sensitive, as we observed in the previous section, to beam imperfections such as initial curvature and eccentricity of axial load. Beams supporting both axial and transverse loads are sometimes known as *beam-columns* or simply as *transversely loaded columns*.

We consider first the case of a pin-ended beam carrying a uniformly distributed load of intensity w per unit length and an axial load P as shown in Fig. 8.11. The bending moment at any section of the beam is

$$M = Pv + \frac{wlz}{2} - \frac{wz^2}{2} = -EI\frac{\mathrm{d}^2v}{\mathrm{d}z^2}$$

giving

$$\frac{\mathrm{d}^2v}{\mathrm{d}z^2} + \frac{P}{EI}v = \frac{w}{2EI}(z^2 - lz) \tag{8.29}$$

The standard solution of Eq. (8.29) is

$$v = A\cos\lambda z + B\sin\lambda z + \frac{w}{2P}\left(z^2 - lz - \frac{2}{\lambda^2}\right)$$

where A and B are unknown constants and $\lambda^2 = P/EI$. Substituting the boundary conditions $v = 0$ at $z = 0$ and l gives

$$A = \frac{w}{\lambda^2 P} \qquad B = \frac{w}{\lambda^2 P\sin\lambda l}(l - \cos\lambda l)$$

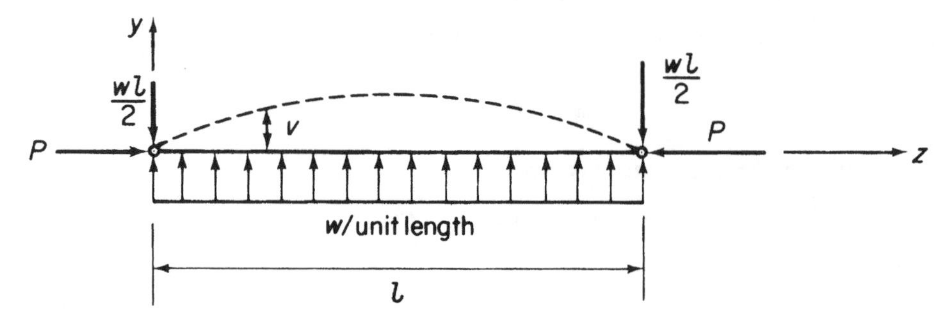

Fig. 8.11 Bending of a uniformly loaded beam-column.

so that the deflection is determinate for any value of w and P and is given by

$$v = \frac{w}{\lambda^2 P}\left[\cos\lambda z + \left(\frac{1-\cos\lambda l}{\sin\lambda l}\right)\sin\lambda z\right] + \frac{w}{2P}\left(z^2 - lz - \frac{2}{\lambda^2}\right) \tag{8.30}$$

In beam-columns, as in beams, we are primarily interested in maximum values of stress and deflection. For this particular case the maximum deflection occurs at the centre of the beam and is, after some transformation of Eq. (8.30)

$$v_{\max} = \frac{w}{\lambda^2 P}\left(\sec\frac{\lambda l}{2} - 1\right) - \frac{wl^2}{8P} \tag{8.31}$$

The corresponding maximum bending moment is

$$M_{\max} = -Pv_{\max} - \frac{wl^2}{8}$$

or, from Eq. (8.31)

$$M_{\max} = \frac{w}{\lambda^2}\left(1 - \sec\frac{\lambda l}{2}\right) \tag{8.32}$$

We may rewrite Eq. (8.32) in terms of the Euler buckling load $P_{\mathrm{CR}} = \pi^2 EI/l^2$ for a pin-ended column. Hence

$$M_{\max} = \frac{wl^2}{\pi^2}\frac{P_{\mathrm{CR}}}{P}\left(1 - \sec\frac{\pi}{2}\sqrt{\frac{P}{P_{\mathrm{CR}}}}\right) \tag{8.33}$$

As P approaches P_{CR} the bending moment (and deflection) becomes infinite. However, the above theory is based on the assumption of small deflections (otherwise $\mathrm{d}^2v/\mathrm{d}z^2$ would not be a close approximation for curvature) so that such a deduction is invalid. The indication is, though, that large deflections will be produced by the presence of a compressive axial load no matter how small the transverse load might be.

Let us consider now the beam-column of Fig. 8.12 with hinged ends carrying a concentrated load W at a distance a from the right-hand support. For

$$z \le l - a \quad EI\frac{\mathrm{d}^2v}{\mathrm{d}z^2} = -M = -Pv - \frac{Waz}{l} \tag{8.34}$$

and for

$$z \ge l - a \quad EI\frac{\mathrm{d}^2v}{\mathrm{d}z^2} = -M = -Pv - \frac{W}{l}(l-a)(l-z) \tag{8.35}$$

Writing

$$\lambda^2 = \frac{P}{EI}$$

Eq. (8.34) becomes

$$\frac{\mathrm{d}^2v}{\mathrm{d}z^2} + \lambda^2 v = -\frac{Wa}{EIl}z$$

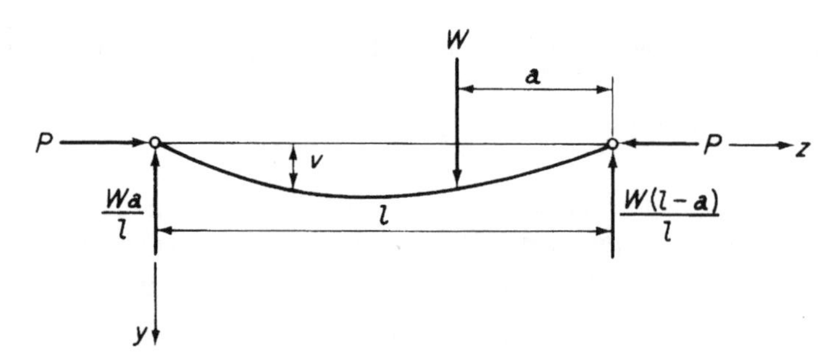

Fig. 8.12 Beam-column supporting a point load.

the general solution of which is

$$v = A\cos\lambda z + B\sin\lambda z - \frac{Wa}{Pl}z \tag{8.36}$$

Similarly, the general solution of Eq. (8.35) is

$$v = C\cos\lambda z + D\sin\lambda z - \frac{W}{Pl}(l-a)(l-z) \tag{8.37}$$

where A, B, C and D are constants which are found from the boundary conditions as follows.

When $z=0$, $v=0$, therefore from Eq. (8.36) $A=0$. *At* $z=l$, $v=0$ giving, from Eq. (8.37), $C=-D\tan\lambda l$. At the point of application of the load the deflection and slope of the beam given by Eqs (8.36) and (8.37) must be the same. Hence, equating deflections

$$B\sin\lambda(l-a) - \frac{Wa}{Pl}(l-a) = D[\sin\lambda(l-a) - \tan\lambda l\cos\lambda(l-a)] - \frac{Wa}{Pl}(l-a)$$

and equating slopes

$$B\lambda\cos\lambda(l-a) - \frac{Wa}{Pl} = D\lambda[\cos\lambda(l-a) - \tan\lambda l\sin\lambda(l-a)] + \frac{W}{Pl}(l-a)$$

Solving the above equations for B and D and substituting for A, B, C and D in Eqs (8.36) and (8.37) we have

$$v = \frac{W\sin\lambda a}{P\lambda\sin\lambda l}\sin\lambda z - \frac{Wa}{Pl}z \quad \text{for} \quad z \leq l-a \tag{8.38}$$

$$v = \frac{W\sin\lambda(l-a)}{P\lambda\sin\lambda l}\sin\lambda(l-z) - \frac{W}{Pl}(l-a)(l-z) \quad \text{for } z \geq l-a \tag{8.39}$$

These equations for the beam-column deflection enable the bending moment and resulting bending stresses to be found at all sections.

A particular case arises when the load is applied at the centre of the span. The deflection curve is then symmetrical with a maximum deflection under the load of

$$v_{\max} = \frac{W}{2P\lambda}\tan\frac{\lambda l}{2} - \frac{Wl}{4p}$$

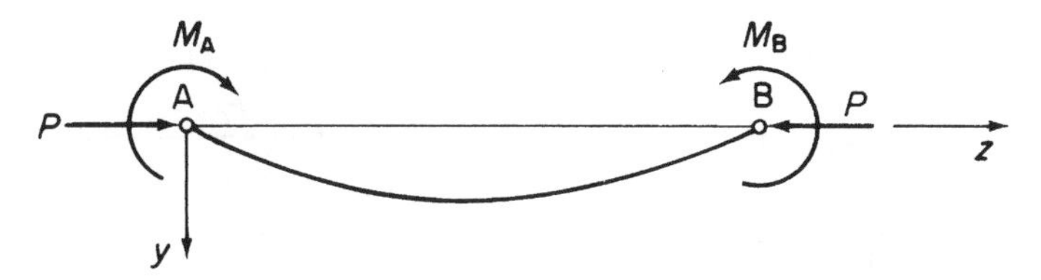

Fig. 8.13 Beam-column supporting end moments.

Finally, we consider a beam-column subjected to end moments M_A and M_B in addition to an axial load P (Fig. 8.13). The deflected form of the beam-column may be found by using the principle of superposition and the results of the previous case. First, we imagine that M_B acts alone with the axial load P. If we assume that the point load W moves towards B and simultaneously increases so that the product $Wa = \text{constant} = M_B$ then, in the limit as a tends to zero, we have the moment M_B applied at B. The deflection curve is then obtained from Eq. (8.38) by substituting λa for $\sin \lambda a$ (since λa is now very small) and M_B for Wa. Thus

$$v = \frac{M_B}{P}\left(\frac{\sin \lambda z}{\sin \lambda l} - \frac{z}{l}\right) \tag{8.40}$$

In a similar way, we find the deflection curve corresponding to M_A acting alone. Suppose that W moves towards A such that the product $W(l-a) = \text{constant} = M_A$. Then as $(l-a)$ tends to zero we have $\sin \lambda(l-a) = \lambda(l-a)$ and Eq. (8.39) becomes

$$v = \frac{M_A}{P}\left[\frac{\sin \lambda(l-z)}{\sin \lambda l} - \frac{(l-z)}{l}\right] \tag{8.41}$$

The effect of the two moments acting simultaneously is obtained by superposition of the results of Eqs (8.40) and (8.41). Hence for the beam-column of Fig. 8.13

$$v = \frac{M_B}{P}\left(\frac{\sin \lambda z}{\sin \lambda l} - \frac{z}{l}\right) + \frac{M_A}{P}\left[\frac{\sin \lambda(l-z)}{\sin \lambda l} - \frac{(l-z)}{l}\right] \tag{8.42}$$

Equation (8.42) is also the deflected form of a beam-column supporting eccentrically applied end loads at A and B. For example, if e_A and e_B are the eccentricities of P at the ends A and B, respectively, then $M_A = Pe_A, M_B = Pe_B$, giving a deflected form of

$$v = e_B\left(\frac{\sin \lambda z}{\sin \lambda l} - \frac{z}{l}\right) + e_A\left[\frac{\sin \lambda(l-z)}{\sin \lambda l} - \frac{(l-z)}{l}\right] \tag{8.43}$$

Other beam-column configurations featuring a variety of end conditions and loading regimes may be analysed by a similar procedure.

8.5 Energy method for the calculation of buckling loads in columns

The fact that the total potential energy of an elastic body possesses a stationary value in an equilibrium state may be used to investigate the neutral equilibrium of a buckled

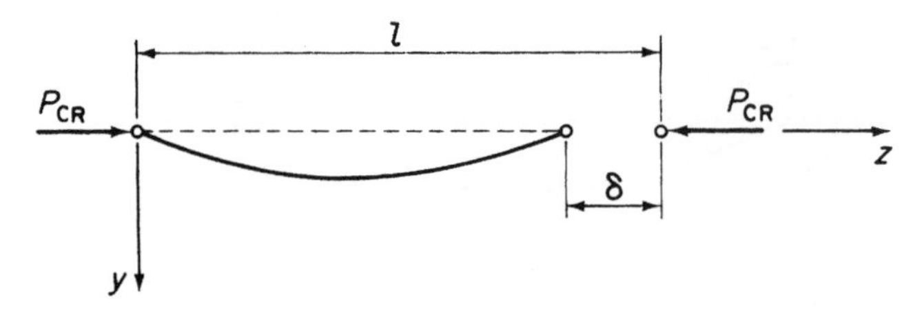

Fig. 8.14 Shortening of a column due to buckling.

column. In particular, the energy method is extremely useful when the deflected form of the buckled column is unknown and has to be 'guessed'.

First, we shall consider the pin-ended column shown in its buckled position in Fig. 8.14. The internal or strain energy U of the column is assumed to be produced by bending action alone and is given by the well known expression

$$U = \int_0^l \frac{M^2}{2EI} \mathrm{d}z \tag{8.44}$$

or alternatively, since $EI\,\mathrm{d}^2v/\mathrm{d}z^2 = -M$

$$U = \frac{EI}{2} \int_0^l \left(\frac{d^2v}{dz^2}\right)^2 \mathrm{d}z \tag{8.45}$$

The potential energy V of the buckling load P_{CR}, referred to the straight position of the column as the datum, is then

$$V = -P_{\mathrm{CR}}\delta$$

where δ is the axial movement of P_{CR} caused by the bending of the column from its initially straight position. By reference to Fig. 7.15(b) and Eq. (7.41) we see that

$$\delta = \frac{1}{2} \int_0^l \left(\frac{\mathrm{d}v}{\mathrm{d}z}\right)^2 \mathrm{d}z$$

giving

$$V = -\frac{P_{\mathrm{CR}}}{2} \int_0^l \left(\frac{\mathrm{d}v}{\mathrm{d}z}\right)^2 \mathrm{d}z \tag{8.46}$$

The total potential energy of the column in the neutral equilibrium of its buckled state is therefore

$$U + V = \int_0^l \frac{M^2}{2EI} \mathrm{d}z - \frac{P_{\mathrm{CR}}}{2} \int_0^l \left(\frac{\mathrm{d}v}{\mathrm{d}z}\right)^2 \mathrm{d}z \tag{8.47}$$

or, using the alternative form of U from Eq. (8.45)

$$U + V = \frac{EI}{2} \int_0^l \left(\frac{\mathrm{d}^2v}{\mathrm{d}z^2}\right)^2 \mathrm{d}z - \frac{P_{\mathrm{CR}}}{2} \int_0^l \left(\frac{\mathrm{d}v}{\mathrm{d}z}\right)^2 \mathrm{d}z \tag{8.48}$$

We have seen in Chapter 7 that exact solutions of plate bending problems are obtainable by energy methods when the deflected shape of the plate is known. An identical situation exists in the determination of critical loads for column and thin plate buckling modes. For the pin-ended column under discussion a deflected form of

$$v = \sum_{n=1}^{\infty} A_n \sin \frac{n\pi z}{l} \tag{8.49}$$

satisfies the boundary conditions of

$$(v)_{z=0} = (v)_{z=l} = 0 \qquad \left(\frac{d^2 v}{dz^2}\right)_{z=0} = \left(\frac{d^2 v}{dz^2}\right)_{z=l} = 0$$

and is capable, within the limits for which it is valid and if suitable values for the constant coefficients A_n are chosen, of representing any continuous curve. We are therefore in a position to find P_{CR} exactly. Substituting Eq. (8.49) into Eq. (8.48) gives

$$\begin{aligned} U + V = {} & \frac{EI}{2}\int_0^l \left(\frac{\pi}{l}\right)^4 \left(\sum_{n=1}^{\infty} n^2 A_n \sin\frac{n\pi z}{l}\right)^2 dz \\ & - \frac{P_{CR}}{2}\int_0^l \left(\frac{\pi}{l}\right)^2 \left(\sum_{n=1}^{\infty} n A_n \cos\frac{n\pi z}{l}\right)^2 dz \end{aligned} \tag{8.50}$$

The product terms in both integrals of Eq. (8.50) disappear on integration, leaving only integrated values of the squared terms. Thus

$$U + V = \frac{\pi^4 EI}{4l^3}\sum_{n=1}^{\infty} n^4 A_n^2 - \frac{\pi^2 P_{CR}}{4l}\sum_{n=1}^{\infty} n^2 A_n^2 \tag{8.51}$$

Assigning a stationary value to the total potential energy of Eq. (8.51) with respect to each coefficient A_n in turn, then taking A_n as being typical, we have

$$\frac{\partial(U+V)}{\partial A_n} = \frac{\pi^4 EIn^4 A_n}{2l^3} - \frac{\pi^2 P_{CR} n^2 A_n}{2l} = 0$$

from which

$$P_{CR} = \frac{\pi^2 EIn^2}{l^2} \quad \text{as before.}$$

We see that each term in Eq. (8.49) represents a particular deflected shape with a corresponding critical load. Hence the first term represents the deflection of the column shown in Fig. 8.14, with $P_{CR} = \pi^2 EI/l^2$. The second and third terms correspond to the shapes shown in Fig. 8.3, having critical loads of $4\pi^2 EI/l^2$ and $9\pi^2 EI/l^2$ and so on. Clearly the column must be constrained to buckle into these more complex forms. In other words the column is being forced into an unnatural shape, is consequently stiffer and offers greater resistance to buckling as we observe from the higher values of critical

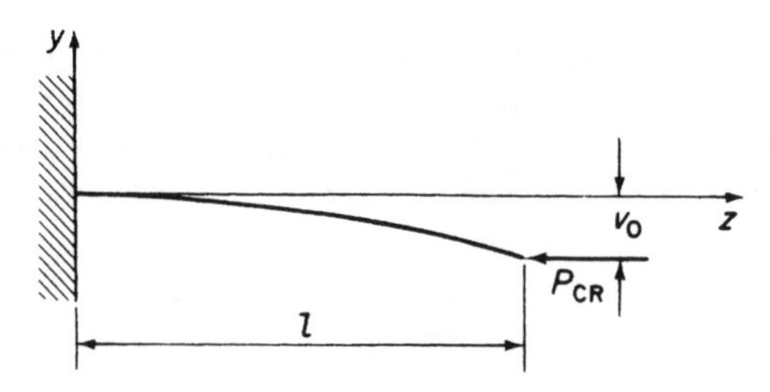

Fig. 8.15 Buckling load for a built-in column by the energy method.

load. Such buckling modes, as stated in Section 8.1, are unstable and are generally of academic interest only.

If the deflected shape of the column is known it is immaterial which of Eqs (8.47) or (8.48) is used for the total potential energy. However, when only an approximate solution is possible Eq. (8.47) is preferable since the integral involving bending moment depends upon the accuracy of the assumed form of v, whereas the corresponding term in Eq. (8.48) depends upon the accuracy of $\mathrm{d}^2v/\mathrm{d}z^2$. Generally, for an assumed deflection curve v is obtained much more accurately than $\mathrm{d}^2v/\mathrm{d}z^2$.

Suppose that the deflection curve of a particular column is unknown or extremely complicated. We then assume a reasonable shape which satisfies, as far as possible, the end conditions of the column and the pattern of the deflected shape (Rayleigh–Ritz method). Generally, the assumed shape is in the form of a finite series involving a series of unknown constants and assumed functions of z. Let us suppose that v is given by

$$v = A_1f_1(z) + A_2f_2(z) + A_3f_3(z)$$

Substitution in Eq. (8.47) results in an expression for total potential energy in terms of the critical load and the coefficients A_1, A_2 and A_3 as the unknowns. Assigning stationary values to the total potential energy with respect to A_1, A_2 and A_3 in turn produces three simultaneous equations from which the ratios A_1/A_2, A_1/A_3 and the critical load are determined. Absolute values of the coefficients are unobtainable since the deflections of the column in its buckled state of neutral equilibrium are indeterminate.

As a simple illustration consider the column shown in its buckled state in Fig. 8.15. An approximate shape may be deduced from the deflected shape of a tip-loaded cantilever. Thus

$$v = \frac{v_0z^2}{2l^3}(3l - z)$$

This expression satisfies the end-conditions of deflection, viz. $v=0$ at $z=0$ and $v=v_0$ at $z=l$. In addition, it satisfies the conditions that the slope of the column is zero at the built-in end and that the bending moment, i.e. $\mathrm{d}^2v/\mathrm{d}z^2$, is zero at the free end. The bending moment at any section is $M = P_{\mathrm{CR}}(v_0 - v)$ so that substitution for M and v in Eq. (8.47) gives

$$U + V = \frac{P_{\mathrm{CR}}^2v_0^2}{2EI}\int_0^l\left(1 - \frac{3z^2}{2l^2} + \frac{z^3}{2l^3}\right)^2\mathrm{d}z - \frac{P_{\mathrm{CR}}}{2}\int_0^l\left(\frac{3v_0}{2l^3}\right)^3 z^2(2l - z)^2\,\mathrm{d}z$$

Integrating and substituting the limits we have

$$U + V = \frac{17}{35}\frac{P_{\mathrm{CR}}^2 v_0^2 l}{2EI} - \frac{3}{5}P_{\mathrm{CR}}\frac{v_0^2}{l}$$

Hence

$$\frac{\partial(U+V)}{\partial v_0} = \frac{17}{35}\frac{P_{\mathrm{CR}}^2 v_0 l}{EI} - \frac{6P_{\mathrm{CR}} v_0}{5l} = 0$$

from which

$$P_{\mathrm{CR}} = \frac{42EI}{17l^2} = 2.471\frac{EI}{l^2}$$

This value of critical load compares with the exact value (see Table 8.1) of $\pi^2 EI/4l^2 = 2.467EI/l^2$; the error, in this case, is seen to be extremely small. Approximate values of critical load obtained by the energy method are always greater than the correct values. The explanation lies in the fact that an assumed deflected shape implies the application of constraints in order to force the column to take up an artificial shape. This, as we have seen, has the effect of stiffening the column with a consequent increase in critical load.

It will be observed that the solution for the above example may be obtained by simply equating the increase in internal energy (U) to the work done by the external critical load ($-V$). This is always the case when the assumed deflected shape contains a single unknown coefficient, such as v_0 in the above example.

8.6 Flexural–torsional buckling of thin-walled columns

It is recommended that the reading of this section be delayed until after Chapter 27 has been studied.

In some instances thin-walled columns of open cross-section do not buckle in bending as predicted by the Euler theory but twist without bending, or bend and twist simultaneously, producing flexural–torsional buckling. The solution of this type of problem relies on the theory presented in Chapter 27 for the torsion of open section beams subjected to warping (axial) restraint. Initially, however, we shall establish a useful analogy between the bending of a beam and the behaviour of a pin-ended column.

The bending equation for a simply supported beam carrying a uniformly distributed load of intensity w_y and having Cx and Cy as principal centroidal axes is

$$EI_{xx}\frac{\mathrm{d}^4 v}{\mathrm{d}z^4} = w_y \quad \text{(see Chapter 16)} \tag{8.52}$$

Also, the equation for the buckling of a pin-ended column about the Cx axis is (see Eq. (8.1))

$$EI_{xx}\frac{\mathrm{d}^2 v}{\mathrm{d}z^2} = -P_{\mathrm{CR}} v \tag{8.53}$$

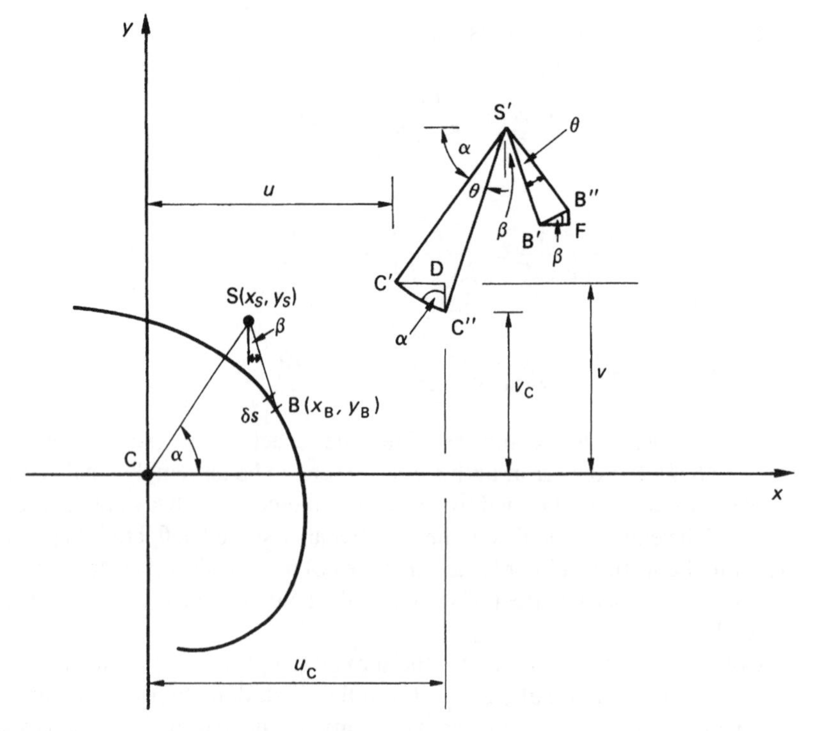

Fig. 8.16 Flexural–torsional buckling of a thin-walled column.

Differentiating Eq. (8.53) twice with respect to z gives

$$EI_{xx}\frac{d^4v}{dz^4} = -P_{CR}\frac{d^2v}{dz^2} \tag{8.54}$$

Comparing Eqs (8.52) and (8.54) we see that the behaviour of the column may be obtained by considering it as a simply supported beam carrying a uniformly distributed load of intensity w_y given by

$$w_y = -P_{CR}\frac{d^2v}{dz^2} \tag{8.55}$$

Similarly, for buckling about the Cy axis

$$w_x = -P_{CR}\frac{d^2u}{dz^2} \tag{8.56}$$

Consider now a thin-walled column having the cross-section shown in Fig. 8.16 and suppose that the centroidal axes Cxy are principal axes (see Chapter 16); S(x_S, y_S) is the shear centre of the column (see Chapter 17) and its cross-sectional area is A. Due to the flexural–torsional buckling produced, say, by a compressive axial load P the cross-section will suffer translations u and v parallel to Cx and Cy, respectively and a rotation θ, positive anticlockwise, about the shear centre S. Thus, due to translation,

C and S move to C′ and S′ and then, due to rotation about S′, C′ moves to C″. The total movement of C, u_C, in the x direction is given by

$$u_c = u + \text{C}'\text{D} = u + \text{C}'\text{C}'' \sin\alpha \quad (\text{S}'\hat{\text{C}}'\text{C}'' \simeq 90°)$$

But

$$\text{C}'\text{C}'' = \text{C}'\text{S}'\theta = \text{CS}\theta$$

Hence

$$u_C = u + \theta\text{CS}\sin\alpha = u + y_S\theta \tag{8.57}$$

Also the total movement of C in the y direction is

$$v_C = v - \text{DC}'' = v - \text{C}'\text{C}''\cos\alpha = v - \theta\text{CS}\cos\alpha$$

so that

$$v_C = v - x_S\theta \tag{8.58}$$

Since at this particular cross-section of the column the centroidal axis has been displaced, the axial load P produces bending moments about the displaced x and y axes given, respectively, by

$$M_x = Pv_C = P(v - x_S\theta) \tag{8.59}$$

and

$$M_y = Pu_C = P(u + y_S\theta) \tag{8.60}$$

From simple beam theory (Chapter 16)

$$EI_{xx}\frac{\text{d}^2v}{\text{d}z^2} = -M_x = -P(v - x_S\theta) \tag{8.61}$$

and

$$EI_{yy}\frac{\text{d}^2u}{\text{d}z^2} = -M_y = -P(u + y_S\theta) \tag{8.62}$$

where I_{xx} and I_{yy} are the second moments of area of the cross-section of the column about the principal centroidal axes, E is Young's modulus for the material of the column and z is measured along the centroidal longitudinal axis.

The axial load P on the column will, at any cross-section, be distributed as a uniform direct stress σ. Thus, the direct load on any element of length δs at a point B(x_B, y_B) is $\sigma t\,\text{d}s$ acting in a direction parallel to the longitudinal axis of the column. In a similar manner to the movement of C to C″ the point B will be displaced to B″. The horizontal movement of B in the x direction is then

$$u_B = u + \text{B}'\text{F} = u + \text{B}'\text{B}''\cos\beta$$

But

$$\text{B}'\text{B}'' = \text{S}'\text{B}'\theta = \text{SB}\theta$$

Hence

$$u_B = u + \theta SB \cos\beta$$

or

$$u_B = u + (y_S - y_B)\theta \tag{8.63}$$

Similarly the movement of B in the y direction is

$$v_B = v - (x_S - x_B)\theta \tag{8.64}$$

Therefore, from Eqs (8.63) and (8.64) and referring to Eqs (8.55) and (8.56), we see that the compressive load on the element δs at B, $\sigma t \delta s$, is equivalent to lateral loads

$$-\sigma t \delta s \frac{d^2}{dz^2}[u + (y_S - y_B)\theta] \quad \text{in the } x \text{ direction}$$

and

$$-\sigma t \delta s \frac{d^2}{dz^2}[v - (x_S - x_B)\theta] \quad \text{in the } y \text{ direction}$$

The lines of action of these equivalent lateral loads do not pass through the displaced position S′ of the shear centre and therefore produce a torque about S′ leading to the rotation θ. Suppose that the element δs at B is of unit length in the longitudinal z direction. The torque per unit length of the column $\delta T(z)$ acting on the element at B is then given by

$$\begin{aligned}\delta T(z) = &-\sigma t \delta s \frac{d^2}{dz^2}[u + (y_S - y_B)\theta](y_S - y_B) \\ &+ \sigma t \delta s \frac{d^2}{dz^2}[v - (x_S - x_B)\theta](x_S - x_B)\end{aligned} \tag{8.65}$$

Integrating Eq. (8.65) over the complete cross-section of the column gives the torque per unit length acting on the column, i.e.

$$\begin{aligned}T(z) = &-\int_{\text{Sect}} \sigma t \frac{d^2 u}{dz^2}(y_S - y_B)ds - \int_{\text{Sect}} \sigma t (y_S - y_B)^2 \frac{d^2\theta}{dz^2}ds \\ &+ \int_{\text{Sect}} \sigma t \frac{d^2 v}{dz^2}(x_S - x_B)ds - \int_{\text{Sect}} \sigma t (x_S - x_B)^2 \frac{d^2\theta}{dz^2}ds\end{aligned} \tag{8.66}$$

Expanding Eq. (8.66) and noting that σ is constant over the cross-section, we obtain

$$
\begin{aligned}
T(z) = &-\sigma\frac{\mathrm{d}^2u}{\mathrm{d}z^2}y_{\mathrm{S}}\int_{\mathrm{Sect}} t\,\mathrm{d}s + \sigma\frac{\mathrm{d}^2u}{\mathrm{d}z^2}\int_{\mathrm{Sect}} ty_{\mathrm{B}}\,\mathrm{d}s - \sigma\frac{\mathrm{d}^2\theta}{\mathrm{d}z^2}y_{\mathrm{S}}^2\int_{\mathrm{Sect}} t\,\mathrm{d}s \\
&+\sigma\frac{\mathrm{d}^2\theta}{\mathrm{d}z^2}2y_{\mathrm{S}}\int_{\mathrm{Sect}} ty_{\mathrm{B}}\,\mathrm{d}s - \sigma\frac{\mathrm{d}^2\theta}{\mathrm{d}z^2}\int_{\mathrm{Sect}} ty_{\mathrm{B}}^2\,\mathrm{d}s + \sigma\frac{\mathrm{d}^2v}{\mathrm{d}z^2}x_{\mathrm{S}}\int_{\mathrm{Sect}} t\,\mathrm{d}s \\
&-\sigma\frac{\mathrm{d}^2v}{\mathrm{d}z^2}\int_{\mathrm{Sect}} tx_{\mathrm{B}}\,\mathrm{d}s - \sigma\frac{\mathrm{d}^2\theta}{\mathrm{d}z^2}x_{\mathrm{S}}^2\int_{\mathrm{Sect}} t\,\mathrm{d}s + \sigma\frac{\mathrm{d}^2\theta}{\mathrm{d}z^2}2x_{\mathrm{S}}\int_{\mathrm{Sect}} tx_{\mathrm{B}}\,\mathrm{d}s \\
&-\sigma\frac{\mathrm{d}^2\theta}{\mathrm{d}z^2}\int_{\mathrm{Sect}} tx_{\mathrm{B}}^2\,\mathrm{d}s
\end{aligned}
\tag{8.67}
$$

Equation (8.67) may be rewritten

$$
T(z) = P\left(x_{\mathrm{S}}\frac{\mathrm{d}^2v}{\mathrm{d}z^2} - y_{\mathrm{S}}\frac{\mathrm{d}^2u}{\mathrm{d}z^2}\right) - \frac{P}{A}\frac{\mathrm{d}^2\theta}{\mathrm{d}z^2}(Ay_{\mathrm{S}}^2 + I_{xx} + Ax_{\mathrm{S}}^2 + I_{yy}) \tag{8.68}
$$

In Eq. (8.68) the term $I_{xx} + I_{yy} + A(x_{\mathrm{S}}^2 + y_{\mathrm{S}}^2)$ is the polar second moment of area I_0 of the column about the shear centre S. Thus Eq. (8.68) becomes

$$
T(z) = P\left(x_{\mathrm{S}}\frac{\mathrm{d}^2v}{\mathrm{d}z^2} - y_{\mathrm{S}}\frac{\mathrm{d}^2u}{\mathrm{d}z^2}\right) - I_0\frac{P}{A}\frac{\mathrm{d}^2\theta}{\mathrm{d}z^2} \tag{8.69}
$$

Substituting for $T(z)$ from Eq. (8.69) in Eq. (27.11), the general equation for the torsion of a thin-walled beam, we have

$$
E\Gamma\frac{\mathrm{d}^4\theta}{\mathrm{d}z^4} - \left(GJ - I_0\frac{P}{A}\right)\frac{\mathrm{d}^2\theta}{\mathrm{d}z^2} - Px_{\mathrm{S}}\frac{\mathrm{d}^2v}{\mathrm{d}z^2} + Py_{\mathrm{S}}\frac{\mathrm{d}^2u}{\mathrm{d}z^2} = 0 \tag{8.70}
$$

Equations (8.61), (8.62) and (8.70) form three simultaneous equations which may be solved to determine the flexural–torsional buckling loads.

As an example, consider the case of a column of length L in which the ends are restrained against rotation about the z axis and against deflection in the x and y directions; the ends are also free to rotate about the x and y axes and are free to warp. Thus $u = v = \theta = 0$ at $z = 0$ and $z = L$. Also, since the column is free to rotate about the x and y axes at its ends, $M_x = M_y = 0$ at $z = 0$ and $z = L$, and from Eqs (8.61) and (8.62)

$$
\frac{\mathrm{d}^2v}{\mathrm{d}z^2} = \frac{\mathrm{d}^2u}{\mathrm{d}z^2} = 0 \text{ at } z = 0 \text{ and } z = L
$$

Further, the ends of the column are free to warp so that

$$
\frac{\mathrm{d}^2\theta}{\mathrm{d}z^2} = 0 \text{ at } z = 0 \text{ and } z = L \text{ (see Eq. (27.1))}
$$

An assumed buckled shape given by

$$
u = A_1 \sin\frac{\pi z}{L} \qquad v = A_2 \sin\frac{\pi z}{L} \qquad \theta = A_3 \sin\frac{\pi z}{L} \tag{8.71}
$$

in which A_1, A_2 and A_3 are unknown constants, satisfies the above boundary conditions. Substituting for u, v and θ from Eqs (8.71) into Eqs (8.61), (8.62) and (8.70), we have

$$\left.\begin{aligned}&\left(P-\frac{\pi^2 EI_{xx}}{L^2}\right)A_2 - Px_S A_3 = 0\\&\left(P-\frac{\pi^2 EI_{yy}}{L^2}\right)A_1 + Py_S A_3 = 0\\&Py_S A_1 - Px_S A_2 - \left(\frac{\pi^2 E\Gamma}{L^2} + GJ - \frac{I_0}{A}P\right)A_3 = 0\end{aligned}\right\} \tag{8.72}$$

For non-zero values of A_1, A_2 and A_3 the determinant of Eqs (8.72) must equal zero, i.e.

$$\begin{vmatrix} 0 & P-\pi^2 EI_{xx}/L^2 & -Px_S \\ P-\pi^2 EI_{yy}/L^2 & 0 & Py_S \\ Py_S & -Px_S & I_0P/A - \pi^2 E\Gamma/L^2 - GJ \end{vmatrix} = 0 \tag{8.73}$$

The roots of the cubic equation formed by the expansion of the determinant give the critical loads for the flexural–torsional buckling of the column; clearly the lowest value is significant.

In the case where the shear centre of the column and the centroid of area coincide, i.e. the column has a doubly symmetrical cross-section, $x_S = y_S = 0$ and Eqs (8.61), (8.62) and (8.70) reduce, respectively, to

$$EI_{xx}\frac{d^2 v}{dz^2} = -Pv \tag{8.74}$$

$$EI_{yy}\frac{d^2 u}{dz^2} = -Pu \tag{8.75}$$

$$E\Gamma\frac{d^4\theta}{dz^4}\left(GJ - I_0\frac{P}{A}\right)\frac{d^2\theta}{dz^2} = 0 \tag{8.76}$$

Equations (8.74), (8.75) and (8.76), unlike Eqs (8.61), (8.62) and (8.70), are uncoupled and provide three separate values of buckling load. Thus, Eqs (8.74) and (8.75) give values for the Euler buckling loads about the x and y axes respectively, while Eq. (8.76) gives the axial load which would produce pure torsional buckling; clearly the buckling load of the column is the lowest of these values. For the column whose buckled shape is defined by Eqs (8.71), substitution for v, u and θ in Eqs (8.74), (8.75) and (8.76), respectively gives

$$P_{CR(xx)} = \frac{\pi^2 EI_{xx}}{L_2} \quad P_{CR(yy)} = \frac{\pi^2 EI_{yy}}{L^2} \quad P_{CR(\theta)} = \frac{A}{I_0}\left(GJ + \frac{\pi^2 E\Gamma}{L^2}\right) \tag{8.77}$$

Example 8.3

A thin-walled pin-ended column is 2 m long and has the cross-section shown in Fig. 8.17. If the ends of the column are free to warp determine the lowest value of axial

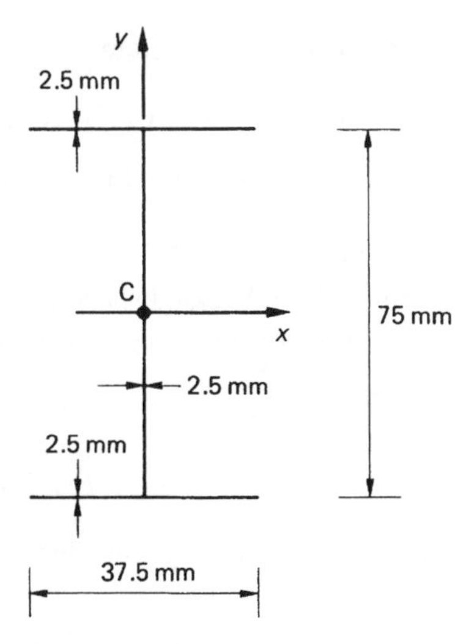

Fig. 8.17 Column section of Example 8.3.

load which will cause buckling and specify the buckling mode. Take $E = 75\,000\ \text{N/mm}^2$ and $G = 21\,000\ \text{N/mm}^2$.

Since the cross-section of the column is doubly-symmetrical, the shear centre coincides with the centroid of area and $x_S = y_S = 0$; Eq. (8.74), (8.75) and (8.76) therefore apply. Further, the boundary conditions are those of the column whose buckled shape is defined by Eqs (8.71) so that the buckling load of the column is the lowest of the three values given by Eqs (8.77).

The cross-sectional area A of the column is

$$A = 2.5(2 \times 37.5 + 75) = 375\ \text{mm}^2$$

The second moments of area of the cross-section about the centroidal axes Cxy are (see Chapter 16), respectively

$$I_{xx} = 2 \times 37.5 \times 2.5 \times 37.5^2 + 2.5 \times 75^3/12 = 3.52 \times 10^5\ \text{mm}^4$$

$$I_{yy} = 2 \times 2.5 \times 37.5^3/12 = 0.22 \times 10^5\ \text{mm}^4$$

The polar second moment of area I_0 is

$$I_0 = I_{xx} + I_{yy} + A(x_S^2 + y_S^2) \quad \text{(see derivation of Eq. (8.69))}$$

i.e.

$$I_0 = 3.52 \times 10^5 + 0.22 \times 10^5 = 3.74 \times 10^5\ \text{mm}^4$$

The torsion constant J is obtained using Eq. (18.11) which gives

$$J = 2 \times 37.5 \times 2.5^3/3 + 75 \times 2.5^3/3 = 781.3\ \text{mm}^4$$

Finally, Γ is found using the method of Section 27.2 and is

$$\Gamma = 2.5 \times 37.5^3 \times 75^2/24 = 30.9 \times 10^6\ \text{mm}^6$$

Substituting the above values in Eqs (8.77) we obtain

$$P_{\text{CR}(xx)} = 6.5 \times 10^4 \text{ N} \quad P_{\text{CR}(yy)} = 0.41 \times 10^4 \text{ N} \quad P_{\text{CR}(\theta)} = 2.22 \times 10^4 \text{ N}$$

The column will therefore buckle in bending about the Cy axis when subjected to an axial load of 0.41×10^4 N.

Equation (8.73) for the column whose buckled shape is defined by Eqs (8.71) may be rewritten in terms of the three separate buckling loads given by Eqs (8.77). Thus

$$\begin{vmatrix} 0 & P - P_{\text{CR}(xx)} & -Px_{\text{S}} \\ P - P_{\text{CR}(yy)} & 0 & Py_{\text{S}} \\ Py_{\text{S}} & -Px_{\text{S}} & I_0(P - P_{\text{CR}(\theta)})/A \end{vmatrix} = 0 \tag{8.78}$$

If the column has, say, Cx as an axis of symmetry, then the shear centre lies on this axis and $y_{\text{S}} = 0$. Equation (8.78) thereby reduces to

$$\begin{vmatrix} P - P_{\text{CR}(xx)} & -Px_{\text{S}} \\ -Px_{\text{S}} & I_0(P - P_{\text{CR}(\theta)})/A \end{vmatrix} = 0 \tag{8.79}$$

The roots of the quadratic equation formed by expanding Eq. (8.79) are the values of axial load which will produce flexural–torsional buckling about the longitudinal and x axes. If $P_{\text{CR}(yy)}$ is less than the smallest of these roots the column will buckle in pure bending about the y axis.

Example 8.4

A column of length 1 m has the cross-section shown in Fig. 8.18. If the ends of the column are pinned and free to warp, calculate its buckling load; $E = 70\,000\text{ N/mm}^2$, $G = 30\,000\text{ N/mm}^2$.

In this case the shear centre S is positioned on the Cx axis so that $y_{\text{S}} = 0$ and Eq. (8.79) applies. The distance $\bar{x}$ of the centroid of area C from the web of the section is found

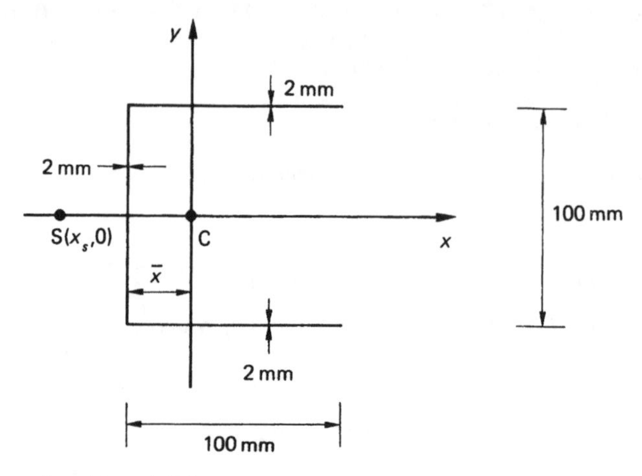

Fig. 8.18 Column section of Example 8.4.

by taking first moments of area about the web. Thus

$$2(100 + 100 + 100)\bar{x} = 2 \times 2 \times 100 \times 50$$

which gives

$$\bar{x} = 33.3\,\text{mm}$$

The position of the shear centre S is found using the method of Example 17.1; this gives $x_S = -76.2$ mm. The remaining section properties are found by the methods specified in Example 8.3 and are listed below

$$A = 600\,\text{mm}^2 \qquad I_{xx} = 1.17 \times 10^6\,\text{mm}^4 \qquad I_{yy} = 0.67 \times 10^6\,\text{mm}^4$$

$$I_0 = 5.32 \times 10^6\,\text{mm}^4 \qquad J = 800\,\text{mm}^4 \qquad \Gamma = 2488 \times 10^6\,\text{mm}^6$$

From Eq. (8.77)

$$P_{\text{CR}(yy)} = 4.63 \times 10^5\,\text{N} \quad P_{\text{CR}(xx)} = 8.08 \times 10^5\,\text{N} \quad P_{\text{CR}(\theta)} = 1.97 \times 10^5\,\text{N}$$

Expanding Eq. (8.79)

$$(P - P_{\text{CR}(xx)})(P - P_{\text{CR}(\theta)})I_0/A - P^2 x_S^2 = 0 \qquad \text{(i)}$$

Rearranging Eq. (i)

$$P^2(1 - Ax_S^2/I_0) - P(P_{\text{CR}(xx)} + P_{\text{CR}(\theta)}) + P_{\text{CR}(xx)}P_{\text{CR}(\theta)} = 0 \qquad \text{(ii)}$$

Substituting the values of the constant terms in Eq. (ii) we obtain

$$P^2 - 29.13 \times 10^5 P + 46.14 \times 10^{10} = 0 \qquad \text{(iii)}$$

The roots of Eq. (iii) give two values of critical load, the lowest of which is

$$P = 1.68 \times 10^5\,\text{N}$$

It can be seen that this value of flexural–torsional buckling load is lower than any of the uncoupled buckling loads $P_{\text{CR}(xx)}$, $P_{\text{CR}(yy)}$ or $P_{\text{CR}(\theta)}$; the reduction is due to the interaction of the bending and torsional buckling modes.

Example 8.5

A thin walled column has the cross-section shown in Fig. 8.19, is of length L and is subjected to an axial load through its shear centre S. If the ends of the column are prevented from warping and twisting determine the value of direct stress when failure occurs due to torsional buckling.

The torsion bending constant Γ is found using the method described in Section 27.2. The position of the shear centre is given but is obvious by inspection. The swept area $2\lambda A_{R,0}$ is determined as a function of s and its distribution is shown in Fig. 8.20. The centre of gravity of the 'wire' is found by taking moments about the s axis.

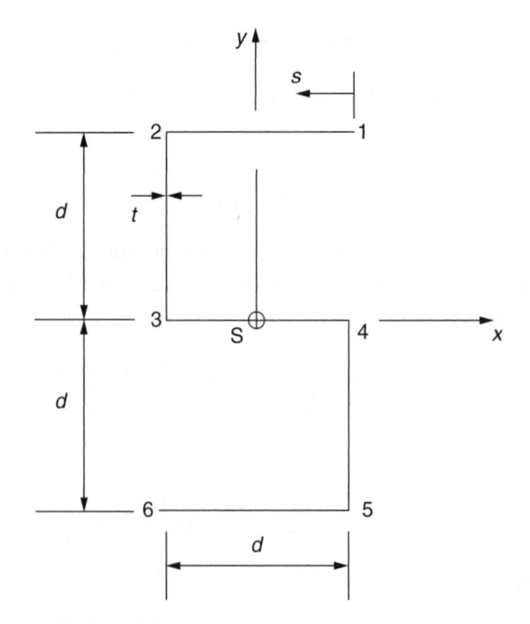

Fig. 8.19 Section of column of Example 8.5.

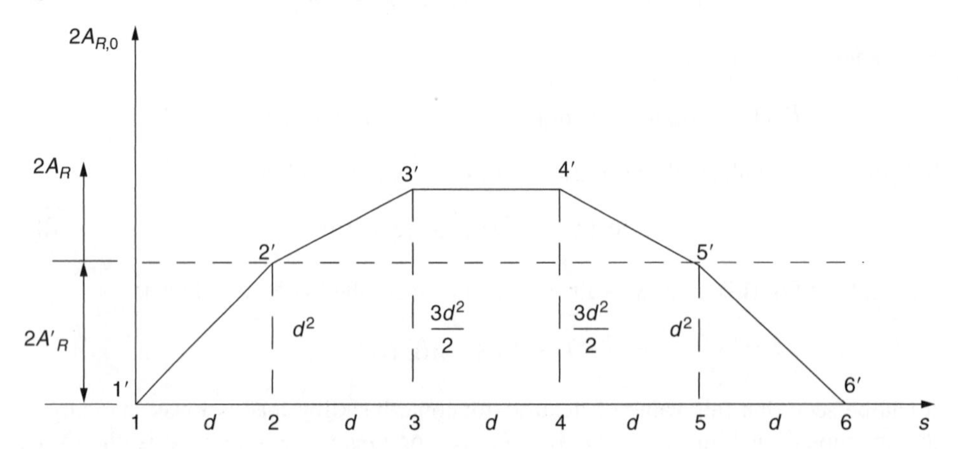

Fig. 8.20 Determination of torsion bending constant for column section of Example 8.5.

Then

$$2A'_R 5td = td\left(\frac{d^2}{2} + \frac{5d^2}{4} + \frac{3d^2}{2} + \frac{5d^2}{4} + \frac{d^2}{2}\right)$$

which gives

$$2A'_R = d^2$$

The torsion bending constant is then the 'moment of inertia' of the 'wire' and is

$$\Gamma = 2td\frac{1}{3}(d^2)^2 + \frac{td}{3}\left(\frac{d^2}{2}\right)^2 \times 2 + td\left(\frac{d^2}{2}\right)^2$$

from which

$$\Gamma = \frac{13}{12} t d^5$$

Also the torsion constant J is given by (see Section 3.4)

$$J = \sum \frac{st^3}{3} = \frac{5dt^3}{3}$$

The shear centre of the section and the centroid of area coincide so that the torsional buckling load is given by Eq. (8.76). Rewriting this equation

$$\frac{d^4\theta}{dz^4} + \mu^2 \frac{d^2\theta}{dz^2} = 0 \qquad \text{(i)}$$

where

$$\mu^2 = (\sigma I_0 - GJ)/E\Gamma \quad (\sigma = P/A)$$

The solution of Eq. (i) is

$$\theta = A \cos \mu z + B \sin \mu z + Cz + D \qquad \text{(ii)}$$

The boundary conditions are $\theta = 0$ when $z = 0$ and $z = L$ and since the warping is suppressed at the ends of the beam

$$\frac{d\theta}{dz} = 0 \quad \text{when } z = 0 \text{ and } z = L \qquad \text{(see Eq. (18.19))}$$

Putting $\theta = 0$ at $z = 0$ in Eq. (ii)

$$0 = A + D$$

or

$$A = -D$$

Also

$$\frac{d\theta}{dz} = -\mu A \sin \mu z + \mu B \cos \mu z + C$$

and since $(d\theta/dz) = 0$ at $z = 0$

$$C = -\mu B$$

When $z = L$, $\theta = 0$ so that, from Eq. (ii)

$$0 = A \cos \mu L + B \sin \mu L + CL + D$$

which may be rewritten

$$0 = B(\sin \mu L - \mu L) + A(\cos \mu L - 1) \qquad \text{(iii)}$$

Then for $(d\theta/dz) = 0$ at $z = L$

$$0 = \mu B \cos \mu L - \mu A \sin \mu L - \mu B$$

or

$$0 = B(\cos \mu L - 1) - A \sin \mu L \tag{iv}$$

Eliminating A from Eqs (iii) and (iv)

$$0 = B[2(1 - \cos \mu L) - \mu L \sin \mu L] \tag{v}$$

Similarly, in terms of the constant C

$$0 = -C[2(1 - \cos \mu L) - \mu L \sin \mu L] \tag{vi}$$

or

$$B = -C$$

But $B = -C/\mu$ so that to satisfy both equations $B = C = 0$ and

$$\theta = A \cos \mu z - A = A(\cos \mu z - 1) \tag{vii}$$

Since $\theta = 0$ at $z = l$

$$\cos \mu L = 1$$

or

$$\mu L = 2n\pi$$

Therefore

$$\mu^2 L^2 = 4n^2\pi^2$$

or

$$\frac{\sigma I_0 - GJ}{E\Gamma} = \frac{4n^2\pi^2}{L^2}$$

The lowest value of torsional buckling load corresponds to $n = 1$ so that, rearranging the above

$$\sigma = \frac{1}{I_0}\left(GJ + \frac{4\pi^2 E\Gamma}{L^2}\right) \tag{viii}$$

The polar second moment of area I_0 is given by

$$I_0 = I_{xx} + I_{yy} \qquad \text{(see Ref. 2)}$$

ie

$$I_0 = 2\left(td\,d^2 + \frac{td^3}{3}\right) + \frac{3td^3}{12} + 2td\frac{d^2}{4}$$

which gives

$$I_0 = \frac{41td^3}{12}$$

Substituting for I_0, J and Γ in Eq. (viii)

$$\sigma = \frac{4}{41d^3}\left(sgt^2 + \frac{13\pi^2 Ed^4}{L^2}\right)$$

References

1 Timoshenko, S. P. and Gere, J. M., *Theory of Elastic Stability*, 2nd edition, McGraw-Hill Book Company, New York, 1961.

2 Megson, T. H. G., *Structural and Stress Analysis*, 2nd edition, Elsevier, Oxford, 2005.

Problems

P.8.1 The system shown in Fig. P.8.1 consists of two bars AB and BC, each of bending stiffness EI elastically hinged together at B by a spring of stiffness K (i.e. bending moment applied by spring $= K \times$ change in slope across B).

Regarding A and C as simple pin-joints, obtain an equation for the first buckling load of the system. What are the lowest buckling loads when (a) $K \to \infty$, (b) $EI \to \infty$. Note that B is free to move vertically.

Ans. $\mu K/\tan \mu l$.

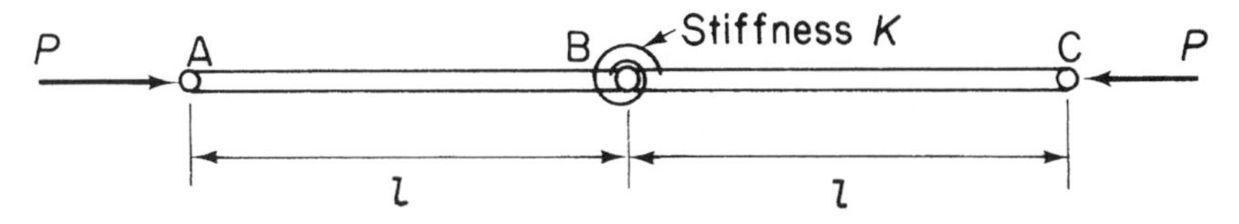

Fig. P.8.1

P.8.2 A pin-ended column of length l and constant flexural stiffness EI is reinforced to give a flexural stiffness $4EI$ over its central half (see Fig. P.8.2).

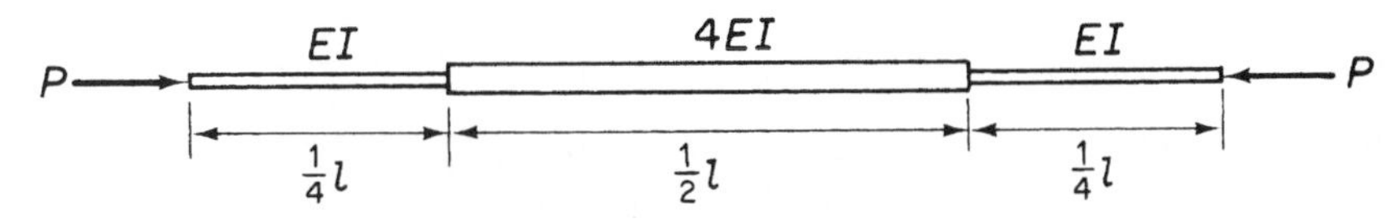

Fig. P.8.2

Considering symmetric modes of buckling only, obtain the equation whose roots yield the flexural buckling loads and solve for the lowest buckling load.

Ans. $\tan \mu l/8 = 1/\sqrt{2}, P = 24.2EI/l^2$

P.8.3 A uniform column of length l and bending stiffness EI is built-in at one end and free at the other and has been designed so that its lowest flexural buckling load is P (see Fig. P.8.3).

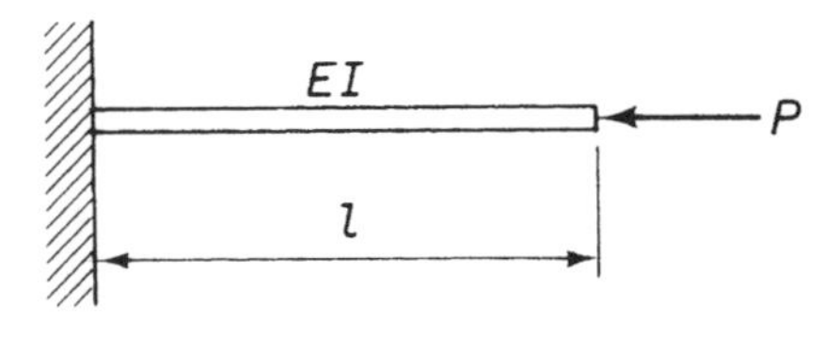

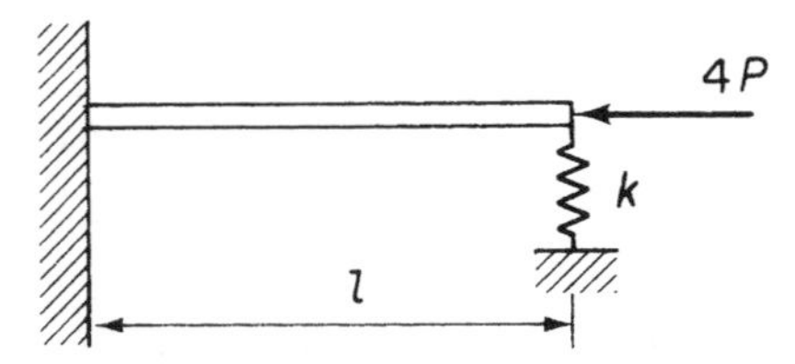

Fig. P.8.3

Subsequently it has to carry an increased load, and for this it is provided with a lateral spring at the free end. Determine the necessary spring stiffness k so that the buckling load becomes $4P$.

Ans. $k = 4P\mu/(\mu l - \tan \mu l)$.

P.8.4 A uniform, pin-ended column of length l and bending stiffness EI has an initial curvature such that the lateral displacement at any point between the column and the straight line joining its ends is given by

$$v_0 = a\frac{4z}{l^2}(l - z) \qquad \text{(see Fig. P.8.4)}$$

Show that the maximum bending moment due to a compressive end load P is given by

$$M_{\max} = -\frac{8aP}{(\lambda l)^2}\left(\sec\frac{\lambda l}{2} - 1\right)$$

where

$$\lambda^2 = P/EI$$

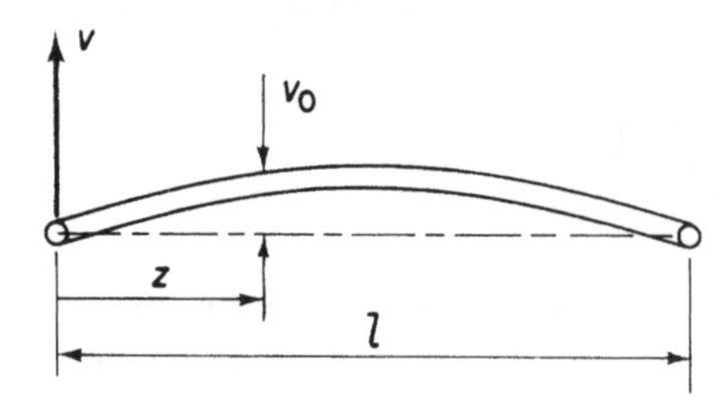

Fig. P.8.4

P.8.5 The uniform pin-ended column shown in Fig. P.8.5 is bent at the centre so that its eccentricity there is δ. If the two halves of the column are otherwise straight and have a flexural stiffness EI, find the value of the maximum bending moment when the column carries a compression load P.

Ans. $-P\frac{2\delta}{l}\sqrt{\frac{EI}{P}}\tan\sqrt{\frac{P}{EI}}\frac{l}{2}$.

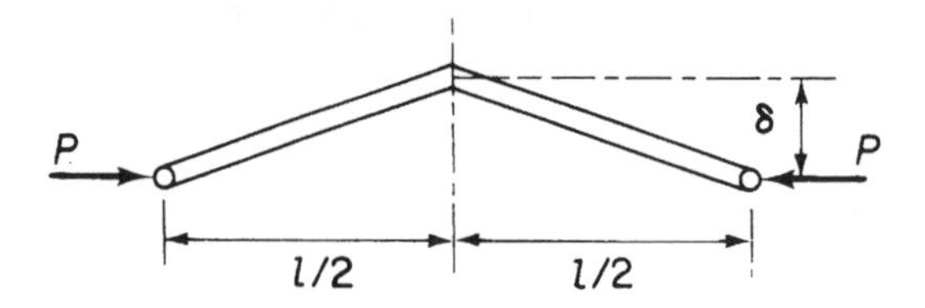

Fig. P.8.5

P.8.6 A straight uniform column of length l and bending stiffness EI is subjected to uniform lateral loading w/unit length. The end attachments do not restrict rotation of the column ends. The longitudinal compressive force P has eccentricity e from the centroids of the end sections and is placed so as to oppose the bending effect of the lateral loading, as shown in Fig. P.8.6. The eccentricity e can be varied and is to be adjusted to the value which, for given values of P and w, will result in the least maximum bending moment on the column. Show that

$$e = (w/P\mu^2)\tan^2 \mu l/4$$

where

$$\mu^2 = P/EI$$

Deduce the end moment which will give the optimum condition when P tends to zero.

Ans. $wl^2/16$.

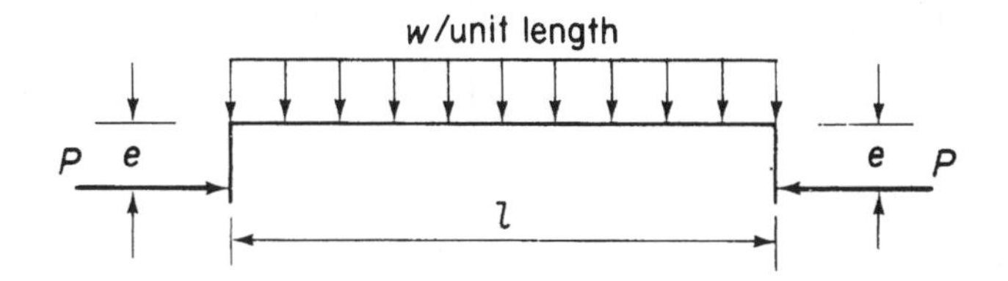

Fig. P.8.6

P.8.7 The relation between stress σ and strain ε in compression for a certain material is

$$10.5 \times 10^6 \varepsilon = \sigma + 21\,000\left(\frac{\sigma}{49\,000}\right)^{16}$$

Assuming the tangent modulus equation to be valid for a uniform strut of this material, plot the graph of σ_b against l/r where σ_b is the flexural buckling stress, l the equivalent pin-ended length and r the least radius of gyration of the cross-section.

Estimate the flexural buckling load for a tubular strut of this material, of 1.5 units outside diameter and 0.08 units wall thickness with effective length 20 units.

Ans. 14 454 force units.

P.8.8 A rectangular portal frame ABCD is rigidly fixed to a foundation at A and D and is subjected to a compression load P applied at each end of the horizontal member BC (see Fig. P.8.8). If the members all have the same bending stiffness EI show that the buckling loads for modes which are symmetrical about the vertical centre line are given by the transcendental equation

$$\frac{\lambda a}{2} = -\frac{1}{2}\left(\frac{a}{b}\right)\tan\left(\frac{\lambda a}{2}\right)$$

where

$$\lambda^2 = P/EI$$

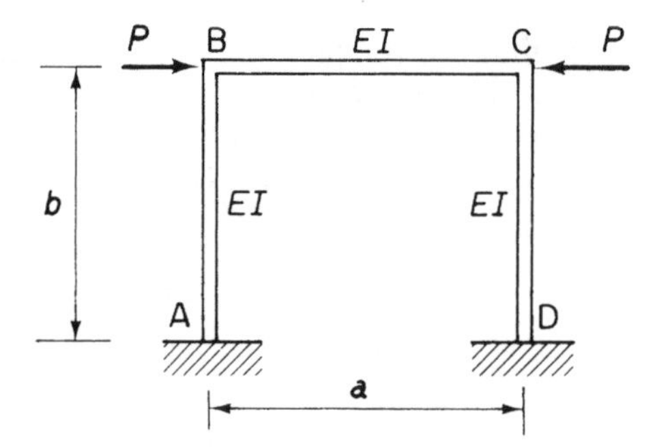

Fig. P.8.8

P.8.9 A compression member (Fig. P.8.9) is made of circular section tube, diameter d, thickness t. The member is not perfectly straight when unloaded, having a slightly bowed shape which may be represented by the expression

$$v = \delta \sin\left(\frac{\pi z}{l}\right)$$

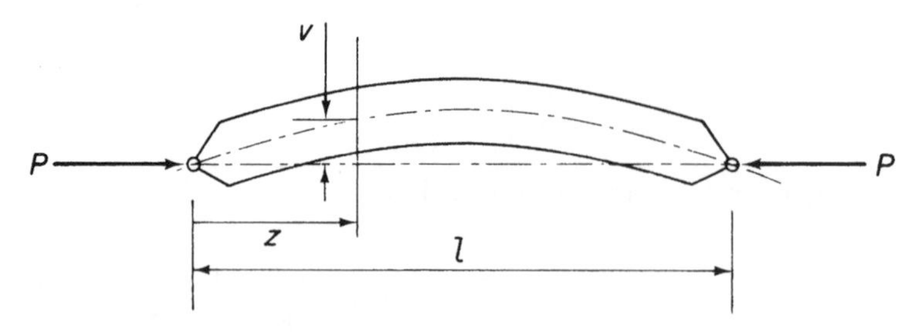

Fig. P.8.9

Show that when the load P is applied, the maximum stress in the member can be expressed as

$$\sigma_{\text{max}} = \frac{P}{\pi dt}\left[1 + \frac{1}{1-\alpha}\frac{4\delta}{d}\right]$$

where

$$\alpha = P/P_{\text{e}}, \quad P_{\text{e}} = \pi^2 EI/l^2$$

Assume t is small compared with d so that the following relationships are applicable:
Cross-sectional area of tube $= \pi dt$.
Second moment of area of tube $= \pi d^3 t/8$.

P.8.10 Figure P.8.10 illustrates an idealized representation of part of an aircraft control circuit. A uniform, straight bar of length a and flexural stiffness EI is built-in at the end A and hinged at B to a link BC, of length b, whose other end C is pinned so that it is free to slide along the line ABC between smooth, rigid guides. A, B and C are initially in a straight line and the system carries a compression force P, as shown.

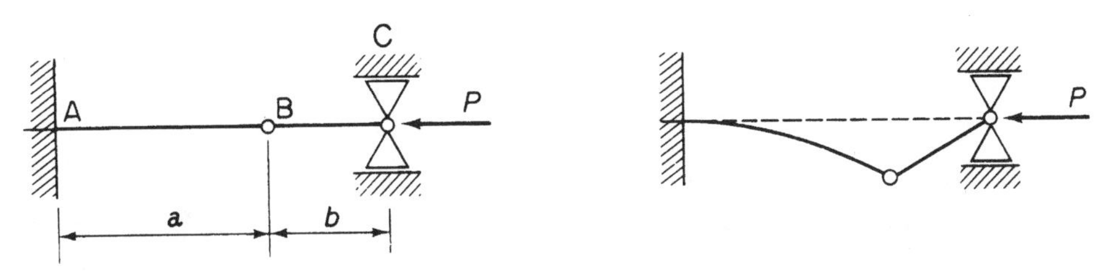

Fig. P.8.10

Assuming that the link BC has a sufficiently high flexural stiffness to prevent its buckling as a pin-ended strut, show, by setting up and solving the differential equation for flexure of AB, that buckling of the system, of the type illustrated in Fig. P.8.10, occurs when P has such a value that

$$\tan \lambda a = \lambda(a + b)$$

where

$$\lambda^2 = P/EI$$

P.8.11 A pin-ended column of length l has its central portion reinforced, the second moment of its area being I_2 while that of the end portions, each of length a, is I_1. Use the energy method to determine the critical load of the column, assuming that its centre-line deflects into the parabola $v = kz(l - z)$ and taking the more accurate of the two expressions for the bending moment.

In the case where $I_2 = 1.6I_1$ and $a = 0.2l$ find the percentage increase in strength due to the reinforcement, and compare it with the percentage increase in weight on the basis that the radius of gyration of the section is not altered.

Ans. $P_{CR} = 14.96EI_1/l^2, 52\%, 36\%$.

P.8.12 A tubular column of length l is tapered in wall-thickness so that the area and the second moment of area of its cross-section decrease uniformly from A_1 and I_1 at its centre to $0.2A_1$ and $0.2I_1$ at its ends.

Assuming a deflected centre-line of parabolic form, and taking the more correct form for the bending moment, use the energy method to estimate its critical load when tested between pin-centres, in terms of the above data and Young's modulus E. Hence show that the saving in weight by using such a column instead of one having the same radius of gyration and constant thickness is about 15%.

Ans. $7.01EI_1/l^2$.

P.8.13 A uniform column (Fig. P.8.13), of length l and bending stiffness EI, is rigidly built-in at the end $z = 0$ and simply supported at the end $z = l$. The column is also attached to an elastic foundation of constant stiffness k/unit length.

Representing the deflected shape of the column by a polynomial

$$v = \sum_{n=0}^{p} a_n \eta^n, \quad \text{where } \eta = z/l$$

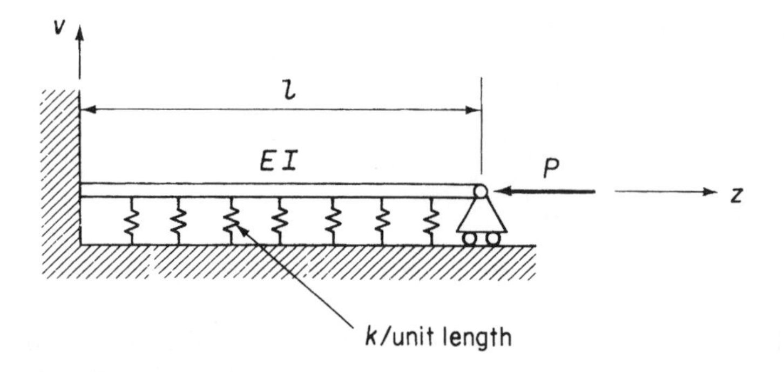

Fig. P.8.13

determine the form of this function by choosing a minimum number of terms p such that all the kinematic (geometric) and static boundary conditions are satisfied, allowing for one arbitrary constant only.

Using the result thus obtained, find an approximation to the lowest flexural buckling load P_{CR} by the Rayleigh–Ritz method.

Ans. $P_{CR} = 21.05EI/l^2 + 0.09kl^2$.

P.8.14 Figure P.8.14 shows the doubly symmetrical cross-section of a thin-walled column with rigidly fixed ends. Find an expression, in terms of the section dimensions and Poisson's ratio, for the column length for which the purely flexural and the purely torsional modes of instability would occur at the same axial load.

In which mode would failure occur if the length were less than the value found? The possibility of local instability is to be ignored.

Ans. $l = (2\pi b^2/t)\sqrt{(1+\nu)/255}$. Torsion.

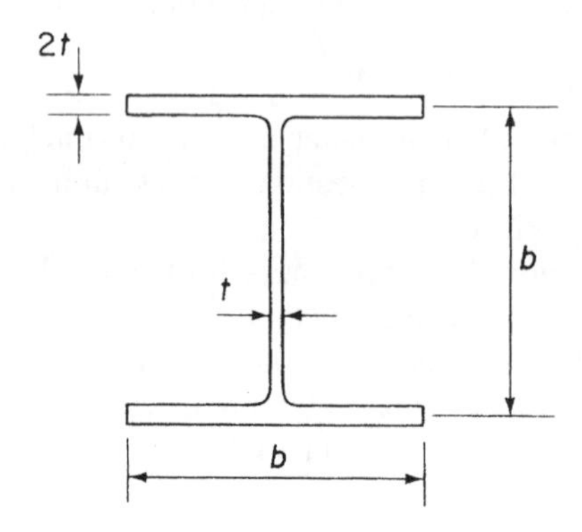

Fig. P.8.14

P.8.15 A column of length $2l$ with the doubly symmetric cross-section shown in Fig. P.8.15 is compressed between the parallel platens of a testing machine which fully prevents twisting and warping of the ends.

Using the data given below, determine the average compressive stress at which the column first buckles in torsion

$$l = 500\,\text{mm},\ b = 25.0\,\text{mm},\ t = 2.5\,\text{mm},\ E = 70\,000\,\text{N/mm}^2,\ E/G = 2.6$$

Ans. $\sigma_{CR} = 282\,\text{N/mm}^2$.

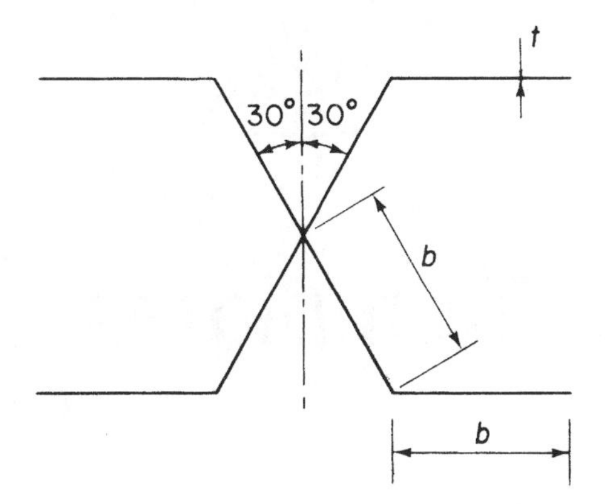

Fig. P.8.15

P.8.16 A pin-ended column of length 1.0 m has the cross-section shown in Fig. P.8.16. If the ends of the column are free to warp determine the lowest value of axial load which will cause the column to buckle, and specify the mode. Take $E = 70\,000\ \text{N/mm}^2$ and $G = 25\,000\ \text{N/mm}^2$.

Ans. 5527 N. Column buckles in bending about an axis in the plane of its web.

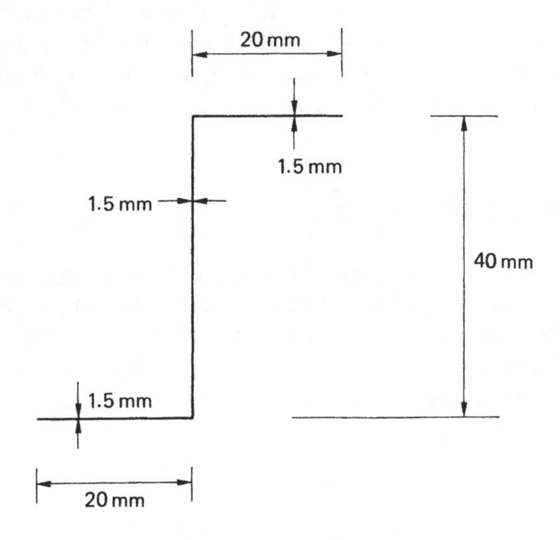

Fig. P.8.16

P.8.17 A pin-ended column of height 3.0 m has a circular cross-section of diameter 80 mm, wall thickness 2.0 mm and is converted to an open section by a narrow longitudinal slit; the ends of the column are free to warp. Determine the values of axial load which would cause the column to buckle in (a) pure bending and (b) pure torsion. Hence determine the value of the flexural–torsional buckling load. Take $E = 70\,000\ \text{N/mm}^2$ and $G = 22\,000\ \text{N/mm}^2$.

Note: the position of the shear centre of the column section may be found using the method described in Chapter 17.

Ans. (a) 3.09×10^4 N, (b) 1.78×10^4 N, 1.19×10^4 N.

9

Thin plates

We shall see in Chapter 12 when we examine the structural components of aircraft that they consist mainly of thin plates stiffened by arrangements of ribs and stringers. Thin plates under relatively small compressive loads are prone to buckle and so must be stiffened to prevent this. The determination of buckling loads for thin plates in isolation is relatively straightforward but when stiffened by ribs and stringers, the problem becomes complex and frequently relies on an empirical solution. In fact it may be the stiffeners which buckle before the plate and these, depending on their geometry, may buckle as a column or suffer local buckling of, say, a flange.

In this chapter we shall present the theory for the determination of buckling loads of flat plates and then examine some of the different empirical approaches which various researchers have suggested. In addition we shall investigate the particular case of flat plates which, when reinforced by horizontal flanges and vertical stiffeners, form the spars of aircraft wing structures; these are known as *tension field beams*.

9.1 Buckling of thin plates

A thin plate may buckle in a variety of modes depending upon its dimensions, the loading and the method of support. Usually, however, buckling loads are much lower than those likely to cause failure in the material of the plate. The simplest form of buckling arises when compressive loads are applied to simply supported opposite edges and the unloaded edges are free, as shown in Fig. 9.1. A thin plate in this configuration

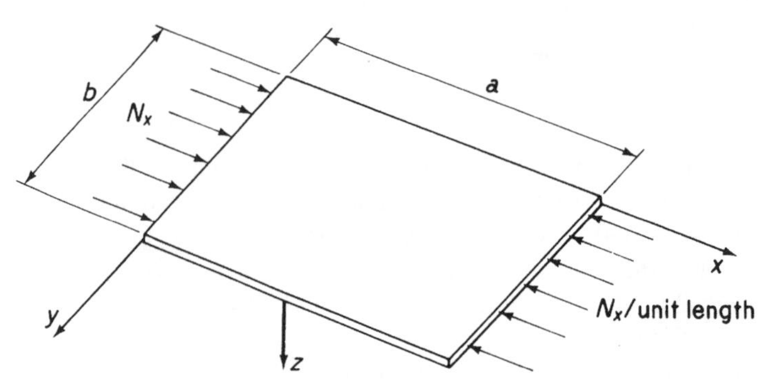

Fig. 9.1 Buckling of a thin flat plate.

behaves in exactly the same way as a pin-ended column so that the critical load is that predicted by the Euler theory. Once this critical load is reached the plate is incapable of supporting any further load. This is not the case, however, when the unloaded edges are supported against displacement out of the xy plane. Buckling, for such plates, takes the form of a bulging displacement of the central region of the plate while the parts adjacent to the supported edges remain straight. These parts enable the plate to resist higher loads; an important factor in aircraft design.

At this stage we are not concerned with this post-buckling behaviour, but rather with the prediction of the critical load which causes the initial bulging of the central area of the plate. For the analysis we may conveniently employ the method of total potential energy since we have already, in Chapter 7, derived expressions for strain and potential energy corresponding to various load and support configurations. In these expressions we assumed that the displacement of the plate comprises bending deflections only and that these are small in comparison with the thickness of the plate. These restrictions therefore apply in the subsequent theory.

First we consider the relatively simple case of the thin plate of Fig. 9.1, loaded as shown, but simply supported along all four edges. We have seen in Chapter 7 that its true deflected shape may be represented by the infinite double trigonometrical series

$$w = \sum_{m=1}^{\infty} \sum_{n=1}^{\infty} A_{mn} \sin \frac{m\pi x}{a} \sin \frac{n\pi y}{b}$$

Also, the total potential energy of the plate is, from Eqs (7.37) and (7.45)

$$\begin{aligned} U + V &= \frac{1}{2} \int_0^a \int_0^b \left[D \left\{ \left(\frac{\partial^2 w}{\partial x^2} + \frac{\partial^2 w}{\partial y^2} \right)^2 \right. \right. \\ & \left. \left. -2(1-\nu) \left[\frac{\partial^2 w}{\partial x^2} \frac{\partial^2 w}{\partial y^2} - \left(\frac{\partial^2 w}{\partial x \, \partial y} \right)^2 \right] \right\} - N_x \left(\frac{\partial w}{\partial x} \right)^2 \right] \mathrm{d}x \, \mathrm{d}y \quad (9.1) \end{aligned}$$

The integration of Eq. (9.1) on substituting for w is similar to those integrations carried out in Chapter 7. Thus, by comparison with Eq. (7.47)

$$U + V = \frac{\pi^4 abD}{8} \sum_{m=1}^{\infty} \sum_{n=1}^{\infty} A_{mn}^2 \left(\frac{m^2}{a^2} + \frac{n^2}{b^2} \right) - \frac{\pi^2 b}{8a} N_x \sum_{m=1}^{\infty} \sum_{n=1}^{\infty} m^2 A_{mn}^2 \quad (9.2)$$

The total potential energy of the plate has a stationary value in the neutral equilibrium of its buckled state (i.e. $N_x = N_{x,\mathrm{CR}}$). Therefore, differentiating Eq. (9.2) with respect to each unknown coefficient A_{mn} we have

$$\frac{\partial(U + V)}{\partial A_{mn}} = \frac{\pi^4 abD}{4} A_{mn} \left(\frac{m^2}{a^2} + \frac{n^2}{b^2} \right)^2 - \frac{\pi^2 b}{4a} N_{x,\mathrm{CR}} m^2 A_{mn} = 0$$

and for a non-trivial solution

$$N_{x,\mathrm{CR}} = \pi^2 a^2 D \frac{1}{m^2} \left(\frac{m^2}{a^2} + \frac{n^2}{b^2} \right)^2 \quad (9.3)$$

Exactly the same result may have been deduced from Eq. (ii) of Example 7.3, where the displacement w would become infinite for a negative (compressive) value of N_x equal to that of Eq. (9.3).

We observe from Eq. (9.3) that each term in the infinite series for displacement corresponds, as in the case of a column, to a different value of critical load (note, the problem is an eigenvalue problem). The lowest value of critical load evolves from some critical combination of integers m and n, i.e. the number of half-waves in the x and y directions, and the plate dimensions. Clearly $n = 1$ gives a minimum value so that no matter what the values of m, a and b the plate buckles into a half sine wave in the y direction. Thus we may write Eq. (9.3) as

$$N_{x,\mathrm{CR}} = \pi^2 a^2 D \frac{1}{m^2}\left(\frac{m^2}{a^2} + \frac{1}{b^2}\right)^2$$

or

$$N_{x,\mathrm{CR}} = \frac{k\pi^2 D}{b^2} \tag{9.4}$$

where the plate *buckling coefficient* k is given by the minimum value of

$$k = \left(\frac{mb}{a} + \frac{a}{mb}\right)^2 \tag{9.5}$$

for a given value of a/b. To determine the minimum value of k for a given value of a/b we plot k as a function of a/b for different values of m as shown by the dotted curves in Fig. 9.2. The minimum value of k is obtained from the lower envelope of the curves shown solid in the figure.

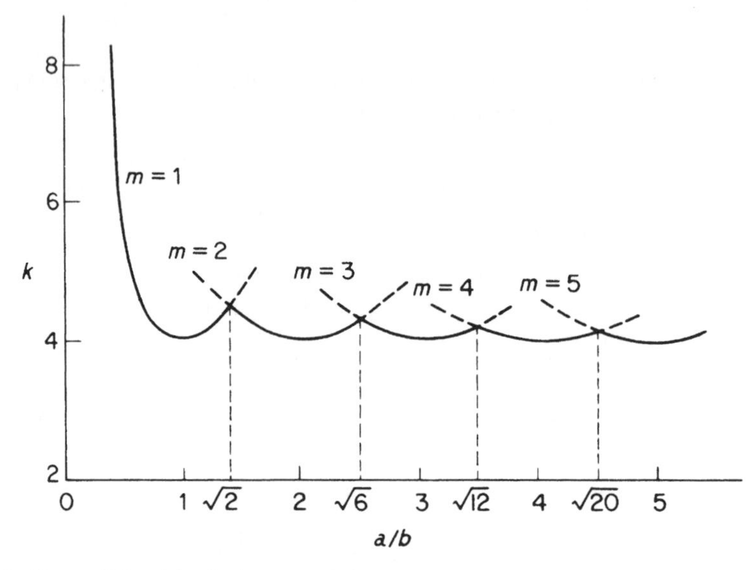

Fig. 9.2 Buckling coefficient k for simply supported plates.

It can be seen that m varies with the ratio a/b and that k and the buckling load are a minimum when $k=4$ at values of $a/b=1,2,3,\ldots$. As a/b becomes large k approaches 4 so that long narrow plates tend to buckle into a series of squares.

The transition from one buckling mode to the next may be found by equating values of k for the m and $m+1$ curves. Hence

$$\frac{mb}{a}+\frac{a}{mb}=\frac{(m+1)b}{a}+\frac{a}{(m+1)b}$$

giving

$$\frac{a}{b}=\sqrt{m(m+1)}$$

Substituting $m=1$, we have $a/b=\sqrt{2}=1.414$, and for $m=2$, $a/b=\sqrt{6}=2.45$ and so on.

For a given value of a/b the critical stress, $\sigma_{CR}=N_{x,CR}/t$, is found from Eqs (9.4) and (7.4), i.e.

$$\sigma_{CR}=\frac{k\pi^2 E}{12(1-\nu^2)}\left(\frac{t}{b}\right)^2 \tag{9.6}$$

In general, the critical stress for a uniform rectangular plate, with various edge supports and loaded by constant or linearly varying in-plane direct forces (N_x, N_y) or constant shear forces (N_{xy}) along its edges, is given by Eq. (9.6). The value of k remains a function of a/b but depends also upon the type of loading and edge support. Solutions for such problems have been obtained by solving the appropriate differential equation or by using the approximate (Rayleigh–Ritz) energy method. Values of k for a variety of loading and support conditions are shown in Fig. 9.3. In Fig. 9.3(c), where k becomes the *shear buckling coefficient*, b is always the smaller dimension of the plate.

We see from Fig. 9.3 that k is very nearly constant for $a/b>3$. This fact is particularly useful in aircraft structures where longitudinal stiffeners are used to divide the skin into narrow panels (having small values of b), thereby increasing the buckling stress of the skin.

9.2 Inelastic buckling of plates

For plates having small values of b/t the critical stress may exceed the elastic limit of the material of the plate. In such a situation, Eq. (9.6) is no longer applicable since, as we saw in the case of columns, E becomes dependent on stress as does Poisson's ratio ν. These effects are usually included in a plasticity correction factor η so that Eq. (9.6) becomes

$$\sigma_{CR}=\frac{\eta k\pi^2 E}{12(1-\nu^2)}\left(\frac{t}{b}\right)^2 \tag{9.7}$$

where E and ν are elastic values of Young's modulus and Poisson's ratio. In the linearly elastic region $\eta=1$, which means that Eq. (9.7) may be applied at all stress levels. The

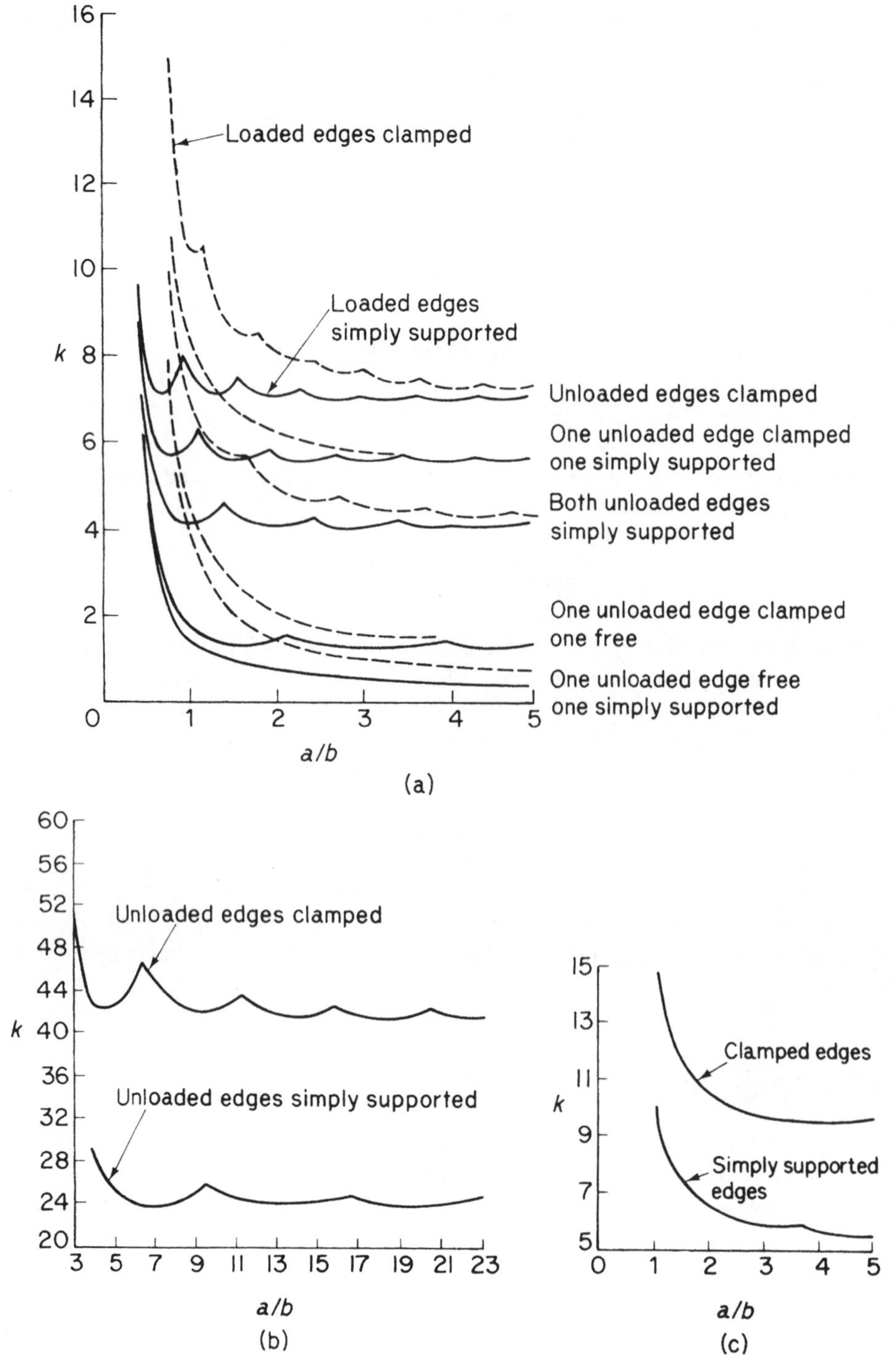

Fig. 9.3 (a) Buckling coefficients for flat plates in compression; (b) buckling coefficients for flat plates in bending; (c) shear buckling coefficients for flat plates.

derivation of a general expression for η is outside the scope of this book but one[1] giving good agreement with experiment is

$$\eta = \frac{1-\nu_e^2}{1-\nu_p^2}\frac{E_s}{E}\left[\frac{1}{2}+\frac{1}{2}\left(\frac{1}{4}+\frac{3}{4}\frac{E_t}{E_s}\right)^{\frac{1}{2}}\right]$$

where E_t and E_s are the tangent modulus and secant modulus (stress/strain) of the plate in the inelastic region and ν_e and ν_p are Poisson's ratio in the elastic and inelastic ranges.

9.3 Experimental determination of critical load for a flat plate

In Section 8.3 we saw that the critical load for a column may be determined experimentally, without actually causing the column to buckle, by means of the Southwell plot. The critical load for an actual, rectangular, thin plate is found in a similar manner.

The displacement of an initially curved plate from the zero load position was found in Section 7.5, to be

$$w_1 = \sum_{m=1}^{\infty}\sum_{n=1}^{\infty} B_{mn} \sin\frac{m\pi x}{a}\sin\frac{n\pi y}{b}$$

where

$$B_{mn} = \frac{A_{mn}N_x}{\frac{\pi^2 D}{a^2}\left(m+\frac{n^2a^2}{mb^2}\right)^2 - N_x}$$

We see that the coefficients B_{mn} increase with an increase of compressive load intensity N_x. It follows that when N_x approaches the critical value, $N_{x,CR}$, the term in the series corresponding to the buckled shape of the plate becomes the most significant. For a square plate $n=1$ and $m=1$ give a minimum value of critical load so that at the centre of the plate

$$w_1 = \frac{A_{11}N_x}{N_{x,CR} - N_x}$$

or, rearranging

$$w_1 = N_{x,CR}\frac{w_1}{N_x} - A_{11}$$

Thus, a graph of w_1 plotted against w_1/N_x will have a slope, in the region of the critical load, equal to $N_{x,CR}$.

9.4 Local instability

We distinguished in the introductory remarks to Chapter 8 between primary and secondary (or local) instability. The latter form of buckling usually occurs in the flanges and webs of thin-walled columns having an effective slenderness ratio, $l_e/r < 20$. For $l_e/r > 80$ this type of column is susceptible to primary instability. In the intermediate range of l_e/r between 20 and 80, buckling occurs by a combination of both primary and secondary modes.

Thin-walled columns are encountered in aircraft structures in the shape of longitudinal stiffeners, which are normally fabricated by extrusion processes or by forming from a flat sheet. A variety of cross-sections are employed although each is usually composed of flat plate elements arranged to form angle, channel, Z- or 'top hat' sections, as shown in Fig. 9.4. We see that the plate elements fall into two distinct categories: flanges which have a free unloaded edge and webs which are supported by the adjacent plate elements on both unloaded edges.

In local instability the flanges and webs buckle like plates with a resulting change in the cross-section of the column. The wavelength of the buckle is of the order of the widths of the plate elements and the corresponding critical stress is generally independent of the length of the column when the length is equal to or greater than three times the width of the largest plate element in the column cross-section.

Buckling occurs when the weakest plate element, usually a flange, reaches its critical stress, although in some cases all the elements reach their critical stresses simultaneously. When this occurs the rotational restraint provided by adjacent elements to each other disappears and the elements behave as though they are simply supported along their common edges. These cases are the simplest to analyse and are found where the cross-section of the column is an equal-legged angle, T-, cruciform or a square tube of constant thickness. Values of local critical stress for columns possessing these types of section may be found using Eq. (9.7) and an appropriate value of k. For example, k for a cruciform section column is obtained from Fig. 9.3(a) for a plate which is simply supported on three sides with one edge free and has $a/b > 3$. Hence $k = 0.43$ and if the section buckles elastically then $\eta = 1$ and

$$\sigma_{\mathrm{CR}} = 0.388E\left(\frac{t}{b}\right)^2 \quad (\nu = 0.3)$$

It must be appreciated that the calculation of local buckling stresses is generally complicated with no particular method gaining universal acceptance, much of the information available being experimental. A detailed investigation of the topic is therefore beyond the scope of this book. Further information may be obtained from all the references listed at the end of this chapter.

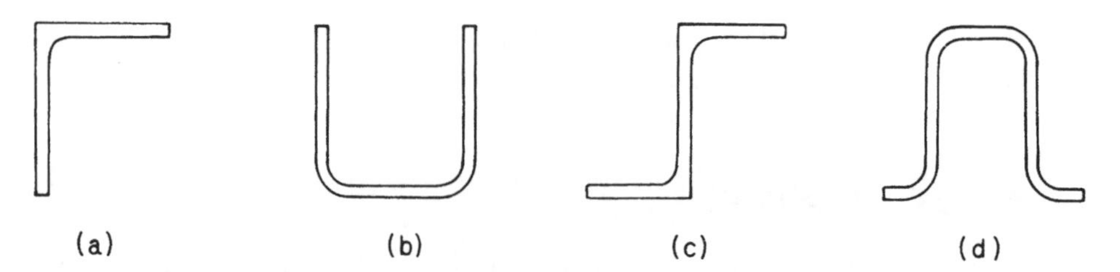

Fig. 9.4 (a) Extruded angle; (b) formed channel; (c) extruded Z; (d) formed 'top hat'.

9.5 Instability of stiffened panels

It is clear from Eq. (9.7) that plates having large values of b/t buckle at low values of critical stress. An effective method of reducing this parameter is to introduce stiffeners along the length of the plate thereby dividing a wide sheet into a number of smaller and more stable plates. Alternatively, the sheet may be divided into a series of wide short columns by stiffeners attached across its width. In the former type of structure the longitudinal stiffeners carry part of the compressive load, while in the latter all the load is supported by the plate. Frequently, both methods of stiffening are combined to form a grid-stiffened structure.

Stiffeners in earlier types of stiffened panel possessed a relatively high degree of strength compared with the thin skin resulting in the skin buckling at a much lower stress level than the stiffeners. Such panels may be analysed by assuming that the stiffeners provide simply supported edge conditions to a series of flat plates.

A more efficient structure is obtained by adjusting the stiffener sections so that buckling occurs in both stiffeners and skin at about the same stress. This is achieved by a construction involving closely spaced stiffeners of comparable thickness to the skin. Since their critical stresses are nearly the same there is an appreciable interaction at buckling between skin and stiffeners so that the complete panel must be considered as a unit. However, caution must be exercised since it is possible for the two simultaneous critical loads to interact and reduce the actual critical load of the structure[2] (see Example 8.4). Various modes of buckling are possible, including primary buckling where the wavelength is of the order of the panel length and local buckling with wavelengths of the order of the width of the plate elements of the skin or stiffeners. A discussion of the various buckling modes of panels having Z-section stiffeners has been given by Argyris and Dunne.[3]

The prediction of critical stresses for panels with a large number of longitudinal stiffeners is difficult and relies heavily on approximate (energy) and semi-empirical methods. Bleich[4] and Timoshenko (see Ref. 1, Chapter 8) give energy solutions for plates with one and two longitudinal stiffeners and also consider plates having a large number of stiffeners. Gerard and Becker[5] have summarized much of the work on stiffened plates and a large amount of theoretical and empirical data is presented by Argyris and Dunne in the *Handbook of Aeronautics*.[3]

For detailed work on stiffened panels, reference should be made to as much as possible of the above work. The literature is, however, extensive so that here we present a relatively simple approach suggested by Gerard[1]. Figure 9.5 represents a panel of width w stiffened by longitudinal members which may be flats (as shown), Z-, I-, channel or

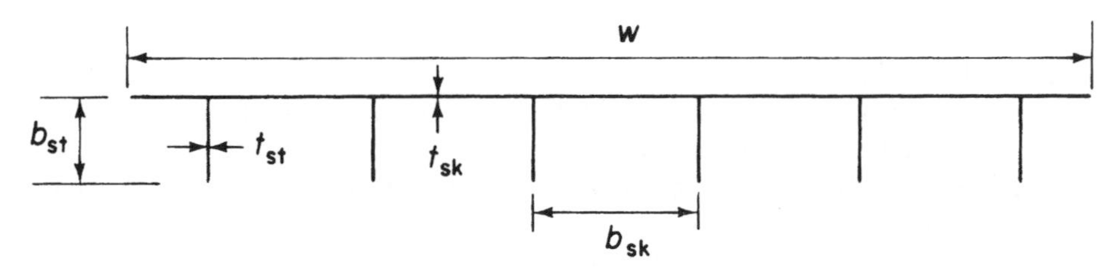

Fig. 9.5 Stiffened panel.

'top hat' sections. It is possible for the panel to behave as an Euler column, its cross-section being that shown in Fig. 9.5. If the equivalent length of the panel acting as a column is l_e then the Euler critical stress is

$$\sigma_{CR,E} = \frac{\pi^2 E}{(l_e/r)^2}$$

as in Eq. (8.8). In addition to the column buckling mode, individual plate elements comprising the panel cross-section may buckle as long plates. The buckling stress is then given by Eq. (9.7), i.e.

$$\sigma_{CR} = \frac{\eta k \pi^2 E}{12(1-\nu^2)}\left(\frac{t}{b}\right)^2$$

where the values of k, t and b depend upon the particular portion of the panel being investigated. For example, the portion of skin between stiffeners may buckle as a plate simply supported on all four sides. Thus, for $a/b > 3$, $k = 4$ from Fig. 9.3(a) and, assuming that buckling takes place in the elastic range

$$\sigma_{CR} = \frac{4\pi^2 E}{12(1-\nu^2)}\left(\frac{t_{sk}}{b_{sk}}\right)^2$$

A further possibility is that the stiffeners may buckle as long plates simply supported on three sides with one edge free. Thus

$$\sigma_{CR} = \frac{0.43\pi^2 E}{12(1-\nu^2)}\left(\frac{t_{st}}{b_{st}}\right)^2$$

Clearly, the minimum value of the above critical stresses is the critical stress for the panel taken as a whole.

The compressive load is applied to the panel over its complete cross-section. To relate this load to an applied compressive stress σ_A acting on each element of the cross-section we divide the load per unit width, say N_x, by an equivalent skin thickness $\bar{t}$, hence

$$\sigma_A = \frac{N_x}{\bar{t}}$$

where

$$\bar{t} = \frac{A_{st}}{b_{sk}} + t_{sk}$$

and A_{st} is the stiffener area.

The above remarks are concerned with the primary instability of stiffened panels. Values of local buckling stress have been determined by Boughan, Baab and Gallaher for idealized web, Z- and T- stiffened panels. The results are reproduced in Rivello[6] together with the assumed geometries.

Further types of instability found in stiffened panels occur where the stiffeners are riveted or spot welded to the skin. Such structures may be susceptible to *interrivet buckling* in which the skin buckles between rivets with a wavelength equal to the rivet pitch, or *wrinkling* where the stiffener forms an elastic line support for the skin. In the latter mode the wavelength of the buckle is greater than the rivet pitch and separation of skin and stiffener does not occur. Methods of estimating the appropriate critical stresses are given in Rivello[6] and the *Handbook of Aeronautics*.[3]

9.6 Failure stress in plates and stiffened panels

The previous discussion on plates and stiffened panels investigated the prediction of buckling stresses. However, as we have seen, plates retain some of their capacity to carry load even though a portion of the plate has buckled. In fact, the ultimate load is not reached until the stress in the majority of the plate exceeds the elastic limit. The theoretical calculation of the ultimate stress is difficult since non-linearity results from both large deflections and the inelastic stress–strain relationship.

Gerard[1] proposes a semi-empirical solution for flat plates supported on all four edges. After elastic buckling occurs theory and experiment indicate that the average compressive stress, $\bar{\sigma}_{\mathrm{a}}$, in the plate and the unloaded edge stress, σ_{e}, are related by the following expression

$$\frac{\bar{\sigma}_{\mathrm{a}}}{\sigma_{\mathrm{CR}}} = \alpha_1 \left(\frac{\sigma_{\mathrm{e}}}{\sigma_{\mathrm{CR}}} \right)^n \tag{9.8}$$

where

$$\sigma_{\mathrm{CR}} = \frac{k\pi^2 E}{12(1-\nu^2)} \left(\frac{t}{b} \right)^2$$

and α_1 is some unknown constant. Theoretical work by Stowell[7] and Mayers and Budiansky[8] shows that failure occurs when the stress along the unloaded edge is approximately equal to the compressive yield strength, σ_{cy}, of the material. Hence substituting σ_{cy} for σ_{e} in Eq. (9.8) and rearranging gives

$$\frac{\bar{\sigma}_{\mathrm{f}}}{\sigma_{\mathrm{cy}}} = \alpha_1 \left(\frac{\sigma_{\mathrm{CR}}}{\sigma_{\mathrm{cy}}} \right)^{1-n} \tag{9.9}$$

where the average compressive stress in the plate has become the average stress at failure $\bar{\sigma}_{\mathrm{f}}$. Substituting for σ_{CR} in Eq. (9.9) and putting

$$\frac{\alpha_1 \pi^{2(1-n)}}{[12(1-\nu^2)]^{1-n}} = \alpha$$

yields

$$\frac{\bar{\sigma}_{\mathrm{f}}}{\sigma_{\mathrm{cy}}} = \alpha k^{1-n} \left[\frac{t}{b} \left(\frac{E}{\sigma_{\mathrm{cy}}} \right)^{\frac{1}{2}} \right]^{2(1-n)} \tag{9.10}$$

or, in a simplified form

$$\frac{\bar{\sigma}_{\mathrm{f}}}{\sigma_{\mathrm{cy}}} = \beta\left[\frac{t}{b}\left(\frac{E}{\sigma_{\mathrm{cy}}}\right)^{\frac{1}{2}}\right]^m \tag{9.11}$$

where $\beta = \alpha k^{m/2}$. The constants β and m are determined by the best fit of Eq. (9.11) to test data.

Experiments on simply supported flat plates and square tubes of various aluminium and magnesium alloys and steel show that $\beta = 1.42$ and $m = 0.85$ fit the results within ± 10 per cent up to the yield strength. Corresponding values for long clamped flat plates are $\beta = 1.80$, $m = 0.85$.

Gerard[9–12] extended the above method to the prediction of local failure stresses for the plate elements of thin-walled columns. Equation (9.11) becomes

$$\frac{\bar{\sigma}_{\mathrm{f}}}{\sigma_{\mathrm{cy}}} = \beta_g\left[\left(\frac{gt^2}{A}\right)\left(\frac{E}{\sigma_{\mathrm{cy}}}\right)^{\frac{1}{2}}\right]^m \tag{9.12}$$

where A is the cross-sectional area of the column, β_g and m are empirical constants and g is the number of cuts required to reduce the cross-section to a series of flanged sections plus the number of flanges that would exist after the cuts are made. Examples of the determination of g are shown in Fig. 9.6.

The local failure stress in longitudinally stiffened panels was determined by Gerard[10,12] using a slightly modified form of Eqs (9.11) and (9.12). Thus, for a section of the panel consisting of a stiffener and a width of skin equal to the stiffener spacing

$$\frac{\bar{\sigma}_{\mathrm{f}}}{\sigma_{\mathrm{cy}}} = \beta_g\left[\frac{g t_{\mathrm{sk}} t_{\mathrm{st}}}{A}\left(\frac{E}{\bar{\sigma}_{\mathrm{cy}}}\right)^{\frac{1}{2}}\right]^m \tag{9.13}$$

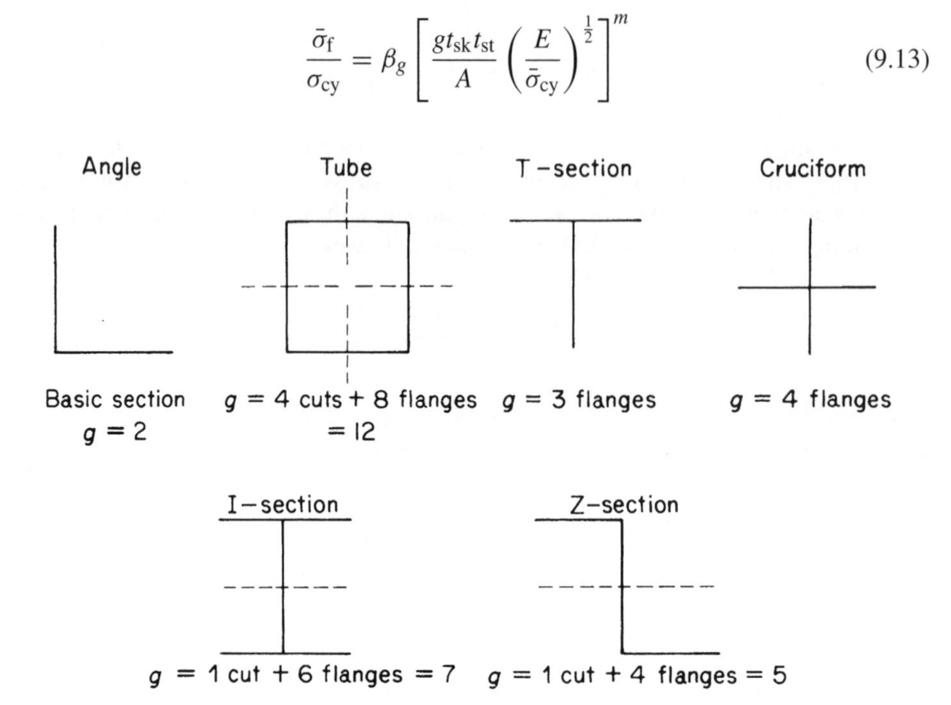

Fig. 9.6 Determination of empirical constant g.

where t_{sk} and t_{st} are the skin and stiffener thicknesses, respectively. A weighted yield stress $\bar{\sigma}_{cy}$ is used for a panel in which the material of the skin and stiffener have different yield stresses, thus

$$\bar{\sigma}_{cy} = \frac{\sigma_{cy} + \sigma_{cy,sk}[(\bar{t}/t_{st}) - 1]}{\bar{t}/t_{st}}$$

where $\bar{t}$ is the average or equivalent skin thickness previously defined. The parameter g is obtained in a similar manner to that for a thin-walled column, except that the number of cuts in the skin and the number of equivalent flanges of the skin are included. A cut to the left of a stiffener is not counted since it is regarded as belonging to the stiffener to the left of that cut. The calculation of g for two types of skin/stiffener combination is illustrated in Fig. 9.7. Equation (9.13) is applicable to either monolithic or built up panels when, in the latter case, interrivet buckling and wrinkling stresses are greater than the local failure stress.

The values of failure stress given by Eqs (9.11), (9.12) and (9.13) are associated with local or secondary instability modes. Consequently, they apply when $l_e/r \leq 20$. In the intermediate range between the local and primary modes, failure occurs through a combination of both. At the moment there is no theory that predicts satisfactorily failure in this range and we rely on test data and empirical methods. The NACA (now NASA) have produced direct reading charts for the failure of 'top hat', Z- and Y-section stiffened panels; a bibliography of the results is given by Gerard.[10]

It must be remembered that research into methods of predicting the instability and post-buckling strength of the thin-walled types of structure associated with aircraft construction is a continuous process. Modern developments include the use of the computer-based finite element technique (see Chapter 6) and the study of the sensitivity of thin-walled structures to imperfections produced during fabrication; much useful information and an extensive bibliography is contained in Murray.[2]

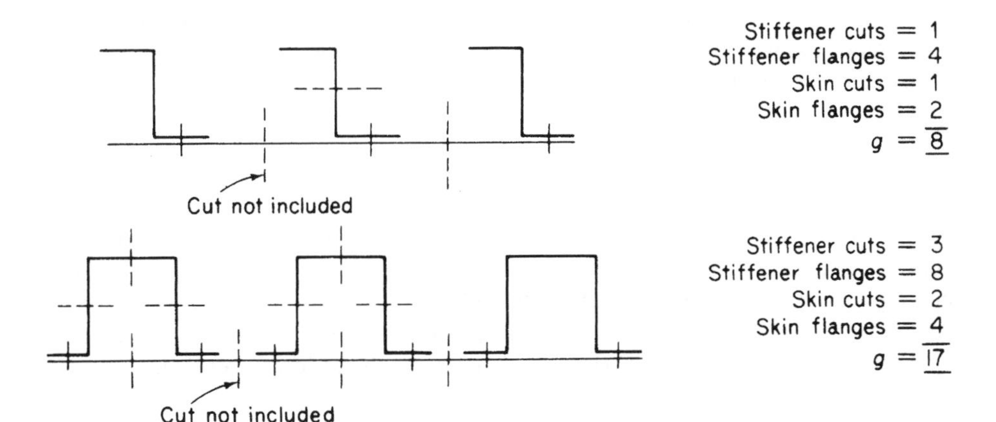

Fig. 9.7 Determination of g for two types of stiffener/skin combination.

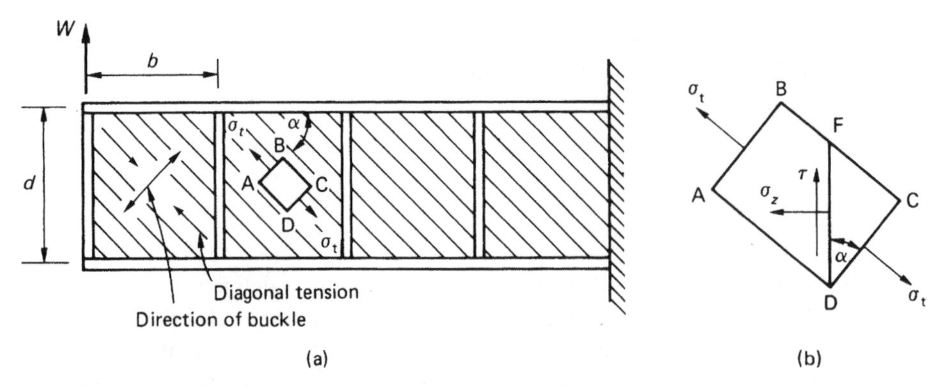

Fig. 9.8 Diagonal tension field beam.

9.7 Tension field beams

The spars of aircraft wings usually comprise an upper and a lower flange connected by thin stiffened webs. These webs are often of such a thickness that they buckle under shear stresses at a fraction of their ultimate load. The form of the buckle is shown in Fig. 9.8(a), where the web of the beam buckles under the action of internal diagonal compressive stresses produced by shear, leaving a wrinkled web capable of supporting diagonal tension only in a direction perpendicular to that of the buckle; the beam is then said to be a *complete tension field beam*.

9.7.1 Complete diagonal tension

The theory presented here is due to H. Wagner.[13]

The beam shown in Fig. 9.8(a) has concentrated flange areas having a depth d between their centroids and vertical stiffeners which are spaced uniformly along the length of the beam. It is assumed that the flanges resist the internal bending moment at any section of the beam while the web, of thickness t, resists the vertical shear force. The effect of this assumption is to produce a uniform shear stress distribution through the depth of the web (see Section 20.3) at any section. Therefore, at a section of the beam where the shear force is S, the shear stress τ is given by

$$\tau = \frac{S}{td} \tag{9.14}$$

Consider now an element ABCD of the web in a panel of the beam, as shown in Fig. 9.8(a). The element is subjected to tensile stresses, σ_t, produced by the diagonal tension on the planes AB and CD; the angle of the diagonal tension is α. On a vertical plane FD in the element the shear stress is τ and the direct stress σ_z. Now considering the equilibrium of the element FCD (Fig. 9.8(b)) and resolving forces vertically, we have (see Section 1.6)

$$\sigma_t \text{CD} t \sin\alpha = \tau \text{FD} t$$

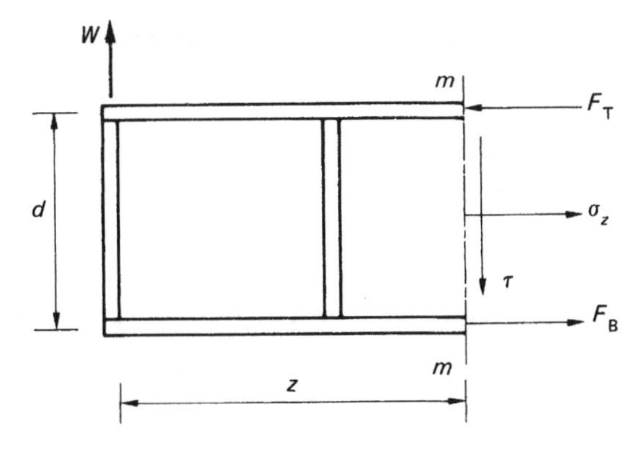

Fig. 9.9 Determination of flange forces.

which gives

$$\sigma_t = \frac{\tau}{\sin\alpha\cos\alpha} = \frac{2\tau}{\sin 2\alpha} \tag{9.15}$$

or, substituting for τ from Eq. (9.14) and noting that in this case $S = W$ at all sections of the beam

$$\sigma_t = \frac{2W}{td\sin 2\alpha} \tag{9.16}$$

Further, resolving forces horizontally for the element FCD

$$\sigma_z \mathrm{FD}t = \sigma_t \mathrm{CD}t\cos\alpha$$

which gives

$$\sigma_z = \sigma_t\cos^2\alpha$$

or, substituting for σ_t from Eq. (9.15)

$$\sigma_z = \frac{\tau}{\tan\alpha} \tag{9.17}$$

or, for this particular beam, from Eq. (9.14)

$$\sigma_z = \frac{W}{td\tan\alpha} \tag{9.18}$$

Since τ and σ_t are constant through the depth of the beam it follows that σ_z is constant through the depth of the beam.

The direct loads in the flanges are found by considering a length z of the beam as shown in Fig. 9.9. On the plane *mm* there are direct and shear stresses σ_z and τ acting in the web, together with direct loads F_T and F_B in the top and bottom flanges respectively. F_T and F_B are produced by a combination of the bending moment Wz at the section plus the compressive action (σ_z) of the diagonal tension. Taking moments about the bottom flange

$$Wz = F_T d - \frac{\sigma_z td^2}{2}$$

Hence, substituting for σ_z from Eq. (9.18) and rearranging

$$F_{\mathrm{T}} = \frac{Wz}{d} + \frac{W}{2\tan\alpha} \tag{9.19}$$

Now resolving forces horizontally

$$F_{\mathrm{B}} - F_{\mathrm{T}} + \sigma_z td = 0$$

which gives, on substituting for σ_z and F_{T} from Eqs (9.18) and (9.19)

$$F_{\mathrm{B}} = \frac{Wz}{d} - \frac{W}{2\tan\alpha} \tag{9.20}$$

The diagonal tension stress σ_{t} induces a direct stress σ_y on horizontal planes at any point in the web. Then, on a horizontal plane HC in the element ABCD of Fig. 9.8 there is a direct stress σ_y and a complementary shear stress τ, as shown in Fig. 9.10.

From a consideration of the vertical equilibrium of the element HDC we have

$$\sigma_y \mathrm{HC}t = \sigma_{\mathrm{t}}\mathrm{CD}t \sin\alpha$$

which gives

$$\sigma_y = \sigma_{\mathrm{t}} \sin^2\alpha$$

Substituting for σ_{t} from Eq. (9.15)

$$\sigma_y = \tau \tan\alpha \tag{9.21}$$

or, from Eq. (9.14) in which $S = W$

$$\sigma_y = \frac{W}{td}\tan\alpha \tag{9.22}$$

The tensile stresses σ_y on horizontal planes in the web of the beam cause compression in the vertical stiffeners. Each stiffener may be assumed to support half of each adjacent panel in the beam so that the compressive load P in a stiffener is given by

$$P = \sigma_y tb$$

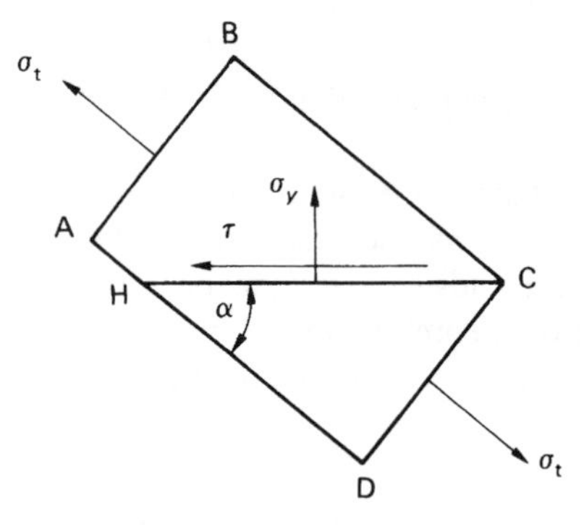

Fig. 9.10 Stress system on a horizontal plane in the beam web.

which becomes, from Eq. (9.22)

$$P = \frac{Wb}{d} \tan \alpha \tag{9.23}$$

If the load P is sufficiently high the stiffeners will buckle. Tests indicate that they buckle as columns of equivalent length

$$\left.\begin{aligned} l_e &= d/\sqrt{4 - 2b/d} \quad \text{for } b < 1.5d \\ \text{or} \quad l_e &= d \quad \text{for } b > 1.5d \end{aligned}\right\} \tag{9.24}$$

In addition to causing compression in the stiffeners the direct stress σ_y produces bending of the beam flanges between the stiffeners as shown in Fig. 9.11. Each flange acts as a continuous beam carrying a uniformly distributed load of intensity $\sigma_y t$. The maximum bending moment in a continuous beam with ends fixed against rotation occurs at a support and is $wL^2/12$ in which w is the load intensity and L the beam span. In this case, therefore, the maximum bending moment M_{max} occurs at a stiffener and is given by

$$M_{max} = \frac{\sigma_y t b^2}{12}$$

or, substituting for σ_y from Eq. (9.22)

$$M_{max} = \frac{Wb^2 \tan \alpha}{12d} \tag{9.25}$$

Midway between the stiffeners this bending moment reduces to $Wb^2 \tan \alpha/24d$.

The angle α adjusts itself such that the total strain energy of the beam is a minimum. If it is assumed that the flanges and stiffeners are rigid then the strain energy comprises the shear strain energy of the web only and $\alpha = 45°$. In practice, both flanges and stiffeners deform so that α is somewhat less than 45°, usually of the order of 40° and, in the type of beam common to aircraft structures, rarely below 38°. For beams having all components made of the same material the condition of minimum strain energy leads to various equivalent expressions for α, one of which is

$$\tan^2 \alpha = \frac{\sigma_t + \sigma_F}{\sigma_t + \sigma_S} \tag{9.26}$$

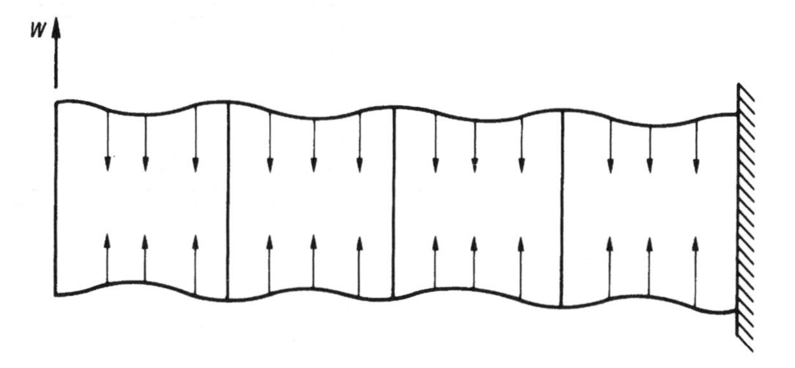

Fig. 9.11 Bending of flanges due to web stress.

in which σ_F and σ_S are the uniform direct *compressive* stresses induced by the diagonal tension in the flanges and stiffeners, respectively. Thus, from the second term on the right-hand side of either of Eqs (9.19) or (9.20)

$$\sigma_F = \frac{W}{2A_F \tan\alpha} \tag{9.27}$$

in which A_F is the cross-sectional area of each flange. Also, from Eq. (9.23)

$$\sigma_S = \frac{Wb}{A_S d}\tan\alpha \tag{9.28}$$

where A_S is the cross-sectional area of a stiffener. Substitution of σ_t from Eq. (9.16) and σ_F and σ_S from Eqs (9.27) and (9.28) into Eq. (9.26), produces an equation which may be solved for α. An alternative expression for α, again derived from a consideration of the total strain energy of the beam, is

$$\tan^4\alpha = \frac{1 + td/2A_F}{1 + tb/A_S} \tag{9.29}$$

Example 9.1

The beam shown in Fig. 9.12 is assumed to have a complete tension field web. If the cross-sectional areas of the flanges and stiffeners are, respectively, 350 mm^2 and 300 mm^2 and the elastic section modulus of each flange is 750 mm^3, determine the maximum stress in a flange and also whether or not the stiffeners will buckle. The thickness of the web is 2 mm and the second moment of area of a stiffener about an axis in the plane of the web is 2000 mm^4; $E = 70\,000$ N/mm^2.

From Eq. (9.29)

$$\tan^4\alpha = \frac{1 + 2 \times 400/(2 \times 350)}{1 + 2 \times 300/300} = 0.7143$$

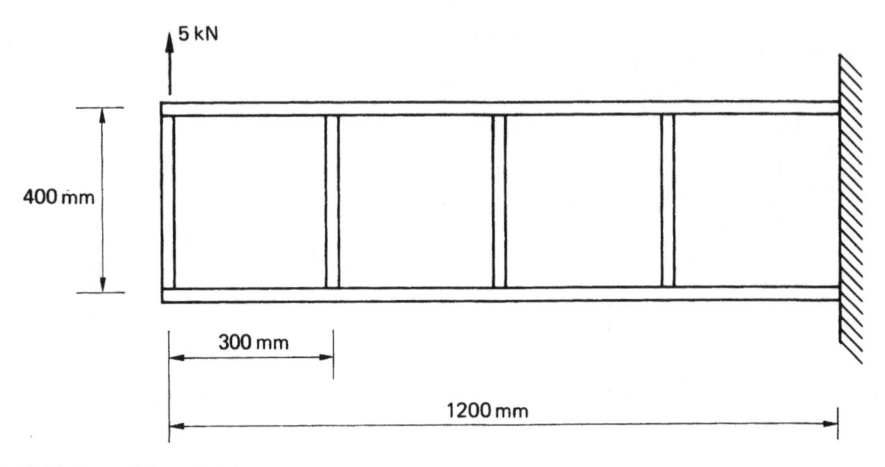

Fig. 9.12 Beam of Example 9.1.

so that

$$\alpha = 42.6^\circ$$

The maximum flange stress will occur in the top flange at the built-in end where the bending moment on the beam is greatest and the stresses due to bending and diagonal tension are additive. Therefore, from Eq. (9.19)

$$F_{\mathrm{T}} = \frac{5 \times 1200}{400} + \frac{5}{2 \tan 42.6^\circ}$$

i.e.

$$F_{\mathrm{T}} = 17.7\,\mathrm{kN}$$

Hence the direct stress in the top flange produced by the externally applied bending moment and the diagonal tension is $17.7 \times 10^3/350 = 50.7\,\mathrm{N/mm^2}$. In addition to this uniform compressive stress, local bending of the type shown in Fig. 9.11 occurs. The local bending moment in the top flange at the built-in end is found using Eq. (9.25), i.e.

$$M_{\max} = \frac{5 \times 10^3 \times 300^2 \tan 42.6^\circ}{12 \times 400} = 8.6 \times 10^4\,\mathrm{N\,mm}$$

The maximum compressive stress corresponding to this bending moment occurs at the lower extremity of the flange and is $8.6 \times 10^4/750 = 114.9\,\mathrm{N/mm^2}$. Thus the maximum stress in a flange occurs on the inside of the top flange at the built-in end of the beam, is compressive and equal to $114.9 + 50.7 = 165.6\,\mathrm{N/mm^2}$.

The compressive load in a stiffener is obtained using Eq. (9.23), i.e.

$$P = \frac{5 \times 300 \tan 42.6^\circ}{400} = 3.4\,\mathrm{kN}$$

Since, in this case, $b < 1.5d$, the equivalent length of a stiffener as a column is given by the first of Eqs (9.24), i.e.

$$l_{\mathrm{e}} = 400/\sqrt{4 - 2 \times 300/400} = 253\,\mathrm{mm}$$

From Eq. (8.7) the buckling load of a stiffener is then

$$P_{\mathrm{CR}} = \frac{\pi^2 \times 70\,000 \times 2000}{253^2} = 22.0\,\mathrm{kN}$$

Clearly the stiffener will not buckle.

In Eqs (9.28) and (9.29) it is implicitly assumed that a stiffener is fully effective in resisting axial load. This will be the case if the centroid of area of the stiffener lies in the plane of the beam web. Such a situation arises when the stiffener consists of two members symmetrically arranged on opposite sides of the web. In the case where the web is stiffened by a single member attached to one side, the compressive load P is offset from the stiffener axis thereby producing bending in addition to axial load. For a stiffener having its centroid a distance e from the centre of the web the combined bending and axial compressive stress, σ_c, at a distance e from the stiffener centroid is

$$\sigma_c = \frac{P}{A_S} + \frac{Pe^2}{A_S r^2}$$

in which r is the radius of gyration of the stiffener cross-section about its neutral axis (note: second moment of area $I = Ar^2$). Then

$$\sigma_c = \frac{P}{A_S}\left[1 + \left(\frac{e}{r}\right)^2\right]$$

or

$$\sigma_c = \frac{P}{A_{S_e}}$$

where

$$A_{S_e} = \frac{A_S}{1 + (e/r)^2} \qquad (9.30)$$

and is termed the effective stiffener area.

9.7.2 Incomplete diagonal tension

In modern aircraft structures, beams having extremely thin webs are rare. They retain, after buckling, some of their ability to support loads so that even near failure they are in a state of stress somewhere between that of pure diagonal tension and the pre-buckling stress. Such a beam is described as an *incomplete diagonal tension field beam* and may be analysed by semi-empirical theory as follows.

It is assumed that the nominal web shear τ ($=S/td$) may be divided into a 'true shear' component τ_S and a diagonal tension component τ_{DT} by writing

$$\tau_{DT} = k\tau, \quad \tau_S = (1-k)\tau \qquad (9.31)$$

where k, the *diagonal tension factor*, is a measure of the degree to which the diagonal tension is developed. A completely unbuckled web has $k=0$ whereas $k=1$ for a web in complete diagonal tension. The value of k corresponding to a web having a critical shear stress τ_{CR} is given by the empirical expression

$$k = \tanh\left(0.5 \log \frac{\tau}{\tau_{CR}}\right) \qquad (9.32)$$

The ratio τ/τ_{CR} is known as the *loading ratio* or *buckling stress ratio.* The buckling stress τ_{CR} may be calculated from the formula

$$\tau_{CR,elastic} = k_{ss}E\left(\frac{t}{b}\right)^2\left[R_d + \frac{1}{2}(R_b - R_d)\left(\frac{b}{d}\right)^3\right] \qquad (9.33)$$

where k_{ss} is the coefficient for a plate with simply supported edges and R_d and R_b are empirical restraint coefficients for the vertical and horizontal edges of the web panel respectively. Graphs giving k_{ss}, R_d and R_b are reproduced in Kuhn.[13]

The stress equations (9.27) and (9.28) are modified in the light of these assumptions and may be rewritten in terms of the applied shear stress τ as

$$\sigma_F = \frac{k\tau \cot\alpha}{(2A_F/td) + 0.5(1-k)} \tag{9.34}$$

$$\sigma_S = \frac{k\tau \tan\alpha}{(A_S/tb) + 0.5(1-k)} \tag{9.35}$$

Further, the web stress σ_t given by Eq. (9.15) becomes two direct stresses: σ_1 along the direction of α given by

$$\sigma_1 = \frac{2k\tau}{\sin 2\alpha} + \tau(1-k)\sin 2\alpha \tag{9.36}$$

and σ_2 perpendicular to this direction given by

$$\sigma_2 = -\tau(1-k)\sin 2\alpha \tag{9.37}$$

The secondary bending moment of Eq. (9.25) is multiplied by the factor k, while the effective lengths for the calculation of stiffener buckling loads become (see Eqs (9.24))

$$l_e = d_s/\sqrt{1 + k^2(3 - 2b/d_s)} \quad \text{for } b < 1.5d$$

or

$$l_e = d_s \quad \text{for } b > 1.5d$$

where d_s is the actual stiffener depth, as opposed to the effective depth d of the web, taken between the web/flange connections as shown in Fig. 9.13. We observe that Eqs (9.34)–(9.37) are applicable to either incomplete or complete diagonal tension field beams since, for the latter case, $k = 1$ giving the results of Eqs (9.27), (9.28) and (9.15).

In some cases beams taper along their lengths, in which case the flange loads are no longer horizontal but have vertical components which reduce the shear load carried by the web. Thus, in Fig. 9.14 where d is the depth of the beam at the section considered, we have, resolving forces vertically

$$W - (F_T + F_B)\sin\beta - \sigma_t(d\cos\alpha)\sin\alpha = 0 \tag{9.38}$$

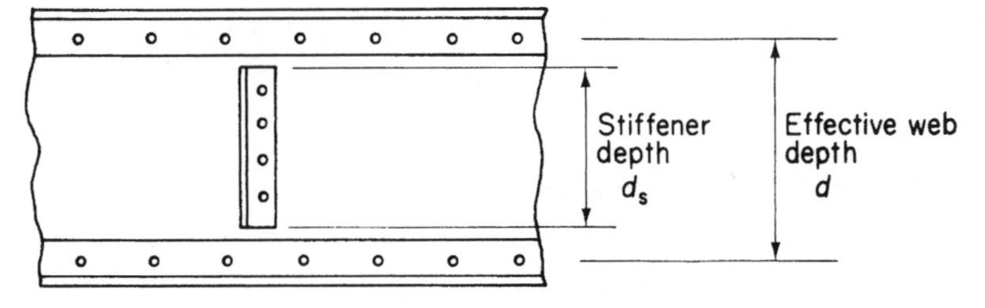

Fig. 9.13 Calculation of stiffener buckling load.

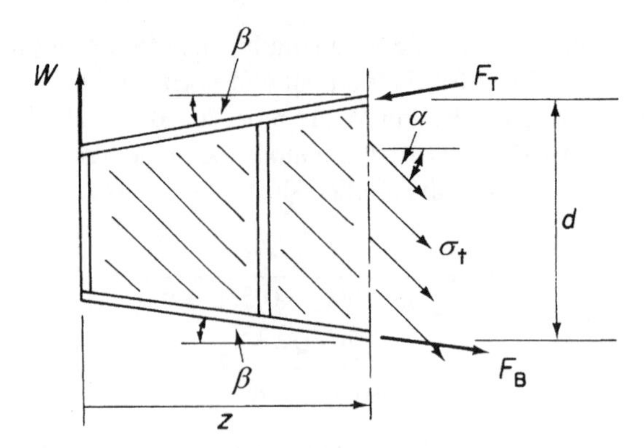

Fig. 9.14 Effect of taper on diagonal tension field beam calculations.

For horizontal equilibrium

$$(F_T - F_B)\cos\beta - \sigma_t td\cos^2\alpha = 0 \tag{9.39}$$

Taking moments about B

$$Wz - F_T d\cos\beta + \tfrac{1}{2}\sigma_t td^2\cos^2\alpha = 0 \tag{9.40}$$

Solving Eqs (9.38), (9.39) and (9.40) for σ_t, F_T and F_B

$$\sigma_t = \frac{2W}{td\sin 2\alpha}\left(1 - \frac{2z}{d}\tan\beta\right) \tag{9.41}$$

$$F_T = \frac{W}{d\cos\beta}\left[z + \frac{d\cot\alpha}{2}\left(1 - \frac{2z}{d}\tan\beta\right)\right] \tag{9.42}$$

$$F_B = \frac{W}{d\cos\beta}\left[z - \frac{d\cot\alpha}{2}\left(1 - \frac{2z}{d}\tan\beta\right)\right] \tag{9.43}$$

Equation (9.23) becomes

$$P = \frac{Wb}{d}\tan\alpha\left(1 - \frac{2z}{d}\tan\beta\right) \tag{9.44}$$

Also the shear force S at any section of the beam is, from Fig. 9.14

$$S = W - (F_T + F_B)\sin\beta$$

or, substituting for F_T and F_B from Eqs (9.42) and (9.43)

$$S = W\left(1 - \frac{2z}{d}\tan\beta\right) \tag{9.45}$$

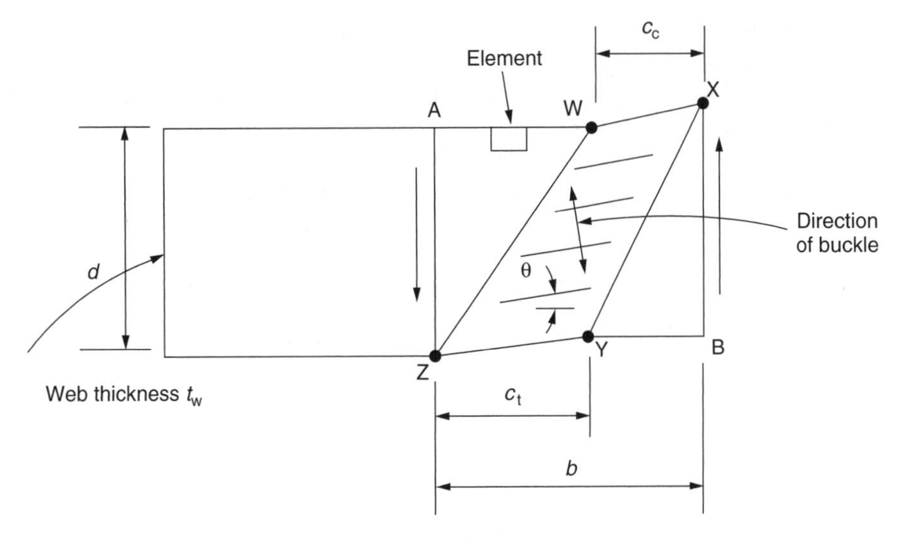

Fig. 9.15 Collapse mechanism of a panel of a tension field beam.

9.7.3 Post buckling behaviour

Sections 9.7.1 and 9.7.2 are concerned with beams in which the thin webs buckle to form tension fields; the beam flanges are then regarded as being subjected to bending action as in Fig. 9.11. It is possible, if the beam flanges are relatively light, for failure due to yielding to occur in the beam flanges after the web has buckled so that plastic hinges form and a failure mechanism of the type shown in Fig. 9.15 exists. This post buckling behaviour was investigated by Evans, Porter and Rockey[15] who developed a design method for beams subjected to bending and shear. It is their method of analysis which is presented here.

Suppose that the panel AXBZ in Fig. 9.15 has collapsed due to a shear load S and a bending moment M; plastic hinges have formed at W, X, Y and Z. In the initial stages of loading the web remains perfectly flat until it reaches its critical stresses i.e., τ_{cr} in shear and σ_{crb} in bending. The values of these stresses may be found approximately from

$$\left(\frac{\sigma_{mb}}{\sigma_{crb}}\right)^2 + \left(\frac{\tau_m}{\tau_{cr}}\right)^2 = 1 \qquad (9.46)$$

where σ_{crb} is the critical value of bending stress with $S = 0$, $M \neq 0$ and τ_{cr} is the critical value of shear stress when $S \neq 0$ and $M = 0$. Once the critical stress is reached the web starts to buckle and cannot carry any increase in compressive stress so that, as we have seen in Section 9.7.1, any additional load is carried by tension field action. It is assumed that the shear and bending stresses remain at their critical values τ_m and σ_{mb} and that there are *additional* stresses σ_t which are inclined at an angle θ to the horizontal and which carry any increases in the applied load. At collapse, i.e. at ultimate load conditions, the additional stress σ_t reaches its maximum value $\sigma_{t(max)}$ and the panel is in the collapsed state shown in Fig. 9.15.

Consider now the small rectangular element on the edge AW of the panel before collapse. The stresses acting on the element are shown in Fig. 9.16(a). The stresses on planes parallel to and perpendicular to the direction of the buckle may be found by considering the equilibrium of triangular elements within this rectangular element. Initially we shall consider the triangular element CDE which is subjected to the stress system shown in Fig. 9.16(b) and is in equilibrium under the action of the forces corresponding to these stresses. Note that the edge CE of the element is parallel to the direction of the buckle in the web.

For equilibrium of the element in a direction perpendicular to CE (see Section 1.6)

$$\sigma_\xi \mathrm{CE} + \sigma_{\mathrm{mb}} \mathrm{ED} \cos\theta - \tau_{\mathrm{m}} \mathrm{ED} \sin\theta - \tau_{\mathrm{m}} \mathrm{DC} \cos\theta = 0$$

Dividing through by CE and rearranging we have

$$\sigma_\xi = -\sigma_{\mathrm{mb}} \cos^2\theta + \tau_{\mathrm{m}} \sin 2\theta \tag{9.47}$$

Similarly, by considering the equilibrium of the element in the direction EC we have

$$\tau_{\eta\xi} = -\frac{\sigma_{\mathrm{mb}}}{2} \sin 2\theta - \tau_{\mathrm{m}} \cos 2\theta \tag{9.48}$$

Further the direct stress σ_η on the plane FD (Fig. 9.16(c)) which is perpendicular to the plane of the buckle is found from the equilibrium of the element FED. Then,

$$\sigma_\eta \mathrm{FD} + \sigma_{\mathrm{mb}} \mathrm{ED} \sin\theta + \tau_{\mathrm{m}} \mathrm{EF} \sin\theta + \tau_{\mathrm{m}} \mathrm{DE} \cos\theta = 0$$

Dividing through by FD and rearranging gives

$$\sigma_\eta = -\sigma_{\mathrm{mb}} \sin^2\theta - \tau_{\mathrm{m}} \sin 2\theta \tag{9.49}$$

Note that the shear stress on this plane forms a complementary shear stress system with $\tau_{\eta\xi}$.

The failure condition is reached by adding $\sigma_{\mathrm{t(max)}}$ to σ_ξ and using the von Mises theory of elastic failure (see Ref. [14]) i.e.

$$\sigma_{\mathrm{y}}^2 = \sigma_1^2 + \sigma_2^2 - \sigma_1\sigma_2 + 3\tau^2 \tag{9.50}$$

where σ_{y} is the yield stress of the material, σ_1 and σ_2 are the direct stresses acting on two mutually perpendicular planes and τ is the shear stress acting on the same two planes. Hence, when the yield stress in the web is σ_{yw} failure occurs when

$$\sigma_{\mathrm{yw}}^2 = (\sigma_\xi + \sigma_{\mathrm{t(max)}})^2 + \sigma_\eta^2 - \sigma_\eta(\sigma_\xi + \sigma_{\mathrm{t(max)}}) + 3\tau_{\eta\xi}^2 \tag{9.51}$$

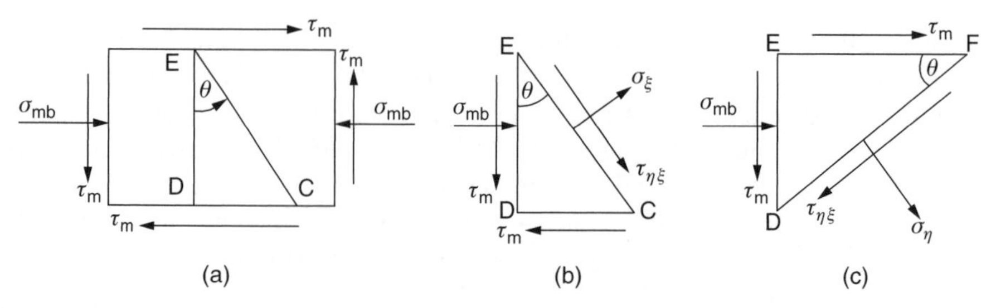

Fig. 9.16 Determination of stresses on planes parallel and perpendicular to the plane of the buckle.

Eqs (9.47), (9.48), (9.49) and (9.51) may be solved for $\sigma_{t(max)}$ which is then given by

$$\sigma_{t(max)} = -\frac{1}{2}A + \frac{1}{2}[A^2 - 4(\sigma_{mb}^2 + 3\tau_m^2 - \sigma_{yw}^2)]^{\frac{1}{2}} \tag{9.52}$$

where

$$A = 3\tau_m \sin 2\theta + \sigma_{mb} \sin^2\theta - 2\sigma_{mb}\cos^2\theta \tag{9.53}$$

These equations have been derived for a point on the edge of the panel but are applicable to any point within its boundary. Therefore the resultant force F_w corresponding to the tension field in the web may be calculated and its line of action determined.

If the average stresses in the compression and tension flanges are σ_{cf} and σ_{tf} and the yield stress of the flanges is σ_{yf} the reduced plastic moments in the flanges are (see Ref. [14])

$$M'_{pc} = M_{pc}\left[1 - \left(\frac{\sigma_{cf}}{\sigma_{yf}}\right)^2\right] \quad \text{(compression flange)} \tag{9.54}$$

$$M'_{pt} = M_{pt}\left[1 - \left(\frac{\sigma_{tf}}{\sigma_{yf}}\right)\right] \quad \text{(tension flange)} \tag{9.55}$$

The position of each plastic hinge may be found by considering the equilibrium of a length of flange and employing the principle of virtual work. In Fig. 9.17 the length WX of the upper flange of the beam is given a virtual displacement ϕ. The work done by the shear force at X is equal to the energy absorbed by the plastic hinges at X and W and the work done *against* the tension field stress $\sigma_{t(max)}$. Suppose the average value of the tension field stress is σ_{tc}, i.e. the stress at the midpoint of WX.

Then

$$S_x c_c \phi = 2M'_{pc}\phi + \sigma_{tc}\, t_w \sin^2\theta \frac{c_c^2}{2}\phi$$

The minimum value of S_x is obtained by differentiating with respect to c_c, i.e.

$$\frac{dS_x}{dc_c} = -2\frac{M'_{pc}}{c_c^2} + \sigma_{tc}\, t_w \frac{\sin^2\theta}{2} = 0$$

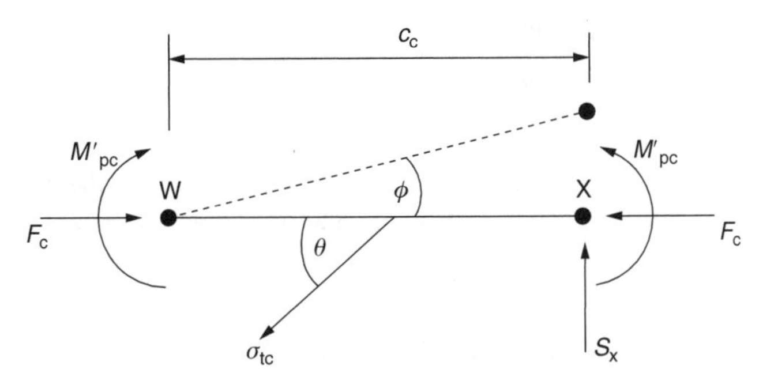

Fig. 9.17 Determination of plastic hinge position.

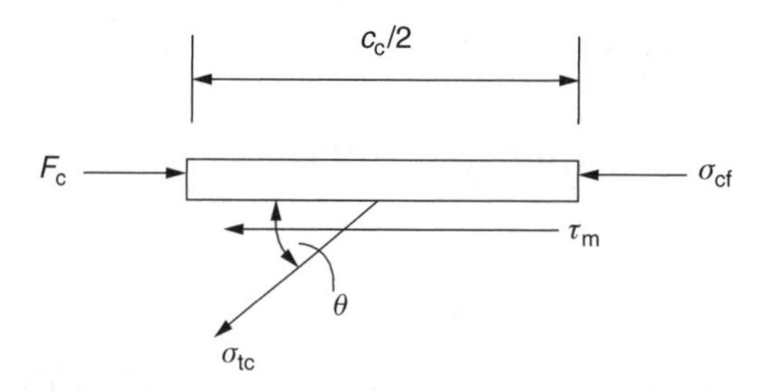

Fig. 9.18 Determination of flange stress.

which gives

$$c_c^2 = \frac{4M'_{pc}}{\sigma_{tc}\, t_w \sin^2\theta} \tag{9.56}$$

Similarly in the tension flange

$$c_t^2 = \frac{4M'_{pt}}{\sigma_{tt}\, t_w \sin^2\theta} \tag{9.57}$$

Clearly for the plastic hinges to occur within a flange both c_c and c_t must be less than b. Therefore from Eq. (9.56)

$$M'_{pc} < \frac{t_w b^2 \sin^2\theta}{4}\sigma_{tc} \tag{9.58}$$

where σ_{tc} is found from Eqs (9.52) and (9.53) at the midpoint of WX.

The average axial stress in the compression flange between W and X is obtained by considering the equilibrium of half of the length of WX (Fig. 9.18).

Then

$$F_c = \sigma_{cf}A_{cf} + \sigma_{tc}t_w\frac{c_c}{2}\sin\theta\cos\theta + \tau_m t_w\frac{c_c}{2}$$

from which

$$\sigma_{cf} = \frac{F_c - \frac{1}{2}(\sigma_{tc}\sin\theta\cos\theta + \tau_m)t_w c_c}{A_{cf}} \tag{9.59}$$

where F_c is the force in the compression flange at W and A_{cf} is the cross-sectional area of the compression flange.

Similarly for the tension flange

$$\sigma_{tf} = \frac{F_t + \frac{1}{2}(\sigma_{tt}\sin\theta\cos\theta + \tau_m)t_w c_t}{A_{tf}} \tag{9.60}$$

The forces F_c and F_t are found by considering the equilibrium of the beam to the right of WY (Fig. 9.19). Then, resolving vertically and noting that $S_{cr} = \tau_m t_w d$

$$S_{ult} = F_w \sin\theta + \tau_m t_w d + \sum W_n \tag{9.61}$$

Resolving horizontally and noting that $H_{cr} = \tau_m t_w\,(b - c_c - c_t)$

$$F_c - F_t = F_w\cos\theta - \tau_m t_w(b - c_c - c_t) \tag{9.62}$$

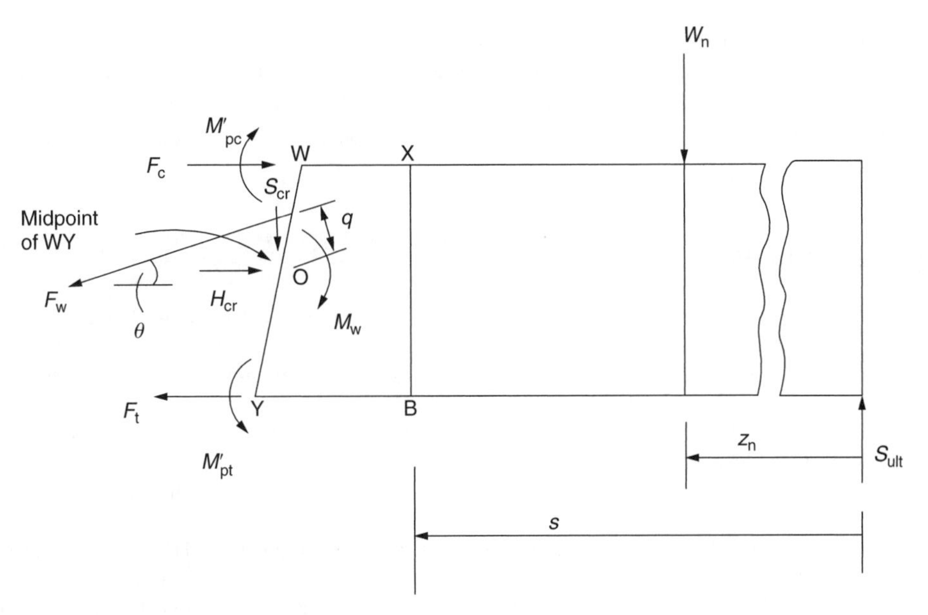

Fig. 9.19 Determination of flange forces.

Taking moments about O we have

$$F_c + F_t = \frac{2}{d}\left[S_{ult}\left(s + \frac{b + c_c - c_t}{2}\right) + M'_{pt} - M'_{pc} + F_w q - M_w - \sum_n W_n z_n\right] \tag{9.63}$$

where W_1 to W_n are external loads applied to the beam to the right of WY and M_w is the bending moment in the web when it has buckled and become a tension field, i.e.

$$M_w = \frac{\sigma_{mb} b d^2}{b}$$

The flange forces are then

$$F_c = \frac{S_{ult}}{2d}(d\cot\theta + 2s + b + c_c - c_t) + \frac{1}{d}\left(M'_{pt} - M'_{pc} + F_w q - M_w - \sum_n W_n z_n\right) - \frac{1}{2}\tau_m t_w (d\cot\theta + b - c_c - c_t) \tag{9.64}$$

$$F_t = \frac{S_{ult}}{2d}(d\cot\theta + 2s + b + c_c - c_t) + \frac{1}{d}\left(M'_{pt} - M'_{pc} - F_w q - M_w - \sum_n W_n z_n\right) + \frac{1}{2}\tau_m t_w (d\cot\theta + b - c_c - c_t) \tag{9.65}$$

Evans, Porter and Rockey adopted an iterative procedure for solving Eqs (9.61)–(9.65) in which an initial value of θ was assumed and σ_{cf} and σ_{tf} were taken to be zero. Then c_c and c_t were calculated and approximate values of F_c and F_t found giving better estimates for σ_{cf} and σ_{tf}. The procedure was then repeated until the required accuracy was obtained.

References

1 Gerard, G., *Introduction to Structural Stability Theory*, McGraw-Hill Book Company, New York, 1962.
2 Murray, N. W., *Introduction to the Theory of Thin-walled Structures*, Oxford Engineering Science Series, Oxford, 1984.
3 *Handbook of Aeronautics No. 1: Structural Principles and Data*, 4th edition, The Royal Aeronautical Society, 1952.
4 Bleich, F., *Buckling Strength of Metal Structures*, McGraw-Hill Book Company, New York, 1952.
5 Gerard, G. and Becker, H., *Handbook of Structural Stability, Pt. I, Buckling of Flat Plates*, NACA Tech. Note 3781, 1957.
6 Rivello, R. M., *Theory and Analysis of Flight Structures*, McGraw-Hill Book Company, New York, 1969.
7 Stowell, E. Z., *Compressive Strength of Flanges*, NACA Tech. Note 1323, 1947.
8 Mayers, J. and Budiansky, B., *Analysis of Behaviour of Simply Supported Flat Plates Compressed Beyond the Buckling Load in the Plastic Range*, NACA Tech. Note 3368, 1955.
9 Gerard, G. and Becker, H., *Handbook of Structural Stability, Pt. IV, Failure of Plates and Composite Elements*, NACA Tech. Note 3784, 1957.
10 Gerard, G., *Handbook of Structural Stability, Pt. V, Compressive Strength of Flat Stiffened Panels*, NACA Tech. Note 3785, 1957.
11 Gerard, G. and Becker, H., *Handbook of Structural Stability, Pt. VII, Strength of Thin Wing Construction*, NACA Tech. Note D-162, 1959.
12 Gerard, G., The crippling strength of compression elements, *J. Aeron. Sci.*, 25(1), 37–52, January 1958.
13 Kuhn, P., *Stresses in Aircraft and Shell Structures, McGraw-Hill Book Company*, New York, 1956.
14 Megson, T. H. G., *Structural and Stress Analysis*, 2nd edition, Elsevier, Oxford, 2005.
15 Evans, H. R., Porter, D. M. and Rockey, K. C. The collapse behaviour of plate girders subjected to shear and bending, Proc. Int. Assn. Bridge and Struct. Eng. P-18/78, 1–20.

Problems

P.9.1 A thin square plate of side a and thickness t is simply supported along each edge, and has a slight initial curvature giving an initial deflected shape.

$$w_0 = \delta \sin\frac{\pi x}{a} \sin\frac{\pi y}{a}$$

If the plate is subjected to a uniform compressive stress σ in the x-direction (see Fig. P.9.1), find an expression for the *elastic* deflection w normal to the plate. Show also that the deflection at the mid-point of the plate can be presented in the form of a Southwell plot and illustrate your answer with a suitable sketch.

Ans. $w = [\sigma t\delta/(4\pi^2 D/a^2 - \sigma t)] \sin \frac{\pi x}{a} \sin \frac{\pi y}{a}$

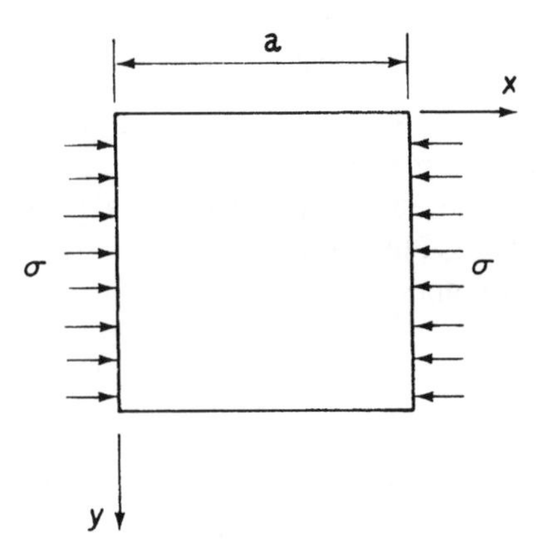

Fig. P.9.1

P.9.2 A uniform flat plate of thickness t has a width b in the y direction and length l in the x direction (see Fig. P.9.2). The edges parallel to the x axis are clamped and those parallel to the y axis are simply supported. A uniform compressive stress σ is applied in the x direction along the edges parallel to the y axis. Using an energy method, find an approximate expression for the magnitude of the stress σ which causes the plate to buckle, assuming that the deflected shape of the plate is given by

$$w = a_{11} \sin \frac{m\pi x}{l} \sin^2 \frac{\pi y}{b}$$

For the particular case $l = 2b$, find the number of half waves m corresponding to the lowest critical stress, expressing the result to the nearest integer. Determine also the lowest critical stress.

Ans. $m = 3,\ \sigma_{CR} = [6E/(1-v^2)](t/b)^2$

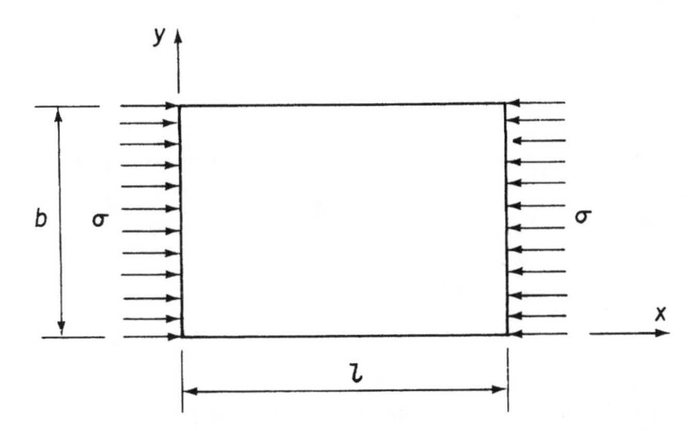

Fig. P.9.2

P.9.3 A panel, comprising flat sheet and uniformly spaced Z-section stringers, a part of whose cross-section is shown in Fig. P.9.3, is to be investigated for strength under uniform compressive loads in a structure in which it is to be stabilized by frames a distance l apart, l being appreciably greater than the spacing b.

(a) State the modes of failure which you would consider and how you would determine appropriate limiting stresses.

(b) Describe a suitable test to verify your calculations, giving particulars of the specimen, the manner of support and the measurements you would take. The latter should enable you to verify the assumptions made, as well as to obtain the load supported.

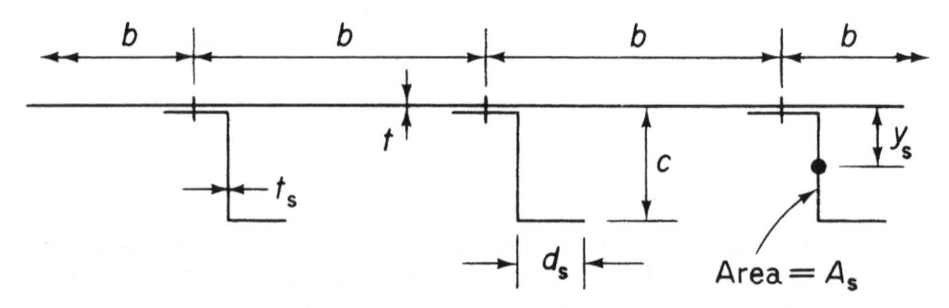

Fig. P.9.3

P.9.4 Part of a compression panel of internal construction is shown in Fig. P.9.4. The equivalent pin-centre length of the panel is 500 mm. The material has a Young's modulus of 70 000 N/mm^2 and its elasticity may be taken as falling catastrophically when a compressive stress of 300 N/mm^2 is reached. Taking coefficients of 3.62 for buckling of a plate with simply supported sides and of 0.385 with one side simply supported and one free, determine (a) the load per mm width of panel when initial buckling may be expected and (b) the load per mm for ultimate failure. Treat the material as thin for calculating section constants and assume that after initial buckling the stress in the plate increases parabolically from its critical value in the centre of sections.

Ans. 613.8 N/mm, 844.7 N/mm.

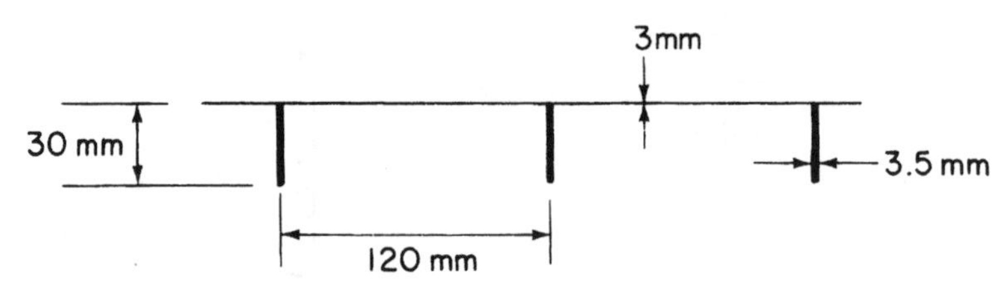

Fig. P.9.4

P.9.5 A simply supported beam has a span of 2.4 m and carries a central concentrated load of 10 kN. The flanges of the beam each have a cross-sectional area of

300 mm^2 while that of the vertical web stiffeners is 280 mm^2. If the depth of the beam, measured between the centroids of area of the flanges, is 350 mm and the stiffeners are symmetrically arranged about the web and spaced at 300 mm intervals, determine the maximum axial load in a flange and the compressive load in a stiffener.

It may be assumed that the beam web, of thickness 1.5 mm, is capable of resisting diagonal tension only.

Ans. 19.9 kN, 3.9 kN.

P.9.6 The spar of an aircraft is to be designed as an incomplete diagonal tension beam, the flanges being parallel. The stiffener spacing will be 250 mm, the effective depth of web will be 750 mm, and the depth between web-to-flange attachments is 725 mm.

The spar is to carry an ultimate shear force of 100 000 N. The maximum permissible shear stress is 165 N/mm^2, but it is also required that the shear stress should not exceed 15 times the critical shear stress for the web panel.

Assuming α to be 40° and using the relationships below:

(i) Select the smallest suitable web thickness from the following range of standard thicknesses. (Take Young's Modulus E as 70 000 N/mm^2.)

0.7 mm, 0.9 mm, 1.2 mm, 1.6 mm

(ii) Calculate the stiffener end load and the secondary bending moment in the flanges (assume stiffeners to be symmetrical about the web).

The shear stress buckling coefficient for the web may be calculated from the expression

$$K = 7.70[1 + 0.75(b/d)^2]$$

b and d having their usual significance.

The relationship between the diagonal tension factor and buckling stress ratio is

τ/τ_{CR}	5	7	9	11	13	15
k	0.37	0.40	0.42	0.48	0.51	0.53

Note that α is the angle of diagonal tension measured from the spanwise axis of the beam, as in the usual notation.

Ans. 1.2 mm, $130A_S/(1 + 0.0113A_S)$, 238 910 N mm.

P. 9.7 The main compressive wing structure of an aircraft consists of stringers, having the section shown in Fig. P.9.7(b), bonded to a thin skin (Fig. P.9.7(a)). Find suitable values for the stringer spacing b and rib spacing L if local instability, skin buckling and panel strut instability all occur at the same stress. Note that in Fig. P.9.7(a) only two of several stringers are shown for diagrammatic clarity. Also the thin skin should be treated as a flat plate since the curvature is small. For a flat plate simply supported along two edges assume a buckling coefficient of 3.62. Take $E = 69\,000\,\text{N/mm}^2$.

Ans. $b = 56.5$ mm, $L = 700$ mm.

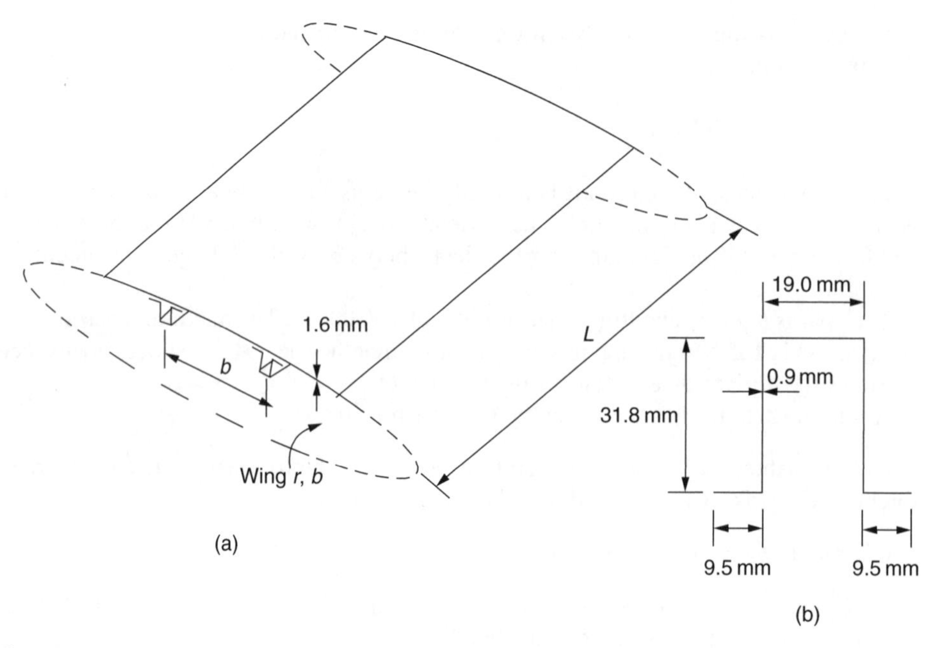

Fig. P.9.7

Section A5 Vibration of Structures

10

Structural vibration

Structures which are subjected to dynamic loading, particularly aircraft, vibrate or oscillate in a frequently complex manner. An aircraft, for example, possesses an infinite number of *natural* or *normal modes* of vibration. Simplifying assumptions, such as breaking down the structure into a number of concentrated masses connected by weightless beams (*lumped mass concept*), are made but whatever method is employed the natural modes and frequencies of vibration of a structure must be known before *flutter* speeds and frequencies can be found. We shall discuss flutter and other dynamic aeroelastic phenomena in Chapter 28 but for the moment we shall concentrate on the calculation of the normal modes and frequencies of vibration of a variety of beam and mass systems.

10.1 Oscillation of mass/spring systems

Let us suppose that the simple mass/spring system shown in Fig. 10.1 is displaced by a small amount x_0 and suddenly released. The equation of the resulting motion in the absence of damping forces is

$$m\ddot{x} + kx = 0 \tag{10.1}$$

where k is the spring stiffness. We see from Eq. (10.1) that the mass, m, oscillates with simple harmonic motion given by

$$x = x_0 \sin(\omega t + \varepsilon) \tag{10.2}$$

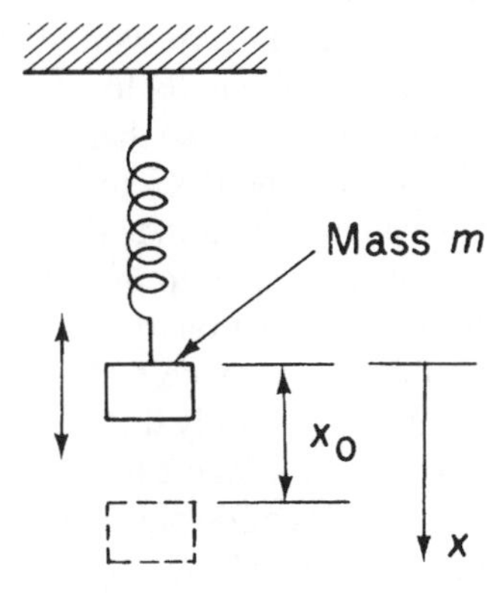

Fig. 10.1 Oscillation of a mass/spring system.

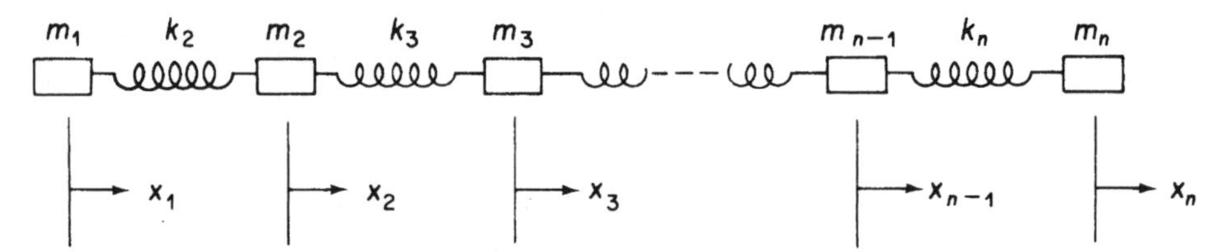

Fig. 10.2 Oscillation of an n mass/spring system.

in which $\omega^2 = k/m$ and ε is a phase angle. The frequency of the oscillation is $\omega/2\pi$ cycles per second and its amplitude x_0. Further, the periodic time of the motion, that is the time taken by one complete oscillation, is $2\pi/\omega$. Both the frequency and periodic time are seen to depend upon the basic physical characteristics of the system, namely the spring stiffness and the magnitude of the mass. Therefore, although the amplitude of the oscillation may be changed by altering the size of the initial disturbance, its frequency is fixed. This frequency is the normal or natural frequency of the system and the vertical simple harmonic motion of the mass is its normal mode of vibration.

Consider now the system of n masses connected by $(n-1)$ springs, as shown in Fig. 10.2. If we specify that motion may only take place in the direction of the spring axes then the system has n degrees of freedom. It is therefore possible to set the system oscillating with simple harmonic motion in n different ways. In each of these n modes of vibration the masses oscillate in phase so that they all attain maximum amplitude at the same time and pass through their zero displacement positions at the same time. The set of amplitudes and the corresponding frequency take up different values in each of the n modes. Again these modes are termed normal or natural modes of vibration and the corresponding frequencies are called normal or natural frequencies.

The determination of normal modes and frequencies for a general spring/mass system involves the solution of a set of n simultaneous second-order differential equations of a type similar to Eq. (10.1). Associated with each solution are two arbitrary constants which determine the phase and amplitude of each mode of vibration. We can therefore relate the vibration of a system to a given set of initial conditions by assigning appropriate values to these constants.

A useful property of the normal modes of a system is their orthogonality, which is demonstrated by the provable fact that the product of the inertia forces in one mode and the displacements in another results in zero work done. In other words displacements in one mode cannot be produced by inertia forces in another. It follows that the normal modes are independent of one another so that the response of each mode to an externally applied force may be found without reference to the other modes. Therefore by considering the response of each mode in turn and adding the resulting motions we can find the response of the complete system to the applied loading. Another useful characteristic of normal modes is their 'stationary property'. It can be shown that if an elastic system is forced to vibrate in a mode that is slightly different from a true normal mode the frequency is only very slightly different to the corresponding natural frequency of the system. Reasonably accurate estimates of natural frequencies may therefore be made from 'guessed' modes of displacement.

We shall proceed to illustrate the general method of solution by determining normal modes and frequencies of some simple beam/mass systems. Two approaches are possible: a *stiffness* or *displacement method* in which spring or elastic forces are expressed in terms of stiffness parameters such as k in Eq. (10.1); and a *flexibility* or *force method* in which elastic forces are expressed in terms of the flexibility δ of the elastic system. In the latter approach δ is defined as the deflection due to unit force; the equation of motion of the spring/mass system of Fig. 10.1 then becomes

$$m\ddot{x} + \frac{x}{\delta} = 0 \tag{10.3}$$

Again the solution takes the form $x = x_0 \sin(\omega t + \varepsilon)$ but in this case $\omega^2 = 1/m\delta$. Clearly by our definitions of k and δ the product $k\delta = 1$. In problems involving rotational oscillations m becomes the moment of inertia of the mass and δ the rotation or displacement produced by unit moment.

Let us consider a spring/mass system having a finite number, n, degrees of freedom. The term spring is used here in a general sense in that the n masses $m_1, m_2, \ldots, m_i, \ldots, m_n$ may be connected by any form of elastic weightless member. Thus, if m_i is the mass at a point i where the displacement is x_i and δ_{ij} is the displacement at the point i due to a unit load at a point j (note from the reciprocal theorem $\delta_{ij} = \delta_{ji}$), the n equations of motion for the system are

$$\left.\begin{array}{l}
m_1\ddot{x}_1\delta_{11} + m_2\ddot{x}_2\delta_{12} + \cdots + m_i\ddot{x}_i\delta_{1i} + \cdots + m_n\ddot{x}_n\delta_{1n} + x_1 = 0 \\
m_1\ddot{x}_1\delta_{21} + m_2\ddot{x}_2\delta_{22} + \cdots + m_i\ddot{x}_i\delta_{2i} + \cdots + m_n\ddot{x}_n\delta_{2n} + x_2 = 0 \\
\cdots\cdots\cdots\cdots\cdots\cdots\cdots\cdots\cdots\cdots\cdots\cdots\cdots\cdots \\
m_1\ddot{x}_1\delta_{i1} + m_2\ddot{x}_2\delta_{i2} + \cdots + m_i\ddot{x}_i\delta_{ii} + \cdots + m_n\ddot{x}_n\delta_{in} + x_i = 0 \\
\cdots\cdots\cdots\cdots\cdots\cdots\cdots\cdots\cdots\cdots\cdots\cdots\cdots\cdots \\
m_1\ddot{x}_1\delta_{n1} + m_2\ddot{x}_2\delta_{n2} + \cdots + m_i\ddot{x}_i\delta_{ni} + \cdots + m_n\ddot{x}_n\delta_{nn} + x_n = 0
\end{array}\right\} \tag{10.4}$$

or

$$\sum_{j=1}^{n} m_j\ddot{x}_j\delta_{ij} + x_i = 0 \quad (i = 1, 2, \ldots, n) \tag{10.5}$$

Since each normal mode of the system oscillates with simple harmonic motion, then the solution for the ith mode takes the form $x = x_i^0 \sin(\omega t + \varepsilon)$ so that $\ddot{x}_i = -\omega^2 x_i^0 \sin(\omega t + \varepsilon) = -\omega^2 x_i$. Equation (10.5) may therefore be written as

$$-\omega^2 \sum_{j=1}^{n} m_j\delta_{ij}x_j + x_i = 0 \quad (i = 1, 2, \ldots, n) \tag{10.6}$$

For a non-trivial solution, that is $x_i \neq 0$, the determinant of Eq. (10.6) must be zero. Hence

$$\begin{vmatrix}
(\omega^2 m_1\delta_{11} - 1) & \omega^2 m_2\delta_{12} & \ldots & \omega^2 m_i\delta_{1i} & \ldots & \omega^2 m_n\delta_{1n} \\
\omega^2 m_1\delta_{21} & (\omega^2 m_2\delta_{22} - 1) & \ldots & \omega^2 m_i\delta_{2i} & \ldots & \omega^2 m_n\delta_{2n} \\
\cdots & \cdots & \cdots & \cdots & \cdots & \cdots \\
\omega^2 m_1\delta_{i1} & \omega^2 m_2\delta_{i2} & \ldots & (\omega^2 m_i\delta_{ii} - 1) & \ldots & \omega^2 m_n\delta_{in} \\
\cdots & \cdots & \cdots & \cdots & \cdots & \cdots \\
\omega^2 m_1\delta_{n1} & \omega^2 m_2\delta_{n2} & \ldots & \omega^2 m_i\delta_{ni} & \ldots & (\omega^2 m_n\delta_{nn} - 1)
\end{vmatrix} = 0 \tag{10.7}$$

The solution of Eq. (10.7) gives the normal frequencies of vibration of the system. The corresponding modes may then be deduced as we shall see in the following examples.

Example 10.1

Determine the normal modes and frequencies of vibration of a weightless cantilever supporting masses *m*/3 and *m* at points 1 and 2 as shown in Fig. 10.3. The flexural rigidity of the cantilever is *EI*.

The equations of motion of the system are

$$(m/3)\ddot{v}_1\delta_{11} + m\ddot{v}_2\delta_{12} + v_1 = 0 \tag{ii}$$

$$(m/3)\ddot{v}_1\delta_{21} + m\ddot{v}_2\delta_{22} + v_2 = 0 \tag{iii}$$

where v_1 and v_2 are the vertical displacements of the masses at any instant of time. In this example, displacements are assumed to be caused by bending strains only; the flexibility coefficients δ_{11}, δ_{22} and $\delta_{12}(=\delta_{21})$ may therefore be found by the unit load method described in Section 5.8. Then

$$\delta_{ij} = \int_L \frac{M_i M_j}{EI}\,dz \tag{iii}$$

where M_i is the bending moment at any section z due to a unit load at the point i and M_j is the bending moment at any section z produced by a unit load at the point j. Therefore, from Fig. 10.3

$$\begin{aligned} M_1 &= 1(l - z) & 0 \le z \le l \\ M_2 &= 1(l/2 - z) & 0 \le z \le l/2 \\ M_2 &= 0 & l/2 \le z \le l \end{aligned}$$

Hence

$$\delta_{11} = \frac{1}{EI}\int_0^l M_1^2\,dz = \frac{1}{EI}\int_0^l (l - z)^2\,dz \tag{iv}$$

$$\delta_{22} = \frac{1}{EI}\int_0^l M_2^2\,dz = \frac{1}{EI}\int_0^{l/2} \left(\frac{l}{2} - z\right)^2\,dz \tag{v}$$

$$\delta_{12} = \delta_{21} = \frac{1}{EI}\int_0^l M_1 M_2\,dz = \frac{1}{EI}\int_0^{l/2} (l - z)\left(\frac{l}{2} - z\right)dz \tag{vi}$$

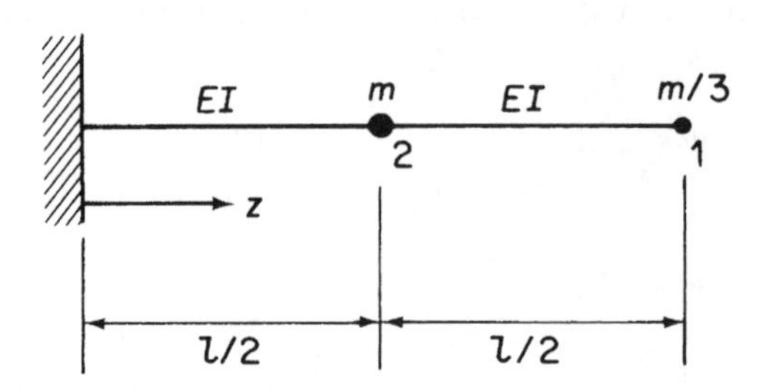

Fig. 10.3 Mass/beam system for Example 10.1.

Integrating Eqs (iv), (v) and (vi) and substituting limits, we obtain

$$\delta_{11} = \frac{l^3}{3EI} \quad \delta_{22} = \frac{l^3}{24EI} \quad \delta_{12} = \delta_{21} = \frac{5l^3}{48EI}$$

Each mass describes simple harmonic motion in the normal modes of oscillation so that $v_1 = v_1^0 \sin(\omega t + \varepsilon)$ and $v_2 = v_2^0 \sin(\omega t + \varepsilon)$. Hence $\ddot{v}_1 = -\omega^2 v_1$ and $\ddot{v}_2 = -\omega^2 v_2$. Substituting for $\ddot{v}_1$, $\ddot{v}_2$, δ_{11}, δ_{22} and $\delta_{12}(=\delta_{21})$ in Eqs (i) and (ii) and writing $\lambda = ml^3/(3 \times 48EI)$, we obtain

$$(1 - 16\lambda\omega^2)v_1 - 15\lambda\omega^2 v_2 = 0 \qquad \text{(vii)}$$

$$5\lambda\omega^2 v_1 - (1 - 6\lambda\omega^2)v_2 = 0 \qquad \text{(viii)}$$

For a non-trivial solution

$$\begin{vmatrix} (1 - 16\lambda\omega^2) & -15\lambda\omega^2 \\ 5\lambda\omega^2 & -(1 - 6\lambda\omega^2) \end{vmatrix} = 0$$

Expanding this determinant we have

$$-(1 - 16\lambda\omega^2)(1 - 6\lambda\omega^2) + 75(\lambda\omega^2)^2 = 0$$

or

$$21(\lambda\omega^2)^2 - 22\lambda\omega^2 + 1 = 0 \qquad \text{(ix)}$$

Inspection of Eq. (ix) shows that

$$\lambda\omega^2 = 1/21 \text{ or } 1$$

Hence

$$\omega^2 = \frac{3 \times 48EI}{21ml^3} \quad \text{or} \quad \frac{3 \times 48EI}{ml^3}$$

The normal or natural frequencies of vibration are therefore

$$f_1 = \frac{\omega_1}{2\pi} = \frac{2}{\pi}\sqrt{\frac{3EI}{7ml^3}}$$

$$f_2 = \frac{\omega_2}{2\pi} = \frac{6}{\pi}\sqrt{\frac{EI}{ml^3}}$$

The system is therefore capable of vibrating at two distinct frequencies. To determine the normal mode corresponding to each frequency we first take the lower frequency f_1 and substitute it in either Eq. (vii) or Eq. (viii). From Eq. (vii)

$$\frac{v_1}{v_2} = \frac{15\lambda\omega^2}{1 - 16\lambda\omega^2} = \frac{15 \times (1/21)}{1 - 16 \times (1/21)}$$

which is a positive quantity. Therefore, at the lowest natural frequency the cantilever oscillates in such a way that the displacement of both masses has the same sign at

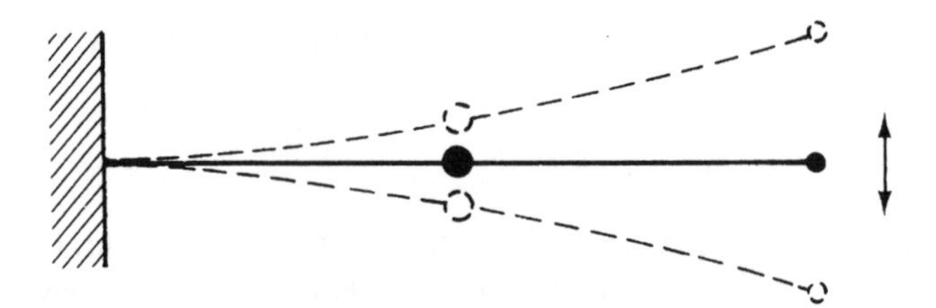

Fig. 10.4 The first natural mode of the mass/beam system of Fig. 10.3.

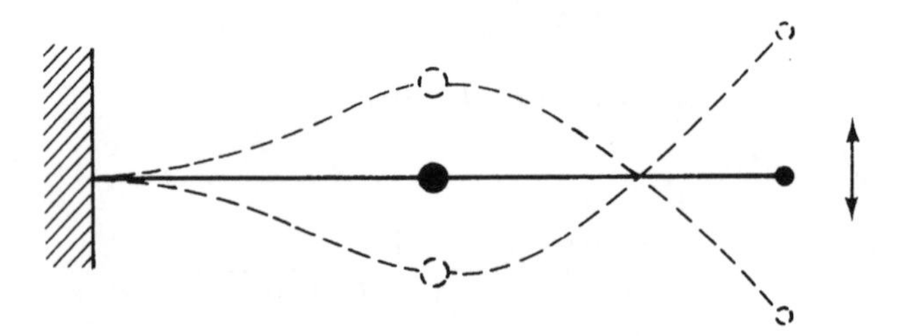

Fig. 10.5 The second natural mode of the mass/beam system of Fig. 10.3.

the same instant of time. Such an oscillation would take the form shown in Fig. 10.4. Substituting the second natural frequency in Eq. (vii) we have

$$\frac{v_1}{v_2} = \frac{15\lambda\omega^2}{1 - 16\lambda\omega^2} = \frac{15}{1 - 16}$$

which is negative so that the masses have displacements of opposite sign at any instant of time as shown in Fig. 10.5.

Example 10.2

Find the lowest natural frequency of the weightless beam/mass system shown in Fig. 10.6. For the beam $GJ = (2/3)EI$.

The equations of motion are

$$m\ddot{v}_1\delta_{11} + 4m\ddot{v}_2\delta_{12} + v_1 = 0 \tag{i}$$

$$m\ddot{v}_1\delta_{21} + 4m\ddot{v}_2\delta_{22} + v_2 = 0 \tag{ii}$$

In this problem displacements are caused by bending and torsion so that

$$\delta_{ij} = \int_L \frac{M_i M_j}{EI}\,ds + \int_L \frac{T_i T_j}{GJ}\,ds \tag{iii}$$

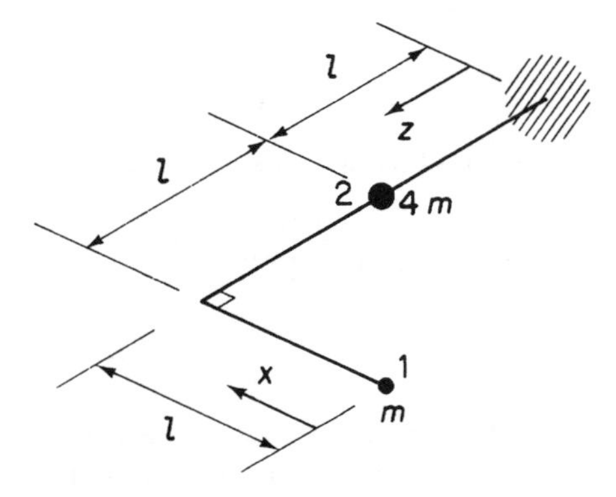

Fig. 10.6 Mass/beam system for Example 10.2.

From Fig. 10.6 we see that

$$
\begin{array}{lll}
M_1 = 1x & 0 \le x \le l & \\
M_1 = 1(2l - z) & 0 \le z \le 2l & \\
M_2 = 1(l - z) & 0 \le z \le l & \\
M_2 = 0 & l \le z \le 2l & 0 \le x \le l \\
T_1 = 1l & 0 \le z \le 2l & \\
T_1 = 0 & 0 \le x \le l & \\
T_2 = 0 & 0 \le z \le 2l & 0 \le x \le l
\end{array}
$$

Hence

$$\delta_{11} = \int_0^l \frac{x^2}{EI}\,dx + \int_0^{2l} \frac{(2l - z)^2}{EI}\,dz + \int_0^{2l} \frac{l^2}{GJ}\,dz \tag{iv}$$

$$\delta_{22} = \int_0^l \frac{(l - z)^2}{EI}\,dz \tag{v}$$

$$\delta_{12} = \delta_{21} = \int_0^l \frac{(2l - z)(l - z)}{EI}\,dz \tag{vi}$$

from which we obtain

$$\delta_{11} = \frac{6l^3}{EI} \quad \delta_{22} = \frac{l^3}{3EI} \quad \delta_{12} = \delta_{21} = \frac{5l^3}{6EI}$$

Writing $\lambda = ml^3/6EI$ and solving Eqs (i) and (ii) in an identical manner to the solution of Eqs (i) and (ii) in Example 10.1 results in a quadratic in $\lambda\omega^2$, namely

$$188(\lambda\omega^2)^2 - 44\lambda\omega^2 + 1 = 0 \tag{vii}$$

Solving Eq. (vii) we obtain

$$\lambda\omega^2 = \frac{44 \pm \sqrt{44^2 - 4 \times 188 \times 1}}{376}$$

which gives

$$\lambda\omega^2 = 0.21 \quad \text{or} \quad 0.027$$

The lowest natural frequency therefore corresponds to $\lambda\omega^2 = 0.027$ and is

$$\frac{1}{2\pi}\sqrt{\frac{0.162EI}{ml^3}}$$

Example 10.3

Determine the natural frequencies of the system shown in Fig. 10.7 and sketch the normal modes. The flexural rigidity EI of the weightless beam is $1.44 \times 10^6\,\text{N}\,\text{m}^2$, $l = 0.76\,\text{m}$, the radius of gyration r of the mass m is $0.152\,\text{m}$ and its weight is 1435 N.

In this problem the mass possesses an inertia about its own centre of gravity (its radius of gyration is not zero) which means that in addition to translational displacements it will experience rotation. The equations of motion are therefore

$$m\ddot{v}\delta_{11} + mr^2\ddot{\theta}\delta_{12} + v = 0 \tag{i}$$

$$m\ddot{v}\delta_{21} + mr^2\ddot{\theta}\delta_{22} + \theta = 0 \tag{ii}$$

where v is the vertical displacement of the mass at any instant of time and θ is the rotation of the mass from its stationary position. Although the beam supports just one mass it is subjected to two moment systems; M_1 at any section z due to the weight of the mass and a constant moment M_2 caused by the inertia couple of the mass as it rotates. Then

$$\begin{aligned} M_1 &= 1z \quad 0 \le z \le l \\ M_1 &= 1l \quad 0 \le y \le l \\ M_2 &= 1 \quad\; 0 \le z \le l \\ M_2 &= 1 \quad\; 0 \le y \le l \end{aligned}$$

Hence

$$\delta_{11} = \int_0^l \frac{z^2}{EI}\,\mathrm{d}z + \int_0^l \frac{l^2}{EI}\,\mathrm{d}y \tag{iii}$$

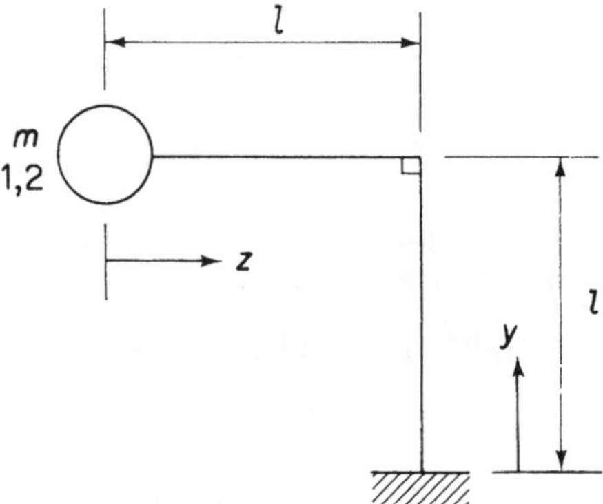

Fig. 10.7 Mass/beam system for Example 10.3.

$$\delta_{22} = \int_0^l \frac{dz}{EI} + \int_0^l \frac{dy}{EI} \quad \text{(iv)}$$

$$\delta_{12} = \delta_{21} = \int_0^l \frac{z\,dz}{EI} + \int_0^l \frac{l}{EI}\,dy \quad \text{(v)}$$

from which

$$\delta_{11} = \frac{4l^3}{3EI} \quad \delta_{22} = \frac{2l}{EI} \quad \delta_{12} = \delta_{21} = \frac{3l^2}{2EI}$$

Each mode will oscillate with simple harmonic motion so that

$$v = v_0 \sin(\omega t + \varepsilon) \quad \theta = \theta_0 \sin(\omega t + \varepsilon)$$

and

$$\ddot{v} = -\omega^2 v \quad \ddot{\theta} = -\omega^2 \theta$$

Substituting in Eqs (i) and (ii) gives

$$\left(1 - \omega^2 m \frac{4l^3}{3EI}\right) v - \omega^2 m r^2 \frac{3l^2}{2EI} \theta = 0 \quad \text{(vi)}$$

$$-\omega^2 m \frac{3l^2}{2EI} v + \left(1 - \omega^2 m r^2 \frac{2l}{EI}\right) \theta = 0 \quad \text{(vii)}$$

Inserting the values of m, r, l and EI we have

$$\left(1 - \frac{1435 \times 4 \times 0.76^3}{9.81 \times 3 \times 1.44 \times 10^6} \omega^2\right) v - \frac{1435 \times 0.152^2 \times 3 \times 0.76^2}{9.81 \times 2 \times 1.44 \times 10^6} \omega^2 \theta = 0 \quad \text{(viii)}$$

$$-\frac{1435 \times 3 \times 0.76^2}{9.81 \times 2 \times 1.44 \times 10^6} \omega^2 v + \left(1 - \frac{1435 \times 0.152^2 \times 2 \times 0.76}{9.81 \times 1.44 \times 10^6} \omega^2\right) \theta = 0 \quad \text{(ix)}$$

or

$$(1 - 6 \times 10^{-5} \omega^2) v - 0.203 \times 10^{-5} \omega^2 \theta = 0 \quad \text{(x)}$$

$$-8.8 \times 10^{-5} \omega^2 v + (1 - 0.36 \times 10^{-5} \omega^2) \theta = 0 \quad \text{(xi)}$$

Solving Eqs (x) and (xi) as before gives

$$\omega = 122 \text{ or } 1300$$

from which the natural frequencies are

$$f_1 = \frac{61}{\pi} \quad f_2 = \frac{650}{\pi}$$

From Eq. (x)

$$\frac{v}{\theta} = \frac{0.203 \times 10^{-5} \omega^2}{1 - 6 \times 10^{-5} \omega^2}$$

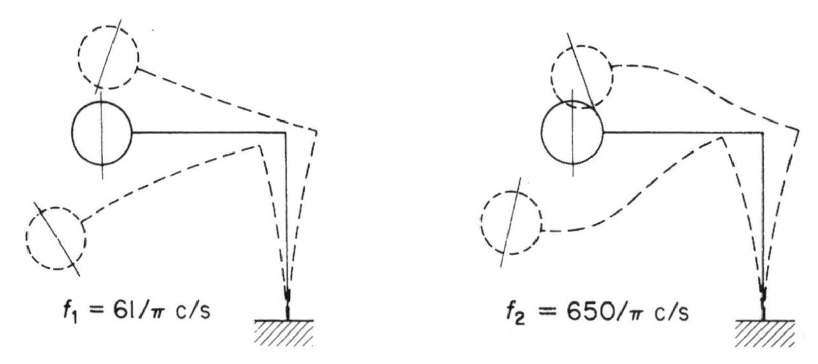

Fig. 10.8 The first two natural modes of vibration of the beam/mass system of Fig. 10.7.

which is positive at the lowest natural frequency, corresponding to $\omega = 122$, and negative for $\omega = 1300$. The modes of vibration are therefore as shown in Fig. 10.8.

10.2 Oscillation of beams

So far we have restricted our discussion to weightless beams supporting concentrated, or otherwise, masses. We shall now investigate methods of determining normal modes and frequencies of vibration of beams possessing weight and therefore inertia. The equations of motion of such beams are derived on the assumption that vibration occurs in one of the principal planes of the beam and that the effects of rotary inertia and shear displacements may be neglected.

Figure 10.9(a) shows a uniform beam of cross-sectional area A vibrating in a principal plane about some axis Oz. The displacement of an element δz of the beam at any instant of time t is v and the moments and forces acting on the element are shown in Fig. 10.9(b). Taking moments about the vertical centre line of the element gives

$$S_y \frac{\delta z}{2} + M_x + \left(S_y + \frac{\partial S_y}{\partial z} \delta z \right) \frac{\delta z}{2} - \left(M_x + \frac{\partial M_x}{\partial z} \delta z \right) = 0$$

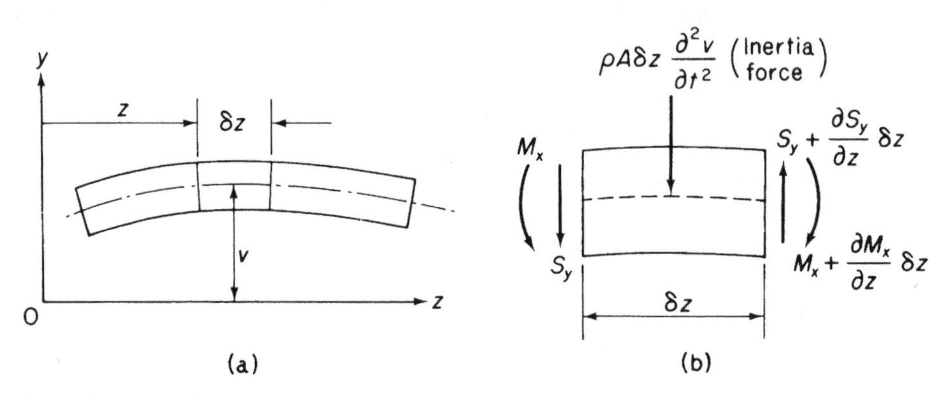

Fig. 10.9 Vibration of a beam possessing mass.

from which, neglecting second-order terms, we obtain

$$S_y = \frac{\partial M_x}{\partial z} \tag{10.8}$$

Considering the vertical equilibrium of the element

$$\left(S_y + \frac{\partial S_y}{\partial z}\delta z\right) - S_y - \rho A \delta z \frac{\partial^2 v}{\partial t^2} = 0$$

so that

$$\frac{\partial S_y}{\partial z} = \rho A \frac{\partial^2 v}{\partial t^2} \tag{10.9}$$

From basic bending theory (Chapter 16)

$$M_x = -EI\frac{\partial^2 v}{\partial z^2} \tag{10.10}$$

It follows from Eqs (10.8), (10.9) and (10.10) that

$$\frac{\partial^2}{\partial z^2}\left(-EI\frac{\partial^2 v}{\partial z^2}\right) = \rho A \frac{\partial^2 v}{\partial t^2} \tag{10.11}$$

Equation (10.11) is applicable to both uniform and non-uniform beams. In the latter case the flexural rigidity, *EI*, and the mass per unit length, ρA, are functions of z. For a beam of uniform section, Eq. (10.11) reduces to

$$EI\frac{\partial^4 v}{\partial z^4} + \rho A \frac{\partial^2 v}{\partial t^2} = 0 \tag{10.12}$$

In the normal modes of vibration each element of the beam describes simple harmonic motion; thus

$$v(z,t) = V(z)\sin(\omega t + \varepsilon) \tag{10.13}$$

where $V(z)$ is the amplitude of the vibration at any section z. Substituting for v from Eq. (10.13) in Eq. (10.12) yields

$$\frac{\mathrm{d}^4 V}{\mathrm{d}z^4} - \frac{\rho A \omega^2}{EI} V = 0 \tag{10.14}$$

Equation (10.14) is a fourth-order differential equation of standard form having the general solution

$$V = B\sin\lambda z + C\cos\lambda z + D\sinh\lambda z + F\cosh\lambda z \tag{10.15}$$

where

$$\lambda^4 = \frac{\rho A \omega^2}{EI}$$

and B, C, D and F are unknown constants which are determined from the boundary conditions of the beam. The ends of the beam may be:

(1) simply supported or pinned, in which case the displacement and bending moment are zero, and therefore in terms of the function $V(z)$ we have $V=0$ and $\mathrm{d}^2V/\mathrm{d}z^2=0$;
(2) fixed, giving zero displacement and slope, that is $V=0$ and $\mathrm{d}V/\mathrm{d}z=0$;
(3) free, for which the bending moment and shear force are zero, hence $\mathrm{d}^2V/\mathrm{d}z^2=0$ and, from Eq. (10.8), $\mathrm{d}^3V/\mathrm{d}z^3=0$.

Example 10.4

Determine the first three normal modes of vibration and the corresponding natural frequencies of the uniform, simply supported beam shown in Fig. 10.10.

Since both ends of the beam are simply supported, $V=0$ and $\mathrm{d}^2V/\mathrm{d}z^2=0$ at $z=0$ and $z=L$. From the first of these conditions and Eq. (10.15) we have

$$0 = C + F \tag{i}$$

and from the second

$$0 = -\lambda^2 C + \lambda^2 F \tag{ii}$$

Hence $C=F=0$. Applying the above boundary conditions at $z=L$ gives

$$0 = B \sin \lambda L + D \sinh \lambda L \tag{iii}$$

and

$$0 = -\lambda^2 B \sin \lambda L + \lambda^2 D \sinh \lambda L \tag{iv}$$

The only non-trivial solution ($\lambda L \neq 0$) of Eqs (iii) and (iv) is $D=0$ and $\sin \lambda L=0$. It follows that

$$\lambda L = n\pi \quad n = 1, 2, 3, \ldots$$

Therefore

$$\omega_\mathrm{n}^2 = \left(\frac{n\pi}{L}\right)^4 \frac{EI}{\rho A} \quad n = 1, 2, 3, \ldots \tag{v}$$

and the normal modes of vibration are given by

$$v(z,t) = B_\mathrm{n} \sin \frac{n\pi z}{L} \sin(\omega_\mathrm{n} t + \varepsilon_\mathrm{n}) \tag{vi}$$

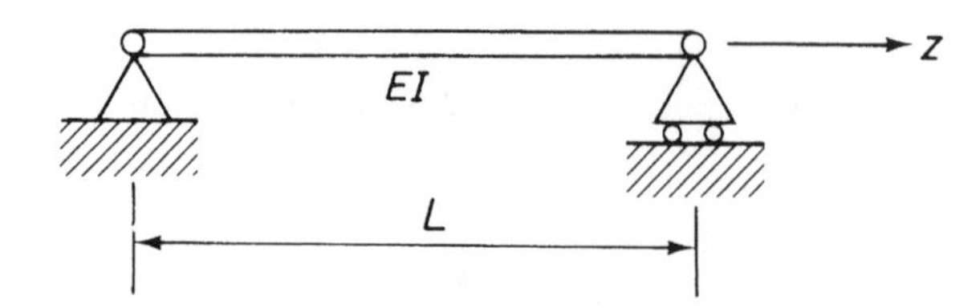

Fig. 10.10 Beam of Example 10.4.

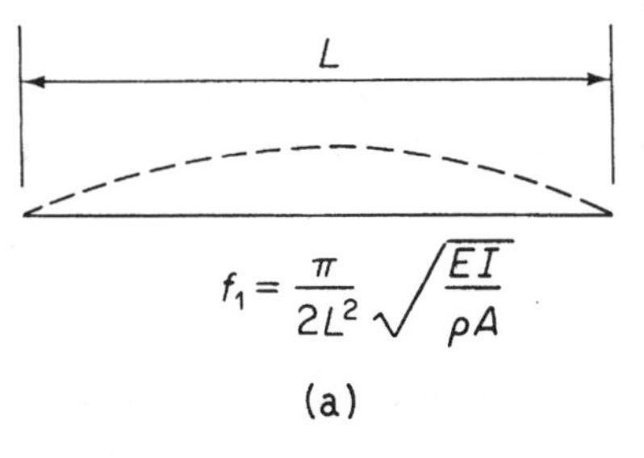

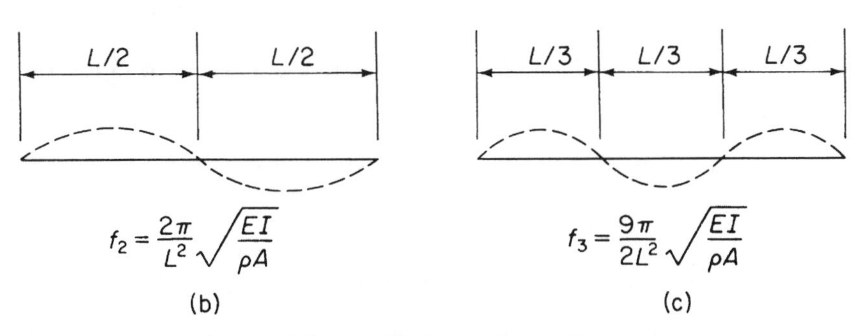

Fig. 10.11 First three normal modes of vibration of the beam of Example 10.4.

with natural frequencies

$$f_{\mathrm{n}} = \frac{\omega_{\mathrm{n}}}{2\pi} = \frac{1}{2\pi}\left(\frac{n\pi}{L}\right)^2\sqrt{\frac{EI}{\rho A}} \tag{vii}$$

The first three normal modes of vibration are shown in Fig. 10.11.

Example 10.5

Find the first three normal modes and corresponding natural frequencies of the uniform cantilever beam shown in Fig. 10.12.

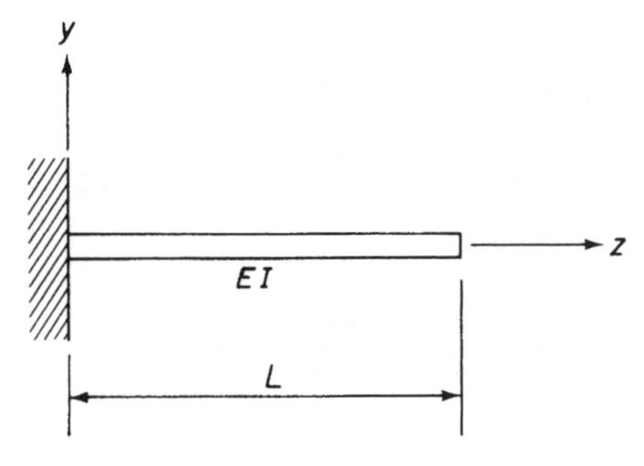

Fig. 10.12 Cantilever beam of Example 10.5.

The boundary conditions in this problem are $V=0$, $\mathrm{d}V/\mathrm{d}z=0$ at $z=0$ and $\mathrm{d}^2V/\mathrm{d}z^2=0$, $\mathrm{d}^3V/\mathrm{d}z^3=0$ at $z=L$. Substituting these in turn in Eq. (10.15) we obtain

$$0 = C + F \quad \text{(i)}$$

$$0 = \lambda B + \lambda D \quad \text{(ii)}$$

$$0 = -\lambda^2 B \sin \lambda L - \lambda^2 C \cos \lambda L + \lambda^2 D \sinh \lambda L + \lambda^2 F \cosh \lambda L \quad \text{(iii)}$$

$$0 = -\lambda^3 B \cos \lambda L + \lambda^3 C \sin \lambda L + \lambda^3 D \cosh \lambda L + \lambda^3 F \sinh \lambda L \quad \text{(iv)}$$

From Eqs (i) and (ii), $C=-F$ and $B=-D$. Thus, replacing F and D in Eqs (iii) and (iv) we obtain

$$B(-\sin \lambda L - \sinh \lambda L) + C(-\cos\lambda L - \cosh \lambda L) = 0 \quad \text{(v)}$$

and

$$B(-\cos\lambda L - \cosh \lambda L) + C(\sin \lambda L - \sinh \lambda L) = 0 \quad \text{(vi)}$$

Eliminating B and C from Eqs (v) and (vi) gives

$$(-\sin\lambda L - \sinh \lambda L)(\sinh \lambda L - \sin \lambda L) + (\cos \lambda L - \cosh \lambda L)^2 = 0$$

Expanding this equation, and noting that $\sin^2 \lambda L + \cos^2 \lambda L = 1$ and $\cosh^2 \lambda L - \sinh^2 \lambda L = 1$, yields the frequency equation

$$\cos \lambda L \cosh \lambda L + 1 = 0 \quad \text{(vii)}$$

Equation (vii) may be solved graphically or by Newton's method. The first three roots λ_1, λ_2 and λ_3 are given by

$$\lambda_1 L = 1.875 \quad \lambda_2 L = 4.694 \quad \lambda_3 L = 7.855$$

from which are found the natural frequencies corresponding to the first three normal modes of vibration. The natural frequency of the rth mode ($r \geq 4$) is obtained from the approximate relationship

$$\lambda_r L \approx (r - \tfrac{1}{2})\pi$$

and its shape in terms of a single arbitrary constant K_r is

$$V_r(z) = K_r[\cosh \lambda_r z - \cos \lambda_r z - k_r(\sinh \lambda_r z - \sin \lambda_r z)]$$

where

$$k_r = \frac{\cos \lambda_r L + \cosh \lambda_r L}{\sin \lambda_r L + \sinh \lambda_r L} \qquad r = 1, 2, 3, \ldots$$

Figure 10.13 shows the first three normal mode shapes of the cantilever and their associated natural frequencies.

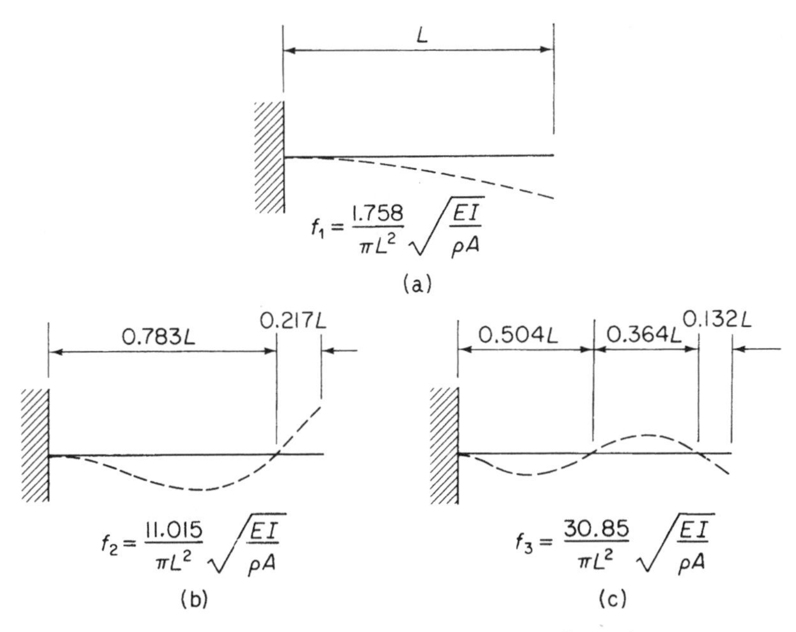

Fig. 10.13 The first three normal modes of vibration of the cantilever beam of Example 10.5.

10.3 Approximate methods for determining natural frequencies

The determination of natural frequencies and normal mode shapes for beams of non-uniform section involves the solution of Eq. (10.11) and fulfilment of the appropriate boundary conditions. However, with the exception of a few special cases, such solutions do not exist and the natural frequencies are obtained by approximate methods such as the Rayleigh and Rayleigh–Ritz methods which are presented here. Rayleigh's method is discussed first.

A beam vibrating in a normal or combination of normal modes possesses kinetic energy by virtue of its motion and strain energy as a result of its displacement from an initial unstrained condition. From the principle of conservation of energy the sum of the kinetic and strain energies is constant with time. In computing the strain energy U of the beam we assume that displacements are due to bending strains only so that

$$U = \int_L \frac{M^2}{2EI}\,\mathrm{d}z \quad \text{(see Chapter 5)} \tag{10.16}$$

where

$$M = -EI\frac{\partial^2 v}{\partial z^2} \quad \text{(see Eq. (10.10))}$$

Substituting for v from Eq. (10.13) gives

$$M = -EI\frac{\mathrm{d}^2 V}{\mathrm{d}z^2}\sin(\omega t + \varepsilon)$$

so that from Eq. (10.16)

$$U = \frac{1}{2}\sin^2(\omega t + \varepsilon)\int_L EI\left(\frac{d^2V}{dz^2}\right)^2 dz \tag{10.17}$$

For a non-uniform beam, having a distributed mass $\rho A(z)$ per unit length and carrying concentrated masses, $m_1, m_2, m_3, \ldots, m_n$ at distances $z_1, z_2, z_3, \ldots, z_n$ from the origin, the kinetic energy K_E may be written as

$$K_E = \frac{1}{2}\int_L \rho A(z)\left(\frac{\partial v}{\partial t}\right)^2 dz + \frac{1}{2}\sum_{r=1}^{n} m_r\left[\left(\frac{\partial v}{\partial t}\right)_{z=z_r}\right]^2$$

Substituting for $v(z)$ from Eq. (10.17) we have

$$K_E = \frac{1}{2}\omega^2\cos^2(\omega t + \varepsilon)\left[\int_L \rho A(z)V^2\,dz + \sum_{r=1}^{n} m_r\{V(z_r)\}^2\right] \tag{10.18}$$

Since $K_E + U = \text{constant}$, say C, then

$$\frac{1}{2}\sin^2(\omega t + \varepsilon)\int_L EI\left(\frac{d^2V}{dz^2}\right)^2 dz + \frac{1}{2}\omega^2\cos^2(\omega t + \varepsilon) \times\left[\int_L \rho A(z)V^2\,dz + \sum_{r=1}^{n} m_r\{V(z_r)\}^2\right] = C \tag{10.19}$$

Inspection of Eq. (10.19) shows that when $(\omega t + \varepsilon) = 0, \pi, 2\pi, \ldots$

$$\frac{1}{2}\omega^2\left[\int_L \rho A(z)V^2\,dz + \sum_{r=1}^{n} m_r\{V(z_r)\}^2\right] = C \tag{10.20}$$

and when

$$(\omega t + \varepsilon) = \pi/2,\ 3\pi/2,\ 5\pi/2, \ldots$$

then

$$\frac{1}{2}\int_L EI\left(\frac{d^2V}{dz^2}\right)^2 dz = C \tag{10.21}$$

In other words the kinetic energy in the mean position is equal to the strain energy in the position of maximum displacement. From Eqs (10.20) and (10.21)

$$\omega^2 = \frac{\int_L EI(d^2V/dz^2)^2\,dz}{\int_L \rho A(z)V^2\,dz + \sum_{r=1}^{n} m_r\{V(z_r)\}^2} \tag{10.22}$$

Equation (10.22) gives the exact value of natural frequency for a particular mode if $V(z)$ is known. In the situation where a mode has to be 'guessed', Rayleigh's principle states that if a mode is assumed which satisfies at least the slope and displacement conditions at the ends of the beam then a good approximation to the true natural frequency will be obtained. We have noted previously that if the assumed normal mode differs only slightly from the actual mode then the stationary property of the normal modes ensures that the approximate natural frequency is only very slightly different to the true value. Furthermore, the approximate frequency will be higher than the actual one since the assumption of an approximate mode implies the presence of some constraints which force the beam to vibrate in a particular fashion; this has the effect of increasing the frequency.

The Rayleigh–Ritz method extends and improves the accuracy of the Rayleigh method by assuming a finite series for $V(z)$, namely

$$V(z) = \sum_{s=1}^{n} B_s V_s(z) \tag{10.23}$$

where each assumed function $V_s(z)$ satisfies the slope and displacement conditions at the ends of the beam and the parameters B_s are arbitrary. Substitution of $V(z)$ in Eq. (10.22) then gives approximate values for the natural frequencies. The parameters B_s are chosen to make these frequencies a minimum, thereby reducing the effects of the implied constraints. Having chosen suitable series, the method of solution is to form a set of equations

$$\frac{\partial \omega^2}{\partial B_s} = 0, \quad s = 1, 2, 3, \ldots, n \tag{10.24}$$

Eliminating the parameter B_s leads to an nth-order determinant in ω^2 whose roots give approximate values for the first n natural frequencies of the beam.

Example 10.6

Determine the first natural frequency of a cantilever beam of length, L, flexural rigidity EI and constant mass per unit length ρA. The cantilever carries a mass $2m$ at the tip, where $m = \rho AL$.

An exact solution to this problem may be found by solving Eq. (10.14) with the appropriate end conditions. Such a solution gives

$$\omega_1 = 1.1582\sqrt{\frac{EI}{mL^3}}$$

and will serve as a comparison for our approximate answer. As an assumed mode shape we shall take the static deflection curve for a cantilever supporting a tip load since, in this particular problem, the tip load $2m$ is greater than the mass ρAL of the cantilever. If the reverse were true we would assume the static deflection curve for a cantilever carrying a uniformly distributed load. Thus

$$V(z) = a(3Lz^2 - z^3) \tag{i}$$

where the origin for z is taken at the built-in end and a is a constant term which includes the tip load and the flexural rigidity of the beam. From Eq. (i)

$$V(L) = 2aL^3 \quad \text{and} \quad \frac{d^2V}{dz^2} = 6a(L - z)$$

Substituting these values in Eq. (10.22) we obtain

$$\omega_1^2 = \frac{36EIa^2 \int_0^L (L - z)^2 \, dz}{\rho Aa^2 \int_0^L (3L - z)^2 z^4 \, dz + 2m(2aL^3)^2} \tag{ii}$$

Evaluating Eq. (ii) and expressing ρA in terms of m we obtain

$$\omega_1 = 1.1584\sqrt{\frac{EI}{mL^3}} \tag{iii}$$

which value is only 0.02 per cent higher than the true value given above. The estimation of higher natural frequencies requires the assumption of further, more complex, shapes for $V(z)$.

It is clear from the previous elementary examples of normal mode and natural frequency calculation that the estimation of such modes and frequencies for a complete aircraft is a complex process. However, the aircraft designer is not restricted to calculation for the solution of such problems, although the advent of the digital computer has widened the scope and accuracy of this approach. Other possible methods are to obtain the natural frequencies and modes by direct measurement from the results of a *resonance test* on the actual aircraft or to carry out a similar test on a simplified scale model. Details of resonance tests are discussed in Section 28.4. Usually a resonance test is impracticable since the designer requires the information before the aircraft is built, although this type of test is carried out on the completed aircraft as a design check. The alternative of building a scale model has found favour for many years. Such models are usually designed to be as light as possible and to represent the stiffness characteristics of the full-scale aircraft. The inertia properties are simulated by a suitable distribution of added masses.

Problems

P.10.1 Figure P.10.1 shows a massless beam ABCD of length $3l$ and uniform bending stiffness EI which carries concentrated masses $2m$ and m at the points B and D, respectively. The beam is built-in at end A and simply supported at C. In addition, there is a hinge at B which allows only shear forces to be transmitted between sections AB and BCD.

Calculate the natural frequencies of free, undamped oscillations of the system and determine the corresponding modes of vibration, illustrating your results by suitably dimensioned sketches.

Ans. $\dfrac{1}{2\pi}\sqrt{\dfrac{3EI}{4ml^3}} \quad \dfrac{1}{2\pi}\sqrt{\dfrac{3EI}{ml^3}}$

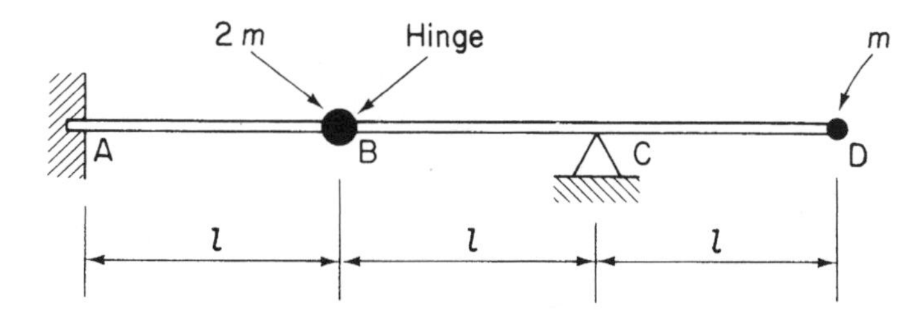

Fig. P.10.1

P.10.2 Three massless beams 12, 23 and 24 each of length l are rigidly joined together in one plane at the point 2, 12 and 23 being in the same straight line with 24 at right angles to them (see Fig. P.10.2). The bending stiffness of 12 is $3EI$ while that of 23 and 24 is EI. The beams carry masses m and $2m$ concentrated at the points 4 and 2, respectively. If the system is simply supported at 1 and 3 determine the natural frequencies of vibration in the plane of the figure.

Ans. $\dfrac{1}{2\pi}\sqrt{\dfrac{2.13EI}{ml^3}}\quad \dfrac{1}{2\pi}\sqrt{\dfrac{5.08EI}{ml^3}}$

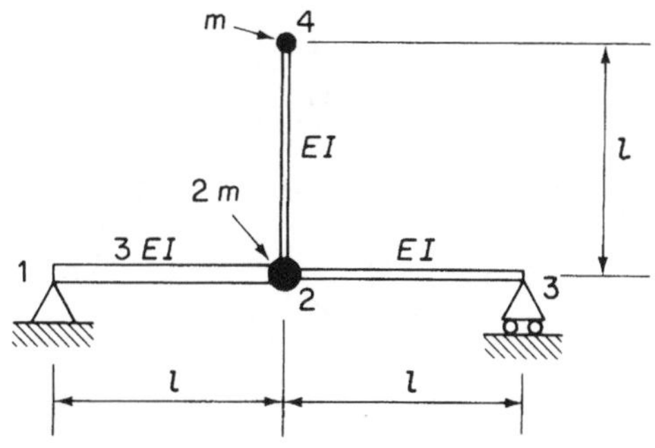

Fig. P.10.2

P.10.3 Two uniform circular tubes AB and BC are rigidly jointed at right angles at B and built-in at A (Fig. P.10.3). The tubes themselves are massless but carry a mass of 20 kg at C which has a polar radius of gyration of $0.25a$ about an axis through its own centre of gravity parallel to AB. Determine the natural frequencies and modes of vibration for small oscillations normal to the plane containing AB and BC. The tube has a mean diameter of 25 mm and wall thickness 1.25 mm. Assume that for the material of the tube $E = 70\,000\ \text{N/mm}^2$, $G = 28\,000\ \text{N/mm}^2$ and $a = 250$ mm.

Ans. 0.09 Hz, 0.62 Hz.

P.10.4 A uniform thin-walled cantilever tube, length L, circular cross-section of radius a and thickness t, carries at its tip two equal masses m. One mass is attached to the tube axis while the other is mounted at the end of a light rigid bar at a distance of $2a$ from the axis (see Fig. P.10.4). Neglecting the mass of the tube and assuming the stresses in the tube are given by basic bending theory and the Bredt–Batho theory of torsion, show that the frequencies ω of the coupled torsion flexure oscillations which

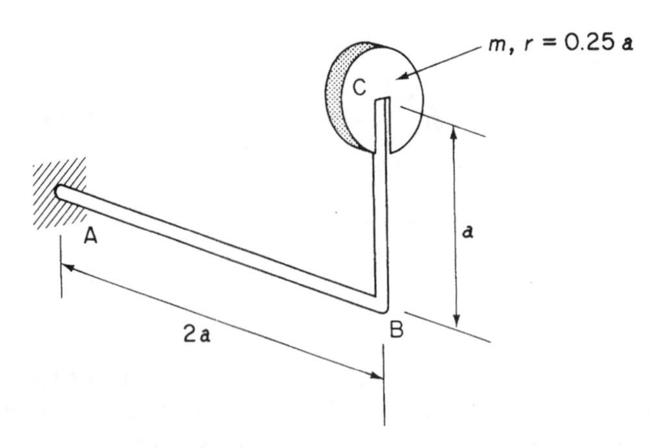

Fig. P.10.3

occur are given by

$$\frac{1}{\omega^2} = \frac{mL^3}{3E\pi a^3 t}[1 + 2\lambda \pm (1 + 2\lambda + 2\lambda^2)^{\frac{1}{2}}]$$

where

$$\lambda = \frac{3E}{G}\frac{a^2}{L^2}$$

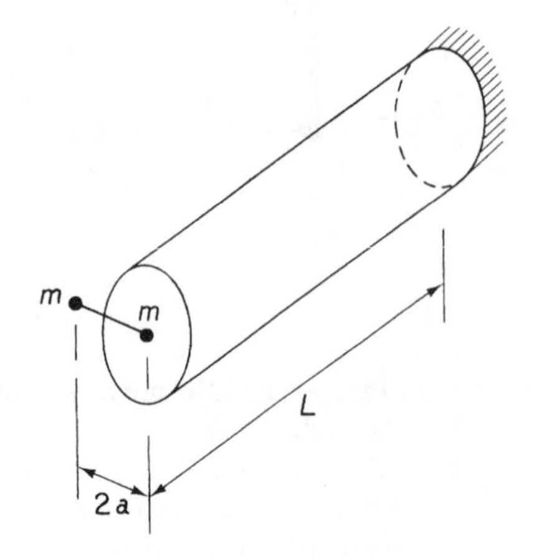

Fig. P.10.4

P.10.5 Figure P.10.5 shows the idealized cross-section of a single cell tube with axis of symmetry xx and length 1525 mm in which the direct stresses due to bending are carried only in the four booms of the cross-section. The walls are assumed to carry only shear stresses. The tube is built-in at the root and carries a weight of 4450 N at its tip; the centre of gravity of the weight coincides with the shear centre of the tube cross-section. Assuming that the direct and shear stresses in the tube are given by basic bending theory, calculate the natural frequency of flexural vibrations of the weight in a vertical direction. The effect of the weight of the tube is to be neglected and it should be noted that it is not necessary to know the position of the shear centre of the cross-section. The effect on the deflections of the shear strains in the tube walls must be included.

$$E = 70\,000\,\text{N/mm}^2 \quad G = 26\,500\,\text{N/mm}^2 \quad \text{boom areas } 970\,\text{mm}^2$$

Ans. 12.1 Hz.

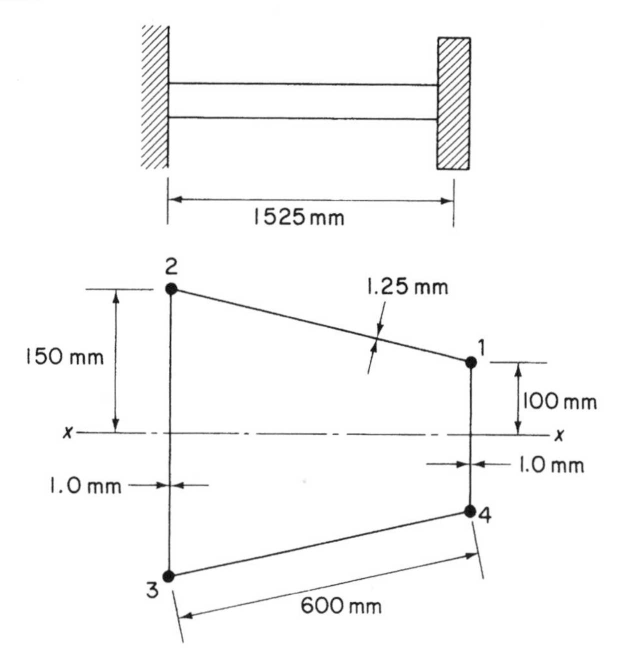

Fig. P.10.5

P.10.6 A straight beam of length l is rigidly built-in at its ends. For one quarter of its length from each end the bending stiffness is $4EI$ and the mass/unit length is $2m$: for the central half the stiffness is EI and the mass m per unit length. In addition, the beam carries three mass concentrations, $\frac{1}{2}ml$ at $\frac{1}{4}l$ from each end and $\frac{1}{4}ml$ at the centre, as shown in Fig. P.10.6.

Use an energy method or other approximation to estimate the lowest frequency of natural flexural vibration. A first approximation solution will suffice if it is accompanied by a brief explanation of a method of obtaining improved accuracy.

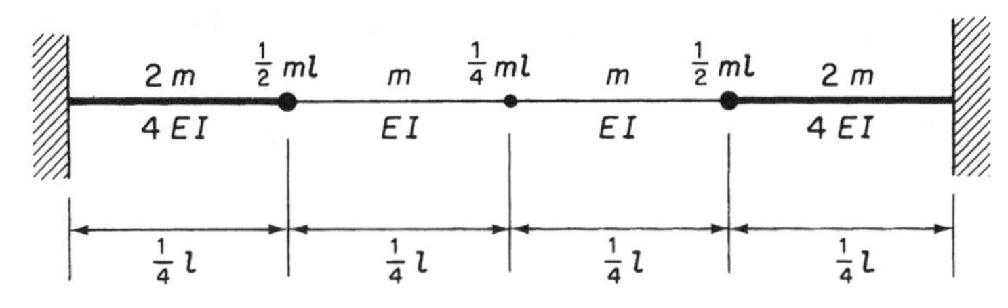

Fig. P.10.6

Ans. $3.7\sqrt{\dfrac{EI}{ml^4}}$

Part B Analysis of Aircraft Structures

Section B1 Principles of Stressed Skin Construction

11

Materials

With the present chapter we begin the purely aeronautical section of the book, where we consider structures peculiar to the field of aeronautical engineering. These structures are typified by arrangements of thin, load-bearing skins, frames and stiffeners, fabricated from lightweight, high strength materials of which aluminium alloys are the most widely used examples.

As a preliminary to the analysis of the basic aircraft structural forms presented in subsequent chapters we shall discuss the materials used in aircraft construction.

Several factors influence the selection of the structural material for an aircraft, but amongst these strength allied to lightness is probably the most important. Other properties having varying, though sometimes critical significance are stiffness, toughness, resistance to corrosion, fatigue and the effects of environmental heating, ease of fabrication, availability and consistency of supply and, not least important, cost.

The main groups of materials used in aircraft construction have been wood, steel, aluminium alloys with, more recently, titanium alloys, and fibre-reinforced composites. In the field of engine design, titanium alloys are used in the early stages of a compressor while nickel-based alloys or steels are used for the hotter later stages. As we are concerned primarily with the materials involved in the construction of the airframe, discussion of materials used in engine manufacture falls outside the scope of this book.

11.1 Aluminium alloys

Pure aluminium is a relatively low strength extremely flexible metal with virtually no structural applications. However, when alloyed with other metals its properties are improved significantly. Three groups of aluminium alloy have been used in the aircraft industry for many years and still play a major role in aircraft construction. In the first of these aluminium is alloyed with copper, magnesium, manganese, silicon and iron, and has a typical composition of 4% copper, 0.5% magnesium, 0.5% manganese, 0.3% silicon and 0.2% iron with the remainder being aluminium. In the wrought, heat-treated, naturally aged condition this alloy possesses a 0.1% proof stress not less than 230 N/mm^2, a tensile strength not less than 390 N/mm^2 and an elongation at fracture of 15%. Artificial ageing at a raised temperature of, for example, 170°C increases the proof stress to not less than 370 N/mm^2 and the tensile strength to not less than 460 N/mm^2 with an elongation of 8%.

The second group of alloys contain, in addition to the above, 1–2% of nickel, a higher content of magnesium and possible variations in the amounts of copper, silicon and iron. The most important property of these alloys is their retention of strength at high temperatures which makes them particularly suitable for aero engine manufacture. A development of these alloys by Rolls-Royce and High Duty Alloys Ltd replaced some of the nickel by iron and reduced the copper content; these RR alloys, as they were called, were used for forgings and extrusions in aero engines and airframes.

The third group of alloys depends upon the inclusion of zinc and magnesium for their high strength and have a typical composition of 2.5% copper, 5% zinc, 3% magnesium and up to 1% nickel with mechanical properties of 0.1% proof stress 510 N/mm^2, tensile strength 585 N/mm^2 and an elongation of 8%. In a modern development of this alloy nickel has been eliminated and provision made for the addition of chromium and further amounts of manganese.

Alloys from each of the above groups have been used extensively for airframes, skins and other stressed components, the choice of alloy being influenced by factors such as strength (proof and ultimate stress), ductility, ease of manufacture (e.g. in extrusion and forging), resistance to corrosion and amenability to protective treatment, fatigue strength, freedom from liability to sudden cracking due to internal stresses and resistance to fast crack propagation under load. Clearly, different types of aircraft have differing requirements. A military aircraft, for instance, having a relatively short life measured in hundreds of hours, does not call for the same degree of fatigue and corrosion resistance as a civil aircraft with a required life of 30 000 hours or more.

Unfortunately, as one particular property of aluminium alloys is improved, other desirable properties are sacrificed. For example, the extremely high static strength of the aluminium–zinc–magnesium alloys was accompanied for many years by a sudden liability to crack in an unloaded condition due to the retention of internal stresses in bars, forgings and sheet after heat treatment. Although variations in composition have eliminated this problem to a considerable extent other deficiencies showed themselves. Early post-war passenger aircraft experienced large numbers of stress-corrosion failures of forgings and extrusions. The problem became so serious that in 1953 it was decided to replace as many aluminium–zinc–manganese components as possible with the aluminium–4 per cent copper Alloy L65 and to prohibit the use of forgings in zinc-bearing alloy in all future designs. However, improvements in the stress-corrosion resistance of the aluminium–zinc–magnesium alloys have resulted in recent years from British, American and German research. Both British and American opinions agree on the benefits of including about 1 per cent copper but disagree on the inclusion of chromium and manganese, while in Germany the addition of silver has been found extremely beneficial. Improved control of casting techniques has brought further improvements in resistance to stress corrosion. The development of aluminium–zinc–magnesium–copper alloys has largely met the requirement for aluminium alloys possessing high strength, good fatigue crack growth resistance and adequate toughness. Further development will concentrate on the production of materials possessing higher specific properties, bringing benefits in relation to weight saving rather than increasing strength and stiffness.

The first group of alloys possess a lower static strength than the above zinc-bearing alloys, but are preferred for portions of the structure where fatigue considerations are of primary importance such as the undersurfaces of wings where tensile fatigue loads predominate. Experience has shown that the naturally aged version of these alloys has important advantages over the fully heat-treated forms in fatigue endurance and resistance to crack propagation. Furthermore, the inclusion of a higher percentage of magnesium was found, in America, to produce, in the naturally aged condition, mechanical properties between those of the normal naturally aged and artificially aged alloy. This alloy, designated 2024 (aluminium–copper alloys form the 2000 series) has the nominal composition: 4.5 per cent copper, 1.5 per cent magnesium, 0.6 per cent manganese, with the remainder aluminium, and appears to be a satisfactory compromise between the various important, but sometimes conflicting, mechanical properties.

Interest in aluminium–magnesium–silicon alloys has recently increased, although they have been in general use in the aerospace industry for decades. The reasons for this renewed interest are that they are potentially cheaper than aluminium–copper alloys and, being weldable, are capable of reducing manufacturing costs. In addition, variants, such as the ISO 6013 alloy, have improved property levels and, generally, possess a similar high fracture toughness and resistance to crack propagation as the 2000 series alloys.

Frequently, a particular form of an alloy is developed for a particular aircraft. An outstanding example of such a development is the use of Hiduminium RR58 as the basis for the main structural material, designated CM001, for Concorde. Hiduminium RR58 is a complex aluminium–copper–magnesium–nickel–iron alloy developed during the 1939–1945 war specifically for the manufacture of forged components in gas turbine aero engines. The chemical composition of the version used in Concorde was decided on the basis of elevated temperature, creep, fatigue and tensile testing programmes and has the detailed specification of:

	%Cu	%Mg	%Si	%Fe	%Ni	%Ti	%Al
Minimum	2.25	1.35	0.18	0.90	1.0	–	Remainder
Maximum	2.70	1.65	0.25	1.20	1.30	0.20	

Generally, CM001 is found to possess better overall strength/fatigue characteristics over a wide range of temperatures than any of the other possible aluminium alloys.

The latest aluminium alloys to find general use in the aerospace industry are the aluminium–lithium alloys. Of these, the aluminium–lithium–copper–manganese alloy, 8090, developed in the UK, is extensively used in the main fuselage structure of GKN Westland Helicopters' design EH101; it has also been qualified for Eurofighter 2000 (now named the Typhoon) but has yet to be embodied. In the USA the aluminium–lithium–copper alloy, 2095, has been used in the fuselage frames of the F16 as a replacement for 2124, resulting in a fivefold increase in fatigue life and a reduction in weight. Aluminium–lithium alloys can be successfully welded, possess a high fracture toughness and exhibit a high resistance to crack propagation.

11.2 Steel

The use of steel for the manufacture of thin-walled, box-section spars in the 1930s has been superseded by the aluminium alloys described in Section 11.1. Clearly, its high specific gravity prevents its widespread use in aircraft construction, but it has retained some value as a material for castings for small components demanding high tensile strengths, high stiffness and high resistance to wear. Such components include undercarriage pivot brackets, wing-root attachments, fasteners and tracks.

Although the attainment of high and ultra-high tensile strengths presents no difficulty with steel, it is found that other properties are sacrificed and that it is difficult to manufacture into finished components. To overcome some of these difficulties types of steel known as *maraging* steels were developed in 1961, from which carbon is either eliminated entirely or present only in very small amounts. Carbon, while producing the necessary hardening of conventional high tensile steels, causes brittleness and distortion; the latter is not easily rectifiable as machining is difficult and cold forming impracticable. Welded fabrication is also almost impossible or very expensive. The hardening of maraging steels is achieved by the addition of other elements such as nickel, cobalt and molybdenum. A typical maraging steel would have these elements present in the proportions: nickel 17–19 per cent, cobalt 8–9 per cent, molybdenum 3–3.5 per cent, with titanium 0.15–0.25 per cent. The carbon content would be a maximum of 0.03 per cent, with traces of manganese, silicon, sulphur, phosphorus, aluminium, boron, calcium and zirconium. Its 0.2 per cent proof stress would be nominally $1400\,\text{N/mm}^2$ and its modulus of elasticity $180\,000\,\text{N/mm}^2$.

The main advantages of maraging steels over conventional low alloy steels are: higher fracture toughness and notched strength, simpler heat treatment, much lower volume change and distortion during hardening, very much simpler to weld, easier to machine and better resistance to stress corrosion/hydrogen embrittlement. On the other hand, the material cost of maraging steels is three or more times greater than the cost of conventional steels, although this may be more than offset by the increased cost of fabricating a complex component from the latter steel.

Maraging steels have been used in: aircraft arrester hooks, rocket motor cases, helicopter undercarriages, gears, ejector seats and various structural forgings.

In addition to the above, steel in its stainless form has found applications primarily in the construction of super- and hypersonic experimental and research aircraft, where temperature effects are considerable. Stainless steel formed the primary structural material in the Bristol 188, built to investigate kinetic heating effects, and also in the American rocket aircraft, the X-15, capable of speeds of the order of Mach 5–6.

11.3 Titanium

The use of titanium alloys increased significantly in the 1980s, particularly in the construction of combat aircraft as opposed to transport aircraft. This increase continued in the 1990s to the stage where, for combat aircraft, the percentage of titanium alloy as a fraction of structural weight is of the same order as that of aluminium alloy. Titanium alloys possess high specific properties, have a good fatigue strength/tensile strength ratio with a distinct fatigue limit, and some retain considerable strength at temperatures up to 400–500°C. Generally, there is also a good resistance to corrosion and corrosion fatigue although properties are adversely affected by exposure to temperature and stress in a salt environment. The latter poses particular problems in the engines of carrier-operated aircraft. Further disadvantages are a relatively high density so that weight penalties are imposed if the alloy is extensively used, coupled with high primary and high fabrication costs, approximately seven times those of aluminium and steel.

In spite of this, titanium alloys were used in the airframe and engines of Concorde, while the Tornado wing carry-through box is fabricated from a weldable medium strength titanium alloy. Titanium alloys are also used extensively in the F15 and F22 American fighter aircraft and are incorporated in the tail assembly of the Boeing 777 civil airliner. Other uses include forged components such as flap and slat tracks and undercarriage parts.

New fabrication processes (e.g. superplastic forming combined with diffusion bonding) enable large and complex components to be produced, resulting in a reduction in production man-hours and weight. Typical savings are 30 per cent in man-hours, 30 per cent in weight and 50 per cent in cost compared with conventional riveted titanium structures. It is predicted that the number of titanium components fabricated in this way for aircraft will increase significantly and include items such as access doors, sheet for areas of hot gas impingement, etc.

11.4 Plastics

Plain plastic materials have specific gravities of approximately unity and are therefore considerably heavier than wood although of comparable strength. On the other hand, their specific gravities are less than half those of the aluminium alloys so that they find uses as windows or lightly stressed parts whose dimensions are established by handling requirements rather than strength. They are also particularly useful as electrical insulators and as energy absorbing shields for delicate instrumentation and even structures where severe vibration, such as in a rocket or space shuttle launch, occurs.

11.5 Glass

The majority of modern aircraft have cabins pressurized for flight at high altitudes. Windscreens and windows are therefore subjected to loads normal to their midplanes. Glass is frequently the material employed for this purpose in the form of plain or laminated plate or heat-strengthened plate. The types of plate glass used in aircraft have a modulus of elasticity between 70 000 and 75 000 N/mm^2 with a modulus of rupture in bending of 45 N/mm^2. Heat-strengthened plate has a modulus of rupture of about four and a half times this figure.

11.6 Composite materials

Composite materials consist of strong fibres such as glass or carbon set in a matrix of plastic or epoxy resin, which is mechanically and chemically protective. The fibres may be continuous or discontinuous but possess a strength very much greater than that of the same bulk materials. For example, carbon fibres have a tensile strength of the order of 2400 N/mm^2 and a modulus of elasticity of 400 000 N/mm^2.

A sheet of fibre-reinforced material is anisotropic, i.e. its properties depend on the direction of the fibres. Generally, therefore, in structural form two or more sheets are sandwiched together to form a *lay-up* so that the fibre directions match those of the major loads.

In the early stages of the development of composite materials glass fibres were used in a matrix of epoxy resin. This glass-reinforced plastic (GRP) was used for radomes and helicopter blades but found limited use in components of fixed wing aircraft due to its low stiffness. In the 1960s, new fibrous reinforcements were introduced; Kevlar, for example, is an aramid material with the same strength as glass but is stiffer. Kevlar composites are tough but poor in compression and difficult to machine, so they were used in secondary structures. Another composite, using boron fibre and developed in the USA, was the first to possess sufficient strength and stiffness for primary structures.

These composites have now been replaced by carbon-fibre-reinforced plastics (CFRP), which have similar properties to boron composites but are very much cheaper. Typically, CFRP has a modulus of the order of three times that of GRP, one and a half times that of a Kevlar composite and twice that of aluminium alloy. Its strength is three times that of aluminium alloy, approximately the same as that of GRP, and slightly less than that of Kevlar composites. CFRP does, however, suffer from some disadvantages. It is a brittle material and therefore does not yield plastically in regions of high stress concentration. Its strength is reduced by impact damage which may not be visible and the epoxy resin matrices can absorb moisture over a long period which reduces its matrix-dependent properties, such as its compressive strength; this effect increases with increase of temperature. Further, the properties of CFRP are subject to more random variation than those of metals. All these factors must be allowed for in design. On the other hand, the stiffness of CFRP is much less affected than its strength by the above and it is less prone to fatigue damage than metals. It is estimated that replacing 40% of an aluminium alloy structure by CFRP would result in a 12% saving in total structural weight.

CFRP is included in the wing, tailplane and forward fuselage of the latest Harrier development, is used in the Tornado taileron and has been used to construct a complete Jaguar wing and engine bay door for testing purposes. The use of CFRP in the fabrication of helicopter blades has led to significant increases in their service life, where fatigue resistance rather than stiffness is of primary importance. Figure 11.1 shows the structural complexity of a Sea King helicopter rotor blade which incorporates CFRP, GRP, stainless steel, a honeycomb core and foam filling. An additional advantage of the use of composites for helicopter rotor blades is that the moulding techniques employed allow variations of cross-section along the span, resulting in substantial aerodynamic benefits. This approach is being employed in the fabrication of the main rotor blades of the GKN Westland Helicopters EH101.

A composite (fibreglass and aluminium) is used in the tail assembly of the Boeing 777 while the leading edge of the Airbus A310–300/A320 fin assembly is of conventional reinforced glass fibre construction, reinforced at the nose to withstand bird strikes. A complete composite airframe was produced for the Beechcraft Starship turboprop executive aircraft which, however, was not a commercial success due to its canard configuration causing drag and weight penalties.

The development of composite materials is continuing with research into the removal of strength-reducing flaws and local imperfections from carbon fibres. Other matrices

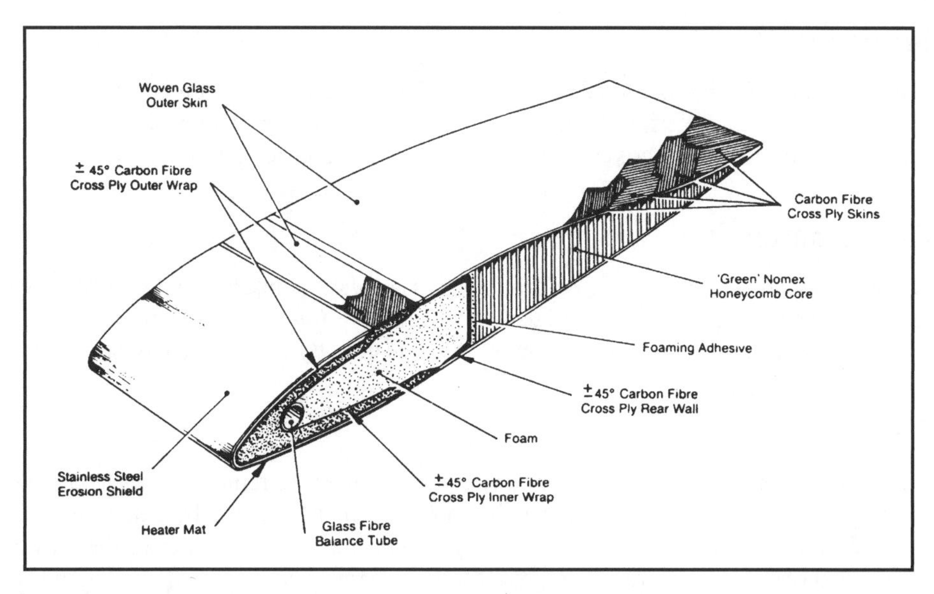

Fig. 11.1 Sectional view of helicopter main rotor blade (courtesy Royal Aeronautical Society, *Aerospace* magazine).

such as polyetheretherketone, which absorbs much less moisture than epoxy resin, has an indefinite shelf life and performs well under impact, are being developed; fabrication, however, requires much higher temperatures. Metal matrix composites such as graphite–aluminium and boron–aluminium are lightweight and retain their strength at higher temperatures than aluminium alloys, but are expensive to produce.

Generally, the use of composites in aircraft construction appears to have reached a plateau, particularly in civil subsonic aircraft where the fraction of the structure comprising composites is approximately 15%. This is due largely to the greater cost of manufacturing composites compared with aluminium alloy structures since composites require hand crafting of the materials and manual construction processes. These increased costs are particularly important in civil aircraft construction and are becoming increasingly important in military aircraft.

11.7 Properties of materials

In Sections 11.1–11.6 we discussed the various materials used in aircraft construction and listed some of their properties. We shall now examine in more detail their behaviour under load and also define different types of material.

Ductility

A material is said to be *ductile* if it is capable of withstanding large strains under load before fracture occurs. These large strains are accompanied by a visible change in cross-sectional dimensions and therefore give warning of impending failure. Materials in this category include mild steel, aluminium and some of its alloys, copper and polymers.

Brittleness

A brittle material exhibits little deformation before fracture, the strain normally being below 5%. Brittle materials therefore may fail suddenly without visible warning. Included in this group are concrete, cast iron, high strength steel, timber and ceramics.

Elastic materials

A material is said to be *elastic* if deformations disappear completely on removal of the load. All known engineering materials are, in addition, *linearly elastic* within certain limits of stress so that strain, within these limits, is directly proportional to stress.

Plasticity

A material is perfectly *plastic* if no strain disappears after the removal of load. Ductile materials are *elastoplastic* and behave in an elastic manner until the *elastic limit* is reached after which they behave plastically. When the stress is relieved the elastic component of the strain is recovered but the plastic strain remains as a *permanent set.*

Isotropic materials

In many materials the elastic properties are the same in all directions at each point in the material although they may vary from point to point, such a material is known as *isotropic*. An isotropic material having the same properties at all points is known as *homogeneous* (e.g. mild steel).

Anisotropic materials

Materials having varying elastic properties in different directions are known as *anisotropic.*

Orthotropic materials

Although a structural material may possess different elastic properties in different directions, this variation may be limited, as in the case of timber which has just two values of Young's modulus, one in the direction of the grain and one perpendicular to the grain. A material whose elastic properties are limited to three different values in three mutually perpendicular directions is known as *orthotropic.*

11.7.1 Testing of engineering materials

The properties of engineering materials are determined mainly by the mechanical testing of specimens machined to prescribed sizes and shapes. The testing may be static or dynamic in nature depending on the particular property being investigated. Possibly the most common mechanical static tests are tensile and compressive tests which are carried out on a wide range of materials. Ferrous and non-ferrous metals are subjected

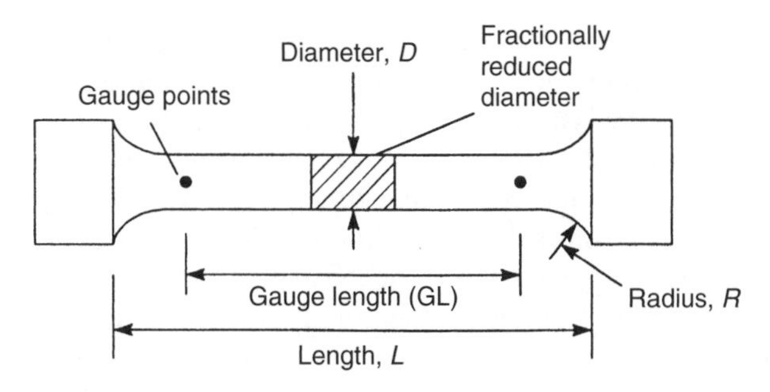

Fig. 11.2 Standard cylindrical test piece.

to both forms of test, while compression tests are usually carried out on many non-metallic materials. Other static tests include bending, shear and hardness tests, while the toughness of a material, in other words its ability to withstand shock loads, is determined by impact tests.

Tensile tests

Tensile tests are normally carried out on metallic materials and, in addition, timber. Test pieces are machined from a batch of material, their dimensions being specified by Codes of Practice. They are commonly circular in cross-section, although flat test pieces having rectangular cross-sections are used when the batch of material is in the form of a plate. A typical test piece would have the dimensions specified in Fig. 11.2. Usually the diameter of a central portion of the test piece is fractionally less than that of the remainder to ensure that the test piece fractures between the gauge points.

Before the test begins, the mean diameter of the test piece is obtained by taking measurements at several sections using a micrometer screw gauge. Gauge points are punched at the required gauge length, the test piece is placed in the testing machine and a suitable strain measuring device, usually an extensometer, is attached to the test piece at the gauge points so that the extension is measured over the given gauge length. Increments of load are applied and the corresponding extensions recorded. This procedure continues until yield occurs, when the extensometer is removed as a precaution against the damage which would be caused if the test piece fractured unexpectedly. Subsequent extensions are measured by dividers placed in the gauge points until, ultimately, the test piece fractures. The final gauge length and the diameter of the test piece in the region of the fracture are measured so that the percentage elongation and percentage reduction in area may be calculated. These two parameters give a measure of the ductility of the material.

A stress–strain curve is drawn (see Figs 11.9 and 11.13), the stress normally being calculated on the basis of the original cross-sectional area of the test piece, i.e. a *nominal stress* as opposed to an *actual stress* (which is based on the actual area of cross-section).

For ductile materials there is a marked difference in the latter stages of the test as a considerable reduction in cross-sectional area occurs between yield and fracture. From the stress–strain curve the ultimate stress, the yield stress and Young's modulus, E, are obtained.

There are a number of variations on the basic tensile test described above. Some of these depend upon the amount of additional information required and some upon the choice of equipment. There is a wide range of strain measuring devices to choose from, extending from different makes of mechanical extensometer, e.g. Huggenberger, Lindley, Cambridge, to the electrical resistance strain gauge. The last would normally be used on flat test pieces, one on each face to eliminate the effects of possible bending. At the same time a strain gauge could be attached in a direction perpendicular to the direction of loading so that lateral strains are measured. The ratio lateral strain/longitudinal strain is Poisson's ratio, ν.

Testing machines are usually driven hydraulically. More sophisticated versions employ load cells to record load and automatically plot load against extension or stress against strain on a pen recorder as the test proceeds, an advantage when investigating the distinctive behaviour of mild steel at yield.

Compression tests

A compression test is similar in operation to a tensile test, with the obvious difference that the load transmitted to the test piece is compressive rather than tensile. This is achieved by placing the test piece between the platens of the testing machine and reversing the direction of loading. Test pieces are normally cylindrical and are limited in length to eliminate the possibility of failure being caused by instability. Again contractions are measured over a given gauge length by a suitable strain measuring device.

Variations in test pieces occur when only the ultimate strength of the material in compression is required. For this purpose concrete test pieces may take the form of cubes having edges approximately 10 cm long, while mild steel test pieces are still cylindrical in section but are of the order of 1 cm long.

Bending tests

Many structural members are subjected primarily to bending moments. Bending tests are therefore carried out on simple beams constructed from the different materials to determine their behaviour under this type of load.

Two forms of loading are employed the choice depending upon the type specified in Codes of Practice for the particular material. In the first a simply supported beam is subjected to a 'two-point' loading system as shown in Fig. 11.3(a). Two concentrated loads are applied symmetrically to the beam, producing zero shear force and constant bending moment in the central span of the beam (Fig. 11.3(b) and (c)). The condition of pure bending is therefore achieved in the central span.

The second form of loading system consists of a single concentrated load at mid-span (Fig. 11.4(a)) which produces the shear force and bending moment diagrams shown in Fig. 11.4(b) and (c).

The loads may be applied manually by hanging weights on the beam or by a testing machine. Deflections are measured by a dial gauge placed underneath the beam. From the recorded results a load–deflection diagram is plotted.

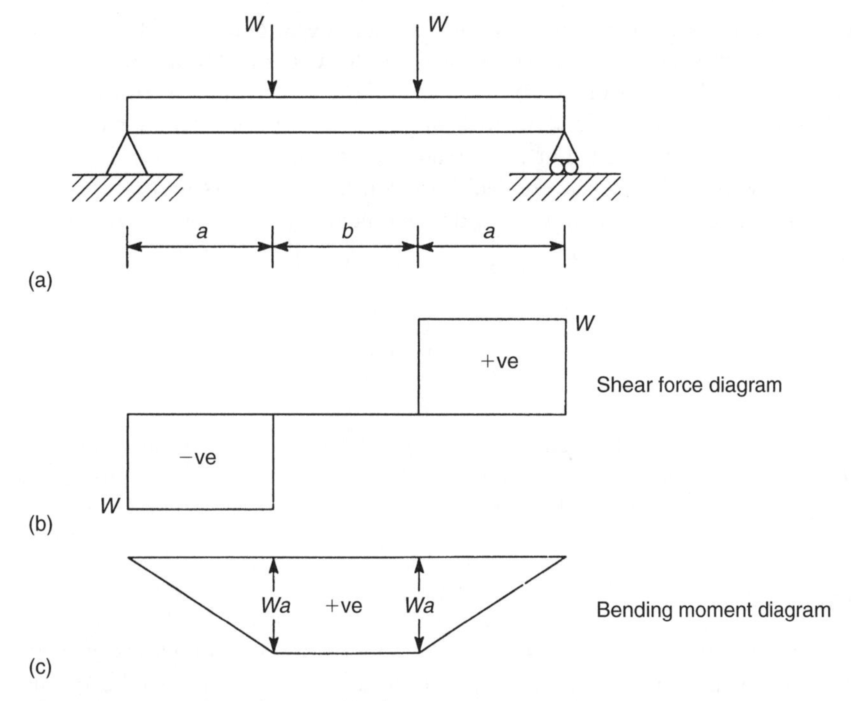

Fig. 11.3 Bending test on a beam, 'two-point' load.

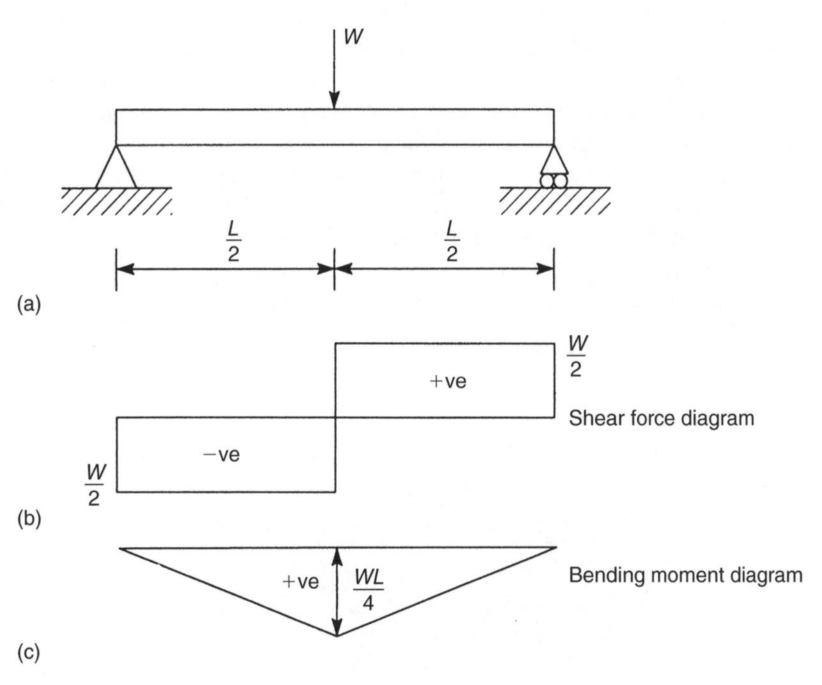

Fig. 11.4 Bending test on a beam, single load.

For most ductile materials the test beams continue to deform without failure and fracture does not occur. Thus plastic properties, for example the ultimate strength in bending, cannot be determined for such materials. In the case of brittle materials, including cast iron, timber and various plastics, failure does occur, so that plastic properties can be evaluated. For such materials the ultimate strength in bending is defined by the *modulus of rupture*. This is taken to be the maximum direct stress in bending, $\sigma_{x,u}$, corresponding to the ultimate moment M_u, and is assumed to be related to M_u by the elastic relationship

$$\sigma_{x,u} = \frac{M_u}{I} y_{max}$$

Other bending tests are designed to measure the ductility of a material and involve the bending of a bar round a pin. The angle of bending at which the bar starts to crack is then taken as an indication of its ductility.

Shear tests

Two main types of shear test are used to determine the shear properties of materials. One type investigates the direct or transverse shear strength of a material and is used in connection with the shear strength of bolts, rivets and beams. A typical arrangement is shown diagrammatically in Fig. 11.5 where the test piece is clamped to a block and the load is applied through the shear tool until failure occurs. In the arrangement shown the test piece is subjected to double shear, whereas if it is extended only partially across the gap in the block it would be subjected to single shear. In either case the average shear strength is taken as the maximum load divided by the shear resisting area.

The other type of shear test is used to evaluate the basic shear properties of a material, such as the shear modulus, G, the shear stress at yield and the ultimate shear stress. In the usual form of test a solid circular-section test piece is placed in a torsion machine and twisted by controlled increments of torque. The corresponding angles of twist are recorded and torque–twist diagrams plotted from which the shear properties of the material are obtained. The method is similar to that used to determine the tensile properties of a material from a tensile test and uses relationships derived in Chapter 3.

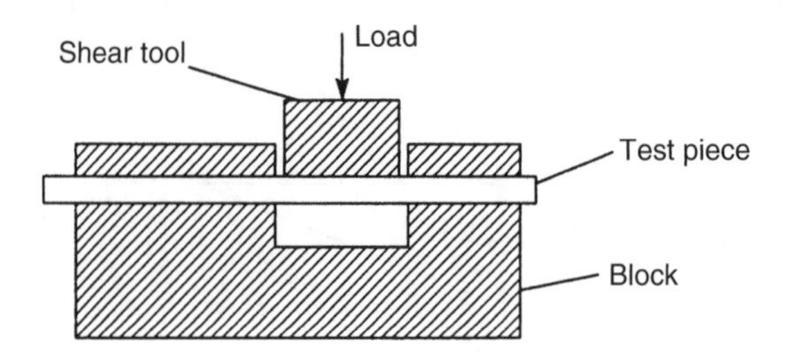

Fig. 11.5 Shear test.

Hardness tests

The machinability of a material and its resistance to scratching or penetration are determined by its 'hardness'. There also appears to be a connection between the hardness of some materials and their tensile strength so that hardness tests may be used to determine the properties of a finished structural member where tensile and other tests would be impracticable. Hardness tests are also used to investigate the effects of heat treatment, hardening and tempering and of cold forming. Two types of hardness test are in common use: *indentation tests* and *scratch and abrasion tests.*

Indentation tests may be subdivided into two classes: static and dynamic. Of the static tests the *Brinell* is the most common. In this a hardened steel ball is pressed into the material under test by a static load acting for a fixed period of time. The load in kg divided by the spherical area of the indentation in mm^2 is called the *Brinell hardness number* (BHN). In Fig. 11.6, if D is the diameter of the ball, F the load in kg, h the depth of the indentation and d the diameter of the indentation, then

$$\text{BHN} = \frac{F}{\pi Dh} = \frac{2F}{\pi D[D - \sqrt{D^2 - d^2}]}$$

In practice, the hardness number of a given material is found to vary with F and D so that for uniformity the test is standardized. For steel and hard materials $F = 3000$ kg and $D = 10$ mm while for soft materials $F = 500$ kg and $D = 10$ mm; in addition the load is usually applied for 15 s.

In the *Brinell* test the dimensions of the indentation are measured by means of a microscope. To avoid this rather tedious procedure, direct reading machines have been devised of which the *Rockwell* is typical. The indenting tool, again a hardened sphere, is first applied under a definite light load. This indenting tool is then replaced by a diamond cone with a rounded point which is then applied under a specified indentation load. The difference between the depth of the indentation under the two loads is taken as a measure of the hardness of the material and is read directly from the scale.

A typical dynamic hardness test is performed by the *Shore Scleroscope* which consists of a small hammer approximately 20 mm long and 6 mm in diameter fitted with a blunt, rounded, diamond point. The hammer is guided by a vertical glass tube and allowed to fall freely from a height of 25 cm onto the specimen, which it indents before rebounding. A certain proportion of the energy of the hammer is expended in forming the indentation so that the height of the rebound, which depends upon the energy still possessed by the hammer, is taken as a measure of the hardness of the material.

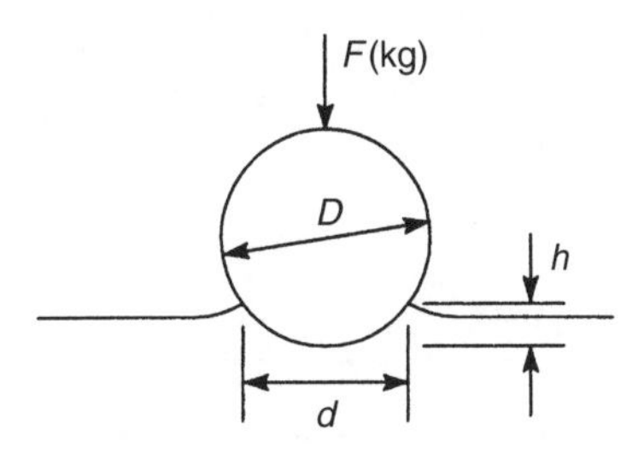

Fig. 11.6 *Brinell* hardness test.

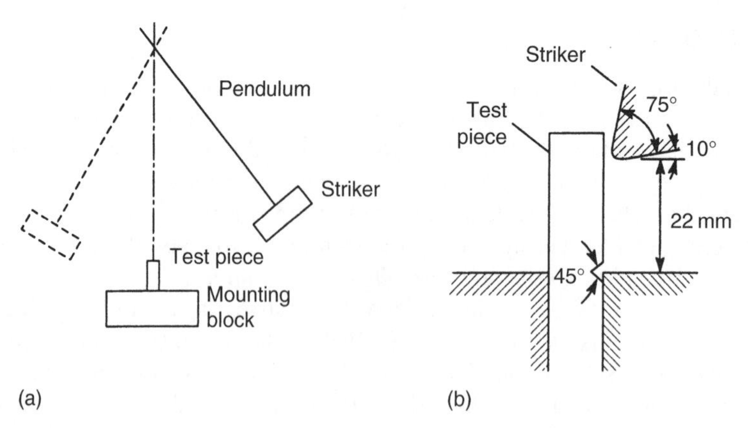

Fig. 11.7 Izod impact test.

A number of tests have been devised to measure the 'scratch hardness' of materials. In one test, the smallest load in grams which, when applied to a diamond point, produces a scratch visible to the naked eye on a polished specimen of material is called its hardness number. In other tests the magnitude of the load required to produce a definite width of scratch is taken as the measure of hardness. Abrasion tests, involving the shaking over a period of time of several specimens placed in a container, measure the resistance to wear of some materials. In some cases, there appears to be a connection between wear and hardness number although the results show no level of consistency.

Impact tests

It has been found that certain materials, particularly heat-treated steels, are susceptible to failure under shock loading whereas an ordinary tensile test on the same material would show no abnormality. Impact tests measure the ability of materials to withstand shock loads and provide an indication of their *toughness*. Two main tests are in use, the *Izod* and the *Charpy*.

Both tests rely on a striker or weight attached to a pendulum. The pendulum is released from a fixed height, the weight strikes a notched test piece and the angle through which the pendulum then swings is a measure of the toughness of the material. The arrangement for the Izod test is shown diagrammatically in Fig. 11.7(a). The specimen and the method of mounting are shown in detail in Fig. 11.7(b). The Charpy test is similar in operation except that the test piece is supported in a different manner as shown in the plan view in Fig. 11.8.

11.7.2 Stress–strain curves

We shall now examine in detail the properties of the different materials from the viewpoint of the results obtained from tensile and compression tests.

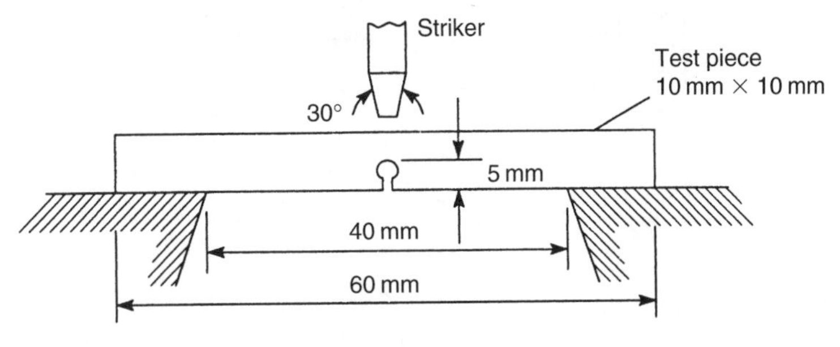

Fig. 11.8 Charpy impact test.

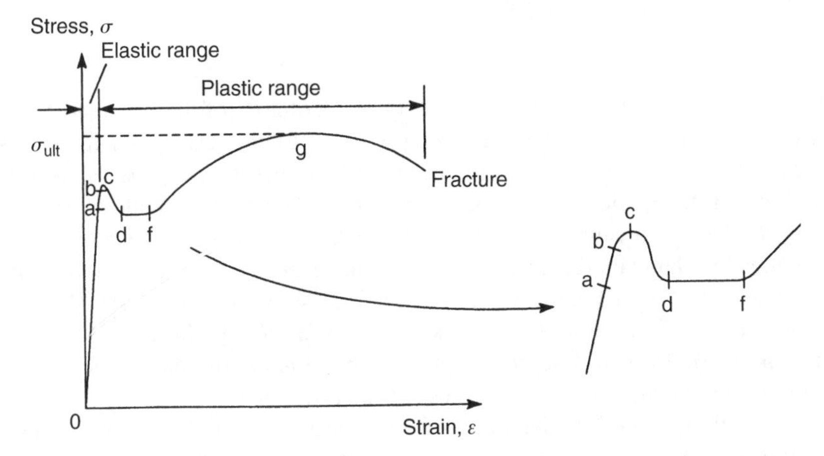

Fig. 11.9 Stress–strain curve for mild steel.

Low carbon steel (mild steel)

A nominal stress–strain curve for mild steel, a ductile material, is shown in Fig. 11.9 From 0 to 'a' the stress–strain curve is linear, the material in this range obeying Hooke's law. Beyond 'a', the *limit of proportionality*, stress is no longer proportional to strain and the stress–strain curve continues to 'b', the *elastic limit*, which is defined as the maximum stress that can be applied to a material without producing a permanent plastic deformation or *permanent set* when the load is removed. In other words, if the material is stressed beyond 'b' and the load then removed, a residual strain exists at zero load. For many materials it is impossible to detect a difference between the limit of proportionality and the elastic limit. From 0 to 'b' the material is said to be in the *elastic range* while from 'b' to fracture the material is in the *plastic range*. The transition from the elastic to the plastic range may be explained by considering the arrangement of crystals in the material. As the load is applied, slipping occurs between the crystals which are aligned most closely to the direction of load. As the load is increased, more and more crystals slip with each equal load increment until appreciable strain increments are produced and the plastic range is reached.

A further increase in stress from 'b' results in the mild steel reaching its *upper yield point* at 'c' followed by a rapid fall in stress to its *lower yield point* at 'd'. The

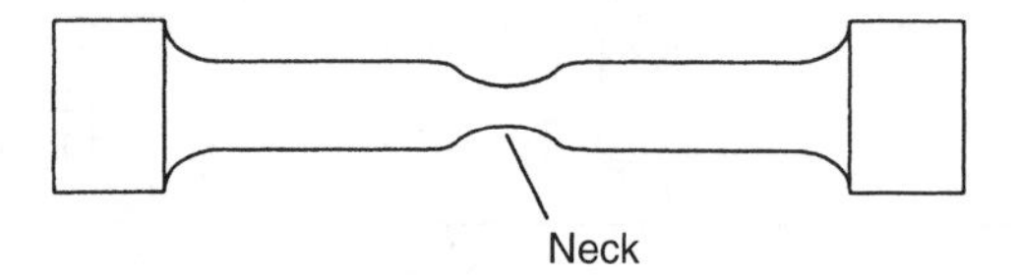

Fig. 11.10 'Necking' of a test piece in the plastic range.

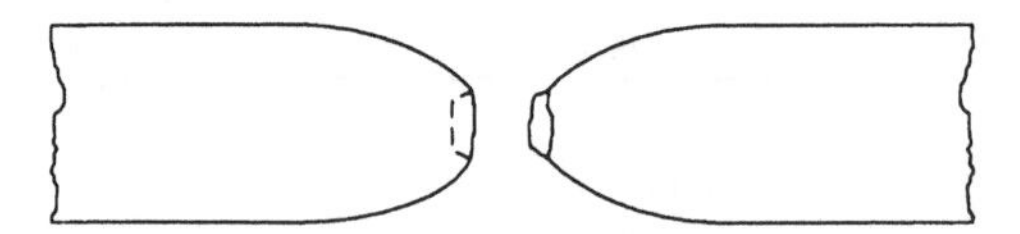

Fig. 11.11 'Cup-and-cone' failure of a mild steel test piece.

existence of a lower yield point for mild steel is a peculiarity of the tensile test wherein the movement of the ends of the test piece produced by the testing machine does not proceed as rapidly as its plastic deformation; the load therefore decreases, as does the stress. From 'd' to 'f' the strain increases at a roughly constant value of stress until *strain hardening* again causes an increase in stress. This increase in stress continues, accompanied by a large increase in strain to 'g', the *ultimate stress*, σ_{ult}, of the material. At this point the test piece begins, visibly, to 'neck' as shown in Fig. 11.10. The material in the test piece in the region of the 'neck' is almost perfectly plastic at this stage and from this point, onwards to fracture, there is a reduction in nominal stress.

For mild steel, yielding occurs at a stress of the order of 300 N/mm^2. At fracture the strain (i.e. the elongation) is of the order of 30%. The gradient of the linear portion of the stress–strain curve gives a value for Young's modulus in the region of 200 000 N/mm^2.

The characteristics of the fracture are worthy of examination. In a cylindrical test piece the two halves of the fractured test piece have ends which form a 'cup and cone' (Fig. 11.11). The actual failure planes in this case are inclined at approximately 45° to the axis of loading and coincide with planes of maximum shear stress. Similarly, if a flat tensile specimen of mild steel is polished and then stressed, a pattern of fine lines appears on the polished surface at yield. These lines, which were first discovered by Lüder in 1854, intersect approximately at right angles and are inclined at 45° to the axis of the specimen, thereby coinciding with planes of maximum shear stress. These forms of yielding and fracture suggest that the crystalline structure of the steel is relatively weak in shear with yielding taking the form of the sliding of one crystal plane over another rather than the tearing apart of two crystal planes.

The behaviour of mild steel in compression is very similar to its behaviour in tension, particularly in the elastic range. In the plastic range it is not possible to obtain ultimate and fracture loads since, due to compression, the area of cross-section increases as the load increases producing a 'barrelling' effect as shown in Fig. 11.12. This increase in cross-sectional area tends to decrease the true stress, thereby increasing the load resistance. Ultimately a flat disc is produced. For design purposes the ultimate stresses of mild steel in tension and compression are assumed to be the same.

Higher grades of steel have greater strengths than mild steel but are not as ductile. They also possess the same Young's modulus so that the higher stresses are accompanied by higher strains.

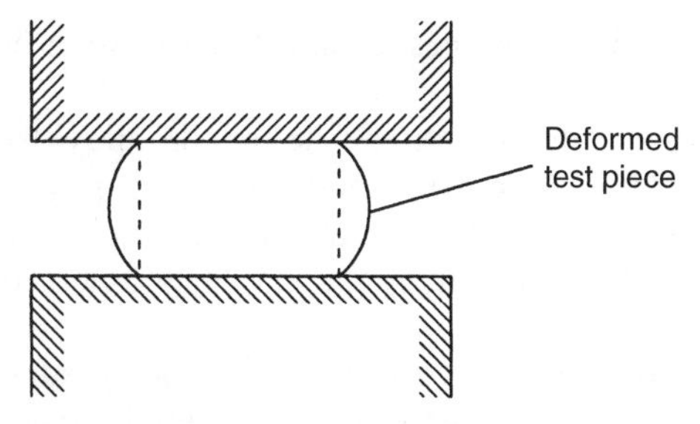

Fig. 11.12 'Barrelling' of a mild steel test piece in compression.

Aluminium

Aluminium and some of its alloys are also ductile materials, although their stress–strain curves do not have the distinct yield stress of mild steel. A typical stress–strain curve is shown in Fig. 11.13. The points 'a' and 'b' again mark the limit of proportionality and elastic limit, respectively, but are difficult to determine experimentally. Instead a *proof stress* is defined which is the stress required to produce a given permanent strain on removal of the load. In Fig. 11.13, a line drawn parallel to the linear portion of the stress–strain curve from a strain of 0.001 (i.e. a strain of 0.1%) intersects the stress–strain curve at the 0.1% proof stress. For elastic design this, or the 0.2% proof stress, is taken as the working stress.

Beyond the limit of proportionality the material extends plastically, reaching its ultimate stress, σ_{ult}, at 'd' before finally fracturing under a reduced nominal stress at 'f'.

A feature of the fracture of aluminium alloy test pieces is the formation of a 'double cup' as shown in Fig. 11.14, implying that failure was initiated in the central portion

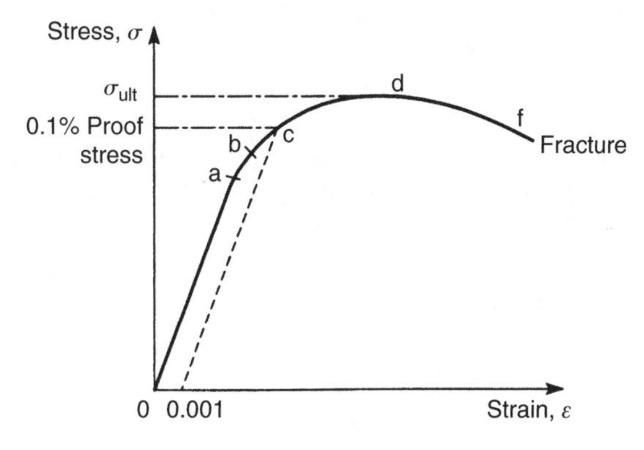

Fig. 11.13 Stress–strain curve for aluminium.

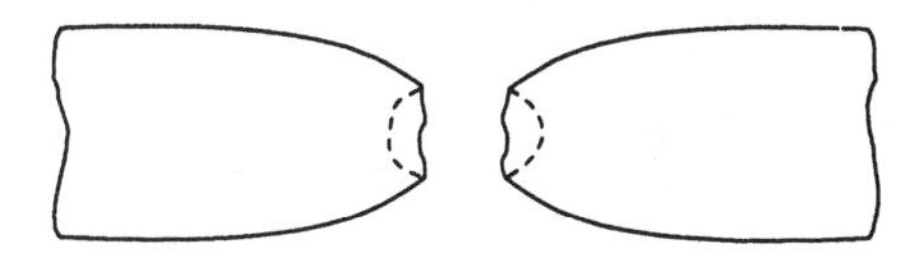

Fig. 11.14 'Double-cup' failure of an aluminium alloy test piece.

of the test piece while the outer surfaces remained intact. Again considerable 'necking' occurs.

In compression tests on aluminium and its ductile alloys similar difficulties are encountered to those experienced with mild steel. The stress–strain curve is very similar in the elastic range to that obtained in a tensile test but the ultimate strength in compression cannot be determined; in design its value is assumed to coincide with that in tension.

Aluminium and its alloys can suffer a form of corrosion particularly in the salt laden atmosphere of coastal regions. The surface becomes pitted and covered by a white furry deposit. This can be prevented by an electrolytic process called *anodizing* which covers the surface with an inert coating. Aluminium alloys will also corrode if they are placed in direct contact with other metals, such as steel. To prevent this, plastic is inserted between the possible areas of contact.

Brittle materials

These include cast iron, high strength steel, concrete, timber, ceramics, glass, etc. The plastic range for brittle materials extends to only small values of strain. A typical stress–strain curve for a brittle material under tension is shown in Fig. 11.15. Little or no yielding occurs and fracture takes place very shortly after the elastic limit is reached.

The fracture of a cylindrical test piece takes the form of a single failure plane approximately perpendicular to the direction of loading with no visible 'necking' and an elongation of the order of 2–3%.

In compression the stress–strain curve for a brittle material is very similar to that in tension except that failure occurs at a much higher value of stress; for concrete the ratio is of the order of 10 : 1. This is thought to be due to the presence of microscopic cracks in the material, giving rise to high stress concentrations which are more likely to have a greater effect in reducing tensile strength than compressive strength.

Composites

Fibre composites have stress–strain characteristics which indicate that they are brittle materials (Fig. 11.16). There is little or no plasticity and the modulus of elasticity is less

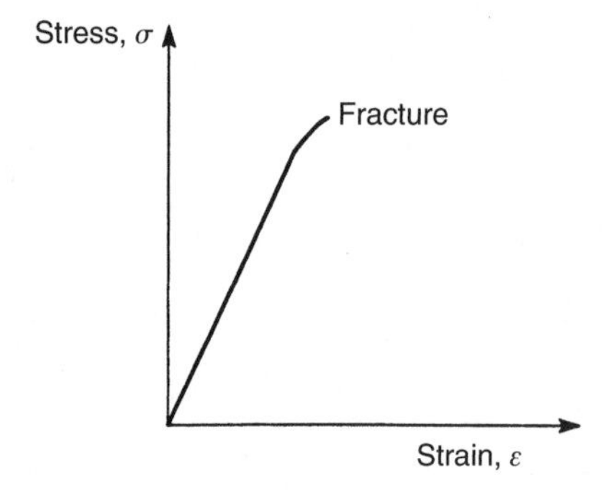

Fig. 11.15 Stress–strain curve for a brittle material.

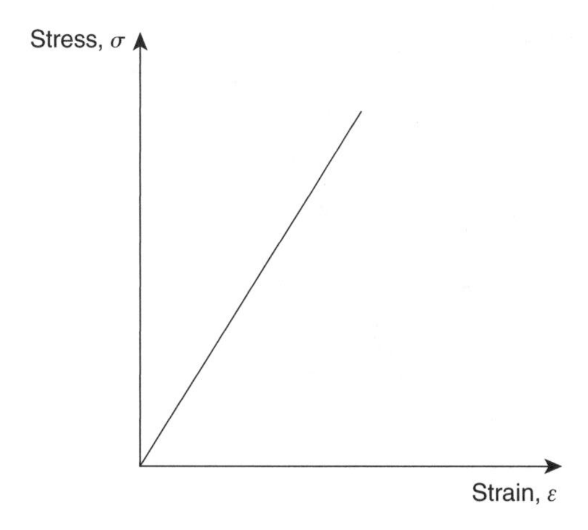

Fig. 11.16 Stress–strain curve for a fibre composite.

than that of steel and aluminium alloy. However, the fibres themselves can have much higher values of strength and modulus of elasticity than the composite. For example, carbon fibres have a tensile strength of the order 2400 N/mm^2 and a modulus of elasticity of 400 000 N/mm^2.

Fibre composites are highly durable, require no maintenance and can be used in hostile chemical and atmospheric environments; vinyls and epoxy resins provide the best resistance.

All the stress–strain curves described in the preceding discussion are those produced in tensile or compression tests in which the strain is applied at a negligible rate. A rapid strain application would result in significant changes in the apparent properties of the materials giving possible variations in yield stress of up to 100%.

11.7.3 Strain hardening

The stress–strain curve for a material is influenced by the *strain history*, or the loading and unloading of the material, within the plastic range. For example, in Fig. 11.17 a test piece is initially stressed in tension beyond the yield stress at, 'a', to a value at 'b'. The material is then unloaded to 'c' and reloaded to 'f' producing an increase in yield stress from the value at 'a' to the value at 'd'. Subsequent unloading to 'g' and loading to 'j' increases the yield stress still further to the value at 'h'. This increase in strength resulting from the loading and unloading is known as *strain hardening*. It can be seen from Fig. 11.17 that the stress–strain curve during the unloading and loading cycles form loops (the shaded areas in Fig. 11.17). These indicate that strain energy is lost during the cycle, the energy being dissipated in the form of heat produced by internal friction. This energy loss is known as *mechanical hysteresis* and the loops as *hysteresis loops*. Although the ultimate stress is increased by strain hardening it is not influenced to the same extent as yield stress. The increase in strength produced by strain hardening is accompanied by decreases in toughness and ductility.

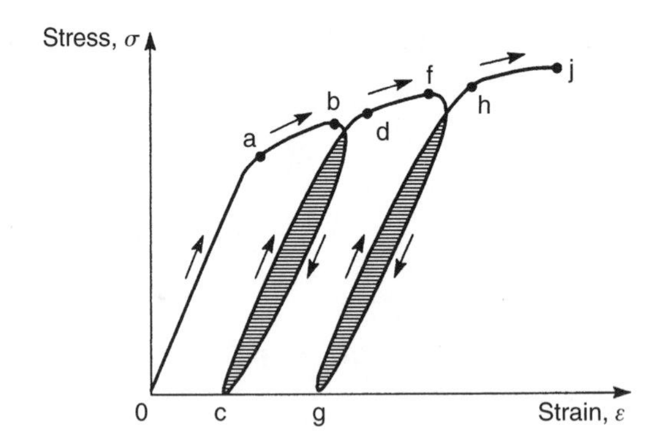

Fig. 11.17 Strain hardening of a material.

11.7.4 Creep and relaxation

We have seen in Chapter 1 that a given load produces a calculable value of stress in a structural member and hence a corresponding value of strain once the full value of the load is transferred to the member. However, after this initial or 'instantaneous' stress and its corresponding value of strain have been attained, a great number of structural materials continue to deform slowly and progressively under load over a period of time. This behaviour is known as *creep*. A typical creep curve is shown in Fig. 11.18.

Some materials, such as plastics and rubber, exhibit creep at room temperatures but most structural materials require high temperatures or long-duration loading at moderate temperatures. In some 'soft' metals, such as zinc and lead, creep occurs over a relatively short period of time, whereas materials such as concrete may be subject to creep over a period of years. Creep occurs in steel to a slight extent at normal temperatures but becomes very important at temperatures above 316°C.

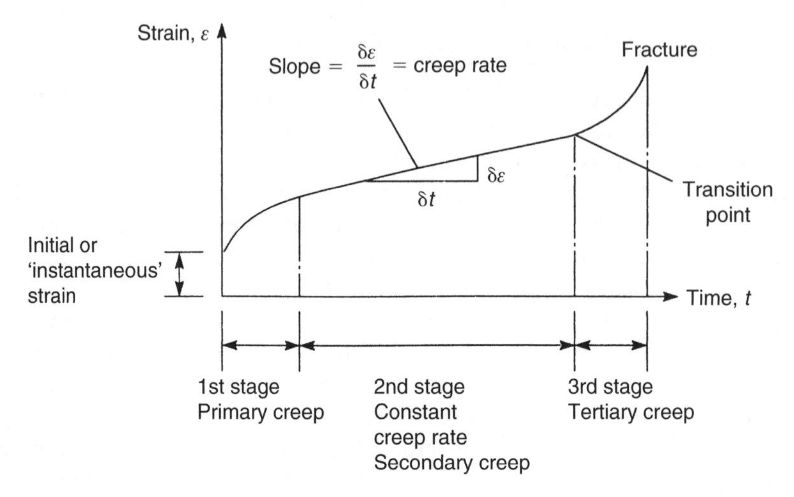

Fig. 11.18 Typical creep curve.

Closely related to creep is *relaxation*. Whereas creep involves an increase in strain under constant stress, relaxation is the decrease in stress experienced over a period of time by a material subjected to a constant strain.

11.7.5 Fatigue

Structural members are frequently subjected to repetitive loading over a long period of time. For example, the members of a bridge structure suffer variations in loading possibly thousands of times a day as traffic moves over the bridge. In these circumstances a structural member may fracture at a level of stress substantially below the ultimate stress for non-repetitive static loads; this phenomenon is known as *fatigue*.

Fatigue cracks are most frequently initiated at sections in a structural member where changes in geometry, e.g. holes, notches or sudden changes in section, cause *stress concentrations*. Designers seek to eliminate such areas by ensuring that rapid changes in section are as smooth as possible. At re-entrant corners for example, fillets are provided as shown in Fig. 11.19.

Other factors which affect the failure of a material under repetitive loading are the type of loading (fatigue is primarily a problem with repeated tensile stresses due, probably, to the fact that microscopic cracks can propagate more easily under tension), temperature, the material, surface finish (machine marks are potential crack propagators), corrosion and residual stresses produced by welding.

Frequently in structural members an alternating stress, σ_{alt}, is superimposed on a static or mean stress, σ_{mean}, as illustrated in Fig. 11.20. The value of σ_{alt} is the most important factor in determining the number of cycles of load that produce failure. The stress σ_{alt}, that can be withstood for a specified number of cycles is called the *fatigue strength* of the material. Some materials, such as mild steel, possess a stress level that can be withstood for an indefinite number of cycles. This stress is known as the *endurance limit* of the material; no such limit has been found for aluminium and its alloys. Fatigue data are frequently presented in the form of an *S–n* curve or stress–endurance curve as shown in Fig. 11.21.

In many practical situations the amplitude of the alternating stress varies and is frequently random in nature. The *S–n* curve does not, therefore, apply directly and an alternative means of predicting failure is required. *Miner's cumulative damage theory*

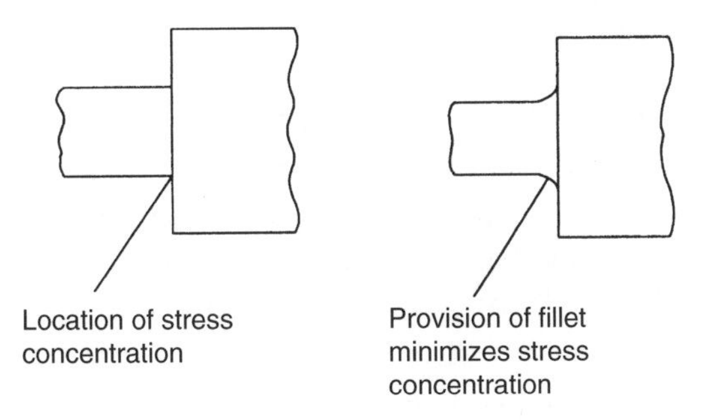

Fig. 11.19 Stress concentration location.

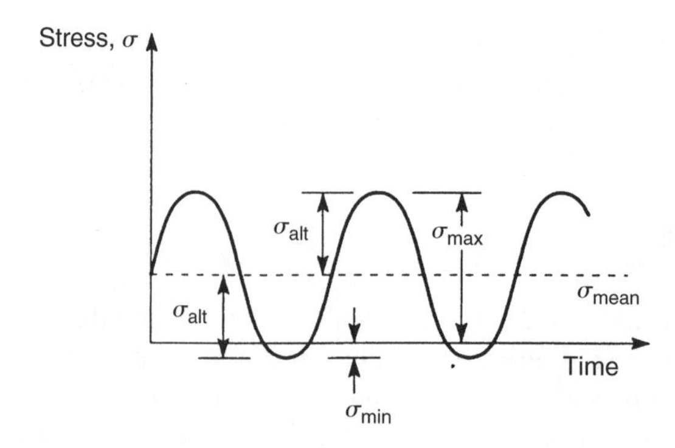

Fig. 11.20 Alternating stress in fatigue loading.

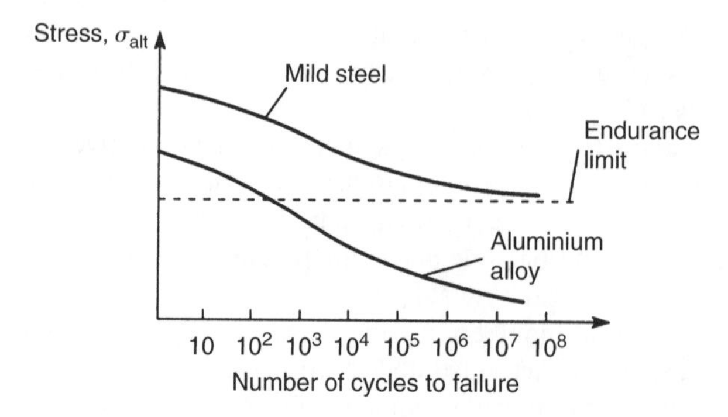

Fig. 11.21 Stress–endurance curves.

suggests that failure will occur when

$$\frac{n_1}{N_1} + \frac{n_2}{N_2} + \cdots + \frac{n_r}{N_r} = 1 \tag{11.1}$$

where $n_1, n_2, \ldots, n_r$ are the number of applications of stresses σ_{alt}, σ_{mean} and N_1, $N_2, \ldots, N_r$ are the number of cycles to failure of stresses σ_{alt}, σ_{mean}.

We shall examine fatigue and its effect on aircraft design in much greater detail in Chapter 15.

Problems

P.11.1 Describe a simple tensile test and show, with the aid of sketches, how measures of the ductility of the material of the specimen may be obtained. Sketch typical stress–strain curves for mild steel and an aluminium alloy showing their important features.

P.11.2 A bar of metal 25 mm in diameter is tested on a length of 250 mm. In tension the following results were recorded (Table P.11.2(a)).

Table P.11.2(a)

Load (kN)	10.4	31.2	52.0	72.8
Extension (mm)	0.036	0.089	0.140	0.191

A torsion test gave the following results (Table P.11.2(b)).

Table P.11.2(b)

Torque (kN m)	0.051	0.152	0.253	0.354
Angle of twist (degrees)	0.24	0.71	1.175	1.642

Represent these results in graphical form and hence determine Young's modulus, E, the modulus of rigidity, G, Poisson's ratio, ν, and the bulk modulus, K, for the metal.

Ans. $E \simeq 205\,000\ \text{N/mm}^2$, $G \simeq 80\,700\ \text{N/mm}^2$, $\nu \simeq 0.27^2$, $K \simeq 148\,500\ \text{N/mm}^2$.

P.11.3 The actual stress–strain curve for a particular material is given by $\sigma = C\varepsilon^n$ where C is a constant. Assuming that the material suffers no change in volume during plastic deformation, derive an expression for the nominal stress–strain curve and show that this has a maximum value when $\varepsilon = n/(1-n)$.

Ans. $\sigma_{\text{nom}} = C\varepsilon^n/(1+\varepsilon)$.

P.11.4 A structural member is to be subjected to a series of cyclic loads which produce different levels of alternating stress as shown in Table P.11.4. Determine whether or not a fatigue failure is probable.

Ans. Not probable $(n_1/N_1 + n_2/N_2 + \cdots = 0.39)$.

Table P.11.4

Loading	Number of cycles	Number of cycles to failure
1	10^4	5×10^4
2	10^5	10^6
3	10^6	24×10^7
4	10^7	12×10^7

12

Structural components of aircraft

Aircraft are generally built up from the basic components of wings, fuselages, tail units and control surfaces. There are variations in particular aircraft, for example, a delta wing aircraft would not necessarily possess a horizontal tail although this is present in a canard configuration such as that of the Eurofighter (Typhoon). Each component has one or more specific functions and must be designed to ensure that it can carry out these functions safely. In this chapter we shall describe the various loads to which aircraft components are subjected, their function and fabrication and also the design of connections.

12.1 Loads on structural components

The structure of an aircraft is required to support two distinct classes of load: the first, termed *ground loads*, includes all loads encountered by the aircraft during movement or transportation on the ground such as taxiing and landing loads, towing and hoisting loads; while the second, *air loads*, comprises lòads imposed on the structure during flight by manoeuvres and gusts. In addition, aircraft designed for a particular role encounter loads peculiar to their sphere of operation. Carrier born aircraft, for instance, are subjected to catapult take-off and arrested landing loads: most large civil and practically all military aircraft have pressurized cabins for high altitude flying; amphibious aircraft must be capable of landing on water and aircraft designed to fly at high speed at low altitude, e.g. the Tornado, require a structure of above average strength to withstand the effects of flight in extremely turbulent air.

The two classes of loads may be further divided into *surface forces* which act upon the surface of the structure, e.g. aerodynamic and hydrostatic pressure, and *body forces* which act over the volume of the structure and are produced by gravitational and inertial effects. Calculation of the distribution of aerodynamic pressure over the various surfaces of an aircraft's structure is presented in numerous texts on aerodynamics and will therefore not be attempted here. We shall, however, discuss the types of load induced by these various effects and their action on the different structural components.

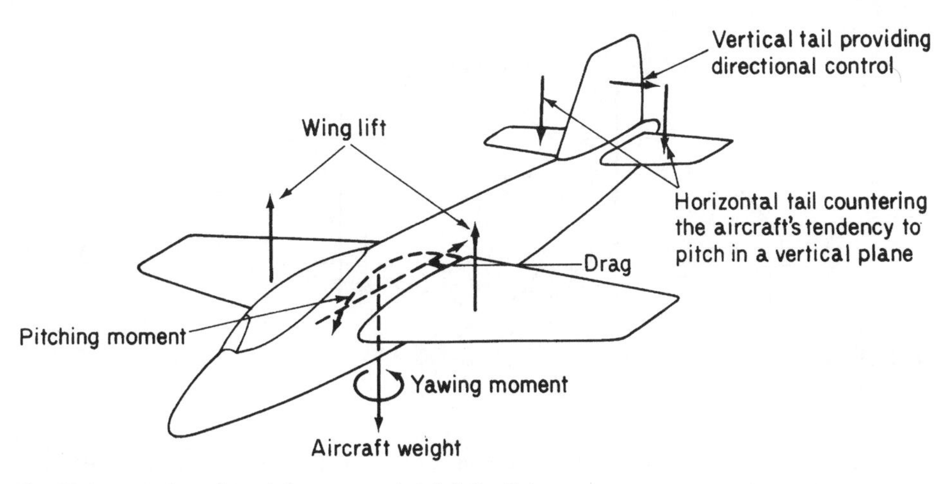

Fig. 12.1 Principal aerodynamic forces on an aircraft during flight.

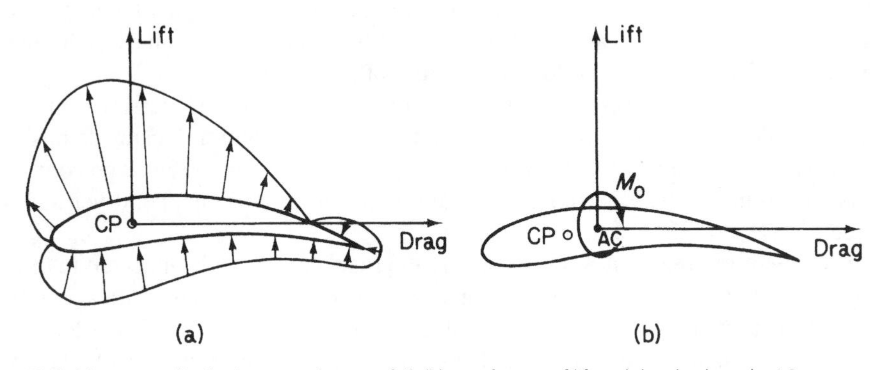

Fig. 12.2 (a) Pressure distribution around an aerofoil; (b) transference of lift and drag loads to the AC.

Basically, all air loads are the resultants of the pressure distribution over the surfaces of the skin produced by steady flight, manoeuvre or gust conditions. Generally, these resultants cause direct loads, bending, shear and torsion in all parts of the structure in addition to local, normal pressure loads imposed on the skin.

Conventional aircraft usually consist of fuselage, wings and tailplane. The fuselage contains crew and payload, the latter being passengers, cargo, weapons plus fuel, depending on the type of aircraft and its function; the wings provide the lift and the tailplane is the main contributor to directional control. In addition, ailerons, elevators and the rudder enable the pilot to manoeuvre the aircraft and maintain its stability in flight, while wing flaps provide the necessary increase of lift for take-off and landing. Figure 12.1 shows typical aerodynamic force resultants experienced by an aircraft in steady flight.

The force on an aerodynamic surface (wing, vertical or horizontal tail) results from a differential pressure distribution caused by incidence, camber or a combination of both. Such a pressure distribution, shown in Fig. 12.2(a), has vertical (lift) and horizontal (drag) resultants acting at a centre of pressure (CP). (In practice, lift and drag are measured perpendicular and parallel to the flight path, respectively.) Clearly the position of the CP changes as the pressure distribution varies with speed or wing incidence.

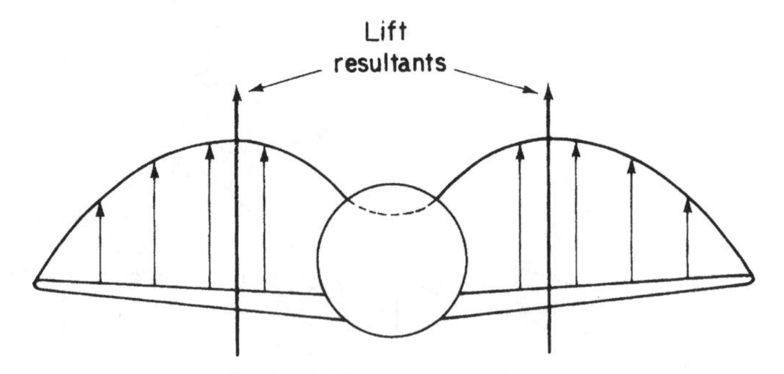

Fig. 12.3 Typical lift distribution for a wing/fuselage combination.

However, there is, conveniently, a point in the aerofoil section about which the moment due to the lift and drag forces remains constant. We therefore replace the lift and drag forces acting at the CP by lift and drag forces acting at the aerodynamic centre (AC) plus a constant moment M_0 as shown in Fig. 12.2(b). (Actually, at high Mach numbers the position of the AC changes due to compressibility effects.)

While the chordwise pressure distribution fixes the position of the resultant aerodynamic load in the wing cross-section, the spanwise distribution locates its position in relation, say, to the wing root. A typical distribution for a wing/fuselage combination is shown in Fig. 12.3. Similar distributions occur on horizontal and vertical tail surfaces.

We see therefore that wings, tailplane and the fuselage are each subjected to direct, bending, shear and torsional loads and must be designed to withstand critical combinations of these. Note that manoeuvres and gusts do not introduce different loads but result only in changes of magnitude and position of the type of existing loads shown in Fig. 12.1. Over and above these basic in-flight loads, fuselages may be pressurized and thereby support hoop stresses, wings may carry weapons and/or extra fuel tanks with resulting additional aerodynamic and body forces contributing to the existing bending, shear and torsion, while the thrust and weight of engines may affect either fuselage or wings depending on their relative positions.

Ground loads encountered in landing and taxiing subject the aircraft to concentrated shock loads through the undercarriage system. The majority of aircraft have their main undercarriage located in the wings, with a nosewheel or tailwheel in the vertical plane of symmetry. Clearly the position of the main undercarriage should be such as to produce minimum loads on the wing structure compatible with the stability of the aircraft during ground manoeuvres. This may be achieved by locating the undercarriage just forward of the flexural axis of the wing and as close to the wing root as possible. In this case the shock landing load produces a given shear, minimum bending plus torsion, with the latter being reduced as far as practicable by offsetting the torque caused by the vertical load in the undercarriage leg by a torque in an opposite sense due to braking.

Other loads include engine thrust on the wings or fuselage which acts in the plane of symmetry but may, in the case of engine failure, cause severe fuselage bending moments, as shown in Fig. 12.4; concentrated shock loads during a catapult launch; and hydrodynamic pressure on the fuselages or floats of seaplanes.

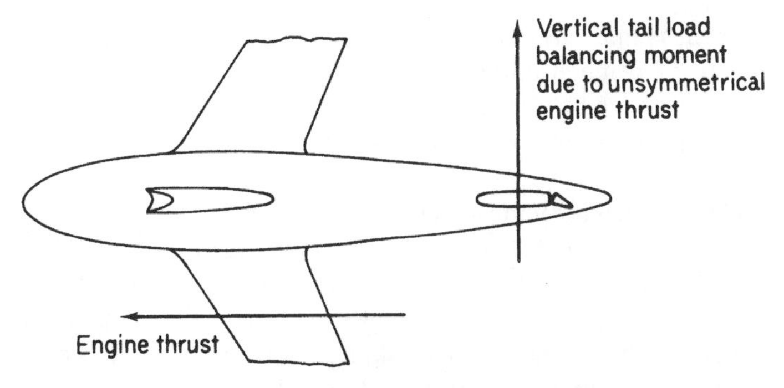

Fig. 12.4 Fuselage and wing bending caused by an unsymmetrical engine load.

In Chapter 13 we shall examine in detail the calculation of ground and air loads for a variety of cases.

12.2 Function of structural components

The basic functions of an aircraft's structure are to transmit and resist the applied loads; to provide an aerodynamic shape and to protect passengers, payload, systems, etc. from the environmental conditions encountered in flight. These requirements, in most aircraft, result in thin shell structures where the outer surface or skin of the shell is usually supported by longitudinal stiffening members and transverse frames to enable it to resist bending, compressive and torsional loads without buckling. Such structures are known as *semi-monocoque*, while thin shells which rely entirely on their skins for their capacity to resist loads are referred to as *monocoque*.

First, we shall consider wing sections which, while performing the same function, can differ widely in their structural complexity, as can be seen by comparing Figs 12.5 and 12.6. In Fig. 12.5, the wing of the small, light passenger aircraft, the De Havilland Canada Twin Otter, comprises a relatively simple arrangement of two spars, ribs, stringers and skin, while the wing of the Harrier in Fig. 12.6 consists of numerous spars, ribs and skin. However, no matter how complex the internal structural arrangement the different components perform the same kind of function. The shape of the cross-section is governed by aerodynamic considerations and clearly must be maintained for all combinations of load; this is one of the functions of the ribs. They also act with the skin in resisting the distributed aerodynamic pressure loads; they distribute concentrated loads (e.g. undercarriage and additional wing store loads) into the structure and redistribute stress around discontinuities, such as undercarriage wells, inspection panels and fuel tanks, in the wing surface. Ribs increase the column buckling stress of the longitudinal stiffeners by providing end restraint and establishing their column length; in a similar manner they increase the plate buckling stress of the skin panels. The dimensions of ribs are governed by their spanwise position in the wing and by the loads they are required to support. In the outer portions of the wing, where the cross-section may be relatively small if the wing is tapered and the loads are light, ribs act primarily as formers for the aerofoil shape. A light structure is sufficient for this purpose whereas at sections closer to the wing root, where the ribs are required to absorb and transmit large concentrated applied loads, such as those from the undercarriage, engine thrust and fuselage attachment point reactions, a much more rugged construction is necessary. Between these two extremes are ribs which support hinge reactions from ailerons, flaps and other control surfaces, plus the many internal loads from fuel, armament and systems installations.

The primary function of the wing skin is to form an impermeable surface for supporting the aerodynamic pressure distribution from which the lifting capability of the wing is derived. These aerodynamic forces are transmitted in turn to the ribs and stringers by the skin through plate and membrane action. Resistance to shear and torsional loads is supplied by shear stresses developed in the skin and spar webs, while axial and bending loads are reacted by the combined action of skin and stringers.

Although the thin skin is efficient for resisting shear and tensile loads, it buckles under comparatively low compressive loads. Rather than increase the skin thickness and suffer a consequent weight penalty, stringers are attached to the skin and ribs, thereby dividing the skin into small panels and increasing the buckling and failing stresses. This stabilizing action of the stringers on the skin is, in fact, reciprocated to some extent although the effect normal to the surface of the skin is minimal. Stringers rely chiefly on rib attachments for preventing column action in this direction. We have noted in the previous paragraph the combined action of stringers and skin in resisting axial and bending loads.

The role of spar webs in developing shear stresses to resist shear and torsional loads has been mentioned previously; they perform a secondary but significant function in stabilizing, with the skin, the spar flanges or caps which are therefore capable of supporting large compressive loads from axial and bending effects. In turn, spar webs exert a stabilizing influence on the skin in a similar manner to the stringers.

While the majority of the above remarks have been directed towards wing structures, they apply, as can be seen by referring to Figs 12.5 and 12.6, to all the aerodynamic surfaces, namely wings, horizontal and vertical tails, except in the obvious cases of undercarriage loading, engine thrust, etc.

Fuselages, while of different shape to the aerodynamic surfaces, comprise members which perform similar functions to their counterparts in the wings and tailplane. However, there are differences in the generation of the various types of load. Aerodynamic forces on the fuselage skin are relatively low; on the other hand, the fuselage supports large concentrated loads such as wing reactions, tailplane reactions, undercarriage reactions and it carries payloads of varying size and weight, which may cause large inertia forces. Furthermore, aircraft designed for high altitude flight must withstand internal pressure. The shape of the fuselage cross-section is determined by operational requirements. For example, the most efficient sectional shape for a pressurized fuselage is circular or a combination of circular elements. Irrespective of shape, the basic fuselage structure is essentially a single cell thin-walled tube comprising skin, transverse frames and stringers; transverse frames which extend completely across the fuselage are known as bulkheads. Three different types of fuselage are shown in Figs 12.5–12.7. In Fig. 12.5 the fuselage is unpressurized so that, in the passenger-carrying area, a more rectangular shape is employed to maximize space. The Harrier fuselage in Fig. 12.6 contains the engine, fuel tanks, etc. so that its cross-sectional shape is, to some extent, predetermined, while in Fig. 12.7 the passenger-carrying fuselage of the British Aerospace 146 is pressurized and therefore circular in cross-section.

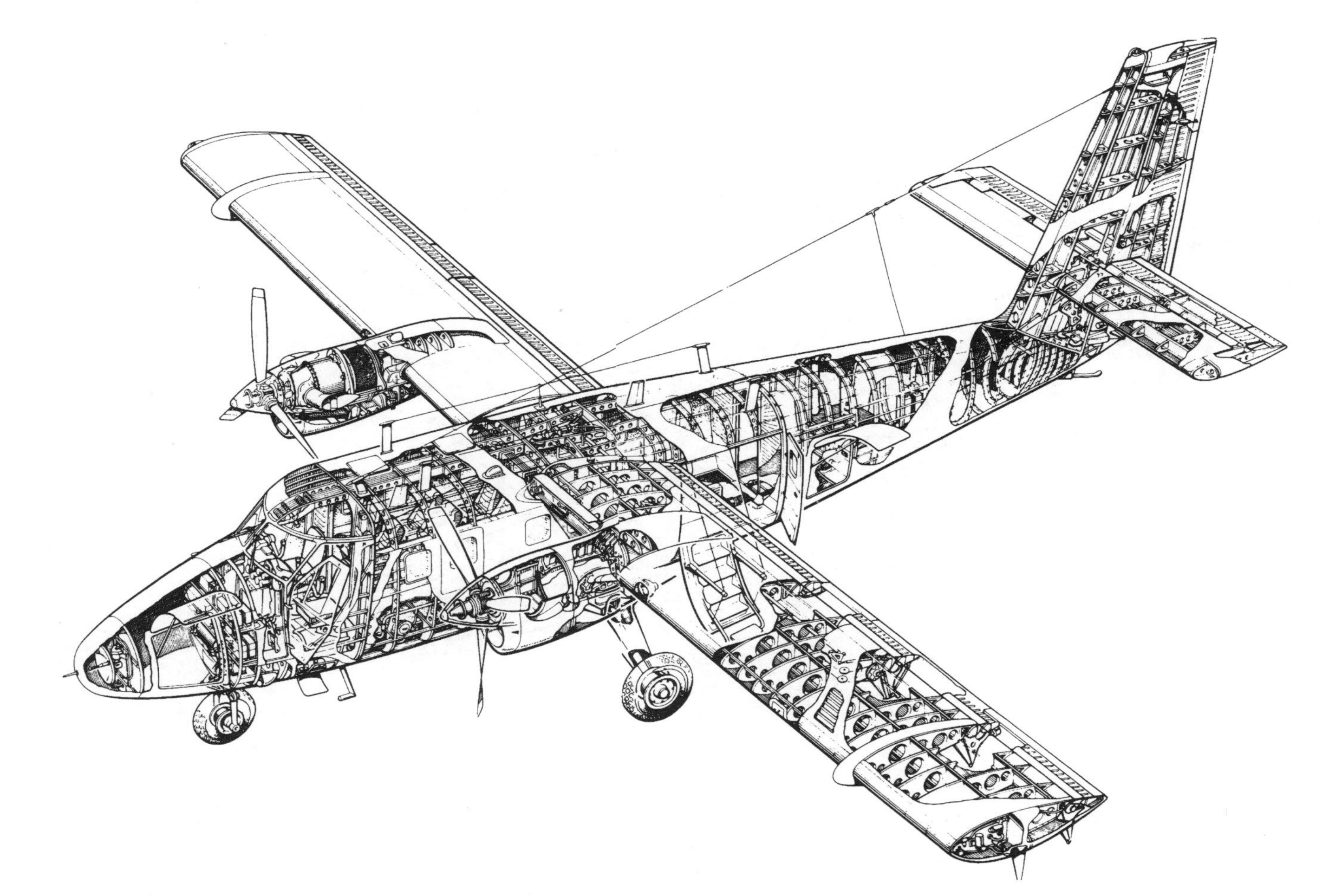

Fig. 12.5 De Havilland Canada Twin Otter (courtesy of De Havilland Aircraft of Canada Ltd.).

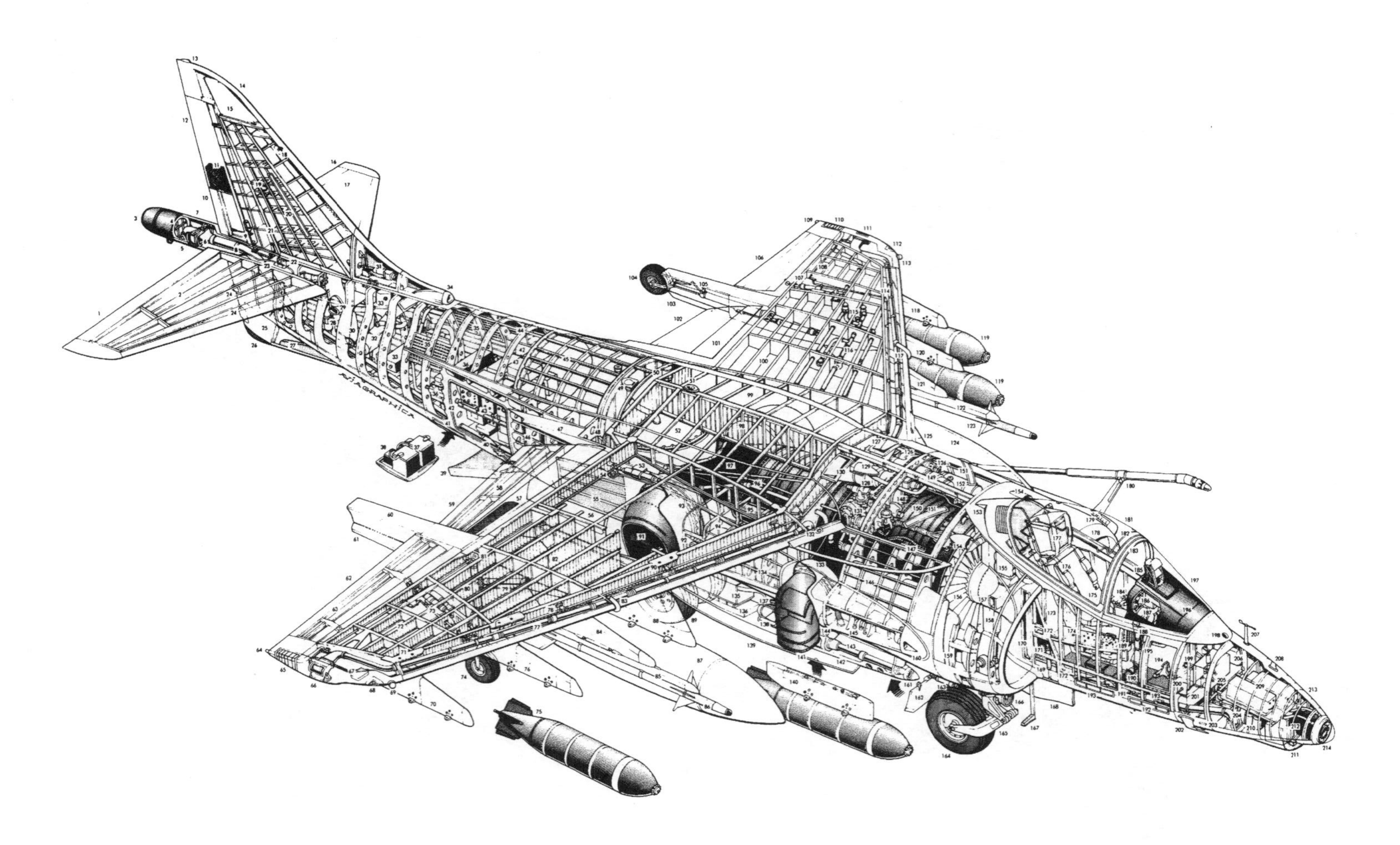

Fig. 12.6 Harrier (courtesy of Pilot Press Ltd.).

1 Starboard all-moving tailplane
2 Tailplane composite construction
3 Tail radome
4 Military equipment
5 Tail pitch control air valve
6 Yaw control air valves
7 Tail 'bullet' fairing
8 Reaction control system air ducting
9 Trim tab actuator
10 Rudder trim tab
11 Rudder composite construction
12 Rudder
13 Antenna
14 Fin tip aerial fairing
15 Upper broad band communications antenna
16 Port tailplane
17 Graphite epoxy tailplane skin
18 Port side temperature probe
19 MAD compensator
20 Formation lighting strip
21 Fin construction
22 Fin attachment joint
23 Tailplane pivot sealing plate
24 Aerials
25 Ventral fin
26 Tail bumper
27 Lower broad band communications antenna
28 Tailplane hydraulic jack
29 Heat exchanger air exhaust
30 Aft fuselage frames
31 Rudder hydraulic actuator
32 Avionics equipment air conditioning plant
33 Avionics equipment racks
34 Heat exchanger ram air intake
35 Electrical system circuit breaker panels, port and starboard
36 Avionic equipment
37 Chaff and flare dispensers
38 Dispenser electronic control units
39 Ventral airbrake
40 Airbrake hydraulic jack
41 Formation lighting strip
42 Avionics bay access door, port and starboard
43 Avionics equipment racks
44 Fuselage frame and stringer construction
45 Rear fuselage fuel tank
46 Main undercarriage wheel bay
47 Wing root fillet
48 Wing spar/fuselage attachment joint
49 Water filler cap
50 Engine fire extinguisher bottle
51 Anti-collision light
52 Water tank
53 Flap hydraulic actuator
54 Flap hinge fitting
55 Nimonic fuselage heat shield
56 Main undercarriage bay doors (closed after cycling of mainwheels)
57 Flap vane composite construction
58 Flap composite construction
59 Starboard slotted flap, lowered
60 Outrigger wheel fairing
61 Outrigger leg doors
62 Starboard aileron
63 Aileron composite construction
64 Fuel jettison
65 Formation lighting panel
66 Roll control airvalve
67 Wing tip fairing
68 Starboard navigation light
69 Radar warning aerial
70 Outboard pylon
71 Pylon attachment joint
72 Graphite epoxy composite wing construction
73 Aileron hydraulic actuator
74 Starboard outrigger wheel
75 BL755 600-lb (272-kg) cluster bomb (CBU)
76 Intermediate pylon
77 Reaction control air ducting
78 Aileron control rod
79 Outrigger hydraulic retraction jack
80 Outrigger leg strut
81 Leg pivot fixing
82 Multi-spar wing construction
83 Leading-edge wing fence
84 Outrigger pylon
85 Missile launch rail
86 AIM-9L Sidewinder air-to-air missile
87 External fuel tank, 300 US gal (1 135 l)
88 Inboard pylon
89 Aft retracting twin mainwheels
90 Inboard pylon attachment joint
91 Rear (hot stream) swivelling exhaust nozzle
92 Position of pressure refuelling connection on port side
93 Rear nozzle bearing
94 Centre fuselage flank fuel tank
95 Hydraulic reservoir
96 Nozzle bearing cooling air duct
97 Engine exhaust divider duct
98 Wing panel centre rib
99 Centre section integral fuel tank
100 Port wing integral fuel tank
101 Flap vane
102 Port slotted flap, lowered
103 Outrigger wheel fairing
104 Port outrigger wheel
105 Torque scissor links
106 Port aileron
107 Aileron hydraulic actuator
108 Aileron/airvalve interconnection
109 Fuel jettison
110 Formation lighting panel
111 Port roll control air valve
112 Port navigation light
113 Radar warning aerial
114 Port wing reaction control air duct
115 Fuel pumps
116 Fuel system piping
117 Port wing leading-edge fence
118 Outboard pylon
119 BL755 cluster bombs (maximum load, seven)
120 Intermediate pylon
121 Port outrigger pylon
122 Missile launch rail
123 AIM-9L Sidewinder air-to-air missile
124 Port leading-edge root extension (LERX)
125 Inboard pylon
126 Hydraulic pumps
127 APU intake
128 Gas turbine starter/auxiliary power unit (APU)
129 Alternator cooling air exhaust
130 APU exhaust
131 Engine fuel control unit
132 Engine bay venting ram air intake
133 Rotary nozzle bearing
134 Nozzle fairing construction
135 Ammunition tank, 100 rounds
136 Cartridge case collector box
137 Ammunition feed chute
138 Fuel vent
139 Gun pack strake
140 Fuselage centreline pylon
141 Zero scarf forward (fan air) nozzle
142 Ventral gun pack (two)
143 Aden 25-mm cannon
144 Engine drain mast
145 Hydraulic system ground connectors
146 Forward fuselage flank fuel tank
147 Engine electronic control units
148 Engine accessory equipment gearbox
149 Gearbox driven alternator
150 Rolls-Royce Pegasus 11 Mk 105 vectored thrust turbofan
151 Formation lighting strips
152 Engine oil tank
153 Bleed air spill duct
154 Air conditioning intake scoops
155 Cockpit air conditioning system heat exchanger
156 Engine compressor/fan face
157 Heat exchanger discharge to intake duct
158 Nose undercarriage hydraulic retraction jack
159 Intake blow-in doors
160 Engine bay venting air scoop
161 Cannon muzzle fairing
162 Lift augmentation retractable cross-dam
163 Cross-dam hydraulic jack
164 Nosewheel
165 Nosewheel forks
166 Landing/taxiing lamp
167 Retractable boarding step
168 Nosewheel doors (closed after cycling of undercarriage)
169 Nosewheel door jack
170 Boundary layer bleed air duct
171 Nose undercarriage wheel bay
172 Kick-in boarding steps
173 Cockpit rear pressure bulkhead
174 Starboard side console panel
175 Martin-Baker Type 12 ejection seat
176 Safety harness
177 Ejection seat headrest
178 Port engine air intake
179 Probe hydraulic jack
180 Retractable in-flight refuelling probe (bolt-on pack)
181 Cockpit canopy cover
182 Miniature detonating cord (MDC) canopy breaker
183 Canopy frame
184 Engine throttle and nozzle angle control levers
185 Pilot's head-up display
186 Instrument panel
187 Moving map display
188 Control column
189 Central warning system panel
190 Cockpit pressure floor
191 Underfloor control runs
192 Formation lighting strips
193 Aileron trim actuator
194 Rudder pedals
195 Cockpit section composite construction
196 Instrument panel shroud
197 One-piece wrap-around windscreen panel
198 Ram air intake (cockpit fresh air)
199 Front pressure bulkhead
200 Incidence vane
201 Air data computer
202 Pitot tube
203 Lower IFF aerial
204 Nose pitch control air valve
205 Pitch trim control actuator
206 Electrical system equipment
207 Yaw vane
208 Upper IFF aerial
209 Avionic equipment
210 ARBS heat exchanger
211 MIRLS sensors
212 Hughes Angle Rate Bombing System (ARBS)
213 Composite construction nose cone
214 ARBS glazed aperture

Fig. 12.7 British Aerospace 146 (courtesy of British Aerospace).

12.3 Fabrication of structural components

The introduction of all-metal, stressed skin aircraft resulted in methods and types of fabrication which remain in use to the present day. However, improvements in engine performance and advances in aerodynamics have led to higher maximum lift, higher speeds and therefore to higher wing loadings so that improved techniques of fabrication are necessary, particularly in the construction of wings. The increase in wing loading from about 350 N/m^2 for 1917–1918 aircraft to around 4800 N/m^2 for modern aircraft, coupled with a drop in the structural percentage of the total weight from 30–40 to 22–25 per cent, gives some indication of the improvements in materials and structural design.

For purposes of construction, aircraft are divided into a number of sub-assemblies. These are built in specially designed jigs, possibly in different parts of the factory or even different factories, before being forwarded to the final assembly shop. A typical break-down into sub-assemblies of a medium-sized civil aircraft is shown in Fig. 12.8. Each

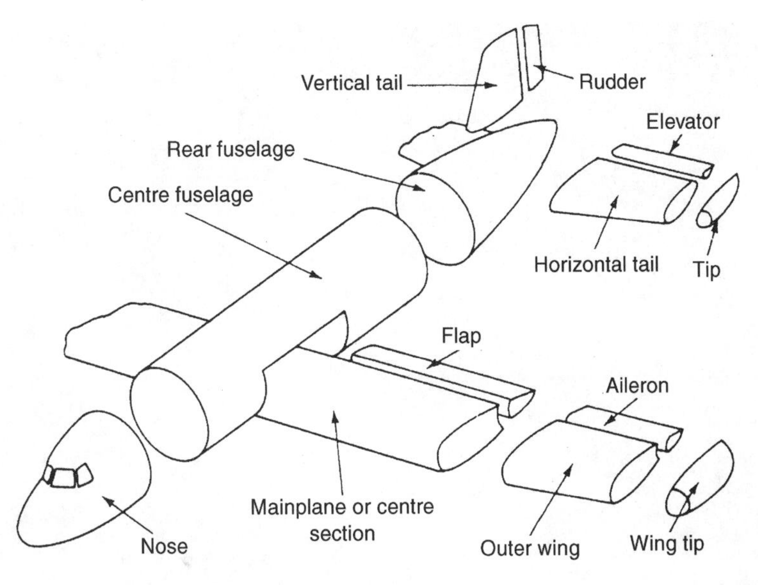

Fig. 12.8 Typical sub-assembly breakdown.

sub-assembly relies on numerous minor assemblies such as spar webs, ribs, frames, and these, in turn, are supplied with individual components from the detail workshop.

Although the wings (and tailsurfaces) of fixed wing aircraft generally consist of spars, ribs, skin and stringers, methods of fabrication and assembly differ. The wing of the aircraft of Fig. 12.5 relies on fabrication techniques that have been employed for many years. In this form of construction the spars comprise thin aluminium alloy webs and flanges, the latter being extruded or machined and are bolted or riveted to the web. The ribs are formed in three parts from sheet metal by large presses and rubber dies and have flanges round their edges so that they can be riveted to the skin and spar webs; cut-outs around their edges allow the passage of spanwise stringers. Holes are cut in the ribs at positions of low stress for lightness and to accommodate control runs, fuel and electrical systems.

Finally, the skin is riveted to the rib flanges and longitudinal stiffeners. Where the curvature of the skin is large, for example at the leading edge, the aluminium alloy sheets are passed through 'rolls' to pre-form them to the correct shape. A further, aerodynamic, requirement is that forward chordwise sections of the wing should be as smooth as possible to delay transition from laminar to turbulent flow. Thus, countersunk rivets are used in these positions as opposed to dome-headed rivets nearer the trailing edge.

The wing is attached to the fuselage through reinforced fuselage frames, frequently by bolts. In some aircraft the wing spars are continuous through the fuselage depending on the demands of space. In a high wing aircraft (Fig. 12.5) deep spars passing through the fuselage would cause obstruction problems. In this case a short third spar provides an additional attachment point. The ideal arrangement is obviously where continuity of the structure is maintained over the entire surface of the wing. In most practical cases this is impossible since cut-outs in the wing surface are required for retracting undercarriages, bomb and gun bays, inspection panels, etc. The last are usually located on the undersurface of the wing and are fastened to stiffeners and rib flanges by screws,

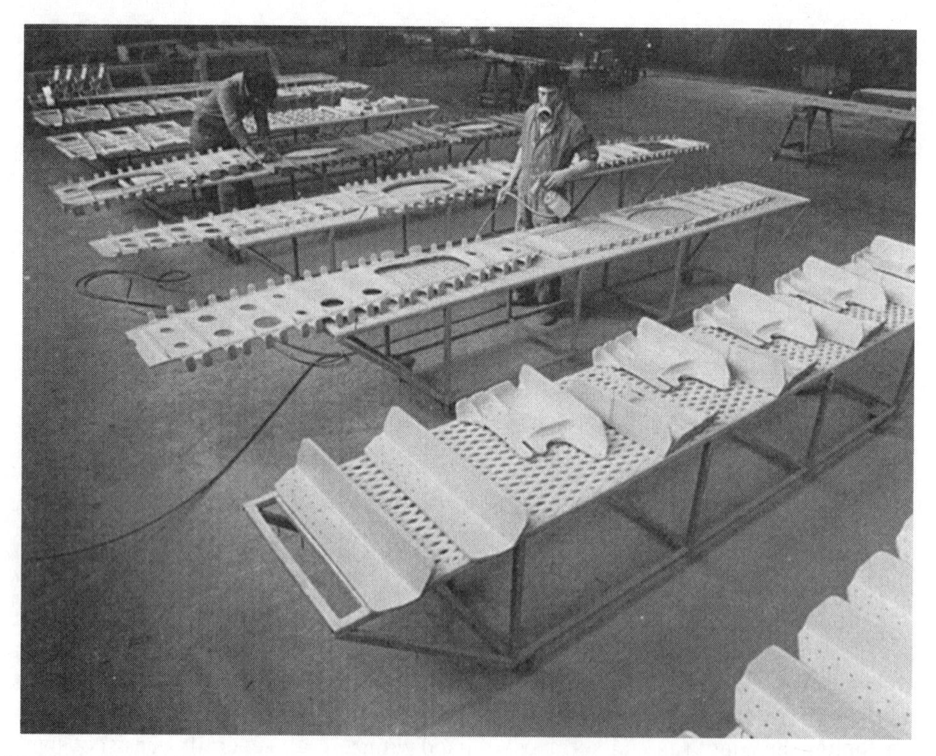

Fig. 12.9 Wing ribs for the European Airbus (courtesy of British Aerospace).

enabling them to resist direct and shear loads. Doors covering undercarriage wells and weapon bays are incapable of resisting wing stresses so that provision must be made for transferring the loads from skin, flanges and shear webs around the cut-out. This may be achieved by inserting strong bulkheads or increasing the spar flange areas, although, no matter the method employed, increased cost and weight result.

The different structural requirements of aircraft designed for differing operational roles lead to a variety of wing constructions. For instance, high-speed aircraft require relatively thin wing sections which support high wing loadings. To withstand the correspondingly high surface pressures and to obtain sufficient strength, much thicker skins are necessary. Wing panels are therefore frequently machined integrally with stringers from solid slabs of material, as are the wing ribs. Figure 12.9 shows wing ribs for the European Airbus in which web stiffeners, flanged lightness holes and skin attachment lugs have been integrally machined from solid. This integral method of construction involves no new design principles and has the advantages of combining a high grade of surface finish, free from irregularities, with a more efficient use of material since skin thicknesses are easily tapered to coincide with the spanwise decrease in bending stresses.

An alternative form of construction is the sandwich panel, which comprises a light honeycomb or corrugated metal core sandwiched between two outer skins of the stress-bearing sheet (see Fig. 12.10). The primary function of the core is to stabilize the outer skins, although it may be stress bearing as well. Sandwich panels are capable of

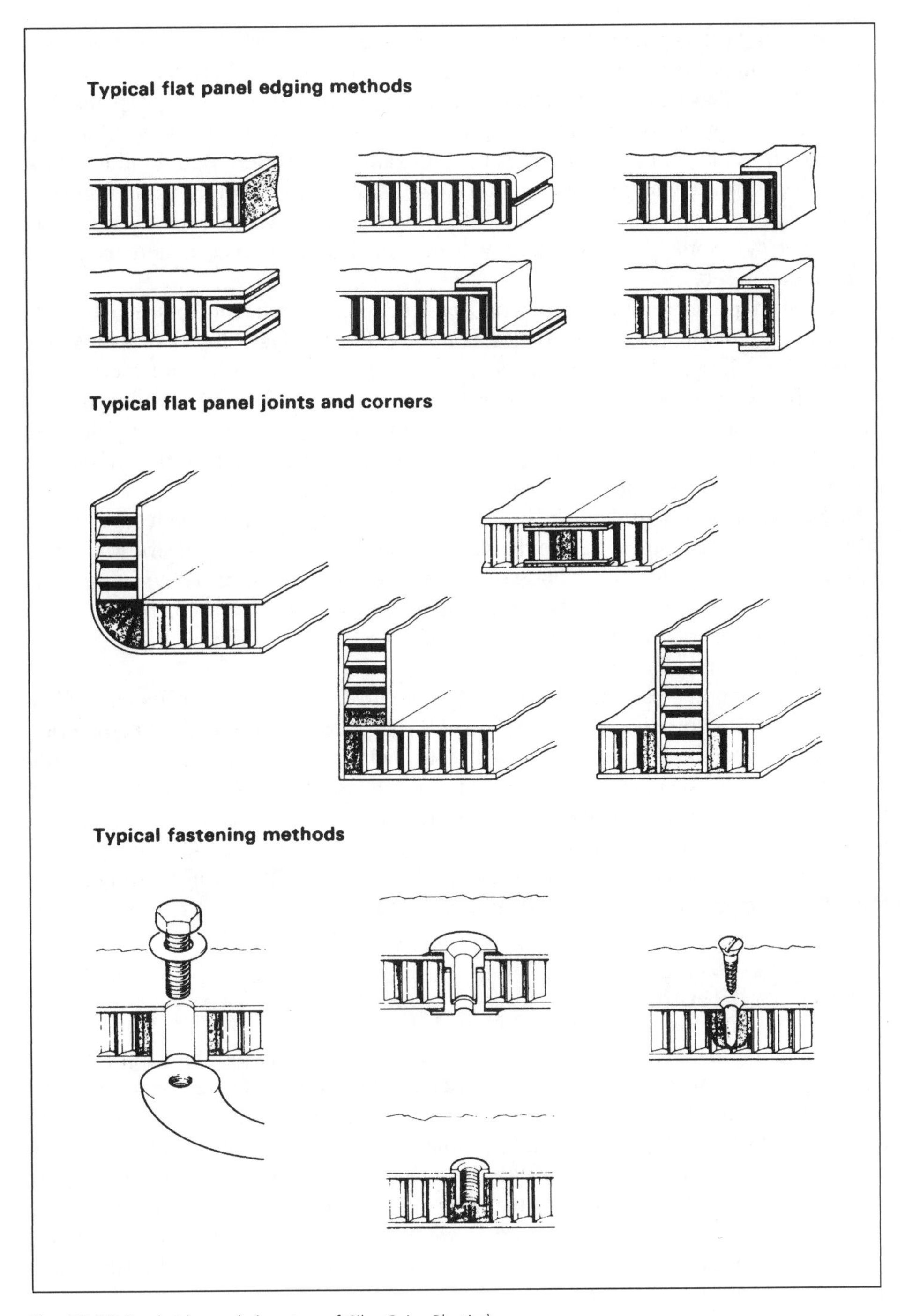

Fig. 12.10 Sandwich panels (courtesy of Ciba-Geigy Plastics).

developing high stresses, have smooth internal and external surfaces and require small numbers of supporting rings or frames. They also possess a high resistance to fatigue from jet efflux. The uses of this method of construction include lightweight 'planks' for cabin furniture, monolithic fairing shells generally having plastic facing skins, and the stiffening of flying control surfaces. Thus, for example, the ailerons and rudder of the British Aerospace Jaguar are fabricated from aluminium honeycomb, while fibreglass and aluminium faced honeycomb are used extensively in the wings and tail surfaces of the Boeing 747. Some problems, mainly disbonding and internal corrosion, have been encountered in service.

The general principles relating to wing construction are applicable to fuselages, with the exception that integral construction is not used in fuselages for obvious reasons. Figures 12.5, 12.6 and 12.7 show that the same basic method of construction is employed in aircraft having widely differing roles. Generally, the fuselage frames that support large concentrated floor loads or loads from wing or tailplane attachment points are heavier than lightly loaded frames and require stiffening, with additional provision for transmitting the concentrated load into the frame and hence the skin.

With the frames in position in the fuselage jig, stringers, passing through cut-outs, are riveted to the frame flanges. Before the skin is riveted to the frames and stringers, other subsidiary frames such as door and window frames are riveted or bolted in position. The areas of the fuselage in the regions of these cut-outs are reinforced by additional stringers, portions of frame and increased skin thickness, to react to the high shear flows and direct stresses developed.

On completion, the various sub-assemblies are brought together for final assembly. Fuselage sections are usually bolted together through flanges around their peripheries, while wings and the tailplane are attached to pick-up points on the relevant fuselage frames. Wing spars on low wing civil aircraft usually pass completely through the fuselage, simplifying wing design and the method of attachment. On smaller, military aircraft, engine installations frequently prevent this so that wing spars are attached directly to and terminate at the fuselage frame. Clearly, at these positions frame/stringer/skin structures require reinforcement.

12.4 Connections

The fabrication of aircraft components generally involves the joining of one part of the component to another. For example, fuselage skins are connected to stringers and frames while wing skins are connected to stringers and wing ribs unless, as in some military aircraft with high wing loadings, the stringers are machined integrally with the wing skin (see Section 12.3). With the advent of all-metal, i.e. aluminium alloy construction, riveted joints became the main form of connection with some welding although aluminium alloys are difficult to weld, and, in the modern era, some glued joints which use epoxy resin. In this section we shall concentrate on the still predominant method of connection, riveting.

In general riveted joints are stressed in complex ways and an accurate analysis is very often difficult to achieve because of the discontinuities in the region of the joint. Fairly crude assumptions as to joint behaviour are made but, when combined with experience, safe designs are produced.

12.4.1 Simple lap joint

Figure 12.11 shows two plates of thickness t connected together by a single line of rivets; this type of joint is termed a lap joint and is one of the simplest used in construction.

Suppose that the plates carry edge loads of P/unit width, that the rivets are of diameter d and are spaced at a distance b apart, and that the distance from the line of rivets to the edge of each plate is a. There are four possible modes of failure which must be considered as follows:

Rivet shear

The rivets may fail by shear across their diameter at the interface of the plates. Then, if the maximum shear stress the rivets will withstand is τ_1 failure will occur when

$$Pb = \tau_1 \left(\frac{\pi d^2}{4}\right)$$

which gives

$$P = \frac{\pi d^2 \tau_1}{4b} \tag{12.1}$$

Bearing pressure

Either the rivet or plate may fail due to bearing pressure. Suppose that p_b is this pressure then failure will occur when

$$\frac{Pb}{td} = p_b$$

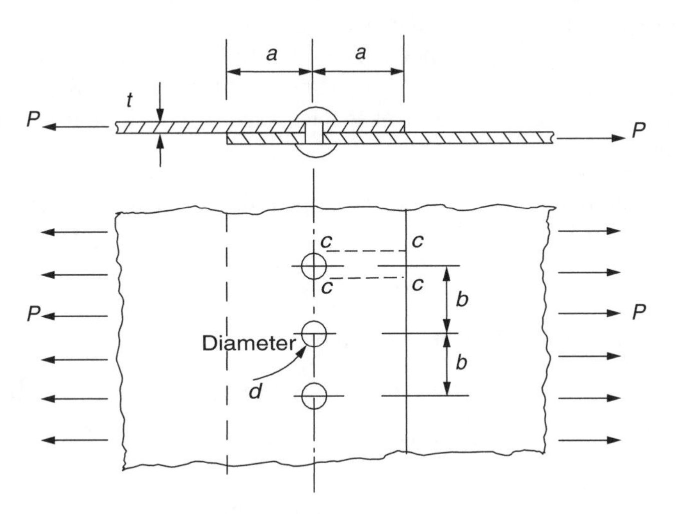

Fig. 12.11 Simple riveted lap joint.

so that

$$P = \frac{p_b t d}{b} \tag{12.2}$$

Plate failure in tension

The area of plate in tension along the line of rivets is reduced due to the presence of rivet holes. Therefore, if the ultimate tensile stress in the plate is σ_{ult} failure will occur when

$$\frac{Pb}{t(b-d)} = \sigma_{ult}$$

from which

$$P = \frac{\sigma_{ult}\, t(b-d)}{b} \tag{12.3}$$

Shear failure in a plate

Shearing of the plates may occur on the planes cc resulting in the rivets being dragged out of the plate. If the maximum shear stress at failure of the material of the plates is τ_2 then a failure of this type will occur when

$$Pb = 2at\,\tau_2$$

which gives

$$P = \frac{2at\,\tau_2}{b} \tag{12.4}$$

Example 12.1

A joint in a fuselage skin is constructed by riveting the abutting skins between two straps as shown in Fig. 12.12. The fuselage skins are 2.5 mm thick and the straps are each 1.2 mm thick; the rivets have a diameter of 4 mm. If the tensile stress in the fuselage skin must not exceed 125 N/mm^2 and the shear stress in the rivets is limited to 120 N/mm^2 determine the maximum allowable rivet spacing such that the joint is equally strong in shear and tension.

A tensile failure in the plate will occur on the reduced plate cross-section along the rivet lines. This area is given by

$$A_p = (b-4) \times 2.5\,\text{mm}^2$$

The failure load/unit width P_f is then given by

$$P_f b = (b-4) \times 2.5 \times 125 \tag{i}$$

The area of cross-section of each rivet is

$$A_r = \frac{\pi \times 4^2}{4} = 12.6\,\text{mm}^2$$

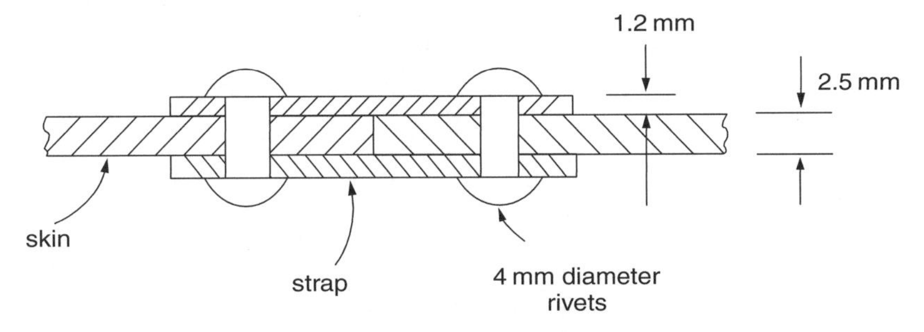

Fig. 12.12 Joint of Example 12.1.

Since each rivet is in double shear (i.e. two failure shear planes) the area of cross-section in shear is

$$2 \times 12.6 = 25.2\,\text{mm}^2$$

Then the failure load/unit width in shear is given by

$$P_f b = 25.2 \times 120 \tag{ii}$$

For failure to occur simultaneously in shear and tension, i.e. equating Eqs (i) and (ii)

$$25.2 \times 120 = (b - 4) \times 2.5 \times 12.5$$

from which

$$b = 13.7\,\text{mm}$$

Say, a rivet spacing of 13 mm.

12.4.2 Joint efficiency

The efficiency of a joint or connection is measured by comparing the actual failure load with that which would apply if there were no rivet holes in the plate. Then, for the joint shown in Fig. 12.11 the joint efficiency η is given by

$$\eta = \frac{\sigma_{\text{ult}}\, t(b - d)/b}{\sigma_{\text{ult}}\, t} = \frac{b - d}{b} \tag{12.5}$$

12.4.3 Group-riveted joints

Rivets may be grouped on each side of a joint such that the efficiency of the joint is a maximum. Suppose that two plates are connected as shown in Fig. 12.13 and that six rivets are required on each side. If it is assumed that each rivet is equally loaded then the single rivet on the line aa will take one-sixth of the total load. The two rivets on the line bb will then share two-sixths of the load while the three rivets on the line cc will share three-sixths of the load. On the line bb the area of cross-section of the

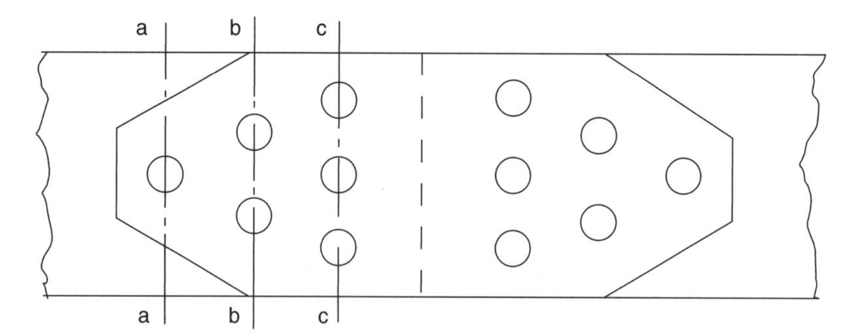

Fig. 12.13 A group-riveted joint.

plate is reduced by two rivet holes and that on the line cc by three rivet holes so that, relatively, the joint is as strong at these sections as at aa. Therefore, a more efficient joint is obtained than if the rivets were arranged in, say, two parallel rows of three.

12.4.4 Eccentrically loaded riveted joints

The bracketed connection shown in Fig. 12.14 carries a load P offset from the centroid of the rivet group. The rivet group is then subjected to a shear load P through its centroid and a moment or torque Pe about its centroid.

It is assumed that the shear load P is distributed equally amongst the rivets causing a shear force in each rivet parallel to the line of action of P. The moment Pe is assumed to produce a shear force S in each rivet where S acts in a direction perpendicular to the line joining a particular rivet to the centroid of the rivet group. Furthermore, the value of S is assumed to be proportional to the distance of the rivet from the centroid of the rivet group. Then

$$Pe = \sum Sr$$

If $S = kr$ where k is a constant for all rivets then

$$Pe = k \sum r^2$$

from which

$$k = Pe/\sum r^2$$

and

$$S = \frac{Pe}{\sum r^2} r \qquad (12.6)$$

The resultant force on a rivet is then the vector sum of the forces due to P and Pe.

Example 12.2

The bracket shown in Fig. 12.15 carries an offset load of 5 kN. Determine the resultant shear forces in the rivets A and B.

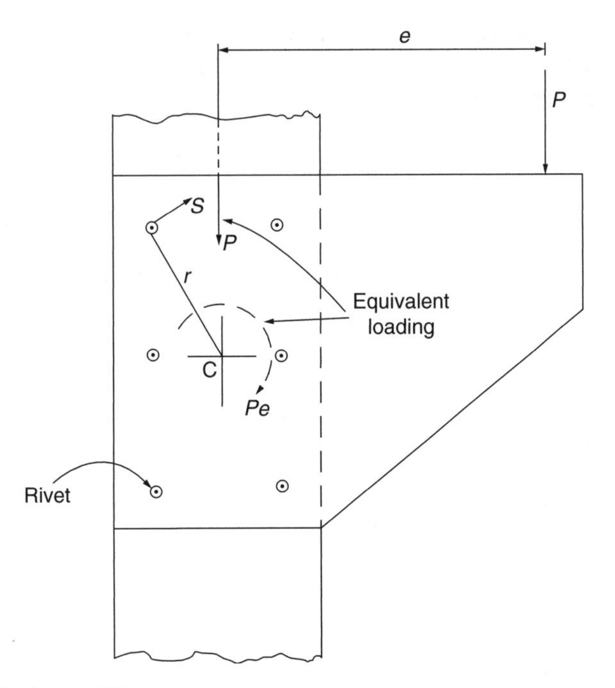

Fig. 12.14 Eccentrically loaded joint.

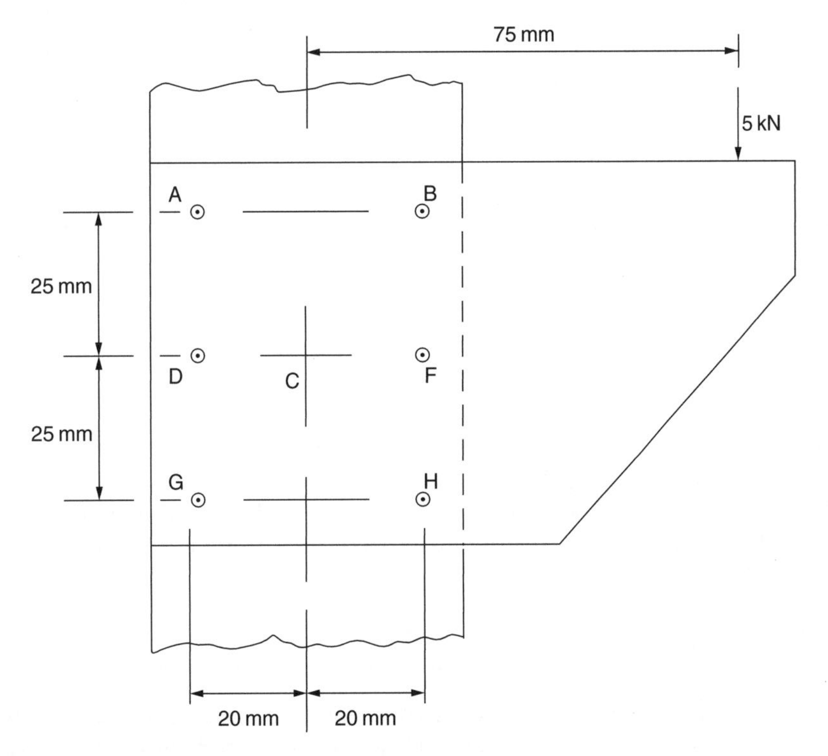

Fig. 12.15 Joint of Example 12.2.

The vertical shear force on each rivet is 5/6 = 0.83 kN. The moment (Pe) on the rivet group is 5 × 75 = 375 kNmm. The distance of rivet A (and B, G and H) from the centroid C of the rivet group is given by

$$r = (20^2 + 25^2)^{1/2} = (1025)^{1/2} = 32.02\,\text{mm}$$

The distance of D (and F) from C is 20 mm. Therefore

$$\sum r^2 = 2 \times 400 + 4 \times 1025 = 4900$$

From Eq. (12.6) the shear forces on rivets A and B due to the moment are

$$S = \frac{375}{4900} \times 32.02 = 2.45\,\text{kN}$$

On rivet A the force system due to P and Pe is that shown in Fig. 12.16(a) while that on B is shown in Fig. 12.16(b).

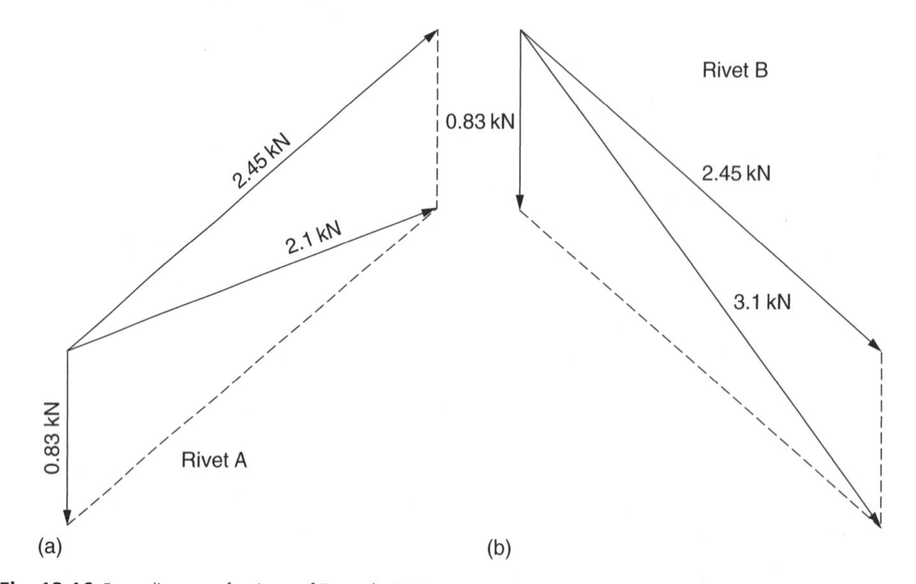

Fig. 12.16 Force diagrams for rivets of Example 12.2.

The resultant forces may then be calculated using the rules of vector addition or determined graphically using the parallelogram of forces.[1]

The design of riveted connections is carried out in the actual design of the rear fuselage of a single-engined trainer/semi-aerobatic aircraft in the Appendix.

12.4.5 Use of adhesives

In addition to riveted connections adhesives have and are being used in aircraft construction although, generally, they are employed in areas of low stress since their application is still a matter of research. Of these adhesives epoxy resins are the most frequently used since they have the advantages over, say polyester resins, of good adhesive properties, low shrinkage during cure so that residual stresses are reduced, good mechanical properties and thermal stability. The modulus and ultimate strength of epoxy resin are, typically, 5000 and 100 N/mm^2. Epoxy resins are now found extensively as the matrix component in fibrous composites.

Reference

1 Megson, T. H. G., *Structural and Stress Analysis*, 2nd edition, Elsevier, Oxford, 2005.

Problems

P.12.1 Examine possible uses of new materials in future aircraft manufacture.

P.12.2 Describe the main features of a stressed skin structure. Discuss the structural functions of the various components with particular reference either to the fuselage or to the wing of a medium-sized transport aircraft.

P.12.3 The double riveted butt joint shown in Fig. P.12.3 connects two plates which are each 2.5 mm thick, the rivets have a diameter of 3 mm. If the failure strength of the rivets in shear is 370 N/mm^2 and the ultimate tensile strength of the plate is 465 N/mm^2 determine the necessary rivet pitch if the joint is to be designed so that failure due to shear in the rivets and failure due to tension in the plate occur simultaneously. Calculate also the joint efficiency.

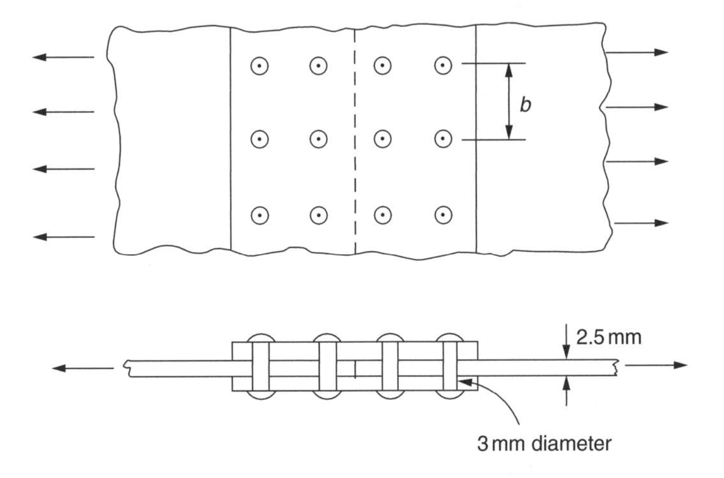

Fig. P.12.3

Ans. Rivet pitch is 12 mm, joint efficiency is 75 per cent.

P.12.4 The rivet group shown in Fig. P.12.4 connects two narrow lengths of plate one of which carries a 15 kN load positioned as shown. If the ultimate shear strength of a rivet is 350 N/mm^2 and its failure strength in compression is 600 N/mm^2 determine the minimum allowable values of rivet diameter and plate thickness.

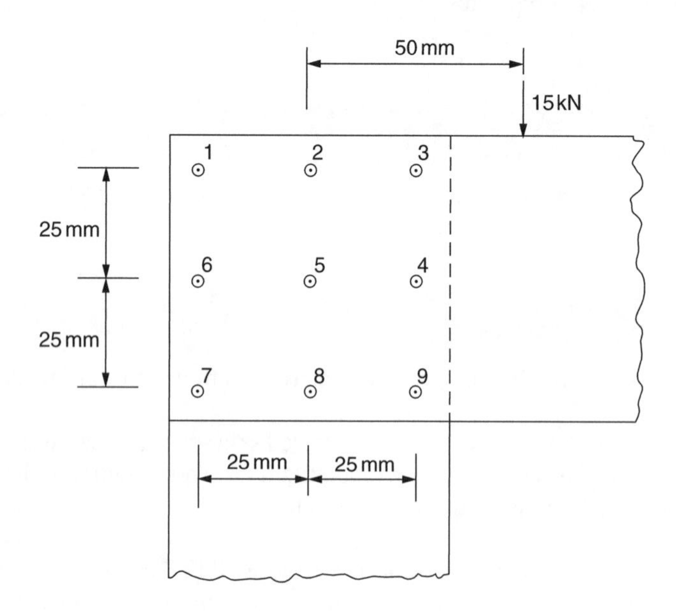

Fig. P.12.4

Ans. Rivet diameter is 4.0 mm, plate thickness is 1.83 mm.

Section B2 Airworthiness and Airframe Loads

13

Airworthiness

The airworthiness of an aircraft is concerned with the standards of safety incorporated in all aspects of its construction. These range from structural strength to the provision of certain safeguards in the event of crash landings, and include design requirements relating to aerodynamics, performance and electrical and hydraulic systems. The selection of minimum standards of safety is largely the concern of 'national and international' airworthiness authorities who prepare handbooks of official requirements. The handbooks include operational requirements, minimum safety requirements, recommended practices and design data, etc.

In this chapter we shall concentrate on the structural aspects of airworthiness which depend chiefly on the strength and stiffness of the aircraft. Stiffness problems may be conveniently grouped under the heading *aeroelasticity* and are discussed in Section B6. Strength problems arise, as we have seen, from ground and air loads, and their magnitudes depend on the selection of manoeuvring and other conditions applicable to the operational requirements of a particular aircraft.

13.1 Factors of safety-flight envelope

The control of weight in aircraft design is of extreme importance. Increases in weight require stronger structures to support them, which in turn lead to further increases in weight and so on. Excesses of structural weight mean lesser amounts of payload, thereby affecting the economic viability of the aircraft. The aircraft designer is therefore constantly seeking to pare his aircraft's weight to the minimum compatible with safety. However, to ensure general minimum standards of strength and safety, airworthiness regulations lay down several factors which the primary structure of the aircraft must satisfy. These are the *limit load*, which is the maximum load that the aircraft is expected to experience in normal operation, the *proof load*, which is the product of the limit load and the *proof factor* (1.0–1.25), and the *ultimate load*, which is the product of the limit load and the *ultimate factor* (usually 1.5). The aircraft's structure must withstand the proof load without detrimental distortion and should not fail until the ultimate load has been achieved. The proof and ultimate factors may be regarded as factors of safety and provide for various contingencies and uncertainties which are discussed in greater detail in Section 13.2.

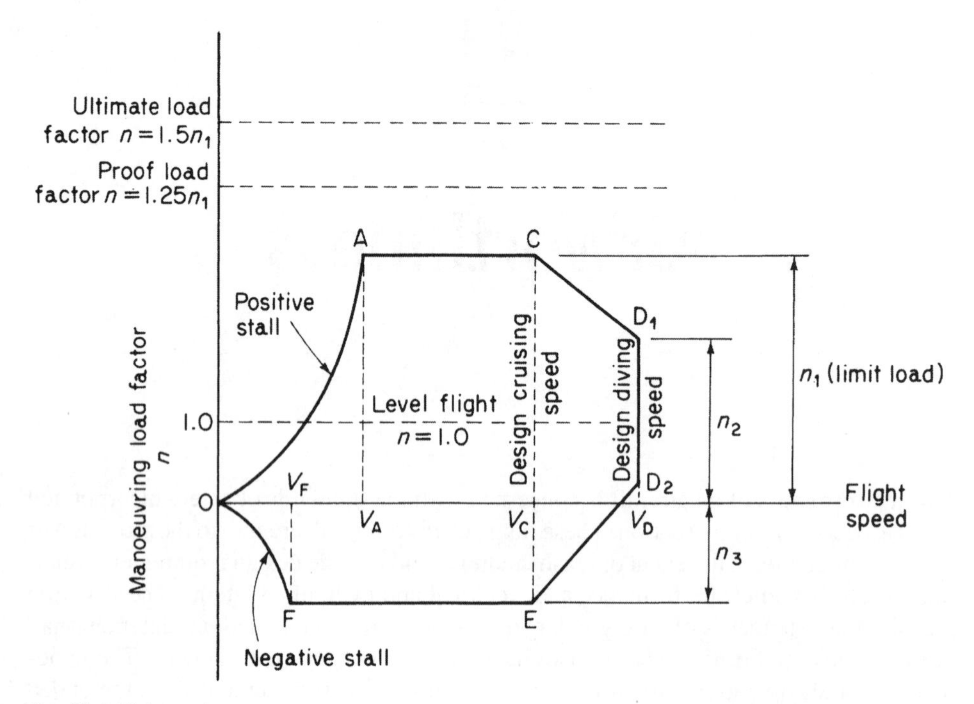

Fig. 13.1 Flight envelope

Table 13.1

Load factor n	Category: Normal	Semi-aerobatic	Aerobatic
n_1	$2.1 + 24\,000/(W + 10\,000)$	4.5	6.0
n_2	$0.75n_1$ but $n_2 \nless 2.0$	3.1	4.5
n_3	1.0	1.8	3.0

The basic strength and flight performance limits for a particular aircraft are selected by the airworthiness authorities and are contained in the *flight envelope* or V–n diagram shown in Fig. 13.1. The curves OA and OF correspond to the stalled condition of the aircraft and are obtained from the well-known aerodynamic relationship

$$\text{Lift} = nW = \tfrac{1}{2}\rho V^2 S C_{\text{L,max}}$$

Therefore, for speeds below V_{A} (positive wing incidence) and V_{F} (negative incidence) the maximum loads which can be applied to the aircraft are governed by $C_{\text{L,max}}$. As the speed increases it is possible to apply the positive and negative limit loads, corresponding to n_1 and n_3, without stalling the aircraft so that AC and FE represent maximum operational load factors for the aircraft. Above the design cruising speed V_{C}, the cut-off lines CD_1 and D_2E relieve the design cases to be covered since it is not expected that the limit loads will be applied at maximum speed. Values of n_1, n_2 and n_3 are specified by the airworthiness authorities for particular aircraft; typical load factors are shown in Table 13.1.

A particular flight envelope is applicable to one altitude only since $C_{L,max}$ is generally reduced with an increase of altitude, and the speed of sound decreases with altitude thereby reducing the critical Mach number and hence the design diving speed V_D. Flight envelopes are therefore drawn for a range of altitudes from sea level to the operational ceiling of the aircraft.

13.2 Load factor determination

Several problems require solution before values for the various load factors in the flight envelope can be determined. The limit load, for example, may be produced by a specified manoeuvre or by an encounter with a particularly severe gust (gust cases and the associated gust envelope are discussed in Section 14.4). Clearly some knowledge of possible gust conditions is required to determine the limiting case. Furthermore, the fixing of the proof and ultimate factors also depends upon the degree of uncertainty of design, variations in structural strength, structural deterioration, etc. We shall now investigate some of these problems to see their comparative influence on load factor values.

13.2.1 Limit load

An aircraft is subjected to a variety of loads during its operational life, the main classes of which are: manoeuvre loads, gust loads, undercarriage loads, cabin pressure loads, buffeting and induced vibrations. Of these, manoeuvre, undercarriage and cabin pressure loads are determined with reasonable simplicity since manoeuvre loads are controlled design cases, undercarriages are designed for given maximum descent rates and cabin pressures are specified. The remaining loads depend to a large extent on the atmospheric conditions encountered during flight. Estimates of the magnitudes of such loads are only possible therefore if in-flight data on these loads is available. It obviously requires a great number of hours of flying if the experimental data are to include possible extremes of atmospheric conditions. In practice, the amount of data required to establish the probable period of flight time before an aircraft encounters, say, a gust load of a given severity, is a great deal more than that available. It therefore becomes a problem in statistics to extrapolate the available data and calculate the probability of an aircraft being subjected to its proof or ultimate load during its operational life. The aim would be for a zero or negligible rate of occurrence of its ultimate load and an extremely low rate of occurrence of its proof load. Having decided on an ultimate load, then the limit load may be fixed as defined in Section 13.1 although the value of the ultimate factor includes, as we have already noted, allowances for uncertainties in design, variation in structural strength and structural deterioration.

13.2.2 Uncertainties in design and structural deterioration

Neither of these presents serious problems in modern aircraft construction and therefore do not require large factors of safety to minimize their effects. Modern methods of aircraft structural analysis are refined and, in any case, tests to determine actual failure loads are carried out on representative full scale components to verify design estimates. The problem of structural deterioration due to corrosion and wear may be largely eliminated by close inspection during service and the application of suitable protective treatments.

13.2.3 Variation in structural strength

To minimize the effect of the variation in structural strength between two apparently identical components, strict controls are employed in the manufacture of materials and in the fabrication of the structure. Material control involves the observance of strict limits in chemical composition and close supervision of manufacturing methods such as machining, heat treatment, rolling, etc. In addition, the inspection of samples by visual, radiographic and other means, and the carrying out of strength tests on specimens, enable below limit batches to be isolated and rejected. Thus, if a sample of a batch of material falls below a specified minimum strength then the batch is rejected. This means of course that an actual structure always comprises materials with properties equal to or better than those assumed for design purposes, an added but unallowed for 'bonus' in considering factors of safety.

Similar precautions are applied to assembled structures with regard to dimension tolerances, quality of assembly, welding, etc. Again, visual and other inspection methods are employed and, in certain cases, strength tests are carried out on sample structures.

13.2.4 Fatigue

Although adequate precautions are taken to ensure that an aircraft's structure possesses sufficient strength to withstand the most severe expected gust or manoeuvre load, there still remains the problem of fatigue. Practically all components of the aircraft's structure are subjected to fluctuating loads which occur a great many times during the life of the aircraft. It has been known for many years that materials fail under fluctuating loads at much lower values of stress than their normal static failure stress. A graph of failure stress against number of repetitions of this stress has the typical form shown in Fig. 13.2. For some materials, such as mild steel, the curve (usually known as an *S–N* curve or diagram) is asymptotic to a certain minimum value, which means that the material has an actual infinite-life stress. Curves for other materials, for example aluminium and its alloys, do not always appear to have asymptotic values so that these materials may not possess an infinite-life stress. We shall discuss the implications of this a little later.

Prior to the mid-1940s little attention had been paid to fatigue considerations in the design of aircraft structures. It was felt that sufficient static strength would eliminate the possibility of fatigue failure. However, evidence began to accumulate that several aircraft crashes had been caused by fatigue failure. The seriousness of the situation was highlighted in the early 1950s by catastrophic fatigue failures of two Comet airliners. These were caused by the once-per-flight cabin pressurization cycle which produced circumferential and longitudinal stresses in the fuselage skin. Although these stresses were well below the allowable stresses for single cycle loading, stress concentrations occurred at the corners of the windows and around rivets which raised local stresses

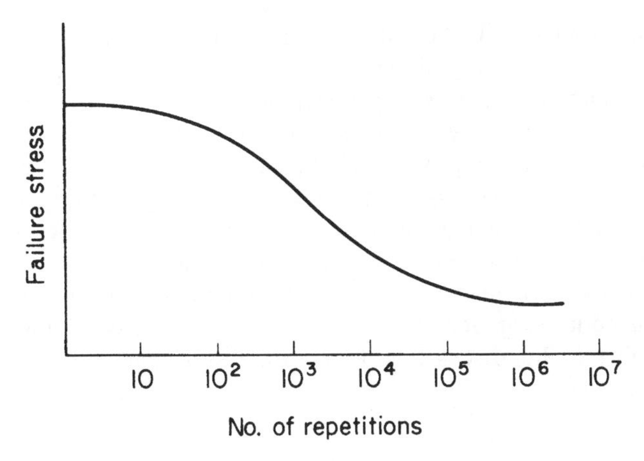

Fig. 13.2 Typical form of *S–N* diagram.

considerably above the general stress level. Repeated cycles of pressurization produced fatigue cracks which propagated disastrously, causing an explosion of the fuselage at high altitude.

Several factors contributed to the emergence of fatigue as a major factor in design. For example, aircraft speeds and sizes increased, calling for higher wing and other loadings. Consequently, the effect of turbulence was magnified and the magnitudes of the fluctuating loads became larger. In civil aviation, airliners had a greater utilization and a longer operational life. The new 'zinc-rich' alloys, used for their high static strength properties, did not show a proportional improvement in fatigue strength, exhibited high crack propagation rates and were extremely notch sensitive.

Despite the fact that the causes of fatigue were reasonably clear at that time its elimination as a threat to aircraft safety was a different matter. The fatigue problem has two major facets: the prediction of the fatigue strength of a structure and a knowledge of the loads causing fatigue. Information was lacking on both counts. The Royal Aircraft Establishment (RAE) and the aircraft industry therefore embarked on an extensive test programme to determine the behaviour of complete components, joints and other detail parts under fluctuating loads. These included fatigue testing by the RAE of some 50 Meteor 4 tailplanes at a range of temperatures, plus research, also by the RAE, into the fatigue behaviour of joints and connections. Further work was undertaken by some universities and by the industry itself into the effects of stress concentrations.

In conjunction with their fatigue strength testing, the RAE initiated research to develop a suitable instrument for counting and recording gust loads over long periods of time. Such an instrument was developed by J. Taylor in 1950 and was designed so that the response fell off rapidly above 10 Hz. Crossings of g thresholds from 0.2 to 1.8 g at 0.1 g intervals were recorded (note that steady level flight is 1 g flight) during experimental flying at the RAE on three different aircraft over 28 000 km, and the best techniques for extracting information from the data established. Civil airlines cooperated by carrying the instruments on their regular air services for a number of years. Eight different types of aircraft were equipped so that by 1961 records had been obtained for regions including Europe, the Atlantic, Africa, India and the Far East, representing 19 000 hours and 8 million km of flying.

Atmospheric turbulence and the cabin pressurization cycle are only two of the many fluctuating loads which cause fatigue damage in aircraft. On the ground the wing is supported on the undercarriage and experiences tensile stresses in its upper surfaces and compressive stresses in its lower surfaces. In flight these stresses are reversed as aerodynamic lift supports the wing. Also, the impact of landing and ground manoeuvring on imperfect surfaces cause stress fluctuations while, during landing and take-off, flaps are lowered and raised, producing additional load cycles in the flap support structure. Engine pylons are subjected to fatigue loading from thrust variations in take-off and landing and also to inertia loads produced by lateral gusts on the complete aircraft.

A more detailed investigation of fatigue and its associated problems is presented in Chapter 15 whilst a fuller discussion of airworthiness as applied to civil jet aircraft is presented in Ref. [1].

Reference

1 Jenkinson, L. R., Simpkin, P. and Rhodes, D., *Civil Jet Aircraft Design*, Arnold, London, 1999.

14

Airframe loads

In Chapter 12, we discussed in general terms the types of load to which aircraft are subjected during their operational life. We shall now examine in more detail the loads which are produced by various manoeuvres and the manner in which they are calculated.

14.1 Aircraft inertia loads

The maximum loads on the components of an aircraft's structure generally occur when the aircraft is undergoing some form of acceleration or deceleration, such as in landings, take-offs and manoeuvres within the flight and gust envelopes. Thus, before a structural component can be designed, the inertia loads corresponding to these accelerations and decelerations must be calculated. For these purposes we shall suppose that an aircraft is a rigid body and represent it by a rigid mass, m, as shown in Fig. 14.1. We shall also, at this stage, consider motion in the plane of the mass which would correspond to pitching of the aircraft without roll or yaw. We shall also suppose that the centre of gravity (CG) of the mass has coordinates $\bar{x}$, $\bar{y}$ referred to x and y axes having an arbitrary origin O; the mass is rotating about an axis through O perpendicular to the xy plane with a constant angular velocity ω.

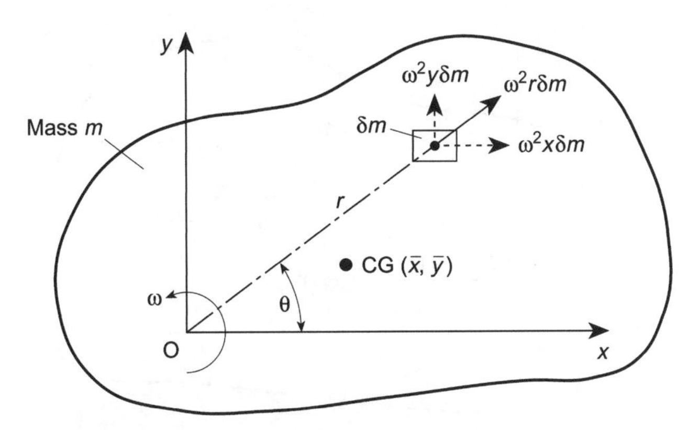

Fig. 14.1 Inertia forces on a rigid mass having a constant angular velocity.

The acceleration of any point, a distance r from O, is $\omega^2 r$ and is directed towards O. Thus, the inertia force acting on the element, δm, is $\omega^2 r\delta m$ in a direction opposite to the acceleration, as shown in Fig. 14.1. The components of this inertia force, parallel to the x and y axes, are $\omega^2 r\delta m\cos\theta$ and $\omega^2 r\delta m\, sin\,\theta$, respectively, or, in terms of x and y, $\omega^2 x\delta m$ and $\omega^2 y\delta m$. The resultant inertia forces, F_x and F_y, are then given by

$$F_x = \int \omega^2 x\,\mathrm{d}m = \omega^2 \int x\,\mathrm{d}m$$

$$F_y = \int \omega^2 y\,\mathrm{d}m = \omega^2 \int y\,\mathrm{d}m$$

in which we note that the angular velocity ω is constant and may therefore be taken outside the integral sign. In the above expressions $\int x\,\mathrm{d}m$ and $\int y\,\mathrm{d}m$ are the moments of the mass, m, about the y and x axes, respectively, so that

$$F_x = \omega^2 \bar{x} m \tag{14.1}$$

and

$$F_y = \omega^2 \bar{y} m \tag{14.2}$$

If the CG lies on the x axis, $\bar{y} = 0$ and $F_y = 0$. Similarly, if the CG lies on the y axis, $F_x = 0$. Clearly, if O coincides with the CG, $\bar{x} = \bar{y} = 0$ and $F_x = F_y = 0$.

Suppose now that the rigid body is subjected to an angular acceleration (or deceleration) α in addition to the constant angular velocity, ω, as shown in Fig. 14.2. An additional inertia force, $\alpha r\delta m$, acts on the element δm in a direction perpendicular to r and in the opposite sense to the angular acceleration. This inertia force has components $\alpha r\delta m\cos\theta$ and $\alpha r\delta m\sin\theta$, i.e. $\alpha x\delta m$ and $\alpha y\delta m$, in the y and x directions, respectively. Thus, the resultant inertia forces, F_x and F_y, are given by

$$F_x = \int \alpha y\,\mathrm{d}m = \alpha \int y\,\mathrm{d}m$$

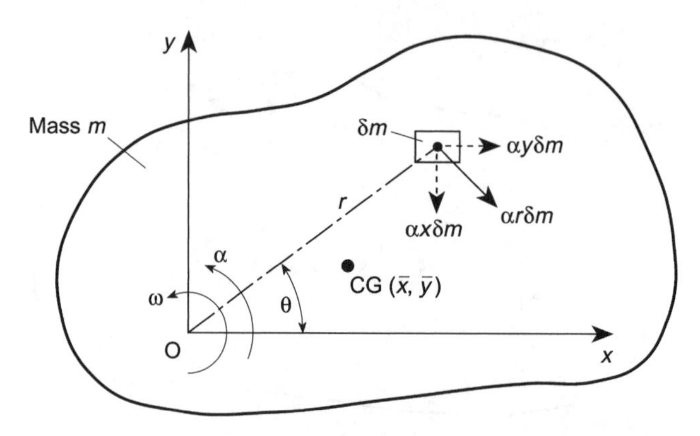

Fig. 14.2 Inertia forces on a rigid mass subjected to an angular acceleration.

and

$$F_y = -\int \alpha x \, dm = -\alpha \int x \, dm$$

for α in the direction shown. Then, as before

$$F_x = \alpha \bar{y} m \tag{14.3}$$

and

$$F_y = \alpha \bar{x} m \tag{14.4}$$

Also, if the CG lies on the x axis, $\bar{y} = 0$ and $F_x = 0$. Similarly, if the CG lies on the y axis, $\bar{x} = 0$ and $F_y = 0$.

The torque about the axis of rotation produced by the inertia force corresponding to the angular acceleration on the element δm is given by

$$\delta T_{\text{O}} = \alpha r^2 \delta m$$

Thus, for the complete mass

$$T_{\text{O}} = \int \alpha r^2 \, dm = \alpha \int r^2 \, dm$$

The integral term in this expression is the moment of inertia, I_{O}, of the mass about the axis of rotation. Thus

$$T_{\text{O}} = \alpha I_{\text{O}} \tag{14.5}$$

Equation (14.5) may be rewritten in terms of I_{CG}, the moment of inertia of the mass about an axis perpendicular to the plane of the mass through the CG. Hence, using the parallel axes theorem

$$I_{\text{O}} = m(\bar{r})^2 + I_{\text{CG}}$$

where $\bar{r}$ is the distance between O and the CG. Then

$$I_{\text{O}} = m[(\bar{x})^2 + (\bar{y})^2] + I_{\text{CG}}$$

and

$$T_{\text{O}} = m[(\bar{x})^2 + (\bar{y})^2]\alpha + I_{\text{CG}}\alpha \tag{14.6}$$

Example 14.1

An aircraft having a total weight of 45 kN lands on the deck of an aircraft carrier and is brought to rest by means of a cable engaged by an arrester hook, as shown in Fig. 14.3. If the deceleration induced by the cable is 3 g determine the tension, T, in the cable, the load on an undercarriage strut and the shear and axial loads in the fuselage at the section AA; the weight of the aircraft aft of AA is 4.5 kN. Calculate also the length of deck covered by the aircraft before it is brought to rest if the touch-down speed is 25 m/s.

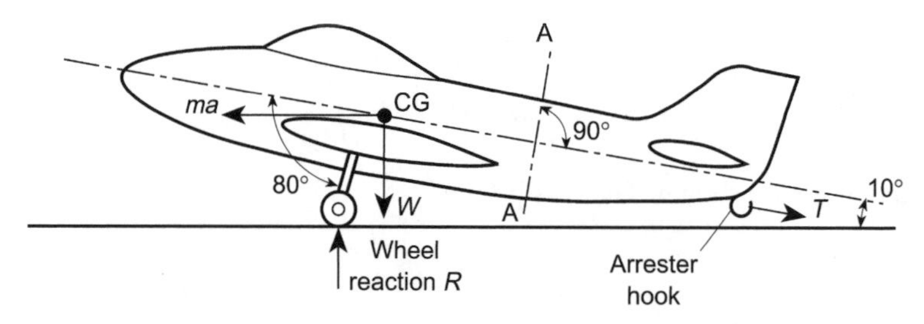

Fig. 14.3 Forces on the aircraft of Example 14.1.

The aircraft is subjected to a horizontal inertia force ma where m is the mass of the aircraft and a its deceleration. Thus, resolving forces horizontally

$$T\cos 10^\circ - ma = 0$$

i.e.

$$T\cos 10^\circ - \frac{45}{g}3\,g = 0$$

which gives

$$T = 137.1\,\text{kN}$$

Now resolving forces vertically

$$R - W - T\sin 10^\circ = 0$$

i.e.

$$R = 45 + 131.1\sin 10^\circ = 68.8\,\text{kN}$$

Assuming two undercarriage struts, the load in each strut will be $(R/2)/\cos 20^\circ = 36.6\,\text{kN}$.

Let N and S be the axial and shear loads at the section AA, as shown in Fig. 14.4. The inertia load acting at the CG of the fuselage aft of AA is $m_1 a$, where m_1 is the mass of the fuselage aft of AA. Then

$$m_1 a = \frac{4.5}{g}3\,g = 13.5\,\text{kN}$$

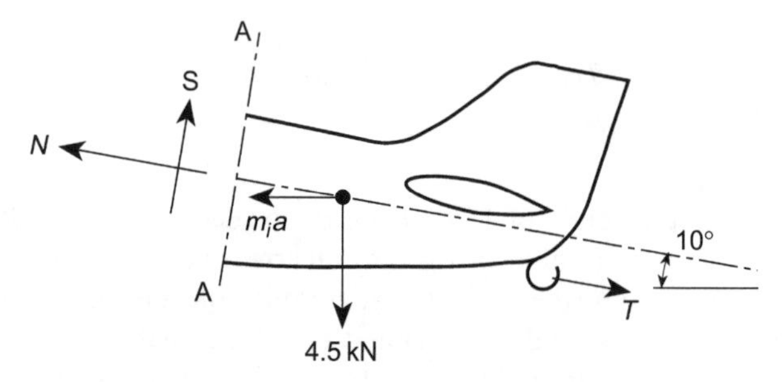

Fig. 14.4 Shear and axial loads at the section AA of the aircraft of Example 14.1.

Resolving forces parallel to the axis of the fuselage

$$N - T + m_1 a \cos 10^\circ - 4.5 \sin 10^\circ = 0$$

i.e.

$$N - 137.1 + 13.5 \cos 10^\circ - 4.5 \sin 10^\circ = 0$$

whence

$$N = 124.6 \, \text{kN}$$

Now resolving forces perpendicular to the axis of the fuselage

$$S - m_1 a \sin 10^\circ - 4.5 \cos 10^\circ = 0$$

i.e.

$$S - 13.5 \sin 10^\circ - 4.5 \cos 10^\circ = 0$$

so that

$$S = 6.8 \, \text{kN}$$

Note that, in addition to the axial load and shear load at the section AA, there will also be a bending moment.

Finally, from elementary dynamics

$$v^2 = v_0^2 + 2as$$

where v_0 is the touchdown speed, v the final speed (=0) and s the length of deck covered. Then

$$v_0^2 = -2as$$

i.e.

$$25^2 = -2(-3 \times 9.81)s$$

which gives

$$s = 10.6 \, \text{m}$$

Example 14.2

An aircraft having a weight of 250 kN and a tricycle undercarriage lands at a vertical velocity of 3.7 m/s, such that the vertical and horizontal reactions on the main wheels are 1200 kN and 400 kN respectively; at this instant the nose wheel is 1.0 m from the ground, as shown in Fig. 14.5. If the moment of inertia of the aircraft about its CG is $5.65 \times 10^8 \, \text{Ns}^2 \, \text{mm}$ determine the inertia forces on the aircraft, the time taken for its vertical velocity to become zero and its angular velocity at this instant.

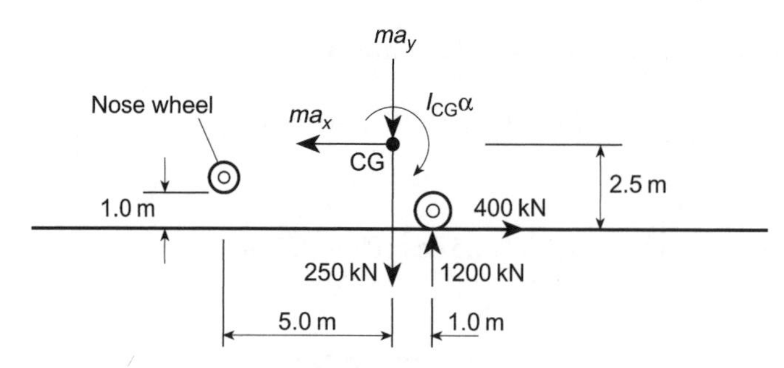

Fig. 14.5 Geometry of the aircraft of Example 14.2.

The horizontal and vertical inertia forces ma_x and ma_y act at the CG, as shown in Fig. 14.5, m is the mass of the aircraft and a_x and a_y its accelerations in the horizontal and vertical directions, respectively. Then, resolving forces horizontally

$$ma_x - 400 = 0$$

whence

$$ma_x = 400\,\text{kN}$$

Now resolving forces vertically

$$ma_y + 250 - 1200 = 0$$

which gives

$$ma_y = 950\,\text{kN}$$

Then

$$a_y = \frac{950}{m} = \frac{950}{250/g} = 3.8\,g \tag{i}$$

Now taking moments about the CG

$$I_{CG}\alpha - 1200 \times 1.0 - 400 \times 2.5 = 0 \tag{ii}$$

from which

$$I_{CG}\alpha = 2200\,\text{m kN}$$

Hence

$$\alpha = \frac{I_{CG}\alpha}{I_{CG}} = \frac{2200 \times 10^6}{5.65 \times 10^8} = 3.9\,\text{rad/s}^2 \tag{iii}$$

From Eq. (i), the aircraft has a vertical deceleration of 3.8 g from an initial vertical velocity of 3.7 m/s. Therefore, from elementary dynamics, the time, t, taken for the vertical velocity to become zero, is given by

$$v = v_0 + a_y t \tag{iv}$$

in which $v = 0$ and $v_0 = 3.7$ m/s. Hence

$$0 = 3.7 - 3.8 \times 9.81t$$

whence

$$t = 0.099\,\text{s}$$

In a similar manner to Eq. (iv) the angular velocity of the aircraft after 0.099 s is given by

$$\omega = \omega_0 + \alpha t$$

in which $\omega_0 = 0$ and $\alpha = 3.9\,\text{rad/s}^2$. Hence

$$\omega = 3.9 \times 0.099$$

i.e.

$$\omega = 0.39\,\text{rad/s}$$

14.2 Symmetric manoeuvre loads

We shall now consider the calculation of aircraft loads corresponding to the flight conditions specified by flight envelopes. There are, in fact, an infinite number of flight conditions within the boundary of the flight envelope although, structurally, those represented by the boundary are the most severe. Furthermore, it is usually found that the corners A, C, D_1, D_2, E and F (see Fig. 13.1) are more critical than points on the boundary between the corners so that, in practice, only the six conditions corresponding to these corner points need be investigated for each flight envelope.

In symmetric manoeuvres we consider the motion of the aircraft initiated by movement of the control surfaces in the plane of symmetry. Examples of such manoeuvres are loops, straight pull-outs and bunts, and the calculations involve the determination of lift, drag and tailplane loads at given flight speeds and altitudes. The effects of atmospheric turbulence and gusts are discussed in Section 14.4.

14.2.1 Level flight

Although steady level flight is not a manoeuvre in the strict sense of the word, it is a useful condition to investigate initially since it establishes points of load application and gives some idea of the equilibrium of an aircraft in the longitudinal plane. The loads acting on an aircraft in steady flight are shown in Fig. 14.6, with the following notation:

L is the lift acting at the aerodynamic centre of the wing.
D is the aircraft drag.
M_0 is the aerodynamic pitching moment of the aircraft *less* its horizontal tail.
P is the horizontal tail load acting at the aerodynamic centre of the tail, usually taken to be at approximately one-third of the tailplane chord.
W is the aircraft weight acting at its CG.
T is the engine thrust, assumed here to act parallel to the direction of flight in order to simplify calculation.

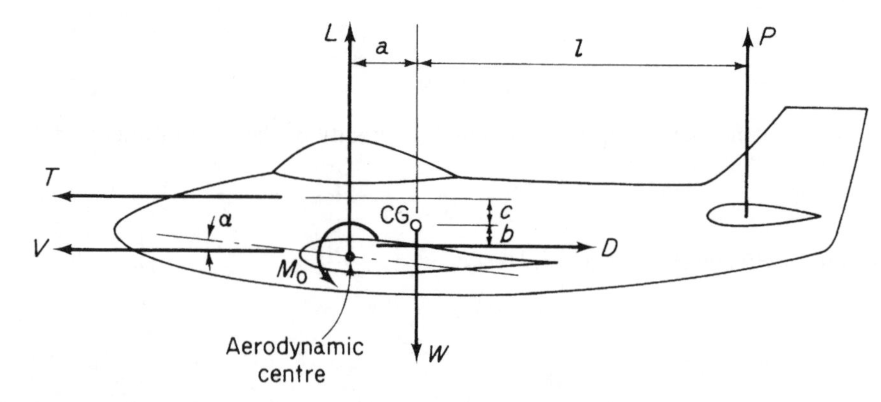

Fig. 14.6 Aircraft loads in level flight.

The loads are in static equilibrium since the aircraft is in a steady, unaccelerated, level flight condition. Thus for vertical equilibrium

$$L + P - W = 0 \tag{14.7}$$

for horizontal equilibrium

$$T - D = 0 \tag{14.8}$$

and taking moments about the aircraft's CG in the plane of symmetry

$$La - Db - Tc - M_0 - Pl = 0 \tag{14.9}$$

For a given aircraft weight, speed and altitude, Eqs (14.7)–(14.9) may be solved for the unknown lift, drag and tail loads. However, other parameters in these equations, such as M_0, depend upon the wing incidence α which in turn is a function of the required wing lift so that, in practice, a method of successive approximation is found to be the most convenient means of solution.

As a first approximation we assume that the tail load P is small compared with the wing lift L so that, from Eq. (14.7), $L \approx W$. From aerodynamic theory with the usual notation

$$L = \tfrac{1}{2}\rho V^2 S C_L$$

Hence

$$\tfrac{1}{2}\rho V^2 S C_L \approx W \tag{14.10}$$

Equation (14.10) gives the approximate lift coefficient C_L and thus (from $C_L{-}\alpha$ curves established by wind tunnel tests) the wing incidence α. The drag load D follows (knowing V and α) and hence we obtain the required engine thrust T from Eq. (14.8). Also M_0, a, b, c and l may be calculated (again since V and α are known) and Eq. (14.9) solved for P. As a second approximation this value of P is substituted in Eq. (14.7) to obtain a more accurate value for L and the procedure is repeated. Usually three approximations are sufficient to produce reasonably accurate values.

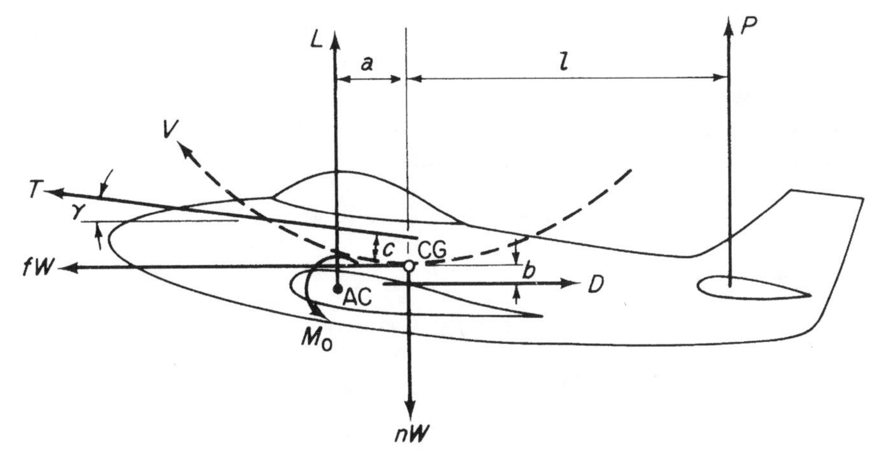

Fig. 14.7 Aircraft loads in a pull-out from a dive.

In most cases P, D and T are small compared with the lift and aircraft weight. Therefore, from Eq. (14.7) $L \approx W$ and substitution in Eq. (14.9) gives, neglecting D and T

$$P \approx W\frac{a}{l} - \frac{M_0}{l} \tag{14.11}$$

We see from Eq. (14.11) that if a is large then P will most likely be positive. In other words the tail load acts upwards when the CG of the aircraft is far aft. When a is small or negative, i.e., a forward CG, then P will probably be negative and act downwards.

14.2.2 General case of a symmetric manoeuvre

In a rapid pull-out from a dive a downward load is applied to the tailplane, causing the aircraft to pitch nose upwards. The downward load is achieved by a backward movement of the control column, thereby applying negative incidence to the elevators, or horizontal tail if the latter is all-moving. If the manoeuvre is carried out rapidly the forward speed of the aircraft remains practically constant so that increases in lift and drag result from the increase in wing incidence only. Since the lift is now greater than that required to balance the aircraft weight the aircraft experiences an upward acceleration normal to its flight path. This normal acceleration combined with the aircraft's speed in the dive results in the curved flight path shown in Fig. 14.7. As the drag load builds up with an increase of incidence the forward speed of the aircraft falls since the thrust is assumed to remain constant during the manoeuvre. It is usual, as we observed in the discussion of the flight envelope, to describe the manoeuvres of an aircraft in terms of a manoeuvring load factor n. For steady level flight $n = 1$, giving 1 g flight, although in fact the acceleration is zero. What is implied in this method of description is that the inertia force on the aircraft in the level flight condition is 1.0 times its weight. It follows that the vertical inertia force on an aircraft carrying out an ng manoeuvre is nW. We may therefore replace the dynamic conditions of the accelerated motion by an equivalent set of static conditions in which the applied loads are in equilibrium with the inertia forces. Thus, in Fig. 14.7, n is the manoeuvre load factor while f is a similar factor giving the horizontal inertia force. Note that the actual normal acceleration in this particular case is $(n - 1)g$.

For vertical equilibrium of the aircraft, we have, referring to Fig. 14.7 where the aircraft is shown at the lowest point of the pull-out

$$L + P + T \sin \gamma - nW = 0 \tag{14.12}$$

For horizontal equilibrium

$$T \cos \gamma + fW - D = 0 \tag{14.13}$$

and for pitching moment equilibrium about the aircraft's CG

$$La - Db - Tc - M_0 - Pl = 0 \tag{14.14}$$

Equation (14.14) contains no terms representing the effect of pitching acceleration of the aircraft; this is assumed to be negligible at this stage.

Again the method of successive approximation is found to be most convenient for the solution of Eqs (14.12)–(14.14). There is, however, a difference to the procedure described for the steady level flight case. The engine thrust T is no longer directly related to the drag D as the latter changes during the manoeuvre. Generally, the thrust is regarded as remaining constant and equal to the value appropriate to conditions before the manoeuvre began.

Example 14.3

The curves C_D, α and $C_{M,CG}$ for a light aircraft are shown in Fig. 14.8(a). The aircraft weight is 8000 N, its wing area 14.5 m^2 and its mean chord 1.35 m. Determine the lift, drag, tail load and forward inertia force for a symmetric manoeuvre corresponding to $n = 4.5$ and a speed of 60 m/s. Assume that engine-off conditions apply and that the air density is 1.223 kg/m^3. Figure 14.8(b) shows the relevant aircraft dimensions.

As a first approximation we neglect the tail load P. Therefore, from Eq. (14.12), since $T = 0$, we have

$$L \approx nW \tag{i}$$

Hence

$$C_L = \frac{L}{\frac{1}{2}\rho V^2 S} \approx \frac{4.5 \times 8000}{\frac{1}{2} \times 1.223 \times 60^2 \times 14.5} = 1.113$$

From Fig. 14.8(a), $\alpha = 13.75°$ and $C_{M,CG} = 0.075$. The tail arm l, from Fig. 14.8(b), is

$$l = 4.18 \cos(\alpha - 2) + 0.31 \sin(\alpha - 2) \tag{ii}$$

Substituting the above value of α gives $l = 4.123$ m. In Eq. (14.14) the terms $La - Db - M_0$ are equivalent to the aircraft pitching moment M_{CG} about its CG. Eq. (14.14) may therefore be written

$$M_{CG} - Pl = 0$$

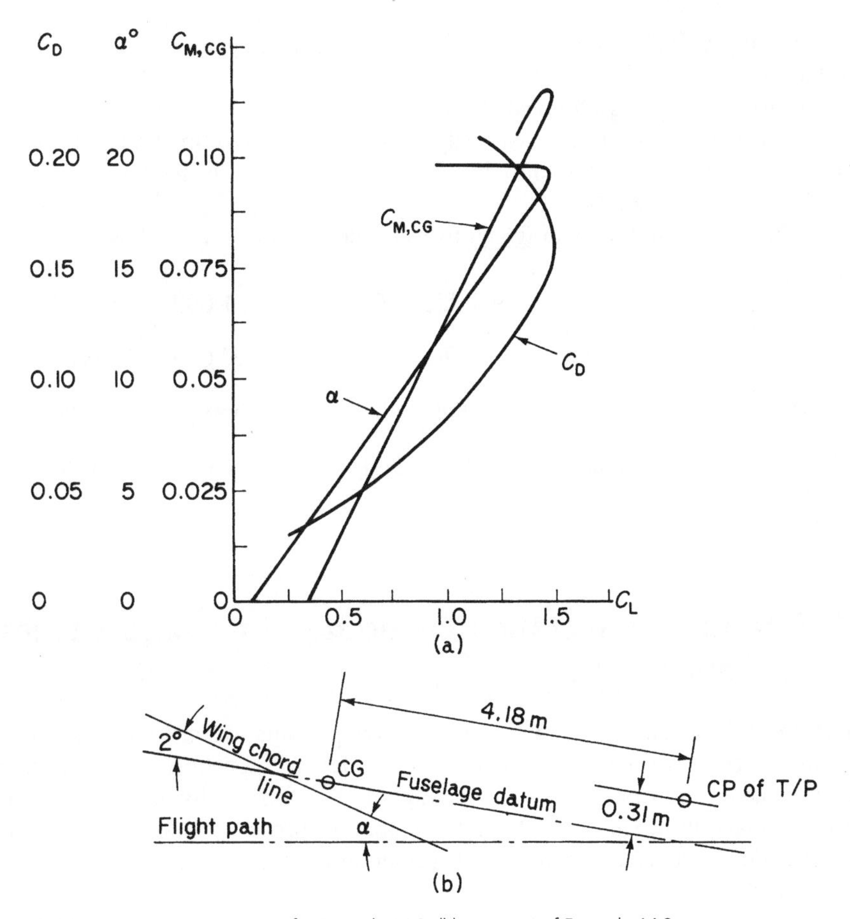

Fig. 14.8 (a) C_D, α, $C_{M,CG} - C_L$ curves for Example 14.3; (b) geometry of Example 14.3.

or

$$Pl = \tfrac{1}{2}\rho V^2 Sc C_{M,CG} \tag{iii}$$

where c = wing mean chord. Substituting P from Eq. (iii) into Eq. (14.12) we have

$$L + \frac{\tfrac{1}{2}\rho V^2 Sc C_{M,CG}}{l} = nW$$

or dividing through by $\tfrac{1}{2}\rho V^2 S$

$$C_L + \frac{c}{l} C_{M,CG} = \frac{nW}{\tfrac{1}{2}\rho V^2 S} \tag{iv}$$

We now obtain a more accurate value for C_L from Eq. (iv)

$$C_L = 1.113 - \frac{1.35}{4.123} \times 0.075 = 1.088$$

giving $\alpha = 13.3°$ and $C_{M,CG} = 0.073$.

Substituting this value of α into Eq. (ii) gives a second approximation for l, namely $l = 4.161$ m.

Equation (iv) now gives a third approximation for C_L, i.e. $C_L = 1.099$. Since the three calculated values of C_L are all extremely close further approximations will not give values of C_L very much different to those above. Therefore, we shall take $C_L = 1.099$. From Fig. 14.8(a) $C_D = 0.0875$.

The values of lift, tail load, drag and forward inertia force then follow:

$$\text{Lift } L = \tfrac{1}{2}\rho V^2 S C_L = \tfrac{1}{2} \times 1.223 \times 60^2 \times 14.5 \times 1.099 = 35\,000\,\text{N}$$

$$\text{Tail load } P = nW - L = 4.5 \times 8000 - 35\,000 = 1000\,\text{N}$$

$$\text{Drag } D = \tfrac{1}{2}\rho V^2 S C_D = \tfrac{1}{2} \times 1.223 \times 60^2 \times 14.5 \times 0.0875 = 2790\,\text{N}$$

$$\text{Forward inertia force } fW = D \text{ (From Eq. (14.13))} = 2790\,\text{N}$$

14.3 Normal accelerations associated with various types of manoeuvre

In Section 14.2 we determined aircraft loads corresponding to a given manoeuvre load factor n. Clearly it is necessary to relate this load factor to given types of manoeuvre. Two cases arise: the first involving a steady pull-out from a dive and the second, a correctly banked turn. Although the latter is not a symmetric manoeuvre in the strict sense of the word, it gives rise to normal accelerations in the plane of symmetry and is therefore included.

14.3.1 Steady pull-out

Let us suppose that the aircraft has just begun its pull-out from a dive so that it is describing a curved flight path but is not yet at its lowest point. The loads acting on the aircraft at this stage of the manoeuvre are shown in Fig. 14.9, where R is the radius of curvature of the flight path. In this case the lift vector must equilibrate the normal (to the flight path) component of the aircraft weight and provide the force producing the centripetal acceleration V^2/R of the aircraft towards the centre of curvature of the flight path. Thus

$$L = \frac{WV^2}{gR} + W\cos\theta$$

or, since $L = nW$ (see Section 14.2)

$$n = \frac{V^2}{gR} + \cos\theta \qquad (14.15)$$

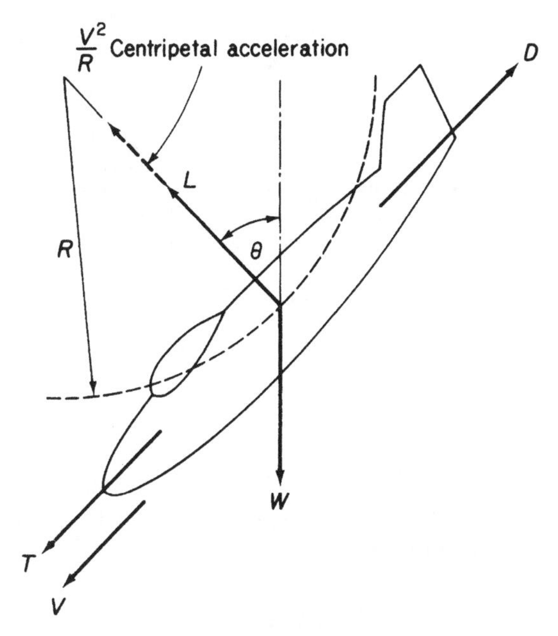

Fig. 14.9 Aircraft loads and acceleration during a steady pull-out.

At the lowest point of the pull-out, $\theta = 0$, and

$$n = \frac{V^2}{gR} + 1 \tag{14.16}$$

We see from either Eq. (14.15) or Eq. (14.16) that the smaller the radius of the flight path, that is the more severe the pull-out, the greater the value of n. It is quite possible therefore for a severe pull-out to overstress the aircraft by subjecting it to loads which lie outside the flight envelope and which may even exceed the proof or ultimate loads. In practice, the control surface movement may be limited by stops incorporated in the control circuit. These stops usually operate only above a certain speed giving the aircraft adequate manoeuvrability at lower speeds. For hydraulically operated controls 'artificial feel' is built in to the system whereby the stick force increases progressively as the speed increases; a necessary precaution in this type of system since the pilot is merely opening and closing valves in the control circuit and therefore receives no direct physical indication of control surface forces.

Alternatively, at low speeds, a severe pull-out or pull-up may stall the aircraft. Again safety precautions are usually incorporated in the form of stall warning devices since, for modern high speed aircraft, a stall can be disastrous, particularly at low altitude.

14.3.2 Correctly banked turn

In this manoeuvre the aircraft flies in a horizontal turn with no sideslip at constant speed. If the radius of the turn is R and the angle of bank ϕ, then the forces acting on

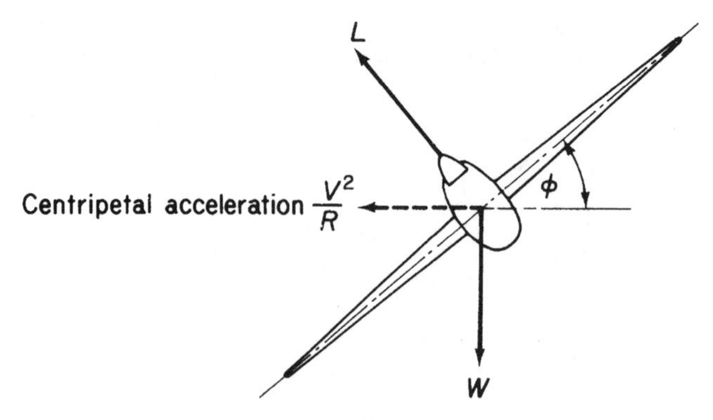

Fig. 14.10 Correctly banked turn.

the aircraft are those shown in Fig. 14.10. The horizontal component of the lift vector in this case provides the force necessary to produce the centripetal acceleration of the aircraft towards the centre of the turn. Then

$$L \sin\phi = \frac{WV^2}{gR} \tag{14.17}$$

and for vertical equilibrium

$$L \cos\phi = W \tag{14.18}$$

or

$$L = W \sec\phi \tag{14.19}$$

From Eq. (14.19) we see that the load factor n in the turn is given by

$$n = \sec\phi \tag{14.20}$$

Also, dividing Eq. (14.17) by Eq. (14.18)

$$\tan\phi = \frac{V^2}{gR} \tag{14.21}$$

Examination of Eq. (14.21) reveals that the tighter the turn the greater the angle of bank required to maintain horizontal flight. Furthermore, we see from Eq. (14.20) that an increase in bank angle results in an increased load factor. Aerodynamic theory shows that for a limiting value of n the minimum time taken to turn through a given angle at a given value of engine thrust occurs when the lift coefficient C_L is a maximum; that is, with the aircraft on the point of stalling.

14.4 Gust loads

In Section 14.2 we considered aircraft loads resulting from prescribed manoeuvres in the longitudinal plane of symmetry. Other types of in-flight load are caused by

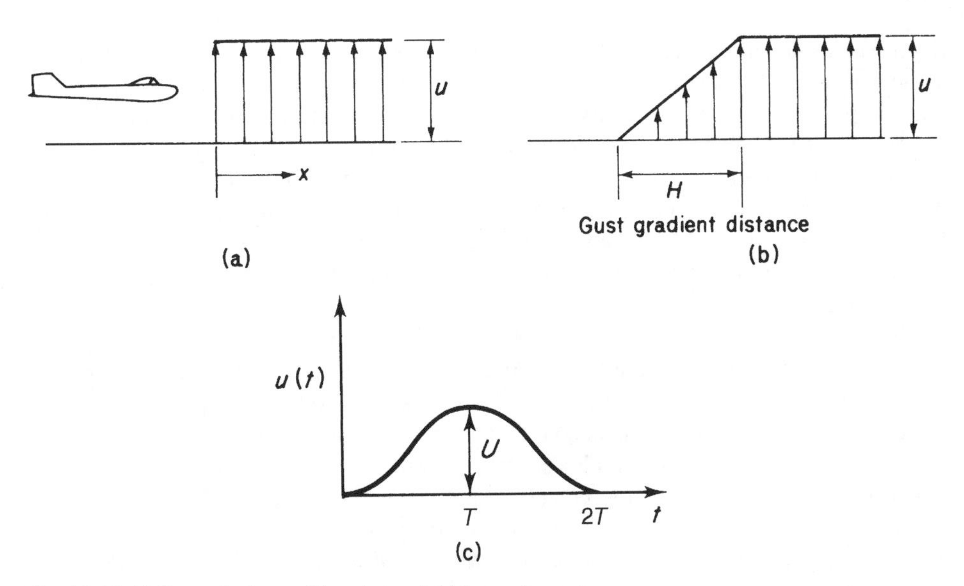

Fig. 14.11 (a) Sharp-edged gust; (b) graded gust; (c) 1 – cosine gust.

air turbulence. The movements of the air in turbulence are generally known as gusts and produce changes in wing incidence, thereby subjecting the aircraft to sudden or gradual increases or decreases in lift from which normal accelerations result. These may be critical for large, high speed aircraft and may possibly cause higher loads than control initiated manoeuvres.

At the present time two approaches are employed in gust analysis. One method, which has been in use for a considerable number of years, determines the aircraft response and loads due to a single or 'discrete' gust of a given profile. This profile is defined as a distribution of vertical gust velocity over a given finite length or given period of time. Examples of these profiles are shown in Fig. 14.11.

Early airworthiness requirements specified an instantaneous application of gust velocity u, resulting in the 'sharp-edged' gust of Fig. 14.11(a). Calculations of normal acceleration and aircraft response were based on the assumptions that the aircraft's flight is undisturbed while the aircraft passes from still air into the moving air of the gust and during the time taken for the gust loads to build up; that the aerodynamic forces on the aircraft are determined by the instantaneous incidence of the particular lifting surface and finally that the aircraft's structure is rigid. The second assumption here relating the aerodynamic force on a lifting surface to its instantaneous incidence neglects the fact that in a disturbance such as a gust there is a gradual growth of circulation and hence of lift to a steady state value (Wagner effect). This in general leads to an overestimation of the upward acceleration of an aircraft and therefore of gust loads.

The 'sharp-edged' gust was replaced when it was realized that the gust velocity built up to a maximum over a period of time. Airworthiness requirements were modified on the assumption that the gust velocity increased linearly to a maximum value over a specified gust gradient distance H. Hence the 'graded' gust of Fig. 14.11(b). In the UK, H is taken as 30.5 m. Since, as far as the aircraft is concerned, the gust velocity builds up to a maximum over a period of time it is no longer allowable to ignore the change of flight path as the aircraft enters the gust. By the time the gust has attained its maximum value the aircraft has developed a vertical component of velocity and, in addition, may be pitching depending on its longitudinal stability characteristics. The effect of the former is to reduce the severity of the gust while the latter may either increase or decrease the loads involved. To evaluate the corresponding gust loads the designer may either calculate the complete motion of the aircraft during the disturbance and hence obtain the gust loads, or replace the 'graded' gust by an equivalent 'sharp-edged' gust producing approximately the same effect. We shall discuss the latter procedure in greater detail later.

The calculation of the complete response of the aircraft to a 'graded' gust may be obtained from its response to a 'sharp-edged' or 'step' gust, by treating the former as comprising a large number of small 'steps' and superimposing the responses to each of these. Such a process is known as convolution or Duhamel integration. This treatment is desirable for large or unorthodox aircraft where aeroelastic (structural flexibility) effects on gust loads may be appreciable or unknown. In such cases the assumption of a rigid aircraft may lead to an underestimation of gust loads. The equations of motion are therefore modified to allow for aeroelastic in addition to aerodynamic effects. For small and medium-sized aircraft having orthodox aerodynamic features the equivalent 'sharp-edged' gust procedure is satisfactory.

While the 'graded' or 'ramp' gust is used as a basis for gust load calculations, other shapes of gust profile are in current use. Typical of these is the '$1 - $cosine' gust of Fig. 14.11(c), where the gust velocity u is given by $u(t) = (U/2)[1 - \cos(\pi t/T)]$. Again the aircraft response is determined by superimposing the responses to each of a large number of small steps.

Although the 'discrete' gust approach still finds widespread use in the calculation of gust loads, alternative methods based on *power spectral* analysis are being investigated. The advantage of the power spectral technique lies in its freedom from arbitrary assumptions of gust shapes and sizes. It is assumed that gust velocity is a random variable which may be regarded for analysis as consisting of a large number of sinusoidal components whose amplitudes vary with frequency. The *power spectrum* of such a function is then defined as the distribution of energy over the frequency range. This may then be related to gust velocity. To establish appropriate amplitude and frequency distributions for a particular random gust profile requires a large amount of experimental data. The collection of such data has been previously referred to in Section 13.2.

Calculations of the complete response of an aircraft and detailed assessments of the 'discrete' gust and power spectral methods of analysis are outside the scope of this book. More information may be found in Refs [1–4] at the end of the chapter. Our present analysis is confined to the 'discrete' gust approach, in which we consider the 'sharp-edged' gust and the equivalent 'sharp-edged' gust derived from the 'graded' gust.

14.4.1 'Sharp-edged' gust

The simplifying assumptions introduced in the determination of gust loads resulting from the 'sharp-edged' gust, have been discussed in the earlier part of this section.

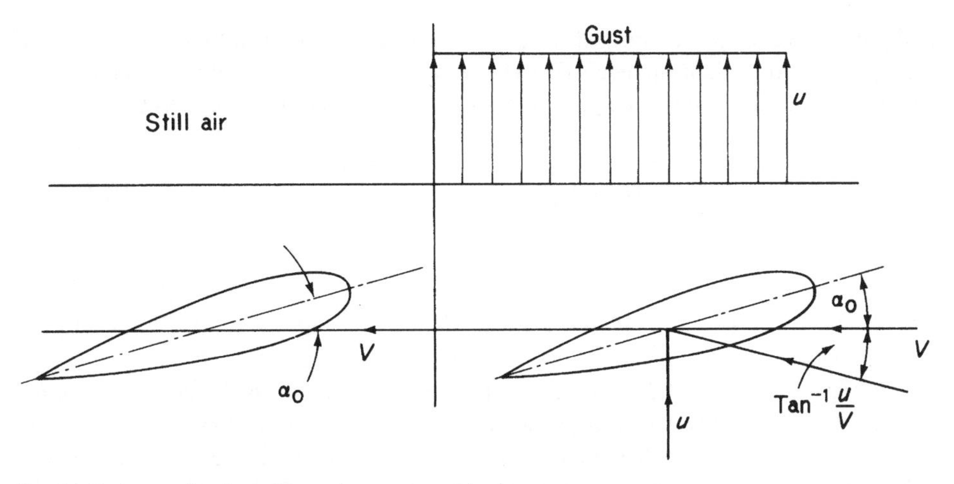

Fig. 14.12 Increase in wing incidence due to a sharp-edged gust.

In Fig. 14.12 the aircraft is flying at a speed V with wing incidence α_0 in still air. After entering the gust of upward velocity u, the incidence increases by an amount $\tan^{-1} u/V$, or since u is usually small compared with V, u/V. This is accompanied by an increase in aircraft speed from V to $(V^2 + u^2)^{\frac{1}{2}}$, but again this increase is neglected since u is small. The increase in wing lift ΔL is then given by

$$\Delta L = \tfrac{1}{2}\rho V^2 S \frac{\partial C_\mathrm{L}}{\partial \alpha}\frac{u}{V} = \frac{1}{2}\rho V S \frac{\partial C_\mathrm{L}}{\partial \alpha} u \tag{14.22}$$

where $\partial C_\mathrm{L}/\partial \alpha$ is the wing lift–curve slope. Neglecting the change of lift on the tailplane as a first approximation, the gust load factor Δn produced by this change of lift is

$$\Delta n = \frac{\frac{1}{2}\rho V S(\partial C_\mathrm{L}/\partial \alpha)u}{W} \tag{14.23}$$

where W is the aircraft weight. Expressing Eq. (14.23) in terms of the wing loading, $w = W/S$, we have

$$\Delta n = \frac{\frac{1}{2}\rho V(\partial C_\mathrm{L}/\partial \alpha)u}{w} \tag{14.24}$$

This increment in gust load factor is additional to the steady level flight value $n = 1$. Therefore, as a result of the gust, the total gust load factor is

$$n = 1 + \frac{\frac{1}{2}\rho V(\partial C_\mathrm{L}/\partial \alpha)u}{w} \tag{14.25}$$

Similarly, for a downgust

$$n = 1 - \frac{\frac{1}{2}\rho V(\partial C_\mathrm{L}/\partial \alpha)u}{w} \tag{14.26}$$

If flight conditions are expressed in terms of equivalent sea-level conditions then V becomes the equivalent airspeed (EAS), V_E, u becomes u_E and the air density ρ is replaced by the sea-level value ρ_0. Equations (14.25) and (14.26) are written

$$n = 1 + \frac{\frac{1}{2}\rho_0 V_E(\partial C_L/\partial\alpha)u_E}{w} \tag{14.27}$$

and

$$n = 1 - \frac{\frac{1}{2}\rho_0 V_E(\partial C_L/\partial\alpha)u_E}{w} \tag{14.28}$$

We observe from Eqs (14.25)–(14.28) that the gust load factor is directly proportional to aircraft speed but inversely proportional to wing loading. It follows that high speed aircraft with low or moderate wing loadings are most likely to be affected by gust loads.

The contribution to normal acceleration of the change in tail load produced by the gust may be calculated using the same assumptions as before. However, the change in tailplane incidence is not equal to the change in wing incidence due to downwash effects at the tail. Thus if ΔP is the increase (or decrease) in tailplane load, then

$$\Delta P = \tfrac{1}{2}\rho_0 V_E^2 S_T \Delta C_{L,T} \tag{14.29}$$

where S_T is the tailplane area and $\Delta C_{L,T}$ the increment of tailplane lift coefficient given by

$$\Delta C_{L,T} = \frac{\partial C_{L,T}}{\partial\alpha}\frac{u_E}{V_E} \tag{14.30}$$

in which $\partial C_{L,T}/\partial\alpha$ is the rate of change of tailplane lift coefficient with wing incidence. From aerodynamic theory

$$\frac{\partial C_{L,T}}{\partial\alpha} = \frac{\partial C_{L,T}}{\partial\alpha_T}\left(1 - \frac{\partial\varepsilon}{\partial\alpha}\right)$$

where $\partial C_{L,T}/\partial\alpha_T$ is the rate of change of $C_{L,T}$ with tailplane incidence and $\partial\varepsilon/\partial\alpha$ the rate of change of downwash angle with wing incidence. Substituting for $\Delta C_{L,T}$ from Eq. (14.30) into Eq. (14.29), we have

$$\Delta P = \tfrac{1}{2}\rho_0 V_E S_T \frac{\partial C_{L,T}}{\partial\alpha} u_E \tag{14.31}$$

For positive increments of wing lift and tailplane load

$$\Delta n W = \Delta L + \Delta P$$

or, from Eqs (14.27) and (14.31)

$$\Delta n = \frac{\frac{1}{2}\rho_0 V_E(\partial C_L/\partial\alpha)u_E}{w}\left(1 + \frac{S_T}{S}\frac{\partial C_{L,T}/\partial\alpha}{\partial C_L/\partial\alpha}\right) \tag{14.32}$$

14.4.2 The 'graded' gust

The 'graded' gust of Fig. 14.11(b) may be converted to an equivalent 'sharp-edged' gust by multiplying the maximum velocity in the gust by a *gust alleviation factor, F*. Equation (14.27) then becomes

$$n = 1 + \frac{\frac{1}{2}\rho_0 V_E(\partial C_L/\partial\alpha)Fu_E}{w} \tag{14.33}$$

Similar modifications are carried out on Eqs (14.25), (14.26), (14.28) and (14.32). The gust alleviation factor allows for some of the dynamic properties of the aircraft, including unsteady lift, and has been calculated taking into account the heaving motion (i.e. the up and down motion with zero rate of pitch) of the aircraft only.[5]

Horizontal gusts cause lateral loads on the vertical tail or fin. Their magnitudes may be calculated in an identical manner to those above, except that areas and values of lift curve slope are referred to the vertical tail. Also, the gust alleviation factor in the 'graded' gust case becomes F_1 and includes allowances for the aerodynamic yawing moment produced by the gust and the yawing inertia of the aircraft.

14.4.3 Gust envelope

Airworthiness requirements usually specify that gust loads shall be calculated at certain combinations of gust and flight speed. The equations for gust load factor in the above analysis show that n is proportional to aircraft speed for a given gust velocity. Therefore, we may plot a gust envelope similar to the flight envelope of Fig. 13.1, as shown in Fig. 14.13. The gust speeds $\pm U_1$, $\pm U_2$ and $\pm U_3$ are high, medium and low velocity gusts, respectively. Cut-offs occur at points where the lines corresponding to each gust

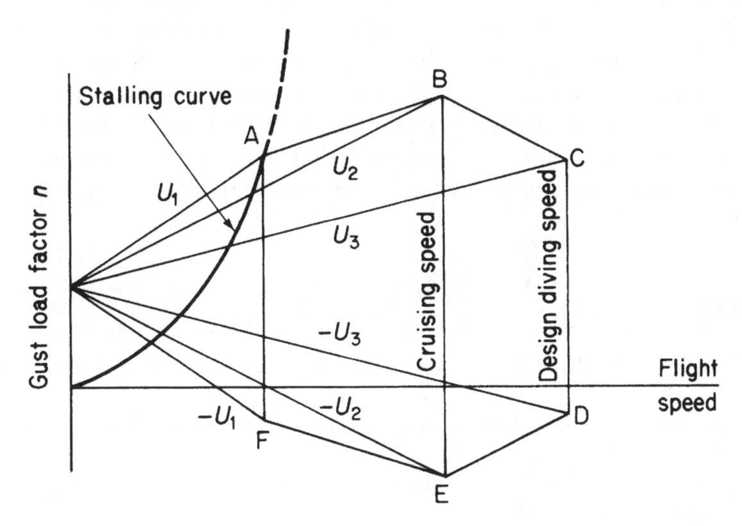

Fig. 14.13 Typical gust envelope.

velocity meet specific aircraft speeds. For example, A and F denote speeds at which a gust of velocity $\pm U_1$ would stall the wing.

The lift coefficient–incidence curve is, as we noted in connection with the flight envelope, affected by compressibility and therefore altitude so that a series of gust envelopes should be drawn for different altitudes. An additional variable in the equations for gust load factor is the wing loading w. Further gust envelopes should therefore be drawn to represent different conditions of aircraft loading.

Typical values of U_1, U_2 and U_3 are 20 m/s, 15.25 m/s and 7.5 m/s. It can be seen from the gust envelope that the maximum gust load factor occurs at the cruising speed V_C. If this value of n exceeds that for the corresponding flight envelope case, that is n_1, then the gust case will be the most critical in the cruise. Let us consider a civil, non-aerobatic aircraft for which $n_1 = 2.5$, $w = 2400\,\text{N/m}^2$ and $\partial C_L/\partial\alpha = 5.0/\text{rad}$. Taking $F = 0.715$ we have, from Eq. (14.33)

$$n = 1 + \frac{\frac{1}{2} \times 1.223\,V_C \times 5.0 \times 0.715 \times 15.25}{2400}$$

giving $n = 1 + 0.0139V_C$, where the cruising speed V_C is expressed as an EAS. For the gust case to be critical

$$1 + 0.0139\,V_C > 2.5$$

or

$$V_C > 108\,\text{m/s}$$

Thus, for civil aircraft of this type having cruising speeds in excess of 108 m/s, the gust case is the most critical. This would, in fact, apply to most modern civil airliners.

Although the same combination of V and n in the flight and gust envelopes will produce the same total lift on an aircraft, the individual wing and tailplane loads will be different, as shown previously (see the derivation of Eq. (14.33)). This situation can be important for aircraft such as the Airbus, which has a large tailplane and a CG forward of the aerodynamic centre. In the flight envelope case the tail load is downwards whereas in the gust case it is upwards; clearly there will be a significant difference in wing load.

The transference of manoeuvre and gust loads into bending, shear and torsional loads on wings, fuselage and tailplanes has been discussed in Section 12.1. Further loads arise from aileron application, in undercarriages during landing, on engine mountings and during crash landings. Analysis and discussion of these may be found in Ref. [6].

References

1 Zbrozek, J. K., Atmospheric gusts – present state of the art and further research, *J. Roy. Aero. Soc.*, January 1965.

2 Cox, R. A., A comparative study of aircraft gust analysis procedures, *J. Roy. Aero. Soc.*, October 1970.

3 Bisplinghoff, R. L., Ashley, H. and Halfman, R. L., *Aeroelasticity*, Addison-Wesley Publishing Co. Inc., Cambridge, Mass., 1955.

4 Babister, A. W., *Aircraft Stability and Control*, Pergamon Press, London, 1961.

5 Zbrozek, J. K., *Gust Alleviation Factor*, R. and M. No. 2970, May 1953.
6 *Handbook of Aeronautics No. 1. Structural Principles and Data*, 4th edition, The Royal Aeronautical Society, 1952.

Problems

P.14.1 The aircraft shown in Fig. P. 14.1(a) weighs 135 kN and has landed such that at the instant of impact the ground reaction on each main undercarriage wheel is 200 kN and its vertical velocity is 3.5 m/s.

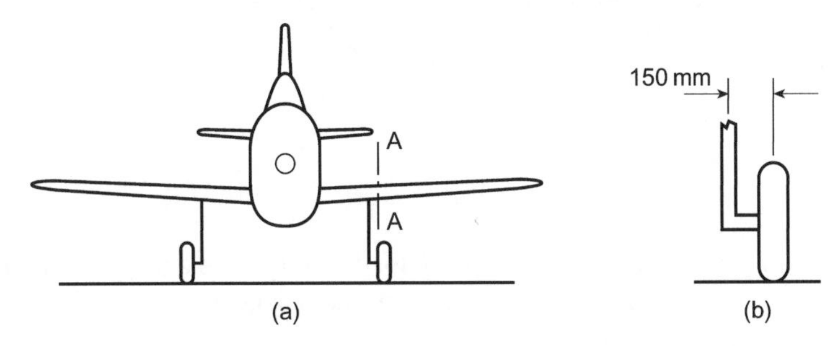

Fig. P.14.1

If each undercarriage wheel weighs 2.25 kN and is attached to an oleo strut, as shown in Fig. P.8.1(b), calculate the axial load and bending moment in the strut; the strut may be assumed to be vertical. Determine also the shortening of the strut when the vertical velocity of the aircraft is zero.

Finally, calculate the shear force and bending moment in the wing at the section AA if the wing, outboard of this section, weighs 6.6 kN and has its CG 3.05 m from AA.

Ans. 193.3 kN, 29.0 kN m (clockwise); 0.32 m; 19.5 kN, 59.6 kN m (anticlockwise).

P.14.2 Determine, for the aircraft of Example 14.2, the vertical velocity of the nose wheel when it hits the ground.

Ans. 3.1 m/s.

P.14.3 Figure P.14.3 shows the flight envelope at sea-level for an aircraft of wing span 27.5 m, average wing chord 3.05 m and total weight 196 000 N. The aerodynamic centre is 0.915 m forward of the CG and the centre of lift for the tail unit is 16.7 m aft of the CG. The pitching moment coefficient is

$$C_{M,0} = -0.0638 \text{ (nose-up positive)}$$

both $C_{M,0}$ and the position of the aerodynamic centre are specified for the complete aircraft less tail unit.

For steady cruising flight at sea-level the fuselage bending moment at the CG is 600 000 Nm. Calculate the maximum value of this bending moment for the given flight envelope. For this purpose it may be assumed that the aerodynamic loadings on the

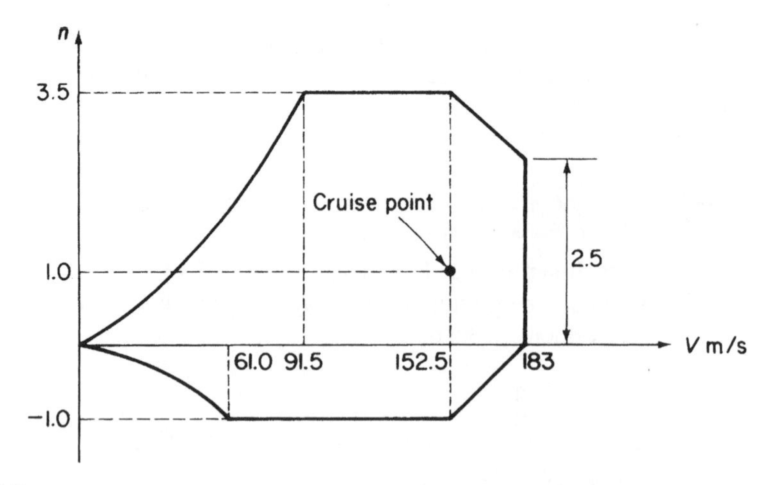

Fig. P.14.3

fuselage itself can be neglected, i.e. the only loads on the fuselage structure aft of the CG are those due to the tail lift and the inertia of the fuselage.

Ans. 1 549 500 N m at $n = 3.5$, $V = 152.5$ m/s.

P.14.4 An aircraft weighing 238 000 N has wings 88.5 m^2 in area for which $C_D = 0.0075 + 0.045C_L^2$ The extra-to-wing drag coefficient based on wing area is 0.0128 and the pitching moment coefficient for all parts excluding the tailplane about an axis through the CG is given by $C_M \cdot c = (0.427C_L - 0.061)$m. The radius from the CG to the line of action of the tail lift may be taken as constant at 12.2 m. The moment of inertia of the aircraft for pitching is 204 000 kg m^2.

During a pull-out from a dive with zero thrust at 215 m/s EAS when the flight path is at 40° to the horizontal with a radius of curvature of 1525 m, the angular velocity of pitch is checked by applying a retardation of 0.25 rad/s^2. Calculate the manoeuvre load factor both at the CG and at the tailplane CP, the forward inertia coefficient and the tail lift.

Ans. $n = 3.78$(CG), $n = 5.19$ at TP, $f = -0.370$, $P = 18\,925$ N.

P.14.5 An aircraft flies at sea level in a correctly banked turn of radius 610 m at a speed of 168 m/s. Figure P.14.5 shows the relative positions of the CG, aerodynamic centre of the complete aircraft less tailplane and the tailplane centre of pressure for the aircraft at zero lift incidence.

Calculate the tail load necessary for equilibrium in the turn. The necessary data are given in the usual notation as follows:

$$\text{Weight } W = 133\,500\text{ N} \qquad dC_L/d\alpha = 4.5/\text{rad}$$
$$\text{Wing area } S = 46.5\text{ m}^2 \qquad C_D = 0.01 + 0.05C_L^2$$
$$\text{Wing mean chord } \bar{c} = 3.0\text{ m} \qquad C_{M,0} = -0.03$$

Ans. 73 160 N

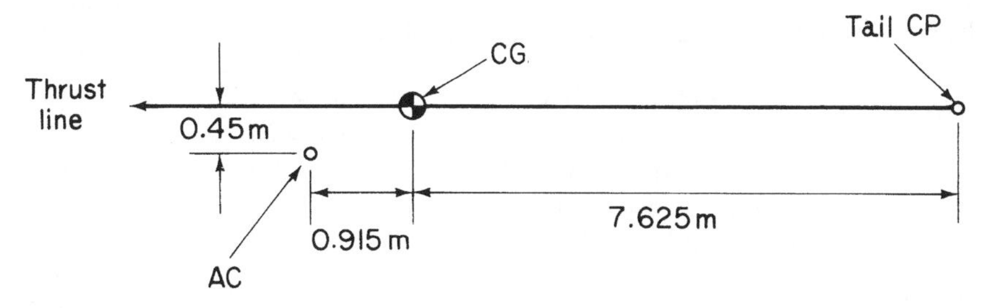

Fig. P.14.5

P.14.6 The aircraft for which the stalling speed V_s in level flight is 46.5 m/s has a maximum allowable manoeuvre load factor n_1 of 4.0. In assessing gyroscopic effects on the engine mounting the following two cases are to be considered:

(a) Pull-out at maximum permissible rate from a dive in symmetric flight, the angle of the flight path to the horizontal being limited to 60° for this aircraft.

(b) Steady, correctly banked turn at the maximum permissible rate in horizontal flight.

Find the corresponding maximum angular velocities in yaw and pitch.

Ans. (a) Pitch, 0.37 rad/s, (b) Pitch, 0.41 rad/s, Yaw, 0.103 rad/s.

P.14.7 A tail-first supersonic airliner, whose essential geometry is shown in Fig. P.14.7, flies at 610 m/s true airspeed at an altitude of 18 300 m. Assuming that thrust and drag forces act in the same straight line, calculate the tail lift in steady straight and level flight.

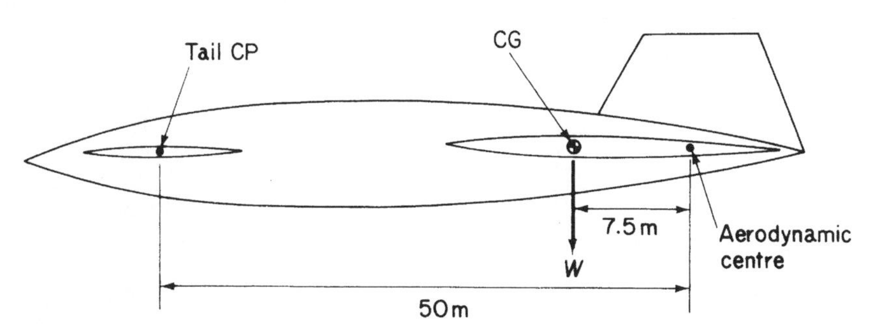

Fig. P.14.7

If, at the same altitude, the aircraft encounters a sharp-edged vertical up-gust of 18 m/s true airspeed, calculate the changes in the lift and tail load and also the resultant load factor n.

The relevant data in the usual notation are as follows:

$$\text{Wing: } S = 280\,\text{m}^2, \qquad \partial C_L/\partial\alpha = 1.5$$
$$\text{Tail: } S_T = 28\,\text{m}^2, \qquad \partial C_{L,T}/\partial\alpha = 2.0$$
$$\text{Weight } W = 1\,600\,000\,\text{N}$$
$$C_{M,0} = -0.01$$
$$\text{Mean chord } \bar{c} = 22.8\,\text{m}$$

At 18 300 m

$$\rho = 0.116\,\text{kg/m}^3$$

Ans. $P = 267\,852\,\text{N}$, $\Delta P = 36\,257\,\text{N}$, $\Delta L = 271\,931\,\text{N}$, $n = 1.19$

P.14.8 An aircraft of all up weight 145 000 N has wings of area 50 m^2 and mean chord 2.5 m. For the whole aircraft $C_D = 0.021 + 0.041C_L^2$, for the wings $\mathrm{d}C_L/\mathrm{d}\alpha = 4.8$, for the tailplane of area 9.0 m^2, $\mathrm{d}C_{L,T}/\mathrm{d}\alpha = 2.2$ allowing for the effects of downwash, and the pitching moment coefficient about the aerodynamic centre (of complete aircraft less tailplane) based on wing area is $C_{M,0} = -0.032$. Geometric data are given in Fig. P.14.8.

During a steady glide with zero thrust at 250 m/s EAS in which $C_L = 0.08$, the aircraft meets a downgust of equivalent 'sharp-edged' speed 6 m/s. Calculate the tail load, the gust load factor and the forward inertia force, $\rho_0 = 1.223\,\text{kg/m}^3$.

Ans. $P = -28\,902\,\text{N}$ (down), $n = -0.64$, forward inertia force $= 40\,703\,\text{N}$.

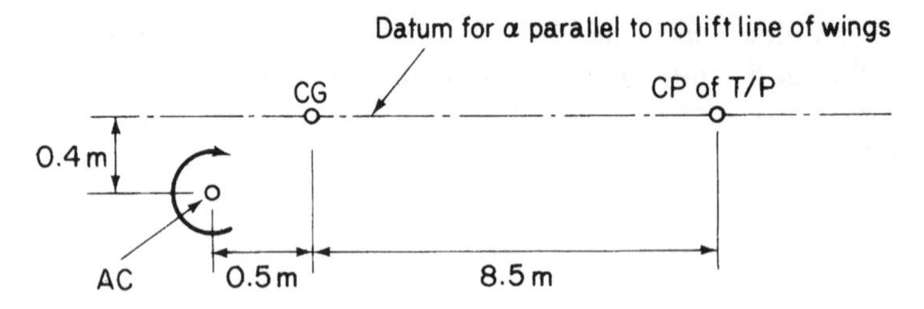

Fig. P.14.8

15

Fatigue

Fatigue has been discussed briefly in Section 11.7 when we examined the properties of materials and also in Section 13.4 as part of the chapter on airworthiness. We shall now look at fatigue in greater detail and consider factors affecting the life of an aircraft including safe life and fail safe structures, designing against fatigue, the fatigue strength of components, the prediction of aircraft fatigue life and crack propagation.

Fatigue is defined as the progressive deterioration of the strength of a material or structural component during service such that failure can occur at much lower stress levels than the ultimate stress level. As we have seen, fatigue is a dynamic phenomenon which initiates small (micro) cracks in the material or component and causes them to grow into large (macro) cracks; these, if not detected, can result in catastrophic failure.

Fatigue damage can be produced in a variety of ways. *Cyclic fatigue* is caused by repeated fluctuating loads. *Corrosion fatigue* is fatigue accelerated by surface corrosion of the material penetrating inwards so that the material strength deteriorates. Small-scale rubbing movements and abrasion of adjacent parts cause *fretting fatigue*, while *thermal fatigue* is produced by stress fluctuations induced by thermal expansions and contractions; the latter does not include the effect on material strength of heat. Finally, high frequency stress fluctuations, due to vibrations excited by jet or propeller noise, cause *sonic* or *acoustic fatigue*.

Clearly an aircraft's structure must be designed so that fatigue does not become a problem. For aircraft in general, the requirements that the strength of an aircraft throughout its operational life shall be such as to ensure that the possibility of a disastrous fatigue failure shall be extremely remote (i.e. the probability of failure is less than 10^{-7}) under the action of the repeated loads of variable magnitude expected in service. Also it is required that the principal parts of the primary structure of the aircraft be subjected to a detailed analysis and to load tests which demonstrate a *safe life*, or that the parts of the primary structure have *fail-safe* characteristics. These requirements do not apply to light aircraft provided that zinc-rich aluminium alloys are not used in their construction and that wing stress levels are kept low, i.e. provided that a 3.05 m/s upgust causes no greater stress than 14 N/mm^2.

15.1 Safe life and fail-safe structures

The danger of a catastrophic fatigue failure in the structure of an aircraft may be eliminated completely or may become extremely remote if the structure is designed to have a safe life or to be fail-safe. In the former approach, the structure is designed to have a minimum life during which it is known that no catastrophic damage will occur. At the end of this life the structure must be replaced even though there may be no detectable signs of fatigue. If a structural component is not economically replaceable when its safe life has been reached the complete structure must be written off. Alternatively, it is possible for easily replaceable components such as undercarriage legs and mechanisms to have a safe life less than that of the complete aircraft since it would probably be more economical to use, say, two lightweight undercarriage systems during the life of the aircraft rather than carry a heavier undercarriage which has the same safe life as the aircraft.

The fail-safe approach relies on the fact that the failure of a member in a redundant structure does not necessarily lead to the collapse of the complete structure, provided that the remaining members are able to carry the load shed by the failed member and can withstand further repeated loads until the presence of the failed member is discovered. Such a structure is called a fail-safe structure or a *damage tolerant* structure.

Generally, it is more economical to design some parts of the structure to be fail-safe rather than to have a long safe life since such components can be lighter. When failure is detected, either through a routine inspection or by some malfunction, such as fuel leakage from a wing crack, the particular aircraft may be taken out of service and repaired. However, the structure must be designed and the inspection intervals arranged such that a failure, for example a crack, too small to be noticed at one inspection must not increase to a catastrophic size before the next. The determination of crack propagation rates is discussed later.

Some components must be designed to have a safe life; these include landing gear, major wing joints, wing–fuselage joints and hinges on all-moving tailplanes or on variable geometry wings. Components which may be designed to be fail-safe include wing skins which are stiffened by stringers and fuselage skins which are stiffened by frames and stringers; the stringers and frames prevent skin cracks spreading disastrously for a sufficient period of time for them to be discovered at a routine inspection.

15.2 Designing against fatigue

Various precautions may be taken to ensure that an aircraft has an adequate fatigue life. We have seen in Chapter 11 that the early aluminium–zinc alloys possessed high ultimate and proof stresses but were susceptible to early failure under fatigue loading; choice of materials is therefore important. The naturally aged aluminium–copper alloys possess good fatigue resistance but with lower static strengths. Modern research is concentrating on alloys which combine high strength with high fatigue resistance.

Attention to detail design is equally important. Stress concentrations can arise at sharp corners and abrupt changes in section. Fillets should therefore be provided at re-entrant corners, and cut-outs, such as windows and access panels, should be reinforced. In machined panels the material thickness should be increased around bolt holes, while holes in primary bolted joints should be reamered to improve surface finish; surface scratches and machine marks are sources of fatigue crack initiation. Joggles in highly stressed members should be avoided while asymmetry can cause additional stresses due to bending.

In addition to sound structural and detail design, an estimation of the number, frequency and magnitude of the fluctuating loads an aircraft encounters is necessary. The *fatigue load spectrum* begins when the aircraft taxis to its take-off position. During taxiing the aircraft may be manoeuvring over uneven ground with a full payload so that wing stresses, for example, are greater than in the static case. Also, during take-off and climb and descent and landing the aircraft is subjected to the greatest load fluctuations. The undercarriage is retracted and lowered; flaps are raised and lowered; there is the impact on landing; the aircraft has to carry out manoeuvres; and, finally, the aircraft, as we shall see, experiences a greater number of gusts than during the cruise.

The loads corresponding to these various phases must be calculated before the associated stresses can be obtained. For example, during take-off, wing bending stresses and shear stresses due to shear and torsion are based on the total weight of the aircraft including full fuel tanks, and maximum payload all factored by 1.2 to allow for a bump during each take-off on a hard runway or by 1.5 for a take-off from grass. The loads produced during level flight and symmetric manoeuvres are calculated using the methods described in Section 14.2. From these values distributions of shear force, bending moment and torque may be found in, say, the wing by integrating the lift distribution. Loads due to gusts are calculated using the methods described in Section 14.4. Thus, due to a single equivalent sharp-edged gust the load factor is given either by Eq (14.25) or Eq (14.26).

Although it is a relatively simple matter to determine the number of load fluctuations during a ground–air–ground cycle caused by standard operations such as raising and lowering flaps, retracting and lowering the undercarriage, etc., it is more difficult to estimate the number and magnitude of gusts an aircraft will encounter. For example, there is a greater number of gusts at low altitude (during take-off, climb and descent) than at high altitude (during cruise). Terrain (sea, flat land, mountains) also affects the number and magnitude of gusts as does weather. The use of radar enables aircraft to avoid cumulus where gusts are prevalent, but has little effect at low altitude in the climb and descent where clouds cannot easily be avoided. The ESDU (Engineering Sciences Data Unit) has produced gust data based on information collected by gust recorders carried by aircraft. These show, in graphical form (l_{10} versus h curves, h is altitude), the average distance flown at various altitudes for a gust having a velocity greater than ± 3.05 m/s to be encountered. In addition, *gust frequency curves* give the number of gusts of a given velocity per 1000 gusts of velocity 3.05 m/s. Combining both sets of data enables the *gust exceedance* to be calculated, i.e. the number of gust cycles having a velocity greater than or equal to a given velocity encountered per kilometre of flight.

Since an aircraft is subjected to the greatest number of load fluctuations during taxi–take-off–climb and descent–standoff–landing while little damage is caused during cruise, the fatigue life of an aircraft does not depend on the number of flying hours but on the number of flights. However, the operational requirements of aircraft differ from class to class. The Airbus is required to have a life free from fatigue cracks of 24 000 flights or 30 000 hours, while its economic repair life is 48 000 flights or 60 000 hours; its landing gear, however, is designed for a safe life of 32 000 flights, after which it must be replaced. On the other hand the BAe 146, with a greater number of shorter flights per day than the Airbus, has a specified crack free life of 40 000 flights and an economic repair life of 80 000 flights. Although the above figures are operational requirements, the nature of fatigue is such that it is unlikely that all of a given type of aircraft will satisfy them. Of the total number of Airbus aircraft, at least 90% will achieve the above values and 50% will be better; clearly, frequent inspections are necessary during an aircraft's life.

15.3 Fatigue strength of components

In Section 13.2.4 we discussed the effect of stress level on the number of cycles to failure of a material such as mild steel. As the stress level is decreased the number of cycles to failure increases, resulting in a fatigue endurance curve (the S–N curve) of the type shown in Fig. 13.2. Such a curve corresponds to the average value of N at each stress amplitude since there will be a wide range of values of N for the given stress; even under carefully controlled conditions the ratio of maximum N to minimum N may be as high as 10 : 1. Two other curves may therefore be drawn, as shown in Fig. 15.1, enveloping all or nearly all the experimental results; these curves are known as the *confidence limits*. If 99.9 per cent of all the results lie between the curves, i.e. only 1 in 1000 falls outside, they represent the 99.9 per cent confidence limits. If 99.99999 per cent of results lie between the curves only 1 in 10^7 results will fall outside them and they represent the 99.99999 per cent confidence limits.

The results from tests on a number of specimens may be represented as a histogram in which the number of specimens failing within certain ranges R of N is plotted against N. Then if N_{av} is the average value of N at a given stress amplitude the probability of failure occurring at N cycles is given by

$$p(N) = \frac{1}{\sigma\sqrt{2\pi}} \exp\left[-\frac{1}{2}\left(\frac{N - N_{av}}{\sigma}\right)^2\right] \tag{15.1}$$

in which σ is the standard deviation of the whole population of N values. The derivation of Eq. (15.1) depends on the histogram approaching the profile of a continuous function close to the *normal distribution*, which it does as the interval N_{av}/R becomes smaller and the number of tests increases. The *cumulative probability*, which gives the probability

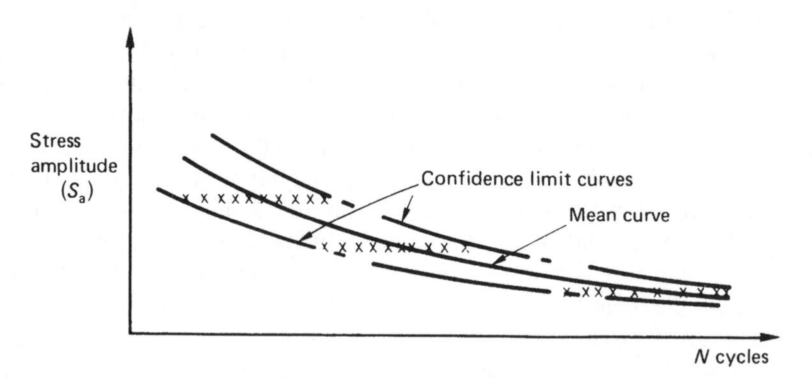

Fig. 15.1 *S–N* diagram.

that a particular specimen will fail at or below N cycles, is defined as

$$P(N) = \int_{-\infty}^{N} p(N)\,\mathrm{d}N \tag{15.2}$$

The probability that a specimen will endure more than N cycles is then $1 - P(N)$. The normal distribution allows negative values of N, which is clearly impossible in a fatigue testing situation. Other distributions, *extreme value distributions*, are more realistic and allow the existence of minimum fatigue endurances and fatigue limits.

The damaging portion of a fluctuating load cycle occurs when the stress is tensile; this causes cracks to open and grow. Therefore, if a steady tensile stress is superimposed on a cyclic stress the maximum tensile stress during the cycle will be increased and the number of cycles to failure will decrease. Conversely, if the steady stress is compressive the maximum tensile stress will decrease and the number of cycles to failure will increase. An approximate method of assessing the effect of a steady mean value of stress is provided by a Goodman diagram, as shown in Fig. 15.2. This shows the cyclic stress amplitudes which can be superimposed upon different mean stress levels to give a constant fatigue life. In Fig. 15.2, S_{a} is the allowable stress amplitude, $S_{\mathrm{a,0}}$ is the stress amplitude required to produce fatigue failure at N cycles with zero mean stress, S_{m} is the mean stress and S_{u} the ultimate tensile stress. If $S_{\mathrm{m}} = S_{\mathrm{u}}$ any cyclic stress will cause failure, while if $S_{\mathrm{m}} = 0$ the allowable stress amplitude is $S_{\mathrm{a,0}}$. The equation of the straight line portion of the diagram is

$$\frac{S_{\mathrm{a}}}{S_{\mathrm{a,0}}} = \left(1 - \frac{S_{\mathrm{m}}}{S_{\mathrm{u}}}\right) \tag{15.3}$$

Experimental evidence suggests a non-linear relationship for particular materials. Equation (15.3) then becomes

$$\frac{S_{\mathrm{a}}}{S_{\mathrm{a,0}}} = \left[1 - \left(\frac{S_{\mathrm{m}}}{S_{\mathrm{u}}}\right)^{m}\right] \tag{15.4}$$

in which m lies between 0.6 and 2.

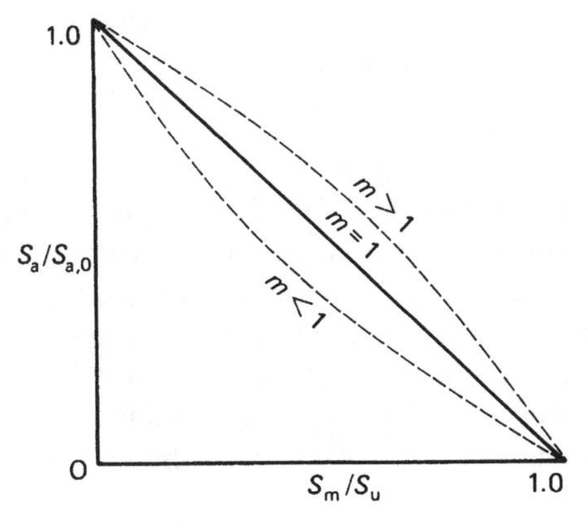

Fig. 15.2 Goodman diagram.

In practical situations, fatigue is not caused by a large number of identical stress cycles but by many different stress amplitude cycles. The prediction of the number of cycles to failure therefore becomes complex. Miner and Palmgren have proposed a *linear cumulative damage law* as follows. If N cycles of stress amplitude S_a cause fatigue failure then 1 cycle produces $1/N$ of the total damage to cause failure. Therefore, if r different cycles are applied in which a stress amplitude S_j $(j = 1, 2, \ldots, r)$ would cause failure in N_j cycles the number of cycles n_j required to cause total fatigue failure is given by

$$\sum_{j=1}^{r} \frac{n_j}{N_j} = 1 \tag{15.5}$$

Although S–N curves may be readily obtained for different materials by testing a large number of small specimens (*coupon tests*), it is not practicable to adopt the same approach for aircraft components since these are expensive to manufacture and the test programme too expensive to run for long periods of time. However, such a programme was initiated in the early 1950s to test the wings and tailplanes of Meteor and Mustang fighters. These were subjected to constant amplitude loading until failure with different specimens being tested at different load levels. Stresses were measured at points where fatigue was expected (and actually occurred) and S–N curves plotted for the complete structure. The curves had the usual appearance and at low stress levels had such large endurances that fatigue did not occur; thus a fatigue limit existed. It was found that the average S–N curve could be approximated to by the equation

$$S_a = 10.3(1 + 1000/\sqrt{N}) \tag{15.6}$$

in which the mean stress was $90\,\text{N/mm}^2$. In general terms, Eq. (15.6) may be written as

$$S_a = S_\infty(1 + C/\sqrt{N}) \tag{15.7}$$

in which S_∞ is the fatigue limit and C is a constant. Thus $S_a \to S_\infty$ as $N \to \infty$. Equation (15.7) may be rearranged to give the endurance directly, i.e.

$$N = C^2 \left(\frac{S_\infty}{S_a - S_\infty} \right)^2 \tag{15.8}$$

which shows clearly that as $S_a \to S_\infty$, $N \to \infty$.

It has been found experimentally that N is inversely proportional to the mean stress as the latter varies in the region of $90\,\text{N/mm}^2$ while C is virtually constant. This suggests a method of determining a 'standard' endurance curve (corresponding to a mean stress level of $90\,\text{N/mm}^2$) from tests carried out on a few specimens at other mean stress levels. Suppose S_m is the mean stress level, not $90\,\text{N/mm}^2$, in tests carried out on a few specimens at an alternating stress level $S_{a,m}$ where failure occurs at a mean number of cycles N_m. Then assuming that the S–N curve has the same form as Eq. (15.7)

$$S_{a,m} = S_{\infty,m}(1 + C/\sqrt{N_m}) \tag{15.9}$$

in which $C = 1000$ and $S_{\infty,\mathrm{m}}$ is the fatigue limit stress corresponding to the mean stress S_{m}. Rearranging Eq. (15.9) we have

$$S_{\infty,\mathrm{m}} = S_{\mathrm{a,m}}/(1 + C/\sqrt{N_m}) \tag{15.10}$$

The number of cycles to failure at a mean stress of 90 N/mm^2 would have been, from the above

$$N' = \frac{S_{\mathrm{m}}}{90} N_{\mathrm{m}} \tag{15.11}$$

The corresponding fatigue limit stress would then have been, from a comparison with Eq. (15.10)

$$S'_{\infty,\mathrm{m}} = S_{\mathrm{a,m}}/(1 + C/\sqrt{N'}) \tag{15.12}$$

The standard endurance curve for the component at a mean stress of 90 N/mm^2 is from Eq. (15.7)

$$S_{\mathrm{a}} = S'_{\infty,\mathrm{m}}/(1 + C/\sqrt{N}) \tag{15.13}$$

Substituting in Eq. (15.13) for $S'_{\infty,\mathrm{m}}$ from Eq. (15.12) we have

$$S_{\mathrm{a}} = \frac{S_{\mathrm{a,m}}}{(1 + C/\sqrt{N'})}(1 + C/\sqrt{N}) \tag{15.14}$$

in which N' is given by Eq. (15.11).

Equation (15.14) will be based on a few test results so that a 'safe' fatigue strength is usually taken to be three standard deviations below the mean fatigue strength. Hence we introduce *a scatter factor* K_n (>1) to allow for this; Eq. (15.14) then becomes

$$S_{\mathrm{a}} \frac{S_{\mathrm{a,m}}}{K_{\mathrm{n}}(1 + C/\sqrt{N'})}(1 + C/\sqrt{N}) \tag{15.15}$$

K_n varies with the number of test results available and for a coefficient of variation of 0.1, $K_n = 1.45$ for 6 specimens, $K_n = 1.445$ for 10 specimens, $K_n = 1.44$ for 20 specimens and for 100 specimens or more $K_n = 1.43$. For typical *S–N* curves a scatter factor of 1.43 is equivalent to a life factor of 3 to 4.

15.4 Prediction of aircraft fatigue life

We have seen that an aircraft suffers fatigue damage during all phases of the ground–air–ground cycle. The various contributions to this damage may be calculated separately and hence the safe life of the aircraft in terms of the number of flights calculated.

In the ground–air–ground cycle the maximum vertical acceleration during take-off is 1.2 g for a take-off from a runway or 1.5 g for a take-off from grass. It is assumed that these accelerations occur at zero lift and therefore produce compressive (negative) stresses, $-S_{\mathrm{TO}}$, in critical components such as the undersurface of wings. The maximum positive stress for the same component occurs in level flight (at 1 g) and is $+S_{1g}$.

The ground–air–ground cycle produces, on the undersurface of the wing, a fluctuating stress $S_{GAG} = (S_{1g} + S_{TO})/2$ about a mean stress $S_{GAG(mean)} = (S_{1g} - S_{TO})/2$. Suppose that tests show that for this stress cycle and mean stress, failure occurs after N_G cycles. For a life factor of 3 the safe life is $N_G/3$ so that the damage done during one cycle is $3/N_G$. This damage is multiplied by a factor of 1.5 to allow for the variability of loading between different aircraft of the same type so that the damage per flight D_{GAG} from the ground–air–ground cycle is given by

$$D_{GAG} = 4.5/N_G \tag{15.16}$$

Fatigue damage is also caused by gusts encountered in flight, particularly during the climb and descent. Suppose that a gust of velocity u_e causes a stress S_u about a mean stress corresponding to level flight, and suppose also that the number of stress cycles of this magnitude required to cause failure is $N(S_u)$; the damage caused by one cycle is then $1/N(S_u)$. Therefore from the Palmgren–Miner hypothesis, when sufficient gusts of this and all other magnitudes together with the effects of all other load cycles produce a cumulative damage of 1.0, fatigue failure will occur. It is therefore necessary to know the number and magnitude of gusts likely to be encountered in flight.

Gust data have been accumulated over a number of years from accelerometer records from aircraft flying over different routes and terrains, at different heights and at different seasons. The ESDU data sheets[1] present the data in two forms, as we have previously noted. First, l_{10} against altitude curves show the distance which must be flown at a given altitude in order that a gust (positive or negative) having a velocity $\geq$3.05 m/s be encountered. It follows that $1/l_{10}$ is the number of gusts encountered in unit distance (1 km) at a particular height. Secondly, gust frequency distribution curves, $r(u_e)$ against u_e, give the number of gusts of velocity u_e for every 1000 gusts of velocity 3.05 m/s.

From these two curves the gust exceedance $E(u_e)$ is obtained; $E(u_e)$ is the number of times a gust of a given magnitude (u_e) will be equalled or exceeded in 1 km of flight. Thus, from the above

$$\text{number of gusts} \geq 3.05\,\text{m/s per km} = 1/l_{10}$$

$$\text{number of gusts equal to } u_e \text{ per 1000 gusts equal to } 3.05\,\text{m/s} = r(u_e)$$

Hence

$$\text{number of gusts equal to } u_e \text{ per single gust equal to } 3.05\,\text{m/s} = r(u_e)/1000$$

It follows that the gust exceedance $E(u_e)$ is given by

$$E(u_e) = \frac{r(u_e)}{1000 l_{10}} \tag{15.17}$$

in which l_{10} is dependent on height. A good approximation for the curve of $r(u_e)$ against u_e in the region $u_e = 3.05$ m/s is

$$r(u_e) = 3.23 \times 10^5 u_e^{-5.26} \tag{15.18}$$

Consider now the typical gust exceedance curve shown in Fig. 15.3. In 1 km of flight there are likely to be $E(u_e)$ gusts exceeding u_e m/s and $E(u_e) - \delta E(u_e)$ gusts exceeding

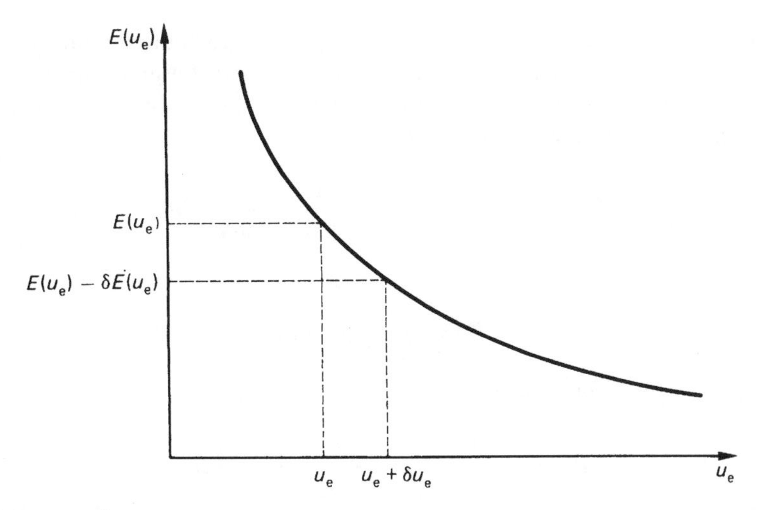

Fig. 15.3 Gust exceedance curve.

$u_e + \delta u_e$ m/s. Thus, there will be $\delta E(u_e)$ *fewer* gusts exceeding $u_e + \delta u_e$ m/s than u_e m/s and the increment in gust speed δu_e corresponds to a number $-\delta E(u_e)$ of gusts at a gust speed close to u_e. Half of these gusts will be positive (upgusts) and half negative (downgusts) so that if it is assumed that each upgust is followed by a downgust of equal magnitude the number of complete gust cycles will be $-\delta E(u_e)/2$. Suppose that each cycle produces a stress $S(u_e)$ and that the number of these cycles required to produce failure is $N(S_{u,e})$. The damage caused by one cycle is then $1/N(S_{u,e})$ and over the gust velocity interval δu_e the total damage δD is given by

$$\delta D = -\frac{\delta E(u_e)}{2N(S_{u,e})} = -\frac{dE(u_e)}{du_e}\frac{\delta u_e}{2N(S_{u,e})} \tag{15.19}$$

Integrating Eq. (15.19) over the whole range of gusts likely to be encountered, we obtain the total damage D_g per km of flight. Thus

$$D_g = -\int_0^\infty \frac{1}{2N(S_{u,e})}\frac{dE(u_e)}{du_e}du_e \tag{15.20}$$

Further, if the average block length journey of an aircraft is R_{av}, the average gust damage per flight is $D_g R_{av}$. Also, some aircraft in a fleet will experience more gusts than others since the distribution of gusts is random. Therefore if, for example, it is found that one particular aircraft encounters 50 per cent more gusts than the average its gust fatigue damage is $1.5\,D_g$/km.

The gust damage predicted by Eq. (15.20) is obtained by integrating over a complete gust velocity range from zero to infinity. Clearly there will be a gust velocity below which no fatigue damage will occur since the cyclic stress produced will be below the fatigue limit stress of the particular component. Equation (15.20) is therefore rewritten

$$D_g = -\int_{u_f}^\infty \frac{1}{2N(S_{u,e})}\frac{dE(u_e)}{du_e}du_e \tag{15.21}$$

in which u_f is the gust velocity required to produce the fatigue limit stress.

We have noted previously that more gusts are encountered during climb and descent than during cruise. Altitude therefore affects the amount of fatigue damage caused by gusts and its effects may be determined as follows. Substituting for the gust exceedance $E(u_e)$ in Eq. (15.21) from Eq. (15.17) we obtain

$$D_g = -\frac{1}{1000 l_{10}} \int_{u_f}^{\infty} \frac{1}{2N(S_{u,e})} \frac{dr(u_e)}{du_e} du_e$$

or

$$D_g = \frac{1}{l_{10}} d_g \text{ per km} \tag{15.22}$$

in which l_{10} is a function of height h and

$$d_g = -\frac{1}{1000} \int_{u_f}^{\infty} \frac{1}{2N(S_{u,e})} \frac{dr(u_e)}{du_e} du_e$$

Suppose that the aircraft is climbing at a speed V with a rate of climb (ROC). The time taken for the aircraft to climb from a height h to a height $h + \delta h$ is δh/ROC during which time it travels a distance $V\delta h$/ROC. Hence, from Eq. (15.22) the fatigue damage experienced by the aircraft in climbing through a height δh is

$$\frac{1}{l_{10}} d_g \frac{V}{\text{ROC}} \delta h$$

The total damage produced during a climb from sea level to an altitude H at a constant speed V and ROC is

$$D_{g,\text{climb}} = d_g \frac{V}{\text{ROC}} \int_0^H \frac{dh}{l_{10}} \tag{15.23}$$

Plotting $1/l_{10}$ against h from ESDU data sheets for aircraft having cloud warning radar and integrating gives

$$\int_0^{3000} \frac{dh}{l_{10}} = 303 \qquad \int_{3000}^{6000} \frac{dh}{l_{10}} = 14 \qquad \int_{6000}^{9000} \frac{dh}{l_{10}} = 3.4$$

From the above $\int_0^{9000} dh/l_{10} = 320.4$, from which it can be seen that approximately 95 per cent of the total damage in the climb occurs in the first 3000 m.

An additional factor influencing the amount of gust damage is forward speed. For example, the change in wing stress produced by a gust may be represented by

$$S_{u,e} = k_1 u_e V_e \quad \text{(see Eq. (14.24))} \tag{15.24}$$

in which the forward speed of the aircraft is in equivalent airspeed (EAS). From Eq. (15.24) we see that the gust velocity u_f required to produce the fatigue limit stress S_∞ is

$$u_f = S_\infty / k_1 V_e \tag{15.25}$$

The gust damage per km at different forward speeds V_e is then found using Eq. (15.21) with the appropriate value of u_f as the lower limit of integration. The integral may be evaluated by using the known approximate forms of $N(S_{u,e})$ and $E(u_e)$ from Eqs (15.15) and (15.17). From Eq. (15.15)

$$S_a = S_{u,e} = \frac{S'_{\infty,m}}{K_n}(1 + C/\sqrt{N(S_{u,e})})$$

from which

$$N(S_{u,e}) = \left(\frac{C}{K_n}\right)^2 \left(\frac{S'_{\infty,m}}{S_{u,e} - S'_{\infty,m}}\right)^2$$

where $S_{u,e} = k_1 V_e u_e$ and $S'_{\infty,m} = k_1 V_e u_f$. Also Eq. (15.17) is

$$E(u_e) = \frac{r(u_e)}{1000 l_{10}}$$

or, substituting for $r(u_e)$ from Eq. (15.18)

$$E(u_e) = \frac{3.23 \times 10^5 u_e^{-5.26}}{1000 l_{10}}$$

Equation (15.21) then becomes

$$D_g = -\int_{u_f}^{\infty} \frac{1}{2} \left(\frac{K_n}{C}\right)^2 \left(\frac{S_{u,e} - S'_{\infty,m}}{S'_{\infty,m}}\right)^2 \left(\frac{-3.23 \times 5.26 \times 10^5 u_e^{-5.26}}{1000 l_{10}}\right) du_e$$

Substituting for $S_{u,e}$ and $S'_{\infty,m}$ we have

$$D_g = \frac{16.99 \times 10^2}{2 l_{10}} \left(\frac{K_n}{C}\right)^2 \int_{u_f}^{\infty} \left(\frac{u_e - u_f}{u_f}\right)^2 u_e^{-6.26} du_e$$

or

$$D_g = \frac{16.99 \times 10^2}{2 l_{10}} \left(\frac{K_n}{C}\right)^2 \int_{u_f}^{\infty} \left(\frac{u_e^{-4.26}}{u_f^2} - \frac{2u_e^{-5.26}}{u_f} + u_e^{-6.26}\right) du_e$$

from which

$$D_g = \frac{46.55}{2 l_{10}} \left(\frac{K_n}{C}\right)^2 u_f^{-5.26}$$

or, in terms of the aircraft speed V_e

$$D_g = \frac{46.55}{2 l_{10}} \left(\frac{K_n}{C}\right)^2 \left(\frac{k_1 V_e}{S'_{\infty,m}}\right)^{5.26} \text{ per km} \qquad (15.26)$$

It can be seen from Eq. (15.26) that gust damage increases in proportion to $V_e^{5.26}$ so that increasing forward speed has a dramatic effect on gust damage.

The total fatigue damage suffered by an aircraft per flight is the sum of the damage caused by the ground–air–ground cycle, the damage produced by gusts and the damage due to other causes such as pilot induced manoeuvres, ground turning and braking, and landing and take-off load fluctuations. The damage produced by these other causes can be determined from load exceedance data. Thus, if this extra damage per flight is D_{extra} the total fractional fatigue damage per flight is

$$D_{\text{total}} = D_{\text{GAG}} + D_{\text{g}}R_{\text{av}} + D_{\text{extra}}$$

or

$$D_{\text{total}} = 4.5/N_{\text{G}} + D_{\text{g}}R_{\text{av}} + D_{\text{extra}} \tag{15.27}$$

and the life of the aircraft in terms of flights is

$$N_{\text{flight}} = 1/D_{\text{total}} \tag{15.28}$$

15.5 Crack propagation

We have seen that the concept of fail-safe structures in aircraft construction relies on a damaged structure being able to retain sufficient of its load-carrying capacity to prevent catastrophic failure, at least until the damage is detected. It is therefore essential that the designer be able to predict how and at what rate a fatigue crack will grow. The ESDU data sheets provide a useful introduction to the study of crack propagation; some of the results are presented here.

The analysis of stresses close to a crack tip using elastic stress concentration factors breaks down since the assumption that the crack tip radius approaches zero results in the stress concentration factor tending to infinity. Instead, linear elastic fracture mechanics analyses the stress field around the crack tip and identifies features of the field common to all cracked elastic bodies.

15.5.1 Stress concentration factor

There are three basic modes of crack growth, as shown in Fig. 15.4. Generally, the stress field in the region of the crack tip is described by a two-dimensional model which may be used as an approximation for many practical three-dimensional loading cases. Thus, the stress system at a distance r ($r \leq a$) from the tip of a crack of length $2a$, shown in Fig. 15.5, can be expressed in the form

$$S_r, S_\theta, S_{r,\theta} = \frac{K}{(2\pi r)^{\frac{1}{2}}} f(\theta) \quad \text{(Ref. [2])} \tag{15.29}$$

in which $f(\theta)$ is a different function for each of the three stresses and K is the *stress intensity factor*; K is a function of the nature and magnitude of the applied stress levels and also of the crack size. The terms $(2\pi r)^{\frac{1}{2}}$ and $f(\theta)$ map the stress field in the

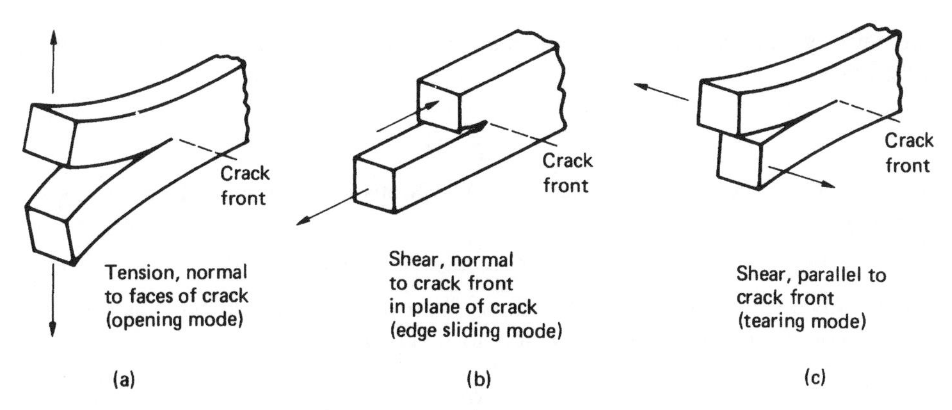

Fig. 15.4 Basic modes of crack growth.

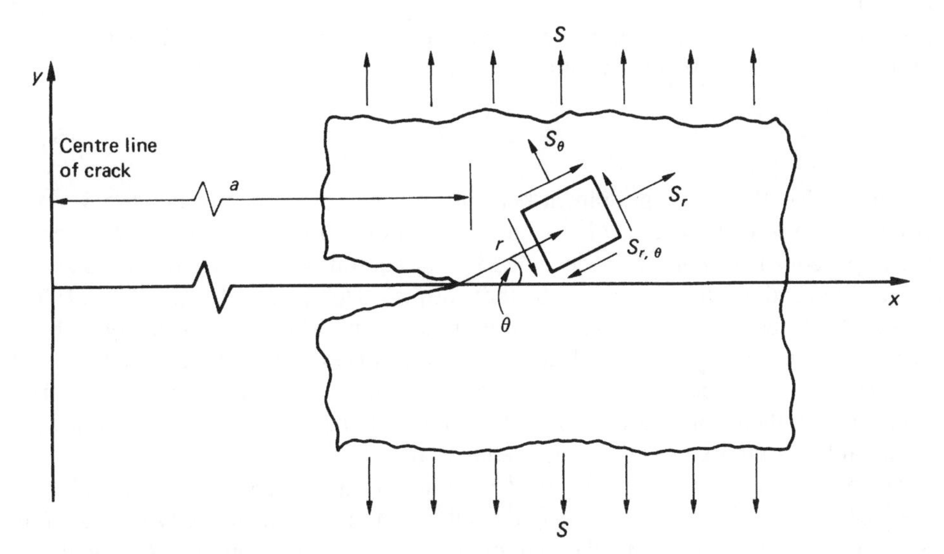

Fig. 15.5 Stress field in the vicinity of a crack.

vicinity of the crack and are the same for all cracks under external loads that cause crack openings of the same type.

Equation (15.29) applies to all modes of crack opening, with K having different values depending on the geometry of the structure, the nature of the applied loads and the type of crack.

Experimental data show that crack growth and residual strength data are better correlated using K than any other parameter. K may be expressed as a function of the nominal applied stress S and the crack length in the form

$$K = S(\pi a)^{\frac{1}{2}}\alpha \tag{15.30}$$

in which α is a non-dimensional coefficient usually expressed as the ratio of crack length to any convenient local dimension in the plane of the component; for a crack in an infinite plate under an applied uniform stress level S remote from the crack, $\alpha = 1.0$.

Alternatively, in cases where opposing loads P are applied at points close to the plane of the crack

$$K = \frac{P\alpha}{(\pi a)^{\frac{1}{2}}} \tag{15.31}$$

in which P is the load/unit thickness. Equations (15.30) and (15.31) may be rewritten as

$$K = K_0 \alpha \tag{15.32}$$

where K_0 is a reference value of the stress intensity factor which depends upon the loading. For the simple case of a remotely loaded plate in tension

$$K_0 = S(\pi a)^{\frac{1}{2}} \tag{15.33}$$

and Eqs (15.32) and (15.30) are identical so that for a given ratio of crack length to plate width α is the same in both formulations. In more complex cases, for example the in-plane bending of a plate of width $2b$ and having a central crack of length $2a$

$$K_0 = \frac{3Ma}{4b^3}(\pi a)^{\frac{1}{2}} \tag{15.34}$$

in which M is the bending moment per unit thickness. Comparing Eqs (15.34) and (15.30), we see that $S = 3Ma/4b^3$ which is the value of direct stress given by basic bending theory at a point a distance $\pm a/2$ from the central axis. However, if S was specified as the bending stress in the outer fibres of the plate, i.e. at $\pm b$, then $S = 3M/2b^2$; clearly the different specifications of S require different values of α. On the other hand the final value of K must be independent of the form of presentation used. Use of Eqs (15.30)–(15.32) depends on the form of the solution for K_0 and care must be taken to ensure that the formula used and the way in which the nominal stress is defined are compatible with those used in the derivation of α.

There are a number of methods available for determining the value of K and α. In one method the solution for a component subjected to more than one type of loading is obtained from available standard solutions using superposition or, if the geometry is not covered, two or more standard solutions may be compounded.[1] Alternatively, a finite element analysis may be used.

The coefficient α in Eq. (15.30) has, as we have noted, different values depending on the plate and crack geometries. Listed below are values of α for some of the more common cases.

(i) A semi-infinite plate having an edge crack of length a; $\alpha = 1.12$.

(ii) An infinite plate having an embedded circular crack or a semi-circular surface crack, each of radius a, lying in a plane normal to the applied stress; $\alpha = 0.64$.

(iii) An infinite plate having an embedded elliptical crack of axes $2a$ and $2b$ or a semi-elliptical crack of width $2b$ in which the depth a is less than half the plate thickness each lying in a plane normal to the applied stress; $\alpha = 1.12\Phi$ in which Φ varies with the ratio a/b as follows:

a/b	0	0.2	0.4	0.6	0.8
Φ	1.0	1.05	1.15	1.28	1.42

For $a/b = 1$ the situation is identical to case (ii).

(iv) A plate of finite width w having a central crack of length $2a$ where $a \leq 0.3w$; $\alpha = [\sec(a\pi/w)]^{1/2}$.
(v) For a plate of finite width w having two symmetrical edge cracks each of depth $2a$, Eq. (15.30) becomes

$$K = S[w \tan(\pi a/w) + (0.1w)\sin(2\pi a/w)]^{1/2}$$

From Eq. (15.29) it can be seen that the stress intensity at a point ahead of a crack can be expressed in terms of the parameter K. Failure will then occur when K reaches a critical value K_c. This is known as the *fracture toughness* of the material and has units $MN/m^{3/2}$ or $N/mm^{3/2}$.

15.5.2 Crack tip plasticity

In certain circumstances it may be necessary to account for the effect of plastic flow in the vicinity of the crack tip. This may be allowed for by estimating the size of the plastic zone and adding this to the actual crack length to form an effective crack length $2a_1$. Thus, if r_p is the radius of the plastic zone, $a_1 = a + r_p$ and Eq. (15.30) becomes

$$K_p = S(\pi a_1)^{\frac{1}{2}}\alpha_1 \tag{15.35}$$

in which K_p is the stress intensity factor corrected for plasticity and α_1 corresponds to a_1. Thus for $r_p/t > 0.5$, i.e. a condition of plane stress

$$r_p = \frac{1}{2\pi}\left(\frac{K}{f_y}\right)^2 \quad \text{or} \quad r_p = \frac{a}{2}\left(\frac{S}{f_y}\right)^2\alpha^2 \quad \text{(Ref. [3])} \tag{15.36}$$

in which f_y is the yield proof stress of the material. For $r_p/t < 0.02$, a condition of plane strain

$$r_p = \frac{1}{6\pi}\left(\frac{K}{f_y}\right)^2 \tag{15.37}$$

For intermediate conditions the correction should be such as to produce a conservative solution.

Dugdale[4] showed that the fracture toughness parameter K_c is highly dependent on plate thickness. In general, since the toughness of a material decreases with decreasing plasticity, it follows that the true fracture toughness is that corresponding to a plane strain condition. This lower limiting value is particularly important to consider in high strength alloys since these are prone to brittle failure. In addition, the assumption that the plastic zone is circular is not representative in plane strain conditions. Rice and Johnson[5] showed that, for a small amount of plane strain yielding, the plastic zone extends as two lobes (Fig. 15.6) each inclined at an angle θ to the axis of the crack where $\theta = 70°$ and the greatest extent L and forward penetration (r_y for $\theta = 0$) of plasticity are given by

$$L = 0.155\,(K/f_y)^2$$

$$r_y = 0.04\,(K/f_y)^2$$

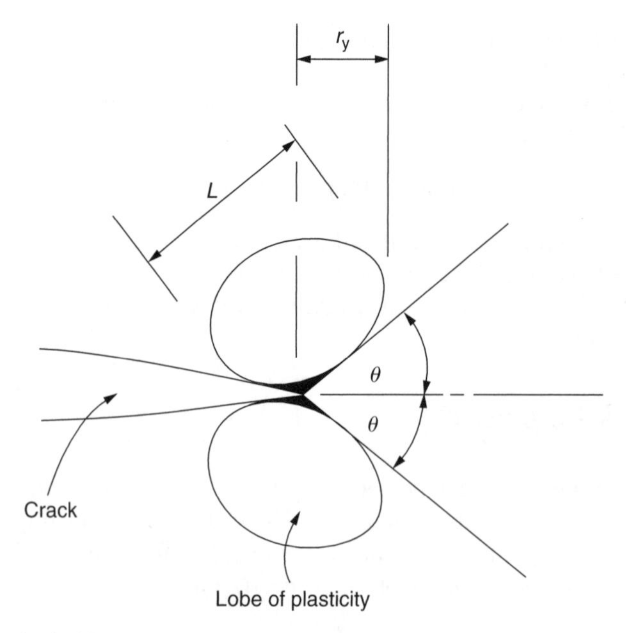

Fig. 15.6 Plane strain plasticity.

15.5.3 Crack propagation rates

Having obtained values of the stress intensity factor and the coefficient α, fatigue crack propagation rates may be estimated. From these, the life of a structure containing cracks or crack-like defects may be determined; alternatively, the loading condition may be modified or inspection periods arranged so that the crack will be detected before failure.

Under constant amplitude loading the rate of crack propagation may be represented graphically by curves described in general terms by the law

$$\frac{da}{dN} = f(R, \Delta K) \quad \text{(Ref. [6])} \tag{15.38}$$

in which ΔK is the stress intensity factor range and $R = S_{min}/S_{max}$. If Eq. (15.30) is used

$$\Delta K = (S_{max} - S_{min})(\pi a)^{\frac{1}{2}}\alpha \tag{15.39}$$

Equation (15.39) may be corrected for plasticity under cyclic loading and becomes

$$\Delta K_p = (S_{max} - S_{min})(\pi a_1)^{\frac{1}{2}}\alpha_1 \tag{15.40}$$

in which $a_1 = a + r_p$, where, for plane stress

$$r_p = \frac{1}{8\pi}\left(\frac{\Delta K}{f_y}\right)^2 \quad \text{(Ref. [7])}$$

The curves represented by Eq. (15.38) may be divided into three regions. The first corresponds to a very slow crack growth rate ($<10^{-8}$ m/cycle) where the curves approach a threshold value of stress intensity factor ΔK^{th} corresponding to 4×10^{-11} m/cycle, i.e. no crack growth. In the second region (10^{-8}–10^{-6} m/cycle) much of the crack life takes place and, for small ranges of ΔK, Eq. (15.38) may be represented by

$$\frac{\mathrm{d}a}{\mathrm{d}N} = C(\Delta K)^n \quad \text{(Ref. [8])} \tag{15.41}$$

in which C and n depend on the material properties; over small ranges of $\mathrm{d}a/\mathrm{d}N$ and ΔK, C and n remain approximately constant. The third region corresponds to crack growth rates $>10^{-6}$ m/cycle, where instability and final failure occur.

An attempt has been made to describe the complete set of curves by the relationship

$$\frac{\mathrm{d}a}{\mathrm{d}N} = \frac{C(\Delta K)^n}{(1-R)K_{\mathrm{c}} - \Delta K} \quad \text{(Ref. [9])} \tag{15.42}$$

in which K_{c} is the fracture toughness of the material obtained from toughness tests. Integration of Eqs (15.41) or (15.42) analytically or graphically gives an estimate of the crack growth life of the structure, i.e. the number of cycles required for a crack to grow from an initial size to an unacceptable length, or the crack growth rate or failure, whichever is the design criterion. Thus, for example, integration of Eq. (15.41) gives, for an infinite width plate for which $\alpha = 1.0$

$$[N]_{N_{\mathrm{i}}}^{N_{\mathrm{f}}} = \frac{1}{C[(S_{\max} - S_{\min})\pi^{\frac{1}{2}}]^n} \left[\frac{a^{(1-n/2)}}{1-n/2}\right]_{a_{\mathrm{i}}}^{a_{\mathrm{f}}} \tag{15.43}$$

for $n > 2$. An analytical integration may only be carried out if n is an integer and α is in the form of a polynomial, otherwise graphical or numerical techniques must be employed.

Substituting the limits in Eq. (15.43) and taking $N_{\mathrm{i}} = 0$, the number of cycles to failure is given by

$$N_{\mathrm{f}} = \frac{2}{C(n-2)[(S_{\max} - S_{\mathrm{m}})\pi^{1/2}]^{\mathrm{n}}} \left[\frac{1}{a_{\mathrm{i}}^{(n-2)/2}} - \frac{1}{a_{\mathrm{f}}^{(n-2)/2}}\right] \tag{15.44}$$

Example 15.1

An infinite plate contains a crack having an initial length of 0.2 mm and is subjected to a cyclic repeated stress range of 175 N/mm^2. If the fracture toughness of the plate is 1708 N/mm$^{3/2}$ and the rate of crack growth is 40×10^{-15} $(\Delta K)^4$ mm/cycle determine the number of cycles to failure.

The crack length at failure is given by Eq. (15.30) in which $\alpha = 1$, $K = 1708$ N/mm$^{3/2}$ and $S = 175$ N/mm^2, i.e.

$$a_{\mathrm{f}} = \frac{1708^2}{\pi \times 175^2} = 30.3\,\text{mm}$$

Also $n = 4$ so that substituting the relevant parameters in Eq. (15.44) gives

$$N_f = \frac{1}{40 \times 10^{-15}[175 \times \pi^{1/2}]^4}\left(\frac{1}{0.1} - \frac{1}{30.3}\right)$$

from which

$$N_f = 26919 \text{ cycles}$$

References

1 ESDU Data Sheets, Fatigue, No. 80036.
2 Knott, J. F., *Fundamentals of Fracture Mechanics*, Butterworths, London, 1973.
3 McClintock, F. A. and Irwin, G. R., Plasticity aspects of fracture mechanics. In: *Fracture Toughness Testing and its Applications*, American Society for Testing Materials, Philadelphia, USA, ASTM STP 381, April, 1965.
4 Dugdale, D. S., *J. Mech. Phys. Solids*, **8**, 1960.
5 Rice, J. R. and Johnson, M. A., *Inelastic Behaviour of Solids*, McGraw Hill, New York, 1970.
6 Paris, P. C. and Erdogan, F., A critical analysis of crack propagation laws, *Trans. Am. Soc. Mech. Engrs.*, **85**, Series D, No. 4, December 1963.
7 Rice, J. R., Mechanics of crack tip deformation and extension by fatigue. In: *Fatigue Crack Propagation*, American Society for Testing Materials, Philadelphia, USA, ASTM STP 415, June, 1967.
8 Paris, P. C., The fracture mechanics approach to fatigue. In: *Fatigue – An Interdisciplinary Approach*, Syracuse University Press, New York, USA, 1964.
9 Forman, R. G., Numerical analysis of crack propagation in cyclic-loaded structures, *Trans. Am. Soc. Mech. Engrs.*, **89**, Series D, No. 3, September 1967.

Further reading

Freudenthal, A. M., *Fatigue in Aircraft Structures*, Academic Press, New York, 1956.

Problems

P.15.1 A material has a fatigue limit of ± 230 N/mm^2 and an ultimate tensile strength of 870 N/mm^2. If the safe range of stress is determined by the Goodman prediction calculate its value.

Ans. 363 N/mm^2.

P.15.2 A more accurate estimate for the safe range of stress for the material of P.15.1 is given by the non-linear form of the Goodman prediction in which $m = 2$. Calculate its value.

Ans. 432 N/mm^2.

P.15.3 A steel component is subjected to a reversed cyclic loading of 100 cycles/day over a period of time in which $\pm 160\,\text{N/mm}^2$ is applied for 200 cycles, $\pm 140\ \text{N/mm}^2$ is applied for 200 cycles and $\pm 100\,\text{N/mm}^2$ is applied for 600 cycles. If the fatigue life of the material at each of these stress levels is 10^4, 10^5 and 2×10^5 cycles, respectively estimate the life of the component using Miner's law.

Ans. 400 days.

P.15.4 An infinite steel plate has a fracture toughness of $3320\,\text{N/mm}^{3/2}$ and contains a 4 mm long crack. Calculate the maximum allowable design stress that could be applied round the boundary of the plate.

Ans. $1324\,\text{N/mm}^2$.

P.15.5 A semi-infinite plate has an edge crack of length 0.4 mm. If the plate is subjected to a cyclic repeated stress loading of $180\,\text{N/mm}^2$, its fracture toughness is $1800\,\text{N/mm}^{3/2}$ and the rate of crack growth is $30 \times 10^{-15}(\Delta K)^4$ mm/cycle determine the crack length at failure and the number of cycles to failure.

Ans. 25.4 mm, 7916 cycles.

P.15.6 An aircraft's cruise speed is increased from 200 m/s to 220 m/s. Determine the percentage increase in gust damage this would cause.

Ans. 65%.

P.15.7 The average block length journey of an executive jet airliner is 1000 km and its cruise speed is 240 m/s. If the damage during the ground–air–ground cycle may be assumed to be 10% of the total damage during a complete flight determine the percentage increase in the life of the aircraft when the cruising speed is reduced to 235 m/s.

Ans. 12%.

Section B3 Bending, Shear and Torsion of Thin-Walled Beams

16

Bending of open and closed, thin-walled beams

In Chapter 12 we discussed the various types of structural component found in aircraft construction and the various loads they support. We saw that an aircraft is basically an assembly of stiffened shell structures ranging from the single cell closed section fuselage to multicellular wings and tail surfaces each subjected to bending, shear, torsional and axial loads. Other, smaller portions of the structure consist of thin-walled channel, T-, Z-, 'top-hat'-or I-sections, which are used to stiffen the thin skins of the cellular components and provide support for internal loads from floors, engine mountings, etc. Structural members such as these are known as *open section* beams, while the cellular components are termed *closed section* beams; clearly, both types of beam are subjected to axial, bending, shear and torsional loads.

In this chapter we shall investigate the stresses and displacements in thin-walled open and single cell closed section beams produced by bending loads.

In Chapter 1 we saw that an axial load applied to a member produces a uniform direct stress across the cross-section of the member. A different situation arises when the applied loads cause a beam to bend which, if the loads are vertical, will take up a sagging '($\smile$)' or hogging shape '($\frown$)'. This means that for loads which cause a beam to sag the upper surface of the beam must be shorter than the lower surface as the upper surface becomes concave and the lower one convex; the reverse is true for loads which cause hogging. The strains in the upper regions of the beam will, therefore, be different to those in the lower regions and since we have established that stress is directly proportional to strain (Eq. (1.40)) it follows that the stress will vary through the depth of the beam.

The truth of this can be demonstrated by a simple experiment. Take a reasonably long rectangular rubber eraser and draw three or four lines on its longer faces as shown in Fig. 16.1(a); the reason for this will become clear a little later. Now hold the eraser between the thumb and forefinger at each end and apply pressure as shown by the direction of the arrows in Fig. 16.1(b). The eraser bends into the shape shown and the lines on the side of the eraser *remain straight* but are now further apart at the top than at the bottom.

Since, in Fig. 16.1(b), the upper fibres have been stretched and the lower fibres compressed there will be fibres somewhere in between which are neither stretched nor compressed; the plane containing these fibres is called the *neutral plane*.

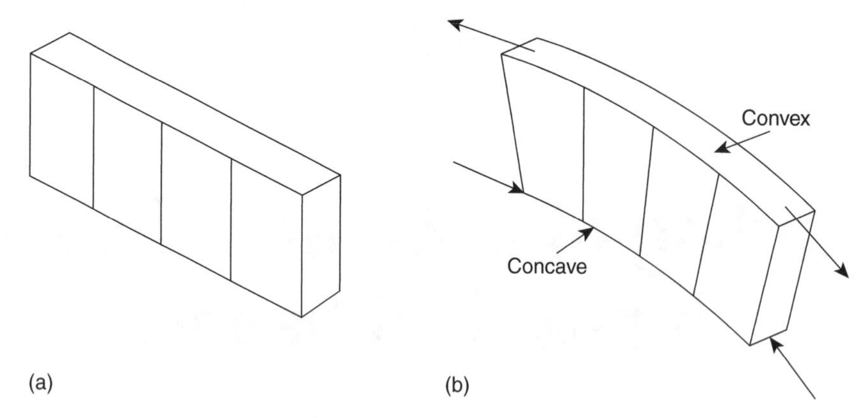

Fig. 16.1 Bending of a rubber eraser.

Now rotate the eraser so that its shorter sides are vertical and apply the same pressure with your fingers. The eraser again bends but now requires much less effort. It follows that the geometry and orientation of a beam section must affect its *bending stiffness*. This is more readily demonstrated with a plastic ruler. When flat it requires hardly any effort to bend it but when held with its width vertical it becomes almost impossible to bend.

16.1 Symmetrical bending

Although symmetrical bending is a special case of the bending of beams of arbitrary cross-section, we shall investigate the former first, so that the more complex general case may be more easily understood.

Symmetrical bending arises in beams which have either singly or doubly symmetrical cross-sections; examples of both types are shown in Fig. 16.2.

Suppose that a length of beam, of rectangular cross-section, say, is subjected to a pure, sagging bending moment, M, applied in a vertical plane. We shall define this later as a negative bending moment. The length of beam will bend into the shape shown in Fig. 16.3(a) in which the upper surface is concave and the lower convex. It can be seen that the upper longitudinal fibres of the beam are compressed while the lower fibres are stretched. It follows that, as in the case of the eraser, between these two extremes there are fibres that remain unchanged in length.

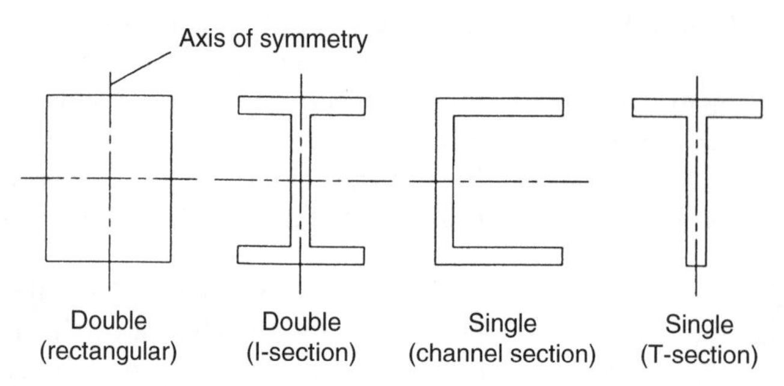

Fig. 16.2 Symmetrical section beams.

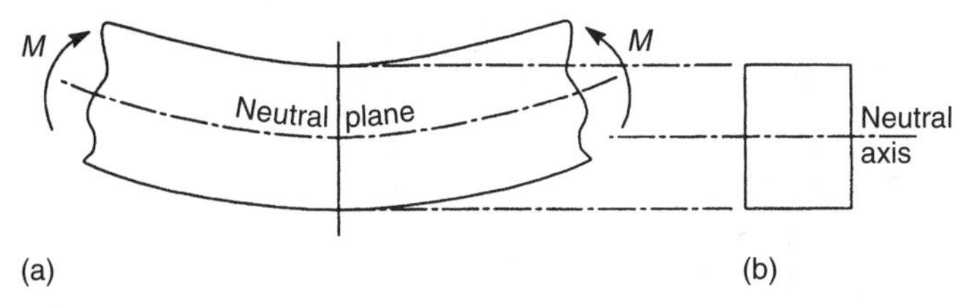

Fig. 16.3 Beam subjected to a pure sagging bending moment.

The direct stress therefore varies through the depth of the beam from compression in the upper fibres to tension in the lower. Clearly the direct stress is zero for the fibres that do not change in length; we have called the plane containing these fibres the *neutral plane*. The line of intersection of the neutral plane and any cross-section of the beam is termed the *neutral axis* (Fig. 16.3(b)).

The problem, therefore, is to determine the variation of direct stress through the depth of the beam, the values of the stresses and subsequently to find the corresponding beam deflection.

16.1.1 Assumptions

The primary assumption made in determining the direct stress distribution produced by pure bending is that plane cross-sections of the beam remain plane and normal to the longitudinal fibres of the beam after bending. Again, we saw this from the lines on the side of the eraser. We shall also assume that the material of the beam is linearly elastic, i.e. it obeys Hooke's law, and that the material of the beam is homogeneous.

16.1.2 Direct stress distribution

Consider a length of beam (Fig. 16.4(a)) that is subjected to a pure, sagging bending moment, M, applied in a vertical plane; the beam cross-section has a vertical axis of symmetry as shown in Fig. 16.4(b). The bending moment will cause the length of beam

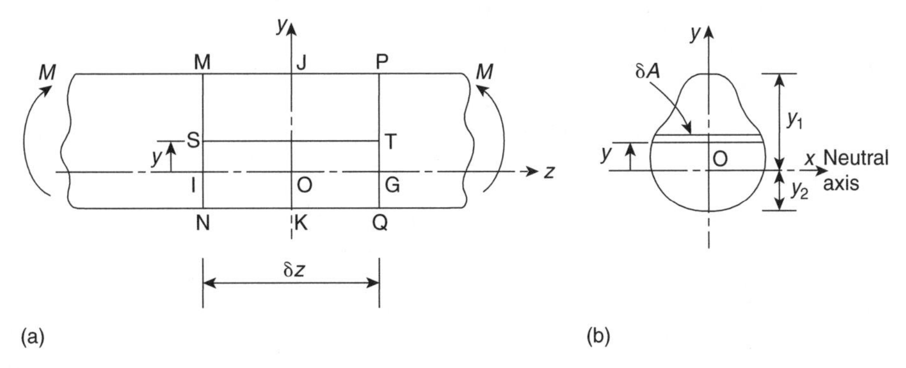

Fig. 16.4 Bending of a symmetrical section beam.

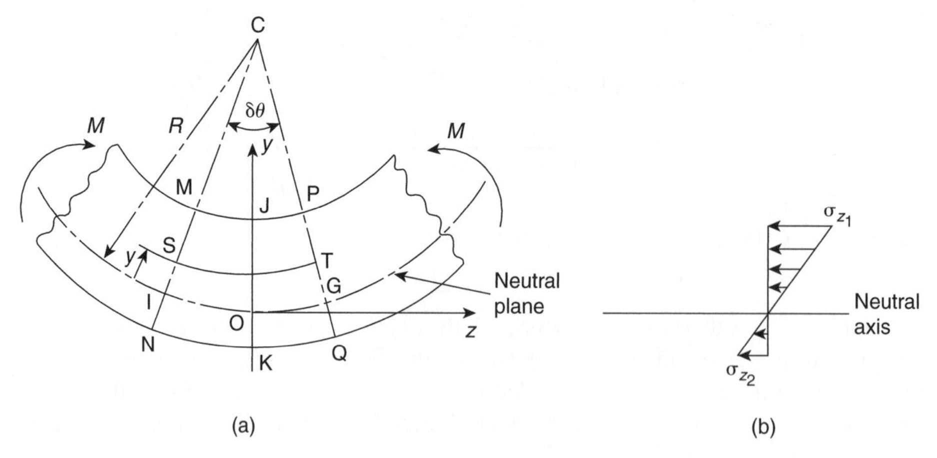

Fig. 16.5 Length of beam subjected to a pure bending moment.

to bend in a similar manner to that shown in Fig. 16.3(a) so that a neutral plane will exist which is, as yet, unknown distances y_1 and y_2 from the top and bottom of the beam, respectively. Coordinates of all points in the beam are referred to axes Oxyz in which the origin O lies in the neutral plane of the beam. We shall now investigate the behaviour of an elemental length, δz, of the beam formed by parallel sections MIN and PGQ (Fig. 16.4(a)) and also the fibre ST of cross-sectional area δA a distance y above the neutral plane. Clearly, before bending takes place MP = IG = ST = NQ = δz.

The bending moment M causes the length of beam to bend about a *centre of curvature* C as shown in Fig. 16.5(a). Since the element is small in length and a pure moment is applied we can take the curved shape of the beam to be circular with a *radius of curvature R* measured to the neutral plane. This is a useful reference point since, as we have seen, strains and stresses are zero in the neutral plane.

The previously parallel plane sections MIN and PGQ remain plane as we have demonstrated but are now inclined at an angle $\delta\theta$ to each other. The length MP is now shorter than δz as is ST while NQ is longer; IG, being in the neutral plane, is still of length δz. Since the fibre ST has changed in length it has suffered a strain ε_z which is given by

$$\varepsilon_z = \frac{\text{change in length}}{\text{original length}}$$

Then

$$\varepsilon_z = \frac{(R - y)\delta\theta - \delta z}{\delta z}$$

i.e.

$$\varepsilon_z = \frac{(R - y)\delta\theta - R\delta\theta}{R\delta\theta}$$

so that

$$\varepsilon_z = -\frac{y}{R} \tag{16.1}$$

The negative sign in Eq. (16.1) indicates that fibres in the region where y is positive will shorten when the bending moment is negative. Then, from Eq. (1.40), the direct stress σ_z in the fibre ST is given by

$$\sigma_z = -E\frac{y}{R} \tag{16.2}$$

The direct or normal force on the cross-section of the fibre ST is $\sigma_z \delta A$. However, since the direct stress in the beam section is due to a pure bending moment, in other words there is no axial load, the resultant normal force on the complete cross-section of the beam must be zero. Then

$$\int_A \sigma_z \, \mathrm{d}A = 0 \tag{16.3}$$

where A is the area of the beam cross-section.

Substituting for σ_z in Eq. (16.3) from (16.2) gives

$$-\frac{E}{R}\int_A y \, \mathrm{d}A = 0 \tag{16.4}$$

in which both E and R are constants for a beam of a given material subjected to a given bending moment. Therefore

$$\int_A y \, \mathrm{d}A = 0 \tag{16.5}$$

Equation (16.5) states that the first moment of the area of the cross-section of the beam with respect to the neutral axis, i.e. the x axis, is equal to zero. Thus we see that *the neutral axis passes through the centroid of area of the cross-section.* Since the y axis in this case is also an axis of symmetry, it must also pass through the centroid of the cross-section. Hence the origin, O, of the coordinate axes, coincides with the centroid of area of the cross-section.

Equation (16.2) shows that for a sagging (i.e. negative) bending moment the direct stress in the beam section is negative (i.e. compressive) when y is positive and positive (i.e. tensile) when y is negative.

Consider now the elemental strip δA in Fig. 16.4(b); this is, in fact, the cross-section of the fibre ST. The strip is above the neutral axis so that there will be a *compressive* force acting on its cross-section of $\sigma_z \delta A$ which is *numerically* equal to $(Ey/R)\delta A$ from Eq. (16.2). Note that this force will act at all sections along the length of ST. At S this force will exert a clockwise moment $(Ey/R)y\delta A$ about the neutral axis while at T the force will exert an identical anticlockwise moment about the neutral axis. Considering either end of ST we see that the moment resultant about the neutral axis of the stresses on all such fibres must be *equivalent* to the applied negative moment M, i.e.

$$M = -\int_A E\frac{y^2}{R}\mathrm{d}A$$

or

$$M = -\frac{E}{R}\int_A y^2 \mathrm{d}A \tag{16.6}$$

The term $\int_A y^2 \, dA$ is known as the *second moment of area* of the cross-section of the beam about the neutral axis and is given the symbol I. Rewriting Eq. (16.6) we have

$$M = -\frac{EI}{R} \tag{16.7}$$

or, combining this expression with Eq. (16.2)

$$\frac{M}{I} = -\frac{E}{R} = \frac{\sigma_z}{y} \tag{16.8}$$

From Eq. (16.8) we see that

$$\sigma_z = \frac{My}{I} \tag{16.9}$$

The direct stress, σ_z, at any point in the cross-section of a beam is therefore directly proportional to the distance of the point from the neutral axis and so varies linearly through the depth of the beam as shown, for the section JK, in Fig. 16.5(b). Clearly, for a positive bending moment σ_z is positive, i.e. tensile, when y is positive and compressive (i.e. negative) when y is negative. Thus in Fig. 16.5(b)

$$\sigma_{z,1} = \frac{My_1}{I} \text{ (compression)} \quad \sigma_{z,2} = \frac{My_2}{I} \text{ (tension)} \tag{16.10}$$

Furthermore, we see from Eq. (16.7) that the curvature, $1/R$, of the beam is given by

$$\frac{1}{R} = \frac{M}{EI} \tag{16.11}$$

and is therefore directly proportional to the applied bending moment and inversely proportional to the product EI which is known as the *flexural rigidity* of the beam.

Example 16.1

The cross-section of a beam has the dimensions shown in Fig. 16.6(a). If the beam is subjected to a negative bending moment of 100 kN m applied in a vertical plane, determine the distribution of direct stress through the depth of the section.

The cross-section of the beam is doubly symmetrical so that the centroid, C, of the section, and therefore the origin of axes, coincides with the mid-point of the web. Furthermore, the bending moment is applied to the beam section in a vertical plane so that the x axis becomes the neutral axis of the beam section; we therefore need to calculate the second moment of area, I_{xx}, about this axis.

$$I_{xx} = \frac{200 \times 300^3}{12} - \frac{175 \times 260^3}{12} = 193.7 \times 10^6 \text{ mm}^4 \text{ (see Section 16.4)}$$

From Eq. (16.9) the distribution of direct stress, σ_z, is given by

$$\sigma_z = -\frac{100 \times 10^6}{193.7 \times 10^6} y = -0.52y \tag{i}$$

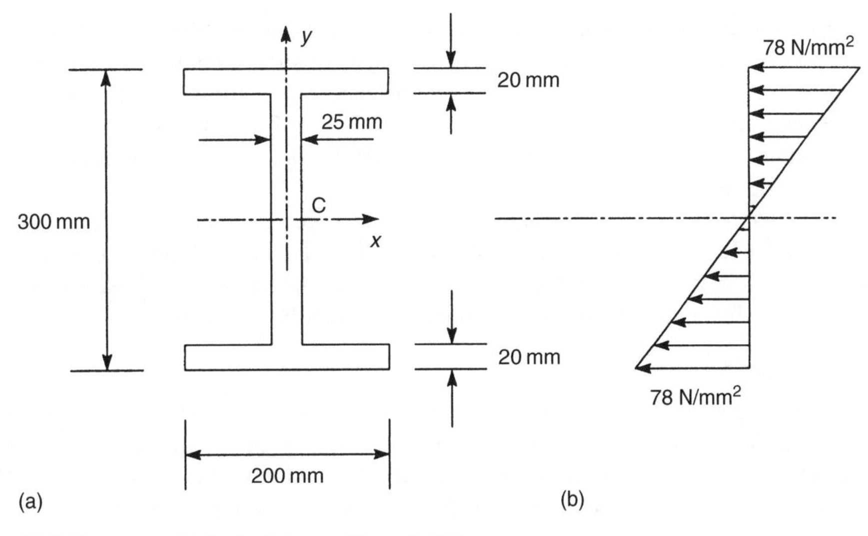

Fig. 16.6 Direct stress distribution in beam of Example 16.1.

The direct stress, therefore, varies linearly through the depth of the section from a value

$$-0.52 \times (+150) = -78\,\text{N/mm}^2 \text{ (compression)}$$

at the top of the beam to

$$-0.52 \times (-150) = +78\,\text{N/mm}^2 \text{ (tension)}$$

at the bottom as shown in Fig. 16.6(b).

Example 16.2

Now determine the distribution of direct stress in the beam of Example 16.1 if the bending moment is applied in a horizontal plane and in a clockwise sense about Cy when viewed in the direction yC.

In this case the beam will bend about the vertical y axis which therefore becomes the neutral axis of the section. Thus Eq. (16.9) becomes

$$\sigma_z = \frac{M}{I_{yy}}x \tag{i}$$

where I_{yy} is the second moment of area of the beam section about the y axis. Again from Section 16.4

$$I_{yy} = 2 \times \frac{20 \times 200^3}{12} + \frac{260 \times 25^3}{12} = 27.0 \times 10^6\,\text{mm}^4$$

Hence, substituting for M and I_{yy} in Eq. (i)

$$\sigma_z = \frac{100 \times 10^6}{27.0 \times 10^6}x = 3.7x$$

We have not specified a sign convention for bending moments applied in a horizontal plane. However, a physical appreciation of the problem shows that the left-hand edges of the beam are in compression while the right-hand edges are in tension. Again the distribution is linear and varies from $3.7 \times (-100) = -370\,\text{N/mm}^2$ (compression) at the left-hand edges of each flange to $3.7 \times (+100) = +370\,\text{N/mm}^2$ (tension) at the right-hand edges.

We note that the maximum stresses in this example are very much greater than those in Example 16.1. This is due to the fact that the bulk of the material in the beam section is concentrated in the region of the neutral axis where the stresses are low. The use of an I-section in this manner would therefore be structurally inefficient.

Example 16.3

The beam section of Example 16.1 is subjected to a bending moment of 100 kN m applied in a plane parallel to the longitudinal axis of the beam but inclined at 30° to the left of vertical. The sense of the bending moment is clockwise when viewed from the left-hand edge of the beam section. Determine the distribution of direct stress.

The bending moment is first resolved into two components, M_x in a vertical plane and M_y in a horizontal plane. Equation (16.9) may then be written in two forms

$$\sigma_z = \frac{M_x}{I_{xx}}y \qquad \sigma_z = \frac{M_y}{I_{yy}}x \tag{i}$$

The separate distributions can then be determined and superimposed. A more direct method is to combine the two equations (i) to give the total direct stress at any point (x, y) in the section. Thus

$$\sigma_z = \frac{M_x}{I_{xx}}y + \frac{M_y}{I_{yy}}x \tag{ii}$$

Now

$$\left.\begin{aligned} M_x &= 100\cos 30° = 86.6\,\text{kN m} \\ M_y &= 100\sin 30° = 50.0\,\text{kN m} \end{aligned}\right\} \tag{iii}$$

M_x is, in this case, a positive bending moment producing tension in the upper half of the beam where y is positive. Also M_y produces tension in the left-hand half of the beam where x is negative; we shall therefore call M_y a negative bending moment. Substituting the values of M_x and M_y from Eq. (iii) but with the appropriate sign in Eq. (ii) together with the values of I_{xx} and I_{yy} from Examples 16.1 and 16.2 we obtain

$$\sigma_z = \frac{86.6 \times 10^6}{193.7 \times 10^6}y - \frac{50.0 \times 10^6}{27.0 \times 10^6}x \tag{iv}$$

or

$$\sigma_z = 0.45y - 1.85x \tag{v}$$

Equation (v) gives the value of direct stress at any point in the cross-section of the beam and may also be used to determine the distribution over any desired portion. Thus on the upper edge of the top flange $y = +150$ mm, $100\,\text{mm} \geq x \geq -100$ mm, so that the direct stress varies linearly with x. At the top left-hand corner of the top flange

$$\sigma_z = 0.45 \times (+150) - 1.85 \times (-100) = +252.5\,\text{N/mm}^2 \text{ (tension)}$$

At the top right-hand corner

$$\sigma_z = 0.45 \times (+150) - 1.85 \times (+100) = -117.5\,\text{N/mm}^2 \text{ (compression)}$$

The distributions of direct stress over the outer edge of each flange and along the vertical axis of symmetry are shown in Fig. 16.7. Note that the neutral axis of the beam section does not in this case coincide with either the x or y axis, although it still passes through the centroid of the section. Its inclination, α, to the x axis, say, can be found by setting $\sigma_z = 0$ in Eq. (v). Then

$$0 = 0.45y - 1.85x$$

or

$$\frac{y}{x} = \frac{1.85}{0.45} = 4.11 = \tan\alpha$$

which gives

$$\alpha = 76.3°$$

Note that α may be found in general terms from Eq. (ii) by again setting $\sigma_z = 0$. Hence

$$\frac{y}{x} = -\frac{M_y I_{xx}}{M_x I_{yy}} = \tan\alpha \tag{16.12}$$

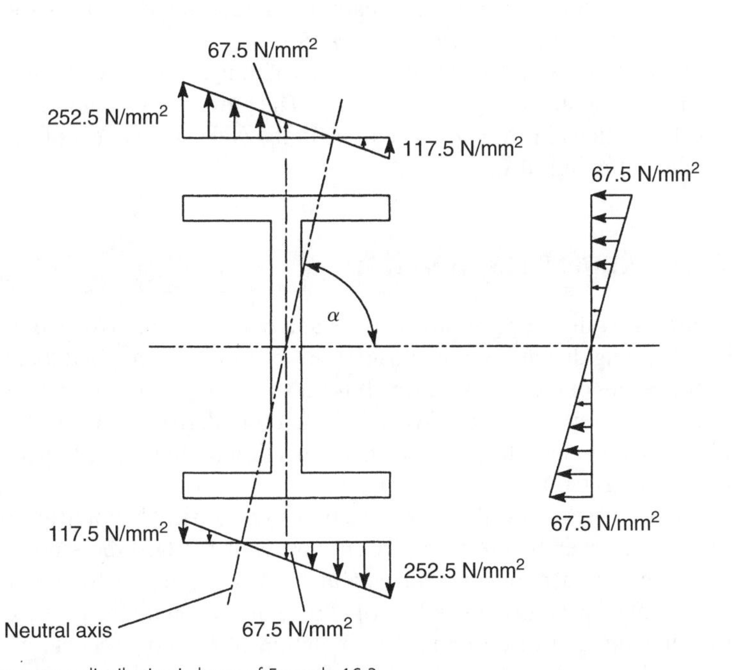

Fig. 16.7 Direct stress distribution in beam of Example 16.3.

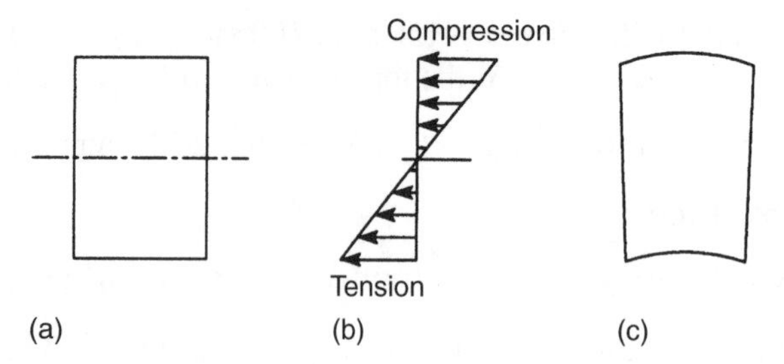

Fig. 16.8 Anticlastic bending of a beam section.

or

$$\tan \alpha = \frac{M_y I_{xx}}{M_x I_{yy}}$$

since y is positive and x is positive for a positive value of α. We shall define in a slightly different way in Section 16.2.4 for beams of unsymmetrical section.

16.1.3 Anticlastic bending

In the rectangular beam section shown in Fig. 16.8(a) the direct stress distribution due to a negative bending moment applied in a vertical plane varies from compression in the upper half of the beam to tension in the lower half (Fig. 16.8(b)). However, due to the Poisson effect the compressive stress produces a lateral elongation of the upper fibres of the beam section while the tensile stress produces a lateral contraction of the lower. The section does not therefore remain rectangular but distorts as shown in Fig. 16.8(c); the effect is known as *anticlastic bending*.

Anticlastic bending is of interest in the analysis of thin-walled box beams in which the cross-sections are maintained by stiffening ribs. The prevention of anticlastic distortion induces local variations in stress distributions in the webs and covers of the box beam and also in the stiffening ribs.

16.2 Unsymmetrical bending

We have shown that the value of direct stress at a point in the cross-section of a beam subjected to bending depends on the position of the point, the applied loading and the geometric properties of the cross-section. It follows that it is of no consequence whether or not the cross-section is open or closed. We therefore derive the theory for a beam of arbitrary cross-section and then discuss its application to thin-walled open and closed section beams subjected to bending moments.

The assumptions are identical to those made for symmetrical bending and are listed in Section 16.1.1. However, before we derive an expression for the direct stress distribution in a beam subjected to bending we shall establish sign conventions for moments, forces and displacements, investigate the effect of choice of section on the positive directions of these parameters and discuss the determination of the components of a bending moment applied in any longitudinal plane.

16.2.1 Sign conventions and notation

Forces, moments and displacements are referred to an arbitrary system of axes Oxyz, of which Oz is parallel to the longitudinal axis of the beam and Oxy are axes in the plane of the cross-section. We assign the symbols M, S, P, T and w to bending moment, shear force, axial or direct load, torque and distributed load intensity, respectively, with suffixes where appropriate to indicate sense or direction. Thus, M_x is a bending moment about the x axis, S_x is a shear force in the x direction and so on. Figure 16.9 shows positive directions and senses for the above loads and moments applied externally to a beam and also the positive directions of the components of displacement u, v and w of any point in the beam cross-section parallel to the x, y and z axes, respectively. A further condition defining the signs of the bending moments M_x and M_y is that they are positive when they induce tension in the positive xy quadrant of the beam cross-section.

If we refer internal forces and moments to that face of a section which is seen when viewed in the direction zO then, as shown in Fig. 16.10, positive internal forces and moments are in the same direction and sense as the externally applied loads whereas on the opposite face they form an opposing system. The former system, which we shall use, has the advantage that direct and shear loads are always positive in the positive directions of the appropriate axes whether they are internal loads or not. It must be realized, though, that internal stress resultants then become equivalent to externally applied forces and moments and are not in equilibrium with them.

16.2.2 Resolution of bending moments

A bending moment M applied in any longitudinal plane parallel to the z axis may be resolved into components M_x and M_y by the normal rules of vectors. However, a visual appreciation of the situation is often helpful. Referring to Fig. 16.11 we see that a bending moment M in a plane at an angle θ to Ox may have components of differing

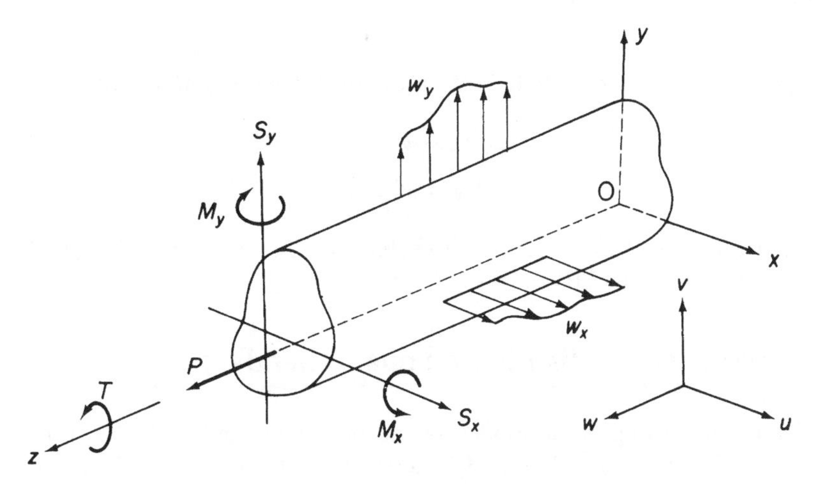

Fig. 16.9 Notation and sign convention for forces, moments and displacements.

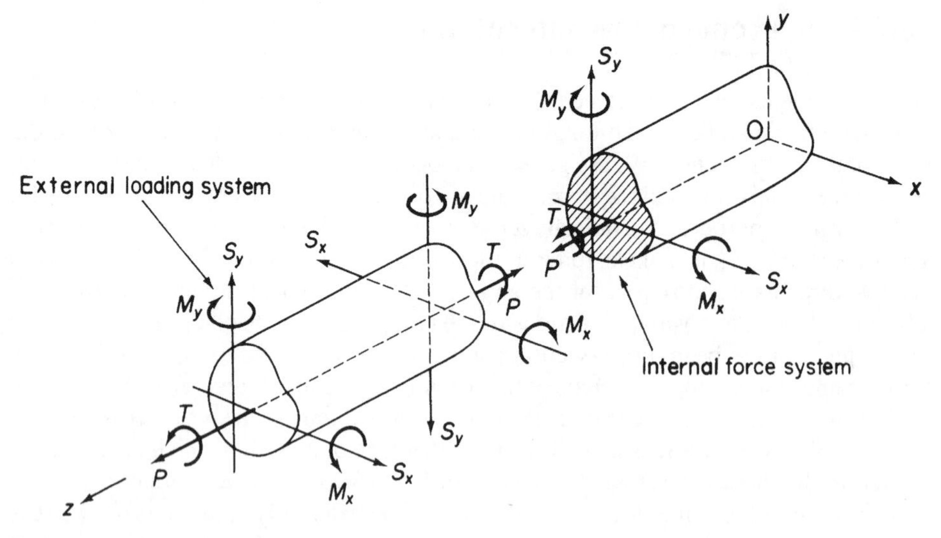

Fig. 16.10 Internal force system.

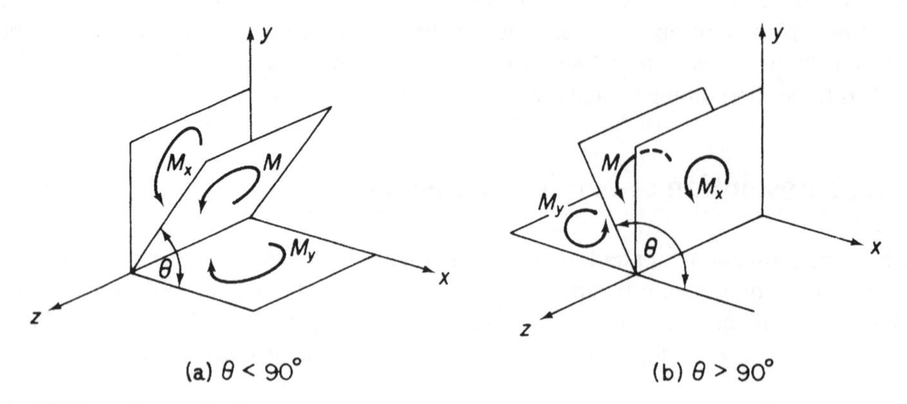

Fig. 16.11 Resolution of bending moments.

sign depending on the size of θ. In both cases, for the sense of M shown

$$M_x = M \sin \theta$$

$$M_y = M \cos \theta$$

which give, for $\theta < \pi/2$, M_x and M_y positive (Fig. 16.11(a)) and for $\theta > \pi/2$, M_x positive and M_y negative (Fig. 16.11(b)).

16.2.3 Direct stress distribution due to bending

Consider a beam having the arbitrary cross-section shown in Fig. 16.12(a). The beam supports bending moments M_x and M_y and bends about some axis in its cross-section which is therefore an axis of zero stress or a *neutral axis* (NA). Let us suppose that the

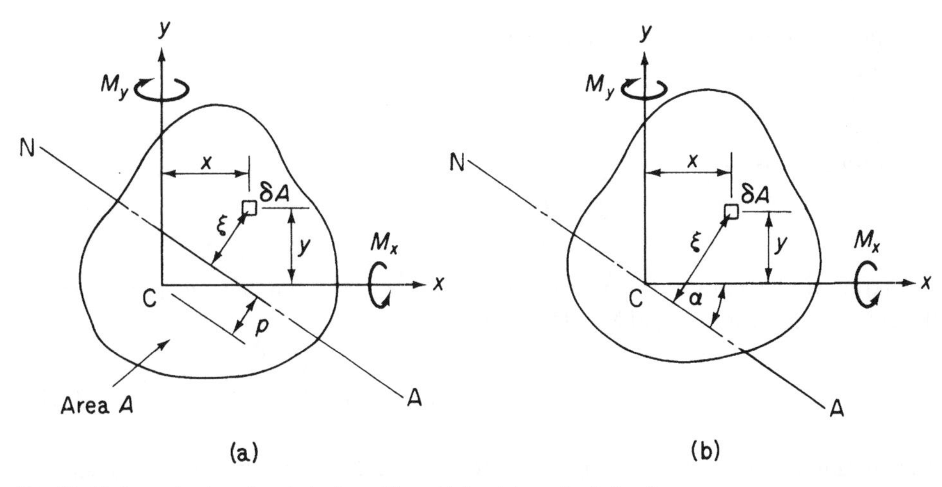

Fig. 16.12 Determination of neutral axis position and direct stress due to bending.

origin of axes coincides with the centroid C of the cross-section and that the neutral axis is a distance p from C. The direct stress σ_z on an element of area δA at a point (x, y) and a distance ξ from the neutral axis is, from the third of Eq. (1.42)

$$\sigma_z = E\varepsilon_z \tag{16.13}$$

If the beam is bent to a radius of curvature ρ about the neutral axis at this particular section then, since plane sections are assumed to remain plane after bending, and by a comparison with symmetrical bending theory

$$\varepsilon_z = \frac{\xi}{\rho}$$

Substituting for ε_z in Eq. (16.13) we have

$$\sigma_z = \frac{E\xi}{\rho} \tag{16.14}$$

The beam supports pure bending moments so that the resultant normal load on any section must be zero. Hence

$$\int_A \sigma_z \, dA = 0$$

Therefore, replacing σ_z in this equation from Eq. (16.14) and cancelling the constant E/ρ gives

$$\int_A \xi \, dA = 0$$

i.e. the first moment of area of the cross-section of the beam about the neutral axis is zero. It follows that the neutral axis passes through the centroid of the cross-section as shown in Fig. 16.12(b) which is the result we obtained for the case of symmetrical bending.

Suppose that the inclination of the neutral axis to Cx is α (measured clockwise from Cx), then

$$\xi = x \sin\alpha + y \cos\alpha \tag{16.15}$$

and from Eq. (16.14)

$$\sigma_z = \frac{E}{\rho}(x \sin\alpha + y \cos\alpha) \tag{16.16}$$

The moment resultants of the internal direct stress distribution have the same sense as the applied moments M_x and M_y. Therefore

$$M_x = \int_A \sigma_z y \, \mathrm{d}A, \quad M_y = \int_A \sigma_z x \, \mathrm{d}A \tag{16.17}$$

Substituting for σ_z from Eq. (16.16) in (16.17) and defining the second moments of area of the section about the axes Cx, Cy as

$$I_{xx} = \int_A y^2 \, \mathrm{d}A, \quad I_{yy} = \int_A x^2 \, \mathrm{d}A, \quad I_{xy} = \int_A xy \, \mathrm{d}A$$

gives

$$M_x = \frac{E \sin\alpha}{\rho} I_{xy} + \frac{E \cos\alpha}{\rho} I_{xx}, \quad M_y = \frac{E \sin\alpha}{\rho} I_{yy} + \frac{E \cos\alpha}{\rho} I_{xy}$$

or, in matrix form

$$\begin{Bmatrix} M_x \\ M_y \end{Bmatrix} = \frac{E}{\rho} \begin{bmatrix} I_{xy} & I_{xx} \\ I_{yy} & I_{xy} \end{bmatrix} \begin{Bmatrix} \sin\alpha \\ \cos\alpha \end{Bmatrix}$$

from which

$$\frac{E}{\rho} \begin{Bmatrix} \sin\alpha \\ \cos\alpha \end{Bmatrix} = \begin{bmatrix} I_{xy} & I_{xx} \\ I_{yy} & I_{xy} \end{bmatrix}^{-1} \begin{Bmatrix} M_x \\ M_y \end{Bmatrix}$$

i.e.

$$\frac{E}{\rho} \begin{Bmatrix} \sin\alpha \\ \cos\alpha \end{Bmatrix} = \frac{1}{I_{xx} I_{yy} - I_{xy}^2} \begin{bmatrix} -I_{xy} & I_{xx} \\ I_{yy} & -I_{xy} \end{bmatrix} \begin{Bmatrix} M_x \\ M_y \end{Bmatrix}$$

so that, from Eq. (16.16)

$$\sigma_z = \left(\frac{M_y I_{xx} - M_x I_{xy}}{I_{xx} I_{yy} - I_{xy}^2} \right) x + \left(\frac{M_x I_{yy} - M_y I_{xy}}{I_{xx} I_{yy} - I_{xy}^2} \right) y \tag{16.18}$$

Alternatively, Eq. (16.18) may be rearranged in the form

$$\sigma_z = \frac{M_x (I_{yy} y - I_{xy} x)}{I_{xx} I_{yy} - I_{xy}^2} + \frac{M_y (I_{xx} x - I_{xy} y)}{I_{xx} I_{yy} - I_{xy}^2} \tag{16.19}$$

From Eq. (16.19) it can be seen that if, say, $M_y = 0$ the moment M_x produces a stress which varies with both x and y; similarly for M_y if $M_x = 0$.

In the case where the beam cross-section has *either* (or both) Cx or Cy as an axis of symmetry the product second moment of area I_{xy} is zero and Cxy are *principal axes.* Equation (16.19) then reduces to

$$\sigma_z = \frac{M_x}{I_{xx}}y + \frac{M_y}{I_{yy}}x \tag{16.20}$$

Further, if either M_y or M_x is zero then

$$\sigma_z = \frac{M_x}{I_{xx}}y \quad \text{or} \quad \sigma_z = \frac{M_y}{I_{yy}}x \tag{16.21}$$

Equations (16.20) and (16.21) are those derived for the bending of beams having at least a singly symmetrical cross-section (see Section 16.1). It may also be noted that in Eq. (16.21) $\sigma_z = 0$ when, for the first equation, $y = 0$ and for the second equation when $x = 0$. Therefore, in symmetrical bending theory the x axis becomes the neutral axis when $M_y = 0$ and the y axis becomes the neutral axis when $M_x = 0$. Thus we see that the position of the neutral axis depends on the form of the applied loading as well as the geometrical properties of the cross-section.

There exists, in any unsymmetrical cross-section, a centroidal set of axes for which the product second moment of area is zero (see Ref. [1]). These axes are then principal axes and the direct stress distribution referred to these axes takes the simplified form of Eqs (16.20) or (16.21). It would therefore appear that the amount of computation can be reduced if these axes are used. This is not the case, however, unless the principal axes are obvious from inspection since the calculation of the position of the principal axes, the principal sectional properties and the coordinates of points at which the stresses are to be determined consumes a greater amount of time than direct use of Eqs (16.18) or (16.19) for an arbitrary, but convenient set of centroidal axes.

16.2.4 Position of the neutral axis

The neutral axis always passes through the centroid of area of a beam's cross-section but its inclination α (see Fig. 16.12(b)) to the x axis depends on the form of the applied loading and the geometrical properties of the beam's cross-section.

At all points on the neutral axis the direct stress is zero. Therefore, from Eq. (16.18)

$$0 = \left(\frac{M_y I_{xx} - M_x I_{xy}}{I_{xx}I_{yy} - I_{xy}^2}\right) x_{\text{NA}} + \left(\frac{M_x I_{yy} - M_y I_{xy}}{I_{xx}I_{yy} - I_{xy}^2}\right) y_{\text{NA}}$$

where x_{NA} and y_{NA} are the coordinates of any point on the neutral axis. Hence

$$\frac{y_{\text{NA}}}{x_{\text{NA}}} = -\frac{M_y I_{xx} - M_x I_{xy}}{M_x I_{yy} - M_y I_{xy}}$$

or, referring to Fig. 16.12(b) and noting that when α is positive x_{NA} and y_{NA} are of opposite sign

$$\tan \alpha = \frac{M_y I_{xx} - M_x I_{xy}}{M_x I_{yy} - M_y I_{xy}} \tag{16.22}$$

Example 16.4

A beam having the cross-section shown in Fig. 16.13 is subjected to a bending moment of 1500 N m in a vertical plane. Calculate the maximum direct stress due to bending stating the point at which it acts.

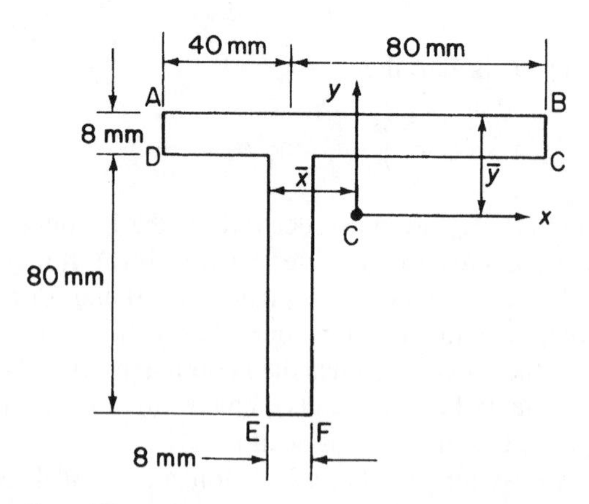

Fig. 16.13 Cross-section of beam in Example 16.4.

The position of the centroid of the section may be found by taking moments of areas about some convenient point. Thus

$$(120 \times 8 + 80 \times 8)\bar{y} = 120 \times 8 \times 4 + 80 \times 8 \times 48$$

giving

$$\bar{y} = 21.6\,\text{mm}$$

and

$$(120 \times 8 + 80 \times 8)\bar{x} = 80 \times 8 \times 4 + 120 \times 8 \times 24$$

giving

$$\bar{x} = 16\,\text{mm}$$

The next step is to calculate the section properties referred to axes Cxy (see Section 16.4)

$$I_{xx} = \frac{120 \times (8)^3}{12} + 120 \times 8 \times (17.6)^2 + \frac{8 \times (80)^3}{12} + 80 \times 8 \times (26.4)^2$$

$$= 1.09 \times 10^6\,\text{mm}^4$$

$$I_{yy} = \frac{8 \times (120)^3}{12} + 120 \times 8 \times (8)^2 + \frac{80 \times (8)^3}{12} + 80 \times 8 \times (12)^2$$

$$= 1.31 \times 10^6\,\text{mm}^4$$

$$I_{xy} = 120 \times 8 \times 8 \times 17.6 + 80 \times 8 \times (-12) \times (-26.4)$$

$$= 0.34 \times 10^6\,\text{mm}^4$$

Since $M_x = 1500\,\text{N m}$ and $M_y = 0$ we have, from Eq. (16.19)

$$\sigma_z = 1.5y - 0.39x \qquad \text{(i)}$$

in which the units are N and mm.

By inspection of Eq. (i) we see that σ_x will be a maximum at F where $x = -8\,\text{mm}$, $y = -66.4\,\text{mm}$. Thus

$$\sigma_{z,\max} = -96\,\text{N/mm}^2 \text{ (compressive)}$$

In some cases the maximum value cannot be obtained by inspection so that values of σ_z at several points must be calculated.

16.2.5 Load intensity, shear force and bending moment relationships, general case

Consider an element of length δz of a beam of unsymmetrical cross-section subjected to shear forces, bending moments and a distributed load of varying intensity, all in the yz plane as shown in Fig. 16.14. The forces and moments are positive in accordance with the sign convention previously adopted. Over the length of the element we may assume that the intensity of the distributed load is constant. Therefore, for equilibrium of the element in the y direction

$$\left(S_y + \frac{\partial S_y}{\partial z}\delta z\right) + w_y \delta z - S_y = 0$$

from which

$$w_y = -\frac{\partial S_y}{\partial z}$$

Taking moments about A we have

$$\left(M_x + \frac{\partial M_x}{\partial z}\delta z\right) - \left(S_y + \frac{\partial S_y}{\partial z}\delta z\right)\delta z - w_y\frac{(\delta z)^2}{2} - M_x = 0$$

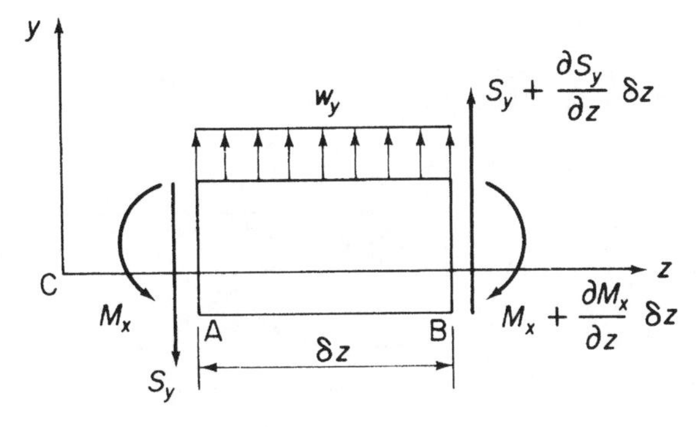

Fig. 16.14 Equilibrium of beam element supporting a general force system in the yz plane.

or, when second-order terms are neglected

$$S_y = \frac{\partial M_x}{\partial z}$$

We may combine these results into a single expression

$$-w_y = \frac{\partial S_y}{\partial z} = \frac{\partial^2 M_x}{\partial z^2} \tag{16.23}$$

Similarly for loads in the xz plane

$$-w_x = \frac{\partial S_x}{\partial z} = \frac{\partial^2 M_y}{\partial z^2} \tag{16.24}$$

16.3 Deflections due to bending

We have noted that a beam bends about its neutral axis whose inclination relative to arbitrary centroidal axes is determined from Eq. (16.22). Suppose that at some section of an unsymmetrical beam the deflection normal to the neutral axis (and therefore an absolute deflection) is ζ, as shown in Fig. 16.15. In other words the centroid C is displaced from its initial position C_I through an amount ζ to its final position C_F. Suppose also that the centre of curvature R of the beam at this particular section is on the opposite side of the neutral axis to the direction of the displacement ζ and that the radius of curvature is ρ. For this position of the centre of curvature and from the usual approximate expression for curvature we have

$$\frac{1}{\rho} = \frac{d^2\zeta}{dz^2} \tag{16.25}$$

The components u and v of ζ are in the negative directions of the x and y axes, respectively, so that

$$u = -\zeta \sin\alpha \quad v = -\zeta \cos\alpha \tag{16.26}$$

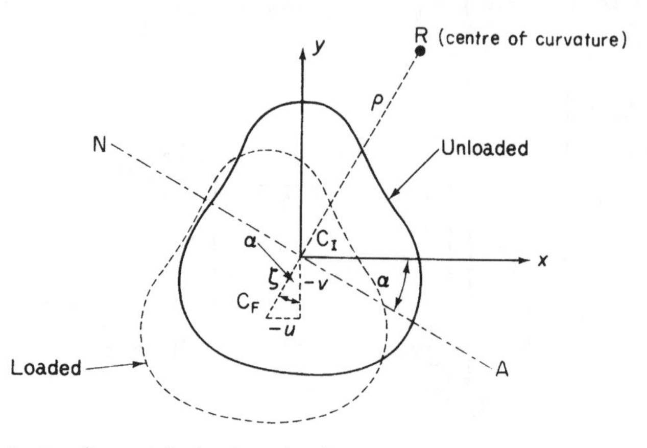

Fig. 16.15 Determination of beam deflection due to bending.

Differentiating Eqs (16.26) twice with respect to z and then substituting for ζ from Eq. (16.25) we obtain

$$\frac{\sin\alpha}{\rho} = -\frac{d^2u}{dz^2}, \quad \frac{\cos\alpha}{\rho} = -\frac{d^2v}{dz^2} \tag{16.27}$$

In the derivation of Eq. (16.18) we see that

$$\frac{1}{\rho}\begin{Bmatrix}\sin\alpha \\ \cos\alpha\end{Bmatrix} = \frac{1}{E(I_{xx}I_{yy} - I_{xy}^2)}\begin{bmatrix}-I_{xy} & I_{xx} \\ I_{yy} & -I_{xy}\end{bmatrix}\begin{Bmatrix}M_x \\ M_y\end{Bmatrix} \tag{16.28}$$

Substituting in Eqs (16.28) for $\sin\alpha/\rho$ and $\cos\alpha/\rho$ from Eqs (16.27) and writing $u'' = d^2u/dz^2$, $v'' = d^2v/dz^2$ we have

$$\begin{Bmatrix}u'' \\ v''\end{Bmatrix} = \frac{-1}{E(I_{xx}I_{yy} - I_{xy}^2)}\begin{bmatrix}-I_{xy} & I_{xx} \\ I_{yy} & -I_{xy}\end{bmatrix}\begin{Bmatrix}M_x \\ M_y\end{Bmatrix} \tag{16.29}$$

It is instructive to rearrange Eq. (16.29) as follows

$$\begin{Bmatrix}M_x \\ M_y\end{Bmatrix} = -E\begin{bmatrix}I_{xy} & I_{xx} \\ I_{yy} & I_{xy}\end{bmatrix}\begin{Bmatrix}u'' \\ v''\end{Bmatrix} \quad \text{(see derivation of Eq. (16.18))} \tag{16.30}$$

i.e.

$$\left.\begin{aligned} M_x &= -EI_{xy}u'' - EI_{xx}v'' \\ M_y &= -EI_{yy}u'' - EI_{xy}v'' \end{aligned}\right\} \tag{16.31}$$

The first of Eqs (16.31) shows that M_x produces curvatures, i.e. deflections, in both the xz and yz planes even though $M_y = 0$; similarly for M_y when $M_x = 0$. Thus, for example, an unsymmetrical beam will deflect both vertically and horizontally even though the loading is entirely in a vertical plane. Similarly, vertical and horizontal components of deflection in an unsymmetrical beam are produced by horizontal loads.

For a beam having either Cx or Cy (or both) as an axis of symmetry, $I_{xy} = 0$ and Eqs (16.29) reduce to

$$u'' = -\frac{M_y}{EI_{yy}}, \quad v'' = -\frac{M_x}{EI_{xx}} \tag{16.32}$$

Example 16.5

Determine the deflection curve and the deflection of the free end of the cantilever shown in Fig. 16.16(a); the flexural rigidity of the cantilever is EI and its section is doubly symmetrical.

The load W causes the cantilever to deflect such that its neutral plane takes up the curved shape shown Fig. 16.16(b); the deflection at any section Z is then v while that at its free end is v_{tip}. The axis system is chosen so that the origin coincides with the built-in end where the deflection is clearly zero.

The bending moment, M, at the section Z is, from Fig. 16.16(a)

$$M = W(L - z) \tag{i}$$

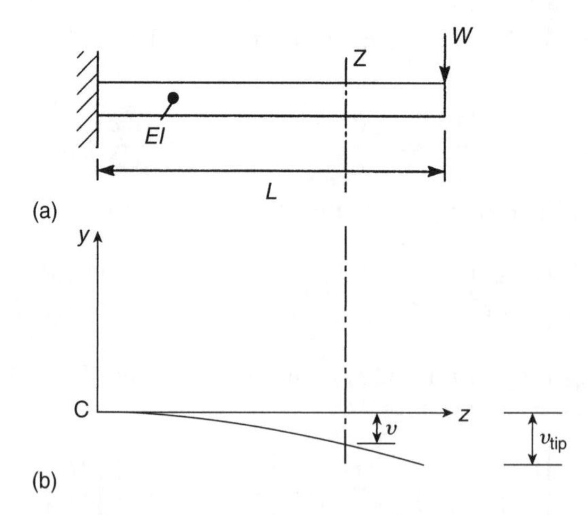

Fig. 16.16 Deflection of a cantilever beam carrying a concentrated load at its free end (Example 16.5).

Substituting for M in the second of Eq. (16.32)

$$v'' = -\frac{W}{EI}(L - z)$$

or in more convenient form

$$EIv'' = -W(L - z) \tag{ii}$$

Integrating Eq. (ii) with respect to z gives

$$EIv'' = -W\left(Lz - \frac{z^2}{2}\right) + C_1$$

where C_1 is a constant of integration which is obtained from the boundary condition that $v' = 0$ at the built-in end where $z = 0$. Hence $C_1 = 0$ and

$$EIv' = -W\left(Lz - \frac{z^2}{2}\right) \tag{iii}$$

Integrating Eq. (iii) we obtain

$$EIv = -W\left(\frac{Lz^2}{2} - \frac{z^3}{6}\right) + C_2$$

in which C_2 is again a constant of integration. At the built-in end $v = 0$ when $z = 0$ so that $C_2 = 0$. Hence the equation of the deflection curve of the cantilever is

$$v = -\frac{W}{6EI}(3Lz^2 - z^3) \tag{iv}$$

The deflection, v_{tip}, at the free end is obtained by setting $z = L$ in Eq. (iv). Then

$$v_{\text{tip}} = -\frac{WL^3}{3EI} \tag{v}$$

and is clearly negative and downwards.

Example 16.6

Determine the deflection curve and the deflection of the free end of the cantilever shown in Fig. 16.17(a). The cantilever has a doubly symmetrical cross-section.

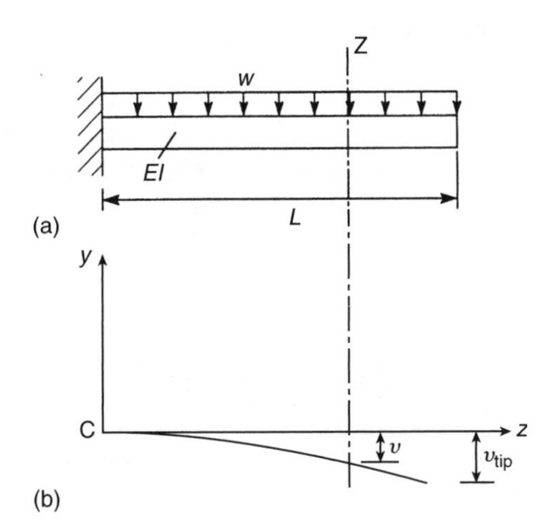

Fig. 16.17 Deflection of a cantilever beam carrying a uniformly distributed load.

The bending moment, M, at any section Z is given by

$$M = \frac{w}{2}(L - z)^2 \tag{i}$$

Substituting for M in the second of Eq. (16.32) and rearranging we have

$$EIv'' = -\frac{w}{2}(L - z)^2 = -\frac{w}{2}(L^2 - 2Lz + z^2) \tag{ii}$$

Integration of Eq. (ii) yields

$$EIv' = -\frac{w}{2}\left(L^2 z - Lz^2 + \frac{z^3}{3}\right) + C_1$$

When $z = 0$ at the built-in end, $v' = 0$ so that $C_1 = 0$ and

$$EIv' = -\frac{w}{2}\left(L^2 z - Lz^2 + \frac{z^3}{3}\right) \tag{iii}$$

Integrating Eq. (iii) we have

$$EIv = -\frac{w}{2}\left(L^2\frac{z^2}{2} - \frac{Lz^3}{3} + \frac{z^4}{12}\right) + C_2$$

and since $v = 0$ when $x = 0$, $C_2 = 0$. The deflection curve of the beam therefore has the equation

$$v = -\frac{w}{24EI}(6L^2z^2 - 4Lz^3 + z^4) \tag{iv}$$

and the deflection at the free end where $x = L$ is

$$v_{\text{tip}} = -\frac{wL^4}{8EI} \tag{v}$$

which is again negative and downwards.

Example 16.7

Determine the deflection curve and the mid-span deflection of the simply supported beam shown in Fig. 16.18(a); the beam has a doubly symmetrical cross-section.

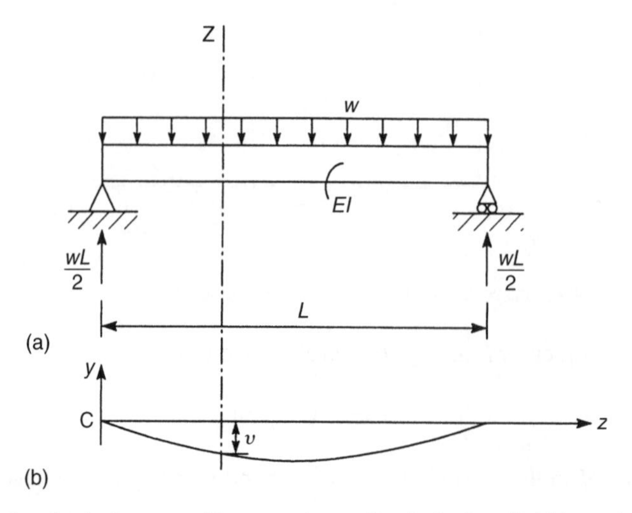

Fig. 16.18 Deflection of a simply supported beam carrying a uniformly distributed load (Example 16.7).

The support reactions are each $wL/2$ and the bending moment, M, at any section Z, a distance z from the left-hand support is

$$M = -\frac{wL}{2}z + \frac{wz^2}{2} \tag{i}$$

Substituting for M in the second of Eq. (16.32) we obtain

$$EIv'' = \frac{w}{2}(Lz - z^2) \tag{ii}$$

Integrating we have

$$EIv' = \frac{w}{2}\left(\frac{Lz^2}{2} - \frac{z^3}{3}\right) + C_1$$

From symmetry it is clear that at the mid-span section the gradient $v' = 0$. Hence

$$0 = \frac{w}{2}\left(\frac{L^3}{8} - \frac{L^3}{24}\right) + C_1$$

which gives

$$C_1 = -\frac{wL^3}{24}$$

Therefore

$$EIv' = \frac{w}{24}(6Lz^2 - 4z^3 - L^3) \tag{iii}$$

Integrating again gives

$$EIv = \frac{w}{24}(2Lz^3 - z^4 - L^3z) + C_2$$

Since $v = 0$ when $z = 0$ (or since $v = 0$ when $z = L$) it follows that $C_2 = 0$ and the deflected shape of the beam has the equation

$$v = \frac{w}{24EI}(2Lz^3 - z^4 - L^3z) \tag{iv}$$

The maximum deflection occurs at mid-span where $z = L/2$ and is

$$v_{\text{mid-span}} = -\frac{5wL^4}{384EI} \tag{v}$$

So far the constants of integration were determined immediately they arose. However, in some cases a relevant boundary condition, say a value of gradient, is not obtainable. The method is then to carry the unknown constant through the succeeding integration and use known values of deflection at two sections of the beam. Thus in the previous example Eq. (ii) is integrated twice to obtain

$$EIv = \frac{w}{2}\left(\frac{Lz^3}{6} - \frac{z^4}{12}\right) + C_1z + C_2$$

The relevant boundary conditions are $v = 0$ at $z = 0$ and $z = L$. The first of these gives $C_2 = 0$ while from the second we have $C_1 = -wL^3/24$. Thus, the equation of the deflected shape of the beam is

$$v = \frac{w}{24EI}(2Lz^3 - z^4 - L^3z)$$

as before.

Example 16.8

Figure 16.19(a) shows a simply supported beam carrying a concentrated load W at mid-span. Determine the deflection curve of the beam and the maximum deflection if the beam section is doubly symmetrical.

The support reactions are each $W/2$ and the bending moment M at a section Z a distance $z(\leq L/2)$ from the left-hand support is

$$M = -\frac{W}{2}z \tag{i}$$

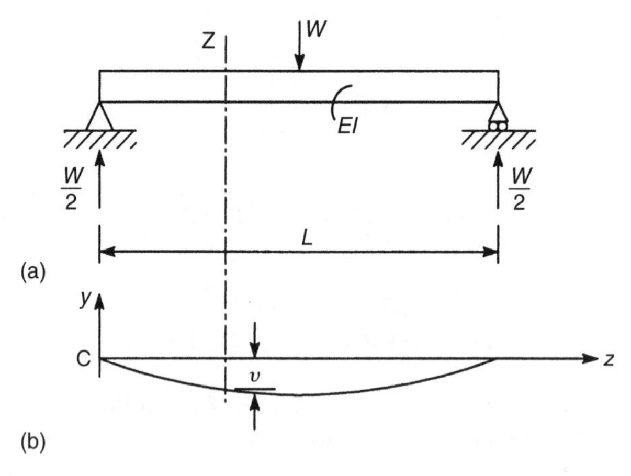

Fig. 16.19 Deflection of a simply supported beam carrying a concentrated load at mid-span (Example 16.8).

From the second of Eq. (16.32) we have

$$EIv'' = \frac{W}{2}z \qquad \text{(ii)}$$

Integrating we obtain

$$EIv' = \frac{W}{2}\frac{z^2}{2} + C_1$$

From symmetry the slope of the beam is zero at mid-span where $z = L/2$. Thus $C_1 = -WL^2/16$ and

$$EIv' = \frac{W}{16}(4z^2 - L^2) \qquad \text{(iii)}$$

Integrating Eq. (iii) we have

$$EIv = \frac{W}{16}\left(\frac{4z^3}{3} - L^2z\right) + C_2$$

and when $z = 0$, $v = 0$ so that $C_2 = 0$. The equation of the deflection curve is, therefore

$$v = \frac{W}{48EI}(4z^3 - 3L^2z) \qquad \text{(iv)}$$

The maximum deflection occurs at mid-span and is

$$v_{\text{mid-span}} = -\frac{WL^3}{48EI} \qquad \text{(v)}$$

Note that in this problem we could not use the boundary condition that $v = 0$ at $z = L$ to determine C_2 since Eq. (i) applies only for $0 \leq z \leq L/2$; it follows that Eqs (iii) and (iv) for slope and deflection apply only for $0 \leq z \leq L/2$ although the deflection curve is clearly symmetrical about mid-span.

Examples 16.5–16.8 are frequently regarded as 'standard' cases of beam deflection.

16.3.1 Singularity functions

The double integration method used in Examples 16.5–16.8 becomes extremely lengthy when even relatively small complications such as the lack of symmetry due to an offset load are introduced. For example, the addition of a second concentrated load on a simply supported beam would result in a total of six equations for slope and deflection producing six arbitrary constants. Clearly the computation involved in determining these constants would be tedious, even though a simply supported beam carrying two concentrated loads is a comparatively simple practical case. An alternative approach is to introduce so-called *singularity* or *half-range* functions. Such functions were first applied to beam deflection problems by Macauley in 1919 and hence the method is frequently known as *Macauley's method.*

We now introduce a quantity $[z-a]$ and define it to be zero if $(z-a)<0$, i.e. $z<a$, and to be simply $(z-a)$ if $z>a$. The quantity $[z-a]$ is known as a singularity or half-range function and is defined to have a value only when the argument is positive in which case the square brackets behave in an identical manner to ordinary parentheses.

Example 16.9

Determine the position and magnitude of the maximum upward and downward deflections of the beam shown in Fig. 16.20.

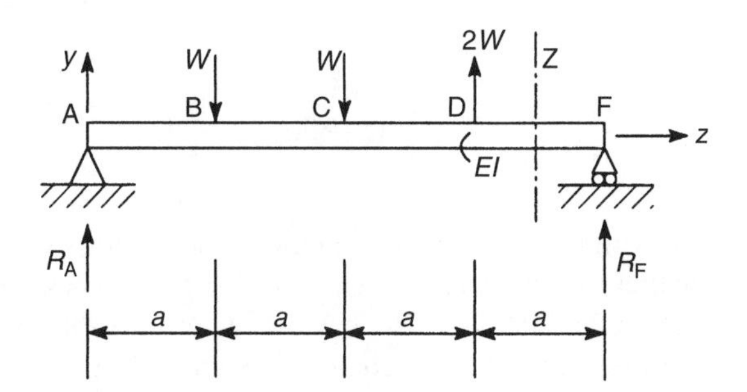

Fig. 16.20 Macauley's method for the deflection of a simply supported beam (Example 16.9).

A consideration of the overall equilibrium of the beam gives the support reactions; thus

$$R_A = \frac{3}{4}W \text{ (upward)} \quad R_F = \frac{3}{4}W \text{ (downward)}$$

Using the method of singularity functions and taking the origin of axes at the left-hand support, we write down an expression for the bending moment, M, at any section Z between D and F, *the region of the beam furthest from the origin.* Thus

$$M = -R_A z + W[z-a] + W[z-2a] - 2W[z-3a] \tag{i}$$

Substituting for M in the second of Eq. (16.32) we have

$$EIv'' = \frac{3}{4}Wz - W[z-a] - W[z-2a] + 2W[z-3a] \tag{ii}$$

Integrating Eq. (ii) and retaining the square brackets we obtain

$$EIv' = \frac{3}{8}Wz^2 - \frac{W}{2}[z-a]^2 - \frac{W}{2}[z-2a]^2 + W[z-3a]^2 + C_1 \qquad \text{(iii)}$$

and

$$EIv = \frac{1}{8}Wz^3 - \frac{W}{6}[z-a]^3 - \frac{W}{6}[z-2a]^3 + \frac{W}{3}[z-3a]^3 + C_1z + C_2 \qquad \text{(iv)}$$

in which C_1 and C_2 are arbitrary constants. When $z=0$ (at A), $v=0$ and hence $C_2=0$. Note that the second, third and fourth terms on the right-hand side of Eq. (iv) disappear for $z<a$. Also $v=0$ at $z=4a$ (F) so that, from Eq. (iv), we have

$$0 = \frac{W}{8}64a^3 - \frac{W}{6}27a^3 - \frac{W}{6}8a^3 + \frac{W}{3}a^3 + 4aC_1$$

which gives

$$C_1 = -\frac{5}{8}Wa^2$$

Equations (iii) and (iv) now become

$$EIv' = \frac{3}{8}Wz^2 - \frac{W}{2}[z-a]^2 - \frac{W}{2}[z-2a]^2 + W[z-3a]^2 - \frac{5}{8}Wa^2 \qquad \text{(v)}$$

and

$$EIv = \frac{1}{8}Wz^3 - \frac{W}{6}[z-a]^3 - \frac{W}{6}[z-2a]^3 + \frac{W}{3}[z-3a]^3 - \frac{5}{8}Wa^2z \qquad \text{(vi)}$$

respectively.

To determine the maximum upward and downward deflections we need to know in which bays $v'=0$ and thereby which terms in Eq. (v) disappear when the exact positions are being located. One method is to select a bay and determine the sign of the slope of the beam at the extremities of the bay. A change of sign will indicate that the slope is zero within the bay.

By inspection of Fig. 16.20 it seems likely that the maximum downward deflection will occur in BC. At B, using Eq. (v)

$$EIv' = \frac{3}{8}Wa^2 - \frac{5}{8}Wa^2$$

which is clearly negative. At C

$$EIv' = \frac{3}{8}W4a^2 - \frac{W}{2}a^2 - \frac{5}{8}Wa^2$$

which is positive. Therefore, the maximum downward deflection does occur in BC and its exact position is located by equating v' to zero for any section in BC. Thus, from Eq. (v)

$$0 = \frac{3}{8}Wz^2 - \frac{W}{2}[z-a]^2 - \frac{5}{8}Wa^2$$

or, simplifying,

$$0 = z^2 - 8az + 9a^2 \tag{vii}$$

Solution of Eq. (vii) gives

$$z = 1.35a$$

so that the maximum downward deflection is, from Eq. (vi)

$$EIv = \frac{1}{8}W(1.35a)^3 - \frac{W}{6}(0.35a)^3 - \frac{5}{8}Wa^2(1.35a)$$

i.e.

$$v_{\max}(\text{downward}) = -\frac{0.54Wa^3}{EI}$$

In a similar manner it can be shown that the maximum upward deflection lies between D and F at $z = 3.42a$ and that its magnitude is

$$v_{\max}(\text{upward}) = \frac{0.04Wa^3}{EI}$$

An alternative method of determining the position of maximum deflection is to select a possible bay, set $v' = 0$ for that bay and solve the resulting equation in z. If the solution gives a value of z that lies within the bay, then the selection is correct, otherwise the procedure must be repeated for a second and possibly a third and a fourth bay. This method is quicker than the former if the correct bay is selected initially; if not, the equation corresponding to each selected bay must be completely solved, a procedure clearly longer than determining the sign of the slope at the extremities of the bay.

Example 16.10

Determine the position and magnitude of the maximum deflection in the beam of Fig. 16.21.

Following the method of Example 16.9 we determine the support reactions and find the bending moment, M, at any section Z in the bay furthest from the origin of the axes.

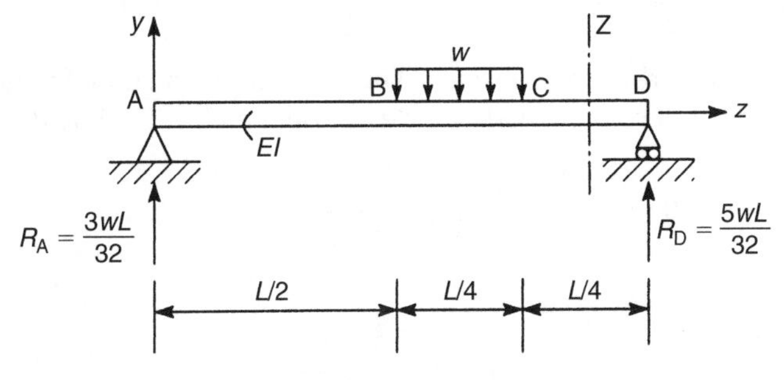

Fig. 16.21 Deflection of a beam carrying a part span uniformly distributed load (Example 16.10).

Then

$$M = -R_A z + w\frac{L}{4}\left[z - \frac{5L}{8}\right] \qquad \text{(i)}$$

Examining Eq. (i) we see that the singularity function $[z - 5L/8]$ does not become zero until $z \leq 5L/8$ although Eq. (i) is only valid for $z \geq 3L/4$. To obviate this difficulty we extend the distributed load to the support D while simultaneously restoring the *status quo* by applying an upward distributed load of the same intensity and length as the additional load (Fig. 16.22).

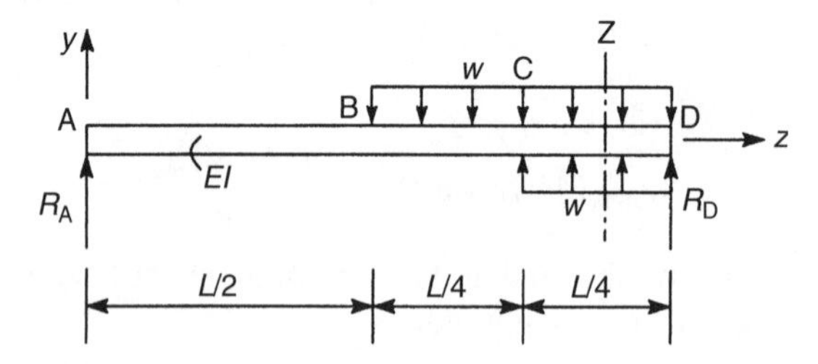

Fig. 16.22 Method of solution for a part span uniformly distributed load.

At the section Z, a distance z from A, the bending moment is now given by

$$M = -R_A z + \frac{w}{2}\left[z - \frac{L}{2}\right]^2 - \frac{w}{2}\left[z - \frac{3L}{4}\right]^2 \qquad \text{(ii)}$$

Equation (ii) is now valid for all sections of the beam if the singularity functions are discarded as they become zero. Substituting Eq. (ii) into the second of Eqs (16.32) we obtain

$$EIv'' = \frac{3}{32}wLz - \frac{w}{2}\left[z - \frac{L}{2}\right]^2 + \frac{w}{2}\left[z - \frac{3L}{4}\right]^2 \qquad \text{(iii)}$$

Integrating Eq. (iii) gives

$$EIv' = \frac{3}{64}wLz^2 - \frac{w}{6}\left[z - \frac{L}{2}\right]^3 + \frac{w}{6}\left[z - \frac{3L}{4}\right]^3 + C_1 \qquad \text{(iv)}$$

$$EIv = \frac{wLz^3}{64} - \frac{w}{24}\left[z - \frac{L}{2}\right]^4 + \frac{w}{24}\left[z - \frac{3L}{4}\right]^4 + C_1 z + C_2 \qquad \text{(v)}$$

where C_1 and C_2 are arbitrary constants. The required boundary conditions are $v = 0$ when $z = 0$ and $z = L$. From the first of these we obtain $C_2 = 0$ while the second gives

$$0 = \frac{wL^4}{64} - \frac{w}{24}\left(\frac{L}{2}\right)^4 + \frac{w}{24}\left(\frac{L}{4}\right)^4 + C_1 L$$

from which

$$C_1 = -\frac{27wL^3}{2048}$$

Equations (iv) and (v) then become

$$EIv' = \frac{3}{64}wLz^2 - \frac{w}{6}\left[z - \frac{L}{2}\right]^3 + \frac{w}{6}\left[z - \frac{3L}{4}\right]^3 - \frac{27wL^3}{2048} \quad \text{(vi)}$$

and

$$EIv = \frac{wLz^3}{64} - \frac{w}{24}\left[z - \frac{L}{2}\right]^4 + \frac{w}{24}\left[z - \frac{3L}{4}\right]^4 - \frac{27wL^3}{2048}z \quad \text{(vii)}$$

In this problem, the maximum deflection clearly occurs in the region BC of the beam. Thus equating the slope to zero for BC we have

$$0 = \frac{3}{64}wLz^2 - \frac{w}{6}\left[z - \frac{L}{2}\right]^3 - \frac{27wL^3}{2048}$$

which simplifies to

$$z^3 - 1.78Lz^2 + 0.75zL^2 - 0.046L^3 = 0 \quad \text{(viii)}$$

Solving Eq. (viii) by trial and error, we see that the slope is zero at $z \simeq 0.6L$. Hence from Eq. (vii) the maximum deflection is

$$v_{\max} = -\frac{4.53 \times 10^{-3}wL^4}{EI}$$

Example 16.11

Determine the deflected shape of the beam shown in Fig. 16.23.

In this problem an external moment M_0 is applied to the beam at B. The support reactions are found in the normal way and are

$$R_A = -\frac{M_0}{L}\text{(downwards)} \quad R_C = \frac{M_0}{L}\text{(upwards)}$$

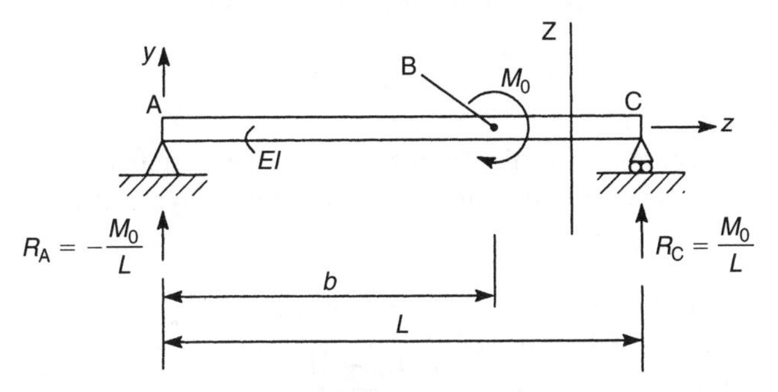

Fig. 16.23 Deflection of a simply supported beam carrying a point moment (Example 16.11).

The bending moment at any section Z between B and C is then given by

$$M = -R_A z - M_0 \tag{i}$$

Equation (i) is valid only for the region BC and clearly does not contain a singularity function which would cause M_0 to vanish for $z \leq b$. We overcome this difficulty by writing

$$M = -R_A z - M_0[z-b]^0 \quad (\text{Note: } [z-b]^0 = 1) \tag{ii}$$

Equation (ii) has the same value as Eq. (i) but is now applicable to all sections of the beam since $[z-b]^0$ disappears when $z \leq b$. Substituting for M from Eq. (ii) in the second of Eq. (16.32) we obtain

$$EIv'' = R_A z + M_0[z-b]^0 \tag{iii}$$

Integration of Eq. (iii) yields

$$EIv' = R_A \frac{z^2}{2} + M_0[z-b] + C_1 \tag{vi}$$

and

$$EIv = R_A \frac{z^3}{6} + \frac{M_0}{2}[z-b]^2 + C_1 z + C_2 \tag{v}$$

where C_1 and C_2 are arbitrary constants. The boundary conditions are $v=0$ when $z=0$ and $z=L$. From the first of these we have $C_2=0$ while the second gives

$$0 = -\frac{M_0}{L}\frac{L^3}{6} + \frac{M_0}{2}[L-b]^2 + C_1 L$$

from which

$$C_1 = -\frac{M_0}{6L}(2L^2 - 6Lb + 3b^2)$$

The equation of the deflection curve of the beam is then

$$v = \frac{M_0}{6EIL}\{z^3 + 3L[z-b]^2 - (2L^2 - 6Lb + 3b^2)z\} \tag{vi}$$

Example 16.12

Determine the horizontal and vertical components of the tip deflection of the cantilever shown in Fig. 16.24. The second moments of area of its unsymmetrical section are I_{xx}, I_{yy} and I_{xy}.

From Eqs (16.29)

$$u'' = \frac{M_x I_{xy} - M_y I_{xx}}{E(I_{xx}I_{yy} - I_{xy}^2)} \tag{i}$$

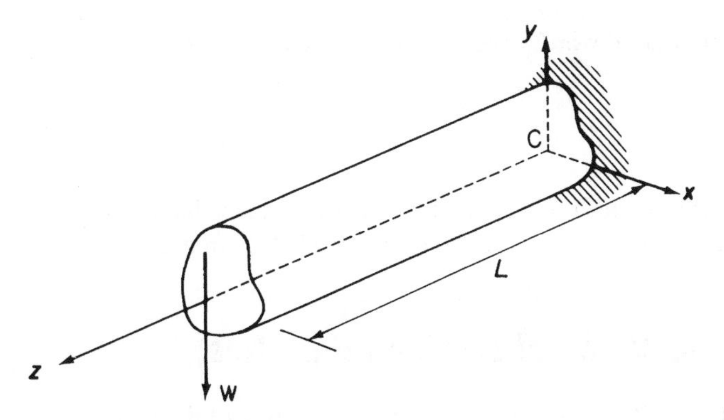

Fig. 16.24 Determination of the deflection of a cantilever.

In this case $M_x = W(L - z)$, $M_y = 0$ so that Eq. (i) simplifies to

$$u'' = \frac{WI_{xy}}{E(I_{xx}I_{yy} - I_{xy}^2)}(L - z) \tag{ii}$$

Integrating Eq. (ii) with respect to z

$$u' = \frac{WI_{xy}}{E(I_{xx}I_{yy} - I_{xy}^2)}\left(Lz - \frac{z^2}{2} + A\right) \tag{iii}$$

and

$$u = \frac{WI_{xy}}{E(I_{xx}I_{yy} - I_{xy}^2)}\left(L\frac{z^2}{2} - \frac{z^3}{6} + Az + B\right) \tag{iv}$$

in which u' denotes du/dz and the constants of integration A and B are found from the boundary conditions, viz. $u' = 0$ and $u = 0$ when $z = 0$. From the first of these and Eq. (iii), $A = 0$, while from the second and Eq. (iv), $B = 0$. Hence the deflected shape of the beam in the xz plane is given by

$$u = \frac{WI_{xy}}{E(I_{xx}I_{yy} - I_{xy}^2)}\left(L\frac{z^2}{2} - \frac{z^3}{6}\right) \tag{v}$$

At the free end of the cantilever ($z = L$) the horizontal component of deflection is

$$u_{\text{f.e.}} = \frac{WI_{xy}L^3}{3E(I_{xx}I_{yy} - I_{xy}^2)} \tag{vi}$$

Similarly, the vertical component of the deflection at the free end of the cantilever is

$$v_{\text{f.e.}} = \frac{-WI_{yy}L^3}{3E(I_{xx}I_{yy} - I_{xy}^2)} \tag{vii}$$

The actual deflection $\delta_{\text{f.e.}}$ at the free end is then given by

$$\delta_{\text{f.e.}} = (u_{\text{f.e.}}^2 + v_{\text{f.e.}}^2)^{\frac{1}{2}}$$

at an angle of $\tan^{-1} u_{\text{f.e.}}/v_{\text{f.e.}}$ to the vertical.

Note that if either Cx or Cy were an axis of symmetry, $I_{xy} = 0$ and Eqs (vi) and (vii) reduce to

$$u_{\text{f.e.}} = 0 \quad v_{\text{f.e.}} = \frac{-WL^3}{3EI_{xx}}$$

the well-known results for the bending of a cantilever having a symmetrical cross-section and carrying a concentrated vertical load at its free end (see Example 16.5).

16.4 Calculation of section properties

It will be helpful at this stage to discuss the calculation of the various section properties required in the analysis of beams subjected to bending. Initially, however, two useful theorems are quoted.

16.4.1 Parallel axes theorem

Consider the beam section shown in Fig. 16.25 and suppose that the second moment of area, I_C, about an axis through its centroid C is known. The second moment of area, I_N, about a parallel axis, NN, a distance b from the centroidal axis is then given by

$$I_\text{N} = I_\text{C} + Ab^2 \tag{16.33}$$

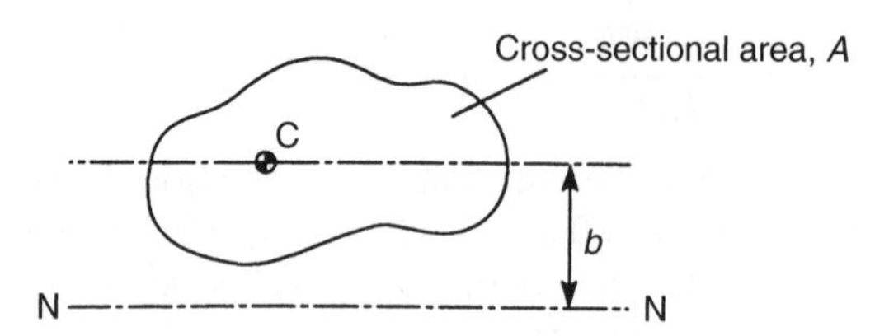

Fig. 16.25 Parallel axes theorem.

16.4.2 Theorem of perpendicular axes

In Fig. 16.26 the second moments of area, I_{xx} and I_{yy}, of the section about Ox and Oy are known. The second moment of area about an axis through O perpendicular to the

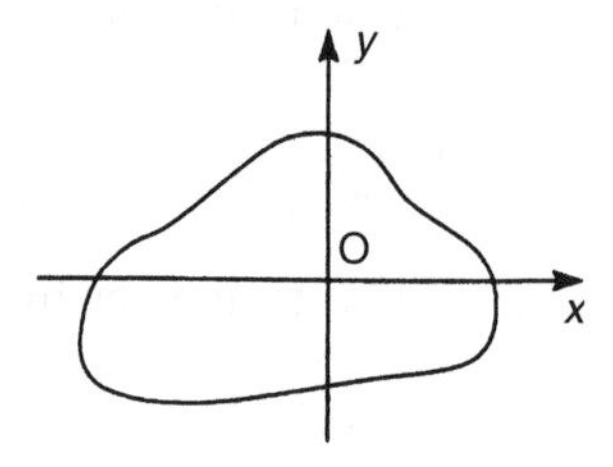

Fig. 16.26 Theorem of perpendicular axes.

plane of the section (i.e. a *polar second moment of area*) is then

$$I_o = I_{xx} + I_{yy} \tag{16.34}$$

16.4.3 Second moments of area of standard sections

Many sections may be regarded as comprising a number of rectangular shapes. The problem of determining the properties of such sections is simplified if the second moments of area of the rectangular components are known and use is made of the parallel axes theorem. Thus, for the rectangular section of Fig. 16.27.

$$I_{xx} = \int_A y^2 \mathrm{d}A = \int_{-d/2}^{d/2} by^2 \mathrm{d}y = b\left[\frac{y^3}{3}\right]_{-d/2}^{d/2}$$

which gives

$$I_{xx} = \frac{bd^3}{12} \tag{16.35}$$

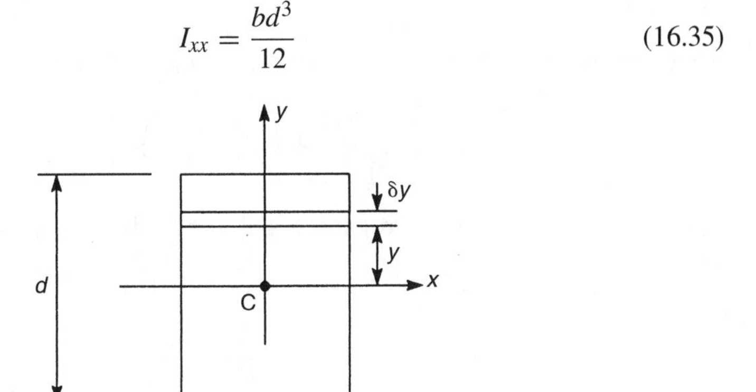

Fig. 16.27 Second moments of area of a rectangular section.

Similarly

$$I_{yy} = \frac{db^3}{12} \tag{16.36}$$

Frequently it is useful to know the second moment of area of a rectangular section about an axis which coincides with one of its edges. Thus in Fig. 16.27, and using the parallel axes theorem

$$I_N = \frac{bd^3}{12} + bd\left(-\frac{d}{2}\right)^2 = \frac{bd^3}{3} \tag{16.37}$$

Example 16.13

Determine the second moments of area I_{xx} and I_{yy} of the I-section shown in Fig. 16.28.

Using Eq. (16.35)

$$I_{xx} = \frac{bd^3}{12} - \frac{(b - t_w)d_w^3}{12}$$

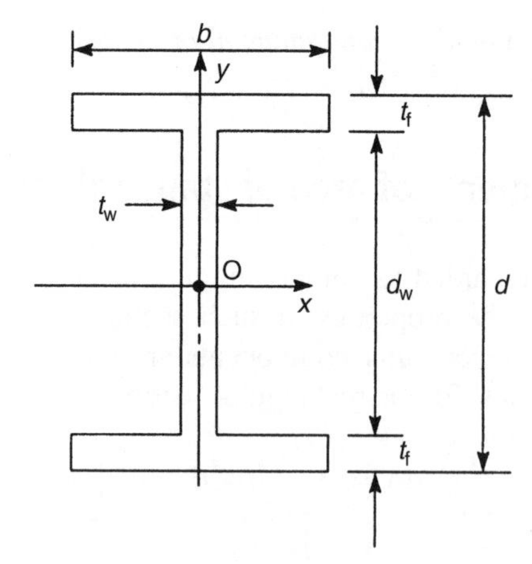

Fig. 16.28 Second moments of area of an I-Section.

Alternatively, using the parallel axes theorem in conjunction with Eq. (16.35)

$$I_{xx} = 2\left[\frac{bt_f^3}{12} + bt_f\left(\frac{d_{w+t_f}}{2}\right)^2\right] + \frac{t_w d_w^3}{12}$$

The equivalence of these two expressions for I_{xx} is most easily demonstrated by a numerical example.

Also, from Eq. (16.36)

$$I_{yy} = 2\frac{t_f b^3}{12} + \frac{d_w t_w^3}{12}$$

It is also useful to determine the second moment of area, about a diameter, of a circular section. In Fig. 16.29 where the x and y axes pass through the centroid of the section

$$I_{xx} = \int_A y^2 \mathrm{d}A = \int_{-d/2}^{d/2} 2\left(\frac{d}{2}\cos\theta\right) y^2 \mathrm{d}y \tag{16.38}$$

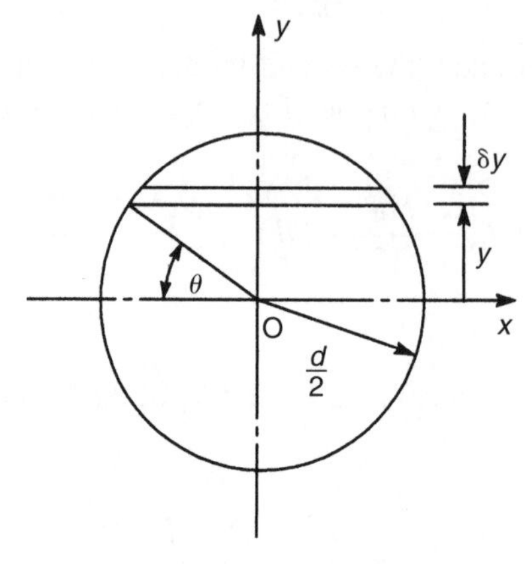

Fig. 16.29 Second moments of area of a circular section.

Integration of Eq. (16.38) is simplified if an angular variable, θ, is used. Thus

$$I_{xx} = \int_{-\pi/2}^{\pi/2} d\cos\theta \left(\frac{d}{2}\sin\theta\right)^2 \frac{d}{2}\cos\theta\,\mathrm{d}\theta$$

i.e.

$$I_{xx} = \frac{d^4}{8}\int_{-\pi/2}^{\pi/2} \cos^2\theta\,\sin^2\theta\,\mathrm{d}\theta$$

which gives

$$I_{xx} = \frac{\pi d^4}{64} \tag{16.39}$$

Clearly from symmetry

$$I_{yy} = \frac{\pi d^4}{64} \tag{16.40}$$

Using the theorem of perpendicular axes, the polar second moment of area, I_o, is given by

$$I_o = I_{xx} + I_{yy} = \frac{\pi d^4}{32} \tag{16.41}$$

16.4.4 Product second moment of area

The product second moment of area, I_{xy}, of a beam section with respect to x and y axes is defined by

$$I_{xy} = \int_A xy\,\mathrm{d}A \tag{16.42}$$

Thus each element of area in the cross-section is multiplied by the product of its coordinates and the integration is taken over the complete area. Although second moments of area are always positive since elements of area are multiplied by the square of one of their coordinates, it is possible for I_{xy} to be negative if the section lies predominantly in the second and fourth quadrants of the axes system. Such a situation would arise in the case of the Z-section of Fig. 16.30(a) where the product second moment of area of each flange is clearly negative.

A special case arises when one (or both) of the coordinate axes is an axis of symmetry so that for any element of area, δA, having the product of its coordinates positive, there is an identical element for which the product of its coordinates is negative (Fig. 16.30 (b)). Summation (i.e. integration) over the entire section of the product second moment of area of all such pairs of elements results in a zero value for I_{xy}.

We have shown previously that the parallel axes theorem may be used to calculate second moments of area of beam sections comprising geometrically simple components. The theorem can be extended to the calculation of product second moments of area. Let us suppose that we wish to calculate the product second moment of area,

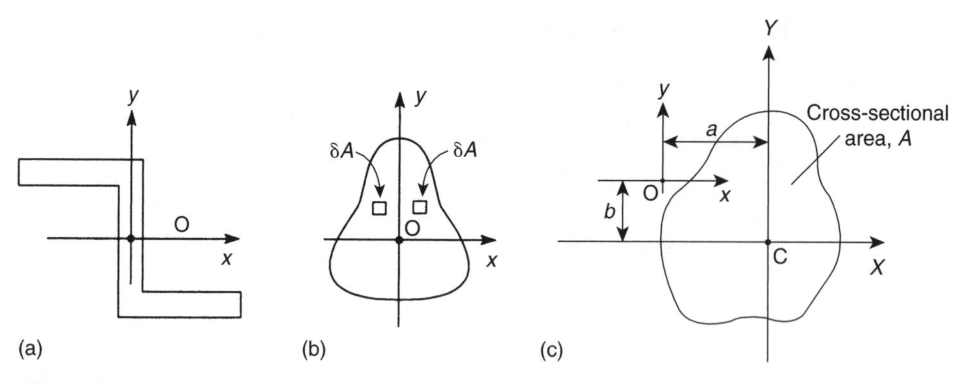

Fig. 16.30 Product second moment of area.

I_{xy}, of the section shown in Fig. 16.30(c) about axes xy when I_{XY} about its own, say centroidal, axes system CXY is known. From Eq. (16.42)

$$I_{xy} = \int_A xy \, \mathrm{d}A$$

or

$$I_{xy} = \int_A (X - a)(Y - b)\mathrm{d}A$$

which, on expanding, gives

$$I_{xy} = \int_A XY \, \mathrm{d}A - b \int_A X\mathrm{d}A - a \int_A Y \, \mathrm{d}A + ab \int_A \mathrm{d}A$$

If X and Y are centroidal axes then $\int_A X \, \mathrm{d}A = \int_A Y \, \mathrm{d}A = 0$. Hence

$$I_{xy} = I_{XY} + abA \tag{16.43}$$

It can be seen from Eq. (16.43) that if either CX or CY is an axis of symmetry, i.e. $I_{XY} = 0$, then

$$I_{xy} = abA \tag{16.44}$$

Therefore for a section component having an axis of symmetry that is parallel to either of the section reference axes the product second moment of area is the product of the coordinates of its centroid multiplied by its area.

16.4.5 Approximations for thin-walled sections

We may exploit the thin-walled nature of aircraft structures to make simplifying assumptions in the determination of stresses and deflections produced by bending. Thus, the thickness t of thin-walled sections is assumed to be small compared with their cross-sectional dimensions so that stresses may be regarded as being constant across the thickness. Furthermore, we neglect squares and higher powers of t in the computation

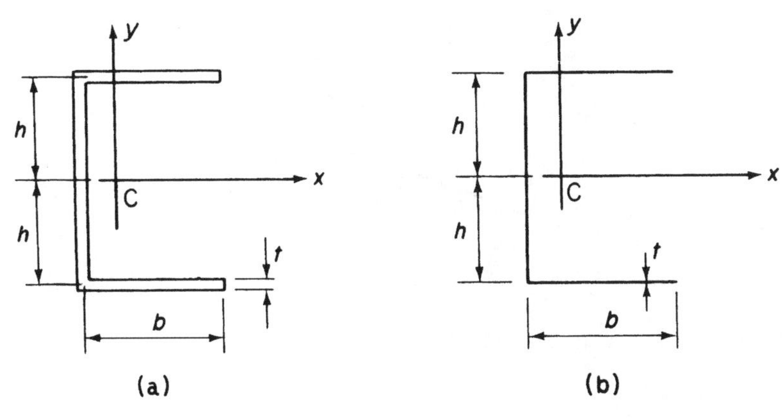

Fig. 16.31 (a) Actual thin-walled channel section; (b) approximate representation of section.

of sectional properties and take the section to be represented by the mid-line of its wall. As an illustration of the procedure we shall consider the channel section of Fig. 16.31(a). The section is singly symmetric about the x axis so that $I_{xy}=0$. The second moment of area I_{xx} is then given by

$$I_{xx}=2\left[\frac{(b+t/2)t^3}{12}+\left(b+\frac{t}{2}\right)th^2\right]+t\frac{[2(h-t/2)]^3}{12}$$

Expanding the cubed term we have

$$I_{xx}=2\left[\frac{(b+t/2)t^3}{12}+\left(b+\frac{t}{2}\right)th^2\right]+\frac{t}{12}\left[(2)^3\left(h^3-3h^2\frac{t}{2}+3h\frac{t^2}{4}-\frac{t^3}{8}\right)\right]$$

which reduces, after powers of t^2 and upwards are ignored, to

$$I_{xx}=2bth^2+t\frac{(2h)^3}{12}$$

The second moment of area of the section about Cy is obtained in a similar manner.

We see, therefore, that for the purpose of calculating section properties we may regard the section as being represented by a single line, as shown in Fig. 16.31(b).

Thin-walled sections frequently have inclined or curved walls which complicate the calculation of section properties. Consider the inclined thin section of Fig. 16.32. Its second moment of area about a horizontal axis through its centroid is given by

$$I_{xx}=2\int_0^{a/2}ty^2\,\mathrm{d}s=2\int_0^{a/2}t(s\sin\beta)^2\,\mathrm{d}s$$

from which

$$I_{xx}=\frac{a^3t\sin^2\beta}{12}$$

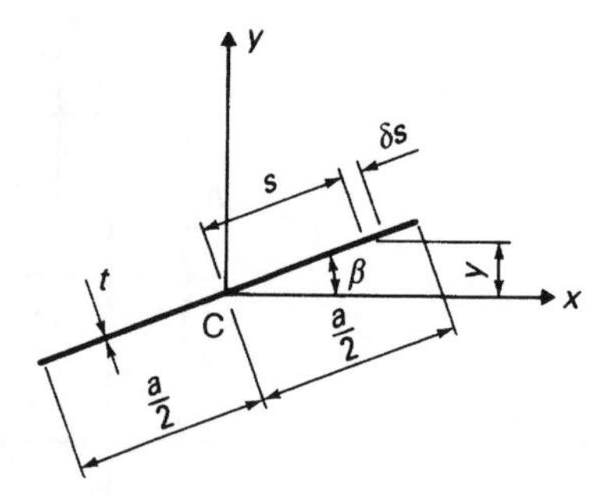

Fig. 16.32 Second moments of area of an inclined thin section.

Similarly

$$I_{yy} = \frac{a^3 t \cos^2 \beta}{12}$$

The product second moment of area is

$$I_{xy} = 2 \int_0^{a/2} txy \, \mathrm{d}s$$

$$= 2 \int_0^{a/2} t(s \cos \beta)(s \sin \beta) \, \mathrm{d}s$$

which gives

$$I_{xy} = \frac{a^3 t \sin 2\beta}{24}$$

We note here that these expressions are approximate in that their derivation neglects powers of t^2 and upwards by ignoring the second moments of area of the element δs about axes through its own centroid.

Properties of thin-walled curved sections are found in a similar manner. Thus, I_{xx} for the semicircular section of Fig. 16.33 is

$$I_{xx} = \int_0^{\pi r} t y^2 \, \mathrm{d}s$$

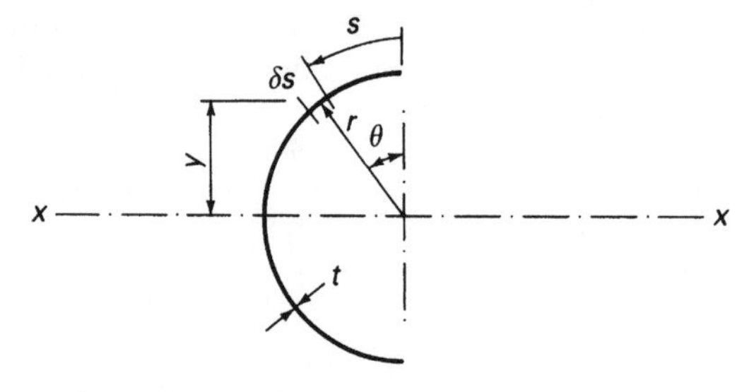

Fig. 16.33 Second moment of area of a semicircular section.

Expressing y and s in terms of a single variable θ simplifies the integration, hence

$$I_{xx} = \int_0^{\pi} t(r\cos\theta)^2 r\,\mathrm{d}\theta$$

from which

$$I_{xx} = \frac{\pi r^3 t}{2}$$

Example 16.14

Determine the direct stress distribution in the thin-walled Z-section shown in Fig. 16.34, produced by a positive bending moment M_x.

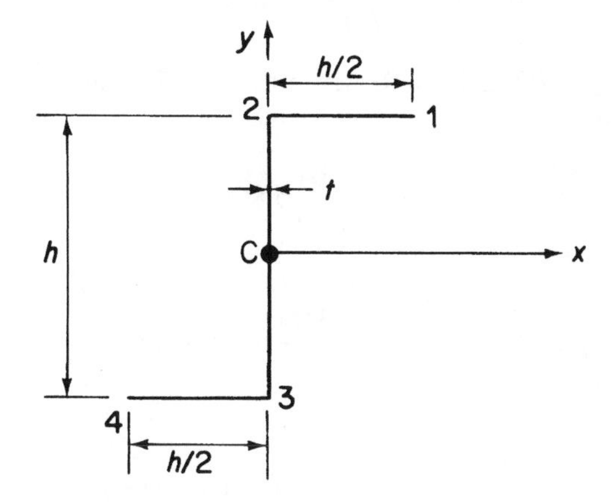

Fig. 16.34 *Z*-section beam of Example 16.14.

The section is antisymmetrical with its centroid at the mid-point of the vertical web. Therefore, the direct stress distribution is given by either of Eq. (16.18) or (16.19) in which $M_y = 0$. From Eq. (16.19)

$$\sigma_z = \frac{M_x(I_{yy}y - I_{xy}x)}{I_{xx}I_{yy} - I_{xy}^2} \tag{i}$$

The section properties are calculated as follows

$$I_{xx} = 2\frac{ht}{2}\left(\frac{h}{2}\right)^2 + \frac{th^3}{12} = \frac{h^3 t}{3}$$

$$I_{yy} = 2\frac{t}{3}\left(\frac{h}{2}\right)^3 = \frac{h^3 t}{12}$$

$$I_{xy} = \frac{ht}{2}\left(\frac{h}{4}\right)\left(\frac{h}{2}\right) + \frac{ht}{2}\left(-\frac{h}{4}\right)\left(-\frac{h}{2}\right) = \frac{h^3 t}{8}$$

Substituting these values in Eq. (i)

$$\sigma_z = \frac{M_x}{h^3 t}(6.86y - 10.30x) \tag{ii}$$

On the top flange $y = h/2$, $0 \leq x \leq h/2$ and the distribution of direct stress is given by

$$\sigma_z = \frac{M_x}{h^3 t}(3.43h - 10.30x)$$

which is linear. Hence

$$\sigma_{z,1} = -\frac{1.72M_x}{h^3 t} \quad \text{(compressive)}$$

$$\sigma_{z,2} = +\frac{3.43M_x}{h^3 t} \quad \text{(tensile)}$$

In the web $h/2 \leq y \leq -h/2$ and $x = 0$. Again the distribution is of linear form and is given by the equation

$$\sigma_z = \frac{M_x}{h^3 t} 6.86y$$

whence

$$\sigma_{z,2} = +\frac{3.43M_x}{h^3 t} \quad \text{(tensile)}$$

and

$$\sigma_{z,3} = -\frac{3.43M_x}{h^3 t} \quad \text{(compressive)}$$

The distribution in the lower flange may be deduced from antisymmetry; the complete distribution is then as shown in Fig. 16.35.

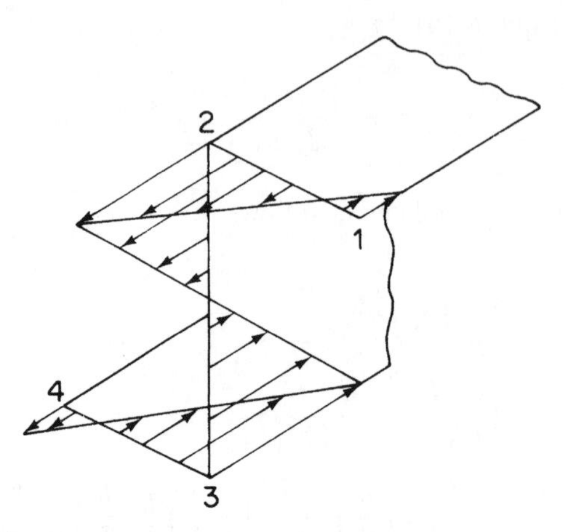

Fig. 16.35 Distribution of direct stress in *Z*-section beam of Example 16.14.

16.5 Applicability of bending theory

The expressions for direct stress and displacement derived in the above theory are based on the assumptions that the beam is of uniform, homogeneous cross-section and that plane sections remain plane after bending. The latter assumption is strictly true only if the bending moments M_x and M_y are constant along the beam. Variation of bending moment implies the presence of shear loads and the effect of these is to deform the beam section into a shallow, inverted 's' (see Section 2.6). However, shear stresses in beams whose cross-sectional dimensions are small in relation to their lengths are comparatively low so that the basic theory of bending may be used with reasonable accuracy.

In thin-walled sections shear stresses produced by shear loads are not small and must be calculated, although the direct stresses may still be obtained from the basic theory of bending so long as axial constraint stresses are absent; this effect is discussed in Chapters 26 and 27. Deflections in thin-walled structures are assumed to result primarily from bending strains; the contribution of shear strains may be calculated separately if required.

16.6 Temperature effects

In Section 1.15.1 we considered the effect of temperature change on stress–strain relationships while in Section 5.11 we examined the effect of a simple temperature gradient on a cantilever beam of rectangular cross-section using an energy approach. However, as we have seen, beam sections in aircraft structures are generally thin walled and do not necessarily have axes of symmetry. We shall now investigate how the effects of temperature on such sections may be determined.

We have seen that the strain produced by a temperature change ΔT is given by

$$\varepsilon = \alpha\,\Delta T \qquad \text{(see Eq. (1.55))}$$

It follows from Eq. (1.40) that the direct stress on an element of cross-sectional area δA is

$$\sigma = E\alpha\,\Delta T\,\delta A \tag{16.45}$$

Consider now the beam section shown in Fig. 16.36 and suppose that a temperature variation ΔT is applied to the complete cross-section, i.e. ΔT is a function of both x and y.

The total normal force due to the temperature change on the beam cross-section is then given by

$$N_T = \int\!\!\int_A E\alpha\,\Delta T\,\mathrm{d}A \tag{16.46}$$

Further, the moments about the x and y axes are

$$M_{xT} = \int\!\!\int_A E\alpha\,\Delta T y\,\mathrm{d}A \tag{16.47}$$

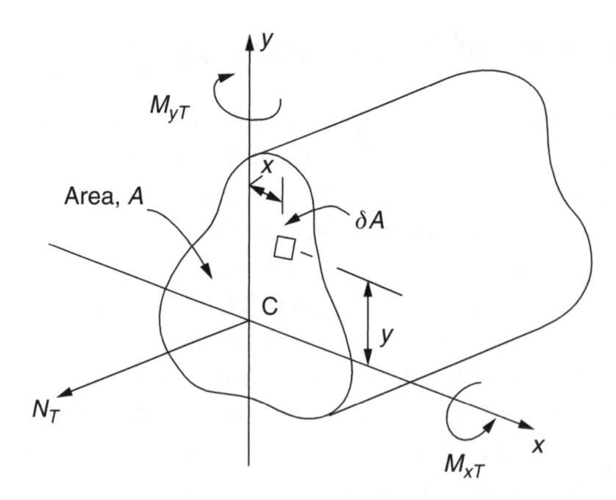

Fig. 16.36 Beam section subjected to a temperature rise.

and

$$M_{yT} = \int \int_A E\alpha \, \Delta T x \, \mathrm{d}A \tag{16.48}$$

respectively.

We have noted that beam sections in aircraft structures are generally thin walled so that Eqs (16.46)–(16.48) may be more easily integrated for such sections by dividing them into thin rectangular components as we did when calculating section properties. We then use the Riemann integration technique in which we calculate the contribution of each component to the normal force and moments and sum them to determine each resultant. Equations (16.46)–(16.48) then become

$$N_T = \Sigma E\alpha \, \Delta T \, A_i \tag{16.49}$$

$$M_{xT} = \Sigma E\alpha \, \Delta T \bar{y}_i A_i \tag{16.50}$$

$$M_{yT} = \Sigma E\alpha \, \Delta T \bar{x}_i A_i \tag{16.51}$$

in which A_i is the cross-sectional area of a component and $\bar{x}_i$ and $\bar{y}_i$ are the coordinates of its centroid.

Example 16.15

The beam section shown in Fig. 16.37 is subjected to a temperature rise of $2T_0$ in its upper flange, a temperature rise of T_0 in its web and zero temperature change in its lower flange. Determine the normal force on the beam section and the moments about the centroidal x and y axes. The beam section has a Young's modulus E and the coefficient of linear expansion of the material of the beam is α.

From Eq. (16.49)

$$N_T = E\alpha(2T_0 \, at + T_0 \, 2at) = 4E\alpha \, at \, T_0$$

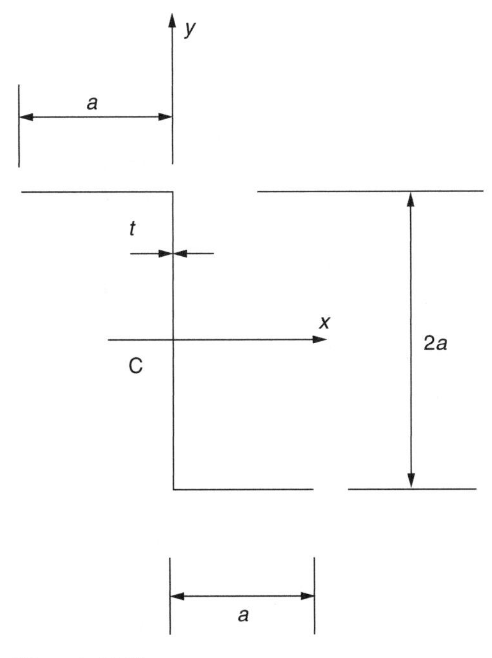

Fig. 16.37 Beam section of Example 16.15.

From Eq. (16.50)

$$M_{xT} = E\alpha[2T_0\, at(a) + T_0\, 2at(0)] = 2E\alpha\, a^2 t\, T_0$$

and from Eq. (16.51)

$$M_{yT} = E\alpha[2T_0\, at(-a/2) + T_0\, 2at(0)] = -E\alpha\, a^2 t\, T_0$$

Note that M_{yT} is negative which means that the upper flange would tend to rotate out of the paper about the web which agrees with a temperature rise for this part of the section. The stresses corresponding to the above stress resultants are calculated in the normal way and are added to those produced by any applied loads.

In some cases the temperature change is not conveniently constant in the components of a beam section and must then be expressed as a function of x and y. Consider the thin-walled beam section shown in Fig. 16.38 and suppose that a temperature change $\Delta T\ (x, y)$ is applied.

The direct stress on an element δs in the wall of the section is then, from Eq. (16.45)

$$\sigma = E\alpha\, \Delta T(x, y)t\, \delta s$$

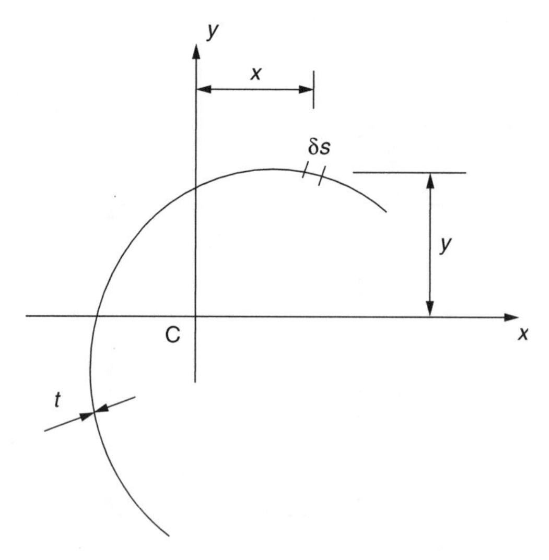

Fig. 16.38 Thin-walled beam section subjected to a varying temperature change.

Equations (16.46)–(16.48) then become

$$N_T = \int_A E\alpha\, \Delta T(x, y)t\, \mathrm{d}s \tag{16.52}$$

$$M_{xT} = \int_A E\alpha\, \Delta T(x, y)ty\, \mathrm{d}s \tag{16.53}$$

$$M_{yT} = \int_A E\alpha\, \Delta T(x, y)tx\, \mathrm{d}s \tag{16.54}$$

Example 16.16

If, in the beam section of Example 16.15, the temperature change in the upper flange is $2T_0$ but in the web varies linearly from $2T_0$ at its junction with the upper flange to zero at its junction with the lower flange determine the values of the stress resultants; the temperature change in the lower flange remains zero.

The temperature change at any point in the web is given by

$$T_{\mathrm{w}} = 2T_0(a + y)/2a = \frac{T_0}{a}(a + y)$$

Then, from Eqs (16.49) and (16.52)

$$N_T = E\alpha\, 2T_0\, at + \int_{-a}^{a} E\alpha \frac{T_0}{a}(a + y)t\, \mathrm{d}s$$

$$\text{i.e.}\quad N_T = E\alpha\, T_0 \left\{ 2at + \frac{1}{a}\left[ay + \frac{y^2}{2} \right]_{-a}^{a} \right\}$$

which gives

$$N_T = 4E\alpha T_0 at$$

Note that, in this case, the answer is identical to that in Example 16.15 which is to be expected since the average temperature change in the web is $(2T_0 + 0)/2 = T_0$ which is equal to the constant temperature change in the web in Example 16.15.

From Eqs (16.50) and (16.53)

$$M_{xT} = E\alpha\, 2T_0 at(a) + \int_{-a}^{a} E\alpha \frac{T_0}{a}(a + y)yt\, \mathrm{d}s$$

i.e.

$$M_{xT} = E\alpha\, T_0 \left\{ 2a^2 t + \frac{1}{a}\left[\frac{ay^2}{2} + \frac{y^3}{3}\right]_{-a}^{a} \right\}$$

from which

$$M_{xT} = \frac{8E\alpha a^2 tT_0}{3}$$

Alternatively, the average temperature change T_0 in the web may be considered to act at the centroid of the temperature change distribution. Then

$$M_{xT} = E\alpha\, 2T_0 at(a) + E\alpha T_0 2at\left(\frac{a}{3}\right)$$

i.e.

$$M_{xT} = \frac{8E\alpha a^2 tT_0}{3} \quad \text{as before}$$

The contribution of the temperature change in the web to M_{yT} remains zero since the section centroid is in the web; the value of M_{yT} is therefore $-E\alpha a^2 tT_0$ as in Example 16.14.

References

1 Megson, T. H. G., *Structures and Stress Analysis*, 2nd edition, Elsevier, Oxford, 2005.

Problems

P.16.1 Figure P.16.1 shows the section of an angle purlin. A bending moment of 3000 N m is applied to the purlin in a plane at an angle of 30° to the vertical y axis. If the sense of the bending moment is such that its components M_x and M_y both produce tension in the positive xy quadrant, calculate the maximum direct stress in the purlin stating clearly the point at which it acts.

Ans. $\sigma_{z,\max} = -63.3\ \text{N/mm}^2$ at C.

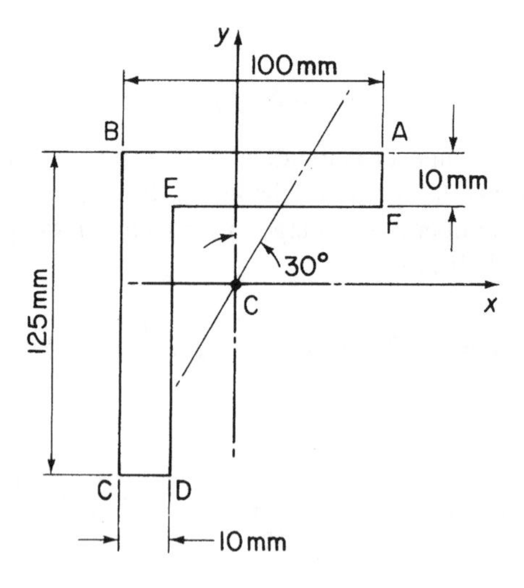

Fig. P.16.1

P.16.2 A thin-walled, cantilever beam of unsymmetrical cross-section supports shear loads at its free end as shown in Fig. P.16.2. Calculate the value of direct stress at the extremity of the lower flange (point A) at a section half-way along the beam if the position of the shear loads is such that no twisting of the beam occurs.

Ans. 194.7 N/mm^2 (tension).

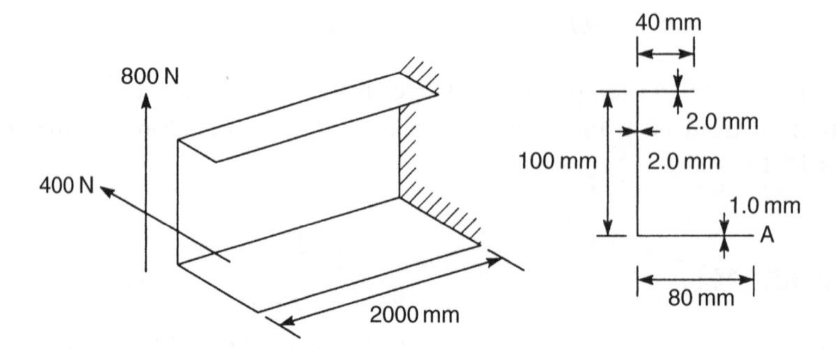

Fig. P.16.2

P.16.3 A beam, simply supported at each end, has a thin-walled cross-section shown in Fig. P.16.3. If a uniformly distributed loading of intensity w/unit length acts on the beam in the plane of the lower, horizontal flange, calculate the maximum direct stress due to bending of the beam and show diagrammatically the distribution of the stress at the section where the maximum occurs.

The thickness t is to be taken as small in comparison with the other cross-sectional dimensions in calculating the section properties I_{xx}, I_{yy} and I_{xy}.

Ans. $\sigma_{z,\max} = \sigma_{z,3} = 13wl^2/384a^2t, \ \sigma_{z,1} = wl^2/96a^2t, \ \sigma_{z,2} = -wl^2/48a^2t.$

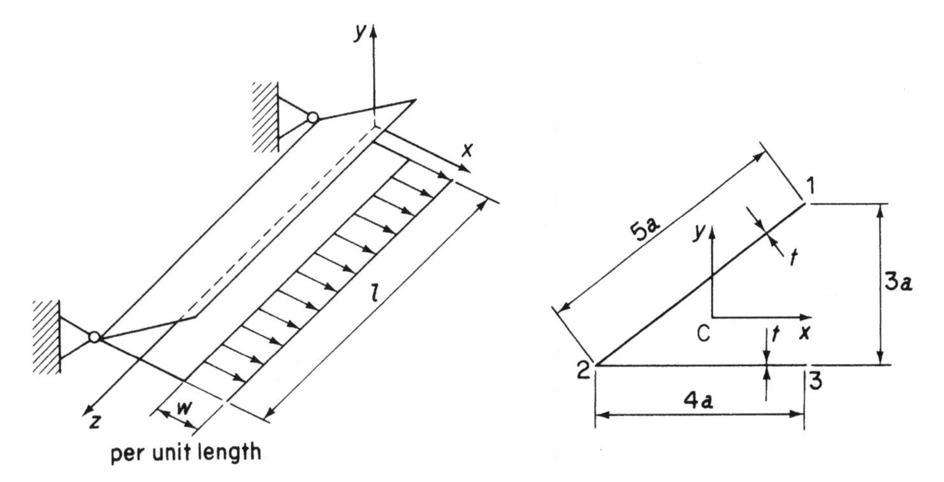

Fig. P.16.3

P.16.4 A thin-walled cantilever with walls of constant thickness t has the cross-section shown in Fig. P.16.4. It is loaded by a vertical force W at the tip and a horizontal force $2W$ at the mid-section, both forces acting through the shear centre. Determine and sketch the distribution of direct stress, according to the basic theory of bending, along the length of the beam for the points 1 and 2 of the cross-section.

The wall thickness t can be taken as very small in comparison with d in calculating the sectional properties I_{xx}, I_{xy}, etc.

Ans. $\sigma_{z,1}$ (mid-point) $= -0.05\ Wl/td^2$, $\sigma_{z,1}$ (built-in end) $= -1.85\ Wl/td^2$
$\sigma_{z,2}$ (mid-point) $= -0.63\ Wl/td^2$, $\sigma_{z,2}$ (built-in end) $= 0.1\ Wl/td^2$.

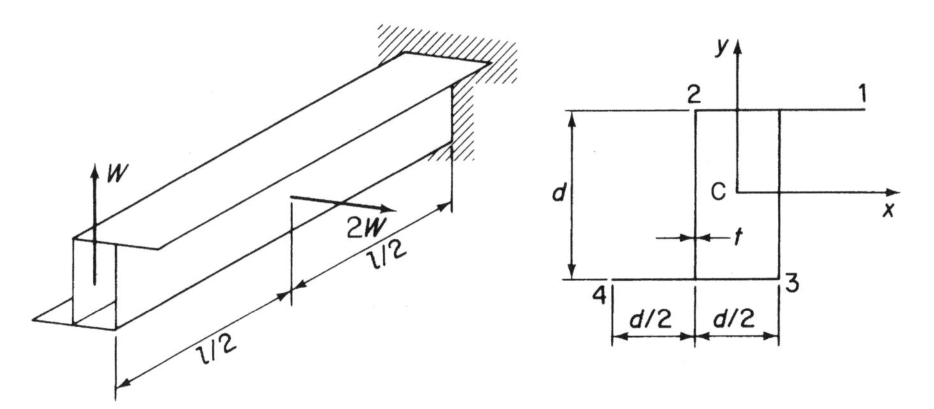

Fig. P.16.4

P. 16.5 A thin-walled beam has the cross-section shown in Fig. P.16.5. If the beam is subjected to a bending moment M_x in the plane of the web 23 calculate and sketch the distribution of direct stress in the beam cross-section.

Ans. At 1, $0.92M_x/th^2$; At 2, $-0.65M_x/th^2$; At 3, $0.65M_x/th^2$;
At 4, $-0.135M_x/th^2$

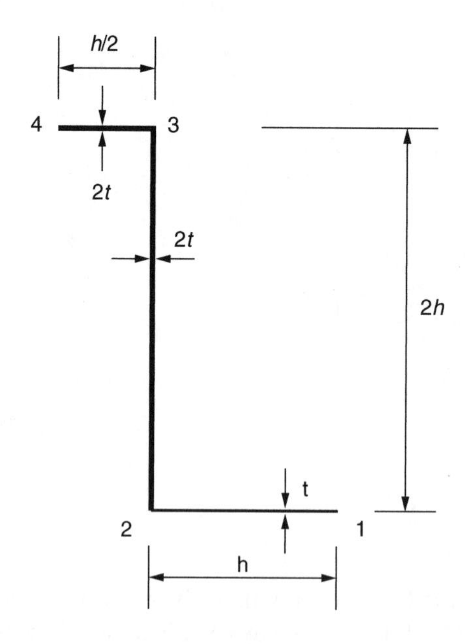

Fig. P.16.5

P.16.6 The thin-walled beam section shown in Fig. P.16.6 is subjected to a bending moment M_x applied in a negative sense. Find the position of the neutral axis and the maximum direct stress in the section.

Ans. NA inclined at 40.9° to Cx. $\pm 0.74\, M_x/ta^2$ at 1 and 2, respectively.

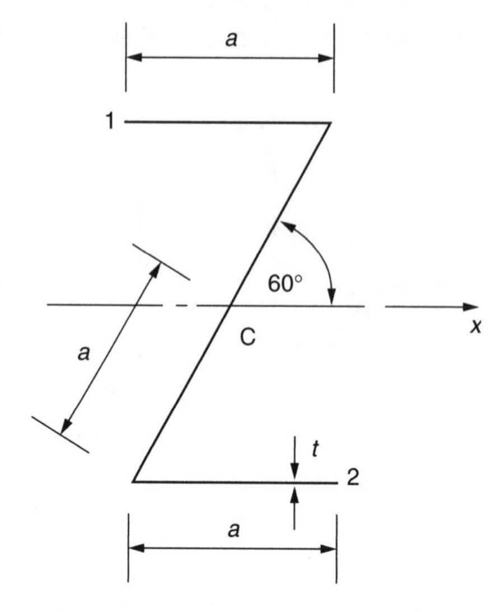

Fig. P.16.6

P.16.7 A thin-walled cantilever has a constant cross-section of uniform thickness with the dimensions shown in Fig. P.16.7. It is subjected to a system of point loads acting in the planes of the walls of the section in the directions shown.

Calculate the direct stresses according to the basic theory of bending at the points 1, 2 and 3 of the cross-section at the built-in end and half-way along the beam. Illustrate your answer by means of a suitable sketch.

The thickness is to be taken as small in comparison with the other cross-sectional dimensions in calculating the section properties I_{xx}, I_{xy}, etc.

Ans. At built-in end, $\sigma_{z,1} = -11.4\,\text{N/mm}^2$, $\sigma_{z,2} = -18.9\,\text{N/mm}^2$, $\sigma_{z,3} = 39.1\,\text{N/mm}^2$
Half-way, $\sigma_{z,1} = -20.3\,\text{N/mm}^2$, $\sigma_{z,2} = -1.1\,\text{N/mm}^2$, $\sigma_{z,3} = 15.4\,\text{N/mm}^2$.

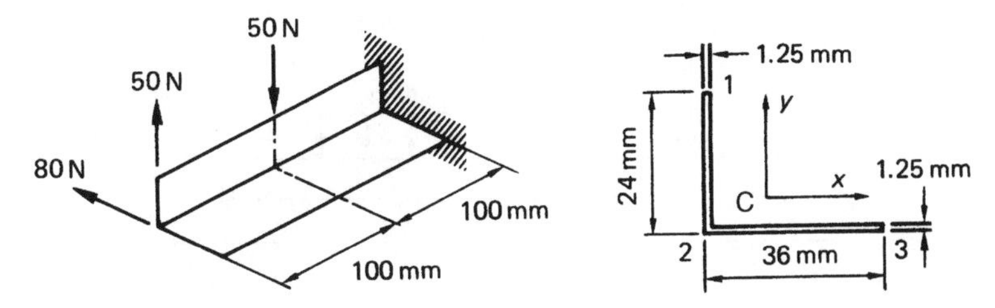

Fig. P.16.7

P.16.8 A uniform thin-walled beam has the open cross-section shown in Fig. P.16.8. The wall thickness t is constant. Calculate the position of the neutral axis and the maximum direct stress for a bending moment $M_x = 3.5\,\text{N m}$ applied about the horizontal axis Cx. Take $r = 5\,\text{mm}$, $t = 0.64\,\text{mm}$.

Ans. $\alpha = 51.9°$, $\sigma_{z,\max} = 101\,\text{N/mm}^2$.

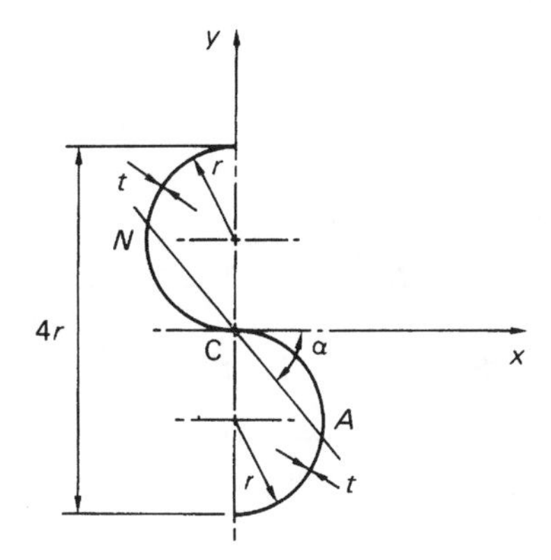

Fig. P.16.8

P.16.9 A uniform beam is simply supported over a span of 6 m. It carries a trapezoidally distributed load with intensity varying from 30 kN/m at the left-hand support to 90 kN/m at the right-hand support. Find the equation of the deflection curve and hence the deflection at the mid-span point. The second moment of area of the cross-section of the beam is $120 \times 10^6\,\text{mm}^4$ and Young's modulus $E = 206\,000\,\text{N/mm}^2$.

Ans. 41 mm (downwards).

P.16.10 A cantilever of length L and having a flexural rigidity EI carries a distributed load that varies in intensity from w/unit length at the built-in end to zero at the free end. Find the deflection of the free end.

Ans. $wL^4/30EI$ (downwards).

P.16.11 Determine the position and magnitude of the maximum deflection of the simply supported beam shown in Fig. P.16.11 in terms of its flexural rigidity EI.

Ans. $38.8/EI$ m downwards at 2.9 m from left-hand support.

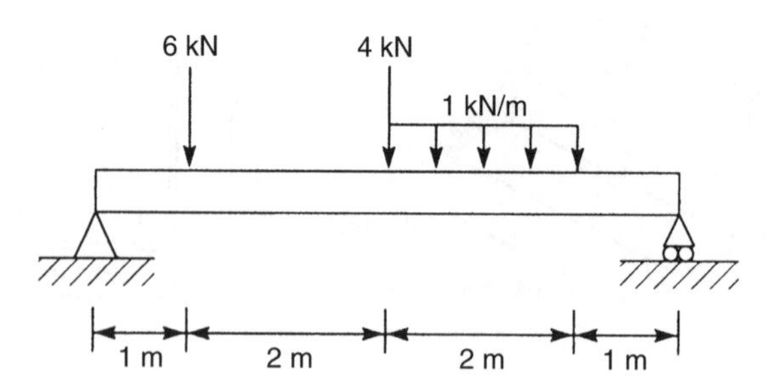

Fig. P.16.11

P.16.12 Determine the equation of the deflection curve of the beam shown in Fig. P.16.12. The flexural rigidity of the beam is EI.

Ans. $$v = -\frac{1}{EI}\left(\frac{125}{6}z^3 - 50[z-1]^2 + \frac{50}{12}[z-2]^4 - \frac{50}{12}[z-4]^4 - \frac{525}{6}[z-4]^3 + 237.5z\right)$$

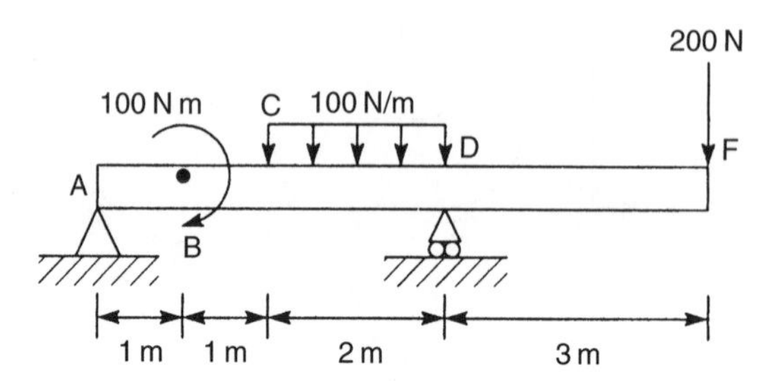

Fig. P.16.12

P.16.13 A uniform thin-walled beam ABD of open cross-section (Fig. P.16.13) is simply supported at points B and D with its web vertical. It carries a downward vertical force W at the end A in the plane of the web.

Derive expressions for the vertical and horizontal components of the deflection of the beam midway between the supports B and D. The wall thickness t and Young's modulus E are constant throughout.

Ans. $u = 0.186Wl^3/Ea^3t$, $v = 0.177Wl^3/Ea^3t$.

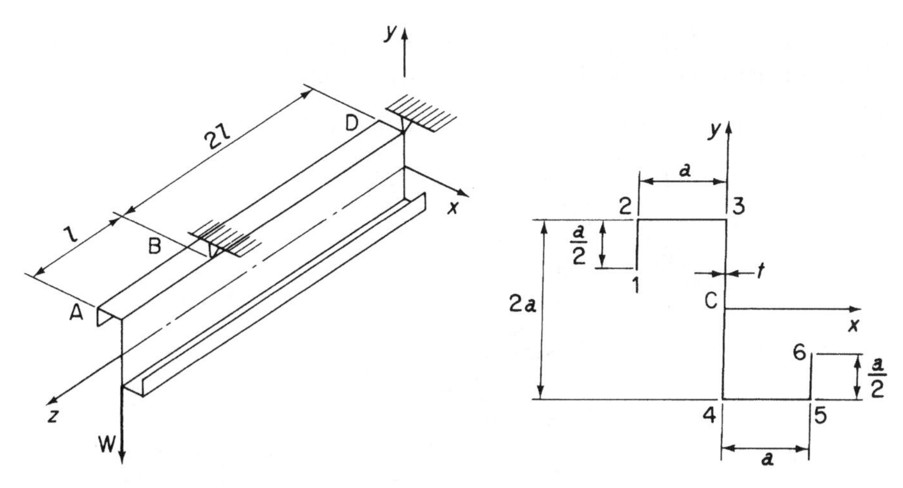

Fig. P.16.13

P.16.14 A uniform cantilever of arbitrary cross-section and length l has section properties I_{xx}, I_{yy} and I_{xy} with respect to the centroidal axes shown in Fig. P.16.14. It is loaded in the vertical (yz) plane with a uniformly distributed load of intensity w/unit length. The tip of the beam is hinged to a horizontal link which constrains it to move in the vertical direction only (provided that the actual deflections are small). Assuming that the link is rigid, and that there are no twisting effects, calculate:

(a) the force in the link;
(b) the deflection of the tip of the beam.

Ans. (a) $3wlI_{xy}/8I_{xx}$; (b) $wl^4/8EI_{xx}$.

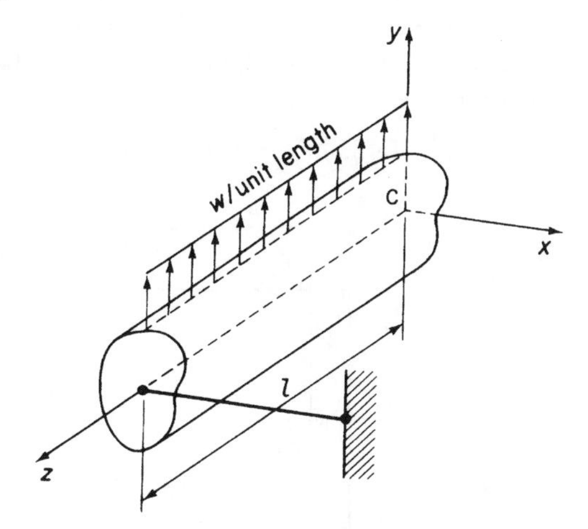

Fig. P.16.14

P.16.15 A uniform beam of arbitrary, unsymmetrical cross-section and length $2l$ is built-in at one end and simply supported in the vertical direction at a point half-way along its length. This support, however, allows the beam to deflect freely in the horizontal x direction (Fig. P.16.15).

For a vertical load W applied at the free end of the beam, calculate and draw the bending moment diagram, putting in the principal values.

Ans. $M_C = 0$, $M_B = Wl$, $M_A = -Wl/2$. Linear distribution.

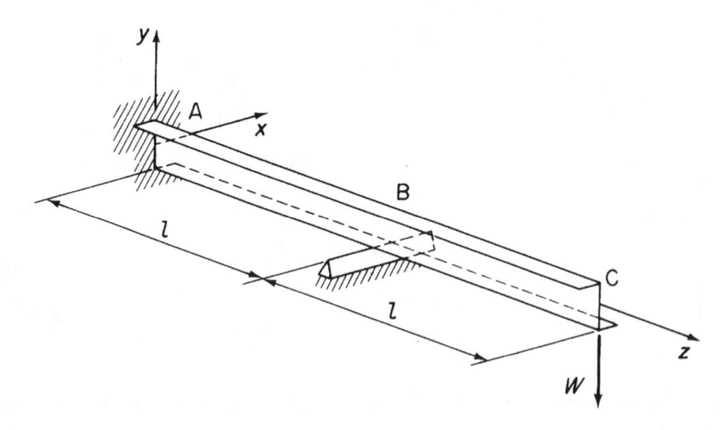

Fig. P.16.15

P.16.16 The beam section of P.16.4 is subjected to a temperature rise of $4T_0$ in its upper flange 12, a temperature rise of $2T_0$ in both vertical webs and a temperature rise of T_0 in its lower flange 34. Determine the changes in axial force and in the bending moments about the x and y axes. Young's modulus for the material of the beam is E and its coefficient of linear expansion is α.

Ans. $N_T = 9E\alpha\, dtT_0$, $M_{xT} = 3E\alpha\, d^2t\, T_0/2$, $M_{yT} = 3E\alpha\, d^2t\, T_0/4$.

P.16.17 The beam section shown in Fig. P.16.17 is subjected to a temperature change which varies with y such that $T = T_0 y/2a$. Determine the corresponding changes in the stress resultants. Young's modulus for the material of the beam is E while its coefficient of linear expansion is α.

Ans. $N_T = 0$, $M_{xT} = 5E\alpha\, a^2t\ T_0/3$, $M_{yT} = E\alpha\, a^2t\ T_0/6$.

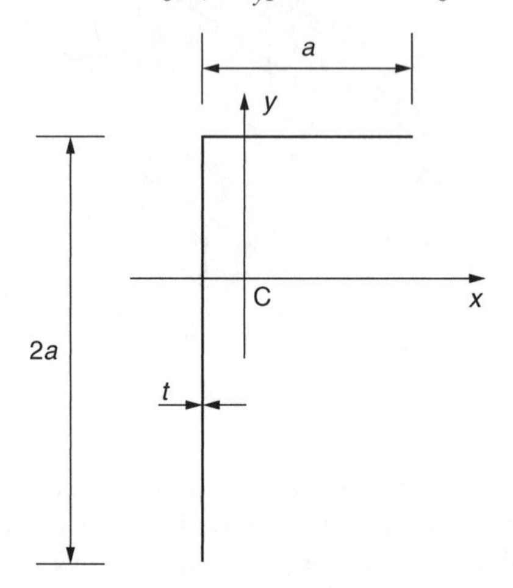

Fig. P.16.17

17

Shear of beams

In Chapter 16 we developed the theory for the bending of beams by considering solid or thick beam sections and then extended the theory to the thin-walled beam sections typical of aircraft structural components. In fact it is only in the calculation of section properties that thin-walled sections subjected to bending are distinguished from solid and thick sections. However, for thin-walled beams subjected to shear, the theory is based on assumptions applicable only to thin-walled sections so that we shall not consider solid and thick sections; the relevant theory for such sections may be found in any text on structural and stress analysis.[1] The relationships between bending moments, shear forces and load intensities derived in Section 16.2.5 still apply.

17.1 General stress, strain and displacement relationships for open and single cell closed section thin-walled beams

We shall establish in this section the equations of equilibrium and expressions for strain which are necessary for the analysis of open section beams supporting shear loads and closed section beams carrying shear and torsional loads. The analysis of open section beams subjected to torsion requires a different approach and is discussed separately in Chapter 18. The relationships are established from first principles for the particular case of thin-walled sections in preference to the adaption of Eqs (1.6), (1.27) and (1.28) which refer to different coordinate axes; the form, however, will be seen to be the same. Generally, in the analysis we assume that axial constraint effects are negligible, that the shear stresses normal to the beam surface may be neglected since they are zero at each surface and the wall is thin, that direct and shear stresses on planes normal to the beam surface are constant across the thickness, and finally that the beam is of uniform section so that the thickness may vary with distance around each section but is constant along the beam. In addition, we ignore squares and higher powers of the thickness t in the calculation of section properties (see Section 16.4.5).

The parameter s in the analysis is distance measured around the cross-section from some convenient origin.

An element $\delta s \times \delta z \times t$ of the beam wall is maintained in equilibrium by a system of direct and shear stresses as shown in Fig. 17.1(a). The direct stress σ_z is produced by bending moments or by the bending action of shear loads while the shear stresses are due

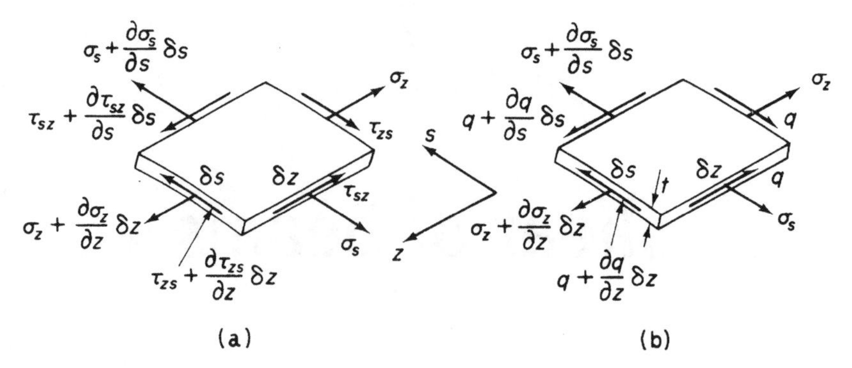

Fig. 17.1 (a) General stress system on element of a closed or open section beam; (b) direct stress and shear flow system on the element.

to shear and/or torsion of a closed section beam or shear of an open section beam. The hoop stress σ_s is usually zero but may be caused, in closed section beams, by internal pressure. Although we have specified that t may vary with s, this variation is small for most thin-walled structures so that we may reasonably make the approximation that t is constant over the length δs. Also, from Eq. (1.4), we deduce that $\tau_{zs} = \tau_{sz} = \tau$ say. However, we shall find it convenient to work in terms of *shear flow* q, i.e. shear force per unit length rather than in terms of shear stress. Hence, in Fig. 17.1(b)

$$q = \tau t \tag{17.1}$$

and is regarded as being positive in the direction of increasing s.

For equilibrium of the element in the z direction and neglecting body forces (see Section 1.2)

$$\left(\sigma_z + \frac{\partial \sigma_z}{\partial z}\delta z\right) t\delta s - \sigma_z t\delta s + \left(q + \frac{\partial q}{\partial s}\delta s\right)\delta_z - q\delta z = 0$$

which reduces to

$$\frac{\partial q}{\partial s} + t\frac{\partial \sigma_z}{\partial z} = 0 \tag{17.2}$$

Similarly for equilibrium in the s direction

$$\frac{\partial q}{\partial z} + t\frac{\partial \sigma_s}{\partial s} = 0 \tag{17.3}$$

The direct stresses σ_z and σ_s produce direct strains ε_z and ε_s, while the shear stress τ induces a shear strain $\gamma(=\gamma_{zs} = \gamma_{sz})$. We shall now proceed to express these strains in terms of the three components of the displacement of a point in the section wall (see Fig. 17.2). Of these components v_t is a tangential displacement in the xy plane and is taken to be positive in the direction of increasing s; v_n is a normal displacement in the xy plane and is positive outwards; and w is an axial displacement which has been defined previously in Section 16.2.1. Immediately, from the third of Eqs (1.18), we have

$$\varepsilon_z = \frac{\partial w}{\partial z} \tag{17.4}$$

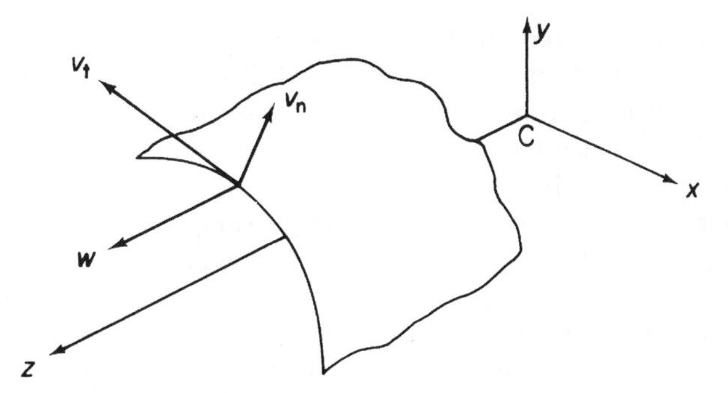

Fig. 17.2 Axial, tangential and normal components of displacement of a point in the beam wall.

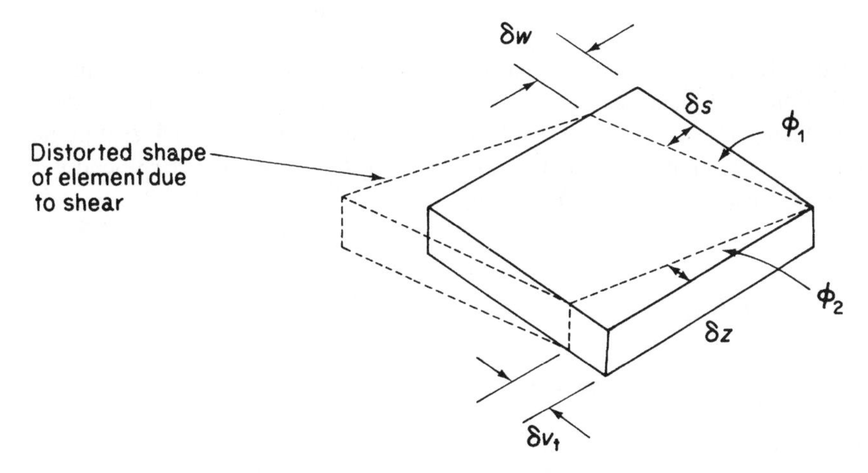

Fig. 17.3 Determination of shear strain γ in terms of tangential and axial components of displacement.

It is possible to derive a simple expression for the direct strain ε_s in terms of v_{t}, v_{n}, s and the curvature $1/r$ in the xy plane of the beam wall. However, as we do not require ε_s in the subsequent analysis we shall, for brevity, merely quote the expression

$$\varepsilon_s = \frac{\partial v_{\mathrm{t}}}{\partial s} + \frac{v_{\mathrm{n}}}{r} \tag{17.5}$$

The shear strain γ is found in terms of the displacements w and v_{t} by considering the shear distortion of an element $\delta s \times \delta z$ of the beam wall. From Fig. 17.3 we see that the shear strain is given by

$$\gamma = \phi_1 + \phi_2$$

or, in the limit as both δs and δz tend to zero

$$\gamma = \frac{\partial w}{\partial s} + \frac{\partial v_{\mathrm{t}}}{\partial z} \tag{17.6}$$

In addition to the assumptions specified in the earlier part of this section, we further assume that during any displacement the shape of the beam cross-section is maintained by a system of closely spaced diaphragms which are rigid in their own plane but are perfectly flexible normal to their own plane (CSRD assumption). There is, therefore, no resistance to axial displacement w and the cross-section moves as a rigid body in its own plane, the displacement of any point being completely specified by translations u and v and a rotation θ (see Fig. 17.4).

At first sight this appears to be a rather sweeping assumption but, for aircraft structures of the thin shell type described in Chapter 12 whose cross-sections are stiffened by ribs or frames positioned at frequent intervals along their lengths, it is a reasonable approximation for the actual behaviour of such sections. The tangential displacement v_t of any point N in the wall of either an open or closed section beam is seen from Fig. 17.4 to be

$$v_t = p\theta + u\cos\psi + v\sin\psi \tag{17.7}$$

where clearly u, v and θ are functions of z only (w may be a function of z and s).

The origin O of the axes in Fig. 17.4 has been chosen arbitrarily and the axes suffer displacements u, v and θ. These displacements, in a loading case such as pure torsion, are equivalent to a pure rotation about some point $R(x_R, y_R)$ in the cross-section where R is the *centre of twist*. Therefore, in Fig. 17.4

$$v_t = p_R\theta \tag{17.8}$$

and

$$p_R = p - x_R\sin\psi + y_R\cos\psi$$

which gives

$$v_t = p\theta - x_R\theta\sin\psi + y_R\theta\cos\psi$$

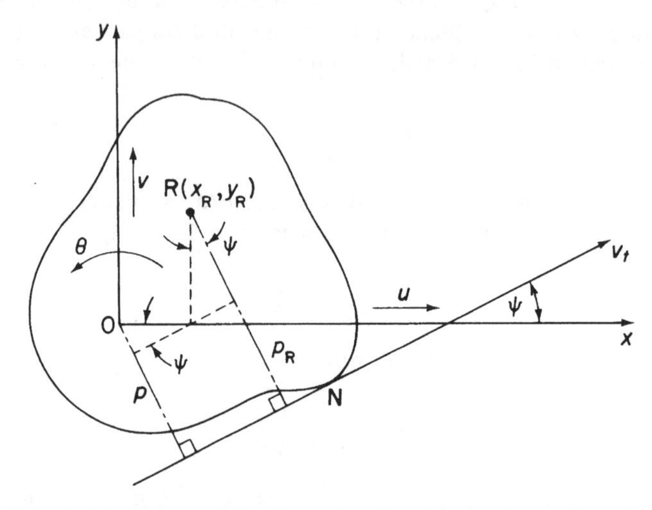

Fig. 17.4 Establishment of displacement relationships and position of centre of twist of beam (open or closed).

and

$$\frac{\partial v_t}{\partial z} = p\frac{d\theta}{dz} - x_R \sin\psi\frac{d\theta}{dz} + y_R \cos\psi\frac{d\theta}{dz} \tag{17.9}$$

Also from Eq. (17.7)

$$\frac{\partial v_t}{\partial z} = p\frac{d\theta}{dz} + \frac{du}{dz}\cos\psi + \frac{dv}{dz}\sin\psi \tag{17.10}$$

Comparing the coefficients of Eqs (17.9) and (17.10) we see that

$$x_R = -\frac{dv/dz}{d\theta/dz} \quad y_R = \frac{du/dz}{d\theta/dz} \tag{17.11}$$

17.2 Shear of open section beams

The open section beam of arbitrary section shown in Fig. 17.5 supports shear loads S_x and S_y such that there is no twisting of the beam cross-section. For this condition to be valid the shear loads must both pass through a particular point in the cross-section known as the *shear centre.*

Since there are no hoop stresses in the beam the shear flows and direct stresses acting on an element of the beam wall are related by Eq. (17.2), i.e.

$$\frac{\partial q}{\partial s} + t\frac{\partial \sigma_z}{\partial z} = 0$$

We assume that the direct stresses are obtained with sufficient accuracy from basic bending theory so that from Eq. (16.18)

$$\frac{\partial \sigma_z}{\partial z} = \frac{[(\partial M_y/\partial z)I_{xx} - (\partial M_x/\partial z)I_{xy}]}{I_{xx}I_{yy} - I_{xy}^2}x + \frac{[(\partial M_x/\partial z)I_{yy} - (\partial M_y/\partial z)I_{xy}]}{I_{xx}I_{yy} - I_{xy}^2}y$$

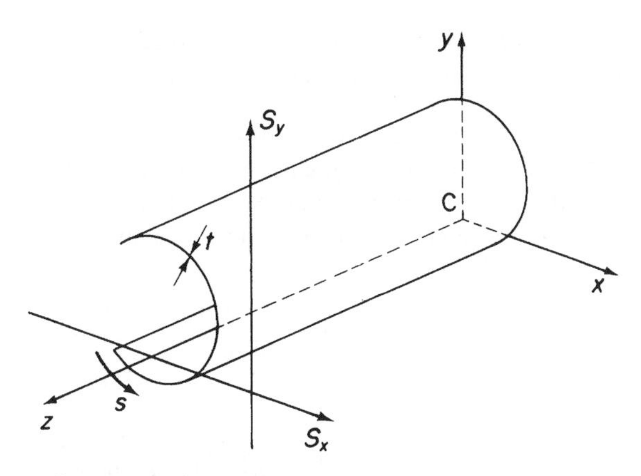

Fig. 17.5 Shear loading of open section beam.

Using the relationships of Eqs (16.23) and (16.24), i.e. $\partial M_y/\partial z = S_x$, etc., this expression becomes

$$\frac{\partial \sigma_z}{\partial z} = \frac{(S_x I_{xx} - S_y I_{xy})}{I_{xx}I_{yy} - I_{xy}^2}x + \frac{(S_y I_{yy} - S_x I_{xy})}{I_{xx}I_{yy} - I_{xy}^2}y$$

Substituting for $\partial \sigma_z/\partial z$ in Eq. (17.2) gives

$$\frac{\partial q}{\partial s} = -\frac{(S_x I_{xx} - S_y I_{xy})}{I_{xx}I_{yy} - I_{xy}^2}tx - \frac{(S_y I_{yy} - S_x I_{xy})}{I_{xx}I_{yy} - I_{xy}^2}ty \tag{17.12}$$

Integrating Eq. (17.12) with respect to s from some origin for s to any point around the cross-section, we obtain

$$\int_0^s \frac{\partial q}{\partial s}\,\mathrm{d}s = -\left(\frac{S_x I_{xx} - S_y I_{xy}}{I_{xx}I_{yy} - I_{xy}^2}\right)\int_0^s tx\,\mathrm{d}s - \left(\frac{S_y I_{yy} - S_x I_{xy}}{I_{xx}I_{yy} - I_{xy}^2}\right)\int_0^s ty\,\mathrm{d}s \tag{17.13}$$

If the origin for s is taken at the open edge of the cross-section, then $q = 0$ when $s = 0$ and Eq. (17.13) becomes

$$q_s = -\left(\frac{S_x I_{xx} - S_y I_{xy}}{I_{xx}I_{yy} - I_{xy}^2}\right)\int_0^s tx\,\mathrm{d}s - \left(\frac{S_y I_{yy} - S_x I_{xy}}{I_{xx}I_{yy} - I_{xy}^2}\right)\int_0^s ty\,\mathrm{d}s \tag{17.14}$$

For a section having either Cx or Cy as an axis of symmetry $I_{xy} = 0$ and Eq. (17.14) reduces to

$$q_s = -\frac{S_x}{I_{yy}}\int_0^s tx\,\mathrm{d}s - \frac{S_y}{I_{xx}}\int_0^s ty\,\mathrm{d}s$$

Example 17.1

Determine the shear flow distribution in the thin-walled Z-section shown in Fig. 17.6 due to a shear load S_y applied through the shear centre of the section.

The origin for our system of reference axes coincides with the centroid of the section at the mid-point of the web. From antisymmetry we also deduce by inspection that the shear centre occupies the same position. Since S_y is applied through the shear centre then no torsion exists and the shear flow distribution is given by Eq. (17.14) in which $S_x = 0$, i.e.

$$q_s = \frac{S_y I_{xy}}{I_{xx}I_{yy} - I_{xy}^2}\int_0^s tx\,\mathrm{d}s - \frac{S_y I_{yy}}{I_{xx}I_{yy} - I_{xy}^2}\int_0^s ty\,\mathrm{d}s$$

or

$$q_s = \frac{S_y}{I_{xx}I_{yy} - I_{xy}^2}\left(I_{xy}\int_0^s tx\,\mathrm{d}s - I_{yy}\int_0^s ty\,\mathrm{d}s\right) \tag{i}$$

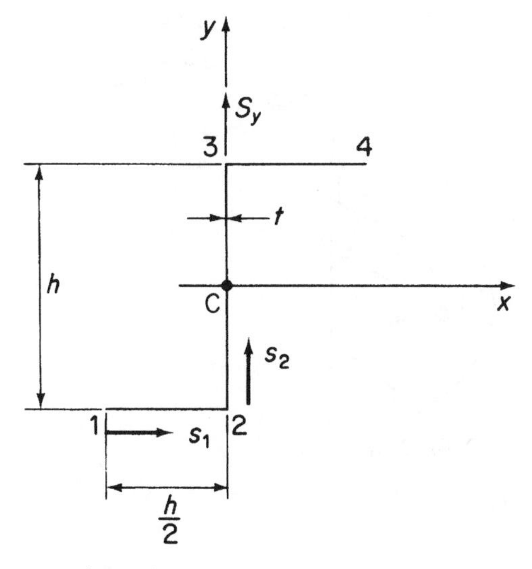

Fig. 17.6 Shear loaded Z-section of Example 17.1.

The second moments of area of the section have previously been determined in Example 16.14 and are

$$I_{xx} = \frac{h^3t}{3}, \quad I_{yy} = \frac{h^3t}{12}, \quad I_{xy} = \frac{h^3t}{8}$$

Substituting these values in Eq. (i) we obtain

$$q_s = \frac{S_y}{h^3}\int_0^s (10.32x - 6.84y)\mathrm{d}s \tag{ii}$$

On the bottom flange 12, $y = -h/2$ and $x = -h/2 + s_1$, where $0 \leq s_1 \leq h/2$. Therefore

$$q_{12} = \frac{S_y}{h^3}\int_0^{s_1} (10.32s_1 - 1.74h)\mathrm{d}s_1$$

giving

$$q_{12} = \frac{S_y}{h^3}(5.16s_1^2 - 1.74hs_1) \tag{iii}$$

Hence at 1 ($s_1 = 0$), $q_1 = 0$ and at 2 ($s_1 = h/2$), $q_2 = 0.42S_y/h$. Further examination of Eq. (iii) shows that the shear flow distribution on the bottom flange is parabolic with a change of sign (i.e. direction) at $s_1 = 0.336h$. For values of $s_1 < 0.336h$, q_{12} is negative and therefore in the opposite direction to s_1.

In the web 23, $y = -h/2 + s_2$, where $0 \leq s_2 \leq h$ and $x = 0$. Then

$$q_{23} = \frac{S_y}{h^3}\int_0^{s_2} (3.42h - 6.84s_2)\mathrm{d}s_2 + q_2 \tag{iv}$$

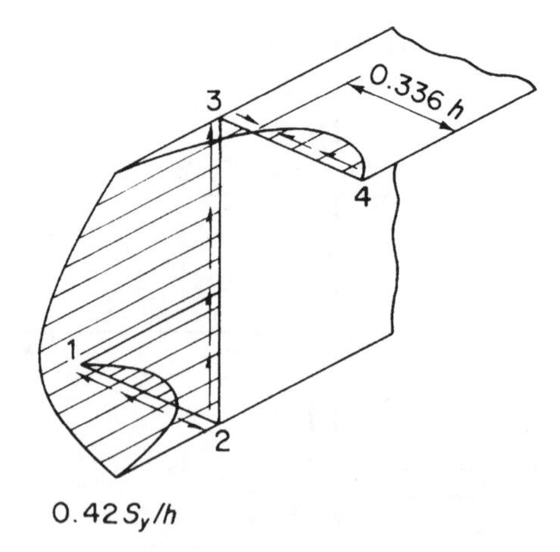

Fig. 17.7 Shear flow distribution in Z-section of Example 17.1.

We note in Eq. (iv) that the shear flow is not zero when $s_2 = 0$ but equal to the value obtained by inserting $s_1 = h/2$ in Eq. (iii), i.e. $q_2 = 0.42S_y/h$. Integration of Eq. (iv) yields

$$q_{23} = \frac{S_y}{h^3}(0.42h^2 + 3.42hs_2 - 3.42s_2^2) \tag{v}$$

This distribution is symmetrical about Cx with a maximum value at $s_2 = h/2 (y = 0)$ and the shear flow is positive at all points in the web.

The shear flow distribution in the upper flange may be deduced from antisymmetry so that the complete distribution is of the form shown in Fig. 17.7.

17.2.1 Shear centre

We have defined the position of the shear centre as that point in the cross-section through which shear loads produce no twisting. It may be shown by use of the reciprocal theorem that this point is also the centre of twist of sections subjected to torsion. There are, however, some important exceptions to this general rule as we shall observe in Section 26.1. Clearly, in the majority of practical cases it is impossible to guarantee that a shear load will act through the shear centre of a section. Equally apparent is the fact that any shear load may be represented by the combination of the shear load applied through the shear centre and a torque. The stresses produced by the separate actions of torsion and shear may then be added by superposition. It is therefore necessary to know the location of the shear centre in all types of section or to calculate its position. Where a cross-section has an axis of symmetry the shear centre must, of course, lie on this axis. For cruciform or angle sections of the type shown in Fig. 17.8 the shear centre is located at the intersection of the sides since the resultant internal shear loads all pass through these points.

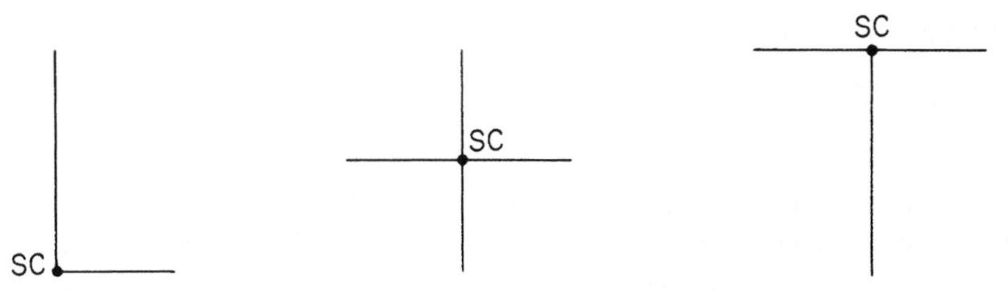

Fig. 17.8 Shear centre position for type of open section beam shown.

Example 17.2

Calculate the position of the shear centre of the thin-walled channel section shown in Fig. 17.9. The thickness t of the walls is constant.

The shear centre S lies on the horizontal axis of symmetry at some distance ξ_S, say, from the web. If we apply an arbitrary shear load S_y through the shear centre then the shear flow distribution is given by Eq. (17.14) and the moment about any point in the cross-section produced by these shear flows is *equivalent* to the moment of the applied shear load. S_y appears on both sides of the resulting equation and may therefore be eliminated to leave ξ_S.

For the channel section, Cx is an axis of symmetry so that $I_{xy} = 0$. Also $S_x = 0$ and therefore Eq. (17.14) simplifies to

$$q_s = -\frac{S_y}{I_{xx}} \int_0^s ty \, ds \tag{i}$$

where

$$I_{xx} = 2bt\left(\frac{h}{2}\right)^2 + \frac{th^3}{12} = \frac{h^3 t}{12}\left(1 + \frac{6b}{h}\right)$$

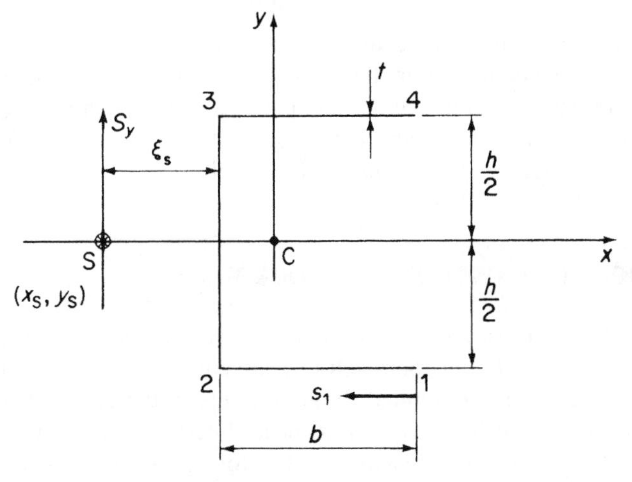

Fig. 17.9 Determination of shear centre position of channel section of Example 17.2.

Substituting for I_{xx} in Eq. (i) we have

$$q_s = \frac{-12S_y}{h^3(1 + 6b/h)} \int_0^s y \, ds \tag{ii}$$

The amount of computation involved may be reduced by giving some thought to the requirements of the problem. In this case we are asked to find the position of the shear centre only, not a complete shear flow distribution. From symmetry it is clear that the moments of the resultant shears on the top and bottom flanges about the mid-point of the web are numerically equal and act in the same rotational sense. Furthermore, the moment of the web shear about the same point is zero. We deduce that it is only necessary to obtain the shear flow distribution on either the top or bottom flange for a solution. Alternatively, choosing a web/flange junction as a moment centre leads to the same conclusion.

On the bottom flange, $y = -h/2$ so that from Eq. (ii) we have

$$q_{12} = \frac{6S_y}{h^2(1 + 6b/h)} s_1 \tag{iii}$$

Equating the clockwise moments of the internal shears about the mid-point of the web to the clockwise moment of the applied shear load about the same point gives

$$S_y \xi_S = 2 \int_0^b q_{12} \frac{h}{2} ds_1$$

or, by substitution from Eq. (iii)

$$S_y \xi_S = 2 \int_0^b \frac{6S_y}{h^2(1 + 6b/h)} \frac{h}{2} s_1 ds_1$$

from which

$$\xi_S = \frac{3b^2}{h(1 + 6b/h)} \tag{iv}$$

In the case of an unsymmetrical section, the coordinates (ξ_S, η_S) of the shear centre referred to some convenient point in the cross-section would be obtained by first determining ξ_S in a similar manner to that of Example 17.2 and then finding η_S by applying a shear load S_x through the shear centre. In both cases the choice of a web/flange junction as a moment centre reduces the amount of computation.

17.3 Shear of closed section beams

The solution for a shear loaded closed section beam follows a similar pattern to that described in Section 17.2 for an open section beam but with two important differences. First, the shear loads may be applied through points in the cross-section other than the shear centre so that torsional as well as shear effects are included. This is possible since, as we shall see, shear stresses produced by torsion in closed section beams have exactly the same form as shear stresses produced by shear, unlike shear stresses due to

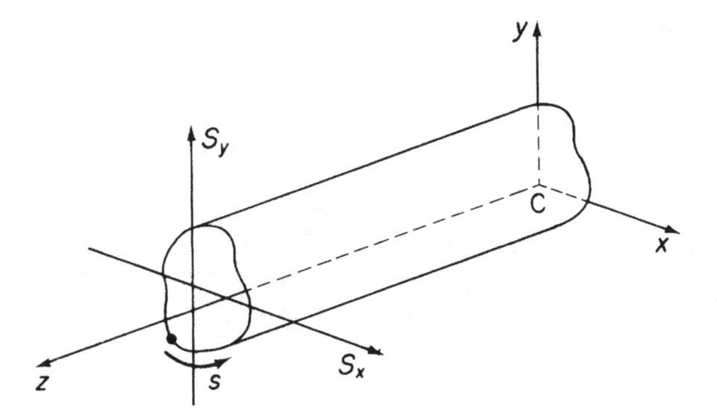

Fig. 17.10 Shear of closed section beams.

shear and torsion in open section beams. Secondly, it is generally not possible to choose an origin for s at which the value of shear flow is known. Consider the closed section beam of arbitrary section shown in Fig. 17.10. The shear loads S_x and S_y are applied through any point in the cross-section and, in general, cause direct bending stresses and shear flows which are related by the equilibrium equation (17.2). We assume that hoop stresses and body forces are absent. Therefore

$$\frac{\partial q}{\partial s} + t\frac{\partial \sigma_z}{\partial z} = 0$$

From this point the analysis is identical to that for a shear loaded open section beam until we reach the stage of integrating Eq. (17.13), namely

$$\int_0^s \frac{\partial q}{\partial s} \mathrm{d}s = -\left(\frac{S_x I_{xx} - S_y I_{xy}}{I_{xx} I_{yy} - I_{xy}^2}\right) \int_0^s tx \, \mathrm{d}s - \left(\frac{S_y I_{yy} - S_x I_{xy}}{I_{xx} I_{yy} - I_{xy}^2}\right) \int_0^s ty \, \mathrm{d}s$$

Let us suppose that we choose an origin for s where the shear flow has the unknown value $q_{s,0}$. Integration of Eq. (17.13) then gives

$$q_s - q_{s,0} = -\left(\frac{S_x I_{xx} - S_y I_{xy}}{I_{xx} I_{yy} - I_{xy}^2}\right) \int_0^s tx \, \mathrm{d}s - \left(\frac{S_y I_{yy} - S_x I_{xy}}{I_{xx} I_{yy} - I_{xy}^2}\right) \int_0^s ty \, \mathrm{d}s$$

or

$$q_s = -\left(\frac{S_x I_{xx} - S_y I_{xy}}{I_{xx} I_{yy} - I_{xy}^2}\right) \int_0^s tx \, \mathrm{d}s - \left(\frac{S_y I_{yy} - S_x I_{xy}}{I_{xx} I_{yy} - I_{xy}^2}\right) \int_0^s ty \, \mathrm{d}s + q_{s,0} \tag{17.15}$$

We observe by comparison of Eqs (17.15) and (17.14) that the first two terms on the right-hand side of Eq. (17.15) represent the shear flow distribution in an open section beam loaded through its shear centre. This fact indicates a method of solution for a shear loaded closed section beam. Representing this 'open' section or 'basic' shear flow by q_{b}, we may write Eq. (17.15) in the form

$$q_s = q_{\mathrm{b}} + q_{s,0} \tag{17.16}$$

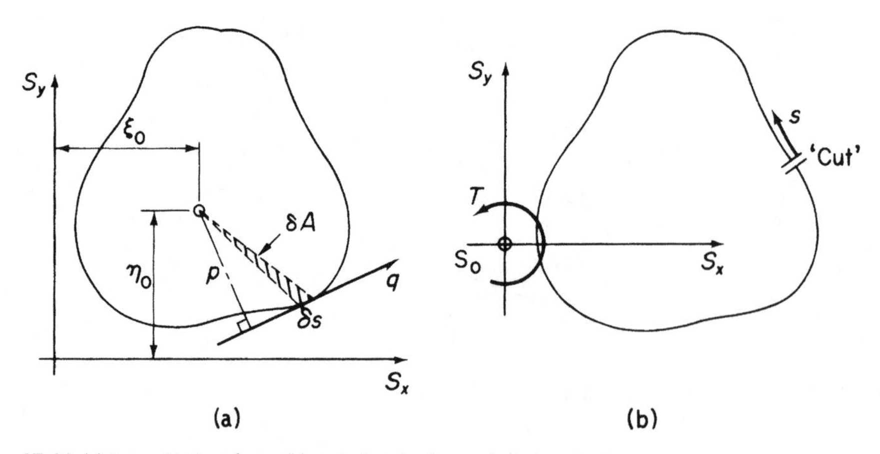

Fig. 17.11 (a) Determination of $q_{s,0}$; (b) equivalent loading on 'open' section beam.

We obtain q_b by supposing that the closed beam section is 'cut' at some convenient point thereby producing an 'open' section (see Fig. 17.11(b)). The shear flow distribution (q_b) around this 'open' section is given by

$$q_b = -\left(\frac{S_x I_{xx} - S_y I_{xy}}{I_{xx} I_{yy} - I_{xy}^2}\right) \int_0^s tx \, ds - \left(\frac{S_y I_{yy} - S_x I_{xy}}{I_{xx} I_{xy} - I_{xy}^2}\right) \int_0^s ty \, ds$$

as in Section 17.2. The value of shear flow at the cut ($s = 0$) is then found by equating applied and internal moments taken about some convenient moment centre. Then, from Fig. 17.11(a)

$$S_x \eta_0 - S_y \xi_0 = \oint pq \, ds = \oint pq_b \, ds + q_{s,0} \oint p \, ds$$

where $\oint$ denotes integration completely around the cross-section. In Fig. 17.11 (a)

$$\delta A = \frac{1}{2} \delta s p$$

so that

$$\oint dA = \frac{1}{2} \oint p \, ds$$

Hence

$$\oint p ds = 2A$$

where A is the area enclosed by the mid-line of the beam section wall. Hence

$$S_x \eta_0 - S_y \xi_0 = \oint pq_b ds + 2A q_{s,0} \tag{17.17}$$

If the moment centre is chosen to coincide with the lines of action of S_x and S_y then Eq. (17.17) reduces to

$$0 = \oint p q_{\mathrm{b}} \, \mathrm{d}s + 2A q_{s,0} \tag{17.18}$$

The unknown shear flow $q_{s,0}$ follows from either of Eqs (17.17) or (17.18).

It is worthwhile to consider some of the implications of the above process. Equation (17.14) represents the shear flow distribution in an open section beam for the condition of zero twist. Therefore, by 'cutting' the closed section beam of Fig. 17.11(a) to determine q_{b}, we are, in effect, replacing the shear loads of Fig. 17.11(a) by shear loads S_x and S_y acting through the shear centre of the resulting 'open' section beam together with a torque T as shown in Fig. 17.11(b). We shall show in Section 18.1 that the application of a torque to a closed section beam results in a constant shear flow. In this case the constant shear flow $q_{s,0}$ corresponds to the torque but will have different values for different positions of the 'cut' since the corresponding various 'open' section beams will have different locations for their shear centres. An additional effect of 'cutting' the beam is to produce a statically determinate structure since the q_{b} shear flows are obtained from statical equilibrium considerations. It follows that a single cell closed section beam supporting shear loads is singly redundant.

17.3.1 Twist and warping of shear loaded closed section beams

Shear loads which are not applied through the shear centre of a closed section beam cause cross-sections to twist and warp; i.e., in addition to rotation, they suffer out of plane axial displacements. Expressions for these quantities may be derived in terms of the shear flow distribution q_s as follows. Since $q = \tau t$ and $\tau = G\gamma$ (see Chapter 1) then we can express q_s in terms of the warping and tangential displacements w and v_{t} of a point in the beam wall by using Eq. (17.6). Thus

$$q_s = Gt \left(\frac{\partial w}{\partial s} + \frac{\partial v_{\mathrm{t}}}{\partial z} \right) \tag{17.19}$$

Substituting for $\partial v_{\mathrm{t}} / \partial z$ from Eq. (17.10) we have

$$\frac{q_s}{Gt} = \frac{\partial w}{\partial s} + p \frac{\mathrm{d}\theta}{\mathrm{d}z} + \frac{\mathrm{d}u}{\mathrm{d}z} \cos \psi + \frac{\mathrm{d}v}{\mathrm{d}z} \sin \psi \tag{17.20}$$

Integrating Eq. (17.20) with respect to s from the chosen origin for s and noting that G may also be a function of s, we obtain

$$\int_0^s \frac{q_s}{Gt} \mathrm{d}s = \int_0^s \frac{\partial w}{\partial s} \mathrm{d}s + \frac{\mathrm{d}\theta}{\mathrm{d}z} \int_0^s p \, \mathrm{d}s + \frac{\mathrm{d}u}{\mathrm{d}z} \int_0^s \cos \psi \, \mathrm{d}s + \frac{\mathrm{d}v}{\mathrm{d}z} \int_0^s \sin \psi \, \mathrm{d}s$$

or

$$\int_0^s \frac{q_s}{Gt} \mathrm{d}s = \int_0^s \frac{\partial w}{\partial s} \mathrm{d}s + \frac{\mathrm{d}\theta}{\mathrm{d}z} \int_0^s p \, \mathrm{d}s + \frac{\mathrm{d}u}{\mathrm{d}z} \int_0^s \mathrm{d}x + \frac{\mathrm{d}v}{\mathrm{d}z} \int_0^s \mathrm{d}y$$

which gives

$$\int_0^s \frac{q_s}{Gt}\mathrm{d}s = (w_s - w_0) + 2A_{Os}\frac{\mathrm{d}\theta}{\mathrm{d}z} + \frac{\mathrm{d}u}{\mathrm{d}z}(x_s - x_0) + \frac{\mathrm{d}v}{\mathrm{d}z}(y_s - y_0) \tag{17.21}$$

where A_{Os} is the area swept out by a generator, centre at the origin of axes, O, from the origin for s to any point s around the cross-section. Continuing the integration completely around the cross-section yields, from Eq. (17.21)

$$\oint \frac{q_s}{Gt}\mathrm{d}s = 2A\frac{\mathrm{d}\theta}{\mathrm{d}z}$$

from which

$$\frac{\mathrm{d}\theta}{\mathrm{d}z} = \frac{1}{2A}\oint \frac{q_s}{Gt}\mathrm{d}s \tag{17.22}$$

Substituting for the rate of twist in Eq. (17.21) from Eq. (17.22) and rearranging, we obtain the warping distribution around the cross-section

$$w_s - w_0 = \int_0^s \frac{q_s}{Gt}\mathrm{d}s - \frac{A_{Os}}{A}\oint \frac{q_s}{Gt}\mathrm{d}s - \frac{\mathrm{d}u}{\mathrm{d}z}(x_s - x_0) - \frac{\mathrm{d}v}{\mathrm{d}z}(y_s - y_0) \tag{17.23}$$

Using Eqs (17.11) to replace $\mathrm{d}u/\mathrm{d}z$ and $\mathrm{d}v/\mathrm{d}z$ in Eq. (17.23) we have

$$w_s - w_0 = \int_0^s \frac{q_s}{Gt}\mathrm{d}s - \frac{A_{Os}}{A}\oint \frac{q_s}{Gt}\mathrm{d}s - y_R\frac{\mathrm{d}\theta}{\mathrm{d}z}(x_s - x_0) + x_R\frac{\mathrm{d}\theta}{\mathrm{d}z}(y_s - y_0) \tag{17.24}$$

The last two terms in Eq. (17.24) represent the effect of relating the warping displacement to an arbitrary origin which itself suffers axial displacement due to warping. In the case where the origin coincides with the centre of twist R of the section then Eq. (17.24) simplifies to

$$w_s - w_0 = \int_0^s \frac{q_s}{Gt}\mathrm{d}s - \frac{A_{Os}}{A}\oint \frac{q_s}{Gt}\mathrm{d}s \tag{17.25}$$

In problems involving singly or doubly symmetrical sections, the origin for s may be taken to coincide with a point of zero warping which will occur where an axis of symmetry and the wall of the section intersect. For unsymmetrical sections the origin for s may be chosen arbitrarily. The resulting warping distribution will have exactly the same form as the actual distribution but will be displaced axially by the unknown warping displacement at the origin for s. This value may be found by referring to the torsion of closed section beams subject to axial constraint (see Section 26.3). In the analysis of such beams it is assumed that the direct stress distribution set up by the constraint is directly proportional to the free warping of the section, i.e.

$$\sigma = \text{constant} \times w$$

Also, since a pure torque is applied the resultant of any internal direct stress system must be zero, in other words it is self-equilibrating. Thus

$$\text{Resultant axial load} = \oint \sigma t \, \mathrm{d}s$$

where σ is the direct stress at any point in the cross-section. Then, from the above assumption

$$0 = \oint wt\,\mathrm{d}s$$

or

$$0 = \oint (w_s - w_0)t\,\mathrm{d}s$$

so that

$$w_0 = \frac{\oint w_s t\,\mathrm{d}s}{\oint t\,\mathrm{d}s} \tag{17.26}$$

17.3.2 Shear centre

The shear centre of a closed section beam is located in a similar manner to that described in Section 17.2.1 for open section beams. Therefore, to determine the coordinate ξ_S (referred to any convenient point in the cross-section) of the shear centre S of the closed section beam shown in Fig. 17.12, we apply an arbitrary shear load S_y through S, calculate the distribution of shear flow q_s due to S_y and then equate internal and external moments. However, a difficulty arises in obtaining $q_{s,0}$ since, at this stage, it is impossible to equate internal and external moments to produce an equation similar to Eq. (17.17) as the position of S_y is unknown. We therefore use the condition that a shear load acting through the shear centre of a section produces zero twist. It follows that $\mathrm{d}\theta/\mathrm{d}z$ in Eq. (17.22) is zero so that

$$0 = \oint \frac{q_s}{Gt}\mathrm{d}s$$

or

$$0 = \oint \frac{1}{Gt}(q_\mathrm{b} + q_{s,0})\mathrm{d}s$$

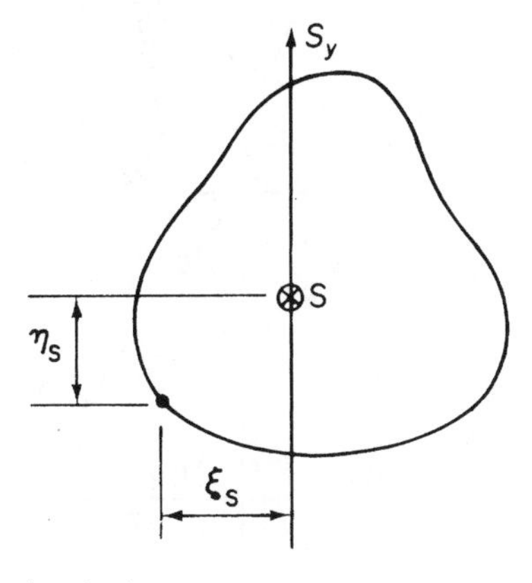

Fig. 17.12 Shear centre of a closed section beam.

which gives

$$q_{s,0} = -\frac{\oint (q_b/Gt)\mathrm{d}s}{\oint \mathrm{d}s/Gt} \tag{17.27}$$

If $Gt = \text{constant}$ then Eq. (17.27) simplifies to

$$q_{s,0} = -\frac{\oint q_b\,\mathrm{d}s}{\oint \mathrm{d}s} \tag{17.28}$$

The coordinate η_S is found in a similar manner by applying S_x through S.

Example 17.3

A thin-walled closed section beam has the singly symmetrical cross-section shown in Fig. 17.13. Each wall of the section is flat and has the same thickness t and shear modulus G. Calculate the distance of the shear centre from point 4.

The shear centre clearly lies on the horizontal axis of symmetry so that it is only necessary to apply a shear load S_y through S and to determine ξ_S. If we take the x reference axis to coincide with the axis of symmetry then $I_{xy} = 0$, and since $S_x = 0$ Eq. (17.15) simplifies to

$$q_s = -\frac{S_y}{I_{xx}}\int_0^s ty\,\mathrm{d}s + q_{s,0} \tag{i}$$

in which

$$I_{xx} = 2\left[\int_0^{10a} t\left(\frac{8}{10}s_1\right)^2 \mathrm{d}s_1 + \int_0^{17a} t\left(\frac{8}{17}s_2\right)^2 \mathrm{d}s_2\right]$$

Evaluating this expression gives $I_{xx} = 1152a^3t$.

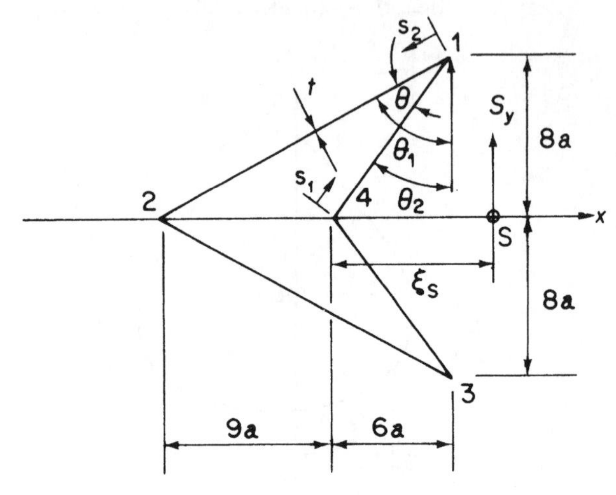

Fig. 17.13 Closed section beam of Example 17.3.

The basic shear flow distribution q_b is obtained from the first term in Eq. (i). Then, for the wall 41

$$q_{b,41} = \frac{-S_y}{1152a^3 t}\int_0^{s_1} t\left(\frac{8}{10}s_1\right)\mathrm{d}s_1 = \frac{-S_y}{1152a^3}\left(\frac{2}{5}s_1^2\right) \tag{ii}$$

In the wall 12

$$q_{b,12} = \frac{-S_y}{1152a^3}\left[\int_0^{s_2}(17a - s_2)\frac{8}{17}\mathrm{d}s_2 + 40a^2\right] \tag{ii}$$

which gives

$$q_{b,12} = \frac{-S_y}{1152a^3}\left(-\frac{4}{17}s_2^2 + 8as_2 + 40a^2\right) \tag{iii}$$

The q_b distributions in the walls 23 and 34 follow from symmetry. Hence from Eq. (17.28)

$$q_{s,0} = \frac{2S_y}{54a \times 1152a^3}\left[\int_0^{10a}\frac{2}{5}s_1^2\,\mathrm{d}s_1 + \int_0^{17a}\left(-\frac{4}{17}s_2^2 + 8as_2 + 40a^2\right)ds_2\right]$$

giving

$$q_{s,0} = \frac{S_y}{1152a^3}(58.7a^2) \tag{iv}$$

Taking moments about the point 2 we have

$$S_y(\xi_S + 9a) = 2\int_0^{10a} q_{41}17a\sin\theta\,\mathrm{d}s_1$$

or

$$S_y(\xi_S + 9a) = \frac{S_y 34a\sin\theta}{1152a^3}\int_0^{10a}\left(-\frac{2}{5}s_1^2 + 58.7a^2\right)\mathrm{d}s_1 \tag{v}$$

We may replace $\sin\theta$ by $\sin(\theta_1 - \theta_2) = \sin\theta_1\cos\theta_2 - \cos\theta_1\sin\theta_2$ where $\sin\theta_1 =$, $15/17$, $\cos\theta_2 = 8/10$, $\cos\theta_1 = 8/17$ and $\sin\theta_2 = 6/10$. Substituting these values and integrating Eq. (v) gives

$$\xi_S = -3.35a$$

which means that the shear centre is inside the beam section.

Reference

1 Megson, T. H. G., *Structural and Stress Analysis,* 2nd edition, Elsevier, Oxford, 2005.

Problems

P.17.1 A beam has the singly symmetrical, thin-walled cross-section shown in Fig. P.17.1. The thickness t of the walls is constant throughout. Show that the distance of the shear centre from the web is given by

$$\xi_S = -d\frac{\rho^2 \sin\alpha\cos\alpha}{1 + 6\rho + 2\rho^3 \sin^2\alpha}$$

where

$$\rho = d/h$$

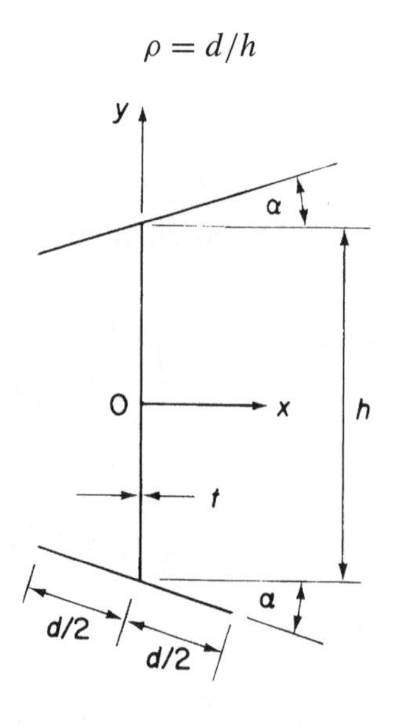

Fig. P.17.1

P.17.2 A beam has the singly symmetrical, thin-walled cross-section shown in Fig. P.17.2. Each wall of the section is flat and has the same length a and thickness t. Calculate the distance of the shear centre from the point 3.

Ans. $5a\cos\alpha/8$

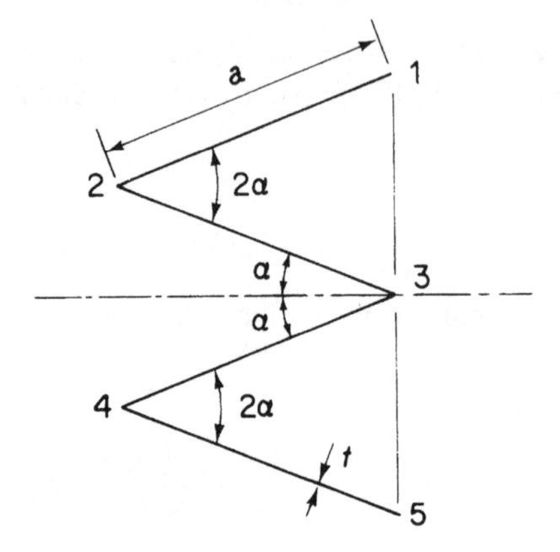

Fig. P.17.2

P.17.3 Determine the position of the shear centre S for the thin-walled, open cross-section shown in Fig. P.17.3. The thickness t is constant.

Ans. $\pi r/3$

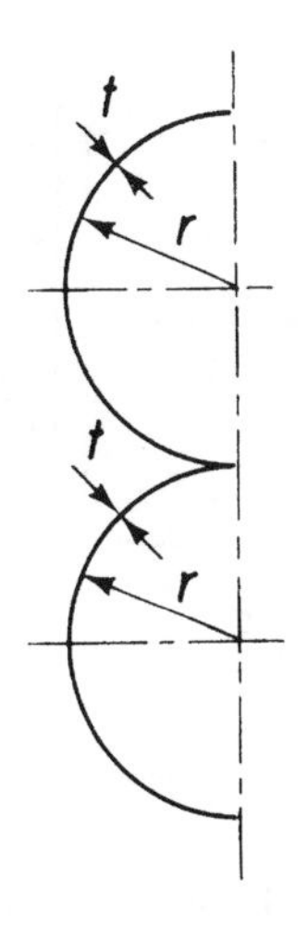

Fig. P.17.3

P.17.4 Figure P.17.4 shows the cross-section of a thin, singly symmetrical I-section. Show that the distance ξ_S of the shear centre from the vertical web is given by

$$\frac{\xi_S}{d} = \frac{3\rho(1-\beta)}{(1+12\rho)}$$

where $\rho = d/h$. The thickness t is taken to be negligibly small in comparison with the other dimensions.

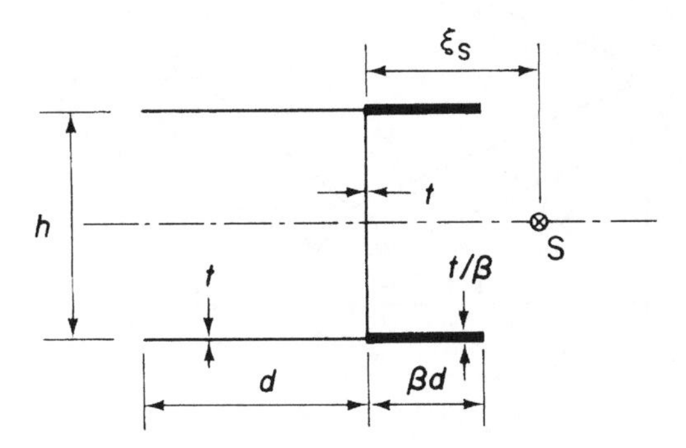

Fig. P.17.4

P.17.5 A thin-walled beam has the cross-section shown in Fig. P.17.5. The thickness of each flange varies linearly from t_1 at the tip to t_2 at the junction with the web. The web itself has a constant thickness t_3. Calculate the distance ξ_S from the web to the shear centre S.

Ans. $d^2(2t_1 + t_2)/[3d(t_1 + t_2) + ht_3]$.

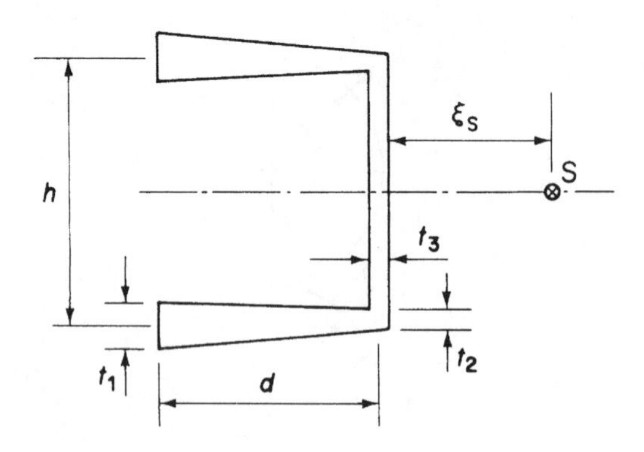

Fig. P.17.5

P.17.6 Figure P.17.6 shows the singly symmetrical cross-section of a thin-walled open section beam of constant wall thickness t, which has a narrow longitudinal slit at the corner 15.

Calculate and sketch the distribution of shear flow due to a vertical shear force S_y acting through the shear centre S and note the principal values. Show also that the distance ξ_S of the shear centre from the nose of the section is $\xi_S = l/2(1 + a/b)$.

Ans. $q_2 = q_4 = 3bS_y/2h(b+a)$, $q_3 = 3S_y/2h$. Parabolic distributions.

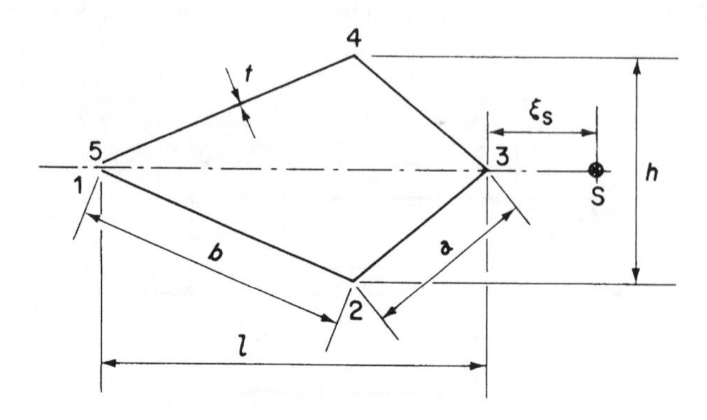

Fig. P.17.6

P.17.7 Show that the position of the shear centre S with respect to the intersection of the web and lower flange of the thin-walled section shown in Fig. P.17.7, is given by

$$\xi_S = -45a/97, \quad \eta_S = 46a/97$$

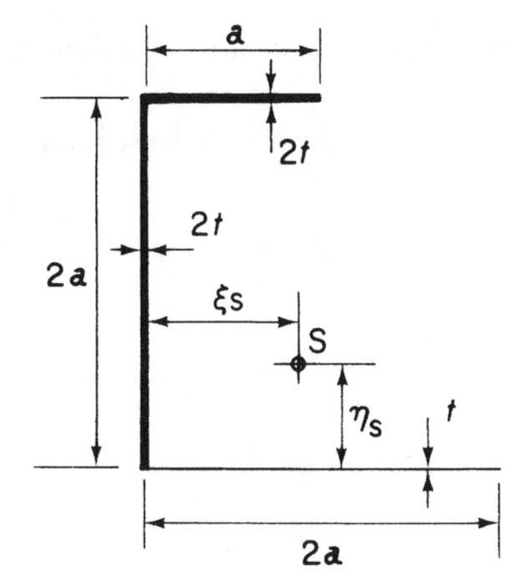

Fig. P.17.7

P.17.8 Define the term 'shear centre' of a thin-walled open section and determine the position of the shear centre of the thin-walled open section shown in Fig. P.17.8.

Ans. $2.66r$ from centre of semicircular wall.

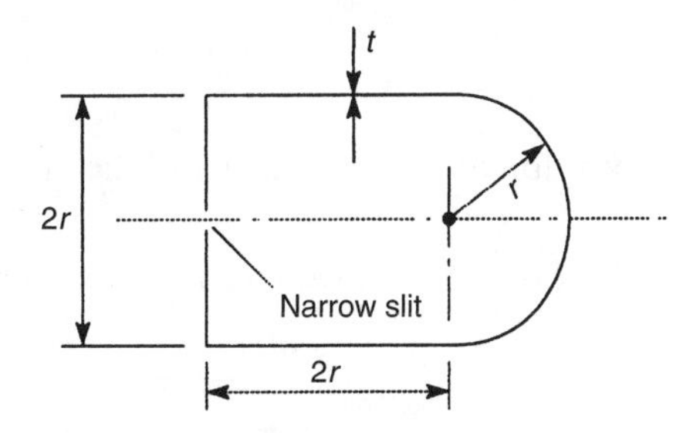

Fig. P.17.8

P.17.9 Determine the position of the shear centre of the cold-formed, thin-walled section shown in Fig. P.17.9. The thickness of the section is constant throughout.

Ans. 87.5 mm above centre of semicircular wall.

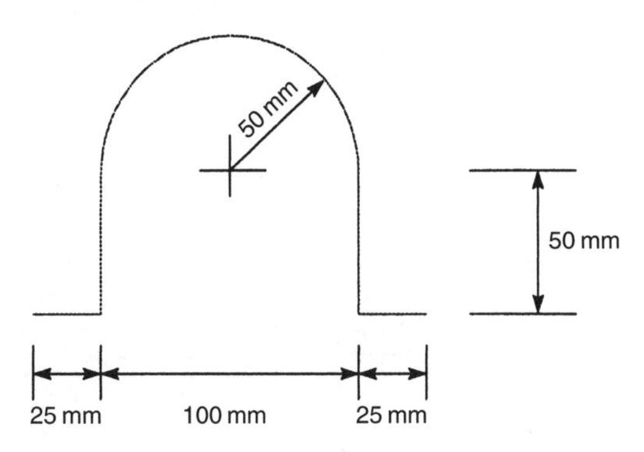

Fig. P.17.9

P.17.10 Find the position of the shear centre of the thin-walled beam section shown in Fig. P.17.10.

Ans. $1.2r$ on axis of symmetry to the left of the section.

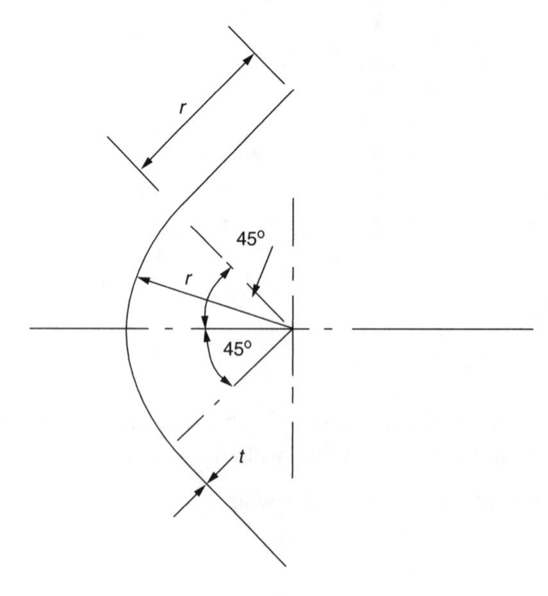

Fig. P.17.10

P.17.11 Calculate the position of the shear centre of the thin-walled section shown in Fig. P.17.11.

Ans. 20.2 mm to the left of the vertical web on axis of symmetry.

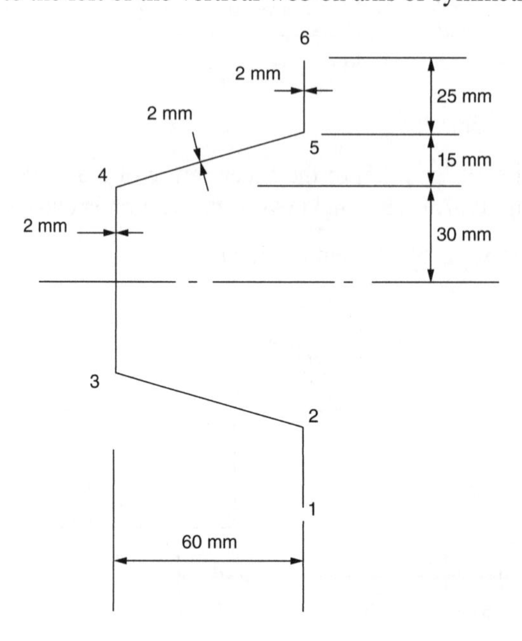

Fig. P.17.11

P.17.12 A thin-walled closed section beam of constant wall thickness t has the cross-section shown in Fig. P.17.12.

Assuming that the direct stresses are distributed according to the basic theory of bending, calculate and sketch the shear flow distribution for a vertical shear force S_y applied tangentially to the curved part of the beam.

Ans. $q_{O1} = S_y(1.61 \cos\theta - 0.80)/r$

$$q_{12} = \frac{S_y}{r^3}(0.57s^2 - 1.14rs + 0.33r^2).$$

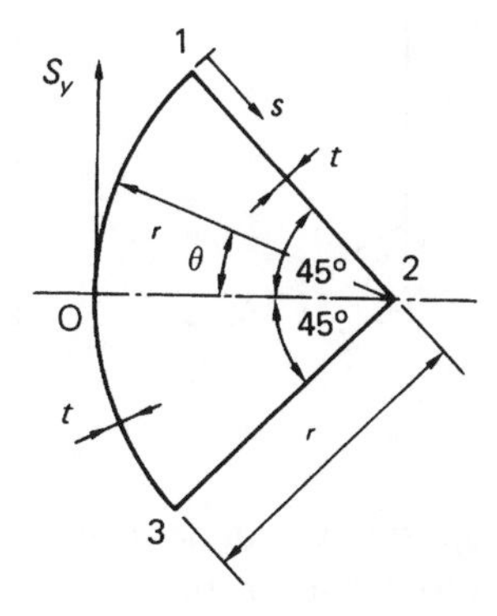

Fig. P.17.12

P.17.13 A uniform thin-walled beam of constant wall thickness t has a cross-section in the shape of an isosceles triangle and is loaded with a vertical shear force S_y applied at the apex. Assuming that the distribution of shear stress is according to the basic theory of bending, calculate the distribution of shear flow over the cross-section.

Illustrate your answer with a suitable sketch, marking in carefully with arrows the direction of the shear flows and noting the principal values.

Ans. $q_{12} = S_y(3s_1^2/d - h - 3d)/h(h + 2d)$
$q_{23} = S_y(-6s_2^2 + 6hs_2 - h^2)/h^2(h + 2d)$

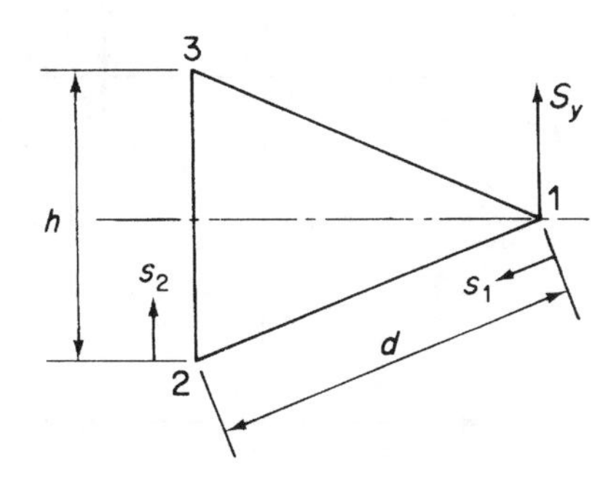

Fig. P.17.13

P.17.14 Figure P.17.14 shows the regular hexagonal cross-section of a thin-walled beam of sides a and constant wall thickness t. The beam is subjected to a transverse shear force S, its line of action being along a side of the hexagon, as shown.

Plot the shear flow distribution around the section, with values in terms of S and a.

Ans. $q_1 = -0.52S/a,\ q_2 = q_8 = -0.47S/a,\ q_3 = q_7 = -0.17S/a,$
$q_4 = q_6 = 0.13S/a,\ q_5 = 0.18S/a$

Parabolic distributions, q positive clockwise.

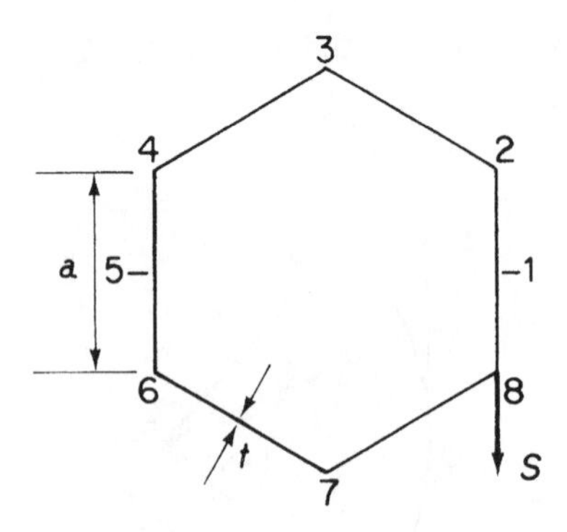

Fig. P.17.14

P.17.15 A box girder has the singly symmetrical trapezoidal cross-section shown in Fig. P.17.15. It supports a vertical shear load of 500 kN applied through its shear centre and in a direction perpendicular to its parallel sides. Calculate the shear flow distribution and the maximum shear stress in the section.

Ans.
$$q_{OA} = 0.25s_A$$
$$q_{AB} = 0.21s_B - 2.14 \times 10^{-4}s_B^2 + 250$$
$$q_{BC} = -0.17s_C + 246$$
$$\tau_{max} = 30.2\ \text{N/mm}^2$$

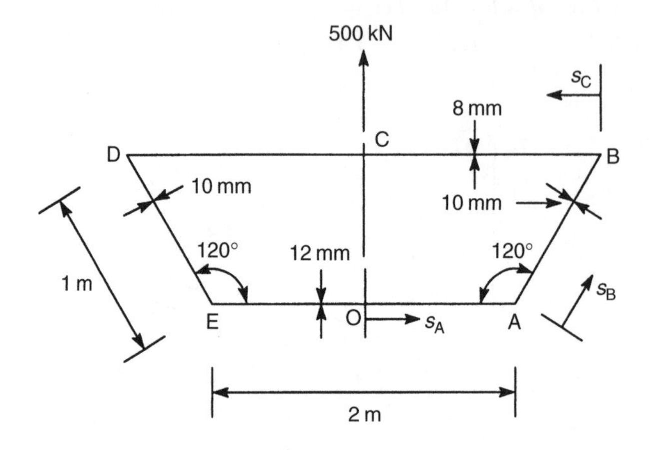

Fig. P.17.15

18

Torsion of beams

In Chapter 3 we developed the theory for the torsion of solid sections using both the Prandtl stress function approach and the St. Venant warping function solution. From that point we looked, via the membrane analogy, at the torsion of a narrow rectangular strip. We shall use the results of this analysis to investigate the torsion of thin-walled open section beams but first we shall examine the torsion of thin-walled closed section beams since the theory for this relies on the general stress, strain and displacement relationships which we established in Chapter 17.

18.1 Torsion of closed section beams

A closed section beam subjected to a pure torque T as shown in Fig. 18.1 does not, in the absence of an axial constraint, develop a direct stress system. It follows that the equilibrium conditions of Eqs (17.2) and (17.3) reduce to $\partial q/\partial s = 0$ and $\partial q/\partial z = 0$, respectively. These relationships may only be satisfied simultaneously by a constant value of q. We deduce, therefore, that the application of a pure torque to a closed section beam results in the development of a constant shear flow in the beam wall. However, the shear stress τ may vary around the cross-section since we allow the wall thickness t to be a function of s. The relationship between the applied torque and this constant shear flow is simply derived by considering the torsional equilibrium of the section shown in Fig. 18.2. The torque produced by the shear flow acting on an element δs of

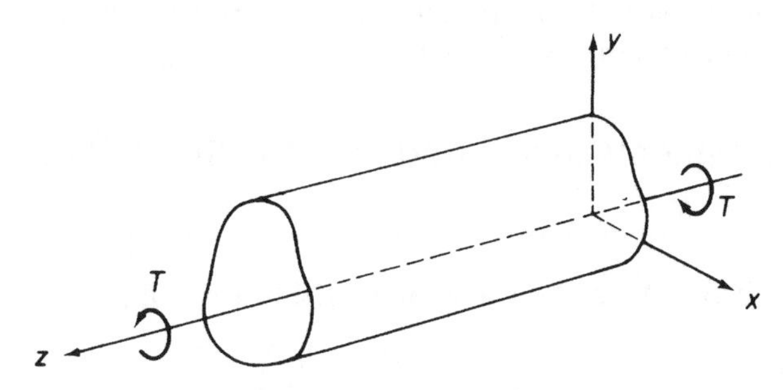

Fig. 18.1 Torsion of a closed section beam.

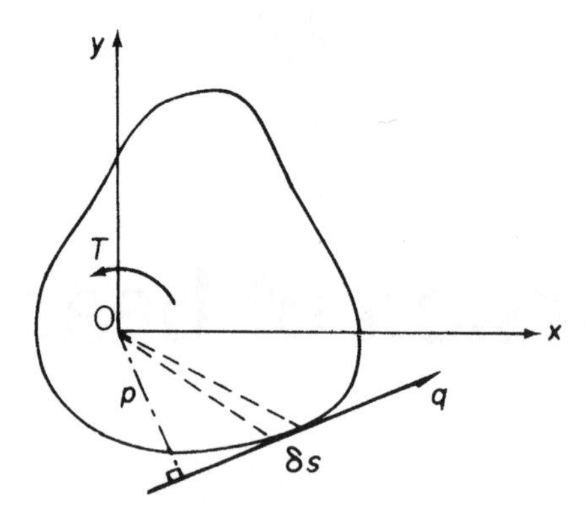

Fig. 18.2 Determination of the shear flow distribution in a closed section beam subjected to torsion.

the beam wall is $pq\delta s$. Hence

$$T = \oint pq \, \mathrm{d}s$$

or, since q is constant and $\oint p \, \mathrm{d}s = 2A$ (see Section 17.3)

$$T = 2Aq \tag{18.1}$$

Note that the origin O of the axes in Fig. 18.2 may be positioned in or outside the cross-section of the beam since the moment of the internal shear flows (whose resultant is a pure torque) is the same about any point in their plane. For an origin outside the cross-section the term $\oint p \, \mathrm{d}s$ will involve the summation of positive and negative areas. The sign of an area is determined by the sign of p which itself is associated with the sign convention for torque as follows. If the movement of the foot of p along the tangent at any point in the positive direction of s leads to an anticlockwise rotation of p about the origin of axes, p is positive. The positive direction of s is in the positive direction of q which is anticlockwise (corresponding to a positive torque). Thus, in Fig. 18.3 a generator OA, rotating about O, will initially sweep out a negative area since p_{A} is negative. At B, however, p_{B} is positive so that the area swept out by the generator has changed sign (at the point where the tangent passes through O and $p = 0$). Positive and negative areas cancel each other out as they overlap so that as the generator moves completely around the section, starting and returning to A say, the resultant area is that enclosed by the profile of the beam.

The theory of the torsion of closed section beams is known as the *Bredt–Batho theory* and Eq. (18.1) is often referred to as the *Bredt–Batho formula*.

18.1.1 Displacements associated with the Bredt–Batho shear flow

The relationship between q and shear strain γ established in Eq. (17.19), namely

$$q = Gt\left(\frac{\partial w}{\partial s} + \frac{\partial v_{\mathrm{t}}}{\partial z}\right)$$

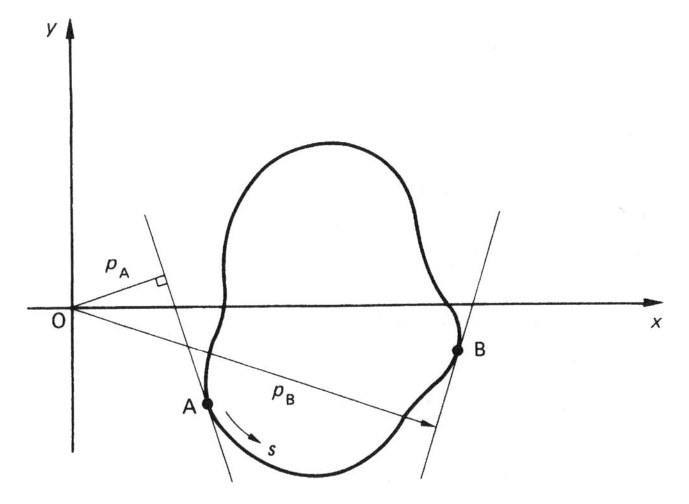

Fig. 18.3 Sign convention for swept areas.

is valid for the pure torsion case where q is constant. Differentiating this expression with respect to z we have

$$\frac{\partial q}{\partial z} = Gt\left(\frac{\partial^2 w}{\partial z\,\partial s} + \frac{\partial^2 v_t}{\partial z^2}\right) = 0$$

or

$$\frac{\partial}{\partial s}\left(\frac{\partial w}{\partial z}\right) + \frac{\partial^2 v_t}{\partial z^2} = 0 \tag{18.2}$$

In the absence of direct stresses the longitudinal strain $\partial w/\partial z(=\varepsilon_z)$ is zero so that

$$\frac{\partial^2 v_t}{\partial z^2} = 0$$

Hence from Eq. (17.7)

$$p\frac{\mathrm{d}^2\theta}{\mathrm{d}z^2} + \frac{\mathrm{d}^2 u}{\mathrm{d}z^2}\cos\psi + \frac{\mathrm{d}^2 v}{\mathrm{d}z^2}\sin\psi = 0 \tag{18.3}$$

For Eq. (18.3) to hold for all points around the section wall, in other words for all values of ψ

$$\frac{\mathrm{d}^2\theta}{\mathrm{d}z^2} = 0, \quad \frac{\mathrm{d}^2 u}{\mathrm{d}z^2} = 0, \quad \frac{\mathrm{d}^2 v}{\mathrm{d}z^2} = 0$$

It follows that $\theta = Az + B$, $u = Cz + D$, $v = Ez + F$, where A, B, C, D, E and F are unknown constants. Thus θ, u and v are all linear functions of z.

Equation (17.22), relating the rate of twist to the variable shear flow q_s developed in a shear loaded closed section beam, is also valid for the case $q_s = q = \text{constant}$. Hence

$$\frac{\mathrm{d}\theta}{\mathrm{d}z} = \frac{q}{2A}\oint \frac{\mathrm{d}s}{Gt}$$

which becomes, on substituting for q from Eq. (18.1)

$$\frac{d\theta}{dz} = \frac{T}{4A^2} \oint \frac{ds}{Gt} \tag{18.4}$$

The warping distribution produced by a varying shear flow, as defined by Eq. (17.25) for axes having their origin at the centre of twist, is also applicable to the case of a constant shear flow. Thus

$$w_s - w_0 = q \int_0^s \frac{ds}{Gt} - \frac{A_{Os}}{A} q \oint \frac{ds}{Gt}$$

Replacing q from Eq. (18.1) we have

$$w_s - w_0 = \frac{T\delta}{2A} \left(\frac{\delta_{Os}}{\delta} - \frac{A_{Os}}{A} \right) \tag{18.5}$$

where

$$\delta = \oint \frac{ds}{Gt} \quad \text{and} \quad \delta_{Os} = \int_0^s \frac{ds}{Gt}$$

The sign of the warping displacement in Eq. (18.5) is governed by the sign of the applied torque T and the signs of the parameters δ_{Os} and A_{Os}. Having specified initially that a positive torque is anticlockwise, the signs of δ_{Os} and A_{Os} are fixed in that δ_{Os} is positive when s is positive, i.e. s is taken as positive in an anticlockwise sense, and A_{Os} is positive when, as before, p (see Fig. 18.3) is positive.

We have noted that the longitudinal strain ε_z is zero in a closed section beam subjected to a pure torque. This means that all sections of the beam must possess identical warping distributions. In other words longitudinal generators of the beam surface remain unchanged in length although subjected to axial displacement.

Example 18.1

A thin-walled circular section beam has a diameter of 200 mm and is 2 m long; it is firmly restrained against rotation at each end. A concentrated torque of 30 kN m is applied to the beam at its mid-span point. If the maximum shear stress in the beam is limited to 200 N/mm^2 and the maximum angle of twist to 2°, calculate the minimum thickness of the beam walls. Take $G = 25\,000$ N/mm^2.

The minimum thickness of the beam corresponding to the maximum allowable shear stress of 200 N/mm^2 is obtained directly using Eq. (18.1) in which $T_{max} = 15$ kN m. Then

$$t_{min} = \frac{15 \times 10^6 \times 4}{2 \times \pi \times 200^2 \times 200} = 1.2\ \text{mm}$$

The rate of twist along the beam is given by Eq. (18.4) in which

$$\oint \frac{ds}{t} = \frac{\pi \times 200}{t_{min}}$$

Hence

$$\frac{\mathrm{d}\theta}{\mathrm{d}z} = \frac{T}{4A^2G} \times \frac{\pi \times 200}{t_{\min}} \tag{i}$$

Taking the origin for z at one of the fixed ends and integrating Eq. (i) for half the length of the beam we obtain

$$\theta = \frac{T}{4A^2G} \times \frac{200\pi}{t_{\min}} z + C_1$$

where C_1 is a constant of integration. At the fixed end where $z = 0$, $\theta = 0$ so that $C_1 = 0$.

Hence

$$\theta = \frac{T}{4A^2G} \times \frac{200\pi}{t_{\min}} z$$

The maximum angle of twist occurs at the mid-span of the beam where $z = 1$ m. Hence

$$t_{\min} = \frac{15 \times 10^6 \times 200 \times \pi \times 1 \times 10^3 \times 180}{4 \times (\pi \times 200^2/4)^2 \times 25\,000 \times 2 \times \pi} = 2.7\,\text{mm}$$

The minimum allowable thickness that satisfies both conditions is therefore 2.7 mm.

Example 18.2

Determine the warping distribution in the doubly symmetrical rectangular, closed section beam, shown in Fig. 18.4, when subjected to an anticlockwise torque T.

From symmetry the centre of twist R will coincide with the mid-point of the cross-section and points of zero warping will lie on the axes of symmetry at the mid-points of the sides. We shall therefore take the origin for s at the mid-point of side 14 and measure s in the positive, anticlockwise, sense around the section. Assuming the shear modulus G to be constant we rewrite Eq. (18.5) in the form

$$w_s - w_0 = \frac{T\delta}{2AG}\left(\frac{\delta_{Os}}{\delta} - \frac{A_{Os}}{A}\right) \tag{i}$$

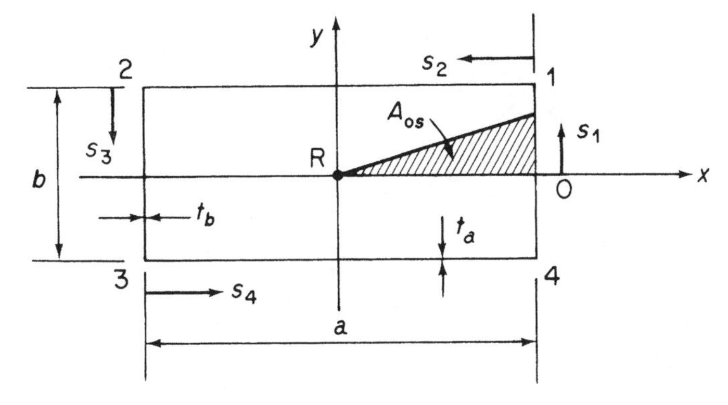

Fig. 18.4 Torsion of a rectangular section beam.

where

$$\delta = \oint \frac{ds}{t} \quad \text{and} \quad \delta_{Os} = \int_0^s \frac{ds}{t}$$

In Eq. (i)

$$w_0 = 0, \quad \delta = 2\left(\frac{b}{t_b} + \frac{a}{t_a}\right) \quad \text{and} \quad A = ab$$

From 0 to 1, $0 \leq s_1 \leq b/2$ and

$$\delta_{Os} = \int_0^{s_1} \frac{ds_1}{t_b} = \frac{s_1}{t_b} \quad A_{Os} = \frac{as_1}{4} \tag{ii}$$

Note that δ_{Os} and A_{Os} are both positive.

Substitution for δ_{Os} and A_{Os} from Eq. (ii) in (i) shows that the warping distribution in the wall 01, w_{01}, is linear. Also

$$w_1 = \frac{T}{2abG} 2\left(\frac{b}{t_b} + \frac{a}{t_a}\right)\left[\frac{b/2t_b}{2(b/t_b + a/t_a)} - \frac{ab/8}{ab}\right]$$

which gives

$$w_1 = \frac{T}{8abG}\left(\frac{b}{t_b} - \frac{a}{t_a}\right) \tag{iii}$$

The remainder of the warping distribution may be deduced from symmetry and the fact that the warping must be zero at points where the axes of symmetry and the walls of the cross-section intersect. It follows that

$$w_2 = -w_1 = -w_3 = w_4$$

giving the distribution shown in Fig. 18.5. Note that the warping distribution will take the form shown in Fig. 18.5 as long as T is positive and $b/t_b > a/t_a$. If *either* of these conditions is reversed w_1 and w_3 will become negative and w_2 and w_4 positive. In the case when $b/t_b = a/t_a$ the warping is zero at all points in the cross-section.

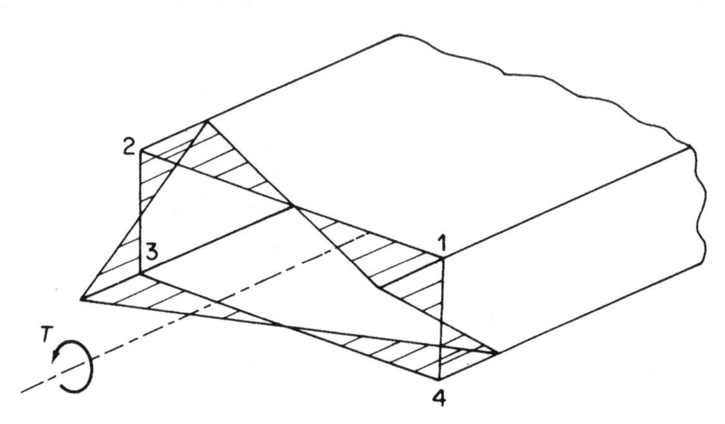

Fig. 18.5 Warping distribution in the rectangular section beam of Example 18.2.

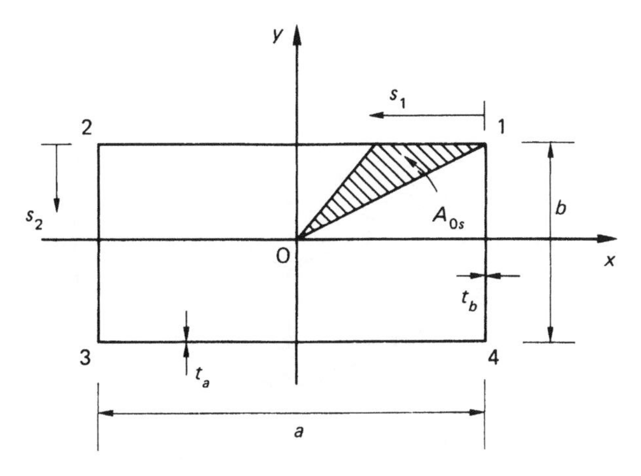

Fig. 18.6 Arbitrary origin for *s*.

Suppose now that the origin for s is chosen arbitrarily at, say, point 1. Then, from Fig. 18.6, δ_{Os} in the wall $12 = s_1/t_a$ and $A_{Os} = \frac{1}{2}s_1 b/2 = s_1 b/4$ and both are positive. Substituting in Eq. (i) and setting $w_O = 0$

$$w'_{12} = \frac{T\delta}{2abG}\left(\frac{s_1}{\delta t_a} - \frac{s_1}{4a}\right) \tag{iv}$$

so that w'_{12} varies linearly from zero at 1 to

$$w'_2 = \frac{T}{2abG}2\left(\frac{b}{t_b} + \frac{a}{t_a}\right)\left[\frac{a}{2(b/t_b + a/t_a)t_a} - \frac{1}{4}\right]$$

at 2. Thus

$$w'_2 = \frac{T}{4abG}\left(\frac{a}{t_a} - \frac{b}{t_b}\right)$$

or

$$w'_2 = -\frac{T}{4abG}\left(\frac{b}{t_b} - \frac{a}{t_a}\right) \tag{v}$$

Similarly

$$w'_{23} = \frac{T\delta}{2abG}\left[\frac{1}{\delta}\left(\frac{a}{t_a} + \frac{s_2}{t_b}\right) - \frac{1}{4b}(b + s_2)\right] \tag{vi}$$

The warping distribution therefore varies linearly from a value $-T(b/t_b - a/t_a)/4abG$ at 2 to zero at 3. The remaining distribution follows from symmetry so that the complete distribution takes the form shown in Fig. 18.7.

Comparing Figs 18.5 and 18.7 it can be seen that the form of the warping distribution is the same but that in the latter case the complete distribution has been displaced axially. The actual value of the warping at the origin for s is found using Eq. (17.26).

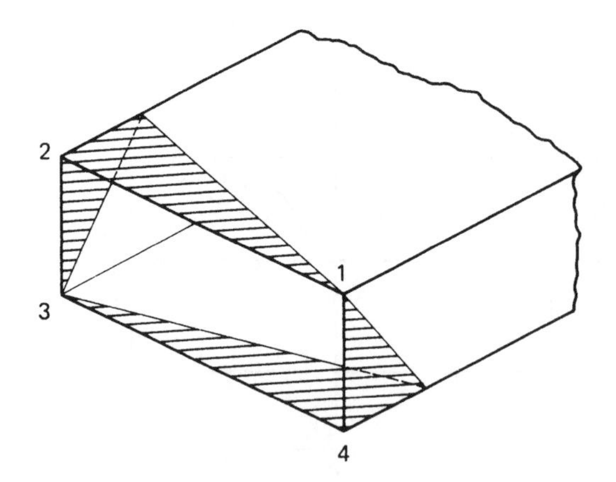

Fig. 18.7 Warping distribution produced by selecting an arbitrary origin for s.

Thus

$$w_0 = \frac{2}{2(at_a + bt_b)} \left(\int_0^a w'_{12} t_a \, \mathrm{d}s_1 + \int_0^b w'_{23} t_b \, \mathrm{d}s_2 \right) \tag{vii}$$

Substituting in Eq. (vii) for w'_{12} and w'_{23} from Eqs (iv) and (vi), respectively, and evaluating gives

$$w_0 = -\frac{T}{8abG} \left(\frac{b}{t_b} - \frac{a}{t_a} \right) \tag{viii}$$

Subtracting this value from the values of $w'_1 (=0)$ and $w'_2 (= -T(b/t_b - a/t_a)/4abG)$ we have

$$w_1 = \frac{T}{8abG} \left(\frac{b}{t_b} - \frac{a}{t_a} \right), \quad w_2 = -\frac{T}{8abG} \left(\frac{b}{t_b} - \frac{a}{t_a} \right)$$

as before. Note that setting $w_0 = 0$ in Eq. (i) implies that w_0, the actual value of warping at the origin for s, has been added to all warping displacements. This value must therefore be *subtracted* from the calculated warping displacements (i.e. those based on an arbitrary choice of origin) to obtain true values.

It is instructive at this stage to examine the mechanics of warping to see how it arises. Suppose that each end of the rectangular section beam of Example 18.2 rotates through opposite angles θ giving a total angle of twist 2θ along its length L. The corner 1 at one end of the beam is displaced by amounts $a\theta/2$ vertically and $b\theta/2$ horizontally as shown in Fig. 18.8. Consider now the displacements of the web and cover of the beam due to rotation. From Figs 18.8 and 18.9 (a) and (b) it can be seen that the angles of rotation of the web and the cover are, respectively

$$\phi_b = (a\theta/2)/(L/2) = a\theta/L$$

and

$$\phi_a = (b\theta/2)/(L/2) = b\theta/L$$

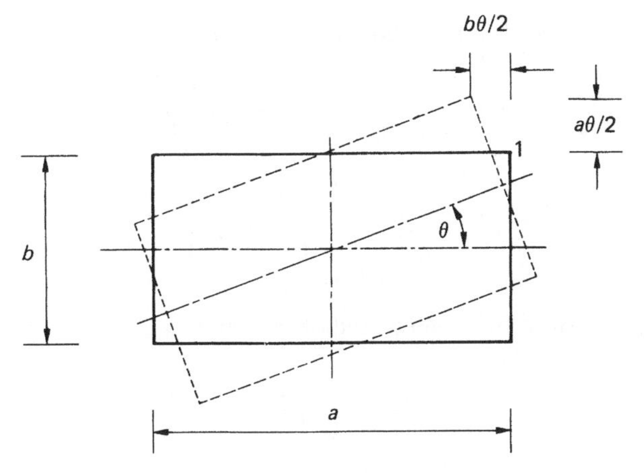

Fig. 18.8 Twisting of a rectangular section beam.

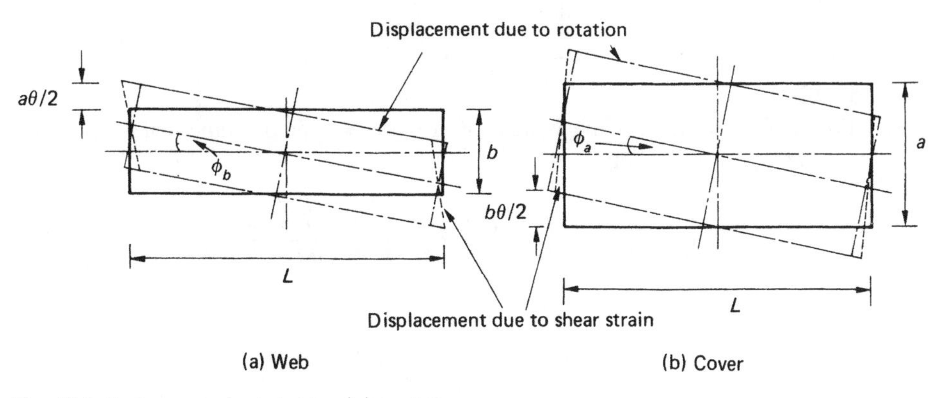

Fig. 18.9 Displacements due to twist and shear strain.

The axial displacements of the corner 1 in the web and cover are then

$$\frac{b}{2}\frac{a\theta}{L}, \quad \frac{a}{2}\frac{b\theta}{L}$$

respectively, as shown in Fig. 18.9(a) and (b). In addition to displacements produced by twisting, the webs and covers are subjected to shear strains γ_b and γ_a corresponding to the shear stress system given by Eq. (18.1). Due to γ_b the axial displacement of corner 1 in the web is $\gamma_b b/2$ in the positive z direction while in the cover the displacement is $\gamma_a a/2$ in the negative z direction. Note that the shear strains γ_b and γ_a correspond to the shear stress system produced by a positive anticlockwise torque. Clearly, the total axial displacement of the point 1 in the web and cover must be the same so that

$$-\frac{b}{2}\frac{a\theta}{L} + \gamma_b\frac{b}{2} = \frac{a}{2}\frac{b\theta}{L} - \gamma_a\frac{a}{2}$$

from which

$$\theta = \frac{L}{2ab}(\gamma_a a + \gamma_b b)$$

The shear strains are obtained from Eq. (18.1) and are

$$\gamma_a = \frac{T}{2abGt_a}, \quad \gamma_b = \frac{T}{2abGt_b}$$

whence

$$\theta = \frac{TL}{4a^2b^2G}\left(\frac{a}{t_a} + \frac{b}{t_b}\right)$$

The total angle of twist from end to end of the beam is 2θ, therefore

$$\frac{2\theta}{L} = \frac{TL}{4a^2b^2G}\left(\frac{2a}{t_a} + \frac{2b}{t_b}\right)$$

or

$$\frac{\mathrm{d}\theta}{\mathrm{d}z} = \frac{T}{4A^2G}\oint \frac{\mathrm{d}s}{t}$$

as in Eq. (18.4).

Substituting for θ in either of the expressions for the axial displacement of the corner 1 gives the warping w_1 at 1. Thus

$$w_1 = \frac{a}{2}\frac{b}{L}\frac{TL}{4a^2b^2G}\left(\frac{a}{t_a} + \frac{b}{t_b}\right) - \frac{T}{2abGt_a}\frac{a}{2}$$

i.e.

$$w_1 = \frac{T}{8abG}\left(\frac{b}{t_b} - \frac{a}{t_a}\right)$$

as before. It can be seen that the warping of the cross-section is produced by a combination of the displacements caused by twisting and the displacements due to the shear strains; these shear strains correspond to the shear stresses whose values are fixed by statics. The angle of twist must therefore be such as to ensure compatibility of displacement between the webs and covers.

18.1.2 Condition for zero warping at a section

The geometry of the cross-section of a closed section beam subjected to torsion may be such that no warping of the cross-section occurs. From Eq. (18.5) we see that this condition arises when

$$\frac{\delta_{Os}}{\delta} = \frac{A_{Os}}{A}$$

or

$$\frac{1}{\delta}\int_0^s \frac{\mathrm{d}s}{Gt} = \frac{1}{2A}\int_0^s p_{\mathrm{R}}\,\mathrm{d}s \qquad (18.6)$$

Differentiating Eq. (18.6) with respect to s gives

$$\frac{1}{\delta G t} = \frac{p_{\mathrm{R}}}{2A}$$

or

$$p_{\mathrm{R}} G t = \frac{2A}{\delta} = \text{constant} \tag{18.7}$$

A closed section beam for which $p_{\mathrm{R}} Gt = \text{constant}$ does not warp and is known as a *Neuber beam*. For closed section beams having a constant shear modulus the condition becomes

$$p_{\mathrm{R}} t = \text{constant} \tag{18.8}$$

Examples of such beams are: a circular section beam of constant thickness; a rectangular section beam for which $at_b = bt_a$ (see Example 18.2); and a triangular section beam of constant thickness. In the last case the shear centre and hence the centre of twist may be shown to coincide with the centre of the inscribed circle so that p_{R} for each side is the radius of the inscribed circle.

18.2 Torsion of open section beams

An approximate solution for the torsion of a thin-walled open section beam may be found by applying the results obtained in Section 3.4 for the torsion of a thin rectangular strip. If such a strip is bent to form an open section beam, as shown in Fig. 18.10(a), and if the distance s measured around the cross-section is large compared with its thickness t then

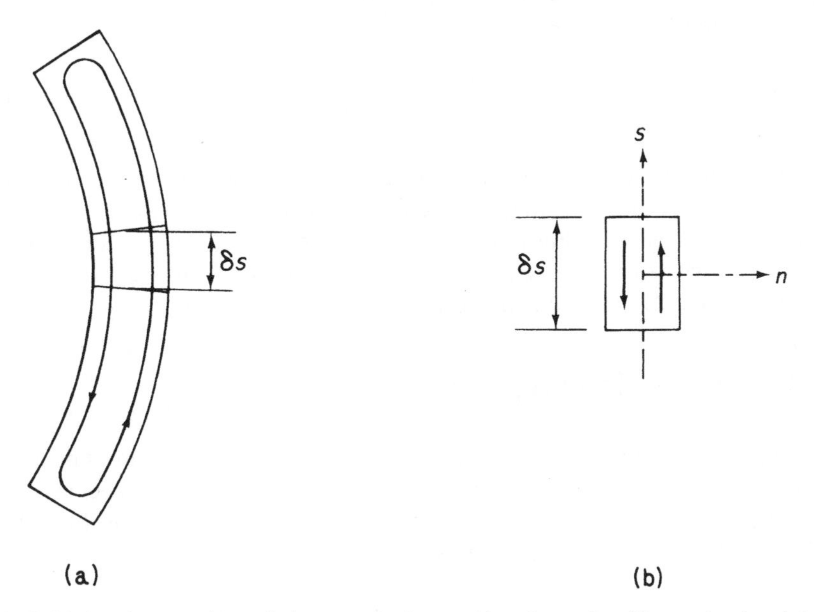

Fig. 18.10 (a) Shear lines in a thin-walled open section beam subjected to torsion; (b) approximation of elemental shear lines to those in a thin rectangular strip.

the contours of the membrane, i.e. lines of shear stress, are still approximately parallel to the inner and outer boundaries. It follows that the shear lines in an element δs of the open section must be nearly the same as those in an element δy of a rectangular strip as demonstrated in Fig. 18.10(b). Equations (3.27)–(3.29) may therefore be applied to the open beam but with reduced accuracy. Referring to Fig. 18.10(b) we observe that Eq. (3.27) becomes

$$\tau_{zs} = 2Gn\frac{\mathrm{d}\theta}{\mathrm{d}z}, \quad \tau_{zn} = 0 \tag{18.9}$$

Equation (3.28) becomes

$$\tau_{zs,\max} = \pm Gt\frac{\mathrm{d}\theta}{\mathrm{d}z} \tag{18.10}$$

and Eq. (3.29) is

$$J = \sum \frac{st^3}{3} \quad \text{or} \quad J = \frac{1}{3}\int_{\text{sect}} t^3\,\mathrm{d}s \tag{18.11}$$

In Eq. (18.11) the second expression for the torsion constant is used if the cross-section has a variable wall thickness. Finally, the rate of twist is expressed in terms of the applied torque by Eq. (3.12), viz.

$$T = GJ\frac{\mathrm{d}\theta}{\mathrm{d}z} \tag{18.12}$$

The shear stress distribution and the maximum shear stress are sometimes more conveniently expressed in terms of the applied torque. Therefore, substituting for $\mathrm{d}\theta/\mathrm{d}z$ in Eqs (18.9) and (18.10) gives

$$\tau_{zs} = \frac{2n}{J}T, \quad \tau_{zs,\max} = \pm\frac{tT}{J} \tag{18.13}$$

We assume in open beam torsion analysis that the cross-section is maintained by the system of closely spaced diaphragms described in Section 17.1 and that the beam is of uniform section. Clearly, in this problem the shear stresses vary across the thickness of the beam wall whereas other stresses such as axial constraint stresses which we shall discuss in Chapter 27 are assumed constant across the thickness.

18.2.1 Warping of the cross-section

We saw in Section 3.4 that a thin rectangular strip suffers warping across its thickness when subjected to torsion. In the same way a thin-walled open section beam will warp across its thickness. This warping, w_{t}, may be deduced by comparing Fig. 18.10(b) with Fig. 3.10 and using Eq. (3.32), thus

$$w_{\mathrm{t}} = ns\frac{\mathrm{d}\theta}{\mathrm{d}z} \tag{18.14}$$

In addition to warping across the thickness, the cross-section of the beam will warp in a similar manner to that of a closed section beam. From Fig. 17.3

$$\gamma_{zs} = \frac{\partial w}{\partial s} + \frac{\partial v_t}{\partial z} \tag{18.15}$$

Referring the tangential displacement v_t to the centre of twist R of the cross-section we have, from Eq. (17.8)

$$\frac{\partial v_t}{\partial z} = p_R \frac{d\theta}{dz} \tag{18.16}$$

Substituting for $\partial v_t/\partial z$ in Eq. (18.15) gives

$$\gamma_{zs} = \frac{\partial w}{\partial s} + p_R \frac{d\theta}{dz}$$

from which

$$\tau_{zs} = G\left(\frac{\partial w}{\partial s} + p_R \frac{d\theta}{dz}\right) \tag{18.17}$$

On the mid-line of the section wall $\tau_{zs} = 0$ (see Eq. (18.9)) so that, from Eq. (18.17)

$$\frac{\partial w}{\partial s} = -p_R \frac{d\theta}{dz}$$

Integrating this expression with respect to s and taking the lower limit of integration to coincide with the point of zero warping, we obtain

$$w_s = -\frac{d\theta}{dz} \int_0^s p_R \, ds \tag{18.18}$$

From Eqs (18.14) and (18.18) it can be seen that two types of warping exist in an open section beam. Equation (18.18) gives the warping of the mid-line of the beam; this is known as *primary warping* and is assumed to be constant across the wall thickness. Equation (18.14) gives the warping of the beam across its wall thickness. This is called *secondary warping*, is very much less than primary warping and is usually ignored in the thin-walled sections common to aircraft structures.

Equation (18.18) may be rewritten in the form

$$w_s = -2A_R \frac{d\theta}{dz} \tag{18.19}$$

or, in terms of the applied torque

$$w_s = -2A_R \frac{T}{GJ} \quad \text{(see Eq. (18.12))} \tag{18.20}$$

in which $A_R = \frac{1}{2}\int_0^s p_R \, ds$ is the area swept out by a generator, rotating about the centre of twist, from the point of zero warping, as shown in Fig. 18.11. The sign of w_s, for a given direction of torque, depends upon the sign of A_R which in turn depends upon the sign of

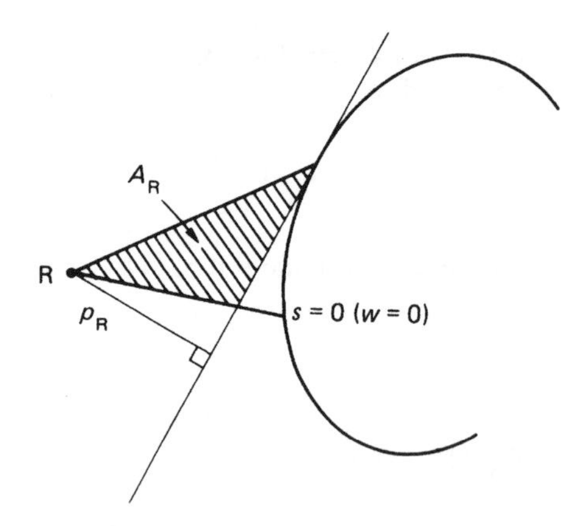

Fig. 18.11 Warping of an open section beam.

p_R, the perpendicular distance from the centre of twist to the tangent at any point. Again, as for closed section beams, the sign of p_R depends upon the assumed direction of a positive torque, in this case anticlockwise. Therefore, p_R (and therefore A_R) is positive if movement of the foot of p_R along the tangent in the assumed direction of s leads to an anticlockwise rotation of p_R about the centre of twist. Note that for open section beams the positive direction of s may be chosen arbitrarily since, for a given torque, the sign of the warping displacement depends only on the sign of the swept area A_R.

Example 18.3

Determine the maximum shear stress and the warping distribution in the channel section shown in Fig. 18.12 when it is subjected to an anticlockwise torque of 10 N m. $G = 25\,000\ \text{N/mm}^2$.

From the second of Eqs (18.13) it can be seen that the maximum shear stress occurs in the web of the section where the thickness is greatest. Also, from the first of Eqs (18.11)

$$J = \frac{1}{3}(2 \times 25 \times 1.5^3 + 50 \times 2.5^3) = 316.7\ \text{mm}^4$$

so that

$$\tau_{\text{max}} = \pm\frac{2.5 \times 10 \times 10^3}{316.7} = \pm 78.9\ \text{N/mm}^2$$

The warping distribution is obtained using Eq. (18.20) in which the origin for s (and hence A_R) is taken at the intersection of the web and the axis of symmetry where the warping is zero. Further, the centre of twist R of the section coincides with its shear centre S whose position is found using the method described in Section 17.2.1, this gives $\xi_S = 8.04$ mm. In the wall O2

$$A_R = \frac{1}{2} \times 8.04 s_1 \quad (p_R \text{ is positive})$$

so that

$$w_{O2} = -2 \times \frac{1}{2} \times 8.04 s_1 \times \frac{10 \times 10^3}{25\,000 \times 316.7} = -0.01 s_1 \tag{i}$$

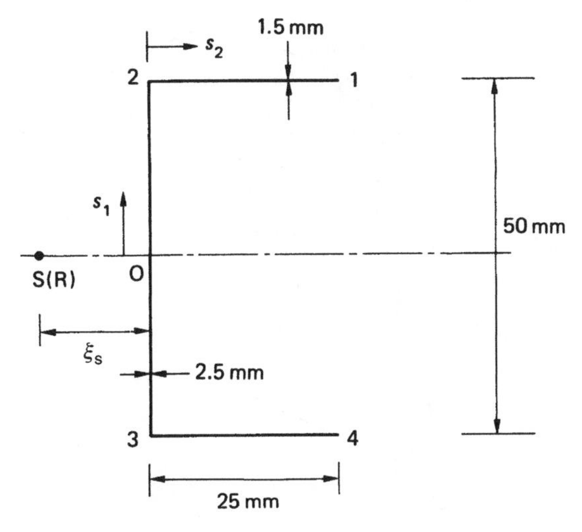

Fig. 18.12 Channel section of Example 18.3.

i.e. the warping distribution is linear in O2 and

$$w_2 = -0.01 \times 25 = -0.25\,\text{mm}$$

In the wall 21

$$A_R = \tfrac{1}{2} \times 8.04 \times 25 - \tfrac{1}{2} \times 25 s_2$$

in which the area swept out by the generator in the wall 21 provides a negative contribution to the total swept area A_R. Thus

$$w_{21} = -25(8.04 - s_2)\frac{10 \times 10^3}{25\,000 \times 316.7}$$

or

$$w_{21} = -0.03(8.04 - s_2) \qquad \text{(ii)}$$

Again the warping distribution is linear and varies from -0.25 mm at 2 to $+0.54$ mm at 1. Examination of Eq. (ii) shows that w_{21} changes sign at $s_2 = 8.04$ mm. The remaining warping distribution follows from symmetry and the complete distribution is shown in Fig. 18.13. In unsymmetrical section beams the position of the point of zero warping is not known but may be found using the method described in Section 27.2 for the restrained warping of an open section beam. From the derivation of Eq. (27.3) we see that

$$2A'_R = \frac{\int_{\text{sect}} 2A_{R,O} t\, ds}{\int_{\text{sect}} t\, ds} \qquad (18.21)$$

in which $A_{R,O}$ is the area swept out by a generator rotating about the centre of twist from some convenient origin and A'_R is the value of $A_{R,O}$ at the point of zero warping. As an illustration we shall apply the method to the beam section of Example 18.3.

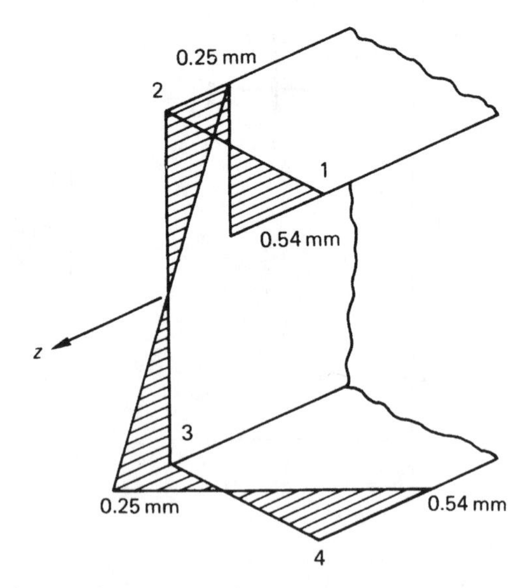

Fig. 18.13 Warping distribution in channel section of Example 18.3.

Suppose that the position of the centre of twist (i.e. the shear centre) has already been calculated and suppose also that we choose the origin for s to be at the point 1. Then, in Fig. 18.14

$$\int_{\text{sect}} t\,\mathrm{d}s = 2 \times 1.5 \times 25 + 2.5 \times 50 = 200\,\text{mm}^2$$

In the wall 12

$$A_{12} = \tfrac{1}{2} \times 25s_1 \quad (A_{\text{R,O}} \text{ for the wall 12}) \tag{i}$$

from which

$$A_2 = \tfrac{1}{2} \times 25 \times 25 = 312.5\,\text{mm}^2$$

Also

$$A_{23} = 312.5 - \tfrac{1}{2} \times 8.04s_2 \tag{ii}$$

and

$$A_3 = 312.5 - \tfrac{1}{2} \times 8.04 \times 50 = 111.5\,\text{mm}^2$$

Finally

$$A_{34} = 111.5 + \tfrac{1}{2} \times 25s_3 \tag{iii}$$

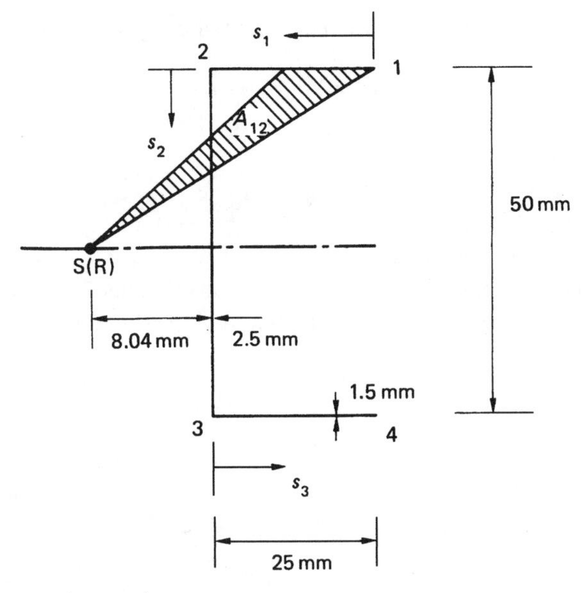

Fig. 18.14 Determination of points of zero warping.

Substituting for A_{12}, A_{23} and A_{34} from Eqs (i)–(iii) in Eq. (18.21) we have

$$2A'_{\rm R} = \frac{1}{200}\left[\int_0^{25} 25 \times 1.15 s_1 \, {\rm d}s_1 + \int_0^{50} 2(312.5 - 4.02 s_2) 2.5 \, {\rm d}s_2 \right.$$
$$\left. + \int_0^{25} 2(111.5 + 12.5 s_3) 1.5 \, {\rm d}s_3 \right] \qquad \text{(iv)}$$

Evaluation of Eq. (iv) gives

$$2A'_{\rm R} = 424\,{\rm mm}^2$$

We now examine each wall of the section in turn to determine points of zero warping. Suppose that in the wall 12 a point of zero warping occurs at a value of s_1 equal to $s_{1,0}$. Then

$$2 \times \tfrac{1}{2} \times 25 s_{1,0} = 424$$

from which

$$s_{1,0} = 16.96\,{\rm mm}$$

so that a point of zero warping occurs in the wall 12 at a distance of 8.04 mm from the point 2 as before. In the web 23 let the point of zero warping occur at $s_2 = s_{2,0}$. Then

$$2 \times \tfrac{1}{2} \times 25 \times 25 - 2 \times \tfrac{1}{2} \times 8.04 s_{2,0} = 424$$

which gives $s_{2,0} = 25$ mm (i.e. on the axis of symmetry). Clearly, from symmetry, a further point of zero warping occurs in the flange 34 at a distance of 8.04 mm from the point 3. The warping distribution is then obtained directly using Eq. (18.20) in which

$$A_R = A_{R,O} - A'_R$$

Problems

P.18.1 A uniform, thin-walled, cantilever beam of closed rectangular cross-section has the dimensions shown in Fig. P.18.1. The shear modulus G of the top and bottom covers of the beam is 18 000 N/mm^2 while that of the vertical webs is 26 000 N/mm^2.

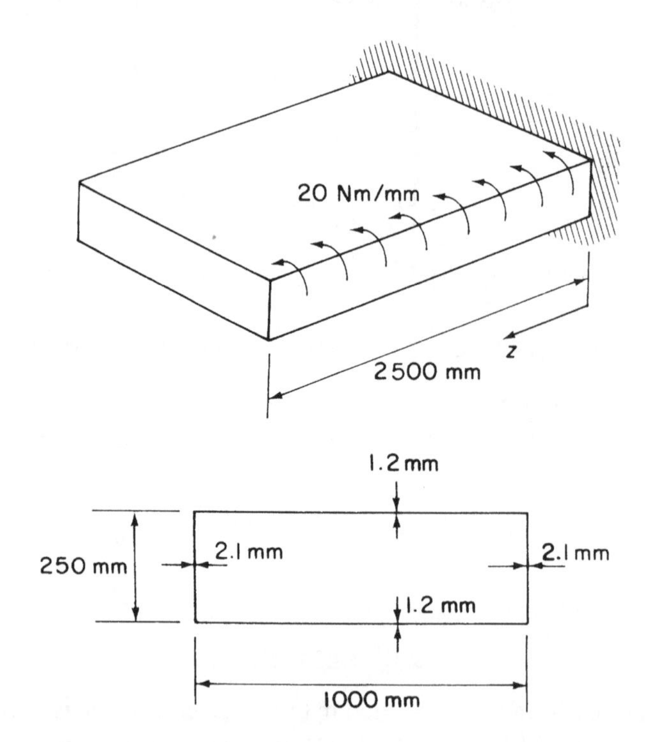

Fig. P.18.1

The beam is subjected to a uniformly distributed torque of 20 N m/mm along its length. Calculate the maximum shear stress according to the Bred–Batho theory of torsion. Calculate also, and sketch, the distribution of twist along the length of the cantilever assuming that axial constraint effects are negligible.

Ans. $\tau_{max} = 83.3\,\text{N/mm}^2$, $\theta = 8.14 \times 10^{-9}\left(2500z - \dfrac{z^2}{2}\right)$ rad.

P.18.2 A single cell, thin-walled beam with the double trapezoidal cross-section shown in Fig. P.18.2, is subjected to a constant torque $T = 90\,500$ N m and is constrained to twist about an axis through the point R. Assuming that the shear stresses are distributed according to the Bredt–Batho theory of torsion, calculate the distribution of warping around the cross-section.

Illustrate your answer clearly by means of a sketch and insert the principal values of the warping displacements.

The shear modulus $G = 27\,500\,\text{N/mm}^2$ and is constant throughout.

Ans. $w_1 = -w_6 = -0.53\,\text{mm}$, $w_2 = -w_5 = 0.05\,\text{mm}$, $w_3 = -w_4 = 0.38\,\text{mm}$.

Linear distribution.

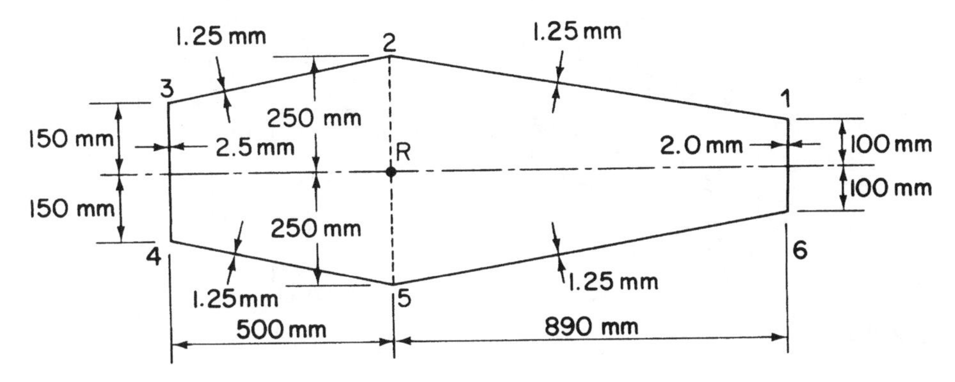

Fig. P.18.2

P.18.3 A uniform thin-walled beam is circular in cross-section and has a constant thickness of 2.5 mm. The beam is 2000 mm long, carrying end torques of 450 N m and, in the same sense, a distributed torque loading of 1.0 N m/mm. The loads are reacted by equal couples R at sections 500 mm distant from each end (Fig. P.18.3).

Calculate the maximum shear stress in the beam and sketch the distribution of twist along its length. Take $G = 30\,000\,\text{N/mm}^2$ and neglect axial constraint effects.

Ans. $\tau_{\max} = 24.2\,\text{N/mm}^2$, $\theta = -0.85 \times 10^{-8} z^2$ rad, $0 \leq z \leq 500\,\text{mm}$,
$\theta = 1.7 \times 10^{-8}(1450z - z^2/2) - 12.33 \times 10^{-3}$ rad, $500 \leq z \leq 1000\,\text{mm}$.

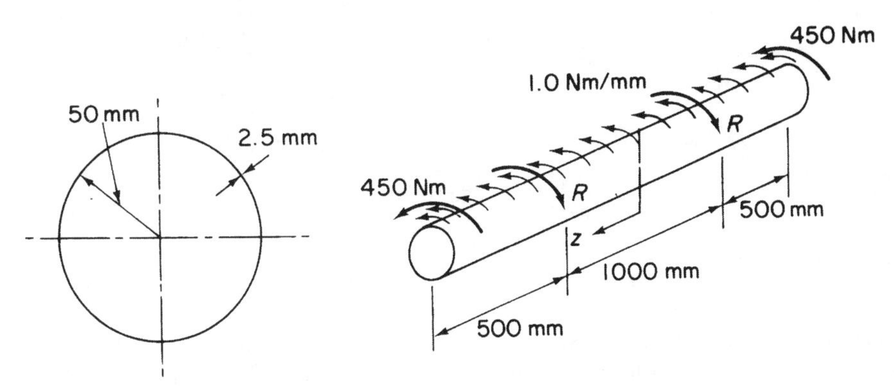

Fig. P.18.3

P.18.4 The thin-walled box section beam ABCD shown in Fig. P.18.4 is attached at each end to supports which allow rotation of the ends of the beam in the longitudinal vertical plane of symmetry but prevent rotation of the ends in vertical planes perpendicular to the longitudinal axis of the beam. The beam is subjected to a uniform torque loading of 20 N m/mm over the portion BC of its span. Calculate the maximum shear stress in the cross-section of the beam and the distribution of angle of twist along its length, $G = 70\,000\ \text{N/mm}^2$.

Ans. $71.4\ \text{N/mm}^2$, $\theta_B = \theta_C = 0.36°$, θ at mid-span $= 0.72°$.

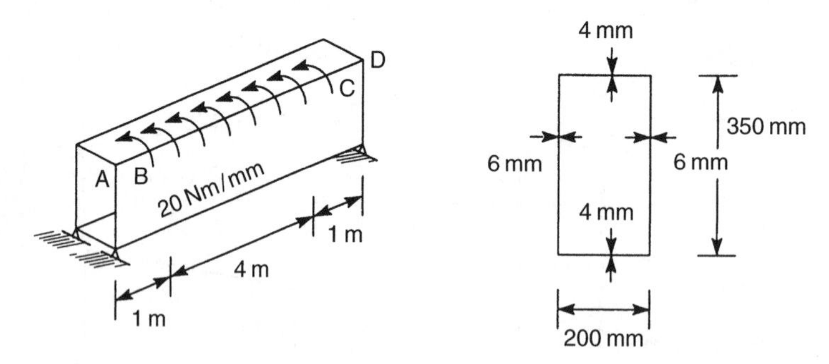

Fig. P.18.4

P.18.5 Figure P.18.5 shows a thin-walled cantilever box beam having a constant width of 50 mm and a depth which decreases linearly from 200 mm at the built-in end to 150 mm at the free end. If the beam is subjected to a torque of 1 kN m at its free end, plot the angle of twist of the beam at 500 mm intervals along its length and determine the maximum shear stress in the beam section. Take $G = 25\,000\ \text{N/mm}^2$.

Ans. $\tau_{max} = 33.3\ \text{N/mm}^2$.

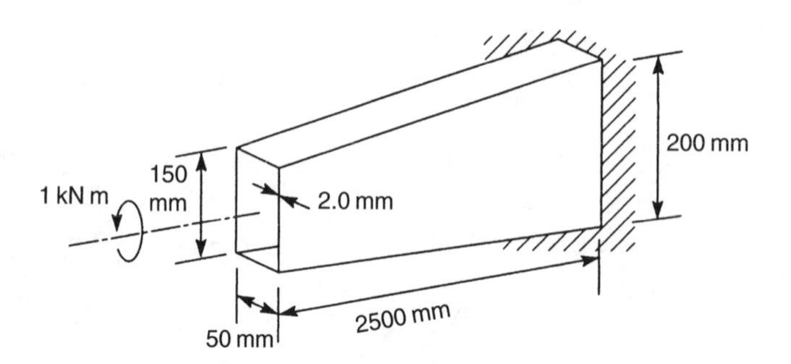

Fig. P.18.5

P.18.6 A uniform closed section beam, of the thin-walled section shown in Fig. P.18.6, is subjected to a twisting couple of 4500 N m. The beam is constrained to twist about a longitudinal axis through the centre C of the semicircular arc 12. For the curved wall 12 the thickness is 2 mm and the shear modulus is $22\,000\ \text{N/mm}^2$. For the plane walls 23, 34 and 41, the corresponding figures are 1.6 mm and $27\,500\ \text{N/mm}^2$. (*Note*: Gt = constant.)

Calculate the rate of twist in rad/mm. Give a sketch illustrating the distribution of warping displacement in the cross-section and quote values at points 1 and 4.

Ans. $d\theta/dz = 29.3 \times 10^{-6}$ rad/mm, $w_3 = -w_4 = -0.19$ mm,
$w_2 = -w_1 = -0.056$ mm.

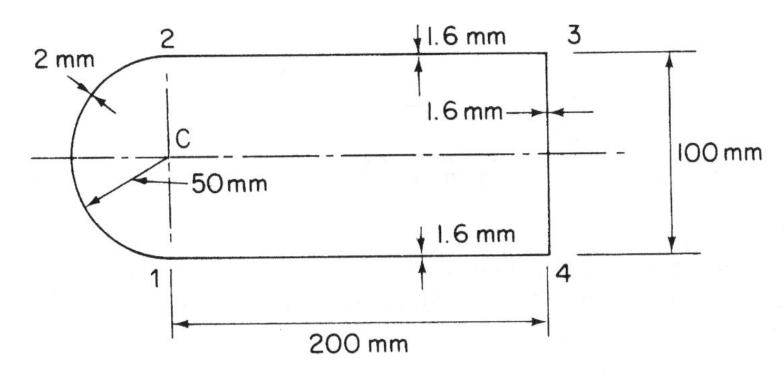

Fig. P.18.6

P.18.7 A uniform beam with the doubly symmetrical cross-section shown in Fig. P.18.7, has horizontal and vertical walls made of different materials which have shear moduli G_a and G_b, respectively. If for any material the ratio mass density/shear modulus is constant find the ratio of the wall thicknesses t_a and t_b, so that for a given torsional stiffness and given dimensions a, b the beam has minimum weight per unit span. Assume the Bredt–Batho theory of torsion is valid.

If this thickness requirement is satisfied find the a/b ratio (previously regarded as fixed), which gives minimum weight for given torsional stiffness.

Ans. $t_b/t_a = G_a/G_b$, $b/a = 1$.

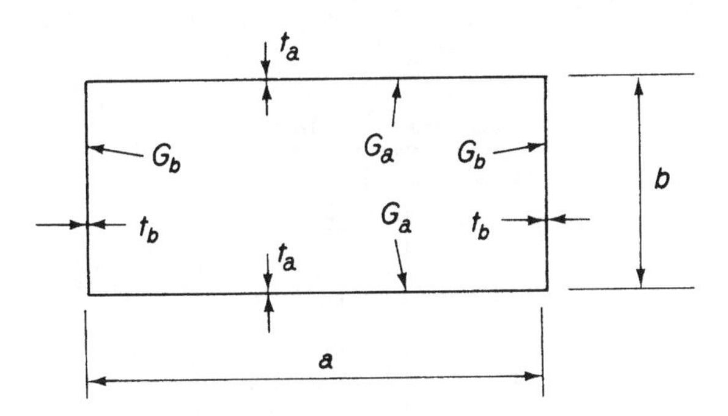

Fig. P.18.7

P.18.8 The cold-formed section shown in Fig. P.18.8 is subjected to a torque of 50 N m. Calculate the maximum shear stress in the section and its rate of twist. $G = 25\,000$ N/mm^2.

Ans. $\tau_{max} = 220.6$ N/mm^2, $d\theta/dz = 0.0044$ rad/mm.

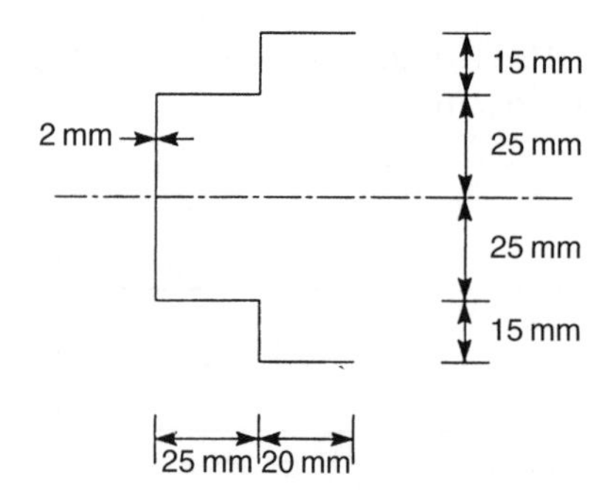

Fig. P.18.8

P.18.9 Determine the rate of twist per unit torque of the beam section shown in Fig. P.17.11 if the shear modulus G is 25 000 N/mm^2. (Note that the shear centre position has been calculated in P.17.11.)

Ans. 6.42×10^{-8} rad/mm.

P.18.10 Figure P.18.10 shows the cross-section of a thin-walled beam in the form of a channel with lipped flanges. The lips are of constant thickness 1.27 mm while the flanges increase linearly in thickness from 1.27 mm where they meet the lips to

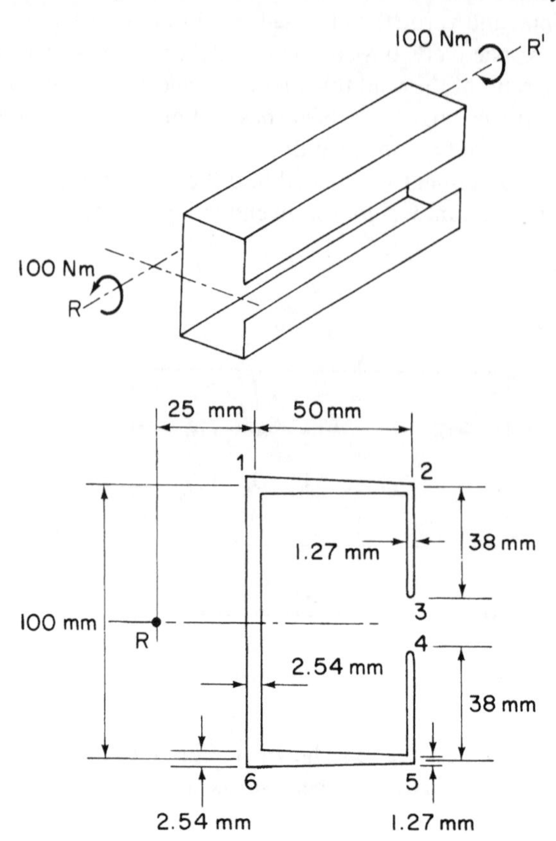

Fig. P.18.10

2.54 mm at their junctions with the web. The web has a constant thickness of 2.54 mm. The shear modulus G is 26 700 N/mm^2 throughout.

The beam has an enforced axis of twist RR$'$ and is supported in such a way that warping occurs freely but is zero at the mid-point of the web. If the beam carries a torque of 100 N m, calculate the maximum shear stress according to the St. Venant theory of torsion for thin-walled sections. Ignore any effects of stress concentration at the corners. Find also the distribution of warping along the middle line of the section, illustrating your results by means of a sketch.

Ans. $\tau_{max} = \pm 297.4\,\text{N/mm}^2$, $w_1 = -5.48\,\text{mm} = -w_6$.
$w_2 = 5.48\,\text{mm} = -w_5$, $w_3 = 17.98\,\text{mm} = -w_4$.

P.18.11 The thin-walled section shown in Fig. P.18.11 is symmetrical about the x axis. The thickness t_0 of the centre web 34 is constant, while the thickness of the other walls varies linearly from t_0 at points 3 and 4 to zero at the open ends 1, 6, 7 and 8.

Determine the St. Venant torsion constant J for the section and also the maximum value of the shear stress due to a torque T. If the section is constrained to twist about an axis through the origin O, plot the relative warping displacements of the section per unit rate of twist.

Ans. $J = 4at_0^3/3$, $\tau_{max} = \pm 3T/4at_0^2$, $w_1 = +a^2(1 + 2\sqrt{2})$.
$w_2 = +\sqrt{2}a^2$, $w_7 = -a^2$, $w_3 = 0$.

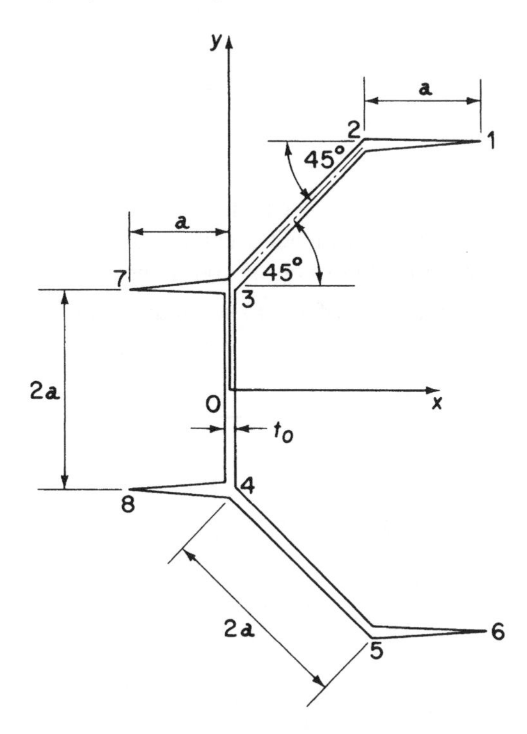

Fig. P.18.11

P.18.12 The thin walled section shown in Fig. P.18.12 is constrained to twist about an axis through R, the centre of the semicircular wall 34. Calculate the maximum shear stress in the section per unit torque and the warping distribution per unit rate of twist. Also compare the value of warping displacement at the point 1 with that corresponding to the section being constrained to twist about an axis through the point O and state what effect this movement has on the maximum shear stress and the torsional stiffness of the section.

Ans. Maximum shear stress is $\pm 0.42/rt^2$ per unit torque.

$$w_{03} = +r^2\theta, \quad w_{32} = +\frac{r}{2}(\pi r + 2s_1), \quad w_{21} = -\frac{r}{2}(2s_2 - 5.142r).$$

With centre of twist at O $_1$ $w_1 = -0.43r^2$. Maximum shear stress is unchanged, torsional stiffness increased since warping reduced.

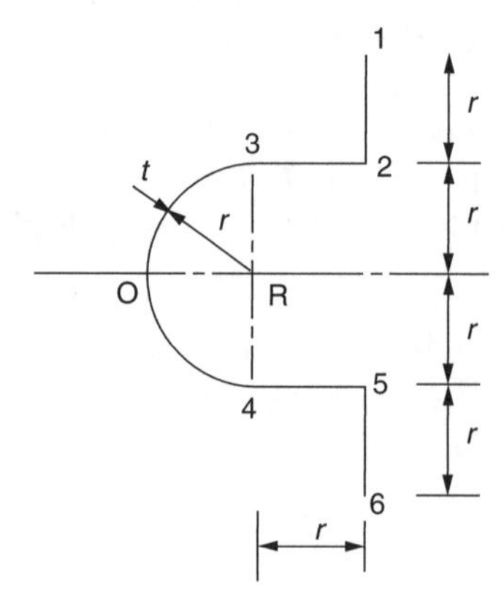

Fig. P.18.12

P.18.13 Determine the maximum shear stress in the beam section shown in Fig. P.18.13 stating clearly the point at which it occurs. Determine also the rate of twist of the beam section if the shear modulus G is 25 000 N/mm^2.

Ans. 70.2 N/mm^2 on underside of 24 at 2 or on upper surface of 32 at 2.
9.0×10^{-4} rad/mm.

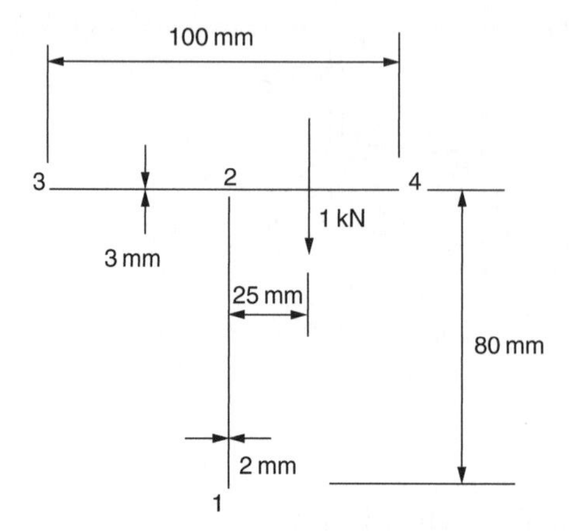

Fig. P.18.13

19

Combined open and closed section beams

So far, in Chapters 16–18, we have analysed thin-walled beams which consist of either completely closed cross-sections or completely open cross-sections. Frequently aircraft components comprise combinations of open and closed section beams. For example the section of a wing in the region of an undercarriage bay could take the form shown in Fig. 19.1. Clearly part of the section is an open channel section while the nose portion is a single cell closed section. We shall now examine the methods of analysis of such sections when subjected to bending, shear and torsional loads.

19.1 Bending

It is immaterial what form the cross-section of a beam takes; the direct stresses due to bending are given by either of Eq. (16.18) or (16.19).

19.2 Shear

The methods described in Sections 17.2 and 17.3 are used to determine the shear stress distribution although, unlike the completely closed section case, shear loads must be applied through the shear centre of the combined section, otherwise shear stresses of the type described in Section 18.2 due to torsion will arise. Where shear loads do not act through the shear centre its position must be found and the loading system replaced

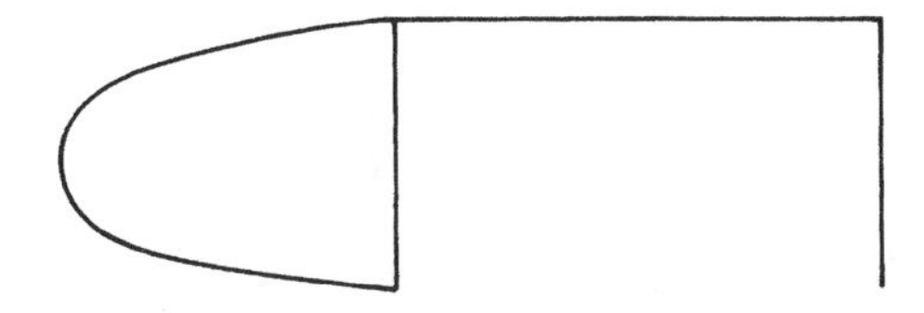

Fig. 19.1 Wing section comprising open and closed components.

by shear loads acting through the shear centre together with a torque; the two loading cases are then analysed separately. Again we assume that the cross-section of the beam remains undistorted by the loading.

Example 19.1

Determine the shear flow distribution in the beam section shown in Fig. 19.2, when it is subjected to a shear load in its vertical plane of symmetry. The thickness of the walls of the section is 2 mm throughout.

The centroid of area C lies on the axis of symmetry at some distance $\bar{y}$ from the upper surface of the beam section. Taking moments of area about this upper surface

$$(4 \times 100 \times 2 + 4 \times 200 \times 2)\bar{y} = 2 \times 100 \times 2 \times 50 + 2 \times 200 \times 2 \times 100 + 200 \times 2 \times 200$$

which gives $\bar{y} = 75$ mm.

The second moment of area of the section about Cx is given by

$$I_{xx} = 2\left(\frac{2 \times 100^3}{12} + 2 \times 100 \times 25^2\right) + 400 \times 2 \times 75^2 + 200 \times 2 \times 125^2 + 2\left(\frac{2 \times 200^3}{12} + 2 \times 200 \times 25^2\right)$$

i.e.

$$I_{xx} = 14.5 \times 10^6 \text{ mm}^4$$

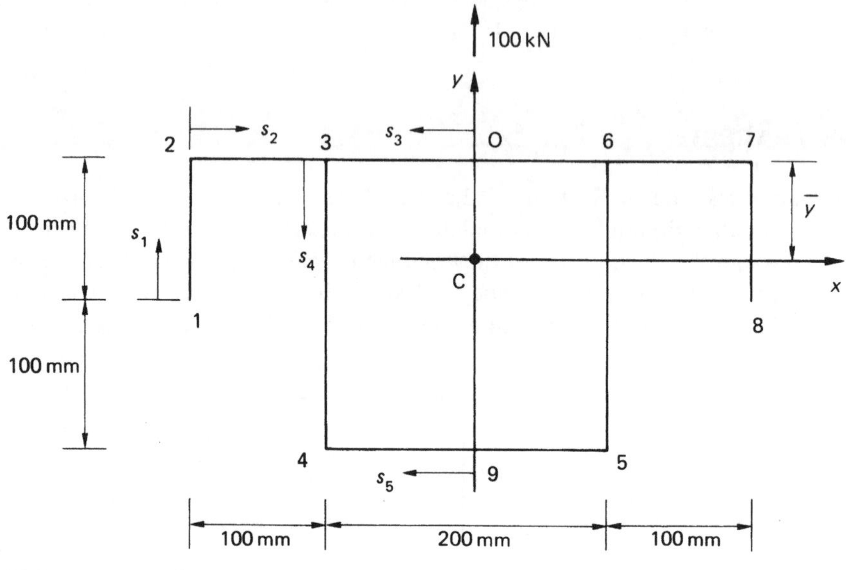

Fig. 19.2 Beam section of Example 19.1.

The section is symmetrical about Cy so that $I_{xy} = 0$ and since $S_x = 0$ the shear flow distribution in the closed section 3456 is, from Eq. (17.15)

$$q_s = -\frac{S_y}{I_{xx}} \int_0^s ty \, ds + q_{s,0} \tag{i}$$

Also the shear load is applied through the shear centre of the complete section, i.e. along the axis of symmetry, so that in the open portions 123 and 678 the shear flow distribution is, from Eq. (17.14)

$$q_s = -\frac{S_y}{I_{xx}} \int_0^s ty \, ds \tag{ii}$$

We note that the shear flow is zero at the points 1 and 8 and therefore the analysis may conveniently, though not necessarily, begin at either of these points. Thus, referring to Fig. 19.2

$$q_{12} = -\frac{100 \times 10^3}{14.5 \times 10^6} \int_0^{s_1} 2(-25 + s_1) \, ds_1$$

i.e.

$$q_{12} = -69.0 \times 10^{-4}(-50s_1 + s_1^2) \tag{iii}$$

whence $q_2 = -34.5$ N/mm.

Examination of Eq. (iii) shows that q_{12} is initially positive and changes sign when $s_1 = 50$ mm. Further, q_{12} has a turning value ($dq_{12}/ds_1 = 0$) at $s_1 = 25$ mm of 4.3 N/mm. In the wall 23

$$q_{23} = -69.0 \times 10^{-4} \int_0^{s_2} 2 \times 75 \, ds_2 - 34.5$$

i.e.

$$q_{23} = -1.04s_2 - 34.5 \tag{iv}$$

Hence q_{23} varies linearly from a value of -34.5 N/mm at 2 to -138.5 N/mm at 3 in the wall 23.

The analysis of the open part of the beam section is now complete since the shear flow distribution in the walls 67 and 78 follows from symmetry. To determine the shear flow distribution in the closed part of the section we must use the method described in Section 17.3 in which the line of action of the shear load is known. Thus we 'cut' the closed part of the section at some convenient point, obtain the q_b or 'open section' shear flows for the complete section and then take moments as in Eqs (17.17) or (17.18). However, in this case, we may use the symmetry of the section and loading to deduce that the final value of shear flow must be zero at the mid-points of the walls 36 and 45, i.e. $q_s = q_{s,0} = 0$ at these points. Hence

$$q_{03} = -69.0 \times 10^{-4} \int_0^{s_3} 2 \times 75 \, ds_3$$

so that

$$q_{03} = -1.04s_3 \tag{v}$$

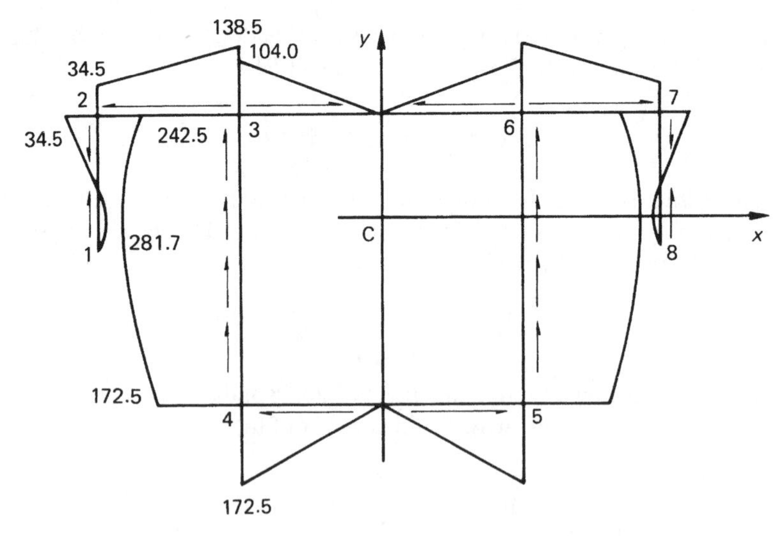

Fig. 19.3 Shear flow distribution in beam of Example 19.1 (all shear flows in N/mm).

and $q_3 = -104$ N/mm in the wall 03. It follows that for equilibrium of shear flows at 3, q_3, in the wall 34, must be equal to $-138.5 - 104 = -242.5$ N/mm. Hence

$$q_{34} = -69.0 \times 10^{-4} \int_0^{s_4} 2(75 - s_4)\, ds_4 - 242.5$$

which gives

$$q_{34} = -1.04 s_4 + 69.0 \times 10^{-4} s_4^2 - 242.5 \tag{vi}$$

Examination of Eq. (vi) shows that q_{34} has a maximum value of -281.7 N/mm at $s_4 = 75$ mm; also $q_4 = -172.5$ N/mm. Finally, the distribution of shear flow in the wall 94 is given by

$$q_{94} = -69.0 \times 10^{-4} \int_0^{s_5} 2(-125)\, ds_5$$

i.e.

$$q_{94} = 1.73 s_5 \tag{vii}$$

The complete distribution is shown in Fig. 19.3.

19.3 Torsion

Generally, in the torsion of composite sections, the closed portion is dominant since its torsional stiffness is far greater than that of the attached open section portion which may

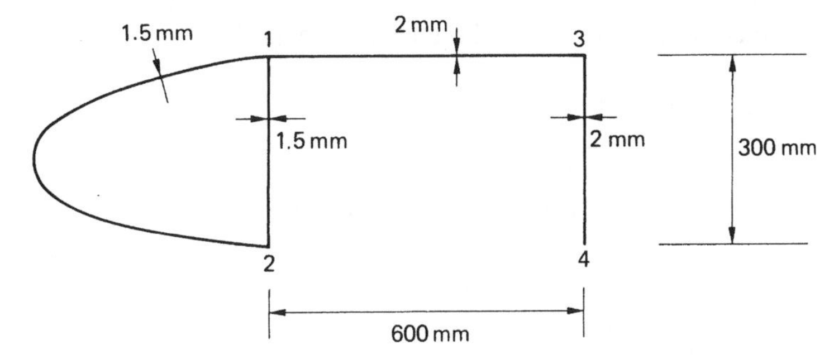

Fig. 19.4 Wing section of Example 19.2.

therefore be frequently ignored in the calculation of torsional stiffness; shear stresses should, however, be checked in this part of the section.

Example 19.2

Find the angle of twist per unit length in the wing whose cross-section is shown in Fig. 19.4 when it is subjected to a torque of 10 kN m. Find also the maximum shear stress in the section. $G = 25\,000\ \text{N/mm}^2$.

Wall 12 (outer) = 900 mm. Nose cell area = $20\,000\ \text{mm}^2$.

It may be assumed, in a simplified approach, that the torsional rigidity GJ of the complete section is the sum of the torsional rigidities of the open and closed portions. For the closed portion the torsional rigidity is, from Eq. (18.4)

$$(GJ)_{\text{cl}} = \frac{4A^2G}{\oint \mathrm{d}s/t} = \frac{4 \times 20\,000^2 \times 25\,000}{(900 + 300)/1.5}$$

which gives

$$(GJ)_{\text{cl}} = 5000 \times 10^7\ \text{N mm}^2$$

The torsional rigidity of the open portion is found using Eq. (18.11), thus

$$(GJ)_{\text{op}} = G \sum \frac{st^3}{3} = \frac{25\,000 \times 900 \times 2^3}{3}$$

i.e.

$$(GJ)_{\text{op}} = 6 \times 10^7\ \text{N mm}^2$$

The torsional rigidity of the complete section is then

$$GJ = 5000 \times 10^7 + 6 \times 10^7 = 5006 \times 10^7\ \text{N mm}^2$$

In all unrestrained torsion problems the torque is related to the rate of twist by the expression

$$T = GJ\frac{d\theta}{dz}$$

The angle of twist per unit length is therefore given by

$$\frac{d\theta}{dz} = \frac{T}{GJ} = \frac{10 \times 10^6}{5006 \times 10^7} = 0.0002 \text{ rad/mm}$$

Substituting for T in Eq. (18.1) from Eq. (18.4), we obtain the shear flow in the closed section. Thus

$$q_{cl} = \frac{(GJ)_{cl}}{2A}\frac{d\theta}{dz} = \frac{5000 \times 10^7}{2 \times 20\,000} \times 0.0002$$

from which

$$q_{cl} = 250 \text{ N/mm}$$

The maximum shear stress in the closed section is then $250/1.5 = 166.7 \text{ N/mm}^2$.

In the open portion of the section the maximum shear stress is obtained directly from Eq. (18.10) and is

$$\tau_{max,op} = 25\,000 \times 2 \times 0.0002 = 10 \text{ N/mm}^2$$

It can be seen from the above that in terms of strength and stiffness the closed portion of the wing section dominates. This dominance may be used to determine the warping distribution. Having first found the position of the centre of twist (the shear centre) the warping of the closed portion is calculated using the method described in Section 18.1. The warping in the walls 13 and 34 is then determined using Eq. (18.19), in which the origin for the swept area A_R is taken at the point 1 and the value of warping is that previously calculated for the closed portion at 1.

Problems

P.19.1 The beam section of Example 19.1 (see Fig. 19.2) is subjected to a bending moment in a vertical plane of 20 kN m. Calculate the maximum direct stress in the cross-section of the beam.

Ans. 172.5 N/mm^2.

P.19.2 A wing box has the cross-section shown diagrammatically in Fig. P.19.2 and supports a shear load of 100 kN in its vertical plane of symmetry. Calculate the shear stress at the mid-point of the web 36 if the thickness of all walls is 2 mm.

Ans. 89.7 N/mm^2.

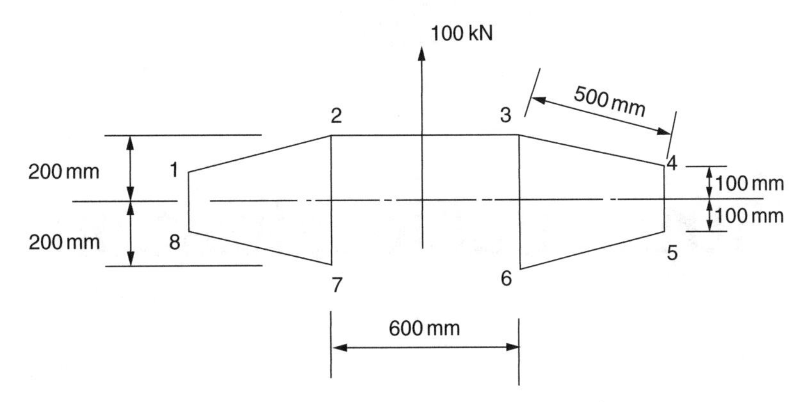

Fig. P.19.2

P.19.3 If the wing box of P.19.2 is subjected to a torque of 100 kN m, calculate the rate of twist of the section and the maximum shear stress. The shear modulus G is 25000 N/mm^2.

Ans. 18.5×10^{-6} rad/mm, 170 N/mm^2.

20

Structural idealization

So far we have been concerned with relatively uncomplicated structural sections which in practice would be formed from thin plate or by the extrusion process. While these sections exist as structural members in their own right they are frequently used, as we saw in Chapter 12, to stiffen more complex structural shapes such as fuselages, wings and tail surfaces. Thus a two spar wing section could take the form shown in Fig. 20.1 in which Z-section stringers are used to stiffen the thin skin while angle sections form the spar flanges. Clearly, the analysis of a section of this type would be complicated and tedious unless some simplifying assumptions are made. Generally, the number and nature of these simplifying assumptions determine the accuracy and the degree of complexity of the analysis; the more complex the analysis the greater the accuracy obtained. The degree of simplification introduced is governed by the particular situation surrounding the problem. For a preliminary investigation, speed and simplicity are often of greater importance than extreme accuracy; on the other hand a final solution must be as exact as circumstances allow.

Complex structural sections may be idealized into simpler 'mechanical model' forms which behave, under given loading conditions, in the same, or very nearly the same, way as the actual structure. We shall see, however, that different models of the same structure are required to simulate actual behaviour under different systems of loading.

20.1 Principle

In the wing section of Fig. 20.1 the stringers and spar flanges have small cross-sectional dimensions compared with the complete section. Therefore, the variation in stress

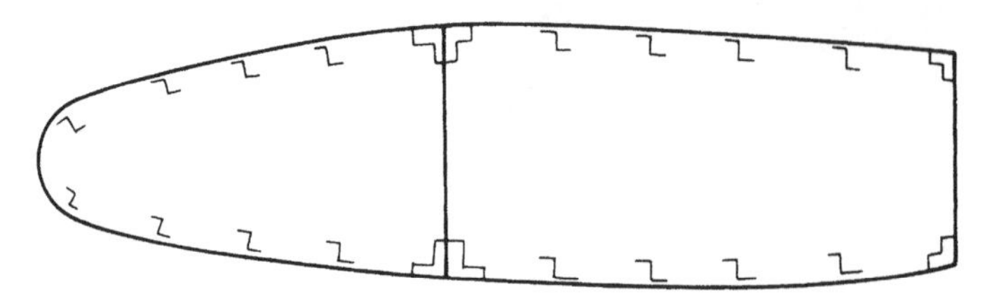

Fig. 20.1 Typical wing section.

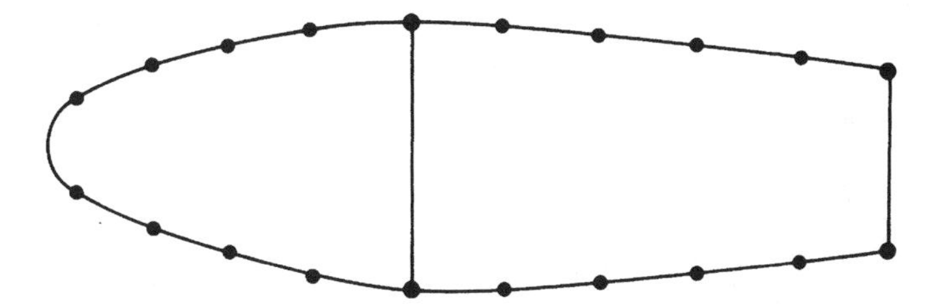

Fig. 20.2 Idealization of a wing section.

over the cross-section of a stringer due to, say, bending of the wing would be small. Furthermore, the difference between the distances of the stringer centroids and the adjacent skin from the wing section axis is small. It would be reasonable to assume therefore that the direct stress is constant over the stringer cross-sections. We could therefore replace the stringers and spar flanges by concentrations of area, known as *booms*, over which the direct stress is constant and which are located along the mid-line of the skin, as shown in Fig. 20.2. In wing and fuselage sections of the type shown in Fig. 20.1, the stringers and spar flanges carry most of the direct stresses while the skin is mainly effective in resisting shear stresses although it also carries some of the direct stresses. The idealization shown in Fig. 20.2 may therefore be taken a stage further by assuming that all direct stresses are carried by the booms while the skin is effective only in shear. The direct stress carrying capacity of the skin may be allowed for by increasing each boom area by an area equivalent to the direct stress carrying capacity of the adjacent skin panels. The calculation of these equivalent areas will generally depend upon an initial assumption as to the form of the distribution of direct stress in a boom/skin panel.

20.2 Idealization of a panel

Suppose that we wish to idealize the panel of Fig. 20.3(a) into a combination of direct stress carrying booms and shear stress only carrying skin as shown in Fig. 20.3(b). In Fig. 20.3(a) the direct stress carrying thickness t_D of the skin is equal to its actual thickness t while in Fig. 20.3(b) $t_D = 0$. Suppose also that the direct stress distribution in the actual panel varies linearly from an unknown value σ_1 to an unknown value σ_2. Clearly the analysis should predict the extremes of stress σ_1 and σ_2 although the distribution of direct stress is obviously lost. Since the loading producing the direct stresses in the actual and idealized panels must be the same we can equate moments to obtain expressions for the boom areas B_1 and B_2. Thus, taking moments about the right-hand edge of each panel

$$\sigma_2 t_D \frac{b^2}{2} + \frac{1}{2}(\sigma_1 - \sigma_2) t_D b \frac{2}{3} b = \sigma_1 B_1 b$$

whence

$$B_1 = \frac{t_D b}{6}\left(2 + \frac{\sigma_2}{\sigma_1}\right) \tag{20.1}$$

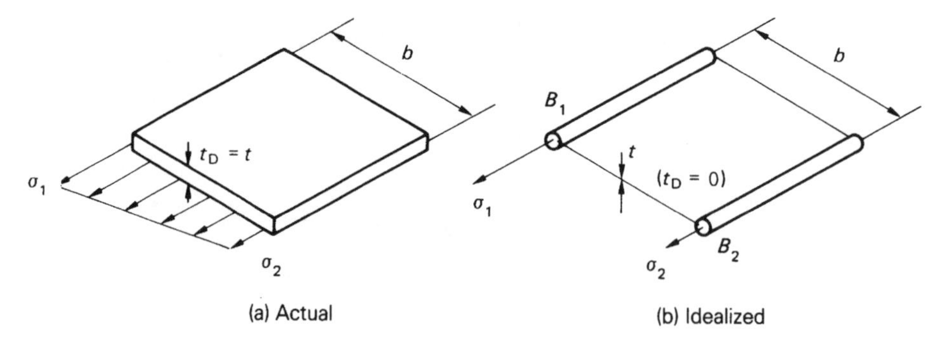

Fig. 20.3 Idealization of a panel.

Similarly

$$B_2 = \frac{t_D b}{6}\left(2 + \frac{\sigma_1}{\sigma_2}\right) \tag{20.2}$$

In Eqs (20.1) and (20.2) the ratio of σ_1 to σ_2, if not known, may frequently be assumed.

The direct stress distribution in Fig. 20.3(a) is caused by a combination of axial load and bending moment. For axial load only $\sigma_1/\sigma_2 = 1$ and $B_1 = B_2 = t_D b/2$; for a pure bending moment $\sigma_1/\sigma_2 = -1$ and $B_1 = B_2 = t_D b/6$. Thus, different idealizations of the same structure are required for different loading conditions.

Example 20.1

Part of a wing section is in the form of the two-cell box shown in Fig. 20.4(a) in which the vertical spars are connected to the wing skin through angle sections all having a cross-sectional area of 300 mm^2. Idealize the section into an arrangement of direct stress carrying booms and shear stress only carrying panels suitable for resisting bending moments in a vertical plane. Position the booms at the spar/skin junctions.

The idealized section is shown in Fig. 20.4(b) in which, from symmetry, $B_1 = B_6$, $B_2 = B_5$, $B_3 = B_4$. Since the section is required to resist bending moments in a vertical plane the direct stress at any point in the actual wing section is directly proportional to its distance from the horizontal axis of symmetry. Further, the distribution of direct stress in all the panels will be linear so that either of Eqs (20.1) or (20.2) may be used. We note that, in addition to contributions from adjacent panels, the boom areas include

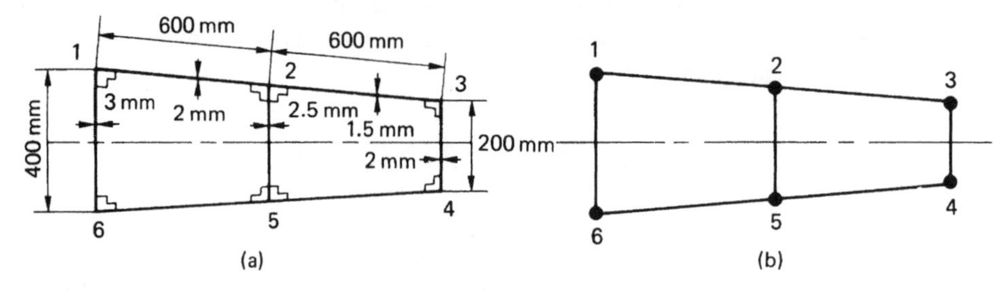

Fig. 20.4 Idealization of a wing section.

the existing spar flanges. Hence

$$B_1 = 300 + \frac{3.0 \times 400}{6}\left(2 + \frac{\sigma_6}{\sigma_1}\right) + \frac{2.0 \times 600}{6}\left(2 + \frac{\sigma_2}{\sigma_1}\right)$$

or

$$B_1 = 300 + \frac{3.0 \times 400}{6}(2 - 1) + \frac{2.0 \times 600}{6}\left(2 + \frac{150}{200}\right)$$

which gives

$$B_1(=B_6) = 1050\,\text{mm}^2$$

Also

$$B_2 = 2 \times 300 + \frac{2.0 \times 600}{6}\left(2 + \frac{\sigma_1}{\sigma_2}\right) + \frac{2.5 \times 300}{6}\left(2 + \frac{\sigma_5}{\sigma_2}\right) + \frac{1.5 \times 600}{6}\left(2 + \frac{\sigma_3}{\sigma_2}\right)$$

i.e.

$$B_2 = 2 \times 300 + \frac{2.0 \times 600}{6}\left(2 + \frac{200}{150}\right) + \frac{2.5 \times 300}{6}(2 - 1) + \frac{1.5 \times 600}{6}\left(2 + \frac{100}{150}\right)$$

from which

$$B_2(= B_5) = 1791.7\,\text{mm}^2$$

Finally

$$B_3 = 300 + \frac{1.5 \times 600}{6}\left(2 + \frac{\sigma_2}{\sigma_3}\right) + \frac{2.0 \times 200}{6}\left(2 + \frac{\sigma_4}{\sigma_3}\right)$$

i.e.

$$B_3 = 300 + \frac{1.5 \times 600}{6}\left(2 + \frac{150}{100}\right) + \frac{2.0 \times 200}{6}(2 - 1)$$

so that

$$B_3(= B_4) = 891.7\,\text{mm}^2$$

20.3 Effect of idealization on the analysis of open and closed section beams

The addition of direct stress carrying booms to open and closed section beams will clearly modify the analyses presented in Chapters 16–18. Before considering individual cases we shall discuss the implications of structural idealization. Generally, in any idealization, different loading conditions require different idealizations of the same structure. In Example 20.1, the loading is applied in a vertical plane. If, however, the loading had been applied in a horizontal plane the assumed stress distribution in the panels of the section would have been different, resulting in different values of boom area.

Suppose that an open or closed section beam is subjected to given bending or shear loads and that the required idealization has been completed. The analysis of such sections usually involves the determination of the neutral axis position and the calculation of sectional properties. The position of the neutral axis is derived from the condition that the resultant load on the beam cross-section is zero, i.e.

$$\int_A \sigma_z \, \mathrm{d}A = 0 \quad \text{(see Eq. (16.3))}$$

The area A in this expression is clearly the direct stress carrying area. It follows that the centroid of the cross-section is the centroid of the direct stress carrying area of the section, depending on the degree and method of idealization. The sectional properties, I_{xx}, etc., must also refer to the direct stress carrying area.

20.3.1 Bending of open and closed section beams

The analysis presented in Sections 16.1 and 16.2 applies and the direct stress distribution is given by any of Eqs (16.9), (16.18) or (16.19), depending on the beam section being investigated. In these equations the coordinates (x, y) of points in the cross-section are referred to axes having their origin at the centroid of the direct stress carrying area. Furthermore, the section properties I_{xx}, I_{yy} and I_{xy} are calculated for the direct stress carrying area only.

In the case where the beam cross-section has been completely idealized into direct stress carrying booms and shear stress only carrying panels, the direct stress distribution consists of a series of direct stresses concentrated at the centroids of the booms.

Example 20.2

The fuselage section shown in Fig. 20.5 is subjected to a bending moment of 100 kN m applied in the vertical plane of symmetry. If the section has been completely idealized into a combination of direct stress carrying booms and shear stress only carrying panels, determine the direct stress in each boom.

The section has Cy as an axis of symmetry and resists a bending moment $M_x = 100$ kN m. Equation (16.18) therefore reduces to

$$\sigma_z = \frac{M_x}{I_{xx}} y \qquad \text{(i)}$$

The origin of axes Cxy coincides with the position of the centroid of the direct stress carrying area which, in this case, is the centroid of the boom areas. Thus, taking moments of area about boom 9

$$(6 \times 640 + 6 \times 600 + 2 \times 620 + 2 \times 850)\bar{y}$$
$$= 640 \times 1200 + 2 \times 600 \times 1140 + 2 \times 600 \times 960 + 2 \times 600 \times 768$$
$$+ 2 \times 620 \times 565 + 2 \times 640 \times 336 + 2 \times 640 \times 144 + 2 \times 850 \times 38$$

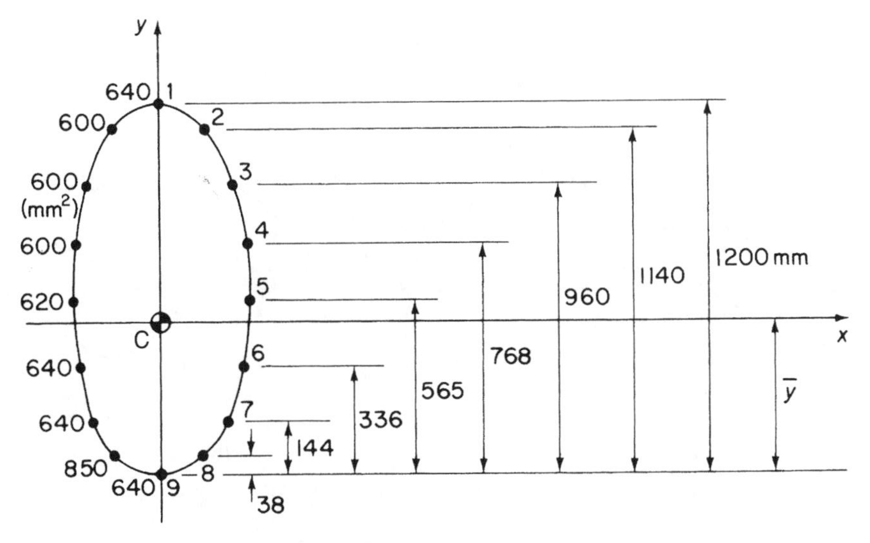

Fig. 20.5 Idealized fuselage section of Example 20.2.

Table 20.1

① Boom	② y (mm)	③ B (mm^2)	④ $\Delta I_{xx} = By^2$ (mm^4)	⑤ σ_z (N/mm^2)
1	+660	640	278×10^6	35.6
2	+600	600	216×10^6	32.3
3	+420	600	106×10^6	22.6
4	+228	600	31×10^6	12.3
5	+25	620	0.4×10^6	1.3
6	−204	640	27×10^6	−11.0
7	−396	640	100×10^6	−21.4
8	−502	850	214×10^6	−27.0
9	−540	640	187×10^6	−29.0

which gives

$$\bar{y} = 540\,\text{mm}$$

The solution is now completed in Table 20.1

From column ④

$$I_{xx} = 1854 \times 10^6\,\text{mm}^4$$

and column ⑤ is completed using Eq. (i).

20.3.2 Shear of open section beams

The derivation of Eq. (17.14) for the shear flow distribution in the cross-section of an open section beam is based on the equilibrium equation (17.2). The thickness t in this

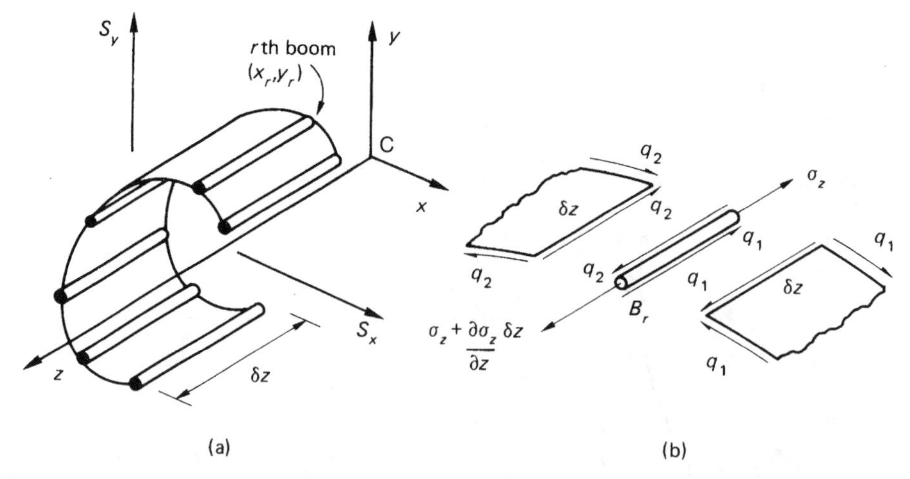

Fig. 20.6 (a) Elemental length of shear loaded open section beam with booms; (b) equilibrium of boom element.

equation refers to the direct stress carrying thickness t_D of the skin. Equation (17.14) may therefore be rewritten

$$q_s = -\left(\frac{S_x I_{xx} - S_y I_{xy}}{I_{xx} I_{yy} - I_{xy}^2}\right) \int_0^s t_D x \, ds - \left(\frac{S_y I_{yy} - S_x I_{xy}}{I_{xx} I_{yy} - I_{xy}^2}\right) \int_0^s t_D y \, ds \qquad (20.3)$$

in which $t_D = t$ if the skin is fully effective in carrying direct stress or $t_D = 0$ if the skin is assumed to carry only shear stresses. Again the section properties in Eq. (20.3) refer to the direct stress carrying area of the section since they are those which feature in Eqs (16.18) and (16.19).

Equation (20.3) makes no provision for the effects of booms which cause discontinuities in the skin and therefore interrupt the shear flow. Consider the equilibrium of the rth boom in the elemental length of beam shown in Fig. 20.6(a) which carries shear loads S_x and S_y acting through its shear centre S. These shear loads produce direct stresses due to bending in the booms and skin and shear stresses in the skin. Suppose that the shear flows in the skin adjacent to the rth boom of cross-sectional area B_r are q_1 and q_2. Then, from Fig. 20.6(b)

$$\left(\sigma_z + \frac{\partial \sigma_z}{\partial z} \delta z\right) B_r - \sigma_z B_r + q_2 \delta z - q_1 \delta z = 0$$

which simplifies to

$$q_2 - q_1 = -\frac{\partial \sigma_z}{\partial z} B_r \qquad (20.4)$$

Substituting for σ_z in Eq. (20.4) from (16.18) we have

$$q_2 - q_1 = -\left[\frac{(\partial M_y/\partial z)I_{xx} - (\partial M_x/\partial z)I_{xy}}{I_{xx}I_{yy} - I_{xy}^2}\right] B_r x_r - \left[\frac{(\partial M_x/\partial z)I_{yy} - (\partial M_y/\partial z)I_{xy}}{I_{xx}I_{yy} - I_{xy}^2}\right] B_r y_r$$

or, using the relationships of Eqs (16.23) and (16.24)

$$q_2 - q_1 = -\left(\frac{S_x I_{xx} - S_y I_{xy}}{I_{xx}I_{yy} - I_{xy}^2}\right) B_r x_r - \left(\frac{S_y I_{yy} - S_x I_{xy}}{I_{xx}I_{yy} - I_{xy}^2}\right) B_r y_r \qquad (20.5)$$

Equation (20.5) gives the change in shear flow induced by a boom which itself is subjected to a direct load ($\sigma_z B_r$). Each time a boom is encountered the shear flow is incremented by this amount so that if, at any distance s around the profile of the section, n booms have been passed, the shear flow at the point is given by

$$q_s = -\left(\frac{S_x I_{xx} - S_y I_{xy}}{I_{xx}I_{yy} - I_{xy}^2}\right)\left(\int_0^s t_D x \, ds + \sum_{r=1}^{n} B_r x_r\right) - \left(\frac{S_y I_{yy} - S_x I_{xy}}{I_{xx}I_{yy} - I_{xy}^2}\right)\left(\int_0^s t_D y \, ds + \sum_{r=1}^{n} B_r y_r\right) \qquad (20.6)$$

Example 20.3

Calculate the shear flow distribution in the channel section shown in Fig. 20.7 produced by a vertical shear load of 4.8 kN acting through its shear centre. Assume that the walls of the section are only effective in resisting shear stresses while the booms, each of area 300 mm^2, carry all the direct stresses.

The effective direct stress carrying thickness t_D of the walls of the section is zero so that the centroid of area and the section properties refer to the boom areas only. Since Cx (and Cy as far as the boom areas are concerned) is an axis of symmetry $I_{xy} = 0$; also $S_x = 0$ and Eq. (20.6) thereby reduces to

$$q_s = -\frac{S_y}{I_{xx}} \sum_{r=1}^{n} B_r y_r \qquad \text{(i)}$$

in which $I_{xx} = 4 \times 300 \times 200^2 = 48 \times 10^6$ mm^4. Substituting the values of S_y and I_{xx} in Eq. (i) gives

$$q_s = -\frac{4.8 \times 10^3}{48 \times 10^6} \sum_{r=1}^{n} B_r y_r = -10^{-4} \sum_{r=1}^{n} B_r y_r \qquad \text{(ii)}$$

At the outside of boom 1, $q_s = 0$. As boom 1 is crossed the shear flow changes by an amount given by

$$\Delta q_1 = -10^{-4} \times 300 \times 200 = -6 \text{ N/mm}$$

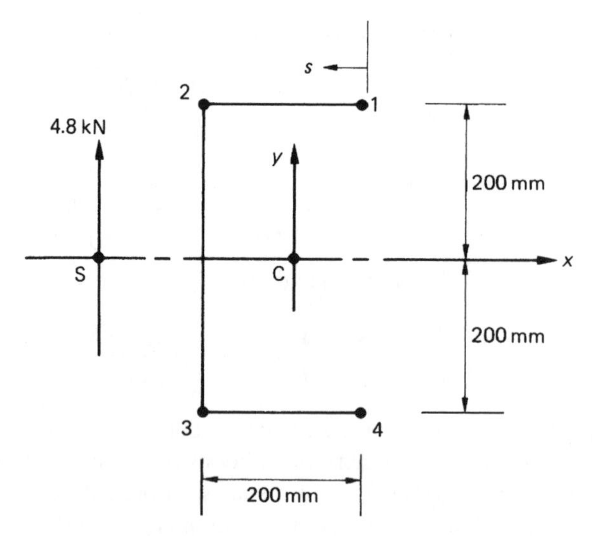

Fig. 20.7 Idealized channel section of Example 20.3.

Hence $q_{12} = -6$ N/mm since, from Eq. (i), it can be seen that no further changes in shear flow occur until the next boom (2) is crossed. Hence

$$q_{23} = -6 - 10^{-4} \times 300 \times 200 = -12\,\text{N/mm}$$

Similarly

$$q_{34} = -12 - 10^{-4} \times 300 \times (-200) = -6\,\text{N/mm}$$

while, finally, at the outside of boom 4 the shear flow is

$$-6 - 10^{-4} \times 300 \times (-200) = 0$$

as expected. The complete shear flow distribution is shown in Fig. 20.8.

It can be seen from Eq. (i) in Example 20.3 that the analysis of a beam section which has been idealized into a combination of direct stress carrying booms and shear

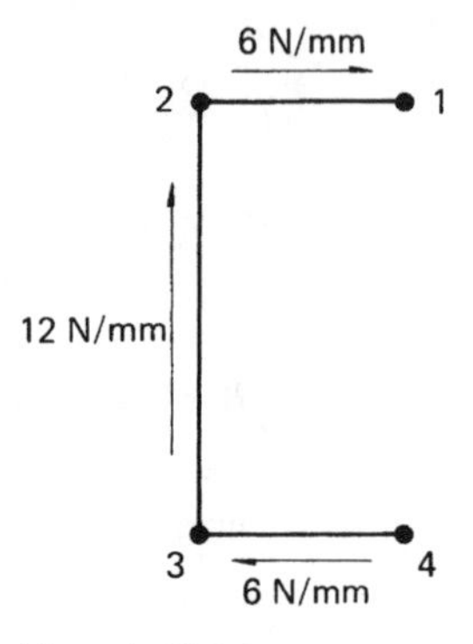

Fig. 20.8 Shear flow in channel section of Example 20.3.

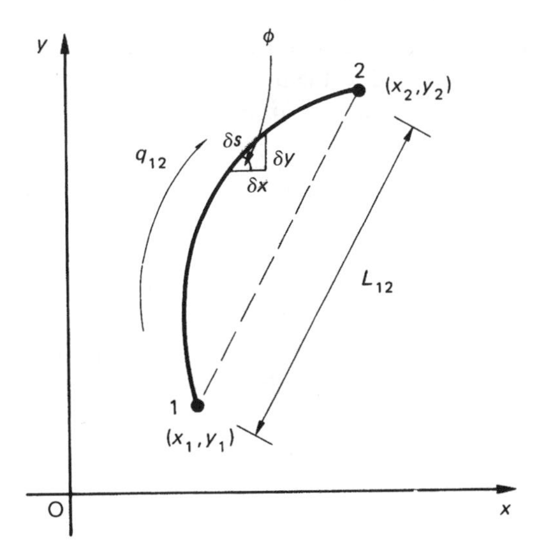

Fig. 20.9 Curved web with constant shear flow.

stress only carrying skin gives constant values of the shear flow in the skin between the booms; the actual distribution of shear flows is therefore lost. What remains is in fact the average of the shear flow, as can be seen by referring to Example 20.3. Analysis of the unidealized channel section would result in a parabolic distribution of shear flow in the web 23 whose resultant is statically equivalent to the externally applied shear load of 4.8 kN. In Fig. 20.8 the resultant of the constant shear flow in the web 23 is $12 \times 400 = 4800\,\text{N} = 4.8\,\text{kN}$. It follows that this constant value of shear flow is the average of the parabolically distributed shear flows in the unidealized section.

The result, from the idealization of a beam section, of a constant shear flow between booms may be used to advantage in parts of the analysis. Suppose that the curved web 12 in Fig. 20.9 has booms at its extremities and that the shear flow q_{12} in the web is constant. The shear force on an element δs of the web is $q_{12}\delta s$, whose components horizontally and vertically are $q_{12}\delta s \cos\phi$ and $q_{12}\delta s \sin\phi$. The resultant, parallel to the x axis, S_x, of q_{12} is therefore given by

$$S_x = \int_1^2 q_{12} \cos\phi \, \mathrm{d}s$$

or

$$S_x = q_{12} \int_1^2 \cos\phi \, \mathrm{d}s$$

which, from Fig. 20.9, may be written

$$S_x = q_{12} \int_1^2 \mathrm{d}x = q_{12}(x_2 - x_1) \tag{20.7}$$

Similarly the resultant of q_{12} parallel to the y axis is

$$S_y = q_{12}(y_2 - y_1) \tag{20.8}$$

Thus the resultant, in a given direction, of a constant shear flow acting on a web is the value of the shear flow multiplied by the projection on that direction of the web.

The resultant shear force S on the web of Fig. 20.9 is

$$S = \sqrt{S_x^2 + S_y^2} = q_{12}\sqrt{(x_2 - x_1)^2 + (y_2 - y_1)^2}$$

i.e.

$$S = q_{12}L_{12} \tag{20.9}$$

Therefore, the resultant shear force acting on the web is the product of the shear flow and the length of the straight line joining the ends of the web; clearly the direction of the resultant is parallel to this line.

The moment M_q produced by the shear flow q_{12} about any point O in the plane of the web is, from Fig. 20.10

$$M_q = \int_1^2 q_{12}p\,\mathrm{d}s = q_{12}\int_1^2 2\,\mathrm{d}A$$

or

$$M_q = 2Aq_{12} \tag{20.10}$$

in which A is the area enclosed by the web and the lines joining the ends of the web to the point O. This result may be used to determine the distance of the line of action of the resultant shear force from any point. From Fig. 20.10

$$Se = 2Aq_{12}$$

from which

$$e = \frac{2A}{S}q_{12}$$

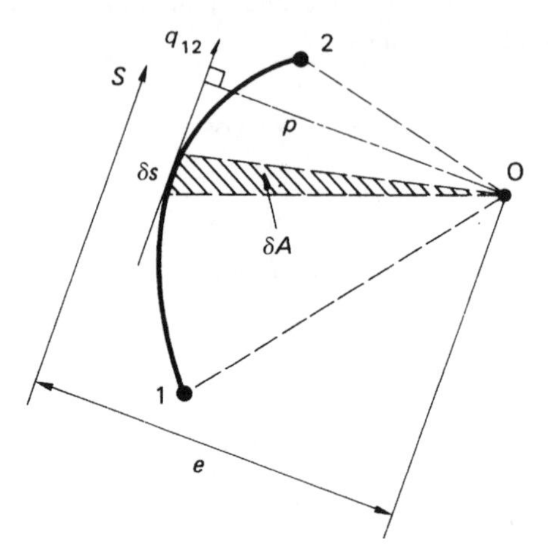

Fig. 20.10 Moment produced by a constant shear flow.

Substituting for q_{12} from Eq. (20.9) gives

$$e = \frac{2A}{L_{12}}$$

20.3.3 Shear loading of closed section beams

Arguments identical to those in the shear of open section beams apply in this case. Thus, the shear flow at any point around the cross-section of a closed section beam comprising booms and skin of direct stress carrying thickness t_D is, by a comparison of Eqs (20.6) and (17.15)

$$q_s = -\left(\frac{S_x I_{xx} - S_y I_{xy}}{I_{xx}I_{yy} - I_{xy}^2}\right)\left(\int_0^s t_D x \, ds + \sum_{r=1}^{n} B_r x_r\right) - \left(\frac{S_y I_{yy} - S_x I_{xy}}{I_{xx}I_{yy} - I_{xy}^2}\right)\left(\int_0^s t_D y \, ds + \sum_{r=1}^{n} B_r y_r\right) + q_{s,0} \qquad (20.11)$$

Note that the zero value of the 'basic' or 'open section' shear flow at the 'cut' in a skin for which $t_D = 0$ extends from the 'cut' to the adjacent booms.

Example 20.4

The thin-walled single cell beam shown in Fig. 20.11 has been idealized into a combination of direct stress carrying booms and shear stress only carrying walls. If the section supports a vertical shear load of 10 kN acting in a vertical plane through booms 3 and 6, calculate the distribution of shear flow around the section.

Boom areas: $B_1 = B_8 = 200\,\text{mm}^2$, $B_2 = B_7 = 250\,\text{mm}^2$, $B_3 = B_6 = 400\,\text{mm}^2$, $B_4 = B_5 = 100\,\text{mm}^2$.

The centroid of the direct stress carrying area lies on the horizontal axis of symmetry so that $I_{xy} = 0$. Also, since $t_D = 0$ and only a vertical shear load is applied,

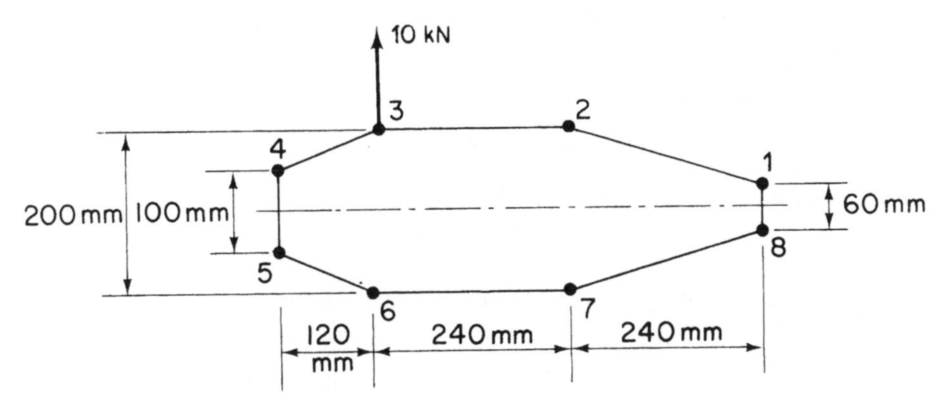

Fig. 20.11 Closed section of beam of Example 20.4.

Eq. (20.11) reduces to

$$q_s = -\frac{S_y}{I_{xx}} \sum_{r=1}^{n} B_r y_r + q_{s,0} \tag{i}$$

in which

$$I_{xx} = 2(200 \times 30^2 + 250 \times 100^2 + 400 \times 100^2 + 100 \times 50^2) = 13.86 \times 10^6 \text{ mm}^4$$

Equation (i) then becomes

$$q_s = -\frac{10 \times 10^3}{13.86 \times 10^6} \sum_{r=1}^{n} B_r y_r + q_{s,0}$$

i.e.

$$q_s = -7.22 \times 10^{-4} \sum_{r=1}^{n} B_r y_r + q_{s,0} \tag{ii}$$

'Cutting' the beam section in the wall 23 (any wall may be chosen) and calculating the 'basic' shear flow distribution q_{b} from the first term on the right-hand side of Eq. (ii) we have

$$q_{\text{b},23} = 0$$

$$q_{\text{b},34} = -7.22 \times 10^{-4}(400 \times 100) = -28.9 \text{ N/mm}$$

$$q_{\text{b},45} = -28.9 - 7.22 \times 10^{-4}(100 \times 50) = -32.5 \text{ N/mm}$$

$$q_{\text{b},56} = q_{\text{b},34} = -28.9 \text{ N/mm (by symmetry)}$$

$$q_{\text{b},67} = q_{\text{b},23} = 0 \text{ (by symmetry)}$$

$$q_{\text{b},21} = -7.22 \times 10^{-4}(250 \times 100) = -18.1 \text{ N/mm}$$

$$q_{\text{b},18} = -18.1 - 7.22 \times 10^{-4}(200 \times 30) = -22.4 \text{ N/mm}$$

$$q_{\text{b},87} = q_{\text{b},21} = -18.1 \text{ N/mm (by symmetry)}$$

Taking moments about the intersection of the line of action of the shear load and the horizontal axis of symmetry and referring to the results of Eqs (20.7) and (20.8) we have, from Eq. (17.18)

$$0 = [q_{\text{b},81} \times 60 \times 480 + 2q_{\text{b},12}(240 \times 100 + 70 \times 240) + 2q_{\text{b},23} \times 240 \times 100$$
$$- 2q_{\text{b},43} \times 120 \times 100 - q_{\text{b},54} \times 100 \times 120] + 2 \times 97\,200 q_{s,0}$$

Substituting the above values of q_b in this equation gives

$$q_{s,0} = -5.4 \text{ N/mm}$$

the negative sign indicating that $q_{s,0}$ acts in a clockwise sense.

In any wall the final shear flow is given by $q_s = q_{\text{b}} + q_{s,0}$ so that

$$q_{21} = -18.1 + 5.4 = -12.7 \text{ N/mm} = q_{87}$$

$$q_{23} = -5.4 \text{ N/mm} = q_{67}$$

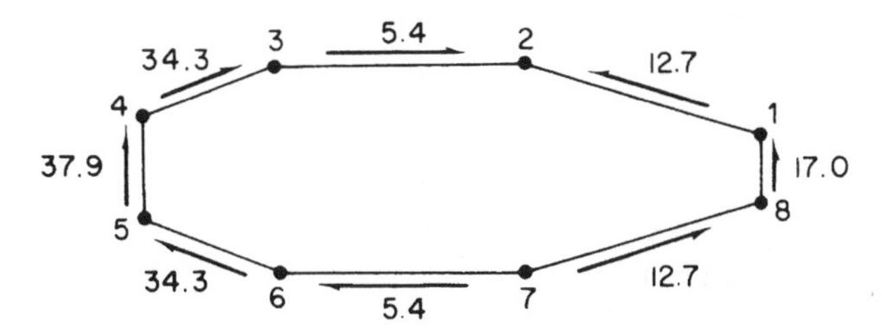

Fig. 20.12 Shear flow distribution N/mm in walls of the beam section of Example 20.4.

$$q_{34} = -34.3\,\mathrm{N/mm} = q_{56}$$

$$q_{45} = -37.9\,\mathrm{N/mm}$$

and

$$q_{81} = 17.0\,\mathrm{N/mm}$$

giving the shear flow distribution shown in Fig. 20.12.

20.3.4 Alternative method for the calculation of shear flow distribution

Equation (20.4) may be rewritten in the form

$$q_2 - q_1 = \frac{\partial P_r}{\partial z} \tag{20.12}$$

in which P_r is the direct load in the rth boom. This form of the equation suggests an alternative approach to the determination of the effect of booms on the calculation of shear flow distributions in open and closed section beams.

Let us suppose that the boom load varies linearly with z. This will be the case for a length of beam over which the shear force is constant. Equation (20.12) then becomes

$$q_2 - q_1 = -\Delta P_r \tag{20.13}$$

in which ΔP_r is the *change* in boom load over unit length of the rth boom. ΔP_r may be calculated by first determining the *change* in bending moment between two sections of a beam a unit distance apart and then calculating the corresponding change in boom stress using either of Eq. (16.18) or (16.19); the change in boom load follows by multiplying the change in boom stress by the boom area B_r. Note that the section properties contained in Eqs (16.18) and (16.19) refer to the direct stress carrying area of the beam section. In cases where the shear force is not constant over the unit length of beam the method is approximate.

We shall illustrate the method by applying it to Example 20.3. In Fig. 20.7 the shear load of 4.8 kN is applied to the face of the section which is seen when a view is taken along the z axis towards the origin. Thus, when considering unit length of the beam, we must ensure that this situation is unchanged. Figure 20.13 shows a unit (1 mm say) length of beam. The change in bending moment between the front and rear faces of the

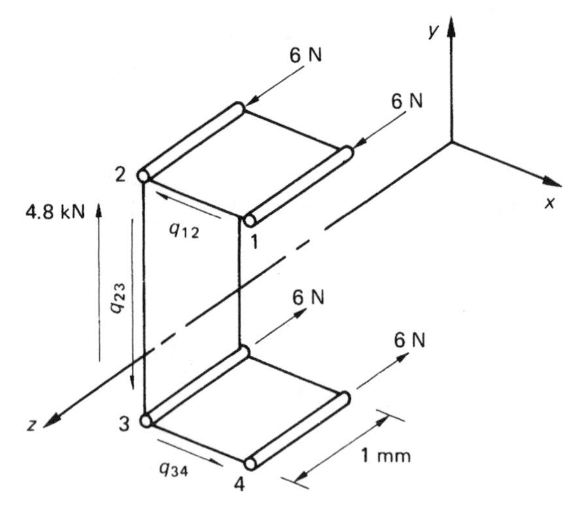

Fig. 20.13 Alternative solution to Example 20.3.

length of beam is 4.8×1 kN mm which produces a change in boom load given by (see Eq. (16.18))

$$\Delta P_r = \frac{4.8 \times 10^3 \times 200}{48 \times 10^6} \times 300 = 6\,\text{N}$$

The change in boom load is compressive in booms 1 and 2 and tensile in booms 3 and 4.

Equation (20.12), and hence Eq. (20.13), is based on the tensile load in a boom increasing with increasing z. If the tensile load had increased with decreasing z the right-hand side of these equations would have been positive. It follows that in the case where a compressive load increases with decreasing z, as for booms 1 and 2 in Fig. 20.13, the right-hand side is negative; similarly for booms 3 and 4 the right-hand side is positive. Thus

$$q_{12} = -6\,\text{N/mm}$$

$$q_{23} = -6 + q_{12} = -12\,\text{N/mm}$$

and

$$q_{34} = +6 + q_{23} = -6\,\text{N/mm}$$

giving the same solution as before. Note that if the unit length of beam had been taken to be 1 m the solution would have been $q_{12} = -6000$ N/m, $q_{23} = -12\,000$ N/m, $q_{34} = -6000$ N/m.

20.3.5 Torsion of open and closed section beams

No direct stresses are developed in either open or closed section beams subjected to a pure torque unless axial constraints are present. The shear stress distribution is therefore unaffected by the presence of booms and the analyses presented in Chapter 18 apply.

20.4 Deflection of open and closed section beams

Bending, shear and torsional deflections of thin-walled beams are readily obtained by application of the unit load method described in Section 5.5.

The displacement in a given direction due to torsion is given directly by the last of Eqs (5.21), thus

$$\Delta_T = \int_L \frac{T_0 T_1}{GJ} \mathrm{d}z \tag{20.14}$$

where J, the torsion constant, depends on the type of beam under consideration. For an open section beam J is given by either of Eqs (18.11) whereas in the case of a closed section beam $J = 4A^2/(\oint \mathrm{d}s/t)$ (Eq. (18.4)) for a constant shear modulus.

Expressions for the bending and shear displacements of unsymmetrical thin-walled beams may also be determined by the unit load method. They are complex for the general case and are most easily derived from first principles by considering the complementary energy of the elastic body in terms of stresses and strains rather than loads and displacements. In Chapter 5 we observed that the theorem of the principle of the stationary value of the total complementary energy of an elastic system is equivalent to the application of the principle of virtual work where virtual forces act through real displacements. We may therefore specify that in our expression for total complementary energy the displacements are the actual displacements produced by the applied loads while the virtual force system is the unit load.

Considering deflections due to bending, we see, from Eq. (5.6), that the increment in total complementary energy due to the application of a virtual unit load is

$$-\int_L \left(\int_A \sigma_{z,1} \varepsilon_{z,0} \, \mathrm{d}A \right) \mathrm{d}z + 1\Delta_M$$

where $\sigma_{z,1}$ is the direct bending stress at any point in the beam cross-section corresponding to the unit load and $\varepsilon_{z,0}$ is the strain at the point produced by the actual loading system. Further, Δ_M is the actual displacement due to bending at the point of application and in the direction of the unit load. Since the system is in equilibrium under the action of the unit load the above expression must equal zero (see Eq. (5.6)). Hence

$$\Delta_M = \int_L \left(\int_A \sigma_{z,1} \varepsilon_{z,0} \, \mathrm{d}A \right) \mathrm{d}z \tag{20.15}$$

From Eq. (16.18) and the third of Eqs (1.42)

$$\sigma_{z,1} = \left(\frac{M_{y,1} I_{xx} - M_{x,1} I_{xy}}{I_{xx} I_{yy} - I_{xy}^2} \right) x + \left(\frac{M_{x,1} I_{yy} - M_{y,1} I_{xy}}{I_{xx} I_{yy} - I_{xy}^2} \right) y$$

$$\varepsilon_{z,0} = \frac{1}{E} \left[\left(\frac{M_{y,0} I_{xx} - M_{x,0} I_{xy}}{I_{xx} I_{yy} - I_{xy}^2} \right) x + \left(\frac{M_{x,0} I_{yy} - M_{y,0} I_{xy}}{I_{xx} I_{yy} - I_{xy}^2} \right) y \right]$$

where the suffixes 1 and 0 refer to the unit and actual loading systems and x, y are the coordinates of any point in the cross-section referred to a centroidal system of axes. Substituting for $\sigma_{z,1}$ and $\varepsilon_{z,0}$ in Eq. (20.15) and remembering that $\int_A x^2\,dA = I_{yy}$, $\int_A y^2\,dA = I_{xx}$, and $\int_A xy\,dA = I_{xy}$, we have

$$\begin{aligned}\Delta_M = \frac{1}{E(I_{xx}I_{yy} - I_{xy}^2)^2} \int_L &\{(M_{y,1}I_{xx} - M_{x,1}I_{xy})(M_{y,0}I_{xx} - M_{x,0}I_{xy})I_{yy} \\ &+ (M_{x,1}I_{yy} - M_{y,1}I_{xy})(M_{x,0}I_{yy} - M_{y,0}I_{xy})I_{xx} \\ &+ [(M_{y,1}I_{xx} - M_{x,1}I_{xy})(M_{x,0}I_{yy} - M_{y,0}I_{xy}) \\ &+ (M_{x,1}I_{yy} - M_{y,1}I_{xy})(M_{y,0}I_{xx} - M_{x,0}I_{xy})]I_{xy}\}\mathrm{d}z \end{aligned} \tag{20.16}$$

For a section having either the x or y axis as an axis of symmetry, $I_{xy} = 0$ and Eq. (20.16) reduces to

$$\Delta_M = \frac{1}{E}\int_L \left(\frac{M_{y,1}M_{y,0}}{I_{yy}} + \frac{M_{x,1}M_{x,0}}{I_{xx}}\right)\mathrm{d}z \tag{20.17}$$

The derivation of an expression for the shear deflection of thin-walled sections by the unit load method is achieved in a similar manner. By comparison with Eq. (20.15) we deduce that the deflection Δ_S, due to shear of a thin-walled open or closed section beam of thickness t, is given by

$$\Delta_S = \int_L \left(\int_{\text{sect}} \tau_1 \gamma_0 t\,\mathrm{d}s\right)\mathrm{d}z \tag{20.18}$$

where τ_1 is the shear stress at an arbitrary point s around the section produced by a unit load applied at the point and in the direction Δ_S, and γ_0 is the shear strain at the arbitrary point corresponding to the actual loading system. The integral in parentheses is taken over all the walls of the beam. In fact, both the applied and unit shear loads must act through the shear centre of the cross-section, otherwise additional torsional displacements occur. Where shear loads act at other points these must be replaced by shear loads at the shear centre plus a torque. The thickness t is the actual skin thickness and may vary around the cross-section but is assumed to be constant along the length of the beam. Rewriting Eq. (20.18) in terms of shear flows q_1 and q_0, we obtain

$$\Delta_S = \int_L \left(\int_{\text{sect}} \frac{q_0 q_1}{Gt}\mathrm{d}s\right)\mathrm{d}z \tag{20.19}$$

where again the suffixes refer to the actual and unit loading systems. In the cases of both open and closed section beams the general expressions for shear flow are long and are best evaluated before substituting in Eq. (20.19). For an open section beam comprising booms and walls of direct stress carrying thickness t_D we have, from Eq. (20.6)

$$\begin{aligned} q_0 = &-\left(\frac{S_{x,0}I_{xx} - S_{y,0}I_{xy}}{I_{xx}I_{yy} - I_{xy}^2}\right)\left(\int_0^s t_D x\,\mathrm{d}s + \sum_{r=1}^n B_r x_r\right) \\ &-\left(\frac{S_{y,0}I_{yy} - S_{x,0}I_{xy}}{I_{xx}I_{yy} - I_{xy}^2}\right)\left(\int_0^s t_D y\,\mathrm{d}s + \sum_{r=1}^n B_r y_r\right) \end{aligned} \tag{20.20}$$

and

$$q_1 = -\left(\frac{S_{x,1}I_{xx} - S_{y,1}I_{xy}}{I_{xx}I_{yy} - I_{xy}^2}\right)\left(\int_0^s t_{\text{D}}x\,\text{d}s + \sum_{r=1}^{n} B_r x_r\right)$$
$$-\left(\frac{S_{y,1}I_{yy} - S_{x,1}I_{xy}}{I_{xx}I_{yy} - I_{xy}^2}\right)\left(\int_0^s t_{\text{D}}y\,\text{d}s + \sum_{r=1}^{n} B_r y_r\right) \tag{20.21}$$

Similar expressions are obtained for a closed section beam from Eq. (20.11).

Example 20.5

Calculate the deflection of the free end of a cantilever 2000 mm long having a channel section identical to that in Example 20.3 and supporting a vertical, upward load of 4.8 kN acting through the shear centre of the section. The effective direct stress carrying thickness of the skin is zero while its actual thickness is 1 mm. Young's modulus E and the shear modulus G are 70 000 and 30 000 N/mm^2, respectively.

The section is doubly symmetrical (i.e. the direct stress carrying area) and supports a vertical load producing a vertical deflection. Thus we apply a unit load through the shear centre of the section at the tip of the cantilever and in the same direction as the applied load. Since the load is applied through the shear centre there is no twisting of the section and the total deflection is given, from Eqs (20.17), (20.19), (20.20) and (20.21), by

$$\Delta = \int_0^L \frac{M_{x,0}M_{x,1}}{EI_{xx}}\text{d}z + \int_0^L \left(\int_{\text{sect}} \frac{q_0 q_1}{Gt}\text{d}s\right)\text{d}z \tag{i}$$

where $M_{x,0} = -4.8 \times 10^3(2000 - z)$, $M_{x,1} = -1(2000 - z)$

$$q_0 = -\frac{4.8 \times 10^3}{I_{xx}}\sum_{r=1}^{n} B_r y_r \quad q_1 = -\frac{1}{I_{xx}}\sum_{r=1}^{n} B_r y_r$$

and z is measured from the built-in end of the cantilever. The actual shear flow distribution has been calculated in Example 20.3. In this case the q_1 shear flows may be deduced from the actual distribution shown in Fig. 20.8, i.e.

$$q_1 = q_0/4.8 \times 10^3$$

Evaluating the bending deflection, we have

$$\Delta_{\text{M}} = \int_0^{2000} \frac{4.8 \times 10^3(2000 - z)^2\text{d}z}{70\,000 \times 48 \times 10^6} = 3.81\text{ mm}$$

The shear deflection Δ_{S} is given by

$$\Delta_{\text{S}} = \int_0^{2000} \frac{1}{30\,000 \times 1}\left[\frac{1}{4.8 \times 10^3}(6^2 \times 200 + 12^2 \times 400 + 6^2 \times 200)\right]\text{d}z$$
$$= 1.0\text{ mm}$$

The total deflection Δ is then $\Delta_{\text{M}} + \Delta_{\text{S}} = 4.81$ mm in a vertical upward direction.

Problems

P.20.1 Idealize the box section shown in Fig. P.20.1 into an arrangement of direct stress carrying booms positioned at the four corners and panels which are assumed to carry only shear stresses. Hence determine the distance of the shear centre from the left-hand web.

Ans. 225 mm.

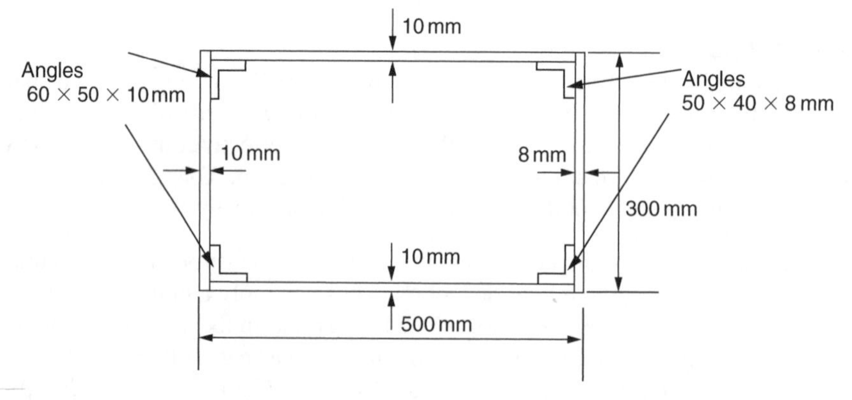

Fig. P.20.1

P.20.2 The beam section shown in Fig. P.20.2 has been idealized into an arrangement of direct stress carrying booms and shear stress only carrying panels. If the beam section is subjected to a vertical shear load of 1495 N through its shear centre, booms 1, 4, 5 and 8 each have an area of 200 mm^2 and booms 2, 3, 6 and 7 each have an area of 250 mm^2 determine the shear flow distribution and the position of the shear centre.

Ans. Wall 12, 1.86 N/mm; 43, 1.49 N/mm; 32, 5.21 N/mm; 27, 10.79 N/mm; remaining distribution follows from symmetry. 122 mm to the left of the web 27.

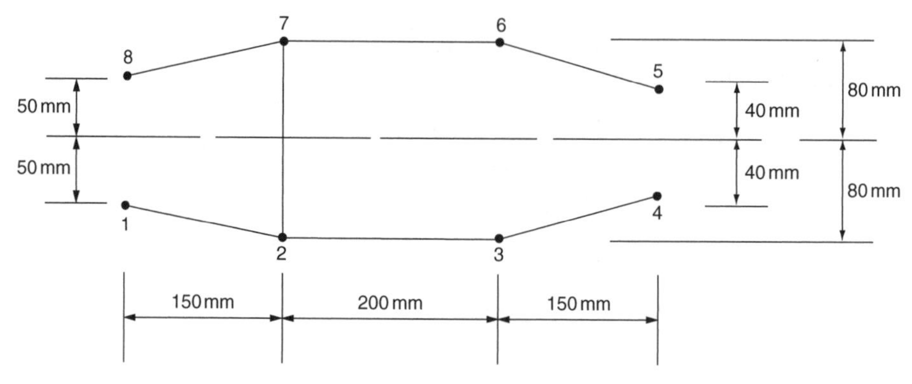

Fig. P.20.2

P.20.3 Figure P.20.3 shows the cross-section of a single cell, thin-walled beam with a horizontal axis of symmetry. The direct stresses are carried by the booms B_1 to B_4, while the walls are effective only in carrying shear stresses. Assuming that the basic theory of bending is applicable, calculate the position of the shear centre S. The shear modulus G is the same for all walls.

Cell area = 135 000 mm^2. Boom areas: $B_1 = B_4 = 450$ mm^2, $B_2 = B_3 = 550$ mm^2.

Ans. 197.2 mm from vertical through booms 2 and 3.

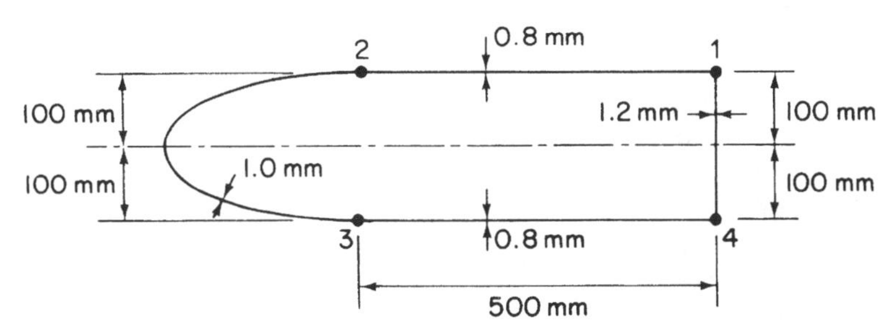

Fig. P.20.3

Wall	Length (mm)	Thickness (mm)
12, 34	500	0.8
23	580	1.0
41	200	1.2

P.20.4 Find the position of the shear centre of the rectangular four boom beam section shown in Fig. P.20.4. The booms carry only direct stresses but the skin is fully effective in carrying both shear and direct stress. The area of each boom is 100 mm^2.

Ans. 142.5 mm from side 23.

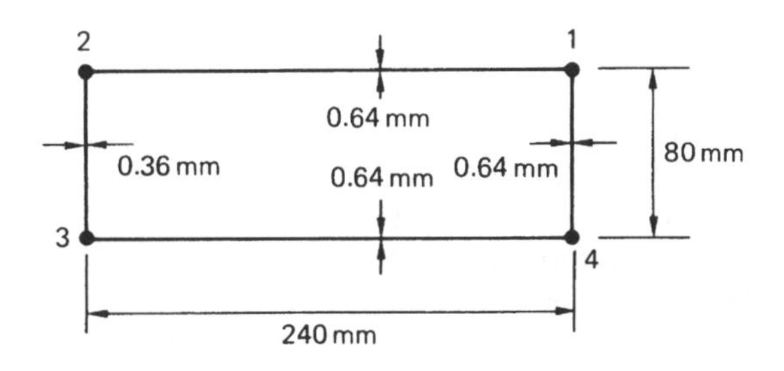

Fig. P.20.4

P.20.5 A uniform beam with the cross-section shown in Fig. P.20.5(a) is supported and loaded as shown in Fig. P.20.5(b). If the direct and shear stresses are given by the basic theory of bending, the direct stresses being carried by the booms and the shear stresses by the walls, calculate the vertical deflection at the ends of the beam when the loads act through the shear centres of the end cross-sections, allowing for the effect of shear strains.

Take $E = 69\,000\,\text{N/mm}^2$ and $G = 26\,700\,\text{N/mm}^2$. Boom areas: $1, 3, 4, 6 = 650\,\text{mm}^2$, $2, 5 = 1300\,\text{mm}^2$.

Ans. 3.4 mm.

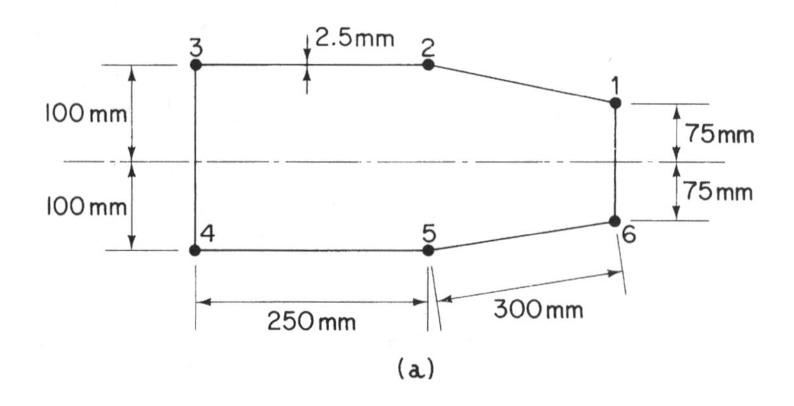

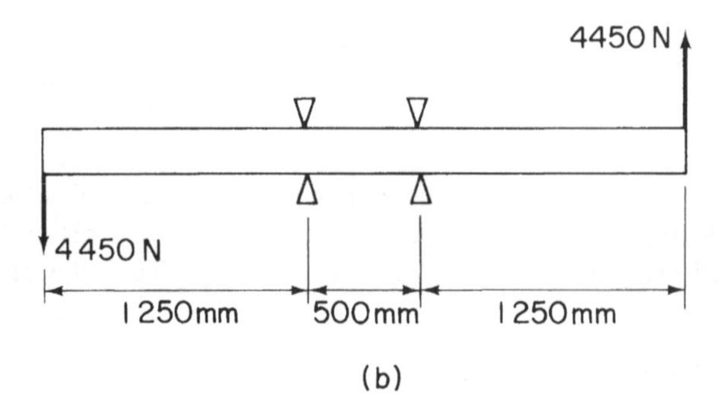

Fig. P.20.5

P.20.6 A cantilever, length L, has a hollow cross-section in the form of a doubly symmetric wedge as shown in Fig. P.20.6. The chord line is of length c, wedge thickness is t, the length of a sloping side is $a/2$ and the wall thickness is constant and equal to t_0. Uniform pressure distributions of magnitudes shown act on the faces of the wedge. Find the vertical deflection of point A due to this given loading. If $G = 0.4E$, $t/c = 0.05$ and $L = 2c$ show that this deflection is approximately $5600 p_0 c^2 / E t_0$.

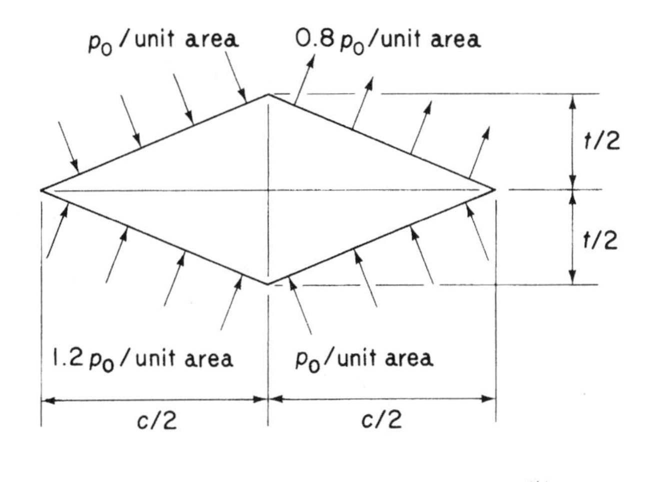

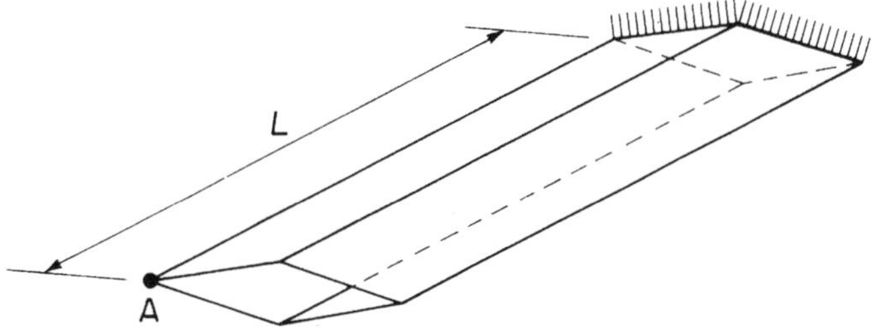

Fig. P.20.6

P.20.7 A rectangular section thin-walled beam of length L and breadth $3b$, depth b and wall thickness t is built in at one end (Fig. P.20.7). The upper surface of the beam is subjected to a pressure which varies linearly across the breadth from a value p_0 at edge AB to zero at edge CD. Thus, at any given value of x the pressure is constant in the z direction. Find the vertical deflection of point A.

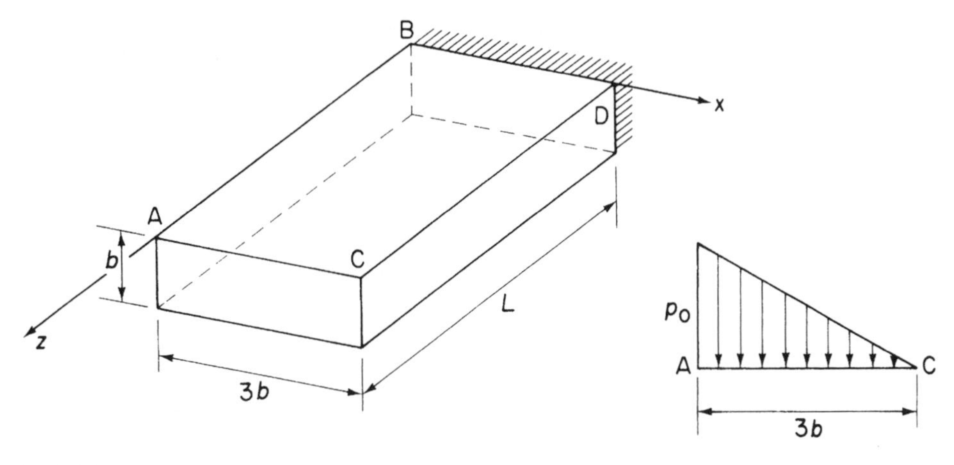

Fig. P.20.7

Ans. $p_0L^2(9L^2/80Eb^2 + 1609/2000G)/t$.

Section B4 Stress Analysis of Aircraft Components

21

Wing spars and box beams

In Chapters 16–18 we established the basic theory for the analysis of open and closed section thin-walled beams subjected to bending, shear and torsional loads. In addition, in Chapter 20, we saw how complex stringer stiffened sections could be idealized into sections more amenable to analysis. We shall now extend this analysis to actual aircraft components including, in this chapter, wing spars and box beams. In subsequent chapters we shall investigate the analysis of fuselages, wings, frames and ribs, and consider the effects of cut-outs in wings and fuselages. Finally, in Chapter 25, an introduction is given to the analysis of components fabricated from composite materials.

Aircraft structural components are, as we saw in Chapter 12, complex, consisting usually of thin sheets of metal stiffened by arrangements of stringers. These structures are highly redundant and require some degree of simplification or idealization before they can be analysed. The analysis presented here is therefore approximate and the degree of accuracy obtained depends on the number of simplifying assumptions made. A further complication arises in that factors such as warping restraint, structural and loading discontinuities and shear lag significantly affect the analysis; we shall investigate these effects in some simple structural components in Chapters 26 and 27. Generally, a high degree of accuracy can only be obtained by using computer-based techniques such as the finite element method (see Chapter 6). However, the simpler, quicker and cheaper approximate methods can be used to advantage in the preliminary stages of design when several possible structural alternatives are being investigated; they also provide an insight into the physical behaviour of structures which computer-based techniques do not.

Major aircraft structural components such as wings and fuselages are usually tapered along their lengths for greater structural efficiency. Thus, wing sections are reduced both chordwise and in depth along the wing span towards the tip and fuselage sections aft of the passenger cabin taper to provide a more efficient aerodynamic and structural shape.

The analysis of open and closed section beams presented in Chapters 16–18 assumes that the beam sections are uniform. The effect of taper on the prediction of direct stresses produced by bending is minimal if the taper is small and the section properties are calculated at the particular section being considered; Eqs (16.18)–(16.22) may therefore be used with reasonable accuracy. On the other hand, the calculation of shear stresses in beam webs can be significantly affected by taper.

21.1 Tapered wing spar

Consider first the simple case of a beam, for example a wing spar, positioned in the yz plane and comprising two flanges and a web: an elemental length δz of the beam is shown in Fig. 21.1. At the section z the beam is subjected to a positive bending moment M_x and a positive shear force S_y. The bending moment resultants $P_{z,1}$ and $P_{z,2}$ are parallel to the z axis of the beam. For a beam in which the flanges are assumed to resist all the direct stresses, $P_{z,1} = M_x/h$ and $P_{z,2} = -M_x/h$. In the case where the web is assumed to be fully effective in resisting direct stress, $P_{z,1}$ and $P_{z,2}$ are determined by multiplying the direct stresses $\sigma_{z,1}$ and $\sigma_{z,2}$ found using Eq. (16.18) or (16.19) by the flange areas B_1 and B_2. $P_{z,1}$ and $P_{z,2}$ are the components in the z direction of the axial loads P_1 and P_2 in the flanges. These have components $P_{y,1}$ and $P_{y,2}$ parallel to the y axis given by

$$P_{y,1} = P_{z,1}\frac{\delta y_1}{\delta z} \quad P_{y,2} = -P_{z,2}\frac{\delta y_2}{\delta z} \tag{21.1}$$

in which, for the direction of taper shown, δy_2 is negative. The axial load in flange ① is given by

$$P_1 = (P_{z,1}^2 + P_{y,1}^2)^{1/2}$$

Substituting for $P_{y,1}$ from Eq. (21.1) we have

$$P_1 = P_{z,1}\frac{(\delta z^2 + \delta y_1^2)^{1/2}}{\delta z} = \frac{P_{z,1}}{\cos\alpha_1} \tag{21.2}$$

Similarly

$$P_2 = \frac{P_{z,2}}{\cos\alpha_2} \tag{21.3}$$

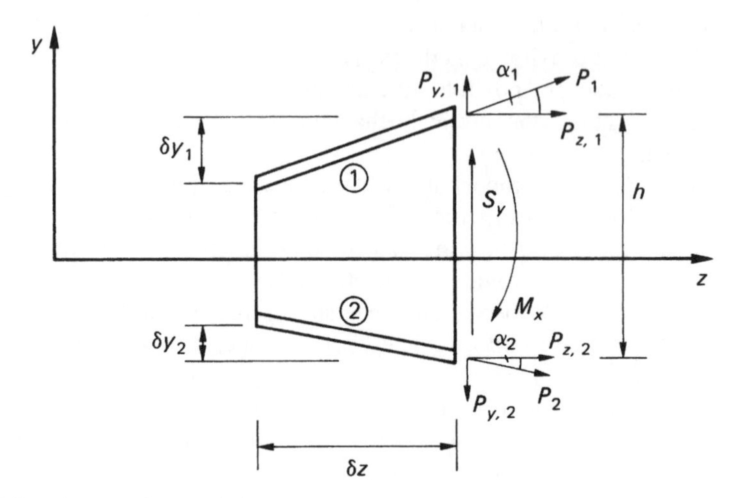

Fig. 21.1 Effect of taper on beam analysis.

The internal shear force S_y comprises the resultant $S_{y,w}$ of the web shear flows together with the vertical components of P_1 and P_2. Thus

$$S_y = S_{y,w} + P_{y,1} - P_{y,2}$$

or

$$S_y = S_{y,w} + P_{z,1}\frac{\delta y_1}{\delta z} + P_{z,2}\frac{\delta y_2}{\delta z} \tag{21.4}$$

so that

$$S_{y,w} = S_y - P_{z,1}\frac{\delta y_1}{\delta z} - P_{z,2}\frac{\delta y_2}{\delta z} \tag{21.5}$$

Again we note that δy_2 in Eqs (21.4) and (21.5) is negative. Equation (21.5) may be used to determine the shear flow distribution in the web. For a completely idealized beam the web shear flow is constant through the depth and is given by $S_{y,w}/h$. For a beam in which the web is fully effective in resisting direct stresses the web shear flow distribution is found using Eq. (20.6) in which S_y is replaced by $S_{y,w}$ and which, for the beam of Fig. 21.1, would simplify to

$$q_s = -\frac{S_{y,w}}{I_{xx}}\left(\int_0^s t_D y \, ds + B_1 y_1\right) \tag{21.6}$$

or

$$q_s = -\frac{S_{y,w}}{I_{xx}}\left(\int_0^s t_D y \, ds + B_2 y_2\right) \tag{21.7}$$

Example 21.1

Determine the shear flow distribution in the web of the tapered beam shown in Fig. 21.2, at a section midway along its length. The web of the beam has a thickness of 2 mm

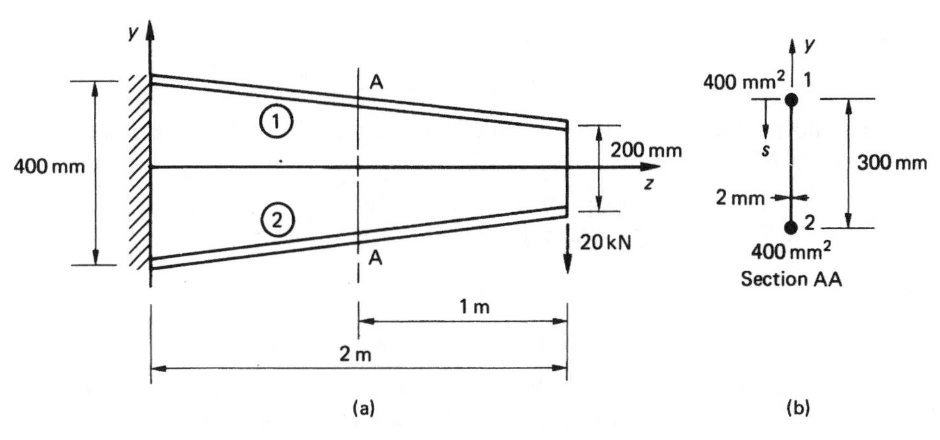

Fig. 21.2 Tapered beam of Example 21.1.

and is fully effective in resisting direct stress. The beam tapers symmetrically about its horizontal centroidal axis and the cross-sectional area of each flange is 400 mm^2.

The internal bending moment and shear load at the section AA produced by the externally applied load are, respectively

$$M_x = 20 \times 1 = 20\,\text{kN m} \quad S_y = -20\,\text{kN}$$

The direct stresses parallel to the z axis in the flanges at this section are obtained either from Eqs (16.18) or (16.19) in which $M_y = 0$ and $I_{xy} = 0$. Thus, from Eq. (16.18)

$$\sigma_z = \frac{M_x y}{I_{xx}} \tag{i}$$

in which

$$I_{xx} = 2 \times 400 \times 150^2 + 2 \times 300^3/12$$

i.e.

$$I_{xx} = 22.5 \times 10^6\,\text{mm}^4$$

Hence

$$\sigma_{z,1} = -\sigma_{z,2} = \frac{20 \times 10^6 \times 150}{22.5 \times 10^6} = 133.3\,\text{N/mm}^2$$

The components parallel to the z axis of the axial loads in the flanges are therefore

$$P_{z,1} = -P_{z,2} = 133.3 \times 400 = 53\,320\,\text{N}$$

The shear load resisted by the beam web is then, from Eq. (21.5)

$$S_{y,w} = -20 \times 10^3 - 53\,320\frac{\delta y_1}{\delta z} + 53\,320\frac{\delta y_2}{\delta z}$$

in which, from Figs 21.1 and 21.2, we see that

$$\frac{\delta y_1}{\delta z} = \frac{-100}{2 \times 10^3} = -0.05 \qquad \frac{\delta y_2}{\delta z} = \frac{100}{2 \times 10^3} = 0.05$$

Hence

$$S_{y,w} = -20 \times 10^3 + 53\,320 \times 0.05 + 53\,320 \times 0.05 = -14\,668\,\text{N}$$

The shear flow distribution in the web follows either from Eq. (21.6) or Eq. (21.7) and is (see Fig. 21.2(b))

$$q_{12} = \frac{14\,668}{22.5 \times 10^6}\left(\int_0^s 2(150 - s)\,\text{d}s + 400 \times 150\right)$$

i.e.

$$q_{12} = 6.52 \times 10^{-4}(-s^2 + 300s + 60\,000) \tag{ii}$$

The maximum value of q_{12} occurs when $s = 150$ mm and q_{12} (max) = 53.8 N/mm. The values of shear flow at points 1 ($s = 0$) and 2 ($s = 300$ mm) are $q_1 = 39.1$ N/mm and $q_2 = 39.1$ N/mm; the complete distribution is shown in Fig. 21.3.

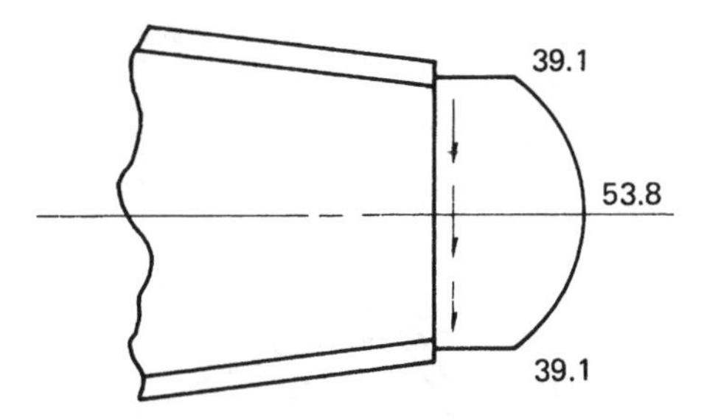

Fig. 21.3 Shear flow (N/mm) distribution at Section AA in Example 21.1.

21.2 Open and closed section beams

We shall now consider the more general case of a beam tapered in two directions along its length and comprising an arrangement of booms and skin. Practical examples of such a beam are complete wings and fuselages. The beam may be of open or closed section; the effects of taper are determined in an identical manner in either case.

Figure 21.4(a) shows a short length δz of a beam carrying shear loads S_x and S_y at the section z; S_x and S_y are positive when acting in the directions shown. Note that if the beam were of open cross-section the shear loads would be applied through its shear centre so that no twisting of the beam occurred. In addition to shear loads the beam is subjected to bending moments M_x and M_y which produce direct stresses σ_z in the booms and skin. Suppose that in the rth boom the direct stress in a direction parallel to the z axis is $\sigma_{z,r}$, which may be found using either Eq. (16.18) or Eq. (16.19). The component $P_{z,r}$ of the axial load P_r in the rth boom is then given by

$$P_{z,r} = \sigma_{z,r} B_r \tag{21.8}$$

where B_r is the cross-sectional area of the rth boom.

From Fig. 21.4(b)

$$P_{y,r} = P_{z,r} \frac{\delta y_r}{\delta z} \tag{21.9}$$

Further, from Fig. 21.4(c)

$$P_{x,r} = P_{y,r} \frac{\delta x_r}{\delta y_r}$$

or, substituting for $P_{y,r}$ from Eq. (21.9)

$$P_{x,r} = P_{z,r} \frac{\delta x_r}{\delta z} \tag{21.10}$$

The axial load P_r is then given by

$$P_r = (P_{x,r}^2 + P_{y,r}^2 + P_{z,r}^2)^{1/2} \tag{21.11}$$

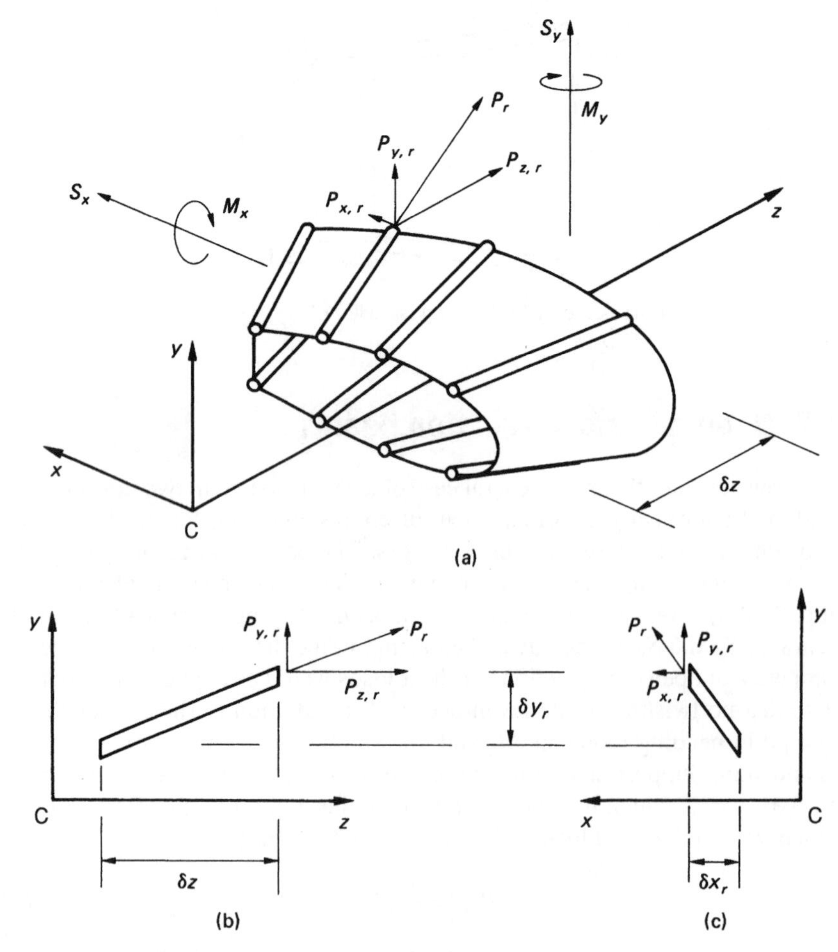

Fig. 21.4 Effect of taper on the analysis of open and closed section beams.

or, alternatively

$$P_r = P_{z,r}\frac{(\delta x_r^2 + \delta y_r^2 + \delta z^2)^{1/2}}{\delta z} \tag{21.12}$$

The applied shear loads S_x and S_y are reacted by the resultants of the shear flows in the skin panels and webs, together with the components $P_{x,r}$ and $P_{y,r}$ of the axial loads in the booms. Therefore, if $S_{x,w}$ and $S_{y,w}$ are the resultants of the skin and web shear flows and there is a total of m booms in the section

$$S_x = S_{x,w} + \sum_{r=1}^{m} P_{x,r} \quad S_y = S_{y,w} + \sum_{r=1}^{m} P_{y,r} \tag{21.13}$$

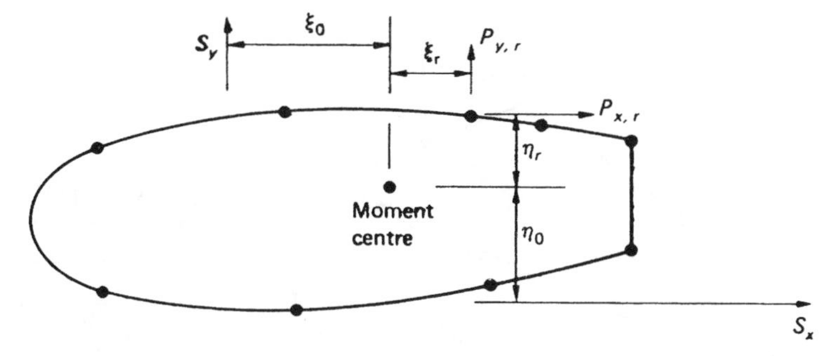

Fig. 21.5 Modification of moment equation in shear of closed section beams due to boom load.

Substituting in Eq. (21.13) for $P_{x,r}$ and $P_{y,r}$ from Eqs (21.10) and (21.9) we have

$$S_x = S_{x,w} + \sum_{r=1}^{m} P_{z,r}\frac{\delta x_r}{\delta z} \quad S_y = S_{y,w} + \sum_{r=1}^{m} P_{z,r}\frac{\delta y_r}{\delta z} \tag{21.14}$$

Hence

$$S_{x,w} = S_x - \sum_{r=1}^{m} P_{z,r}\frac{\delta x_r}{\delta z} \quad S_{y,w} = S_y - \sum_{r=1}^{m} P_{z,r}\frac{\delta y_r}{\delta z} \tag{21.15}$$

The shear flow distribution in an open section beam is now obtained using Eq. (20.6) in which S_x is replaced by $S_{x,w}$ and S_y by $S_{y,w}$ from Eq. (21.15). Similarly for a closed section beam, S_x and S_y in Eq. (20.11) are replaced by $S_{x,w}$ and $S_{y,w}$. In the latter case the moment equation (Eq. (17.17)) requires modification due to the presence of the boom load components $P_{x,r}$ and $P_{y,r}$. Thus from Fig. 21.5 we see that Eq. (17.17) becomes

$$S_x\eta_0 - S_y\xi_0 = \oint q_b p \, ds + 2Aq_{s,0} - \sum_{r=1}^{m} P_{x,r}\eta_r + \sum_{r=1}^{m} P_{y,r}\xi_r \tag{21.16}$$

Equation (21.16) is directly applicable to a tapered beam subjected to forces positioned in relation to the moment centre as shown. Care must be taken in a particular problem to ensure that the moments of the forces are given the correct sign.

Example 21.2

The cantilever beam shown in Fig. 21.6 is uniformly tapered along its length in both x and y directions and carries a load of 100 kN at its free end. Calculate the forces in the booms and the shear flow distribution in the walls at a section 2 m from the built-in end if the booms resist all the direct stresses while the walls are effective only in shear. Each corner boom has a cross-sectional area of 900 mm^2 while both central booms have cross-sectional areas of 1200 mm^2.

The internal force system at a section 2 m from the built-in end of the beam is

$$S_y = 100\,\text{kN} \quad S_x = 0 \quad M_x = -100 \times 2 = -200\,\text{kN m} \quad M_y = 0$$

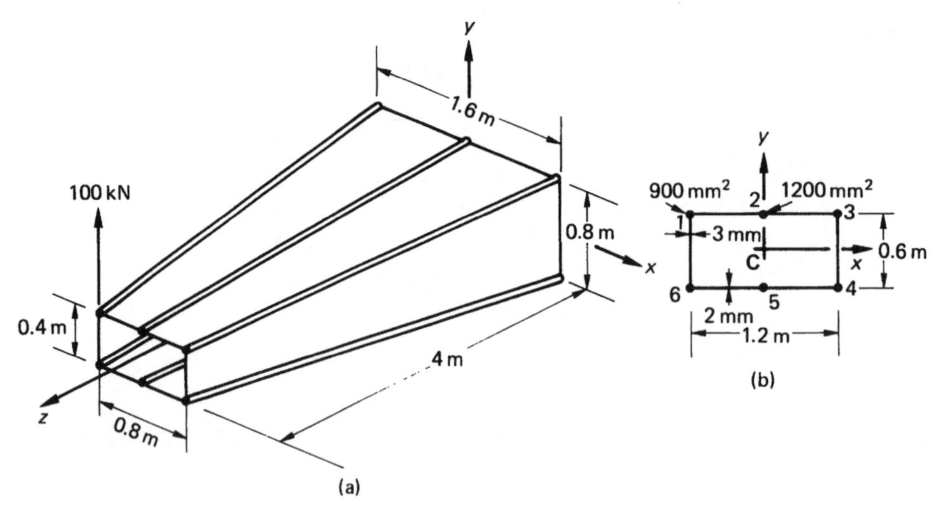

Fig. 21.6 (a) Beam of Example 21.2; (b) section 2 m from built-in end.

The beam has a doubly symmetrical cross-section so that $I_{xy} = 0$ and Eq. (16.18) reduces to

$$\sigma_z = \frac{M_x y}{I_{xx}} \tag{i}$$

in which, for the beam section shown in Fig. 21.6(b)

$$I_{xx} = 4 \times 900 \times 300^2 + 2 \times 1200 \times 300^2 = 5.4 \times 10^8 \text{ mm}^4$$

Then

$$\sigma_{z,r} = \frac{-200 \times 10^6}{5.4 \times 10^8} y_r$$

or

$$\sigma_{z,r} = -0.37 y_r \tag{ii}$$

Hence

$$P_{z,r} = -0.37 y_r B_r \tag{iii}$$

The value of $P_{z,r}$ is calculated from Eq. (iii) in column ② in Table 21.1; $P_{x,r}$ and $P_{y,r}$ follow from Eqs (21.10) and (21.9), respectively in columns ⑤ and ⑥. The axial load P_r, column ⑦, is given by $[②^2 + ⑤^2 + ⑥^2]^{1/2}$ and has the same sign as $P_{z,r}$ (see Eq. (21.12)). The moments of $P_{x,r}$ and $P_{y,r}$ are calculated for a moment centre at the centre of symmetry with anticlockwise moments taken as positive. Note that in Table 21.1, $P_{x,r}$ and $P_{y,r}$ are positive when they act in the positive directions of the section x and y axes, respectively; the distances η_r and ξ_r of the lines of action of $P_{x,r}$ and

Table 21.1

① Boom	② $P_{z,r}$ (kN)	③ $\delta x_r/\delta z$	④ $\delta y_r/\delta z$	⑤ $P_{x,r}$ (kN)	⑥ $P_{y,r}$ (kN)	⑦ P_r (kN)	⑧ ξ_r (m)	⑨ η_r (m)	⑩ $P_{x,r}\eta_r$ (kN m)	⑪ $P_{y,r}\xi_r$ (kN m)
1	−100	0.1	−0.05	−10	5	−101.3	0.6	0.3	3	−3
2	−133	0	−0.05	0	6.7	−177.3	0	0.3	0	0
3	−100	−0.1	−0.05	10	5	−101.3	0.6	0.3	−3	3
4	100	−0.1	0.05	−10	5	101.3	0.6	0.3	−3	3
5	133	0	0.05	0	6.7	177.3	0	0.3	0	0
6	100	0.1	0.05	10	5	101.3	0.6	0.3	3	−3

$P_{y,r}$ from the moment centre are not given signs since it is simpler to determine the sign of each moment, $P_{x,r}\eta_r$ and $P_{y,r}\xi_r$, by referring to the directions of $P_{x,r}$ and $P_{y,r}$ individually.

From column ⑥

$$\sum_{r=1}^{6} P_{y,r} = 33.4\,\text{kN}$$

From column ⑩

$$\sum_{r=1}^{6} P_{x,r}\eta_r = 0$$

From column ⑪

$$\sum_{r=1}^{6} P_{y,r}\xi_r = 0$$

From Eq. (21.15)

$$S_{x,w} = 0 \quad S_{y,w} = 100 - 33.4 = 66.6\,\text{kN}$$

The shear flow distribution in the walls of the beam is now found using the method described in Section 20.3. Since, for this beam, $I_{xy} = 0$ and $S_x = S_{x,w} = 0$, Eq. (20.11) reduces to

$$q_s = \frac{-S_{y,w}}{I_{xx}} \sum_{r=1}^{n} B_r y_r + q_{s,0} \tag{iv}$$

We now 'cut' one of the walls, say 16. The resulting 'open section' shear flow is given by

$$q_b = -\frac{66.6 \times 10^3}{5.4 \times 10^8} \sum_{r=1}^{n} B_r y_r$$

or

$$q_{\rm b} = -1.23 \times 10^{-4} \sum_{r=1}^{n} B_r y_r \qquad \text{(v)}$$

Thus

$$\begin{aligned}
q_{\rm b,16} &= 0 \\
q_{\rm b,12} &= 0 - 1.23 \times 10^{-4} \times 900 \times 300 = -33.2\,\text{N/mm} \\
q_{\rm b,23} &= -33.2 - 1.23 \times 10^{-4} \times 1200 \times 300 = -77.5\,\text{N/mm} \\
q_{\rm b,34} &= -77.5 - 1.23 \times 10^{-4} \times 900 \times 300 = -110.7\,\text{N/mm} \\
q_{\rm b,45} &= -77.5\,\text{N/mm (from symmetry)} \\
q_{\rm b,56} &= -33.2\,\text{N/mm (from symmetry)}
\end{aligned}$$

giving the distribution shown in Fig. 21.7. Taking moments about the centre of symmetry we have, from Eq. (21.16)

$$\begin{aligned}
-100 \times 10^3 \times 600 = {}& 2 \times 33.2 \times 600 \times 300 + 2 \times 77.5 \times 600 \times 300 \\
& + 110.7 \times 600 \times 600 + 2 \times 1200 \times 600 q_{s,0}
\end{aligned}$$

from which $q_{s,0} = -97.0\,\text{N/mm}$ (i.e. clockwise). The complete shear flow distribution is found by adding the value of $q_{s,0}$ to the $q_{\rm b}$ shear flow distribution of Fig. 21.7 and is shown in Fig. 21.8.

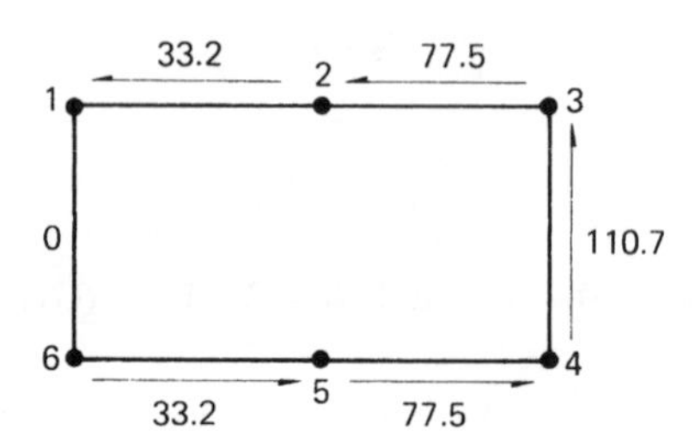

Fig. 21.7 'Open section' shear flow (N/mm) distribution in beam section of Example 21.2.

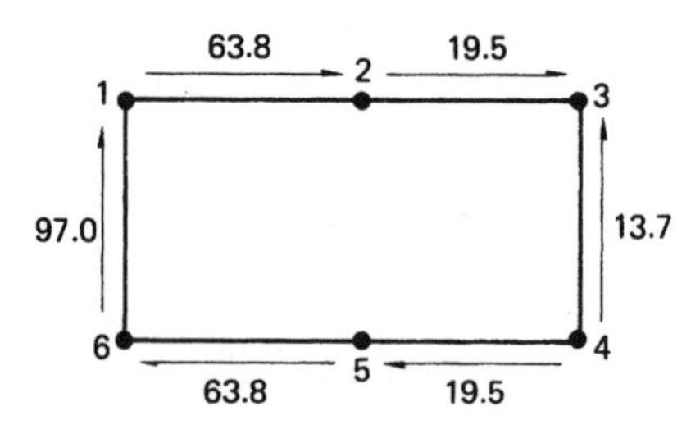

Fig. 21.8 Shear flow (N/mm) distribution in beam section of Example 21.2.

21.3 Beams having variable stringer areas

In many aircraft, structural beams, such as wings, have stringers whose cross-sectional areas vary in the spanwise direction. The effects of this variation on the determination of shear flow distribution cannot therefore be found by the methods described in Section 20.3 which assume constant boom areas. In fact, as we noted in Section 20.3, if the stringer stress is made constant by varying the area of cross-section there is no change in shear flow as the stringer/boom is crossed.

The calculation of shear flow distributions in beams having variable stringer areas is based on the alternative method for the calculation of shear flow distributions described in Section 20.3 and illustrated in the alternative solution of Example 20.3. The stringer loads $P_{z,1}$ and $P_{z,2}$ are calculated at two sections z_1 and z_2 of the beam a convenient distance apart. We assume that the stringer load varies linearly along its length so that the change in stringer load per unit length of beam is given by

$$\Delta P = \frac{P_{z,1} - P_{z,2}}{z_1 - z_2}$$

The shear flow distribution follows as previously described.

Example 21.3

Solve Example 21.2 by considering the differences in boom load at sections of the beam either side of the specified section.

In this example the stringer areas do not vary along the length of the beam but the method of solution is identical.

We are required to find the shear flow distribution at a section 2 m from the built-in end of the beam. We therefore calculate the boom loads at sections, say 0.1 m either side of this section. Thus, at a distance 2.1 m from the built-in end

$$M_x = -100 \times 1.9 = -190 \text{ kN m}$$

The dimensions of this section are easily found by proportion and are width = 1.18 m, depth = 0.59 m. Thus the second moment of area is

$$I_{xx} = 4 \times 900 \times 295^2 + 2 \times 1200 \times 295^2 = 5.22 \times 10^8 \text{ mm}^4$$

and

$$\sigma_{z,r} = \frac{-190 \times 10^6}{5.22 \times 10^8} y_r = -0.364 y_r$$

Hence

$$P_1 = P_3 = -P_4 = -P_6 = -0.364 \times 295 \times 900 = -96\,642 \text{ N}$$

and

$$P_2 = -P_5 = -0.364 \times 295 \times 1200 = -128\,856 \text{ N}$$

At a section 1.9 m from the built-in end

$$M_x = -100 \times 2.1 = -210\,\text{kN m}$$

and the section dimensions are width = 1.22 m, depth = 0.61 m so that

$$I_{xx} = 4 \times 900 \times 305^2 + 2 \times 1200 \times 305^2 = 5.58 \times 10^8\,\text{mm}^4$$

and

$$\sigma_{z,r} = \frac{-210 \times 10^6}{5.58 \times 10^8} y_r = -0.376 y_r$$

Hence

$$P_1 = P_3 = -P_4 = -P_6 = -0.376 \times 305 \times 900 = -103\,212\,\text{N}$$

and

$$P_2 = -P_5 = -0.376 \times 305 \times 1200 = -137\,616\,\text{N}$$

Thus, there is an increase in compressive load of 103 212 − 96 642 = 6570 N in booms 1 and 3 and an increase in tensile load of 6570 N in booms 4 and 6 between the two sections. Also, the compressive load in boom 2 increases by 137 616 − 128 856 = 8760 N while the tensile load in boom 5 increases by 8760 N. Therefore, the change in boom load per unit length is given by

$$\Delta P_1 = \Delta P_3 = -\Delta P_4 = -\Delta P_6 = \frac{6570}{200} = 32.85\,\text{N}$$

and

$$\Delta P_2 = -\Delta P_5 = \frac{8760}{200} = 43.8\,\text{N}$$

The situation is illustrated in Fig. 21.9. Suppose now that the shear flows in the panels 12, 23, 34, etc. are q_{12}, q_{23}, q_{34}, etc. and consider the equilibrium of boom 2, as shown in Fig. 21.10, with adjacent portions of the panels 12 and 23. Thus

$$q_{23} + 43.8 - q_{12} = 0$$

or

$$q_{23} = q_{12} - 43.8$$

Similarly

$$q_{34} = q_{23} - 32.85 = q_{12} - 76.65$$

$$q_{45} = q_{34} + 32.85 = q_{12} - 43.8$$

$$q_{56} = q_{45} + 43.8 = q_{12}$$

$$q_{61} = q_{45} + 32.85 = q_{12} + 32.85$$

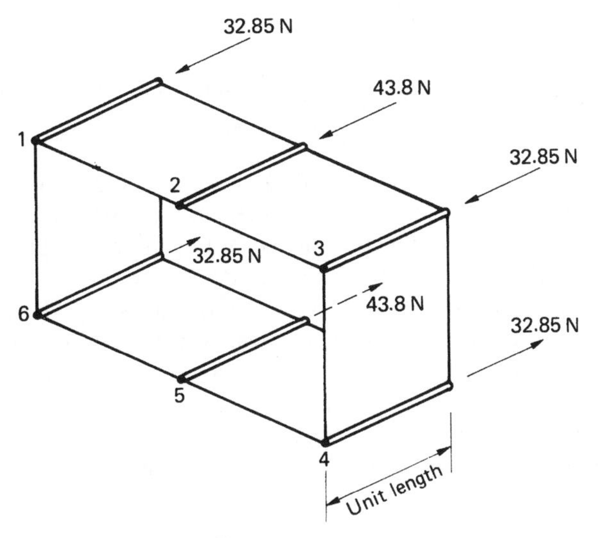

Fig. 21.9 Change in boom loads/unit length of beam.

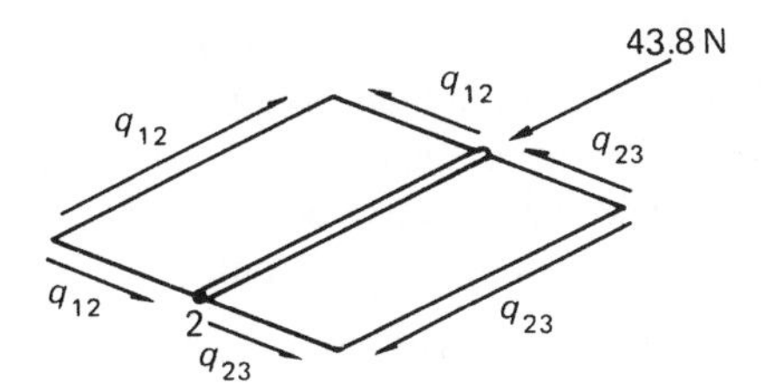

Fig. 21.10 Equilibrium of boom.

The moment resultant of the internal shear flows, together with the moments of the components $P_{y,r}$ of the boom loads about any point in the cross-section, is equivalent to the moment of the externally applied load about the same point. We note from Example 21.2 that for moments about the centre of symmetry

$$\sum_{r=1}^{6} P_{x,r}\eta_r = 0 \quad \sum_{r=1}^{6} P_{y,r}\xi_r = 0$$

Therefore, taking moments about the centre of symmetry

$$\begin{aligned}100 \times 10^3 \times 600 = 2q_{12} \times 600 \times 300 + 2(q_{12} - 43.8)600 \times 300 \\ + (q_{12} - 76.65)600 \times 600 + (q_{12} + 32.85)600 \times 600\end{aligned}$$

from which

$$q_{12} = 62.5\,\text{N/mm}$$

whence

$$q_{23} = 19.7\,\text{N/mm} \quad q_{34} = -13.2\,\text{N/mm} \quad q_{45} = 19.7\,\text{N/mm},$$

$$q_{56} = 63.5\,\text{N/mm} \quad q_{61} = 96.4\,\text{N/mm}$$

so that the solution is almost identical to the longer exact solution of Example 21.2.

The shear flows q_{12}, q_{23}, etc. induce complementary shear flows q_{12}, q_{23}, etc. in the panels in the longitudinal direction of the beam; these are, in fact, the average shear flows between the two sections considered. For a complete beam analysis the above procedure is applied to a series of sections along the span. The distance between adjacent sections may be taken to be any convenient value; for actual wings distances of the order of 350–700 mm are usually chosen. However, for very small values small percentage errors in $P_{z,1}$ and $P_{z,2}$ result in large percentage errors in ΔP. On the other hand, if the distance is too large the average shear flow between two adjacent sections may not be quite equal to the shear flow midway between the sections.

Problems

P.21.1 A wing spar has the dimensions shown in Fig. P.21.1 and carries a uniformly distributed load of 15 kN/m along its complete length. Each flange has a cross-sectional area of 500 mm^2 with the top flange being horizontal. If the flanges are assumed to resist all direct loads while the spar web is effective only in shear, determine the flange loads and the shear flows in the web at sections 1 and 2 m from the free end.

Ans. 1 m from free end: $P_U = 25$ kN (tension), $P_L = 25.1$ kN (compression), q = 41.7 N/mm.

2 m from free end: $P_U = 75$ kN (tension), $P_L = 75.4$ kN (compression), $q = 56.3$ N/mm.

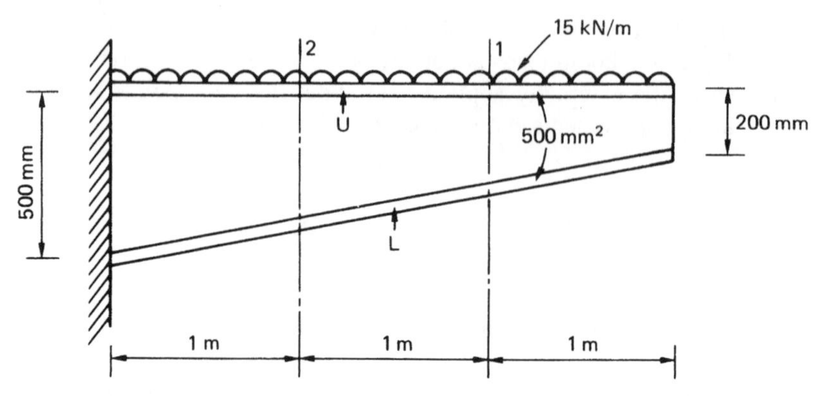

Fig. P.21.1

P.21.2 If the web in the wing spar of P.21.1 has a thickness of 2 mm and is fully effective in resisting direct stresses, calculate the maximum value of shear flow in the web at a section 1 m from the free end of the beam.

Ans. 46.8 N/mm.

P.21.3 Calculate the shear flow distribution and the stringer and flange loads in the beam shown in Fig. P.21.3 at a section 1.5 m from the built-in end. Assume that the skin and web panels are effective in resisting shear stress only; the beam tapers symmetrically in a vertical direction about its longitudinal axis.

Ans. $q_{13} = q_{42} = 36.9\,\text{N/mm}$, $q_{35} = q_{64} = 7.3\,\text{N/mm}$, $q_{21} = 96.2\,\text{N/mm}$, $q_{65} = 22.3\,\text{N/mm}$.

$$P_2 = -P_1 = 133.3\,\text{kN},\ P_4 = P_6 = -P_3 = -P_5 = 66.7\,\text{kN}.$$

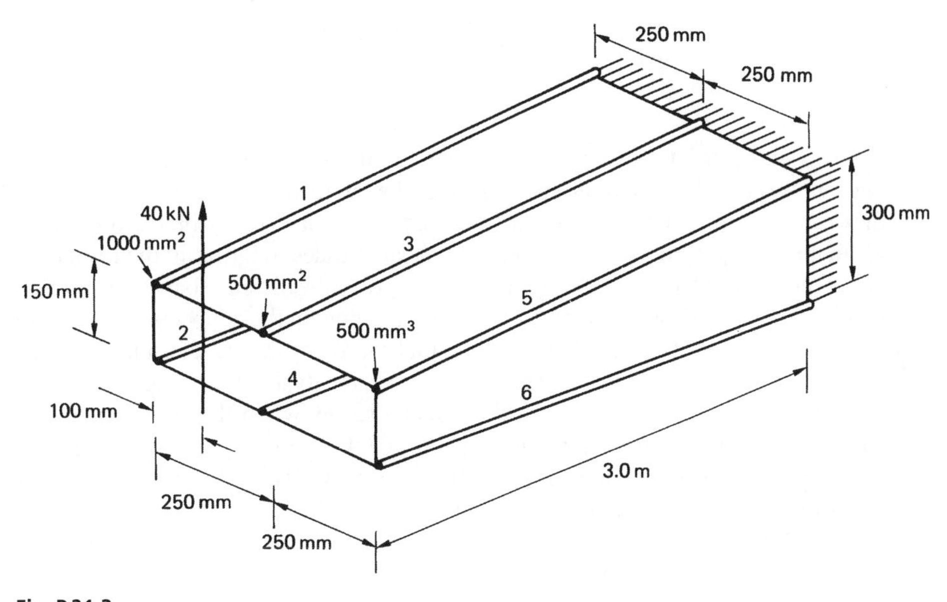

Fig. P.21.3

22

Fuselages

Aircraft fuselages consist, as we saw in Chapter 12, of thin sheets of material stiffened by large numbers of longitudinal stringers together with transverse frames. Generally, they carry bending moments, shear forces and torsional loads which induce axial stresses in the stringers and skin together with shear stresses in the skin; the resistance of the stringers to shear forces is generally ignored. Also, the distance between adjacent stringers is usually small so that the variation in shear flow in the connecting panel will be small. It is therefore reasonable to assume that the shear flow is constant between adjacent stringers so that the analysis simplifies to the analysis of an idealized section in which the stringers/booms carry all the direct stresses while the skin is effective only in shear. The direct stress carrying capacity of the skin may be allowed for by increasing the stringer/boom areas as described in Section 20.3. The analysis of fuselages therefore involves the calculation of direct stresses in the stringers and the shear stress distributions in the skin; the latter are also required in the analysis of transverse frames, as we shall see in Chapter 24.

22.1 Bending

The skin/stringer arrangement is idealized into one comprising booms and skin as described in Section 20.3. The direct stress in each boom is then calculated using either Eqs (16.18) or (16.19) in which the reference axes and the section properties refer to the direct stress carrying areas of the cross-section.

Example 22.1

The fuselage of a light passenger carrying aircraft has the circular cross-section shown in Fig. 22.1(a). The cross-sectional area of each stringer is $100\,\text{mm}^2$ and the vertical distances given in Fig. 22.1(a) are to the mid-line of the section wall at the corresponding stringer position. If the fuselage is subjected to a bending moment of 200 kN m applied in the vertical plane of symmetry, at this section, calculate the direct stress distribution.

The section is first idealized using the method described in Section 20.3. As an approximation we shall assume that the skin between adjacent stringers is flat so that

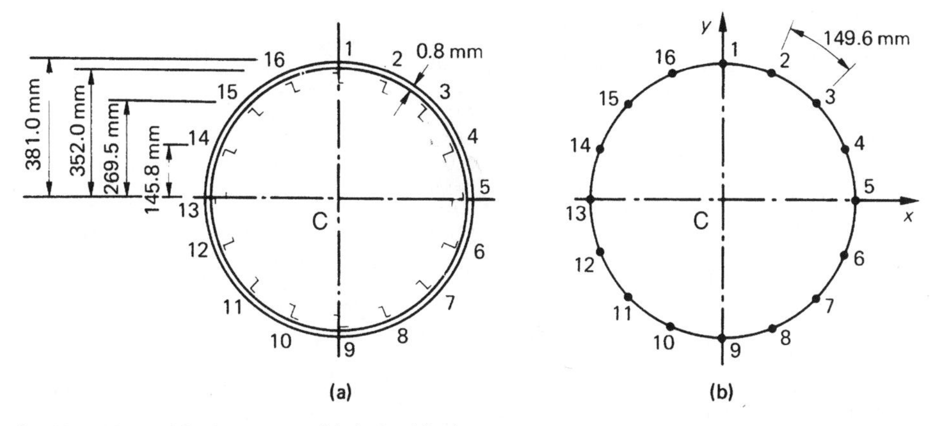

Fig. 22.1 (a) Actual fuselage section; (b) idealized fuselage section.

we may use either Eq. (20.1) or Eq. (20.2) to determine the boom areas. From symmetry $B_1 = B_9$, $B_2 = B_8 = B_{10} = B_{16}$, $B_3 = B_7 = B_{11} = B_{15}$, $B_4 = B_6 = B_{12} = B_{14}$ and $B_5 = B_{13}$. From Eq. (20.1)

$$B_1 = 100 + \frac{0.8 \times 149.6}{6}\left(2 + \frac{\sigma_2}{\sigma_1}\right) + \frac{0.8 \times 149.6}{6}\left(2 + \frac{\sigma_{16}}{\sigma_1}\right)$$

i.e.

$$B_1 = 100 + \frac{0.8 \times 149.6}{6}\left(2 + \frac{352.0}{381.0}\right) \times 2 = 216.6\,\text{mm}^2$$

Similarly $B_2 = 216.6\,\text{mm}^2$, $B_3 = 216.6\,\text{mm}^2$, $B_4 = 216.7\,\text{mm}^2$. We note that stringers 5 and 13 lie on the neutral axis of the section and are therefore unstressed; the calculation of boom areas B_5 and B_{13} does not then arise. For this particular section $I_{xy} = 0$ since Cx (and Cy) is an axis of symmetry. Further, $M_y = 0$ so that Eq. (16.18) reduces to

$$\sigma_z = \frac{M_x y}{I_{xx}}$$

in which

$$I_{xx} = 2 \times 216.6 \times 381.0^2 + 4 \times 216.6 \times 352.0^2 + 4 \times 216.6 \times 2695^2$$
$$+ 4 \times 216.7 \times 145.8^2 = 2.52 \times 10^8\,\text{mm}^4$$

The solution is completed in Table 22.1.

Table 22.1

Stringer/boom	y (mm)	σ_z (N/mm^2)
1	381.0	302.4
2, 16	352.0	279.4
3, 15	269.5	213.9
4, 14	145.8	115.7
5, 13	0	0
6, 12	−145.8	−115.7
7, 11	−269.5	−213.9
8, 10	−352.0	−279.4
9	−381.0	−302.4

22.2 Shear

For a fuselage having a cross-section of the type shown in Fig. 22.1(a), the determination of the shear flow distribution in the skin produced by shear is basically the analysis of an idealized single cell closed section beam. The shear flow distribution is therefore given by Eq. (20.11) in which the direct stress carrying capacity of the skin is assumed to be zero, i.e. $t_D = 0$, thus

$$q_s = -\left(\frac{S_x I_{xx} - S_y I_{xy}}{I_{xx} I_{yy} - I_{xy}^2}\right) \sum_{r=1}^{n} B_r x_r - \left(\frac{S_y I_{yy} - S_x I_{xy}}{I_{xx} I_{yy} - I_{xy}^2}\right) \sum_{r=1}^{n} B_r y_r + q_{s,0} \qquad (22.1)$$

Equation (22.1) is applicable to loading cases in which the shear loads are not applied through the section shear centre so that the effects of shear and torsion are included simultaneously. Alternatively, if the position of the shear centre is known, the loading system may be replaced by shear loads acting through the shear centre together with a pure torque, and the corresponding shear flow distributions may be calculated separately and then superimposed to obtain the final distribution.

Example 22.2

The fuselage of Example 22.1 is subjected to a vertical shear load of 100 kN applied at a distance of 150 mm from the vertical axis of symmetry as shown, for the idealized section, in Fig. 22.2. Calculate the distribution of shear flow in the section.

As in Example 22.1, $I_{xy} = 0$ and, since $S_x = 0$, Eq. (22.1) reduces to

$$q_s = -\frac{S_y}{I_{xx}} \sum_{r=1}^{n} B_r y_r + q_{s,0} \qquad \text{(i)}$$

in which $I_{xx} = 2.52 \times 10^8$ mm^4 as before. Then

$$q_s = \frac{-100 \times 10^3}{2.52 \times 10^8} \sum_{r=1}^{n} B_r y_r + q_{s,0}$$

or

$$q_s = -3.97 \times 10^{-4} \sum_{r=1}^{n} B_r y_r + q_{s,0} \qquad \text{(ii)}$$

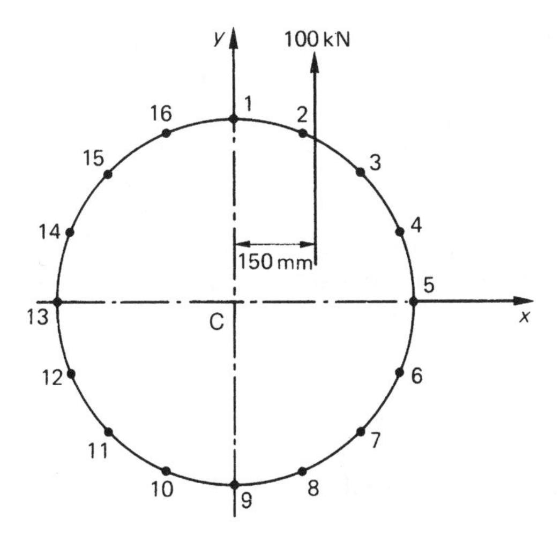

Fig. 22.2 Idealized fuselage section of Example 22.2.

Table 22.2

Skin panel	Boom	B_r (mm^2)	y_r (mm)	q_b (N/mm)
1 2	–	–	–	0
2 3	2	216.6	352.0	−30.3
3 4	3	216.6	269.5	−53.5
4 5	4	216.7	145.8	−66.0
5 6	5	–	0	−66.0
6 7	6	216.7	−145.8	−53.5
7 8	7	216.6	−269.5	−30.3
8 9	8	216.6	−352.0	0
1 16	1	216.6	381.0	−32.8
16 15	16	216.6	352.0	−63.1
15 14	15	216.6	269.5	−86.3
14 13	14	216.6	145.8	−98.8
13 12	13	–	0	−98.8
12 11	12	216.7	−145.8	−86.3
11 10	11	216.6	−269.5	−63.1
10 9	10	216.6	−352.0	−32.8

The first term on the right-hand side of Eq. (ii) is the 'open section' shear flow q_b. We therefore 'cut' one of the skin panels, say 12, and calculate q_b. The results are presented in Table 22.2.

Note that in Table 22.2, the column headed Boom indicates the boom that is crossed when the analysis moves from one panel to the next. Note also that, as would be expected, the q_b shear flow distribution is symmetrical about the Cx axis. The shear flow $q_{s,0}$ in the panel 12 is now found by taking moments about a convenient moment centre, say C. Therefore from Eq. (17.17)

$$100 \times 10^3 \times 150 = \oint q_b p \mathrm{d}s + 2Aq_{s,0} \tag{iii}$$

in which $A = \pi \times 381.0^2 = 4.56 \times 10^5\ \text{mm}^2$. Since the q_b shear flows are constant between the booms, Eq. (iii) may be rewritten in the form (see Eq. (20.10))

$$100 \times 10^3 \times 150 = -2A_{12}q_{b,12} - 2A_{23}q_{b,23} - \cdots - 2A_{161}q_{b,161} + 2Aq_{s,0} \qquad \text{(iv)}$$

in which $A_{12}, A_{23}, \ldots, A_{161}$ are the areas subtended by the skin panels 12, 23, ..., 161 at the centre C of the circular cross-section and anticlockwise moments are taken as positive. Clearly $A_{12} = A_{23} = \cdots = A_{161} = 4.56 \times 10^5/16 = 28\,500\ \text{mm}^2$. Equation (iv) then becomes

$$100 \times 10^3 \times 150 = 2 \times 28\,500(-q_{b_{12}} - q_{b_{23}} - \cdots - q_{b_{161}}) + 2 \times 4.56 \times 10^5 q_{s,0} \quad \text{(v)}$$

Substituting the values of q_b from Table 22.2 in Eq. (v), we obtain

$$100 \times 10^3 \times 150 = 2 \times 28\,500(-262.4) + 2 \times 4.56 \times 10^5 q_{s,0}$$

from which

$$q_{s,0} = 32.8\ \text{N/mm (acting in an anticlockwise sense)}$$

The complete shear flow distribution follows by adding the value of $q_{s,0}$ to the q_b shear flow distribution, giving the final distribution shown in Fig. 22.3. The solution may be checked by calculating the resultant of the shear flow distribution parallel to the Cy axis. Thus

$$2[(98.8 + 66.0)145.8 + (86.3 + 53.5)123.7 + (63.1 + 30.3)82.5$$
$$+ (32.8 - 0)29.0] \times 10^{-3} = 99.96\ \text{kN}$$

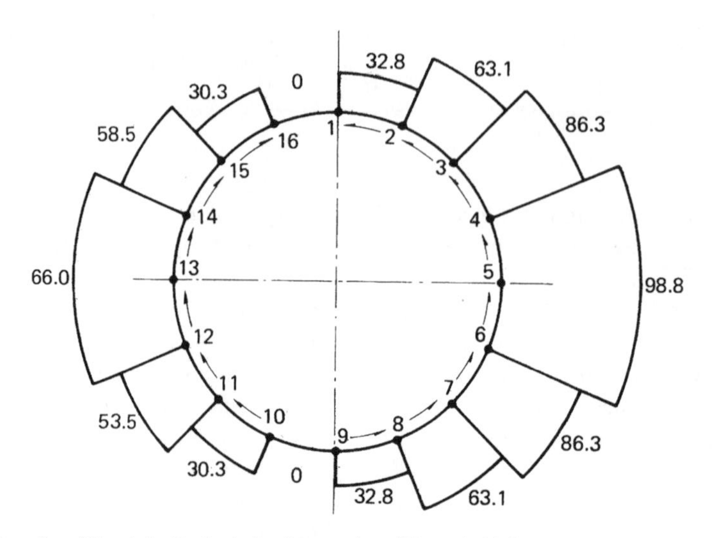

Fig. 22.3 Shear flow (N/mm) distribution in fuselage section of Example 22.2.

which agrees with the applied shear load of 100 kN. The analysis of a fuselage which is tapered along its length is carried out using the method described in Section 21.2 and illustrated in Example 21.2.

22.3 Torsion

A fuselage section is basically a single cell closed section beam. The shear flow distribution produced by a pure torque is therefore given by Eq. (18.1) and is

$$q = \frac{T}{2A} \tag{22.2}$$

It is immaterial whether or not the section has been idealized since, in both cases, the booms are assumed not to carry shear stresses.

Equation (22.2) provides an alternative approach to that illustrated in Example 22.2 for the solution of shear loaded sections in which the position of the shear centre is known. In Fig. 22.1 the shear centre coincides with the centre of symmetry so that the loading system may be replaced by the shear load of 100 kN acting through the shear centre together with a pure torque equal to $100 \times 10^3 \times 150 = 15 \times 10^6$ N mm as shown in Fig. 22.4. The shear flow distribution due to the shear load may be found using the method of Example 22.2 but with the left-hand side of the moment equation (iii) equal to zero for moments about the centre of symmetry. Alternatively, use may be made of the symmetry of the section and the fact that the shear flow is constant between adjacent booms. Suppose that the shear flow in the panel 21 is q_{21}. Then from symmetry and using the results of Table 22.2

$$q_{98} = q_{910} = q_{161} = q_{21}$$

$$q_{32} = q_{87} = q_{1011} = q_{1516} = 30.3 + q_{21}$$

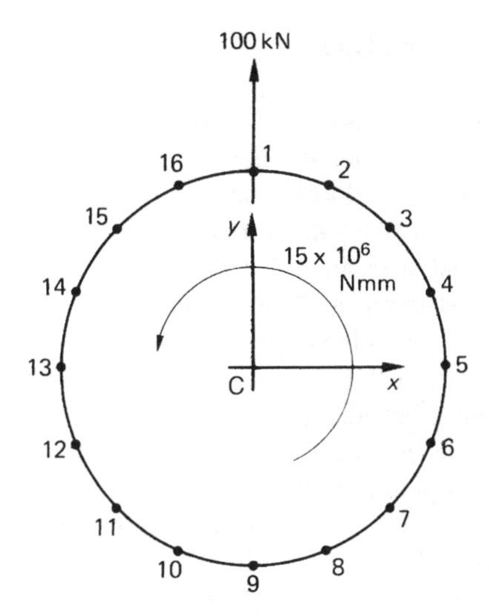

Fig. 22.4 Alternative solution of Example 22.2.

$$q_{43} = q_{76} = q_{11\,12} = q_{14\,15} = 53.5 + q_{21}$$

$$q_{54} = q_{65} = q_{12\,13} = q_{13\,14} = 66.0 + q_{21}$$

The resultant of these shear flows is statically equivalent to the applied shear load so that

$$4(29.0q_{21} + 82.5q_{32} + 123.7q_{43} + 145.8q_{54}) = 100 \times 10^3$$

Substituting for q_{32}, q_{43} and q_{54} from the above we obtain

$$4(381q_{21} + 18\,740.5) = 100 \times 10^3$$

whence

$$q_{21} = 16.4\,\text{N/mm}$$

and

$$q_{32} = 46.7\,\text{N/mm}, \quad q_{43} = 69.9\,\text{N/mm}, \quad q_{54} = 83.4\,\text{N/mm etc.}$$

The shear flow distribution due to the applied torque is, from Eq. (22.2)

$$q = \frac{15 \times 10^6}{2 \times 4.56 \times 10^5} = 16.4\,\text{N/mm}$$

acting in an anticlockwise sense completely around the section. This value of shear flow is now superimposed on the shear flows produced by the shear load; this gives the solution shown in Fig. 22.3, i.e.

$$q_{21} = 16.4 + 16.4 = 32.8\,\text{N/mm}$$

$$q_{161} = 16.4 - 16.4 = 0 \text{ etc.}$$

22.4 Cut-outs in fuselages

So far we have considered fuselages to be closed sections stiffened by transverse frames and longitudinal stringers. In practice it is necessary to provide openings in these closed stiffened shells for, for example, doors, cockpits, bomb bays, windows in passenger cabins, etc. These openings or 'cut-outs' produce discontinuities in the otherwise continuous shell structure so that loads are redistributed in the vicinity of the cut-out thereby affecting loads in the skin, stringers and frames. Frequently these regions must be heavily reinforced resulting in unavoidable weight increases. In some cases, for example door openings in passenger aircraft, it is not possible to provide rigid fuselage frames on each side of the opening because the cabin space must not be restricted. In such situations a rigid frame is placed around the opening to resist shear loads and to transmit loads from one side of the opening to the other.

The effects of smaller cut-outs, such as those required for rows of windows in passenger aircraft, may be found approximately as follows. Figure 22.5 shows a fuselage panel provided with cut-outs for windows which are spaced a distance l apart. The panel is subjected to an average shear flow q_{av} which would be the value of the shear

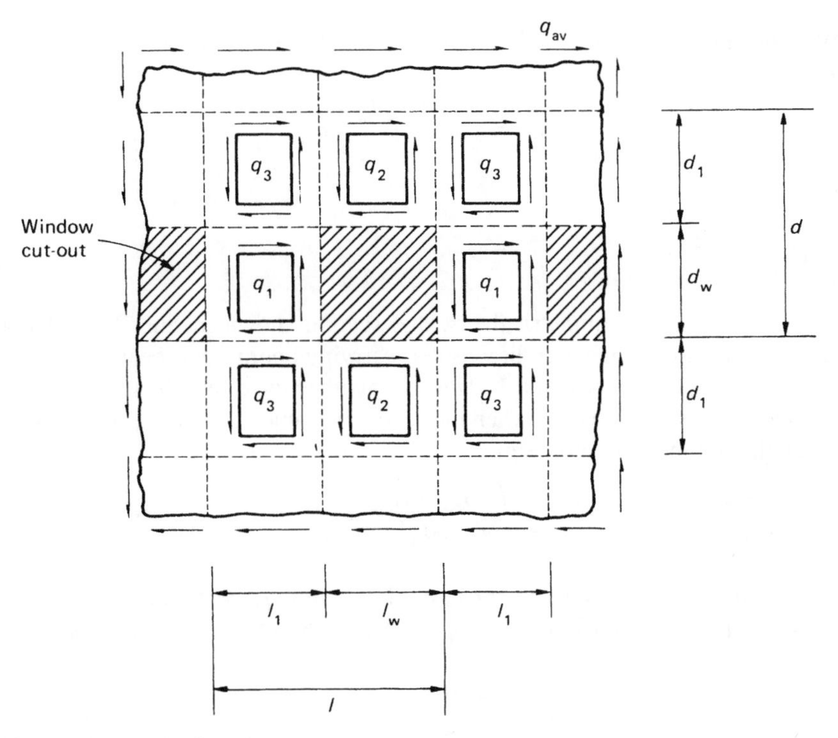

Fig. 22.5 Fuselage panel with windows.

flow in the panel without cut-outs. Considering a horizontal length of the panel through the cut-outs we see that

$$q_1 l_1 = q_{\mathrm{av}} l$$

or

$$q_1 = \frac{l}{l_1} q_{\mathrm{av}} \tag{22.3}$$

Now considering a vertical length of the panel through the cut-outs

$$q_2 d_1 = q_{\mathrm{av}} d$$

or

$$q_2 = \frac{d}{d_1} q_{\mathrm{av}} \tag{22.4}$$

The shear flows q_3 may be obtained by considering either vertical or horizontal sections not containing the cut-out. Thus

$$q_3 l_1 + q_2 l_{\mathrm{w}} = q_{\mathrm{av}} l$$

Substituting for q_2 from Eq. (22.3) and noting that $l = l_1 + l_w$ and $d = d_1 + d_w$, we obtain

$$q_3 = \left(1 - \frac{d_w}{d_l}\frac{l_w}{l_l}\right) q_{av} \tag{22.5}$$

Problems

P.22.1 The doubly symmetrical fuselage section shown in Fig. P.22.1 has been idealized into an arrangement of direct stress carrying booms and shear stress carrying skin panels; the boom areas are all 150 mm^2. Calculate the direct stresses in the booms and the shear flows in the panels when the section is subjected to a shear load of 50 kN and a bending moment of 100 kN m.

Ans. $\sigma_{z,1} = -\sigma_{z,6} = 180\ \text{N/mm}^2$, $\sigma_{z,2} = \sigma_{z,10} = -\sigma_{z,5} = -\sigma_{z,7} = 144.9\ \text{N/mm}^2$, $\sigma_{z,3} = \sigma_{z,9} = -\sigma_{z,4} = -\sigma_{z,8} = 60\ \text{N/mm}^2$.

$q_{21} = q_{65} = 1.9$ N/mm, $q_{32} = q_{54} = 12.8$ N/mm, $q_{43} = 17.3$ N/mm, $q_{67} = q_{101} = 11.6$ N/mm, $q_{78} = q_{910} = 22.5$ N/mm, $q_{89} = 27.0$ N/mm.

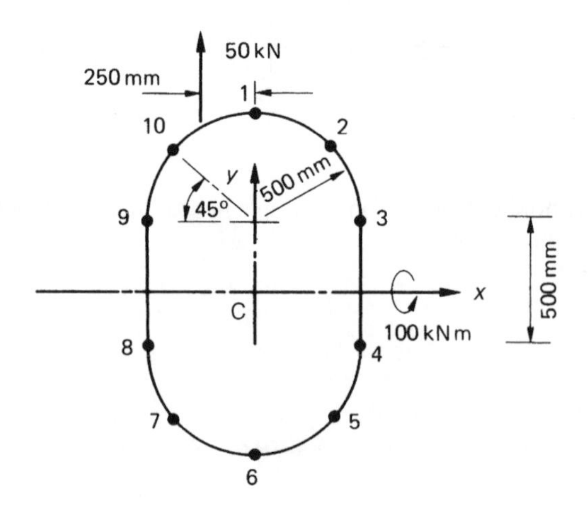

Fig. P.22.1

P.22.2 Determine the shear flow distribution in the fuselage section of P.22.1 by replacing the applied load by a shear load through the shear centre together with a pure torque.

23

Wings

We have seen in Chapters 12 and 20 that wing sections consist of thin skins stiffened by combinations of stringers, spar webs, and caps and ribs. The resulting structure frequently comprises one, two or more cells, and is highly redundant. However, as in the case of fuselage sections, the large number of closely spaced stringers allows the assumption of a constant shear flow in the skin between adjacent stringers so that a wing section may be analysed as though it were completely idealized as long as the direct stress carrying capacity of the skin is allowed for by additions to the existing stringer/boom areas. We shall investigate the analysis of multicellular wing sections subjected to bending, torsional and shear loads, although, initially, it will be instructive to examine the special case of an idealized three-boom shell.

23.1 Three-boom shell

The wing section shown in Fig. 23.1 has been idealized into an arrangement of direct stress carrying booms and shear–stress-only carrying skin panels. The part of the wing section aft of the vertical spar 31 performs an aerodynamic role only and is therefore

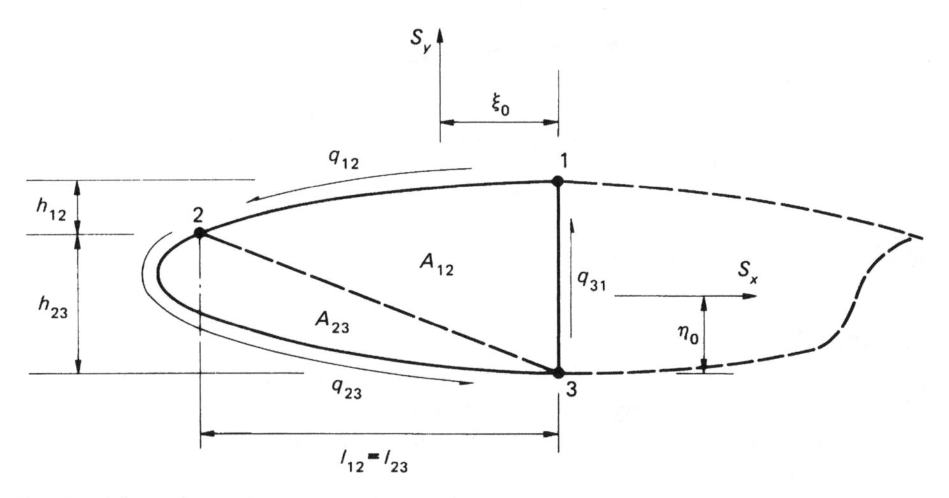

Fig. 23.1 Three-boom wing section.

unstressed. Lift and drag loads, S_y and S_x, induce shear flows in the skin panels which are constant between adjacent booms since the section has been completely idealized. Therefore, resolving horizontally and noting that the resultant of the internal shear flows is equivalent to the applied load, we have

$$S_x = -q_{12}l_{12} + q_{23}l_{23} \tag{23.1}$$

Now resolving vertically

$$S_y = q_{31}(h_{12} + h_{23}) - q_{12}h_{12} - q_{23}h_{23} \tag{23.2}$$

Finally, taking moments about, say, boom 3

$$S_x\eta_0 + S_y\xi_0 = -2A_{12}q_{12} - 2A_{23}q_{23} \tag{23.3}$$

(see Eqs (20.9) and (20.10)). In the above there are three unknown values of shear flow, q_{12}, q_{23}, q_{31} and three equations of statical equilibrium. We conclude therefore that a three-boom idealized shell is statically determinate.

We shall return to the simple case of a three-boom wing section when we examine the distributions of direct load and shear flows in wing ribs. Meanwhile, we shall consider the bending, torsion and shear of multicellular wing sections.

23.2 Bending

Bending moments at any section of a wing are usually produced by shear loads at other sections of the wing. The direct stress system for such a wing section (Fig. 23.2) is given by either Eqs (16.18) or (16.19) in which the coordinates (x, y) of any point in the cross-section and the sectional properties are referred to axes Cxy in which the origin C coincides with the centroid of the direct stress carrying area.

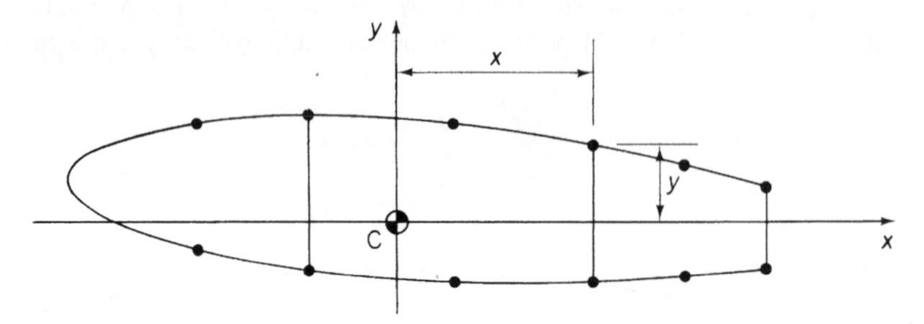

Fig. 23.2 Idealized section of a multicell wing.

Example 23.1

The wing section shown in Fig. 23.3 has been idealized such that the booms carry all the direct stresses. If the wing section is subjected to a bending moment of 300 kN m applied in a vertical plane, calculate the direct stresses in the booms.

Boom areas: $B_1 = B_6 = 2580\,\text{mm}^2$ $\quad B_2 = B_5 = 3880\,\text{mm}^2$ $\quad B_3 = B_4 = 3230\,\text{mm}^2$

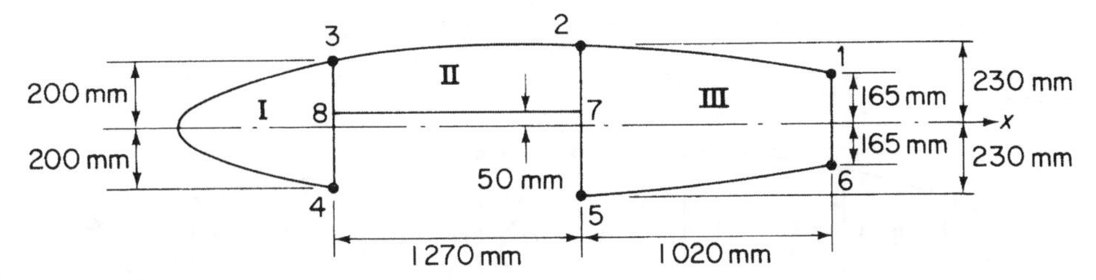

Fig. 23.3 Wing section of Example 23.1.

Table 23.1

Boom	y (mm)	σ_z (N/mm^2)
1	165	61.2
2	230	85.3
3	200	74.2
4	−200	−74.2
5	−230	−85.3
6	−165	−61.2

We note that the distribution of the boom areas is symmetrical about the horizontal x axis. Hence, in Eq. (16.18), $I_{xy}=0$. Further, $M_x=300$ kN m and $M_y=0$ so that Eq. (16.18) reduces to

$$\sigma_z = \frac{M_x y}{I_{xx}} \tag{i}$$

in which

$$I_{xy} = 2(2580 \times 165^2 + 3880 \times 230^2 + 3230 \times 200^2) = 809 \times 10^6 \text{ mm}^4$$

Hence

$$\sigma_z = \frac{300 \times 10^6}{809 \times 10^6} y = 0.371y \tag{ii}$$

The solution is now completed in Table 23.1 in which positive direct stresses are tensile and negative direct stresses compressive.

23.3 Torsion

The chordwise pressure distribution on an aerodynamic surface may be represented by shear loads (lift and drag loads) acting through the aerodynamic centre together with a pitching moment M_0 (see Section 12.1). This system of shear loads may be transferred to the shear centre of the section in the form of shear loads S_x and S_y together with a torque T. It is the pure torsion case that is considered here. In the analysis we assume that no axial constraint effects are present and that the shape of the wing section remains unchanged by the load application. In the absence of axial constraint there is no development of direct stress in the wing section so that only shear

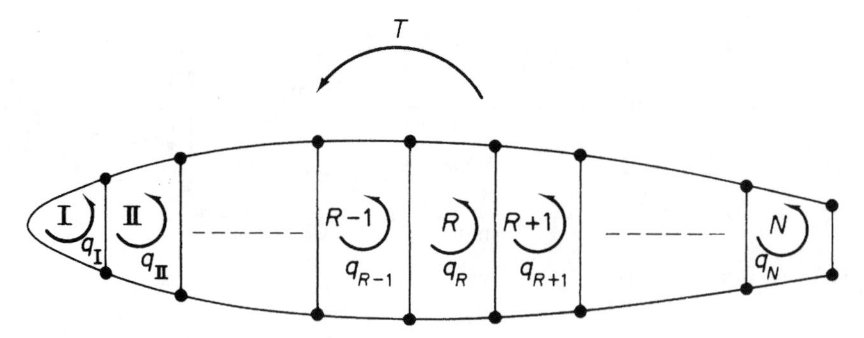

Fig. 23.4 Multicell wing section subjected to torsion.

stresses are present. It follows that the presence of booms does not affect the analysis in the pure torsion case.

The wing section shown in Fig. 23.4 comprises N cells and carries a torque T which generates individual but unknown torques in each of the N cells. Each cell therefore develops a constant shear flow $q_{\mathrm{I}}, q_{\mathrm{II}}, \ldots, q_R, \ldots, q_N$ given by Eq. (18.1).

The total is therefore

$$T = \sum_{R=1}^{N} 2A_R q_R \tag{23.4}$$

Although Eq. (23.4) is sufficient for the solution of the special case of a single cell section, which is therefore statically determinate, additional equations are required for an N-cell section. These are obtained by considering the rate of twist in each cell and the compatibility of displacement condition that all N cells possess the same rate of twist $\mathrm{d}\theta/\mathrm{d}z$; this arises directly from the assumption of an undistorted cross-section.

Consider the Rth cell of the wing section shown in Fig. 23.5. The rate of twist in the cell is, from Eq. (17.22)

$$\frac{\mathrm{d}\theta}{\mathrm{d}z} = \frac{1}{2A_R G} \oint_R q \frac{\mathrm{d}s}{t} \tag{23.5}$$

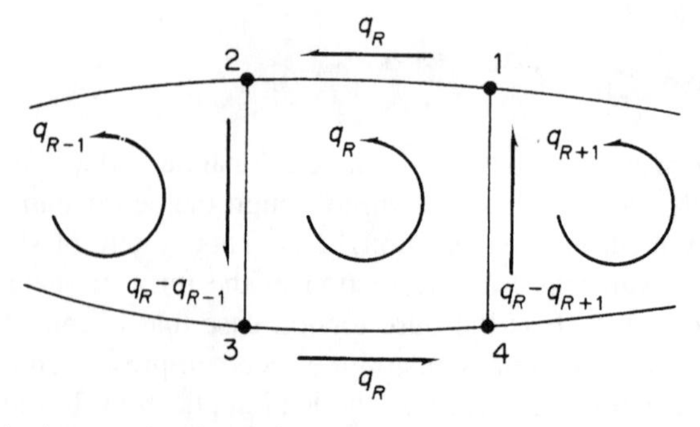

Fig. 23.5 Shear flow distribution in the Rth cell of an N-cell wing section.

The shear flow in Eq. (23.5) is constant along each wall of the cell and has the values shown in Fig. 23.5. Writing $\int ds/t$ for each wall as δ, Eq. (23.5) becomes

$$\frac{d\theta}{dz} = \frac{1}{2A_R G}[q_R\delta_{12} + (q_R - q_{R-1})\delta_{23} + q_R\delta_{34} + (q_R - q_{R+1})\delta_{41}]$$

or, rearranging the terms in square brackets

$$\frac{d\theta}{dz} = \frac{1}{2A_R G}[-q_{R-1}\delta_{23} + q_R(\delta_{12} + \delta_{23} + \delta_{34} + \delta_{41}) - q_{R+1}\delta_{41}]$$

In general terms, this equation may be rewritten in the form

$$\frac{d\theta}{dz} = \frac{1}{2A_R G}(-q_{R-1}\delta_{R-1,R} + q_R\delta_R - q_{R+1}\delta_{R+1,R}) \tag{23.6}$$

in which $\delta_{R-1,R}$ is $\int ds/t$ for the wall common to the Rth and $(R-1)$th cells, δ_R is $\int ds/t$ for all the walls enclosing the Rth cell and $\delta_{R+1,R}$ is $\int ds/t$ for the wall common to the Rth and $(R+1)$th cells.

The general form of Eq. (23.6) is applicable to multicell sections in which the cells are connected consecutively, i.e. cell I is connected to cell II, cell II to cells I and III and so on. In some cases, cell I may be connected to cells II and III, etc. (see problem P.23.4) so that Eq. (23.6) cannot be used in its general form. For this type of section the term $\oint q(ds/t)$ should be computed by considering $\int q(ds/t)$ for each wall of a particular cell in turn.

There are N equations of the type (23.6) which, with Eq. (23.4), comprise the $N+1$ equations required to solve for the N unknown values of shear flow and the one unknown value of $d\theta/dz$.

Frequently, in practice, the skin panels and spar webs are fabricated from materials possessing different properties such that the shear modulus G is not constant. The analysis of such sections is simplified if the actual thickness t of a wall is converted to a modulus-weighted thickness t^* as follows. For the Rth cell of an N-cell wing section in which G varies from wall to wall, Eq. (23.5) takes the form

$$\frac{d\theta}{dz} = \frac{1}{2A_R}\oint_R q\frac{ds}{Gt}$$

This equation may be rewritten as

$$\frac{d\theta}{dz} = \frac{1}{2A_R G_{REF}}\oint_R q\frac{ds}{(G/G_{REF})t} \tag{23.7}$$

in which G_{REF} is a convenient reference value of the shear modulus. Equation (23.7) is now rewritten as

$$\frac{d\theta}{dz} = \frac{1}{2A_R G_{REF}}\oint_R q\frac{ds}{t^*} \tag{23.8}$$

in which the modulus-weighted thickness t^* is given by

$$t^* = \frac{G}{G_{REF}}t \tag{23.9}$$

Then, in Eq. (23.6), δ becomes $\int ds/t^*$.

Example 23.2

Calculate the shear stress distribution in the walls of the three-cell wing section shown in Fig. 23.6, when it is subjected to an anticlockwise torque of 11.3 kN m.

Wall	Length (mm)	Thickness (mm)	G (N/mm^2)	Cell area (mm^2)
12°	1650	1.22	24 200	$A_I = 258\,000$
12^i	508	2.03	27 600	$A_{II} = 355\,000$
13, 24	775	1.22	24 200	$A_{III} = 161\,000$
34	380	1.63	27 600	
35, 46	508	0.92	20 700	
56	254	0.92	20 700	

Note: The superscript symbols o and i are used to distinguish between outer and inner walls connecting the same two booms.

Since the wing section is loaded by a pure torque the presence of the booms has no effect on the analysis.

Choosing $G_{REF} = 27\,600$ N/mm^2 then, from Eq. (23.9)

$$t^*_{12^o} = \frac{24\,200}{27\,600} \times 1.22 = 1.07 \text{ mm}$$

Similarly

$$t^*_{13} = t^*_{24} = 1.07 \text{ mm} \quad t^*_{35} = t^*_{46} = t^*_{56} = 0.69 \text{ mm}$$

Hence

$$\delta_{12^o} = \int_{12^o} \frac{ds}{t^*} = \frac{1650}{1.07} = 1542$$

Similarly

$$\delta_{12^i} = 250 \quad \delta_{13} = \delta_{24} = 725 \quad \delta_{34} = 233 \quad \delta_{35} = \delta_{46} = 736 \quad \delta_{56} = 368$$

Substituting the appropriate values of δ in Eq. (23.6) for each cell in turn gives the following:

- For cell I

$$\frac{d\theta}{dz} = \frac{1}{2 \times 258\,000 G_{REF}} [q_I(1542 + 250) - 250 q_{II}] \tag{i}$$

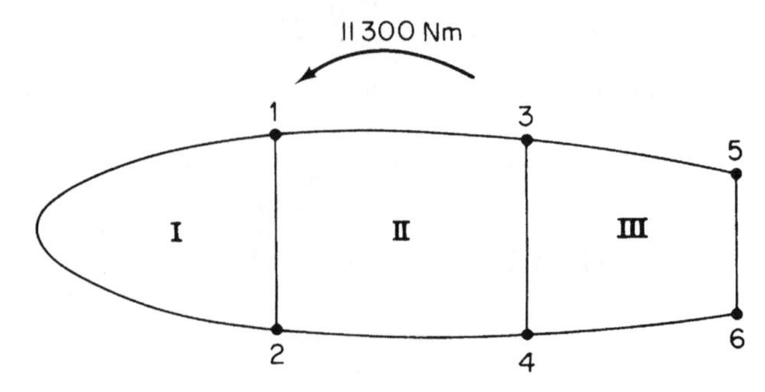

Fig. 23.6 Wing section of Example 23.2

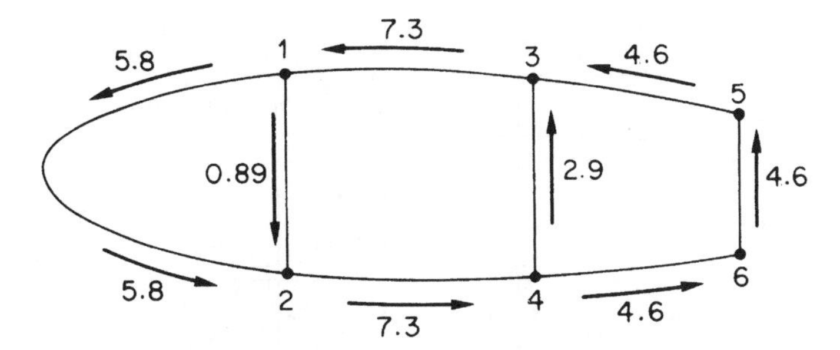

Fig. 23.7 Shear stress (N/mm²) distribution in wing section of Example 23.2.

- For cell II

$$\frac{d\theta}{dz} = \frac{1}{2 \times 355\,000 G_{REF}}[-250q_I + q_{II}(250 + 725 + 233 + 725) - 233q_{III}] \quad \text{(ii)}$$

- For cell III

$$\frac{d\theta}{dz} = \frac{1}{2 \times 161\,000 G_{REF}}[-233q_{II} + q_{III}(736 + 233 + 736 + 368)] \quad \text{(iii)}$$

In addition, from Eq. (23.4)

$$11.3 \times 10^6 = 2(258\,000q_I + 355\,000q_{II} + 161\,000q_{III}) \quad \text{(iv)}$$

Solving Eqs (i)–(iv) simultaneously gives

$$q_I = 7.1\,\text{N/mm} \quad q_{II} = 8.9\,\text{N/mm} \quad q_{III} = 4.2\,\text{N/mm}$$

The shear stress in any wall is obtained by dividing the shear flow by the *actual* wall thickness. Hence the shear stress distribution is as shown in Fig. 23.7.

23.4 Shear

Initially we shall consider the general case of an N-cell wing section comprising booms and skin panels, the latter being capable of resisting both direct and shear stresses. The wing section is subjected to shear loads S_x and S_y whose lines of action do not necessarily pass through the shear centre S (see Fig. 23.8); the resulting shear flow distribution is therefore due to the combined effects of shear and torsion.

The method for determining the shear flow distribution and the rate of twist is based on a simple extension of the analysis of a single cell beam subjected to shear loads (Sections 17.3 and 20.3). Such a beam is statically indeterminate, the single redundancy being selected as the value of shear flow at an arbitrarily positioned 'cut'. Thus, the N-cell wing section of Fig. 23.8 may be made statically determinate by 'cutting' a skin panel in each cell as shown. While the actual position of these 'cuts' is theoretically immaterial there are advantages to be gained from a numerical point of view if the 'cuts' are made near the centre of the top or bottom skin panel in each cell. Generally,

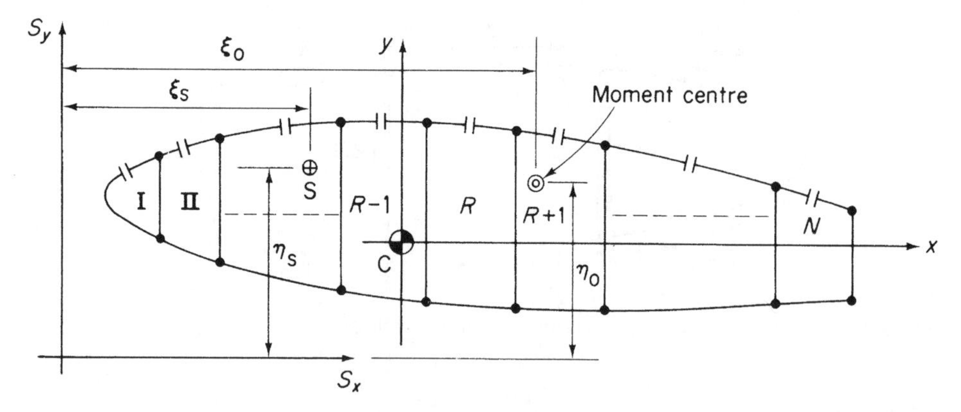

Fig. 23.8 *N*-cell wing section subjected to shear loads.

at these points, the redundant shear flows ($q_{s,0}$) are small so that the final shear flows differ only slightly from those of the determinate structure. The system of simultaneous equations from which the final shear flows are found will then be 'well conditioned' and will produce reliable results. The solution of an 'ill conditioned' system of equations would probably involve the subtraction of large numbers of a similar size which would therefore need to be expressed to a large number of significant figures for reasonable accuracy. Although this reasoning does not apply to a completely idealized wing section since the calculated values of shear flow are constant between the booms, it is again advantageous to 'cut' either the top or bottom skin panels for, in the special case of a wing section having a horizontal axis of symmetry, a 'cut' in, say, the top skin panels will result in the 'open section' shear flows (q_b) being zero in the bottom skin panels. This decreases the arithmetical labour and simplifies the derivation of the moment equation, as will become obvious in Example 23.4.

The 'open section' shear flow q_b in the wing section of Fig. 23.8 is given by Eq. (20.6), i.e.

$$q_\text{b} = -\left(\frac{S_x I_{xx} - S_y I_{xy}}{I_{xx}I_{yy} - I_{xy}^2}\right)\left(\int_0^s t_\text{D} x \,\text{d}s + \sum_{r=1}^{n} B_r x_r\right) - \left(\frac{S_y I_{yy} - S_x I_{xy}}{I_{xx}I_{yy} - I_{xy}^2}\right)\left(\int_0^s t_\text{D} y \,\text{d}s + \sum_{r=1}^{n} B_r y_r\right)$$

We are left with an unknown value of shear flow at each of the 'cuts', i.e. $q_{s,0,\text{I}}$, $q_{s,0,\text{II}}, \ldots, q_{s,0,N}$ plus the unknown rate of twist $\text{d}\theta/\text{d}z$ which, from the assumption of an undistorted cross-section, is the same for each cell. Therefore, as in the torsion case, there are $N+1$ unknowns requiring $N+1$ equations for a solution.

Consider the Rth cell shown in Fig. 23.9. The complete distribution of shear flow around the cell is given by the summation of the 'open section' shear flow q_b and the value of shear flow at the 'cut', $q_{s,0,R}$. We may therefore regard $q_{s,0,R}$ as a constant shear flow acting around the cell. The rate of twist is again given by Eq. (17.22); thus

$$\frac{\text{d}\theta}{\text{d}z} = \frac{1}{2A_R G}\oint_R q\frac{\text{d}s}{t} = \frac{1}{2A_R G}\oint_R (q_\text{b} + q_{s,0,R})\frac{\text{d}s}{t}$$

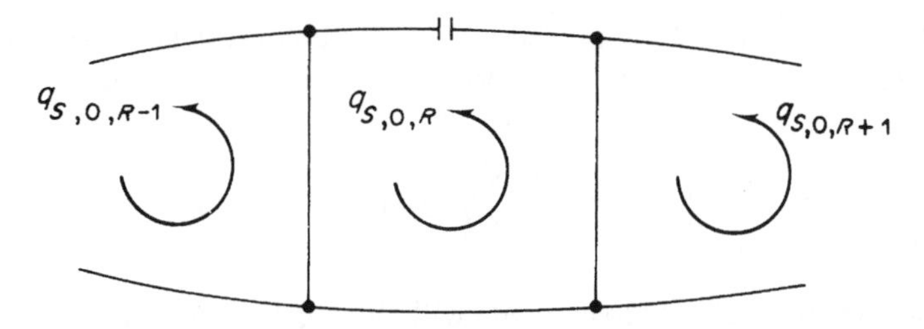

Fig. 23.9 Redundant shear flow in the *R*th cell of an *N*-cell wing section subjected to shear.

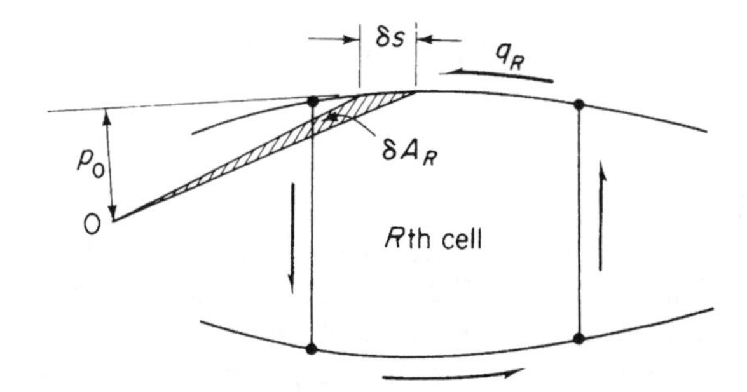

Fig. 23.10 Moment equilibrium of *R*th cell.

By comparison with the pure torsion case we deduce that

$$\frac{\mathrm{d}\theta}{\mathrm{d}z} = \frac{1}{2A_R G}\left(-q_{s,0,R-1}\delta_{R-1,R} + q_{s,0,R}\delta_R - q_{s,0,R+1}\delta_{R+1,R} + \oint_R q_b \frac{\mathrm{d}s}{t}\right) \tag{23.10}$$

in which q_b has previously been determined. There are N equations of the type (23.10) so that a further equation is required to solve for the $N+1$ unknowns. This is obtained by considering the moment equilibrium of the *R*th cell in Fig. 23.10.

The moment $M_{q,R}$ produced by the total shear flow about any convenient moment centre O is given by

$$M_{q,R} = \oint q_R p_0 \, \mathrm{d}s \quad \text{(see Section 18.1)}$$

Substituting for q_R in terms of the 'open section' shear flow q_b and the redundant shear flow $q_{s,0,R}$, we have

$$M_{q,R} = \oint_R q_b p_0 \, \mathrm{d}s + q_{s,0,R} \oint_R p_0 \, \mathrm{d}s$$

or

$$M_{q,R} = \oint_R q_b p_0 \, \mathrm{d}s + 2A_R q_{s,0,R}$$

The sum of the moments from the individual cells is equivalent to the moment of the externally applied loads about the same point. Thus, for the wing section of Fig. 23.8

$$S_x\eta_0 - S_y\xi_0 = \sum_{R=1}^{N} M_{q,R} = \sum_{R=1}^{N} \oint_R q_b p_0 \, \mathrm{d}s + \sum_{R=1}^{N} 2A_R q_{s,0,R} \tag{23.11}$$

If the moment centre is chosen to coincide with the point of intersection of the lines of action of S_x and S_y, Eq. (23.11) becomes

$$0 = \sum_{R=1}^{N} \oint_R q_b p_0 \, \mathrm{d}s + \sum_{R=1}^{N} 2A_R q_{s,0,R} \tag{23.12}$$

Example 23.3

The wing section of Example 23.1 (Fig. 23.3) carries a vertically upward shear load of 86.8 kN in the plane of the web 572. The section has been idealized such that the booms resist all the direct stresses while the walls are effective only in shear. If the shear modulus of all walls is 27 600 N/mm^2 except for the wall 78 for which it is three times this value, calculate the shear flow distribution in the section and the rate of twist. Additional data are given below.

Wall	Length (mm)	Thickness (mm)	Cell area (mm^2)
12, 56	1023	1.22	$A_I = 265\,000$
23	1274	1.63	$A_{II} = 213\,000$
34	2200	2.03	$A_{III} = 413\,000$
483	400	2.64	
572	460	2.64	
61	330	1.63	
78	1270	1.22	

Choosing G_{REF} as 27 600 N/mm^2 then, from Eq. (23.9)

$$t_{78}^* = \frac{3 \times 27\,600}{27\,600} \times 1.22 = 3.66\,\mathrm{mm}$$

Hence

$$\delta_{78} = \frac{1270}{3.66} = 347$$

Also

$$\delta_{12} = \delta_{56} = 840 \quad \delta_{23} = 783 \quad \delta_{34} = 1083 \quad \delta_{38} = 57 \quad \delta_{84} = 95 \quad \delta_{87} = 347$$
$$\delta_{27} = 68 \quad \delta_{75} = 106 \quad \delta_{16} = 202$$

We now 'cut' the top skin panels in each cell and calculate the 'open section' shear flows using Eq. (20.6) which, since the wing section is idealized, singly symmetrical (as far as the direct stress carrying area is concerned) and is subjected to a vertical shear load only, reduces to

$$q_b = \frac{-S_y}{I_{xx}} \sum_{r=1}^{n} B_r y_r \qquad \text{(i)}$$

where, from Example 23.1, $I_{xx} = 809 \times 10^6\,\text{mm}^4$. Thus, from Eq. (i)

$$q_b = -\frac{86.8 \times 10^3}{809 \times 10^6} \sum_{r=1}^{n} B_r y_r = -1.07 \times 10^{-4} \sum_{r=1}^{n} B_r y_r \qquad \text{(ii)}$$

Since $q_b = 0$ at each 'cut', then $q_b = 0$ for the skin panels 12, 23 and 34. The remaining q_b shear flows are now calculated using Eq. (ii). Note that the order of the numerals in the subscript of q_b indicates the direction of movement from boom to boom.

$$q_{b,27} = -1.07 \times 10^{-4} \times 3880 \times 230 = -95.5\,\text{N/mm}$$

$$q_{b,16} = -1.07 \times 10^{-4} \times 2580 \times 165 = -45.5\,\text{N/mm}$$

$$q_{b,65} = -45.5 - 1.07 \times 10^{-4} \times 2580 \times (-165) = 0$$

$$q_{b,57} = -1.07 \times 10^{-4} \times 3880 \times (-230) = 95.5\,\text{N/mm}$$

$$q_{b,38} = -1.07 \times 10^{-4} \times 3230 \times 200 = -69.0\,\text{N/mm}$$

$$q_{b,48} = -1.07 \times 10^{-4} \times 3230 \times (-200) = 69.0\,\text{N/mm}$$

Therefore, as $q_{b,83} = q_{b,48}$ (or $q_{b,72} = q_{b,57}$), $q_{b,78} = 0$. The distribution of the q_b shear flows is shown in Fig. 23.11. The values of δ and q_b are now substituted in Eq. (23.10) for each cell in turn.

- For cell I

$$\frac{d\theta}{dz} = \frac{1}{2 \times 265\,000 G_{\text{REF}}} [q_{s,0,\text{I}}(1083+95+57) - 57q_{s,0,\text{II}} + 69 \times 95 + 69 \times 57] \qquad \text{(iii)}$$

- For cell II

$$\frac{d\theta}{dz} = \frac{1}{2 \times 213\,000 G_{\text{REF}}} [-57q_{s,0,\text{I}} + q_{s,0,\text{II}}(783 + 57 + 347 + 68) - 68q_{s,0,\text{III}} + 95.5 \times 68 - 69 \times 57] \qquad \text{(iv)}$$

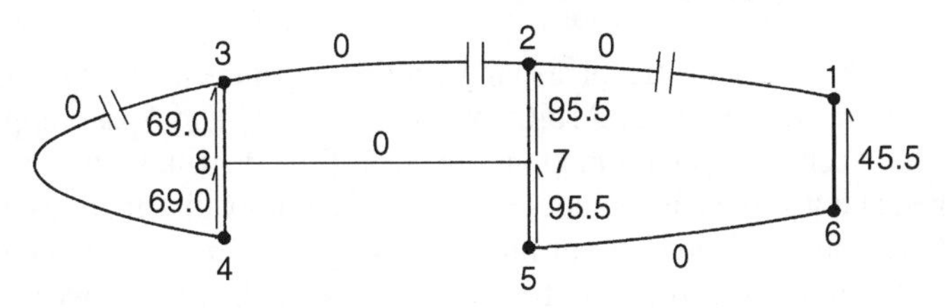

Fig. 23.11 q_b distribution (N/mm).

- For cell III

$$\frac{d\theta}{dz} = \frac{1}{2 \times 413\,000 G_{REF}}[-68q_{s,0,II} + q_{s,0,III}(840 + 68 + 106 + 840 + 202) + 45.5 \times 202 - 95.5 \times 68 - 95.5 \times 106] \quad \text{(v)}$$

The solely numerical terms in Eqs (iii)–(v) represent $\oint_R q_b(ds/t)$ for each cell. Care must be taken to ensure that the contribution of each q_b value to this term is interpreted correctly. The path of the integration follows the positive direction of $q_{s,0}$ in each cell, i.e. anticlockwise. Thus, the positive contribution of $q_{b,83}$ to $\oint_I q_b(ds/t)$ becomes a negative contribution to $\oint_{II} q_b(ds/t)$ and so on.

The fourth equation required for a solution is obtained from Eq. (23.12) by taking moments about the intersection of the x axis and the web 572. Thus

$$0 = -69.0 \times 250 \times 1270 - 69.0 \times 150 \times 1270 + 45.5 \times 330 \times 1020 + 2 \times 265\,000 q_{s,0,I} + 2 \times 213\,000 q_{s,0,II} + 2 \times 413\,000 q_{s,0,III} \quad \text{(vi)}$$

Simultaneous solution of Eqs (iii)–(vi) gives

$$q_{s,0,I} = 5.5\,\text{N/mm} \quad q_{s,0,II} = 10.2\,\text{N/mm} \quad q_{s,0,III} = 16.5\,\text{N/mm}$$

Superimposing these shear flows on the q_b distribution of Fig. 23.11, we obtain the final shear flow distribution. Thus

$$q_{34} = 5.5\,\text{N/mm} \quad q_{23} = q_{87} = 10.2\,\text{N/mm} \quad q_{12} = q_{56} = 16.5\,\text{N/mm}$$

$$q_{61} = 62.0\,\text{N/mm} \quad q_{57} = 79.0\,\text{N/mm} \quad q_{72} = 89.2\,\text{N/mm}$$

$$q_{48} = 74.5\,\text{N/mm} \quad q_{83} = 64.3\,\text{N/mm}$$

Finally, from any of Eqs (iii)–(v)

$$\frac{d\theta}{dz} = 1.16 \times 10^{-6}\,\text{rad/mm}$$

23.5 Shear centre

The position of the shear centre of a wing section is found in an identical manner to that described in Section 17.3. Arbitrary shear loads S_x and S_y are applied in turn through the shear centre S, the corresponding shear flow distributions determined and moments taken about some convenient point. The shear flow distributions are obtained as described previously in the shear of multicell wing sections except that the N equations of the type (23.10) are sufficient for a solution since the rate of twist $d\theta/dz$ is zero for shear loads applied through the shear centre.

23.6 Tapered wings

Wings are generally tapered in both spanwise and chordwise directions. The effects on the analysis of taper in a single cell beam have been discussed in Section 21.2. In a multicell wing section the effects are dealt with in an identical manner except that the moment equation (21.16) becomes, for an N-cell wing section (see Figs 21.5 and 23.8)

$$S_x\eta_0 - S_y\xi_0 = \sum_{R=1}^{N}\oint_R q_b p_0\,\mathrm{d}s + \sum_{R=1}^{N} 2A_R q_{s,0,R} - \sum_{r=1}^{m} P_{x,r}\eta_r + \sum_{r=1}^{m} P_{y,r}\xi_r \qquad (23.13)$$

Example 23.4

A two-cell beam has singly symmetrical cross-sections 1.2 m apart and tapers symmetrically in the y direction about a longitudinal axis (Fig. 23.12). The beam supports loads which produce a shear force $S_y = 10$ kN and a bending moment $M_x = 1.65$ kN m at the larger cross-section; the shear load is applied in the plane of the internal spar web. If booms 1 and 6 lie in a plane which is parallel to the yz plane calculate the forces in the booms and the shear flow distribution in the walls at the larger cross-section. The booms are assumed to resist all the direct stresses while the walls are effective only in shear. The shear modulus is constant throughout, the vertical webs are all 1.0 mm thick while the remaining walls are all 0.8 mm thick:

$$\text{Boom areas: } B_1 = B_3 = B_4 = B_6 = 600\,\text{mm}^2 \quad B_2 = B_5 = 900\,\text{mm}^2$$

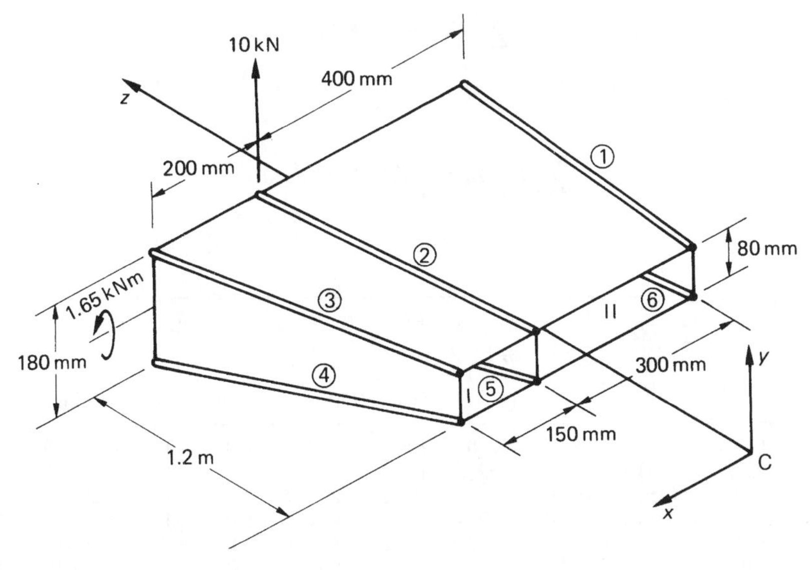

Fig. 23.12 Tapered beam of Example 23.4.

At the larger cross-section

$$I_{xx} = 4 \times 600 \times 90^2 + 2 \times 900 \times 90^2 = 34.02 \times 10^6\,\text{mm}^4$$

The direct stress in a boom is given by Eq. (16.18) in which $I_{xy} = 0$ and $M_y = 0$, i.e.

$$\sigma_{z,r} = \frac{M_x y_r}{I_{xx}}$$

whence

$$P_{z,r} = \frac{M_x y_r}{I_{xx}} B_r$$

or

$$P_{z,r} = \frac{1.65 \times 10^6 y_r B_r}{34.02 \times 10^6} = 0.08 y_r B_r \qquad \text{(i)}$$

The value of $P_{z,r}$ is calculated from Eq. (i) in column ② of Table 23.2; $P_{x,r}$ and $P_{y,r}$ follow from Eqs (21.10) and (21.9), respectively in columns ⑤ and ⑥. The axial load P_r is given by $[②^2 + ⑤^2 + ⑥^2]^{1/2}$ in column ⑦ and has the same sign as $P_{z,r}$ (see Eq. (21.12)). The moments of $P_{x,r}$ and $P_{y,r}$, columns ⑩ and ⑪, are calculated for a moment centre at the mid-point of the internal web taking anticlockwise moments as positive.
From column ⑤

$$\sum_{r=1}^{6} P_{x,r} = 0$$

(as would be expected from symmetry).
From column ⑥

$$\sum_{r=1}^{6} P_{y,r} = 764.4\,\text{N}$$

From column ⑩

$$\sum_{r=1}^{6} P_{x,r}\eta_r = -117\,846\,\text{N mm}$$

Table 23.2

①	②	③	④	⑤	⑥	⑦	⑧	⑨	⑩	⑪
Boom	$P_{z,r}$ (N)	$\frac{\delta x_r}{\delta z}$	$\frac{\delta y_r}{\delta z}$	$P_{x,r}$ (N)	$P_{y,r}$ (N)	P_r (N)	ξ_r (mm)	η_r (mm)	$P_{x,r}\eta_r$ (N mm)	$P_{y,r}\xi_r$ (N mm)
1	2619.0	0	0.0417	0	109.2	2621.3	400	90	0	43 680
2	3928.6	0.0833	0.0417	327.3	163.8	3945.6	0	90	−29 457	0
3	2619.0	0.1250	0.0417	327.4	109.2	2641.6	200	90	−29 466	21 840
4	−2619.0	0.1250	−0.0417	−327.4	109.2	−2641.6	200	90	−29 466	21 840
5	−3928.6	0.0833	−0.0417	−327.3	163.8	−3945.6	0	90	−29 457	0
6	−2619.0	0	−0.0417	0	109.2	−2621.3	400	90	0	−43 680

From column ⑪

$$\sum_{r=1}^{6} P_{y,r}\xi_r = -43\,680\,\text{N mm}$$

From Eq. (21.15)

$$S_{x,w} = 0 \quad S_{y,w} = 10 \times 10^3 - 764.4 = 9235.6\,\text{N}$$

Also, since Cx is an axis of symmetry, $I_{xy} = 0$ and Eq. (20.6) for the 'open section' shear flow reduces to

$$q_\text{b} = -\frac{S_{y,w}}{I_{xx}} \sum_{r=1}^{n} B_r y_r$$

or

$$q_\text{b} = -\frac{9235.6}{34.02 \times 10^6} \sum_{r=1}^{n} B_r y_r = -2.715 \times 10^{-4} \sum_{r=1}^{n} B_r y_r \qquad \text{(ii)}$$

'Cutting' the top walls of each cell and using Eq. (ii), we obtain the q_b distribution shown in Fig. 23.13. Evaluating δ for each wall and substituting in Eq. (23.10) gives for cell I

$$\frac{d\theta}{dz} = \frac{1}{2 \times 36\,000G}(760q_{s,0,\text{I}} - 180q_{s,0,\text{II}} - 1314) \qquad \text{(iii)}$$

for cell II

$$\frac{d\theta}{dz} = \frac{1}{2 \times 72\,000G}(-180q_{s,0,\text{I}} + 1160q_{s,0,\text{II}} + 1314) \qquad \text{(iv)}$$

Taking moments about the mid-point of web 25 we have, using Eq. (23.13)

$$0 = -14.7 \times 180 \times 400 + 14.7 \times 180 \times 200 + 2 \times 36\,000q_{s,0,\text{I}} + 2 \times 72\,000q_{s,0,\text{II}} - 117\,846 - 43\,680$$

or

$$0 = -690\,726 + 72\,000q_{s,0,\text{I}} + 144\,000q_{s,0,\text{II}} \qquad \text{(v)}$$

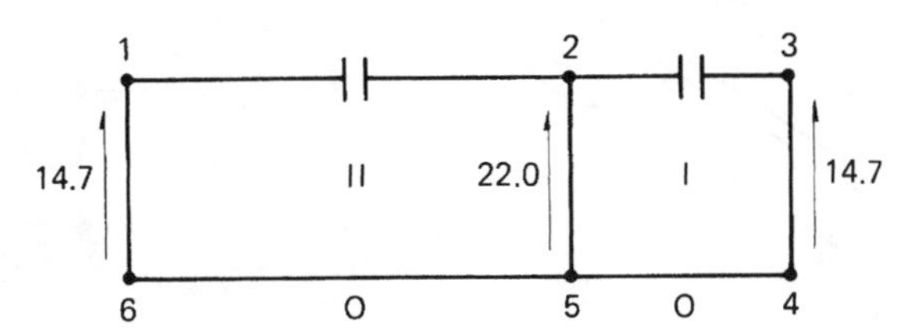

Fig. 23.13 q_b (N/mm) distribution in beam section of Example 23.4 (view along z axis towards C).

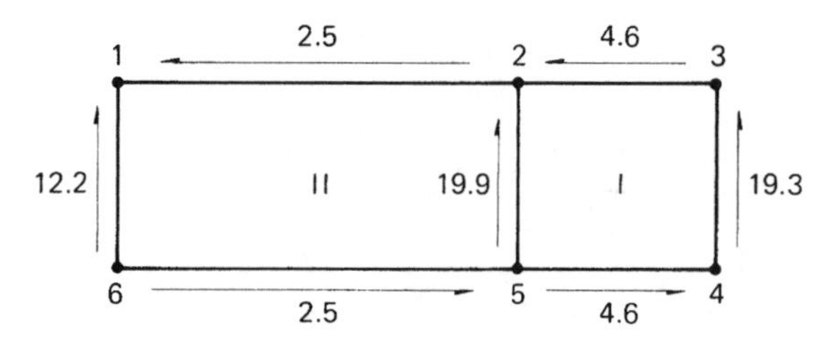

Fig. 23.14 Shear flow (N/mm) distribution in tapered beam of Example 23.4.

Solving Eqs (iii)–(v) gives

$$q_{s,0,\mathrm{I}} = 4.6\,\mathrm{N/mm} \quad q_{s,0,\mathrm{II}} = 2.5\,\mathrm{N/mm}$$

and the resulting shear flow distribution is shown in Fig. 23.14.

23.7 Deflections

Deflections of multicell wings may be calculated by the unit load method in an identical manner to that described in Section 20.4 for open and single cell beams.

Example 23.5

Calculate the deflection at the free end of the two-cell beam shown in Fig. 23.15 allowing for both bending and shear effects. The booms carry all the direct stresses while the skin panels, of constant thickness throughout, are effective only in shear.

$$\text{Take } E = 69\,000\,\mathrm{N/mm^2} \quad \text{and} \quad G = 25\,900\,\mathrm{N/mm^2}$$

$$\text{Boom areas: } B_1 = B_3 = B_4 = B_6 = 650\,\mathrm{mm^2} \quad B_2 = B_5 = 1300\,\mathrm{mm^2}$$

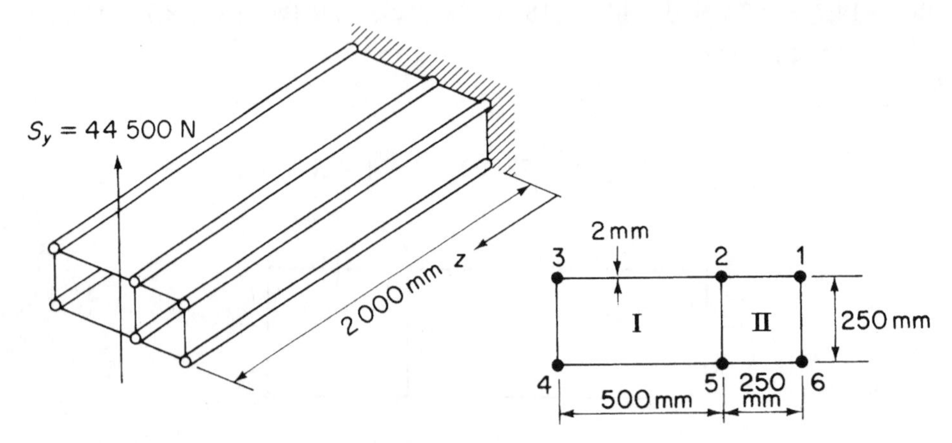

Fig. 23.15 Deflection of two-cell wing section.

The beam cross-section is symmetrical about a horizontal axis and carries a vertical load at its free end through the shear centre. The deflection Δ at the free end is then, from Eqs (20.17) and (20.19)

$$\Delta = \int_0^{2000} \frac{M_{x,0}M_{x,1}}{EI_{xx}}\mathrm{d}z + \int_0^{2000} \left(\int_{\text{section}} \frac{q_0 q_1}{Gt}\mathrm{d}s\right)\mathrm{d}z \tag{i}$$

where

$$M_{x,0} = -44.5 \times 10^3(2000 - z) \quad M_{x,1} = -(2000 - z)$$

and

$$I_{xx} = 4 \times 650 \times 125^2 + 2 \times 1300 \times 125^2 = 81.3 \times 10^6\,\text{mm}^4$$

also

$$S_{y,0} = 44.5 \times 10^3\,\text{N} \quad S_{y,1} = 1$$

The q_0 and q_1 shear flow distributions are obtained as previously described (note $\mathrm{d}\theta/\mathrm{d}z = 0$ for a shear load through the shear centre) and are

$$q_{0,12} = 9.6\,\text{N/mm} \quad q_{0,23} = -5.8\,\text{N/mm} \quad q_{0,43} = 50.3\,\text{N/mm}$$

$$q_{0,45} = -5.8\,\text{N/mm} \quad q_{0,56} = 9.6\,\text{N/mm} \quad q_{0,61} = 54.1\,\text{N/mm}$$

$$q_{0,52} = 73.6\,\text{N/mm at all sections of the beam}$$

The q_1 shear flows in this case are given by $q_0/44.5 \times 10^3$. Thus

$$\begin{aligned}\int_{\text{section}} \frac{q_0 q_1}{Gt}\mathrm{d}s &= \frac{1}{25\,900 \times 2 \times 44.5 \times 10^3}(9.6^2 \times 250 \times 2 + 5.8^2 \times 500 \times 2 \\ &\quad + 50.3^2 \times 250 + 54.1^2 \times 250 + 73.6^2 \times 250) \\ &= 1.22 \times 10^{-3}\end{aligned}$$

Hence, from Eq. (i)

$$\Delta = \int_0^{2000} \frac{44.5 \times 10^3(2000 - z)^2}{69\,000 \times 81.3 \times 10^6}\mathrm{d}z + \int_0^{2000} 1.22 \times 10^{-3}\mathrm{d}z$$

giving

$$\Delta = 23.5\,\text{mm}$$

23.8 Cut-outs in wings

Wings, as well as fuselages, have openings in their surfaces to accommodate undercarriages, engine nacelles and weapons installations, etc. In addition inspection panels are required at specific positions so that, as for fuselages, the loads in adjacent portions of the wing structure are modified.

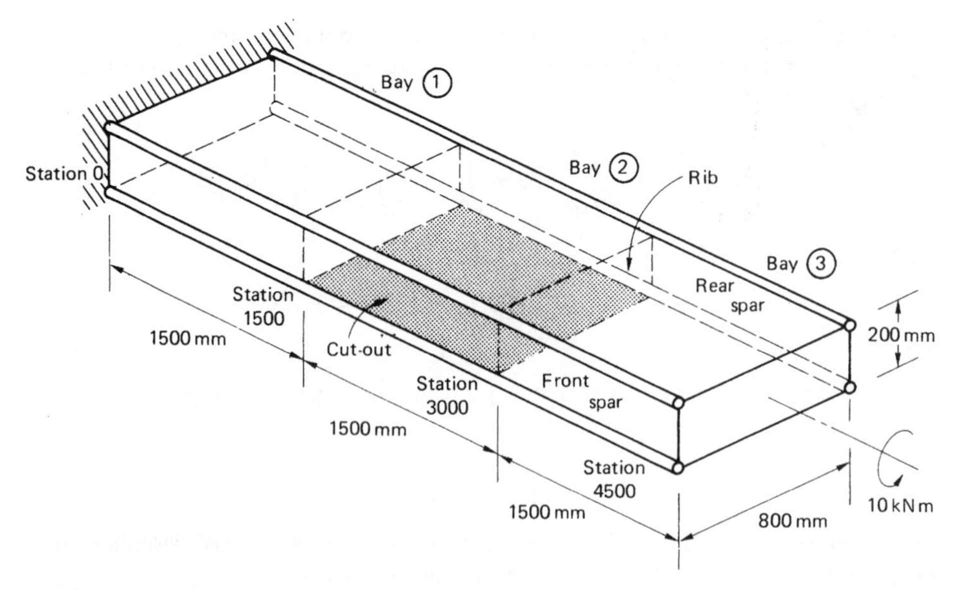

Fig. 23.16 Three-bay wing structure with cut-out of Example 23.6.

Initially we shall consider the case of a wing subjected to a pure torque in which one bay of the wing has the skin on its undersurface removed. The method is best illustrated by a numerical example.

Example 23.6

The structural portion of a wing consists of a three-bay rectangular section box which may be assumed to be firmly attached at all points around its periphery to the aircraft fuselage at its inboard end. The skin on the undersurface of the central bay has been removed and the wing is subjected to a torque of 10 kN m at its tip (Fig. 23.16). Calculate the shear flows in the skin panels and spar webs, the loads in the corner flanges and the forces in the ribs on each side of the cut-out assuming that the spar flanges carry all the direct loads while the skin panels and spar webs are effective only in shear.

If the wing structure were continuous and the effects of restrained warping at the built-in end ignored, the shear flows in the skin panels would be given by Eq. (18.1), i.e.

$$q = \frac{T}{2A} = \frac{10 \times 10^6}{2 \times 200 \times 800} = 31.3\,\text{N/mm}$$

and the flanges would be unloaded. However, the removal of the lower skin panel in bay ② results in a torsionally weak channel section for the length of bay ② which must in any case still transmit the applied torque to bay ① and subsequently to the wing support points. Although open section beams are inherently weak in torsion (see Section 18.2), the channel section in this case is attached at its inboard and outboard ends to torsionally stiff closed boxes so that, in effect, it is built-in at both ends. We shall examine the effect of axial constraint on open section beams subjected to torsion in Chapter 27. An alternative approach is to assume that the torque is transmitted

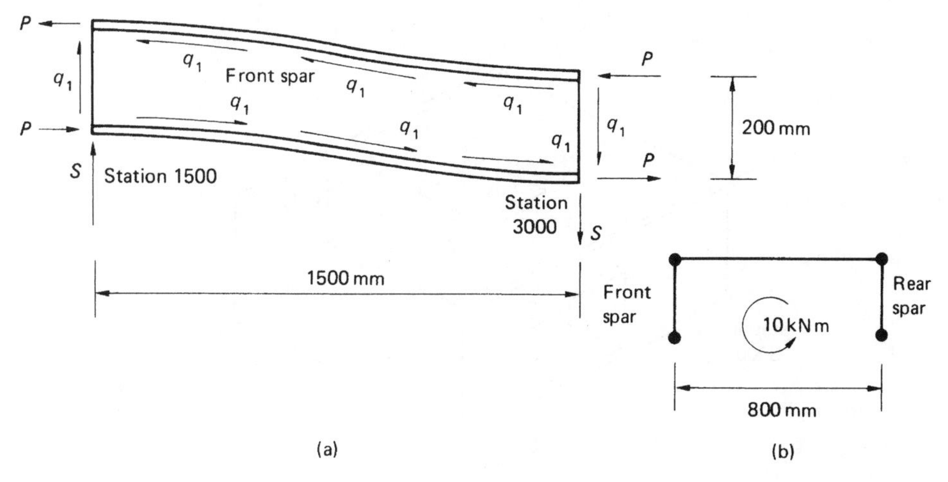

Fig. 23.17 Differential bending of front spar.

across bay ② by the differential bending of the front and rear spars. The bending moment in each spar is resisted by the flange loads P as shown, for the front spar, in Fig. 23.17(a). The shear loads in the front and rear spars form a couple at any station in bay ② which is equivalent to the applied torque. Thus, from Fig. 23.17(b)

$$800S = 10 \times 10^6 \text{ N mm}$$

i.e.

$$S = 12\,500 \text{ N}$$

The shear flow q_1 in Fig. 23.17(a) is given by

$$q_1 = \frac{12\,500}{200} = 62.5 \text{ N/mm}$$

Midway between stations 1500 and 3000 a point of contraflexure occurs in the front and rear spars so that at this point the bending moment is zero. Hence

$$200P = 12\,500 \times 750 \text{ N mm}$$

so that

$$P = 46\,875 \text{ N}$$

Alternatively, P may be found by considering the equilibrium of either of the spar flanges. Thus

$$2P = 1500q_1 = 1500 \times 62.5 \text{ N}$$

whence

$$P = 46\,875 \text{ N}$$

The flange loads P are reacted by loads in the flanges of bays ① and ③. These flange loads are transmitted to the adjacent spar webs and skin panels as shown

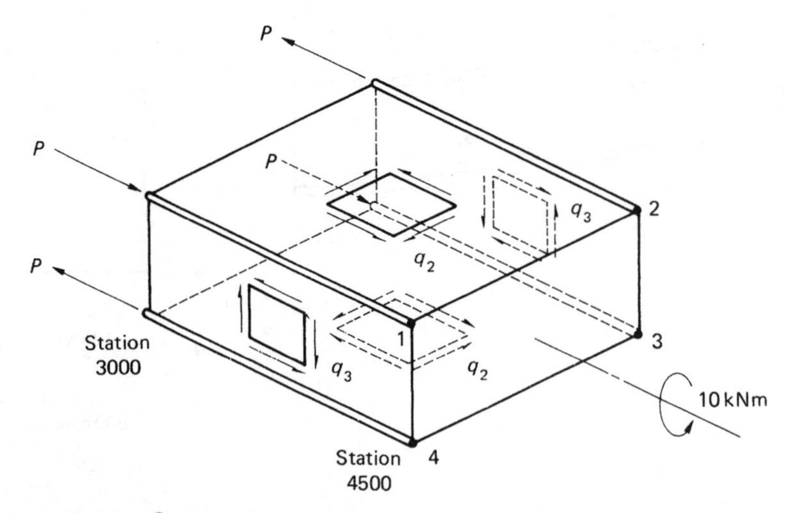

Fig. 23.18 Loads on bay ③ of the wing of Example 23.6.

in Fig. 23.18 for bay ③ and modify the shear flow distribution given by Eq. (18.1). For equilibrium of flange 1

$$1500q_2 - 1500q_3 = P = 46\,875\ \text{N}$$

or

$$q_2 - q_3 = 31.3 \qquad \text{(i)}$$

The resultant of the shear flows q_2 and q_3 must be equivalent to the applied torque. Hence, for moments about the centre of symmetry at any section in bay ③ and using Eq. (20.10)

$$200 \times 800q_2 + 200 \times 800q_3 = 10 \times 10^6\ \text{N mm}$$

or

$$q_2 + q_3 = 62.5 \qquad \text{(ii)}$$

Solving Eqs (i) and (ii) we obtain

$$q_2 = 46.9\ \text{N/mm} \quad q_3 = 15.6\ \text{N/mm}$$

Comparison with the results of Eq. (18.1) shows that the shear flows are increased by a factor of 1.5 in the upper and lower skin panels and decreased by a factor of 0.5 in the spar webs.

The flange loads are in equilibrium with the resultants of the shear flows in the adjacent skin panels and spar webs. Thus, for example, in the top flange of the front spar

$$P(\text{st}.4500) = 0$$

$$P(\text{st}.3000) = 1500q_2 - 1500q_3 = 46\,875\ \text{N (compression)}$$

$$P(\text{st}.2250) = 1500q_2 - 1500q_3 - 750q_1 = 0$$

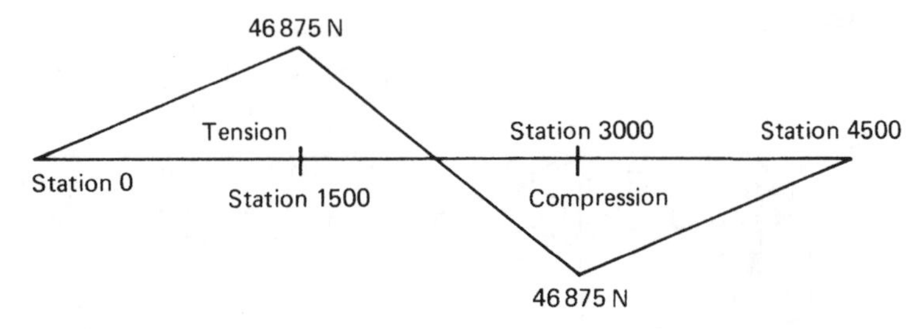

Fig. 23.19 Distribution of load in the top flange of the front spar of the wing of Example 23.6.

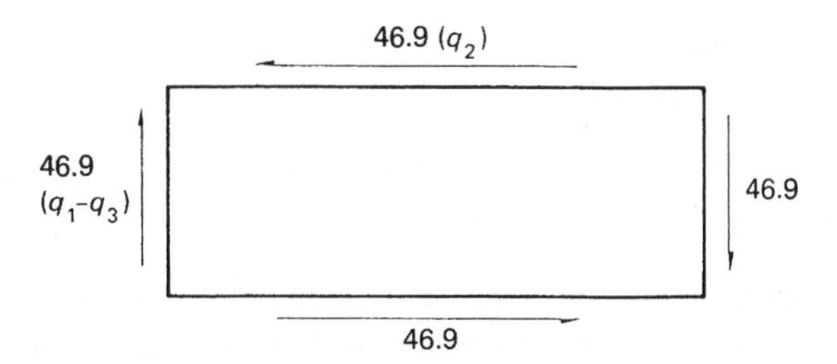

Fig. 23.20 Shear flows (N/mm) on wing rib at station 3000 in the wing of Example 23.6.

The loads along the remainder of the flange follow from antisymmetry giving the distribution shown in Fig. 23.19. The load distribution in the bottom flange of the rear spar will be identical to that shown in Fig. 23.19 while the distributions in the bottom flange of the front spar and the top flange of the rear spar will be reversed. We note that the flange loads are zero at the built-in end of the wing (station 0). Generally, however, additional stresses are induced by the warping restraint at the built-in end; these are investigated in Chapter 26. The loads on the wing ribs on either the inboard or outboard end of the cut-out are found by considering the shear flows in the skin panels and spar webs immediately inboard and outboard of the rib. Thus, for the rib at station 3000 we obtain the shear flow distribution shown in Fig. 23.20.

In Example 23.6 we implicitly assumed in the analysis that the local effects of the cut-out were completely dissipated within the length of the adjoining bays which were equal in length to the cut-out bay. The validity of this assumption relies on St. Venant's principle (Section 2.4). It may generally be assumed therefore that the effects of a cut-out are restricted to spanwise lengths of the wing equal to the length of the cut-out on both inboard and outboard ends of the cut-out bay.

We shall now consider the more complex case of a wing having a cut-out and subjected to shear loads which produce both bending and torsion. Again the method is illustrated by a numerical example.

Example 23.7

A wing box has the skin panel on its undersurface removed between stations 2000 and 3000 and carries lift and drag loads which are constant between stations 1000 and 4000 as shown in Fig. 23.21(a). Determine the shear flows in the skin panels and spar webs and also the loads in the wing ribs at the inboard and outboard ends of the cut-out bay.

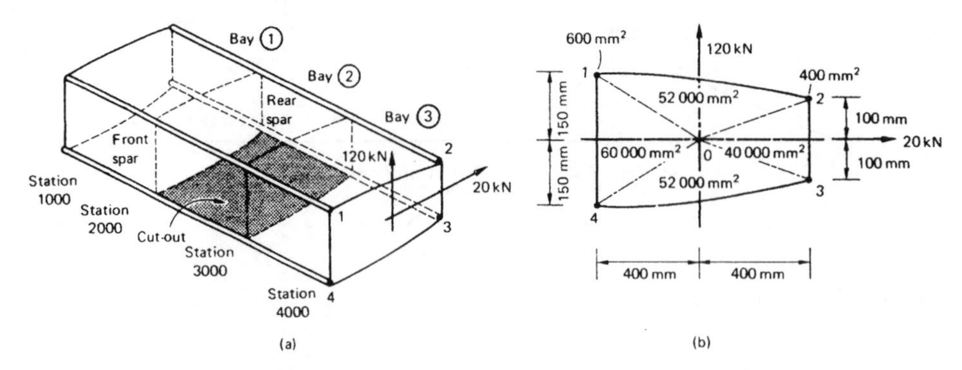

Fig. 23.21 Wing box of Example 23.7.

Assume that all bending moments are resisted by the spar flanges while the skin panels and spar webs are effective only in shear.

The simplest approach is first to determine the shear flows in the skin panels and spar webs as though the wing box were continuous and then to apply an equal and opposite shear flow to that calculated around the edges of the cut-out. The shear flows in the wing box without the cut-out will be the same in each bay and are calculated using the method described in Section 20.3 and illustrated in Example 20.4. This gives the shear flow distribution shown in Fig. 23.22.

We now consider bay ② and apply a shear flow of 75.9 N/mm in the wall 34 in the opposite sense to that shown in Fig. 23.22. This reduces the shear flow in the wall 34 to zero and, in effect, restores the cut-out to bay ②. The shear flows in the remaining walls of the cut-out bay will no longer be equivalent to the externally applied shear loads so that corrections are required. Consider the cut-out bay (Fig. 23.23) with the shear flow of 75.9 N/mm applied in the opposite sense to that shown in Fig. 23.22. The correction shear flows q'_{12}, q'_{32} and q'_{14} may be found using statics. Thus, resolving forces horizontally we have

$$800q'_{12} = 800 \times 75.9\,\text{N}$$

whence

$$q'_{12} = 75.9\,\text{N/mm}$$

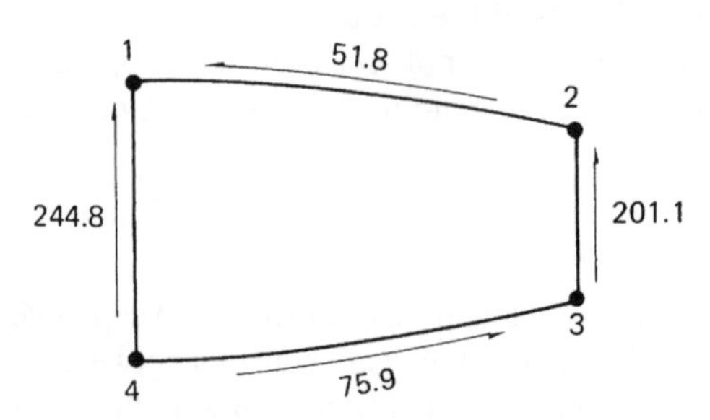

Fig. 23.22 Shear flow (N/mm) distribution at any station in the wing box of Example 23.7 without cut-out.

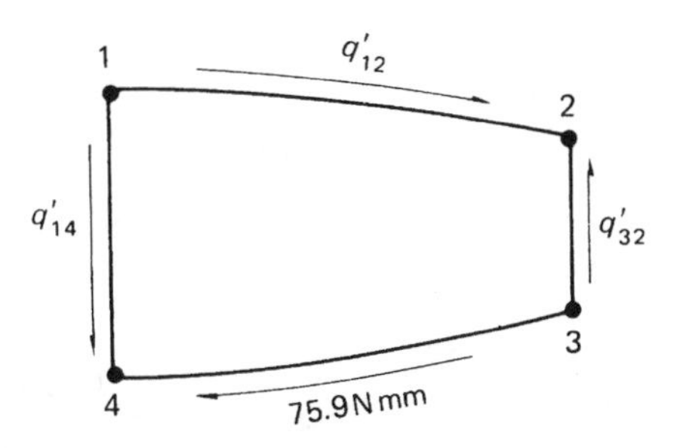

Fig. 23.23 Correction shear flows in the cut-out bay of the wing box of Example 23.7.

Resolving forces vertically

$$200q'_{32} = 50q'_{12} - 50 \times 75.9 - 300q'_{14} = 0 \tag{i}$$

and taking moments about O in Fig. 23.21(b) we obtain

$$2 \times 52\,000q'_{12} - 2 \times 40\,000q'_{32} + 2 \times 52\,000 \times 75.9 - 2 \times 60\,000q'_{14} = 0 \tag{ii}$$

Solving Eqs (i) and (ii) gives

$$q'_{32} = 117.6\,\text{N/mm} \quad q'_{14} = 53.1\,\text{N/mm}$$

The final shear flows in bay ② are found by superimposing q'_{12}, q'_{32} and q'_{14} on the shear flows in Fig. 23.22, giving the distribution shown in Fig. 23.24. Alternatively, these shear flows could have been found directly by considering the equilibrium of the cut-out bay under the action of the applied shear loads.

The correction shear flows in bay ② (Fig. 23.23) will also modify the shear flow distributions in bays ① and ③. The correction shear flows to be applied to those shown in Fig. 23.22 for bay ③ (those in bay ① will be identical) may be found by determining the flange loads corresponding to the correction shear flows in bay ②.

It can be seen from the magnitudes and directions of these correction shear flows (Fig. 23.23) that at any section in bay ② the loads in the upper and lower flanges of the front spar are equal in magnitude but opposite in direction; similarly for the rear spar. Thus, the correction shear flows in bay ② produce an identical system of flange loads to that shown in Fig. 23.17 for the cut-out bays in the wing structure of Example 23.6.

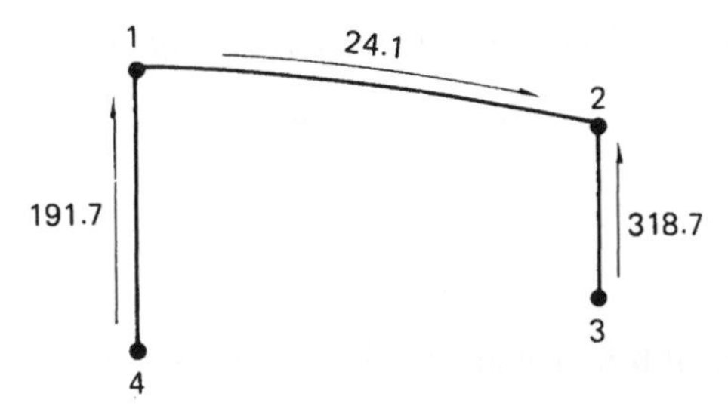

Fig. 23.24 Final shear flows (N/mm) in the cut-out bay of the wing box of Example 23.7.

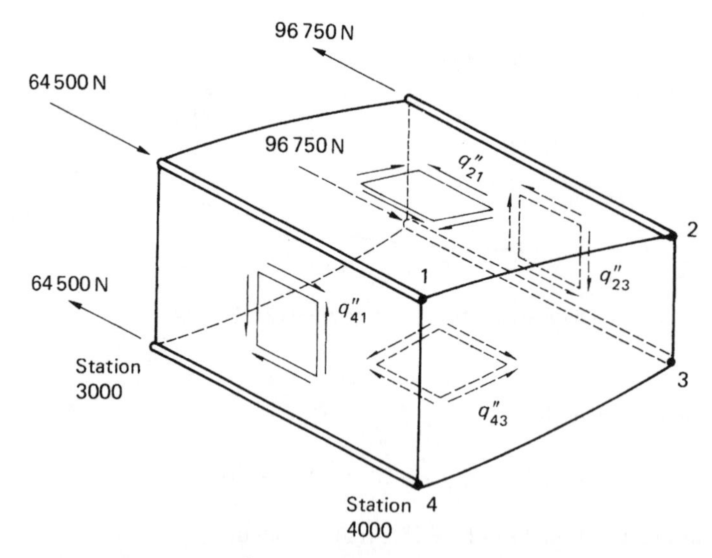

Fig. 23.25 Correction shear flows in bay ③ of the wing box of Example 23.7.

It follows that these correction shear flows produce differential bending of the front and rear spars in bay ② and that the spar bending moments and hence the flange loads are zero at the mid-bay points. Therefore, at station 3000 the flange loads are

$$P_1 = (75.9 + 53.1) \times 500 = 64\,500\,\text{N (compression)}$$

$$P_4 = 64\,500\,\text{N (tension)}$$

$$P_2 = (75.9 + 117.6) \times 500 = 96\,750\,\text{N (tension)}$$

$$P_3 = 96\,750\,\text{N (tension)}$$

These flange loads produce correction shear flows q''_{21}, q''_{43}, q''_{23} and q''_{41} in the skin panels and spar webs of bay ③ as shown in Fig. 23.25. Thus for equilibrium of flange 1

$$1000q''_{41} + 1000q''_{21} = 64\,500\,\text{N} \tag{iii}$$

and for equilibrium of flange 2

$$1000q''_{21} + 1000q''_{23} = 96\,750\,\text{N} \tag{iv}$$

For equilibrium in the chordwise direction at any section in bay ③

$$800q''_{21} = 800q''_{43}$$

or

$$q''_{21} = q''_{43} \tag{v}$$

Finally, for vertical equilibrium at any section in bay ③

$$300q''_{41} + 50q''_{43} + 50q''_{21} - 200q''_{23} = 0 \tag{vi}$$

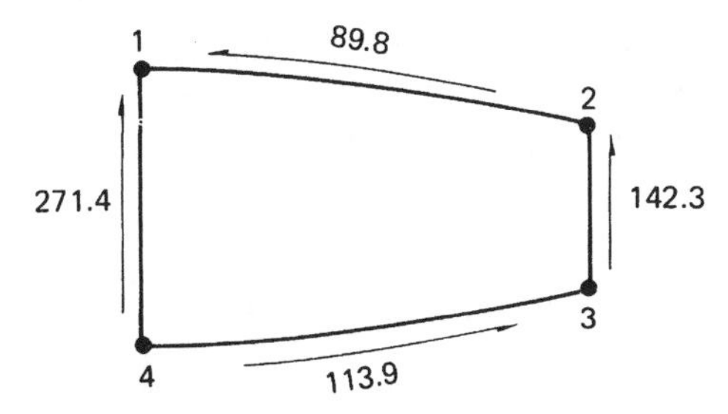

Fig. 23.26 Final shear flows in bay ③ (and bay ①) of the wing box of Example 23.7.

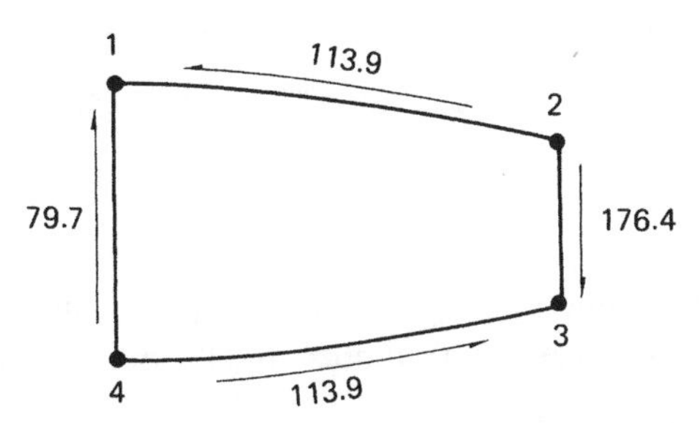

Fig. 23.27 Shear flows (N/mm) applied to the wing rib at station 3000 in the wing box of Example 23.7.

Simultaneous solution of Eqs (iii)–(vi) gives

$$q''_{21} = q''_{43} = 38.0\,\text{N/mm} \quad q''_{23} = 58.8\,\text{N/mm} \quad q''_{41} = 26.6\,\text{N/mm}$$

Superimposing these correction shear flows on those shown in Fig. 23.22 gives the final shear flow distribution in bay ③ as shown in Fig. 23.26. The rib loads at stations 2000 and 3000 are found as before by adding algebraically the shear flows in the skin panels and spar webs on each side of the rib. Thus, at station 3000 we obtain the shear flows acting around the periphery of the rib as shown in Fig. 23.27. The shear flows applied to the rib at the inboard end of the cut-out bay will be equal in magnitude but opposite in direction.

Note that in this example only the shear loads on the wing box between stations 1000 and 4000 are given. We cannot therefore determine the final values of the loads in the spar flanges since we do not know the values of the bending moments at these positions caused by loads acting on other parts of the wing.

Problems

P.23.1 The central cell of a wing has the idealized section shown in Fig. P.23.1. If the lift and drag loads on the wing produce bending moments of $-120\,000$ N m and $-30\,000$ N m, respectively at the section shown, calculate the direct stresses in the booms. Neglect axial constraint effects and assume that the lift and drag vectors are in vertical and horizontal planes.

$$\text{Boom areas: } B_1 = B_4 = B_5 = B_8 = 1000\,\text{mm}^2$$
$$B_2 = B_3 = B_6 = B_7 = 600\,\text{mm}^2$$

Ans. $\sigma_1 = -190.7\,\text{N/mm}^2$ $\sigma_2 = -181.7\,\text{N/mm}^2$ $\sigma_3 = -172.8\,\text{N/mm}^2$
$\sigma_4 = -163.8\,\text{N/mm}^2$ $\sigma_5 = 140\,\text{N/mm}^2$ $\sigma_6 = 164.8\,\text{N/mm}^2$
$\sigma_7 = 189.6\,\text{N/mm}^2$ $\sigma_8 = 214.4\,\text{N/mm}^2$.

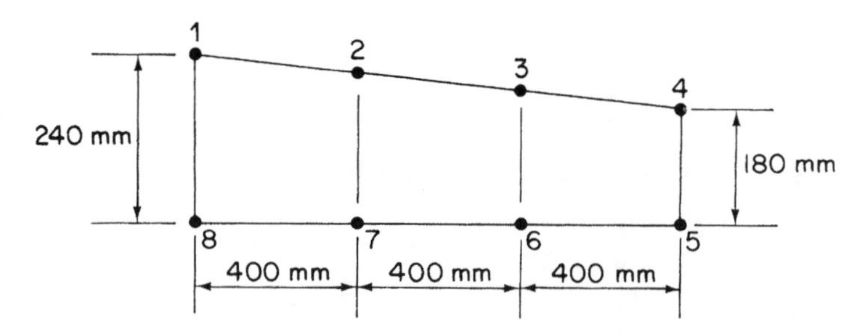

Fig. P.23.1

P.23.2 Figure P.23.2 shows the cross-section of a two-cell torque box. If the shear stress in any wall must not exceed $140\,\text{N/mm}^2$, find the maximum torque which can be applied to the box.

If this torque were applied at one end and resisted at the other end of such a box of span 2500 mm, find the twist in degrees of one end relative to the other and the torsional rigidity of the box. The shear modulus $G = 26\,600\,\text{N/mm}^2$ for all walls.

Data:

Shaded areas:	$A_{34} = 6450\,\text{mm}^2,\ A_{16} = 7750\,\text{mm}^2$
Wall lengths:	$s_{34} = 250\,\text{mm},\ s_{16} = 300\,\text{mm}$
Wall thickness:	$t_{12} = 1.63\,\text{mm},\ t_{34} = 0.56\,\text{mm}$
	$t_{23} = t_{45} = t_{56} = 0.92\,\text{mm}$
	$t_{61} = 2.03\,\text{mm}$
	$t_{25} = 2.54\,\text{mm}$

Ans. $T = 102\,417\,\text{N m}$, $\theta = 1.46°$, $GJ = 10 \times 10^{12}\,\text{N mm}^2/\text{rad}$.

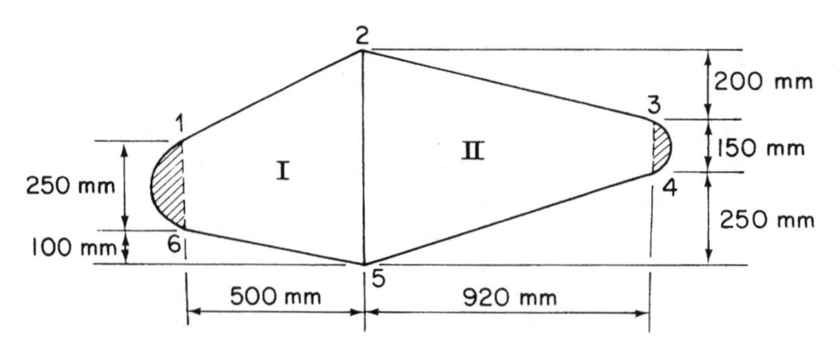

Fig. P.23.2

P.23.3 Determine the torsional stiffness of the four-cell wing section shown in Fig. P.23.3.

Data:

Wall	12	23	34					
	78	67	56	45°	45^i	36	27	18
Peripheral length (mm)	762	812	812	1525	356	406	356	254
Thickness (mm)	0.915	0.915	0.915	0.711	1.220	1.625	1.220	0.915

Cell areas (mm^2) $A_I = 161\,500$ $A_{II} = 291\,000$
$A_{III} = 291\,000$ $A_{IV} = 226\,000$

Ans. $522.5 \times 10^6 G\,N\,mm^2/rad$.

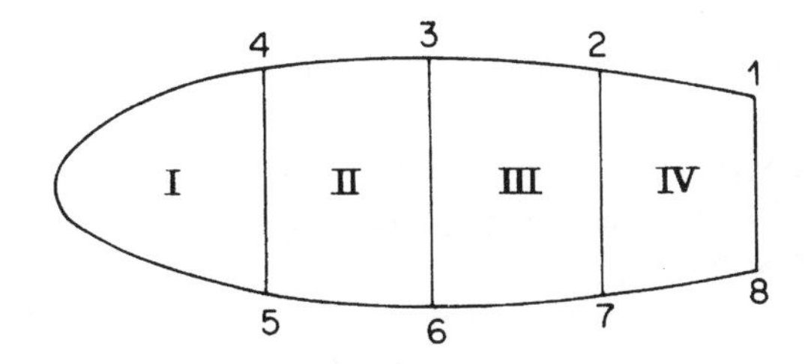

Fig. P.23.3

P.23.4 Determine the shear flow distribution for a torque of 56 500 N m for the three cell section shown in Fig. P.23.4. The section has a constant shear modulus throughout.

Wall	Length (mm)	Thickness (mm)	Cell	Area (mm^2)
12^U	1084	1.220	I	108 400
12^L	2160	1.625	II	202 500
14, 23	127	0.915	III	528 000
34^U	797	0.915		
34^L	797	0.915		

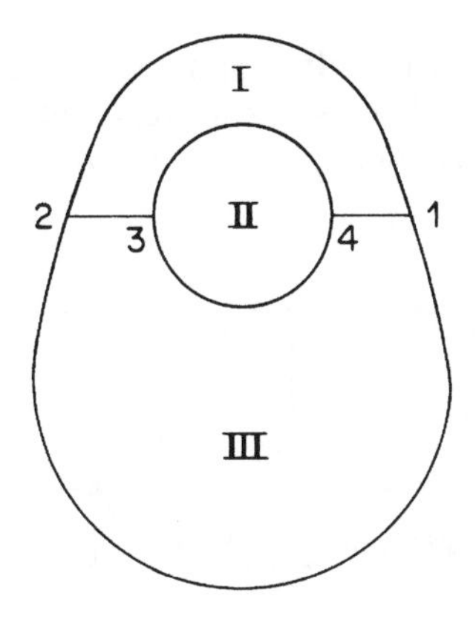

Fig. P.23.4

Ans. $q_{12\text{U}} = 25.4\,\text{N/mm}$ $q_{21\text{L}} = 33.5\,\text{N/mm}$ $q_{14} = q_{32} = 8.1\,\text{N/mm}$
$q_{43\text{U}} = 13.4\,\text{N/mm}$ $q_{34\text{L}} = 5.3\,\text{N/mm}$.

P.23.5 The idealized cross-section of a two-cell thin-walled wing box is shown in Fig. P.23.5. If the wing box supports a load of 44 500 N acting along the web 25, calculate the shear flow distribution. The shear modulus G is the same for all walls of the wing box.

Wall	Length (mm)	Thickness (mm)	Boom	Area (mm^2)
16	254	1.625	1, 6	1290
25	406	2.032	2, 5	1936
34	202	1.220	3, 4	645
12, 56	647	0.915		
23, 45	775	0.559		

Cell areas: $A_{\text{I}} = 232\,000\,\text{mm}^2$, $A_{\text{II}} = 258\,000\,\text{mm}^2$

Ans. $q_{16} = 33.9\,\text{N/mm}$ $q_{65} = q_{21} = 1.1\,\text{N/mm}$
$q_{45} = q_{23} = 7.2\,\text{N/mm}$ $q_{34} = 20.8\,\text{N/mm}$
$q_{25} = 73.4\,\text{N/mm}$.

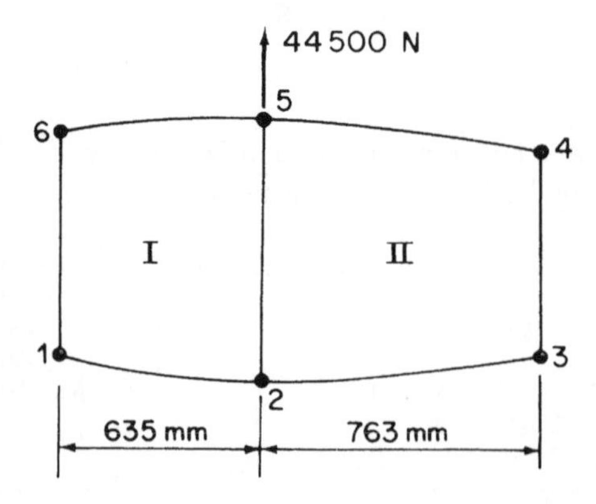

Fig. P.23.5

P.23.6 Figure P.23.6 shows a singly symmetric, two-cell wing section in which all direct stresses are carried by the booms, shear stresses alone being carried by the walls. All walls are flat with the exception of the nose portion 45. Find the position of the

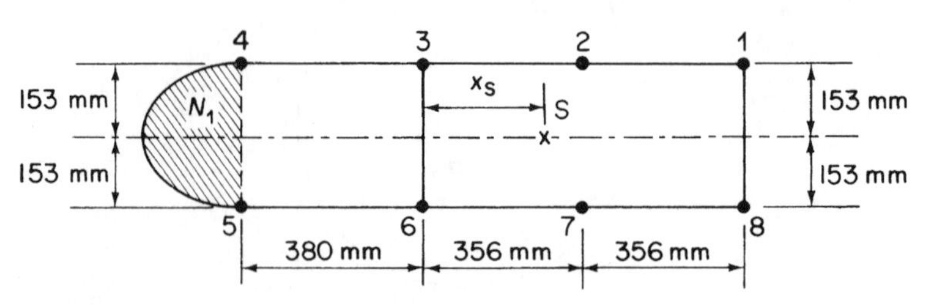

Fig. P.23.6

shear centre S and the shear flow distribution for a load of $S_y = 66\,750$ N through S. Tabulated below are lengths, thicknesses and shear moduli of the shear carrying walls. Note that dotted line 45 is not a wall.

Wall	Length (mm)	Thickness (mm)	G(N/mm^2)	Boom	Area (mm^2)
34, 56	380	0.915	20 700	1, 3, 6, 8	1290
12, 23, 67, 78	356	0.915	24 200	2, 4, 5, 7	645
36, 81	306	1.220	24 800		
45	610	1.220	24 800		

Nose area $N_I = 51\,500$ mm^2

Ans. $x_S = 160.1$ mm $q_{12} = q_{78} = 17.8$ N/mm $q_{32} = q_{76} = 18.5$ N/mm
$q_{63} = 88.2$ N/mm $q_{43} = q_{65} = 2.9$ N/mm $q_{54} = 39.2$ N/mm
$q_{81} = 90.4$ N/mm.

P.23.7 A singly symmetric wing section consists of two closed cells and one open cell (see Fig. P.23.7). The webs 25, 34 and the walls 12, 56 are straight, while all other walls are curved. All walls of the section are assumed to be effective in carrying shear stresses only, direct stresses being carried by booms 1–6. Calculate the distance x_S of the shear centre S aft of the web 34. The shear modulus G is the same for all walls.

Wall	Length (mm)	Thickness (mm)	Boom	Area (mm^2)	Cell	Area (mm^2)
12, 56	510	0.559	1, 6	645	I	93 000
23, 45	765	0.915	2, 5	1290	II	258 000
34°	1015	0.559	3, 4	1935		
34^i	304	2.030				
25	304	1.625				

Ans. 241.4 mm.

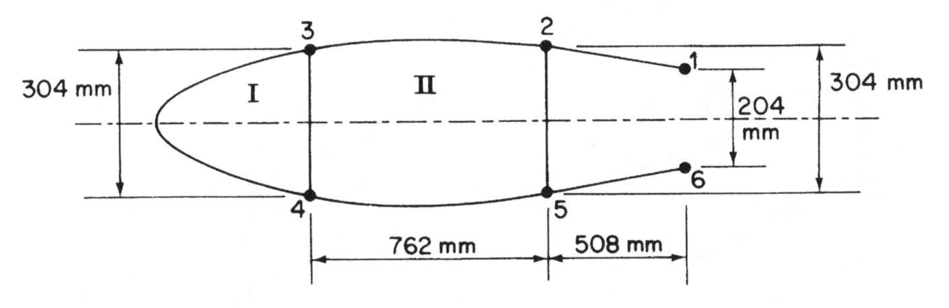

Fig. P.23.7

P.23.8 A portion of a tapered, three-cell wing has singly symmetrical idealized cross-sections 1000 mm apart as shown in Fig. P.23.8. A bending moment $M_x = 1800$ N m and a shear load $S_y = 12\,000$ N in the plane of the web 52 are applied at the larger cross-section. Calculate the forces in the booms and the shear flow distribution at this cross-section. The modulus G is constant throughout. Section dimensions at the larger cross-section are given below.

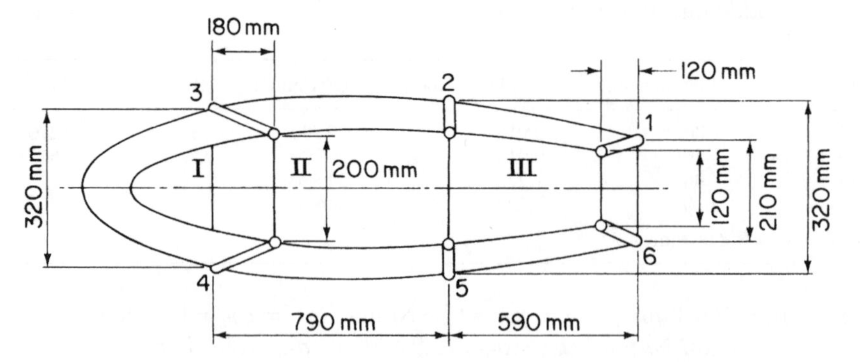

Fig. P.23.8

Wall	Length (mm)	Thickness (mm)	Boom	Area (mm^2)	Cell	Area (mm^2)
12, 56	600	1.0	1, 6	600	I	100 000
23, 45	800	1.0	2, 5	800	II	260 000
34°	1200	0.6	3, 4	800	III	180 000
34^i	320	2.0				
25	320	2.0				
16	210	1.5				

Ans. $P_1 = -P_6 = 1200\,\text{N}$ $P_2 = -P_5 = 2424\,\text{N}$ $P_3 = -P_4 = 2462\,\text{N}$
$q_{12} = q_{56} = 3.74\,\text{N/mm}$ $q_{23} = q_{45} = 3.11\,\text{N/mm}$ $q_{34^o} = 0.06\,\text{N/mm}$
$q_{43^i} = 12.16\,\text{N/mm}$ $q_{52} = 14.58\,\text{N/mm}$ $q_{61} = 11.22\,\text{N/mm}$.

P.23.9 A portion of a wing box is built-in at one end and carries a shear load of 2000 N through its shear centre and a torque of 1000 N m as shown in Fig. P.23.9. If

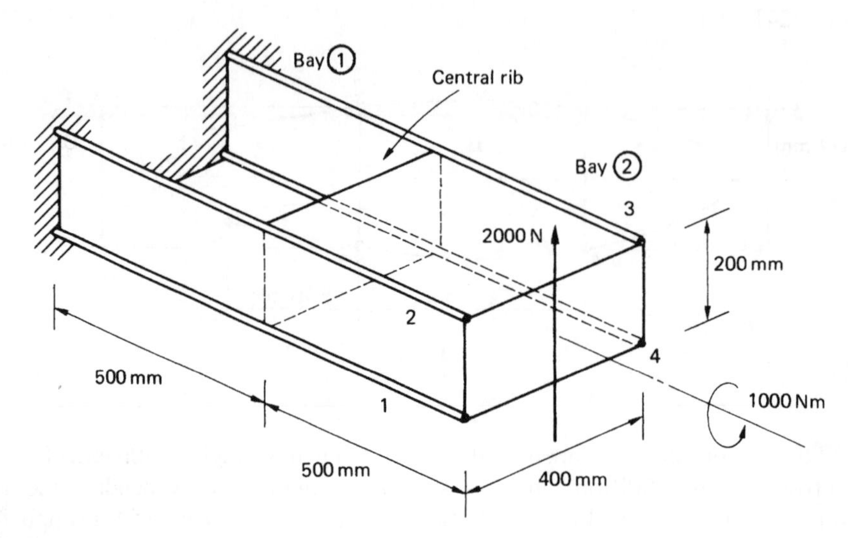

Fig. P.23.9

the skin panel in the upper surface of the inboard bay is removed, calculate the shear flows in the spar webs and remaining skin panels, the distribution of load in the spar flanges and the loading on the central rib. Assume that the spar webs and skin panels are effective in resisting shear stresses only.

Ans. Bay ①: q in spar webs = 7.5 N/mm
Bay ②: q in spar webs = 1.9 N/mm, in skin panels = 9.4 N/mm
Flange loads (2): at built-in end = 1875 N (compression)
at central rib = 5625 N (compression)
Rib loads: q (horizontal edges) = 9.4 N/mm
q (vertical edges) = 9.4 N/mm.

24

Fuselage frames and wing ribs

Aircraft are constructed primarily from thin metal skins which are capable of resisting in-plane tension and shear loads but buckle under comparatively low values of in-plane compressive loads. The skins are therefore stiffened by longitudinal stringers which resist the in-plane compressive loads and, at the same time, resist small distributed loads normal to the plane of the skin. The effective length in compression of the stringers is reduced, in the case of fuselages, by transverse frames or bulkheads or, in the case of wings, by ribs. In addition, the frames and ribs resist concentrated loads in transverse planes and transmit them to the stringers and the plane of the skin. Thus, cantilever wings may be bolted to fuselage frames at the spar caps while undercarriage loads are transmitted to the wing through spar and rib attachment points.

24.1 Principles of stiffener/web construction

Generally, frames and ribs are themselves fabricated from thin sheets of metal and therefore require stiffening members to distribute the concentrated loads to the thin webs. If the load is applied in the plane of a web the stiffeners must be aligned with the direction of the load. Alternatively, if this is not possible, the load should be applied at the intersection of two stiffeners so that each stiffener resists the component of load in its direction. The basic principles of stiffener/web construction are illustrated in Example 24.1.

Example 24.1

A cantilever beam (Fig. 24.1) carries concentrated loads as shown. Calculate the distribution of stiffener loads and the shear flow distribution in the web panels assuming that the latter are effective only in shear.

We note that stiffeners HKD and JK are required at the point of application of the 4000 N load to resist its vertical and horizontal components. A further transverse stiffener GJC is positioned at the unloaded end J of the stiffener JK since stress concentrations are produced if a stiffener ends in the centre of a web panel. We note also that the web panels are only effective in shear so that the shear flow is constant throughout a particular web panel; the assumed directions of the shear flows are shown in Fig. 24.1.

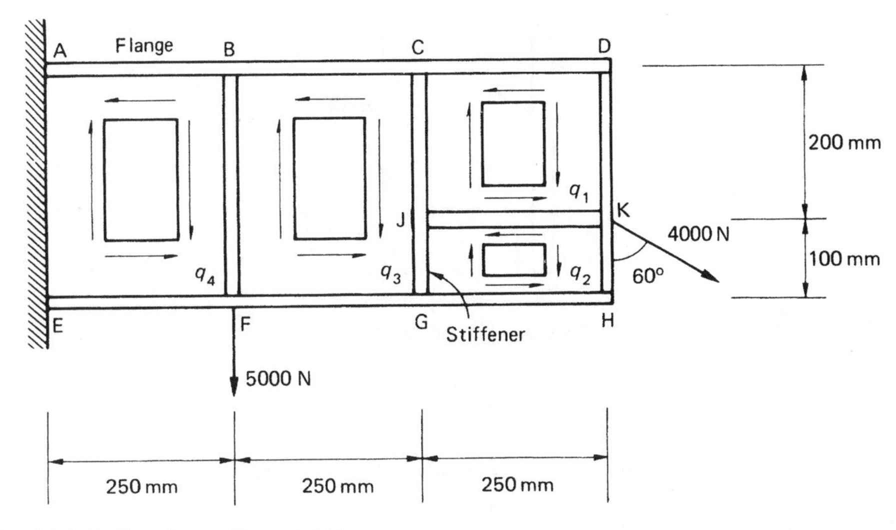

Fig. 24.1 Cantilever beam of Example 24.1.

It is instructive at this stage to examine the physical role of the different structural components in supporting the applied loads. Generally, stiffeners are assumed to withstand axial forces only so that the horizontal component of the load at K is equilibrated locally by the axial load in the stiffener JK and not by the bending of stiffener HKD. By the same argument the vertical component of the load at K is resisted by the axial load in the stiffener HKD. These axial stiffener loads are equilibrated in turn by the resultants of the shear flows q_1 and q_2 in the web panels CDKJ and JKHG. Thus we see that the web panels resist the shear component of the externally applied load and at the same time transmit the bending and axial load of the externally applied load to the beam flanges; subsequently, the flange loads are reacted at the support points A and E.

Consider the free body diagrams of the stiffeners JK and HKD shown in Figs. 24.2(a) and (b).

From the equilibrium of stiffener JK we have

$$(q_1 - q_2) \times 250 = 4000 \sin 60^\circ = 3464.1 \text{ N} \qquad \text{(i)}$$

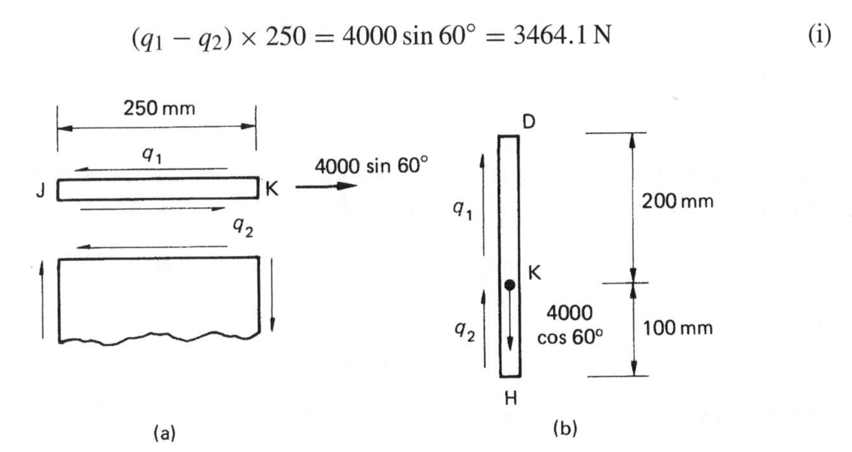

Fig. 24.2 Free body diagrams of stiffeners JK and HKD in the beam of Example 24.1.

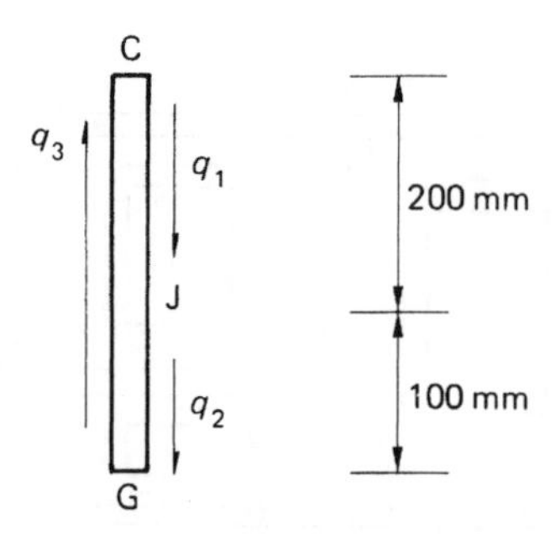

Fig. 24.3 Equilibrium of stiffener CJG in the beam of Example 24.1.

and from the equilibrium of stiffener HKD

$$200q_1 + 100q_2 = 4000 \cos 60^\circ = 2000\,\text{N} \qquad \text{(ii)}$$

Solving Eqs (i) and (ii) we obtain

$$q_1 = 11.3\,\text{N/mm} \quad q_2 = -2.6\,\text{N/mm}$$

The vertical shear force in the panel BCGF is equilibrated by the vertical resultant of the shear flow q_3. Thus

$$300q_3 = 4000 \cos 60^\circ = 2000\,\text{N}$$

whence

$$q_3 = 6.7\,\text{N/mm}$$

Alternatively, q_3 may be found by considering the equilibrium of the stiffener CJG. From Fig. 24.3

$$300q_3 = 200q_1 + 100q_2$$

or

$$300q_3 = 200 \times 11.3 - 100 \times 2.6$$

from which

$$q_3 = 6.7\,\text{N/mm}$$

The shear flow q_4 in the panel ABFE may be found using either of the above methods. Thus, considering the vertical shear force in the panel

$$300q_4 = 4000 \cos 60^\circ + 5000 = 7000\,\text{N}$$

whence

$$q_4 = 23.3\,\text{N/mm}$$

Alternatively, from the equilibrium of stiffener BF

$$300q_4 - 300q_3 = 5000\,\text{N}$$

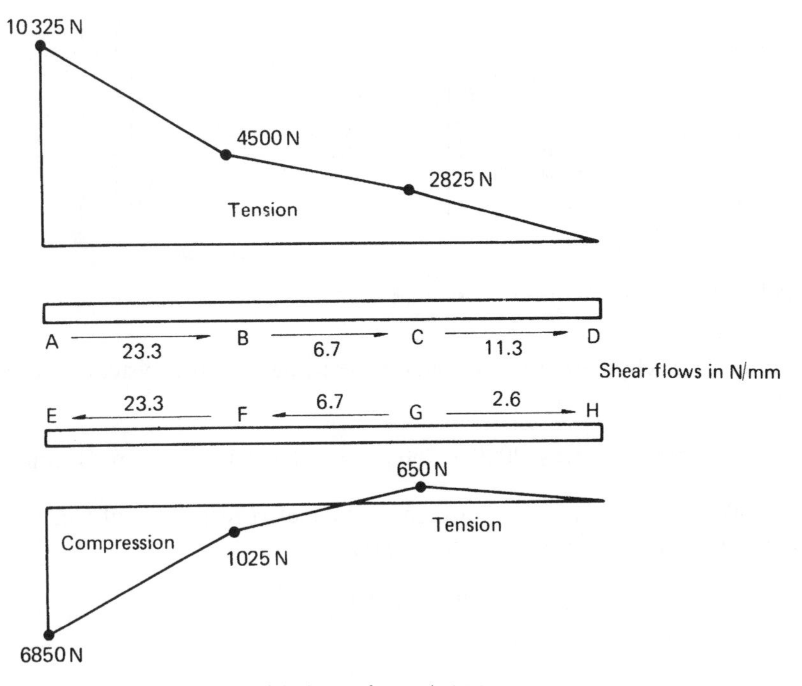

Fig. 24.4 Load distributions in flanges of the beam of Example 24.1.

whence

$$q_4 = 23.3\,\text{N/mm}$$

The flange and stiffener load distributions are calculated in the same way and are obtained from the algebraic summation of the shear flows along their lengths. For example, the axial load P_A at A in the flange ABCD is given by

$$P_\text{A} = 250q_1 + 250q_3 + 250q_4$$

or

$$P_\text{A} = 250 \times 11.3 + 250 \times 6.7 + 250 \times 23.3 = 10\,325\,\text{N (tension)}$$

Similarly

$$P_\text{E} = -250q_2 - 250q_3 - 250q_4$$

i.e.

$$P_\text{E} = 250 \times 2.6 - 250 \times 6.7 - 250 \times 23.3 = -6850\,\text{N (compression)}$$

The complete load distribution in each flange is shown in Fig. 24.4. The stiffener load distributions are calculated in the same way and are shown in Fig. 24.5.

The distribution of flange load in the bays ABFE and BCGF could have been obtained by considering the bending and axial loads on the beam at any section. For

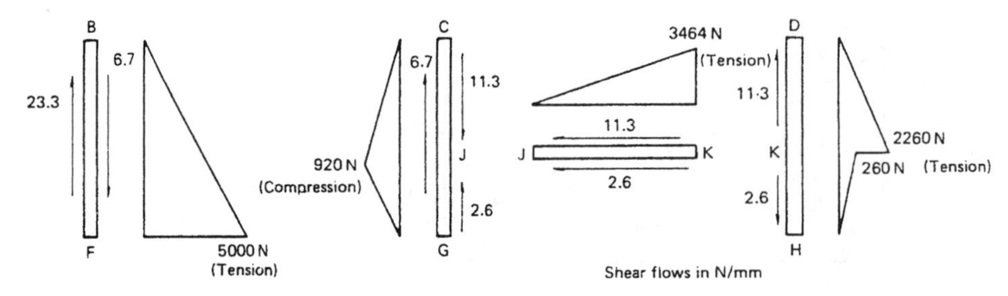

Fig. 24.5 Load distributions in stiffeners of the beam of Example 24.1.

example, at the section AE we can replace the actual loading system by a bending moment

$$M_{AE} = 5000 \times 250 + 2000 \times 750 - 3464.1 \times 50 = 2\,576\,800 \text{ N mm}$$

and an axial load acting midway between the flanges (irrespective of whether or not the flange areas are symmetrical about this point) of

$$P = 3464.1 \text{ N}$$

Thus

$$P_A = \frac{2\,576\,800}{300} + \frac{3464.1}{2} = 10\,321 \text{ N (tension)}$$

and

$$P_E = \frac{-2\,576\,800}{300} + \frac{3464.1}{2} = -6857 \text{ N (compression)}$$

This approach cannot be used in the bay CDHG except at the section CJG since the axial load in the stiffener JK introduces an additional unknown.

The above analysis assumes that the web panels in beams of the type shown in Fig. 24.1 resist pure shear along their boundaries. In Chapter 9 we saw that thin webs may buckle under the action of such shear loads producing tension field stresses which, in turn, induce additional loads in the stiffeners and flanges of beams. The tension field stresses may be calculated separately by the methods described in Chapter 9 and then superimposed on the stresses determined as described above.

So far we have been concerned with web/stiffener arrangements in which the loads have been applied in the plane of the web so that two stiffeners are sufficient to resist the components of a concentrated load. Frequently, loads have an out-of-plane component in which case the structure should be arranged so that two webs meet at the point of load application with stiffeners aligned with the three component directions (Fig. 24.6). In some situations it is not practicable to have two webs meeting at the point of load application so that a component normal to a web exists. If this component is small it may be resisted in bending by an in-plane stiffener, otherwise an additional member must be provided spanning between adjacent frames or ribs, as shown in Fig. 24.7. In general, no normal loads should be applied to an unsupported web no matter how small their magnitude.

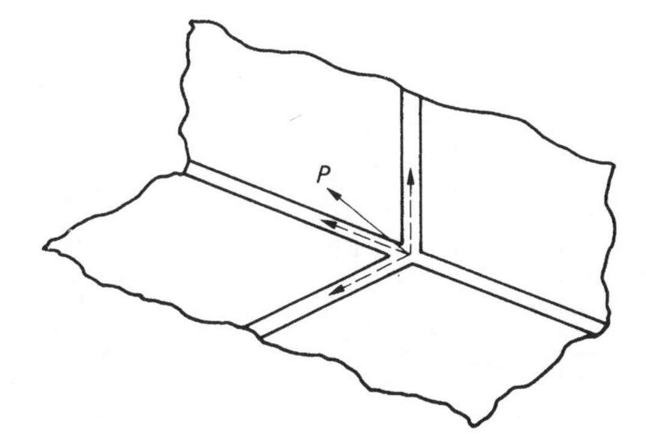

Fig. 24.6 Structural arrangement for an out of plane load.

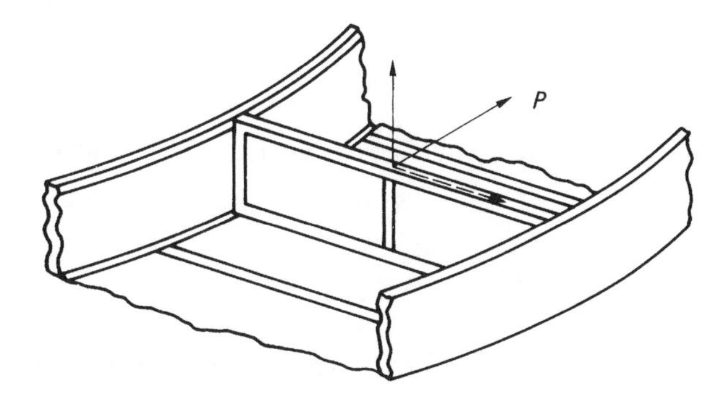

Fig. 24.7 Support of load having a component normal to a web.

24.2 Fuselage frames

We have noted that fuselage frames transfer loads to the fuselage shell and provide column support for the longitudinal stringers. The frames generally take the form of open rings so that the interior of the fuselage is not obstructed. They are connected continuously around their peripheries to the fuselage shell and are not necessarily circular in form but will usually be symmetrical about a vertical axis.

A fuselage frame is in equilibrium under the action of any external loads and the reaction shear flows from the fuselage shell. Suppose that a fuselage frame has a vertical axis of symmetry and carries a vertical external load W, as shown in Fig. 24.8(a) and (b). The fuselage shell/stringer section has been idealized such that the fuselage skin is effective only in shear. Suppose also that the shear force in the fuselage immediately to the left of the frame is $S_{y,1}$ and that the shear force in the fuselage immediately to the right of the frame is $S_{y,2}$; clearly, $S_{y,2} = S_{y,1} - W$. $S_{y,1}$ and $S_{y,2}$ generate shear flow distributions q_1 and q_2, respectively in the fuselage skin, each given by Eq. (22.1) in which $S_{x,1} = S_{x,2} = 0$ and $I_{xy} = 0$ (Cy is an axis of symmetry). The shear flow q_f transmitted to the periphery of the frame is equal to the algebraic sum of q_1 and q_2, i.e.

$$q_f = q_1 - q_2$$

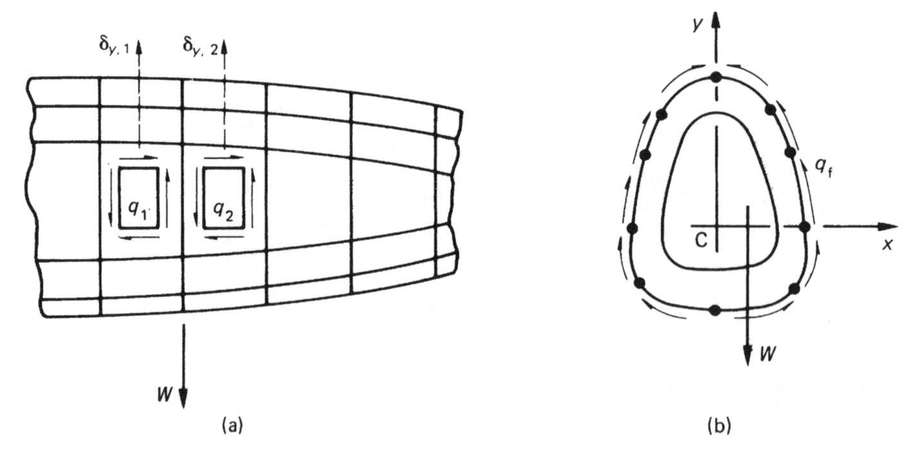

Fig. 24.8 Loads on a fuselage frame.

Thus, substituting for q_1 and q_2 obtained from Eq. (22.1) and noting that $S_{y,2} = S_{y,1} - W$, we have

$$q_{\mathrm{f}} = \frac{-W}{I_{xx}} \sum_{r=1}^{n} B_r y_r + q_{s,0}$$

in which $q_{s,0}$ is calculated using Eq. (17.17) where the shear load is W and

$$q_{\mathrm{b}} = \frac{-W}{I_{xx}} \sum_{r=1}^{n} B_r y_r$$

The method of determining the shear flow distribution applied to the periphery of a fuselage frame is identical to the method of solution (or the alternative method) of Example 22.2.

Having determined the shear flow distribution around the periphery of the frame, the frame itself may be analysed for distributions of bending moment, shear force and normal force, as described in Section 5.4.

24.3 Wing ribs

Wing ribs perform similar functions to those performed by fuselage frames. They maintain the shape of the wing section, assist in transmitting external loads to the wing skin and reduce the column length of the stringers. Their geometry, however, is usually different in that they are frequently of unsymmetrical shape and possess webs which are continuous except for lightness holes and openings for control runs.

Wing ribs are subjected to loading systems which are similar to those applied to fuselage frames. External loads applied in the plane of the rib produce a change in shear force in the wing across the rib; this induces reaction shear flows around its periphery. These are calculated using the methods described in Chapter 17 and in Chapter 23.

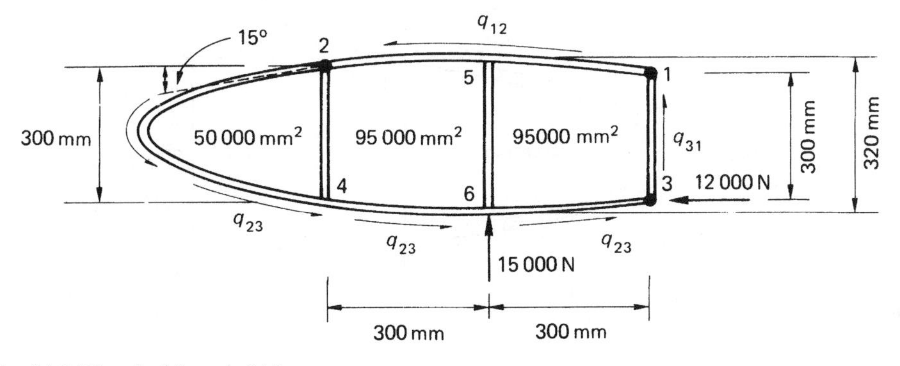

Fig. 24.9 Wing rib of Example 24.2.

To illustrate the method of rib analysis we shall use the example of a three-flange wing section in which, as we noted in Section 23.1, the shear flow distribution is statically determinate.

Example 24.2

Calculate the shear flows in the web panels and the axial loads in the flanges of the wing rib shown in Fig. 24.9. Assume that the web of the rib is effective only in shear while the resistance of the wing to bending moments is provided entirely by the three flanges 1, 2 and 3.

Since the wing bending moments are resisted entirely by the flanges 1, 2 and 3, the shear flows developed in the wing skin are constant between the flanges. Using the method described in Section 23.1 for a three-flange wing section we have, resolving forces horizontally

$$600q_{12} - 600q_{23} = 12\,000\,\text{N} \tag{i}$$

Resolving vertically

$$300q_{31} - 300q_{23} = 15\,000\,\text{N} \tag{ii}$$

Taking moments about flange 3

$$2(50\,000 + 95\,000)q_{23} + 2 \times 95\,000q_{12} = -15\,000 \times 300\,\text{N mm} \tag{iii}$$

Solution of Eqs (i)–(iii) gives

$$q_{12} = 13.0\,\text{N/mm} \quad q_{23} = -7.0\,\text{N/mm} \quad q_{31} = 43.0\,\text{N/mm}$$

Consider now the nose portion of the rib shown in Fig. 24.10 and suppose that the shear flow in the web immediately to the left of the stiffener 24 is q_1. The total vertical shear force $S_{y,1}$ at this section is given by

$$S_{y,1} = 7.0 \times 300 = 2100\,\text{N}$$

The horizontal components of the rib flange loads resist the bending moment at this section. Thus

$$P_{x,4} = P_{x,2} = \frac{2 \times 50\,000 \times 7.0}{300} = 2333.3\,\text{N}$$

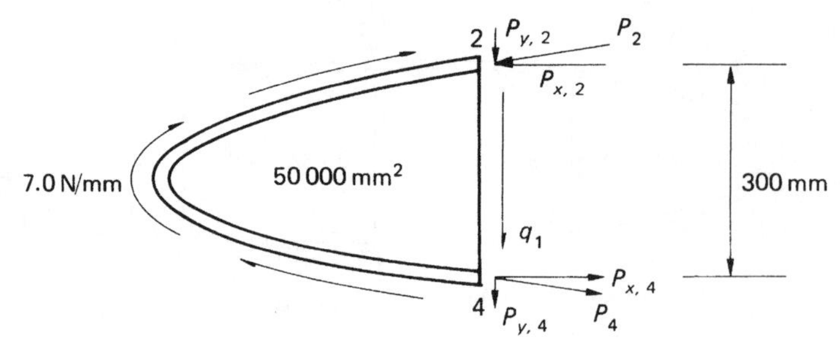

Fig. 24.10 Equilibrium of nose portion of the rib.

The corresponding vertical components are then

$$P_{y,2} = P_{y,4} = 2333.3 \tan 15° = 625.2\,\text{N}$$

Thus the shear force carried by the web is $2100 - 2 \times 625.2 = 849.6\,\text{N}$. Hence

$$q_1 = \frac{849.6}{300} = 2.8\,\text{N/mm}$$

The axial loads in the rib flanges at this section are given by

$$P_2 = P_4 = (2333.3^2 + 625.2^2)^{1/2} = 2415.6\,\text{N}$$

The rib flange loads and web panel shear flows, at a vertical section immediately to the left of the intermediate web stiffener 56, are found by considering the free body diagram shown in Fig. 24.11. At this section the rib flanges have zero slope so that the flange loads P_5 and P_6 are obtained directly from the value of bending moment at this section. Thus

$$P_5 = P_6 = 2[(50\,000 + 46\,000) \times 7.0 - 49\,000 \times 13.0]/320 = 218.8\,\text{N}$$

The shear force at this section is resisted solely by the web. Hence

$$320q_2 = 7.0 \times 300 + 7.0 \times 10 - 13.0 \times 10 = 2040\,\text{N}$$

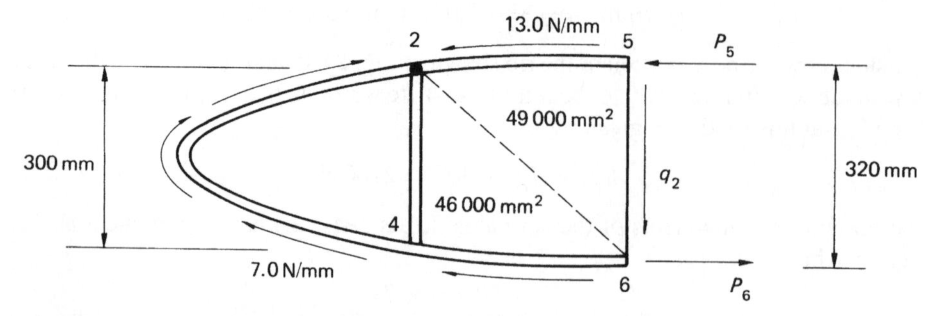

Fig. 24.11 Equilibrium of rib forward of intermediate stiffener 56.

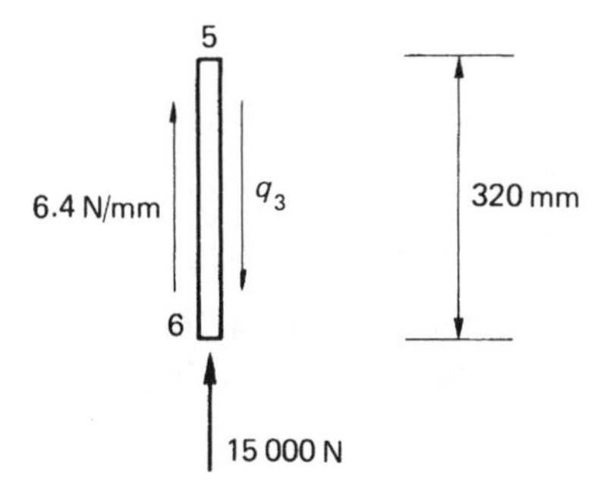

Fig. 24.12 Equilibrium of stiffener 56.

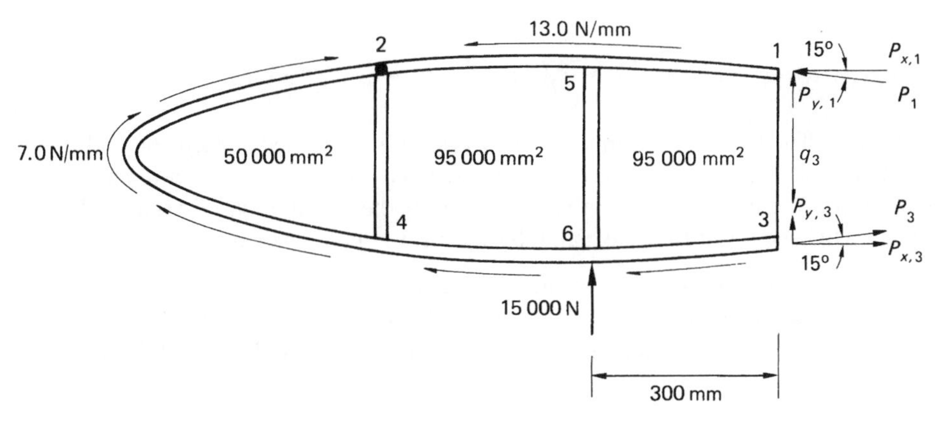

Fig. 24.13 Equilibrium of the rib forward of stiffener 31.

so that

$$q_2 = 6.4\,\text{N/mm}$$

The shear flow in the rib immediately to the right of stiffener 56 is found most simply by considering the vertical equilibrium of stiffener 56 as shown in Fig. 24.12. Thus

$$320q_3 = 6.4 \times 320 + 15\,000$$

which gives

$$q_3 = 53.3\,\text{N/mm}$$

Finally, we shall consider the rib flange loads and the web shear flow at a section immediately forward of stiffener 31. From Fig. 24.13, in which we take moments about the point 3

$$M_3 = 2[(50\,000 + 95\,000) \times 7.0 - 95\,000 \times 13.0] + 15\,000 \times 300 = 4.06 \times 10^6\,\text{N mm}$$

The horizontal components of the flange loads at this section are then

$$P_{x,1} = P_{x,3} = \frac{4.06 \times 10^6}{300} = 13\,533.3\,\text{N}$$

and the vertical components are

$$P_{y,1} = P_{y,3} = 3626.2\,\text{N}$$

Hence

$$P_1 = P_3 = \sqrt{13\,533.3^2 + 3626.2^2} = 14\,010.7\,\text{N}$$

The total shear force at this section is $15\,000 + 300 \times 7.0 = 17\,100$ N. Therefore, the shear force resisted by the web is $17\,100 - 2 \times 3626.2 = 9847.6$ N so that the shear flow q_3 in the web at this section is

$$q_3 = \frac{9847.6}{300} = 32.8\,\text{N/mm}$$

Problems

P.24.1 The beam shown in Fig. P.24.1 is simply supported at each end and carries a load of 6000 N. If all direct stresses are resisted by the flanges and stiffeners and the web panels are effective only in shear, calculate the distribution of axial load in the flange ABC and the stiffener BE and the shear flows in the panels.

Ans: $q(\text{ABEF}) = 4$ N/mm, $q(\text{BCDE}) = 2$ N/mm
P_{BE} increases linearly from zero at B to 6000 N (tension) at E
P_{AB} and P_{CB} increase linearly from zero at A and C to 4000 N (compression) at B.

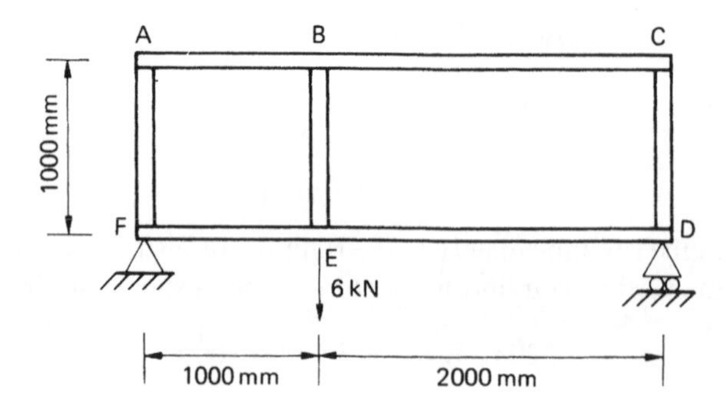

Fig. P.24.1

P.24.2 Calculate the shear flows in the web panels and direct load in the flanges and stiffeners of the beam shown in Fig. P.24.2 if the web panels resist shear stresses only.

Ans. $q_1 = 21.6$ N/mm $\quad q_2 = -1.6$ N/mm $\quad q_3 = 10$ N/mm
$P_{\text{C}} = 0 \quad P_{\text{B}} = 6480$ N (tension) $\quad P_{\text{A}} = 9480$ N (tension)
$P_{\text{F}} = 0 \quad P_{\text{G}} = 480$ N (tension) $\quad P_{\text{H}} = 2520$ N (compression)
P_{E} in BEG $= 2320$ N (compression) $\quad P_{\text{D}}$ in ED $= 6928$ N (tension)
P_{D} in CD $= 4320$ N (tension) $\quad P_{\text{D}}$ in DF $= 320$ N (tension).

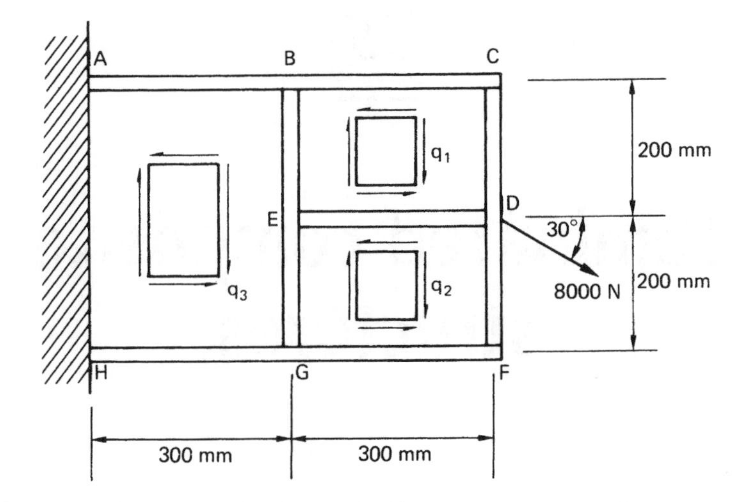

Fig. P.24.2

P.24.3 A three-flange wing section is stiffened by the wing rib shown in Fig. P.24.3. If the rib flanges and stiffeners carry all the direct loads while the rib panels are effective only in shear, calculate the shear flows in the panels and the direct loads in the rib flanges and stiffeners.

Ans. $q_1 = 4.0\,\text{N/mm}$ $\quad q_2 = 26.0\,\text{N/mm}$ $\quad q_3 = 6.0\,\text{N/mm}$

P_2 in $12 = -P_3$ in $43 = 1200\,\text{N}$ (tension) $\quad P_5$ in $154 = 2000\,\text{N}$ (tension)

P_3 in $263 = 8000\,\text{N}$ (compression) $\quad P_5$ in $56 = 12\,000\,\text{N}$ (tension)

P_6 in $263 = 6000\,\text{N}$ (compression).

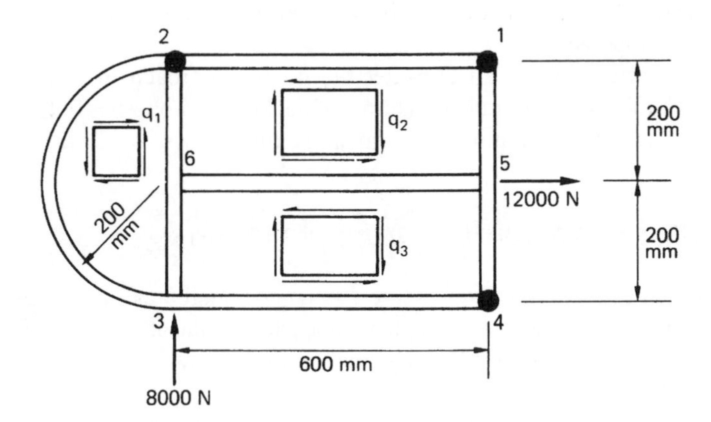

Fig. P.24.3

25

Laminated composite structures

An increasingly large proportion of the structures of many modern aircraft are fabricated from composite materials. These, as we saw in Chapter 12 consist of laminas in which a stiff, high strength filament, for example carbon fibre, is embedded in a matrix such as epoxy, polyester, etc. The use of composites can lead to considerable savings in weight over conventional metallic structures. They also have the advantage that the direction of the filaments in a multi-lamina structure may be aligned with the direction of the major loads at a particular point resulting in a more efficient design.

There are two approaches to the analysis of composite materials. In the first, micromechanics, the constituent materials, i.e. the fibres and resin (the matrix) are considered separately. The properties of the composite will then change from point to point in a particular direction depending on whether the fibre or the resin is being examined. In the second approach, macromechanics, the composite material is regarded as a whole so that the properties will not change from point to point in a particular direction. Generally, the design and analysis of composite materials are based on the macro- rather than the micro- approach.

Initially, but briefly, we shall consider the micro- approach in which the elastic constants of a lamina are determined in terms of the known properties of the constituent materials; we shall then determine the corresponding stresses.

25.1 Elastic constants of a simple lamina

A simple lamina of a composite structure can be considered as orthotropic with two principal material directions in its own plane: one parallel, the other perpendicular to the direction of the filaments; we shall designate the former the longitudinal direction (1), the latter the transverse direction (t).

In Fig. 25.1 a portion of a lamina containing a single filament is subjected to a stress, σ_1, in the longitudinal direction which produces an extension Δl. If it is assumed that plane sections remain plane during deformation then the strain ε_1 corresponding to σ_1 is given by

$$\varepsilon_1 = \frac{\Delta l}{l} \qquad (25.1)$$

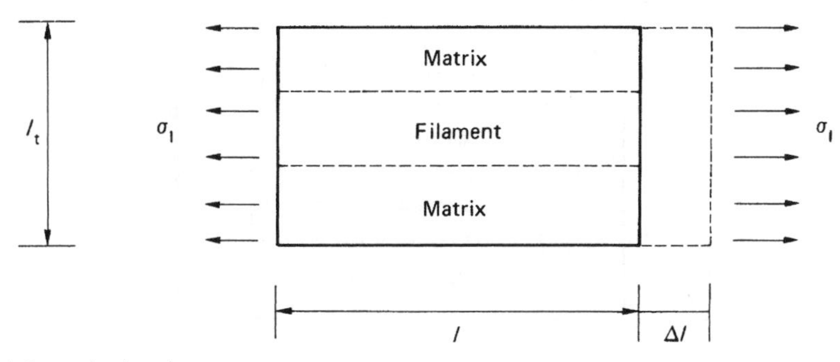

Fig. 25.1 Determination of E_l.

and

$$\sigma_l = E_l \varepsilon_l \tag{25.2}$$

where E_l is the modulus of elasticity of the lamina in the direction of the filament. Also, using the suffixes f and m to designate filament and matrix parameters, we have

$$\sigma_f = E_f \varepsilon_l \quad \sigma_m = E_m \varepsilon_l \tag{25.3}$$

Further, if A is the total area of cross-section of the lamina in Fig. 25.1, A_f is the cross-sectional area of the filament and A_m the cross-sectional area of the matrix then, for equilibrium in the direction of the filament

$$\sigma_l A = \sigma_f A_f + \sigma_m A_m$$

or, substituting for σ_l, σ_f and σ_m from Eqs (25.2) and (25.3)

$$E_l \varepsilon_l A = E_f \varepsilon_f A_f + E_m \varepsilon_l A_m$$

so that

$$E_l = E_f \frac{A_f}{A} + E_m \frac{A_m}{A} \tag{25.4}$$

Writing $A_f/A = v_f$ and $A_m/A = v_m$, Eq. (25.4) becomes

$$E_l = v_f E_f + v_m E_m \tag{25.5}$$

Equation (25.5) is generally referred to as the *law of mixtures*.

A similar approach may be used to determine the modulus of elasticity in the transverse direction (E_t). In Fig. 25.2 the total extension in the transverse direction is produced by σ_t and is given by

$$\varepsilon_t l_t = \varepsilon_m l_m + \varepsilon_f l_f$$

or

$$\frac{\sigma_t}{E_t} l_t = \frac{\sigma_t}{E_m} l_m + \frac{\sigma_t}{E_f} l_f$$

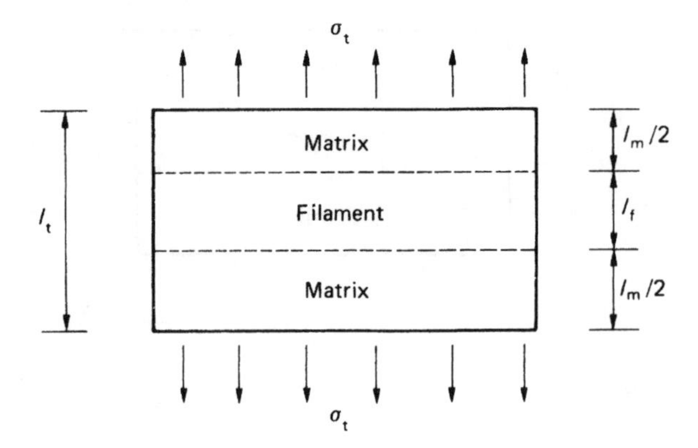

Fig. 25.2 Determination of E_t.

which gives

$$\frac{1}{E_t} = \frac{v_m}{E_m} + \frac{v_f}{E_f}$$

Rearranging this we obtain

$$E_t = \frac{E_m E_f}{v_m E_f + v_f E_m} \tag{25.6}$$

The major Poisson's ratio ν_{lt} may be found by referring to the stress system of Fig. 25.1 and the dimensions given in Fig. 25.2. The total displacement in the transverse direction produced by σ_l is given by

$$\Delta_t = \nu_{lt}\varepsilon_l l_t$$

i.e.

$$\Delta_t = \nu_{lt}\varepsilon_l l_t = \nu_m \varepsilon_l l_m + \nu_f \varepsilon_l l_f$$

from which

$$\nu_{lt} = v_m \nu_m + v_f \nu_f \tag{25.7}$$

The minor Poisson's ratio ν_{tl} is found by referring to Fig. 25.2. The strain in the longitudinal direction produced by the transverse stress σ_t is given by

$$\nu_{tl}\frac{\sigma_t}{E_t} = \nu_m \frac{\sigma_t}{E_t} = \nu_f \frac{\sigma_t}{E_f} \tag{25.8}$$

From the last two of Eqs (25.8)

$$\nu_f = \frac{E_f}{E_m}\nu_m$$

Substituting in Eq. (25.7)

$$\nu_{lt} = \nu_m\left(v_m + \frac{E_f}{E_m}v_f\right) = \frac{\nu_m}{E_m}(v_m E_m + v_f E_f)$$

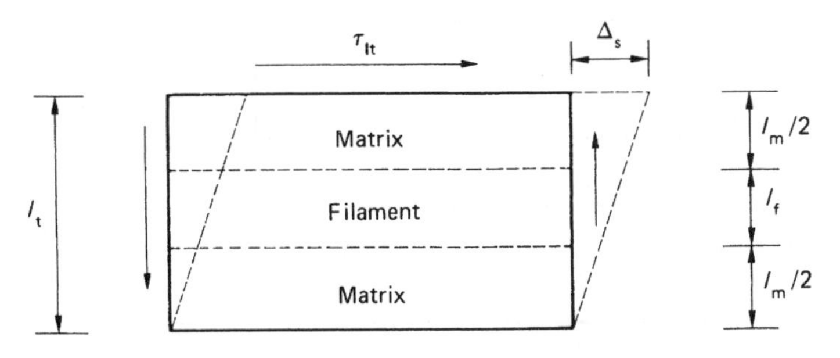

Fig. 25.3 Determination of G_{lt}.

or, from Eq. (25.5)

$$\nu_{lt} = \nu_m \frac{E_l}{E_m}$$

Now substituting for ν_m in the first two of Eqs (25.8)

$$\frac{\nu_{tl}}{E_t} = \frac{\nu_{lt}}{E_t}$$

or

$$\nu_{tl} = \frac{E_t}{E_l}\nu_{lt} = \frac{E_t}{E_l}(v_m \nu_m + v_f \nu_f) \tag{25.9}$$

Finally, the shear modulus $G_{lt}(=G_{tl})$ is determined by assuming that the constituent materials are subjected to the same shear stress τ_{lt} as shown in Fig. 25.3. The displacement Δ_s produced by shear is

$$\Delta_s = \frac{\tau_{lt}}{G_{lt}} l_t = \frac{\tau_{lt}}{G_m} l_m + \frac{\tau_{lt}}{G_f} l_f$$

in which G_m and G_f are the shear moduli of the matrix and filament, respectively. Then

$$\frac{l_t}{G_{lt}} = \frac{l_m}{G_m} + \frac{l_f}{G_f}$$

whence

$$G_{lt} = \frac{G_m G_f}{v_m G_f + v_f G_m} \tag{25.10}$$

Example 25.1

A laminated bar whose cross-section is shown in Fig. 25.4 is 500 mm long and comprises an epoxy resin matrix reinforced by a carbon filament having moduli equal to 5000 N/mm^2 and 200 000 N/mm^2, respectively; the corresponding values of Poisson's ratio are 0.2 and 0.3. If the bar is subjected to an axial tensile load of 100 kN, determine the lengthening of the bar and the reduction in its thickness. Calculate also the stresses in the epoxy resin and the carbon filament.

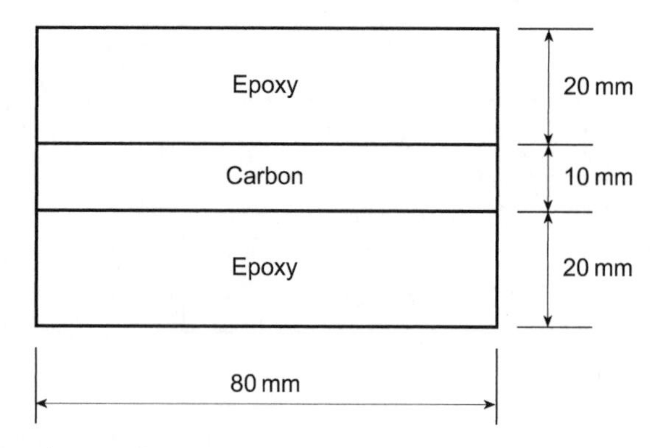

Fig. 25.4 Cross-section of the bar of Example 25.1.

From Eq. (25.5) the modulus of the bar is given by

$$E_l = 200\,000 \times \frac{80 \times 10}{80 \times 50} + 5000 \times \frac{80 \times 40}{80 \times 50}$$

i.e.

$$E_l = 44\,000\,\text{N/mm}^2$$

The direct stress, σ_l, in the longitudinal direction is given by

$$\sigma_l = \frac{100 \times 10^3}{80 \times 50} = 25.0\,\text{N/mm}^2$$

Therefore, from Eq. (25.2), the longitudinal strain in the bar is

$$\varepsilon_l = \frac{25.0}{44\,000} = 5.68 \times 10^{-4}$$

The lengthening, Δ_l, of the bar is then

$$\Delta_l = 5.68 \times 10^{-4} \times 500$$

i.e.

$$\Delta_l = 0.284\,\text{mm}$$

The major Poisson's ratio for the bar is found from Eq. (25.7), i.e.

$$\nu_{lt} = \frac{80 \times 40}{80 \times 50} \times 0.2 + \frac{80 \times 10}{80 \times 50} \times 0.3 = 0.22$$

The strain in the bar across its thickness is then

$$\varepsilon_t = -0.22 \times 5.68 \times 10^{-4} = -1.25 \times 10^{-4}$$

The reduction in thickness, Δ_t, of the bar is then

$$\Delta_t = 1.25 \times 10^{-4} \times 50$$

i.e.

$$\Delta_t = 0.006\,\text{mm}$$

The stresses in the epoxy and the carbon are found using Eq. (25.3). Thus

$$\sigma_m(\text{epoxy}) = 5000 \times 5.68 \times 10^{-4} = 2.84\,\text{N/mm}^2$$

$$\sigma_f(\text{carbon}) = 200\,000 \times 5.68 \times 10^{-4} = 113.6\,\text{N/mm}^2$$

25.2 Stress–strain relationships for an orthotropic ply (macro- approach)

A single sheet of composite material in which the sheet has been preimpregnated with resin (a prepreg) and with the fibres aligned with one particular direction is called a *unidirectional ply* or *lamina* (Fig. 25.5(a)). On the other hand a woven ply has the fibres placed in two perpendicular directions (Fig. 25.5(b)); generally the fibre reinforcement will be the same in both directions. In plies where this is not the case so that the material properties are different in the two mutually perpendicular directions the ply is said to be orthotropic (see Section 11.7). Two cases of orthotropic plies arise. In the first, the directions of the applied loads coincide with directions of the plies; these are known as *specially orthotropic plies*. In the second the applied loads are applied in any direction; these are termed *generally orthotropic plies*.

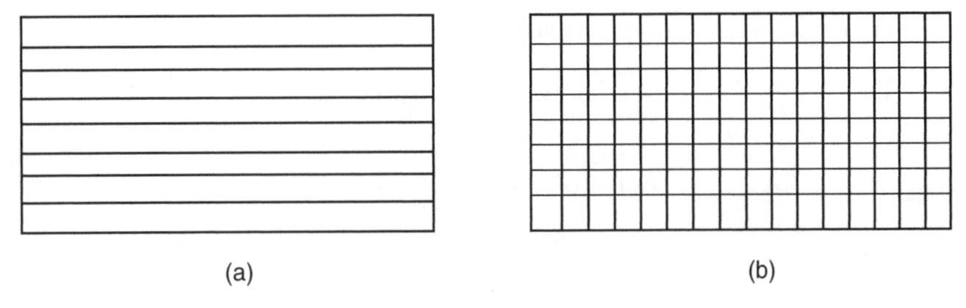

Fig. 25.5 Types of ply. (a) Unidirectional ply; (b) Woven ply.

25.2.1 Specially orthotropic ply

Figure 25.6 shows an element of a specially orthotropic ply. The ply reference axes are the same as in Section 25.1, i.e. longitudinal (suffix l) and transverse (suffix t). Of course these axes do not have the same significance for a woven ply as they do for a unidirectional ply but reference axes must be specified and these are as convenient as any. We also specify loading axes, x and y, which, for a specially orthotropic ply, coincide with the ply reference axes.

Suppose that the ply is subjected to direct stresses σ_x and σ_y, shear and complementary shear stresses τ_{xy} and that the elastic constants for the ply are E_l, E_t, G_{lt} ($=G_{tl}$), ν_{lt} and ν_{tl} (see Eqs (25.5)–(25.10)). Note that, unlike an isotropic material, the shear modulus

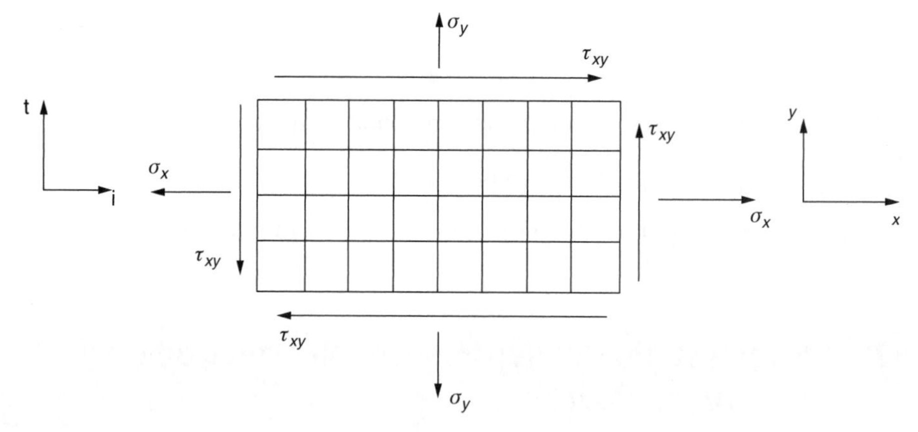

Fig. 25.6 Reference axes for a specially orthotropic ply.

G_{lt} is not related to the other elastic constants. From Section 1.15 (see Eqs (1.52)) the strains in the longitudinal and transverse directions are given by

$$\left.\begin{aligned}\varepsilon_{\mathrm{l}} &= \frac{\sigma_x}{E_{\mathrm{l}}} - \frac{\nu_{\mathrm{tl}}\sigma_y}{E_{\mathrm{t}}}\\ \varepsilon_{\mathrm{t}} &= \frac{\sigma_y}{E_{\mathrm{t}}} - \frac{\nu_{\mathrm{lt}}\sigma_x}{E_{\mathrm{l}}}\end{aligned}\right\} \tag{25.11}$$

Eqs (25.11) may be written in matrix form, i.e.

$$\begin{Bmatrix} E_{\mathrm{l}} \\ E_{\mathrm{t}} \end{Bmatrix} = \begin{bmatrix} \dfrac{1}{E_{\mathrm{l}}} & -\dfrac{\nu_{\mathrm{tl}}}{E_{\mathrm{t}}} \\ -\dfrac{\nu_{\mathrm{lt}}}{E_{\mathrm{l}}} & \dfrac{1}{E_{\mathrm{t}}} \end{bmatrix} \begin{Bmatrix} \sigma_{\mathrm{x}} \\ \sigma_{\mathrm{y}} \end{Bmatrix} \tag{25.12}$$

or, in general terms

$$[\varepsilon] = [K][\sigma] \tag{25.13}$$

It may be shown, using an energy approach, that the stiffness matrix $[K]$ must be symmetric about the leading diagonal. Therefore

$$-\frac{\nu_{\mathrm{tl}}}{E_{\mathrm{t}}} = -\frac{\nu_{\mathrm{lt}}}{E_{\mathrm{l}}}$$

giving

$$\frac{\nu_{\mathrm{tl}}}{E_{\mathrm{t}}} = \frac{\nu_{\mathrm{lt}}}{E_{\mathrm{l}}} \tag{25.14}$$

so that, of the four elastic constants E_{l}, E_{t}, ν_{lt} and ν_{tl} only three are independent.

Eqs (25.11) may be transposed (as in Section 1.15) to give stress–strain relationships. Then

$$\left.\begin{aligned}\sigma_x &= \frac{E_\mathrm{l}}{1-\nu_\mathrm{lt}\nu_\mathrm{tl}}\varepsilon_\mathrm{l} + \frac{\nu_\mathrm{tl}E_\mathrm{l}}{1-\nu_\mathrm{lt}\nu_\mathrm{tl}}\varepsilon_\mathrm{t}\\ \sigma_y &= \frac{E_\mathrm{t}}{1-\nu_\mathrm{lt}\nu_\mathrm{tl}}\varepsilon_\mathrm{t} + \frac{\nu_\mathrm{lt}E_\mathrm{t}}{1-\nu_\mathrm{lt}\nu_\mathrm{tl}}\varepsilon_\mathrm{l}\end{aligned}\right\} \tag{25.15}$$

From the last of Eqs (1.52)

$$\gamma_\mathrm{lt} = \frac{\tau_{xy}}{G_\mathrm{lt}}$$

$$\tau_\mathrm{xy} = \gamma_\mathrm{lt}G_\mathrm{lt} \tag{25.16}$$

Eqs (25.15) and (25.16) may be written in matrix form, i.e.

$$\begin{Bmatrix}\sigma_x\\ \sigma_y\\ \tau_{xy}\end{Bmatrix} = \begin{bmatrix}\dfrac{E_\mathrm{l}}{1-\nu_\mathrm{lt}\nu_\mathrm{tl}} & \dfrac{\nu_\mathrm{tl}E_\mathrm{l}}{1-\nu_\mathrm{lt}\nu_\mathrm{tl}} & 0\\ \dfrac{\nu_\mathrm{lt}E_\mathrm{t}}{1-\nu_\mathrm{lt}\nu_\mathrm{tl}} & \dfrac{E_\mathrm{t}}{1-\nu_\mathrm{lt}\nu_\mathrm{tl}} & 0\\ 0 & 0 & G_\mathrm{lt}\end{bmatrix}\begin{Bmatrix}\varepsilon_\mathrm{l}\\ \varepsilon_\mathrm{t}\\ \gamma_\mathrm{lt}\end{Bmatrix} \tag{25.17}$$

Example 25.2

A single sheet of woven ply is subjected to longitudinal and transverse direct stresses of 50 and 25 N/mm^2, respectively together with a shear stress of 40 N/mm^2. The elastic constants for the ply are $E_\mathrm{l} = 120\,000$ N/mm^2, $E_\mathrm{t} = 80\,000$ N/mm^2, $G_\mathrm{lt} = 5000$ N/mm^2 and $\nu_\mathrm{lt} = 0.3$. Calculate the direct strains in the longitudinal and transverse directions and the shear strain in the ply.

The value of the minor Poisson's ratio, ν_tl, is not given and must be calculated first. From Eq. (25.14)

$$\nu_\mathrm{tl} = \nu_\mathrm{lt}\frac{E_\mathrm{t}}{E_\mathrm{l}} = 0.3 \times \frac{80\,000}{120\,000} = 0.2$$

Therefore, from Eqs (25.11)

$$\varepsilon_\mathrm{l} = \frac{50}{120\,000} - \frac{0.2\times 25}{80\,000} = 3.54\times 10^{-4}$$

$$\varepsilon_\mathrm{t} = \frac{25}{80\,000} - \frac{0.3\times 50}{120\,000} = 1.88\times 10^{-4}$$

and from Eq. (25.16)

$$\gamma_\mathrm{lt} = \frac{40}{5000} = 80.0\times 10^{-4}$$

25.2.2 Generally orthotropic ply

In Fig. 25.7 the direction of the fibres in the ply does not coincide with the loading axes x and y. We shall specify that the longitudinal fibres of the ply are inclined at an angle θ to the x axis; θ is positive when the fibres are rotated in an anticlockwise sense from the x axis.

Suppose that an element of the ply is subjected to stresses σ_x, σ_y and τ_{xy} as shown in Fig. 25.8. The stresses on an element of the ply in the directions of the fibres may be found in terms of the applied stresses using the method described in Section 1.6. Therefore, by comparison with Eq. (1.8)

$$\sigma_{\mathrm{l}} = \sigma_x \cos^2\theta + \sigma_y \sin^2\theta + 2\tau_{xy}\cos\theta\sin\theta \tag{25.18}$$

Similarly

$$\sigma_{\mathrm{t}} = \sigma_x \sin^2\theta + \sigma_y \cos^2\theta - 2\tau_{xy}\cos\theta\sin\theta \tag{25.19}$$

and by comparison with Eq. (1.9) but noting that τ_{lt} is in the opposite sense to τ

$$\tau_{\mathrm{lt}} = -\sigma_x \cos\theta\sin\theta + \sigma_y \cos\theta\sin\theta + \tau_{xy}(\cos^2\theta - \sin^2\theta) \tag{25.20}$$

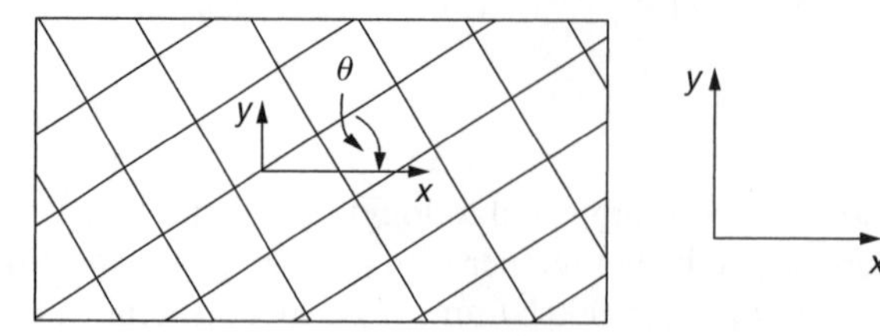

Fig. 25.7 Generally orthotropic ply.

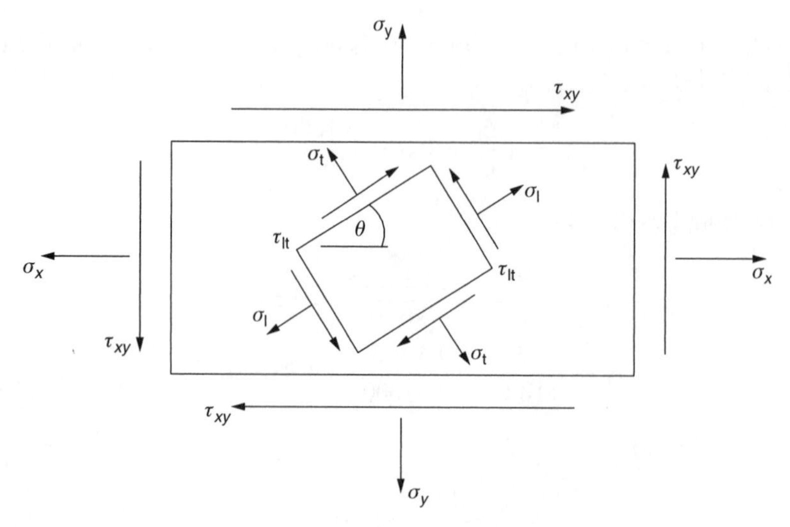

Fig. 25.8 Stresses in a generally orthotropic ply.

If we write $m = \cos\theta$ and $n = \sin\theta$, Eqs (25.18)–(25.20) become

$$\sigma_{\mathrm{l}} = m^2\sigma_x + n^2\sigma_y + 2mn\tau_{xy} \tag{25.21}$$

$$\sigma_{\mathrm{t}} = n^2\sigma_x + m^2\sigma_y - 2mn\tau_{xy} \tag{25.22}$$

$$\tau_{\mathrm{lt}} = -mn\sigma_x + mn\sigma_y + (m^2 - n^2)\tau_{xy} \tag{25.23}$$

Writing Eqs (25.21)–(25.23) in matrix form we have

$$\begin{Bmatrix} \sigma_{\mathrm{l}} \\ \sigma_{\mathrm{t}} \\ \tau_{\mathrm{lt}} \end{Bmatrix} = \begin{bmatrix} m^2 & n^2 & 2mn \\ n^2 & m^2 & -2mn \\ -mn & mn & m^2 - n^2 \end{bmatrix} \begin{Bmatrix} \sigma_x \\ \sigma_y \\ \tau_{xy} \end{Bmatrix} \tag{25.24}$$

Similarly, from Eqs (1.31) and (1.34)

$$\begin{Bmatrix} \varepsilon_{\mathrm{l}} \\ \varepsilon_{\mathrm{t}} \\ \gamma_{\mathrm{lt}} \end{Bmatrix} = \begin{bmatrix} m^2 & n^2 & mn \\ n^2 & m^2 & -mn \\ -2mn & 2mn & m^2 - n^2 \end{bmatrix} \begin{Bmatrix} \varepsilon_x \\ \varepsilon_y \\ \gamma_{xy} \end{Bmatrix} \tag{25.25}$$

Eqs (25.24) may be transposed so that the applied stresses are expressed in terms of the ply stresses. Then

$$\begin{Bmatrix} \sigma_x \\ \sigma_y \\ \tau_{xy} \end{Bmatrix} = \begin{bmatrix} m^2 & n^2 & -2mn \\ n^2 & m^2 & 2mn \\ mn & -mn & m^2 - n^2 \end{bmatrix} \begin{Bmatrix} \sigma_{\mathrm{l}} \\ \sigma_{\mathrm{lt}} \\ \tau_{\mathrm{lt}} \end{Bmatrix} \tag{25.26}$$

In Eqs (25.17) for a specially orthotropic ply the ply stresses and loading stresses are identical so that we may use this equation to relate the ply stresses in a generally orthotropic ply to the ply strains. Then

$$\begin{Bmatrix} \sigma_{\mathrm{l}} \\ \sigma_{\mathrm{t}} \\ \tau_{\mathrm{lt}} \end{Bmatrix} = \begin{bmatrix} \dfrac{E_{\mathrm{l}}}{1 - \nu_{\mathrm{lt}}\nu_{\mathrm{tl}}} & \dfrac{\nu_{\mathrm{tl}}E_{\mathrm{l}}}{1 - \nu_{\mathrm{lt}}\nu_{\mathrm{tl}}} & 0 \\ \dfrac{\nu_{\mathrm{lt}}E_{\mathrm{t}}}{1 - \nu_{\mathrm{lt}}\nu_{\mathrm{tl}}} & \dfrac{E_{\mathrm{t}}}{1 - \nu_{\mathrm{lt}}\nu_{\mathrm{tl}}} & 0 \\ 0 & 0 & G_{\mathrm{lt}} \end{bmatrix} \begin{Bmatrix} \varepsilon_{\mathrm{l}} \\ \varepsilon_{\mathrm{t}} \\ \gamma_{\mathrm{lt}} \end{Bmatrix} \tag{25.27}$$

Substituting for the ply stresses in Eqs (25.26) from Eqs (25.27) we express the applied stresses in terms of the ply strains, i.e.

$$\begin{Bmatrix} \sigma_x \\ \sigma_y \\ \tau_{xy} \end{Bmatrix} = \begin{bmatrix} m^2 & n^2 & -2mn \\ n^2 & m^2 & 2mn \\ mn & -mn & m^2 - n^2 \end{bmatrix} \begin{bmatrix} \dfrac{E_{\mathrm{l}}}{1 - \nu_{\mathrm{lt}}\nu_{\mathrm{tl}}} & \dfrac{\nu_{\mathrm{tl}}E_{\mathrm{l}}}{1 - \nu_{\mathrm{lt}}\nu_{\mathrm{tl}}} & 0 \\ \dfrac{\nu_{\mathrm{lt}}E_{\mathrm{t}}}{1 - \nu_{\mathrm{lt}}\nu_{\mathrm{tl}}} & \dfrac{E_{\mathrm{t}}}{1 - \nu_{\mathrm{lt}}\nu_{\mathrm{tl}}} & 0 \\ 0 & 0 & G_{\mathrm{lt}} \end{bmatrix} \begin{Bmatrix} \varepsilon_{\mathrm{l}} \\ \varepsilon_{\mathrm{t}} \\ \gamma_{\mathrm{lt}} \end{Bmatrix} \tag{25.28}$$

Finally, by substituting for the ply strains in Eqs (25.28) from Eqs (25.25) we obtain the applied stresses in terms of the strains referred to the *xy* axes, i.e.

$$\begin{Bmatrix}\sigma_x\\ \sigma_y\\ \tau_{xy}\end{Bmatrix} = \begin{bmatrix} m^2 & n^2 & -2mn\\ n^2 & m^2 & 2mn\\ mn & -mn & m^2-n^2\end{bmatrix}\begin{bmatrix}\dfrac{E_{\mathrm{l}}}{1-\nu_{\mathrm{lt}}\nu_{\mathrm{tl}}} & \dfrac{\nu_{\mathrm{tl}}E_{\mathrm{l}}}{1-\nu_{\mathrm{lt}}\nu_{\mathrm{tl}}} & 0\\ \dfrac{\nu_{\mathrm{lt}}E_{\mathrm{t}}}{1-\nu_{\mathrm{lt}}\nu_{\mathrm{tl}}} & \dfrac{E_{\mathrm{t}}}{1-\nu_{\mathrm{lt}}\nu_{\mathrm{tl}}} & 0\\ 0 & 0 & G_{\mathrm{lt}}\end{bmatrix}$$

$$\times\begin{bmatrix} m^2 & n^2 & mn\\ n^2 & m^2 & -mn\\ -2mn & 2mn & m^2-n^2\end{bmatrix}\begin{Bmatrix}\varepsilon_x\\ \varepsilon_y\\ \gamma_{xy}\end{Bmatrix} \tag{25.29}$$

Writing the individual terms of the central matrix as

$$k_{11} = \frac{E_{\mathrm{l}}}{1-\nu_{\mathrm{lt}}\nu_{\mathrm{tl}}} \quad k_{12} = \frac{\nu_{\mathrm{tl}}E_{\mathrm{l}}}{1-\nu_{\mathrm{lt}}\nu_{\mathrm{tl}}} = k_{21} \text{ (from Eq. (25.14))}$$

$$k_{22} = \frac{E_{\mathrm{t}}}{1-\nu_{\mathrm{lt}}\nu_{\mathrm{tl}}} \quad k_{33} = G_{\mathrm{lt}}$$

Eqs (25.29) become

$$\begin{Bmatrix}\sigma_x\\ \sigma_y\\ \tau_{xy}\end{Bmatrix} = \begin{bmatrix} m^2 & n^2 & -2mn\\ n^2 & m^2 & 2mn\\ mn & -mn & m^2-n^2\end{bmatrix}\begin{bmatrix} k_{11} & k_{12} & 0\\ k_{12} & k_{22} & 0\\ 0 & 0 & k_{33}\end{bmatrix}$$

$$\times\begin{bmatrix} m^2 & n^2 & mn\\ n^2 & m^2 & -mn\\ -2mn & 2mn & m^2-n^2\end{bmatrix}\begin{Bmatrix}\varepsilon_x\\ \varepsilon_y\\ \gamma_{xy}\end{Bmatrix} \tag{25.30}$$

Carrying out the matrix multiplication in Eqs (25.30) we obtain

$$\begin{Bmatrix}\sigma_x\\ \sigma_y\\ \tau_{xy}\end{Bmatrix} = \begin{bmatrix} \begin{matrix} m^4k_{11}+m^2n^2(2k_{12}\\ +4k_{33})+n^4k_{22}\end{matrix} & \begin{matrix} m^2n^2(k_{11}+k_{22}-4k_{33})\\ +(m^4+n^4)k_{12}\end{matrix} & \begin{matrix} m^3n(k_{11}-k_{12}-2k_{33})\\ +mn^3(k_{12}-k_{22}+2k_{33})\end{matrix}\\ \begin{matrix} m^2n^2(k_{11}+k_{22}-4k_{33})\\ +(m^4+n^4)k_{12}\end{matrix} & \begin{matrix} n^4k_{11}+m^2n^2(2k_{12}\\ +4k_{33})+m^4k_{22}\end{matrix} & \begin{matrix} mn^3(k_{11}-k_{12}-2k_{33})\\ +m^3n(k_{12}-k_{22}+2k_{33})\end{matrix}\\ \begin{matrix} m^3n(k_{11}-k_{12}-2k_{33})\\ +mn^3(k_{12}-k_{22}+2k_{33})\end{matrix} & \begin{matrix} mn^3(k_{11}-k_{12}-2k_{33})\\ +m^3n(k_{12}-k_{22}+2k_{33})\end{matrix} & \begin{matrix} m^2n^2(k_{11}-k_{22}-2k_{12}\\ -2k_{33})+(m^4+n^4)k_{33}\end{matrix}\end{bmatrix}\begin{Bmatrix}\varepsilon_x\\ \varepsilon_y\\ \gamma_{xy}\end{Bmatrix} \tag{25.31}$$

It can be seen that for a specially orthotropic ply where $\theta = 0$, Eqs (25.31) reduce to

$$\begin{Bmatrix} \sigma_x \\ \sigma_y \\ \tau_{xy} \end{Bmatrix} = \begin{bmatrix} k_{11} & k_{12} & 0 \\ k_{12} & k_{22} & 0 \\ 0 & 0 & k_{33} \end{bmatrix} \begin{Bmatrix} \varepsilon_x \\ \varepsilon_y \\ \gamma_{xy} \end{Bmatrix} \tag{25.32}$$

which are identical to Eqs (25.17).

Having expressed the applied stresses in terms of the xy strains Eqs (25.31) may be transposed to obtain the xy strains in terms of the applied stresses. This may be shown to be

$$\begin{Bmatrix} \varepsilon_x \\ \varepsilon_y \\ \gamma_{xy} \end{Bmatrix} = \begin{bmatrix} \begin{matrix} m^4 s_{11} + n^4 s_{22} \\ +2m^2n^2 s_{12} + m^2n^2 s_{33} \end{matrix} & \begin{matrix} m^2n^2 s_{11} + m^2n^2 s_{22} \\ +(m^4 + n^4) s_{12} \\ -m^2n^2 s_{33} \end{matrix} & \begin{matrix} 2m^3n s_{11} - 2mn^3 s_{22} \\ +2(mn^3 - m^3n) s_{12} \\ +(mn^3 - m^3n) s_{33} \end{matrix} \\ \begin{matrix} m^2n^2 s_{11} + m^2n^2 s_{22} \\ +(m^4 + n^4) s_{12} \\ -m^2n^2 s_{33} \end{matrix} & \begin{matrix} n^4 s_{11} + m^4 s_{22} \\ +2m^2n^2 s_{12} + m^2n^2 s_{33} \end{matrix} & \begin{matrix} 2mn^3 s_{11} - 2m^3n s_{22} \\ +2(m^3n - mn^3) s_{12} \\ +(m^3n - mn^3) s_{33} \end{matrix} \\ \begin{matrix} 2m^3n s_{11} - 2mn^3 s_{22} \\ +2(mn^3 - m^3n) s_{12} \\ +(mn^3 - m^3n) s_{33} \end{matrix} & \begin{matrix} 2mn^3 s_{11} - 2m^3n s_{22} \\ +2(m^3n - mn^3) s_{12} \\ +(m^3n - mn^3) s_{33} \end{matrix} & \begin{matrix} 4m^2n^2 s_{11} + 4m^2n^2 s_{22} \\ -8m^2n^2 s_{12} \\ +(m^2 - n^2)^2 s_{33} \end{matrix} \end{bmatrix} \begin{Bmatrix} \sigma_x \\ \sigma_y \\ \tau_{xy} \end{Bmatrix} \tag{25.33}$$

in which

$$s_{11} = 1/E_{\mathrm{l}}, \quad s_{12} = -\nu_{\mathrm{tl}}/E_{\mathrm{t}}, \quad s_{22} = 1/E_{\mathrm{t}}, \quad s_{33} = 1/G_{\mathrm{lt}}$$

For a specially orthotropic ply in which only direct stresses σ_x and σ_y are applied, Eqs (25.33) reduce to

$$\begin{Bmatrix} \varepsilon_x \\ \varepsilon_y \end{Bmatrix} = \begin{bmatrix} s_{11} & s_{12} \\ s_{12} & s_{22} \end{bmatrix} \begin{Bmatrix} \sigma_x \\ \sigma_y \end{Bmatrix}$$

i.e.

$$\begin{Bmatrix} \varepsilon_x \\ \varepsilon_y \end{Bmatrix} = \begin{bmatrix} \dfrac{1}{E_{\mathrm{l}}} & -\dfrac{\nu_{\mathrm{tl}}}{E_{\mathrm{t}}} \\ -\dfrac{\nu_{\mathrm{lt}}}{E_{\mathrm{l}}} & \dfrac{1}{E_{\mathrm{t}}} \end{bmatrix} \begin{Bmatrix} \sigma_x \\ \sigma_y \end{Bmatrix}$$

which are identical to Eqs (25.12).

Example 25.3

A generally orthotropic ply is subjected to direct stresses of $60\,\mathrm{N/mm^2}$ parallel to the x reference axis and $40\,\mathrm{N/mm^2}$ perpendicular to the x reference axis. If the longitudinal plies are inclined at an angle of 45° to the x axis and the elastic constants are

$E_l = 150\,000\,\text{N/mm}^2$, $E_t = 90\,000\,\text{N/mm}^2$, $G_{lt} = 5000\,\text{N/mm}^2$ and $\nu_{lt} = 0.3$, calculate the direct strains parallel to the x and y directions and the shear strain referred to the xy axes.

We note that there is no applied shear stress so that it is unnecessary to calculate the terms in the third column of the matrix of Eqs (25.33). Then

$$s_{11} = \frac{1}{E_l} = \frac{1}{150\,000} = 6.7 \times 10^{-6}$$

$$s_{22} = \frac{1}{E_t} = \frac{1}{90\,000} = 11.1 \times 10^{-6}$$

$$s_{12} = -\frac{\nu_{lt}}{E_l} = -\frac{0.3}{150\,000} = -2.0 \times 10^{-6}$$

$$s_{33} = \frac{1}{G_{lt}} = \frac{1}{5000} = 200 \times 10^{-6}$$

Also

$$\cos\theta = \sin\theta = \cos 45° = 1/\sqrt{2}$$

so that

$$m^2 = 0.5 = n^2, \quad m^4 = n^4 = 0.25, \quad m^2n^2 = 0.25, \text{ etc.}$$

Substituting these values in Eqs (25.33) we have

$$\begin{Bmatrix} \varepsilon_x \\ \varepsilon_y \\ \gamma_{xy} \end{Bmatrix} = \begin{bmatrix} 53.45 & -46.55 & - \\ -46.55 & 53.45 & - \\ -2.2 & -2.2 & - \end{bmatrix} \begin{Bmatrix} 60 \\ 40 \\ 0 \end{Bmatrix}$$

which gives

$$\varepsilon_x = 1345 \times 10^{-6}$$

$$\varepsilon_y = -655 \times 10^{-6}$$

$$\gamma_{xy} = -220 \times 10^{-6}$$

It should be noted that the above is an introduction into the analysis and design of composite materials. Complete texts[1,2] are devoted to the subject in which multi-ply laminates, laminate failure, residual thermal stresses, etc. are considered.

25.3 Thin-walled composite beams

We noted in Chapter 11 that some structural components in many modern aircraft are fabricated from composite materials. These components are generally in the form of laminates which are stacks of plies bonded together. The orientation of each ply will be different to that of its immediate neighbour so that the required strength and stiffness in a

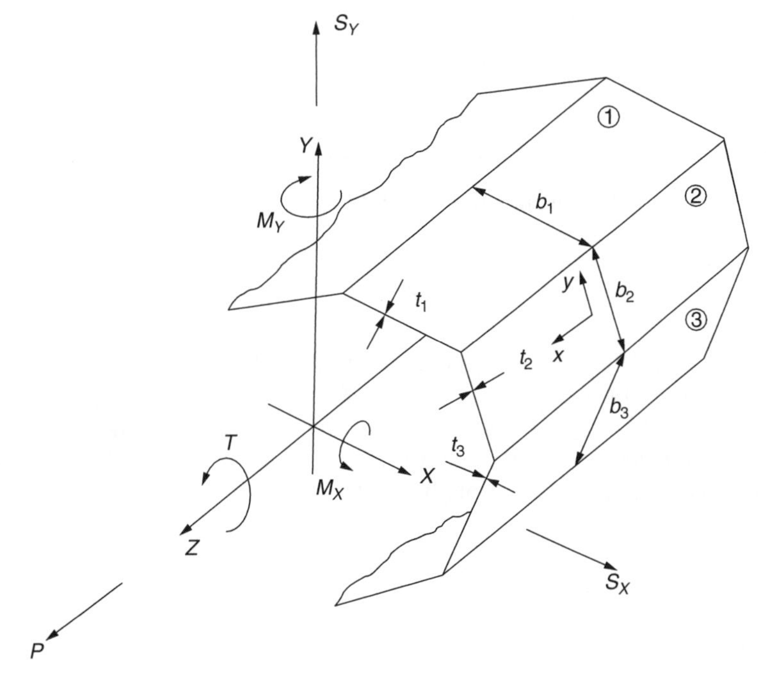

Fig. 25.9 Composite thin-walled section.

particular direction is obtained. The determination of the elastic properties of a laminate is discussed in Refs [1] and [2] and is lengthy so that we shall assume that these are known and concentrate on the effects of the composite construction on the analysis.

In Chapters 16–18 we determined stresses and displacements in open and closed section thin-walled beams subjected to bending, shear and torsional loads; the effect of axial load was considered in Chapter 1. We shall now re-examine these cases to determine the effect of composite construction.

Figure 25.9 shows a thin-walled beam which may be of either open or closed section and which is fabricated from laminates ①, ②, ③, ... The dimensions of each laminate are different as are their elastic properties.

The beam is subjected to axial, bending, shear and torsional loads which are positive in the directions shown (see also Fig 16.9). The beam axes *XYZ* are now in upper case letters to avoid confusion with the laminate axes *xy*.

25.3.1 Axial load

Suppose that the portion of the axial load P taken by the ith laminate is P_i. The longitudinal strain $\varepsilon_{x,i}$ in the laminate is equal to the longitudinal strain ε_z in the beam since one of the basic assumptions of our analysis, except in the case of torsion, is that plane sections remain plane after the load is applied. Then, from Eq. (1.40)

$$\frac{P_i}{b_i t_i} = \varepsilon_{x,i} E_{x,i}$$

Therefore

$$P_i = b_i t_i \, \varepsilon_{x,i} \, E_{x,i}$$

i.e.

$$P_i = \varepsilon_z \, b_i t_i \, E_{x,i} \tag{25.34}$$

The total axial load on the beam is then given by

$$P = \varepsilon_Z \sum_{i=1}^{n} b_i t_i \, E_{x,i} \tag{25.35}$$

Note that in Eq. (25.35) ε_z is the longitudinal strain in the beam section and is therefore the same for every laminate, it may therefore be taken outside the summation. Further, the value of Young's modulus for a particular laminate is the same whether referred to the laminate x axis or the beam Z axis; we shall therefore refer it to the beam Z axis. Equation (25.35) may therefore be written

$$P = \varepsilon_Z \sum_{i=1}^{n} b_i t_i \, E_{z,i} \tag{25.36}$$

from which

$$\varepsilon_Z = \frac{P}{\sum_{i=1}^{n} b_i t_i \, E_{z,i}} \tag{25.37}$$

Example 25.4

A beam has the singly symmetrical composite section shown in Fig. 25.10. The flange laminates are identical and have a Young's modulus, E_z, of 60 000 N/mm^2 while the

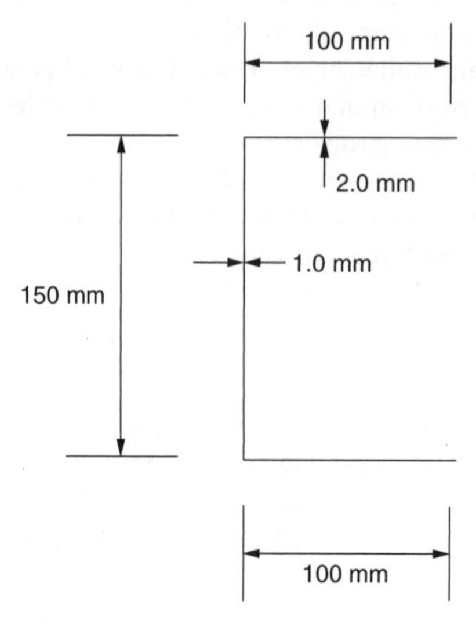

Fig. 25.10 Beam section of Example 25.4.

vertical web has a Young's modulus, E_z, of 20 000 N/mm^2. If the beam is subjected to an axial load of 40 kN determine the axial load in each laminate.

For each flange

$$b_i t_i E_{z,i} = 100 \times 2.0 \times 60\,000 = 12 \times 10^6$$

and for the web

$$b_i t_i E_{z,i} = 150 \times 1.0 \times 20\,000 = 3 \times 10^6$$

Therefore

$$\sum_{i=1}^{n} b_i t_i E_{z,i} = 2 \times 12 \times 10^6 + 3 \times 10^6 = 27 \times 10^6$$

Then, from Eq. (25.37)

$$\varepsilon_z = \frac{40 \times 10^3}{27 \times 10^6} = 1.48 \times 10^{-3}$$

Therefore, from Eq. (25.34)

$$P(\text{flanges}) = 1.48 \times 10^{-3} \times 12 \times 10^6 = 17\,760\,\text{N} = 17.76\,\text{kN}$$

$$P(\text{web}) = 1.48 \times 10^{-3} \times 3 \times 10^6 = 4440\,\text{N} = 4.44\,\text{kN}$$

Note that $2 \times 17.76 + 4.44 = 39.96$ kN, the discrepancy, 0.04 kN, is due to rounding off errors.

25.3.2 Bending

In Section 16.2 we derived an expression for the direct stress distribution in a beam of unsymmetrical cross-section (Eqs (16.18) or (16.19)). In this derivation the direct stress on an element of the beam cross-section was expressed in terms of Young's modulus, the radius of curvature of the beam, the coordinates of the element and the inclination of the neutral axis to the section x axis (see Eq. (16.16)). The beam was assumed to be comprised of homogenous material so that Young's modulus was a constant. This, as we have seen, is not necessarily the case for a composite beam where E can vary from laminate to laminate. We therefore rewrite Eq. (16.17) in the form

$$M_x = \int_A \frac{E_{z,i}}{\rho}(x \sin\alpha + y \cos\alpha) y \, dA, \quad M_y = \int_A \frac{E_{z,i}}{\rho}(x \sin\alpha + y \cos\alpha) x \, dA,$$

or

$$M_x = \frac{\sin\alpha}{\rho}\int_A E_{z,i}\, xy \, dA + \frac{\cos\alpha}{\rho}\int_A E_{z,i}\, y^2 \, dA,$$

$$M_y = \frac{\sin\alpha}{\rho}\int_A E_{z,i}\, x^2 \, dA + \frac{\cos\alpha}{\rho}\int_A E_{z,i}\, xy \, dA.$$

We therefore define modified second moments of area which include the laminate value of Young's modulus, $E_{Z,i}$, and which are referred to the XYZ axes of Fig. 25.10. Then

$$I'_{XX} = \int_A E_{Z,i}\, Y^2\, \mathrm{d}A, \quad I'_{YY} = \int_A E_{Z,i}\, X^2\, \mathrm{d}A, \quad I'_{XY} = \int_A E_{Z,i}\, XY\, \mathrm{d}A \tag{25.38}$$

so that

$$M_X = \frac{\sin\alpha}{\rho} I'_{XY} + \frac{\cos\alpha}{\rho} I'_{XX}$$

$$M_Y = \frac{\sin\alpha}{\rho} I'_{YY} + \frac{\cos\alpha}{\rho} I'_{XY}$$

Solving, we obtain

$$\frac{\sin\alpha}{\rho} = \frac{M_Y I'_{XX} - M_X I'_{XY}}{I'_{XX} I'_{YY} - I'^{2}_{XY}}$$

$$\frac{\cos\alpha}{\rho} = \frac{M_X I'_{YY} - M_Y I'_{XY}}{I'_{XX} I'_{YY} - I'^{2}_{XY}}$$

Then, from Eq. (16.16)

$$\sigma_Z = E_{Z,i}\left[\left(\frac{M_Y I'_{XX} - M_X I'_{XY}}{I'_{XX} I'_{YY} - I'^{2}_{XY}}\right)x + \left(\frac{M_X I'_{YY} - M_Y I'_{XY}}{I'_{XX} I'_{YY} - I'^{2}_{XY}}\right)y\right] \tag{25.39}$$

Note that the above applies equally to open or closed section thin-walled beams.

Example 25.5

A thin-walled beam has the composite cross-section shown in Fig. 25.11 and is subjected to a bending moment of 1 kN m applied in a vertical plane. If the values of Young's modulus for the flange laminates are each 50 000 N/mm^2 and that of the web is 15 000 N/mm^2 determine the maximum value of direct stress in the cross-section of the beam.

From Section 16.4.5 and Eqs (25.38)

$$I'_{XX} = 2 \times 50\,000 \times 50 \times 2.0 \times 50^2 + 15\,000 \times 1.0 \times \frac{100^3}{12} = 2.63 \times 10^{10}\ \mathrm{N\,mm^2}$$

$$I'_{YY} = 50\,000 \times 2.0 \times \frac{100^3}{12} = 0.83 \times 10^{10}\ \mathrm{N\,mm^2}$$

$$I'_{XY} = 50\,000 \times 50 \times 2.0(+50)(+50) + 50\,000 \times 50 \times 2.0(-50)(-50)$$

$$= 2.50 \times 10^{10}\ \mathrm{N\,mm^2}$$

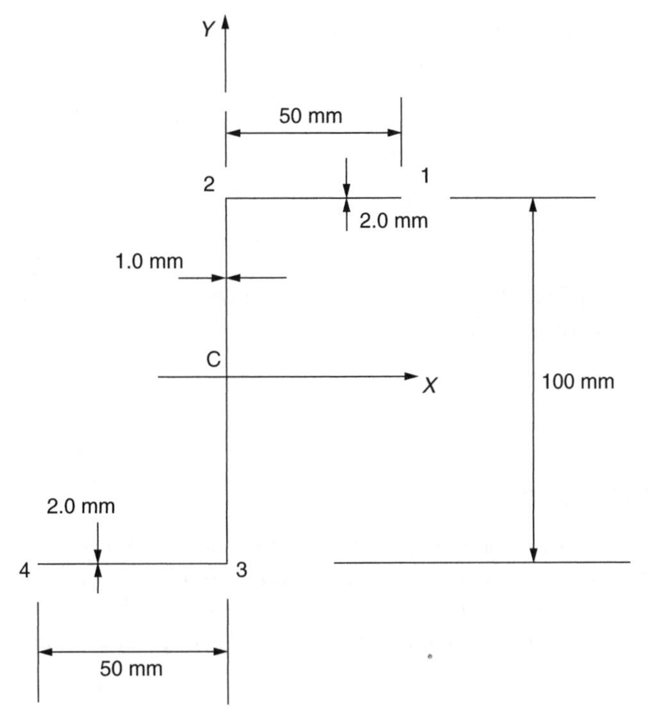

Fig. 25.11 Beam section of Example 25.5.

Also since $M_X = 1\,\text{kNm}$ and $M_Y = 0$, Eq. (25.39) becomes

$$\sigma_Z = E_{Z,i}\left[\frac{-1 \times 10^6 \times 2.5 \times 10^{10}}{10^{20}(2.63 \times 0.83 - 2.5^2)}X + \frac{1 \times 10^6 \times 0.83 \times 10^{10}}{10^{20}(2.63 \times 0.83 - 2.5^2)}Y\right]$$

i.e.

$$\sigma_Z = E_{Z,i}(6.15 \times 10^{-5}X - 2.04 \times 10^{-5}Y) \qquad \text{(i)}$$

On the top flange 12, $E_{Z,i} = 50\,000\,\text{N/mm}^2$ and $Y = 50\,\text{mm}$ so that Eq. (i) becomes

$$\sigma_Z = 3.08X - 51.0$$

Then

$$\sigma_{Z,1} = 3.08 \times 50 - 51.0 = 103.0\,\text{N/mm}^2$$

and

$$\sigma_{Z,2} = -51.0\,\text{N/mm}^2$$

In the web 23, $E_{Z,i} = 15\,000\,\text{N/mm}^2$ and $X = 0$. Equation (i) then becomes

$$\sigma_Z = -0.31Y$$

and

$$\sigma_{Z,2} = -15.5\,\text{N/mm}^2$$

The remaining distribution follows from antisymmetry so that the maximum direct stress in the beam cross-section is $\pm 103\,\text{N/mm}^2$.

25.3.3 Shear

Open section beams

In Section 17.2, we derived an expression for the shear flow distribution in an open section thin-walled beam subjected to shear loads (Eq. (17.14)). This is related to the direct stress distribution in the section (Eq. (17.2)) so that the arguments applied to composite section beams subjected to bending apply to the case of composite beams subjected to shear. Equation (17.14) then becomes

$$q_s = -E_{Z,i}\left[\left(\frac{S_X I'_{XX} - S_Y I'_{XY}}{I'_{XX} I'_{YY} - {I'_{XY}}^2}\right)\int_0^s t_i\, x\, \text{d}s + \left(\frac{S_Y I'_{YY} - S_X I'_{XX}}{I'_{XX} I'_{YY} - {I'_{XY}}^2}\right)\int_0^s t_i\, Y\, \text{d}s\right] \tag{25.40}$$

Note that in Eq. (25.40) s is measured from an open edge in the beam section and the second moments of area are those defined in Eq. (25.38).

Closed section beams

Again the same arguments apply to the composite case as before and Eq. (17.15) becomes

$$q_s = -E_{Z,i}\left[\left(\frac{S_X I'_{XX} - S_Y I'_{XY}}{I'_{XX} I'_{YY} - {I'_{XY}}^2}\right)\int_0^s t_i\, x\, \text{d}s + \left(\frac{S_Y I'_{YY} - S_X I'_{XX}}{I'_{XX} I'_{YY} - {I'_{XY}}^2}\right)\int_0^s t_i\, y\, \text{d}s\right] + q_{s,0} \tag{25.41}$$

In Eq. (25.41) the value of the shear flow, $q_{s,0}$, at the origin for s is found using either of Eqs (17.17) or (17.18).

Example 25.6

The composite triangular section thin-walled beam shown in Fig. 25.12 carries a vertical shear load of 2 kN applied at the apex. If the walls 12 and 13 have a laminate Young's modulus of $45\,000\,\text{N/mm}^2$ while that of the vertical web 23 is $20\,000\,\text{N/mm}^2$ determine the shear flow distribution in the section.

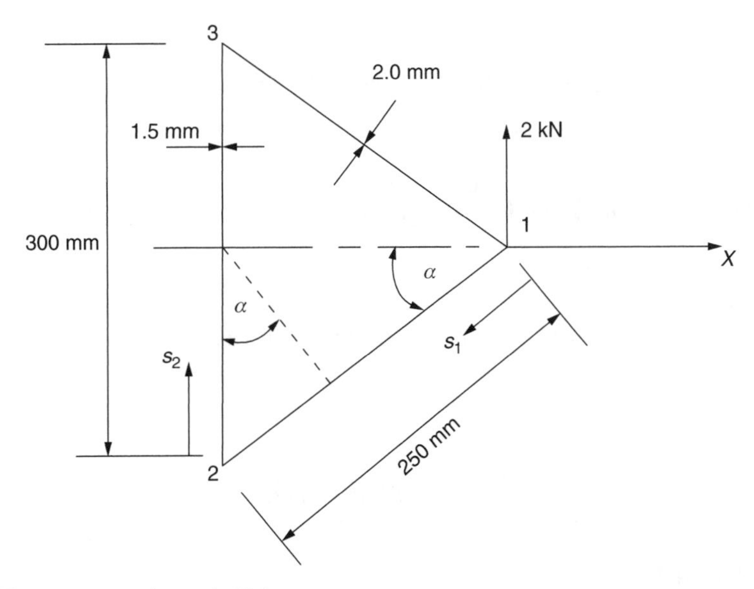

Fig. 25.12 Beam section of Example 25.6.

The X axis is an axis of symmetry so that $I'_{XY} = 0$ and, since $S_X = 0$, Eq. (25.41) reduces to

$$q_s = -E_{Z,i}\frac{S_Y}{I'_{XX}}\int_0^s tY\,\mathrm{d}s + q_{s,0} \tag{i}$$

From Section 16.4.5 and Eq. (25.38)

$$I'_{XX} = \frac{2 \times 45\,000 \times 2.0 \times 250^3(150/250)^2}{12} + \frac{20\,000 \times 1.5 \times 300^3}{12}$$
$$= 15.2 \times 10^{10}\,\mathrm{N\,mm^2}$$

'Cut' the section at 1. Then, from the first term on the right-hand side of Eq. (i)

$$q_{b,12} = -\frac{45\,000 \times 2 \times 10^3}{15.2 \times 10^{10}}\int_0^{s_1} 2.0(-s_1 \sin\alpha)\mathrm{d}s_1$$

in which $\sin\alpha = 150/250 = 0.6$. Therefore

$$q_{b,12} = 3.6 \times 10^{-4} s_1^2 \tag{ii}$$

so that

$$q_{b,2} = 22.2\,\mathrm{N/mm}$$

Also

$$q_{b,23} = -\frac{20\,000 \times 2 \times 10^3}{15.2 \times 10^{10}}\int_0^{s_2} 1.5(-150 + s_2)\mathrm{d}s_2 + 22.2$$

from which

$$q_{b,23} = 0.06s_2 - 1.95 \times 10^{-4}s_2^2 + 22.2 \qquad \text{(iii)}$$

Taking moments about the mid-point of the wall 23 (or about point 1) we have

$$2 \times 10^3 \times 250 \cos\alpha = -2\int_0^{250} q_{b,12} 150 \cos\alpha \, \mathrm{d}s_2 + 2 \times \frac{300}{2} \times (250\cos\alpha) q_{s,0}$$

which gives

$$q_{s,0} = 14.2 \text{ N/mm (in an anticlockwise sense)}$$

The shear flow distribution is then

$$q_{12} = 3.6 \times 10^{-4}s_1^2 - 14.2$$

$$q_{23} = -1.95 \times 10^{-4}s_2^2 + 0.06s_2 + 8.0$$

25.3.4 Torsion

Closed section beams

We shall consider composite closed section beams first since, as we saw in Chapters 17 and 18, the strain–displacement relationships derived for open and closed section beams subjected to shear loads apply to the torsion of closed section beams so that the analysis follows logically on.

The shear flow distribution in a closed section thin-walled beam subjected to a torque in which the warping is unrestrained is given by Eq. (18.1), i.e.

$$T = 2Aq$$

or

$$q = \frac{T}{2A} \qquad (25.42)$$

The derivation of Eq. (25.42) is based purely on equilibrium considerations and does not, therefore, rely on the elastic properties of the beam section. Equation (25.42) therefore applies equally to composite as well as to isotropic beam sections.

The rate of twist of a closed section beam subjected to a torque is given by Eq. (18.4), i.e.

$$\frac{\mathrm{d}\theta}{\mathrm{d}z} = \frac{T}{4A^2}\oint \frac{\mathrm{d}s}{Gt}$$

This expression also applies to a composite closed section beam provided that the shear modulus G remains within the integration and that the laminate shear modulus $G_{XY,i}$ is used as appropriate. Equation (18.4) then becomes

$$\frac{\mathrm{d}\theta}{\mathrm{d}Z} = \frac{T}{4A^2}\oint \frac{\mathrm{d}s}{G_{XY,i}\,t_i} \qquad (25.43)$$

Rearranging

$$T = \frac{4A^2}{\oint \frac{\mathrm{d}s}{G_{XY,i}\,t_i}} \frac{\mathrm{d}\theta}{\mathrm{d}Z} \tag{25.44}$$

We saw in Chapter 3, Eq. (3.12), that the torque and rate of twist in a beam are related by the torsional stiffness GJ. Therefore, from Eq. (25.44), we see that the torsional stiffness of a composite closed section beam is given by

$$GJ = \frac{4A^2}{\oint \frac{\mathrm{d}s}{G_{XY,i}\,t_i}} \tag{25.45}$$

The above arguments apply to the determination of the warping distribution in a closed section composite beam. This is then given by (see the derivation of Eq. (18.5))

$$W_s - W_0 = q \int_0^s \frac{\mathrm{d}s}{G_{XY,i}\,t_i} - \frac{A_{0s}}{A} q \oint \frac{\mathrm{d}s}{G_{XY,i}\,t_i} \tag{25.46}$$

or, from Eq. (25.42) in terms of the applied torque

$$W_s - W_0 = \frac{T}{2A}\left(\int_0^s \frac{\mathrm{d}s}{G_{XY,i}\,t_i} - \frac{A_{0s}}{A} \oint \frac{\mathrm{d}s}{G_{XY,i}\,t_i} \right) \tag{25.47}$$

Example 25.7

The rectangular section, thin-walled, composite beam shown in Fig. 25.13 is subjected to a torque of 10 kN m. If the laminate shear modulus of the covers is 20 000 N/mm^2 and that of the webs is 35 000 N/mm^2 determine the shear flow distribution in the section and the distribution of warping.

The shear flow distribution is obtained from Eq. (25.42) and is

$$q = \frac{10 \times 10^6}{2 \times 200 \times 100} = 250\,\mathrm{N/mm}$$

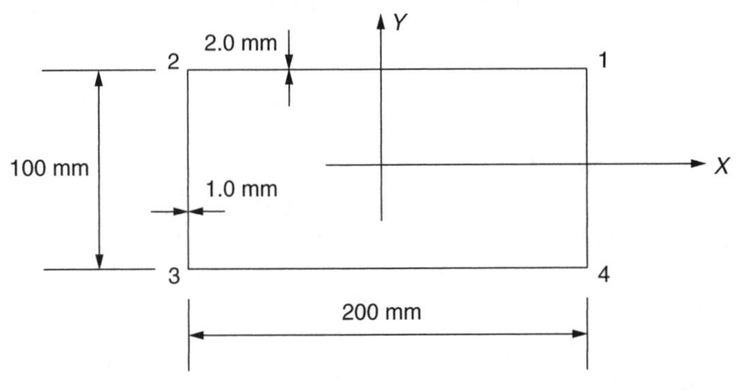

Fig. 25.13 Beam section of Example 25.7.

The warping distribution is given by Eq. (25.47) in which

$$\oint \frac{\mathrm{d}s}{G_{XY,i}\,t_i} = \frac{2 \times 200}{20\,000 \times 2.0} + \frac{2 \times 100}{35\,000 \times 1.0} = 0.0157$$

Eq. (25.47) then becomes

$$W_s - W_0 = 250\left(\int_0^s \frac{\mathrm{d}s}{G_{XY,i}\,t_i} - \frac{A_{Os}}{200 \times 100} \times 0.0157\right)$$

or

$$W_2 - W_0 = 250\left(\int_0^s \frac{\mathrm{d}s}{G_{XY,i}\,t_i} - 0.785 \times 10^{-6} A_{Os}\right) \qquad \text{(i)}$$

We saw in Example 18.2 that the warping distribution in a rectangular section thin-walled beam is linear with zero values at the mid-points of the webs and covers. The same situation applies in this example so that it is only necessary to calculate the value of warping at, say, corner 1. Then, from Eq. (i)

$$W_1 = 250\left(\frac{50}{35\,000 \times 1.0} - 0.785 \times 10^{-6} \times 100 \times 50\right)$$

which gives

$$W_1 = -0.62\,\mathrm{mm}$$

The remaining distribution follows from symmetry.

Open section beams

The torsional stiffness of an open section thin-walled beam is, as for a closed section beam, GJ, but in which the torsion constant, J, is given by either of Eqs (18.11). However, for a composite beam section the shear modulus must be taken inside the summation or integral and will be the laminate shear modulus $G_{XY,i}$. Then

$$GJ = \sum_{i=1}^{n} G_{XY,i} \frac{s t_i^3}{3} \quad \text{or} \quad GJ = \frac{1}{3}\int_{\text{sect}} G_{XY,i}\, t_i^3\, \mathrm{d}s \qquad (25.48)$$

The rate of twist of a beam is related to the applied torque by Eq. (3.12). For a composite open section beam the relationship holds but the torsional stiffness is given by either of Eqs (25.48), i.e.

$$T = \left(\sum_{i=1}^{n} G_{XY,i} \frac{s t_i^3}{3}\right)\frac{\mathrm{d}\theta}{\mathrm{d}Z} \quad \text{or} \quad T = \left(\frac{1}{3}\int_{\text{sect}} G_{XY,i}\, t_i^3\, \mathrm{d}s\right)\frac{\mathrm{d}\theta}{\mathrm{d}Z} \qquad (25.49)$$

Having obtained the rate of twist Eq. (18.9) gives the shear stress distribution across the thickness at any point round the beam section, i.e.

$$\tau = 2G_{XY,i}\, n \frac{d\theta}{dZ} \qquad (25.50)$$

Again the maximum shear stress will occur at the surface of the beam section where $n = \pm t/2$.

The primary warping distribution follows from Eq. (18.19) in which the rate of twist is found from either of Eqs (25.49).

Example 25.8

A composite channel section has the dimensions shown in Fig. 18.12 and is subjected to a torque of 10 Nm. If the flanges have a laminate shear modulus of 20 000 N/mm^2 and that of the web is 15 000 N/mm^2 determine the maximum shear stress in the beam section and the distribution of warping assuming that the beam is constrained to twist about an axis through the mid-point of the web.

The torsional stiffness of the section is obtained from the first of Eqs (25.48) and is

$$GJ = 2 \times 20\,000 \times 25 \times \frac{1.5^3}{3} + 15\,000 \times 50 \times \frac{2.5^3}{3} = 5.03 \times 10^6\,\text{N mm}^2$$

Then, from Eq. (25,49)

$$\frac{d\theta}{dZ} = \frac{10 \times 10^3}{5.03 \times 10^6} = 1.99 \times 10^{-3}$$

and from Eq. (25.50)

$$\tau_{max}(12) = 2 \times 20\,000 \times (1.5/2) \times 1.99 \times 10^{-3} = 59.7\,\text{N/mm}^2$$

$$\tau_{max}(23) = 2 \times 15\,000 \times (2.5/2) \times 1.99 \times 10^{-3} = 74.6\,\text{N/mm}^2$$

The maximum therefore occurs in the web and is 74.6 N/mm^2.

The section is constrained to twist about an axis through the mid-point of the web so that W is zero everywhere in the web. Then, from Eq. (18.19)

$$W_1 = -2 \times \frac{1}{2} \times 25 \times 25 \times 1.99 \times 10^{-3} = -1.24\,\text{mm}$$

The warping is linear along the flange 12, the warping along the flange 34 follows from symmetry.

Note that if the axis of twist had not been specified the position of the shear centre of the section would have had to have been found using the method previously described.

References

1 Calcote, L. R., *The Analysis of Laminated Composite Structures*, Van Nostrand Reinhold Co., New York, 1969.
2 Datoo, M. H., *Mechanics of Fibrous Composites*, Elsevier Applied Science, London, 1991.

Problems

P.25.1 A bar, whose cross-section is shown in Fig. P.25.1, comprises a polyester matrix and Kevlar filaments; the respective moduli are 3000 and 140 000 N/mm^2 with corresponding Poisson's ratios of 0.16 and 0.28. If the bar is 1 m long and is subjected to a compressive axial load of 500 kN, determine the shortening of the bar, the increase in its thickness and the stresses in the polyester and Kevlar.

Ans. 3.26 mm, 0.032 mm, 9.78 N/mm^2, 456.4 N/mm^2.

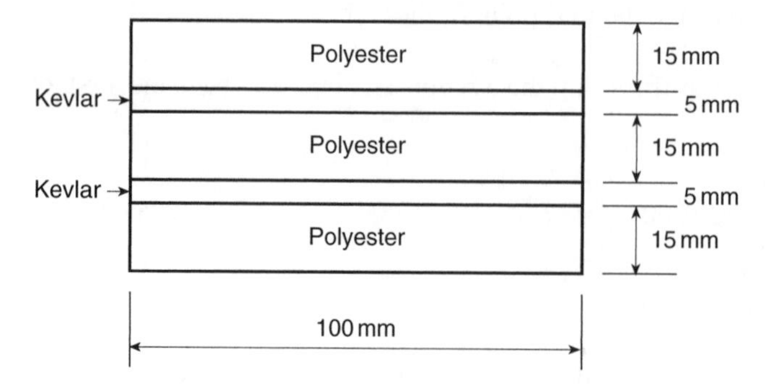

Fig. P.25.1

P.25.2 A box beam has the thin-walled composite cross-section shown in Fig. P.25.2. The cover laminates are identical and have a Young's modulus of 20 000 N/mm^2 while that of the vertical webs is 60 000 N/mm^2. If the beam is subjected to an axial load of 40 kN determine the axial force in each laminate.

Ans. Covers, 4 kN; webs,16 kN.

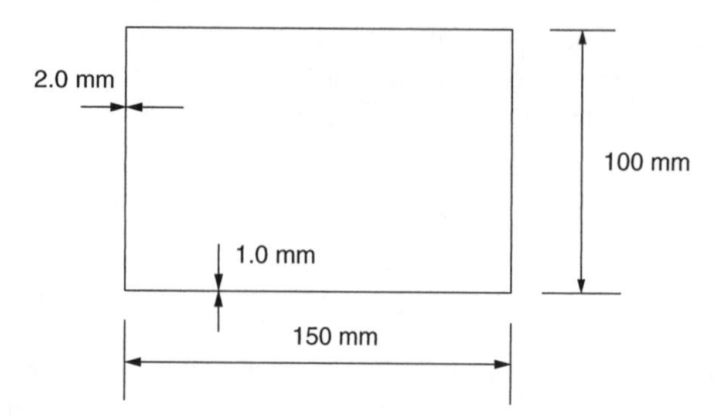

Fig. P.25.2

P.25.3 If the thin-walled box beam of Fig. P.25.2 carries a bending moment of 1 kN m applied in a vertical plane, determine the maximum direct stress in the cross-section of the beam.

Ans. 85.8 N/mm^2.

P.25.4 If the thin-walled composite beam of Example 25.5 is subjected to a bending moment of 0.5 kN m applied in a horizontal plane calculate the maximum value of direct stress in the beam section.

Ans. 76.8 N/mm^2.

P.25.5 The thin-walled composite beam section of Example 25.5 carries a vertical shear load of 2 kN applied in the plane of the web. Determine the shear flow distribution.

Ans. $q_{12} = 0.00575s_1^2 - 0.385s_1$
$q_{23} = 0.0287s_2 - 2.865 \times 10^{-4}s_2^2 - 4.875$.

P.25.6 The closed, composite section, thin-walled beam shown in Fig. P.25.6 is subjected to a vertical shear load of 20 kN applied through its centre of symmetry. If the laminate elastic properties are: for the covers, $E_{Z,i} = 54\,100$ N/mm^2; for the webs $E_{Z,i} = 17\,700$ N/mm^2, determine the distribution of shear flow round the cross-section.

Ans. $q_{01} = -1.98s_1$, $q_{12} = 6.5 \times 10^{-3}s_2^2 - 0.325s_2 - 198$.

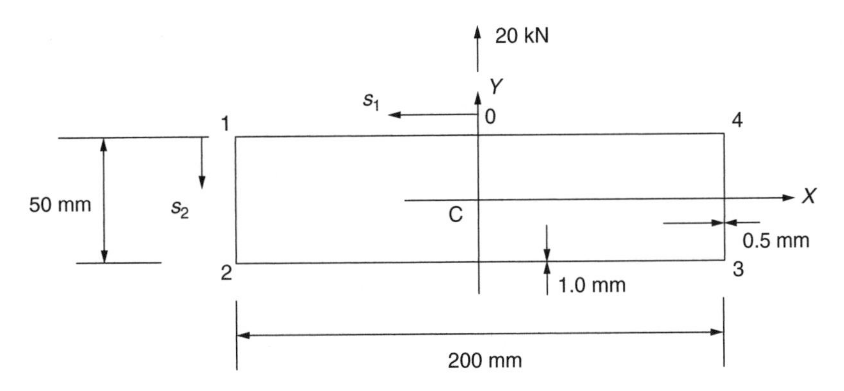

Fig. P.25.6

P.25.7 The beam section shown in Fig. P.25.6 is subjected to an anticlockwise torque of 1 kN m. If the laminate shear modulus of the covers is 20 700 N/mm^2 and that of the webs is 36 400 N/mm^2 determine the maximum shear stress in the section, its rate of twist and the distribution of warping.

Ans. 100 N/mm^2, 6.25×10^{-5} rad/mm, -0.086 mm (at 4, zero at 0).

P.25.8 The thin-walled, composite beam section shown in Fig. P.25.8 has laminate shear moduli of 16 300 N/mm^2 for the flanges and 20 900 N/mm^2 for the web. If the beam is subjected to a torque of 0.5 kN mm determine the rate of twist in the section, the maximum shear stress and the value of warping at the point 1.

Ans. 0.8×10^{-3} rad/mm, ± 13 N/mm^2 (in flanges), 2.0 mm.

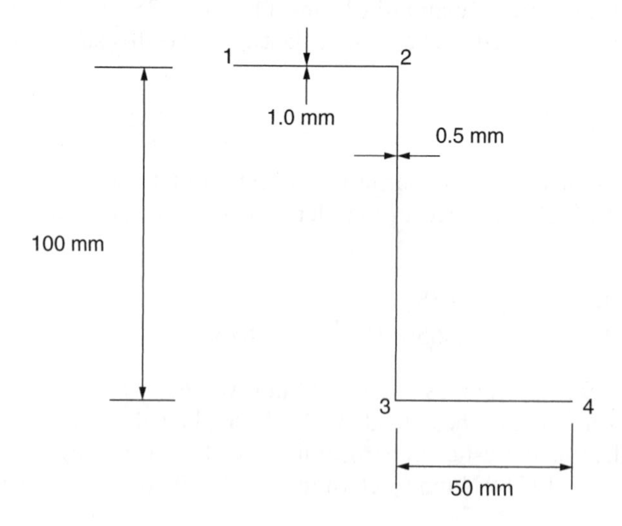

Fig. P.25.8

Section B5 Structural and Loading Discontinuities

26

Closed section beams

The analysis presented in Chapters 16–20 relies on elementary theory for the determination of stresses and displacements produced by axial loads, shear forces and bending moments and torsion. No allowance is made for the effects of restrained warping produced by structural or loading discontinuities in the torsion of open or closed section beams, or for the effects of shear strains on the calculation of direct and shear stresses in beams subjected to bending and shear.

In this chapter we shall examine some relatively simple examples of the above effects; more complex cases require analysis by computer-based techniques such as the finite element method.

26.1 General aspects

Structural constraint stresses in either closed or open beams result from a restriction on the freedom of any section of the beam to assume its normal displaced shape under load. Such a restriction arises when one end of the beam is built-in although the same effect may be produced practically, in a variety of ways. For example, the root section of a beam subjected to torsion is completely restrained from warping into the displaced shape indicated by Eq. (18.5) and a longitudinal stress system is induced which, in a special case discussed later, is proportional to the free warping of the beam.

A slightly different situation arises when the beam supports shear loads. The stress system predicted by elementary bending theory relies on the basic assumption of plane sections remaining plane after bending. However, for a box beam comprising thin skins and booms, the shear strains in the skins are of sufficient magnitude to cause a measurable redistribution of direct load in the booms and hence previously plane sections warp. We shall discuss the phenomenon of load redistribution resulting from shear, known as *shear lag*, in detail later in the chapter. The prevention of this warping by some form of axial constraint modifies the stress system still further.

The most comprehensive analysis yet published of multi-cell and single cell beams under arbitrary loading and support conditions is that by Argyris and Dunne.[1] Their work concentrates in the main on beams of idealized cross-section and while the theory they present is in advance of that required here, it is beneficial to examine some of the results of their analysis. We shall limit the present discussion to closed beams of idealized cross-section.

The problem of axial constraint may be conveniently divided into two parts. In the first, the shear stress distribution due to an arbitrary loading is calculated exclusively at the built-in end of the beam. In the second, the stress (and/or load) distributions are calculated along the length of the beam for the separate loading cases of torsion and shear. Obviously the shear stress systems predicted by each portion of theory must be compatible at the built-in end.

Argyris and Dunne showed that the calculation of the shear stress distribution at a built-in end is a relatively simple problem, the solution being obtained for any loading and beam cross-section by statics. More complex is the determination of the stress distributions at sections along the beam. These stresses, for the torsion case, are shown to be the sum of the stresses predicted by elementary theory and stresses caused by systems of self-equilibrating end loads. For a beam supporting shear loads the total stresses are again the sum of those corresponding to elementary bending theory and stresses due to systems of self-equilibrating end loads.

For an n-boom, idealized beam, Argyris and Dunne found that there are $n-3$ self-equilibrating end load, or *eigenload*, systems required to nullify $n-3$ possible modes of warping displacement. These eigenloads are analogous to, say, the buckling loads corresponding to the different buckled shapes of an elastic strut. The fact that, generally, there are a number of warping displacements possible in an idealized beam invalidates the use of the shear centre or flexural axis as a means of separating torsion and shear loads. For, associated with each warping displacement is an axis of twist that is different for each warping mode. In practice, a good approximation is obtained if the torsion loads are referred to the axis of twist corresponding to the lowest eigenload. Transverse loads through this axis, the *zero warping axis* produce no warping due to twist, although axial constraint stresses due to shear will still be present.

In the special case of a doubly symmetrical section the problem of separating the torsion and bending loads does not arise since it is obvious that the torsion loads may be referred to the axis of symmetry. Double symmetry has the further effect of dividing the eigenloads into four separate groups corresponding to $(n/4)-1$ pure flexural modes in each of the xz and yz planes, $(n/4)$ pure twisting modes about the centre of symmetry and $(n/4)-1$ pure warping modes which involve neither flexure nor twisting. A doubly symmetrical six boom beam supporting a single shear load has therefore just one eigenload system if the centre boom in the top and bottom panels is regarded as being divided equally on either side of the axis of symmetry thereby converting it, in effect, into an eight boom beam.

It will be obvious from the above that, generally, the self-equilibrating stress systems cannot be proportional to the free warping of the beam unless the free warping can be nullified by just one eigenload system. This is true only for the four boom beam which, from the above, has one possible warping displacement. If, in addition, the beam is doubly symmetrical then its axis of twist will pass through the centre of symmetry. We note that only in cases of doubly symmetrical beams do the zero warping and flexural axes coincide.

A further special case arises when the beam possesses the properties of a Neuber beam (Section 18.1.2) which does not warp under torsion. The stresses in this case are the elementary torsion theory stresses since no constraint effects are present. When bending loads predominate, however, it is generally impossible to design an efficient structure which does not warp.

In this chapter the calculation of spanwise stress distributions in closed section beams is limited to simple cases of beams having doubly symmetrical cross-sections. It should be noted that simplifications of this type can be misleading in that some of the essential characteristics of beam analysis, for example the existence of the $n-3$ self-equilibrating end load systems, vanish.

26.2 Shear stress distribution at a built-in end of a closed section beam

This special case of structural constraint is of interest due to the fact that the shear stress distribution at the built-in end of a closed section beam is statically determinate. Figure 26.1 represents the cross-section of a thin-walled closed section beam at its built-in end. It is immaterial for this analysis whether or not the section is idealized since the expression for shear flow in Eq. (17.19), on which the solution is based, is applicable to either case. The beam supports shear loads S_x and S_y which generally will produce torsion in addition to shear. We again assume that the cross-section of the beam remains undistorted by the applied loads so that the displacement of the beam cross-section is completely defined by the displacements u, v, w and the rotation θ referred to an arbitrary system of axes Oxy. The shear flow q at any section of the beam is then given by Eq. (17.20), that is

$$q = Gt\left(p\frac{\mathrm{d}\theta}{\mathrm{d}z} + \frac{\mathrm{d}u}{\mathrm{d}z}\cos\psi + \frac{\mathrm{d}v}{\mathrm{d}z}\sin\psi + \frac{\partial w}{\partial s}\right)$$

At the built-in end, $\partial w/\partial s$ is zero and hence

$$q = Gt\left(p\frac{\mathrm{d}\theta}{\mathrm{d}z} + \frac{\mathrm{d}u}{\mathrm{d}z}\cos\psi + \frac{\mathrm{d}v}{\mathrm{d}z}\sin\psi\right) \tag{26.1}$$

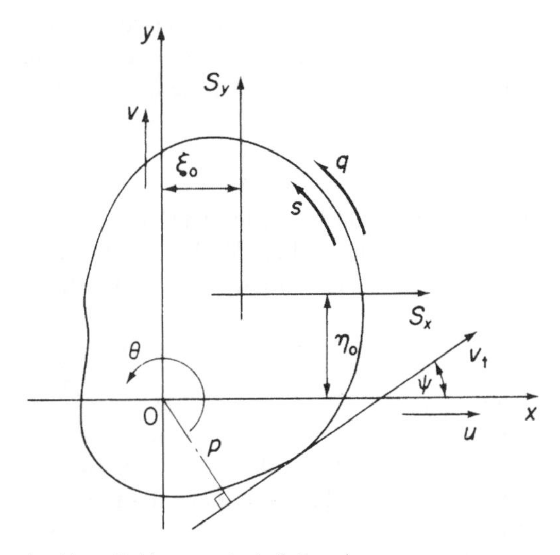

Fig. 26.1 Cross-section of a thin-walled beam at the built-in end.

in which $\mathrm{d}\theta/\mathrm{d}z$, $\mathrm{d}u/\mathrm{d}z$ and $\mathrm{d}v/\mathrm{d}z$ are the unknowns, the remaining terms being functions of the section geometry.

The resultants of the internal shear flows q must be statically equivalent to the applied loading, so that

$$\left.\begin{aligned} \oint q\cos\psi\,\mathrm{d}s &= S_x \\ \oint q\sin\psi\,\mathrm{d}s &= S_y \\ \oint qp\,\mathrm{d}s &= S_y\xi_0 - S_x\eta_0 \end{aligned}\right\} \tag{26.2}$$

Substitution for q from Eq. (26.1) in Eqs (26.2) yields

$$\left.\begin{aligned} \frac{\mathrm{d}\theta}{\mathrm{d}z}\oint tp\cos\psi\,\mathrm{d}s + \frac{\mathrm{d}u}{\mathrm{d}z}\oint t\cos^2\psi\,\mathrm{d}s + \frac{\mathrm{d}v}{\mathrm{d}z}\oint t\cos\psi\sin\psi\,\mathrm{d}s &= \frac{S_x}{G} \\ \frac{\mathrm{d}\theta}{\mathrm{d}z}\oint tp\sin\psi\,\mathrm{d}s + \frac{\mathrm{d}u}{\mathrm{d}z}\oint t\sin\psi\cos\psi\,\mathrm{d}s + \frac{\mathrm{d}v}{\mathrm{d}z}\oint t\sin^2\psi\,\mathrm{d}s &= \frac{S_y}{G} \\ \frac{\mathrm{d}\theta}{\mathrm{d}z}\oint tp^2\,\mathrm{d}s + \frac{\mathrm{d}u}{\mathrm{d}z}\oint tp\cos\psi\,\mathrm{d}s + \frac{\mathrm{d}v}{\mathrm{d}z}\oint tp\sin\psi\,\mathrm{d}s &= \frac{(S_y\xi_0 - S_x\eta_0)}{G} \end{aligned}\right\} \tag{26.3}$$

Equations (26.3) are solved simultaneously for $\mathrm{d}\theta/\mathrm{d}z$, $\mathrm{d}u/\mathrm{d}z$ and $\mathrm{d}v/\mathrm{d}z$. These values are then substituted in Eq. (26.1) to obtain the shear flow, and hence the shear stress distribution.

Attention must be paid to the signs of ψ, p and q in Eqs (26.3). Positive directions for each parameter are suggested in Fig. 26.1 although alternative conventions may be adopted. In general, however, there are rules which must be obeyed, these having special importance in the solution of multicell beams. Briefly, these are as follows. The positive directions of q and s are the same but may be assigned arbitrarily in each wall. Then p is positive if movement of the foot of the perpendicular along the positive direction of the tangent leads to an anticlockwise rotation of p about O. ψ is the clockwise rotation of the tangent vector necessary to bring it into coincidence with the positive direction of the x axis.

Example 26.1

Calculate the shear stress distribution at the built-in end of the beam shown in Fig. 26.2(a) when, at this section, it carries a shear load of 22 000 N acting at a distance of 100 mm from and parallel to side 12. The modulus of rigidity G is constant throughout the section:

Wall	12	34	23
Length (mm)	375	125	500

It is helpful at the start of the problem to sketch the notation and sign convention as shown in Fig. 26.2(b). The walls of the beam are flat and therefore p and ψ are constant along each wall. Also the thickness of each wall is constant so that the shear flow q is independent of s in each wall. Let point 1 be the origin of the axes, then, writing

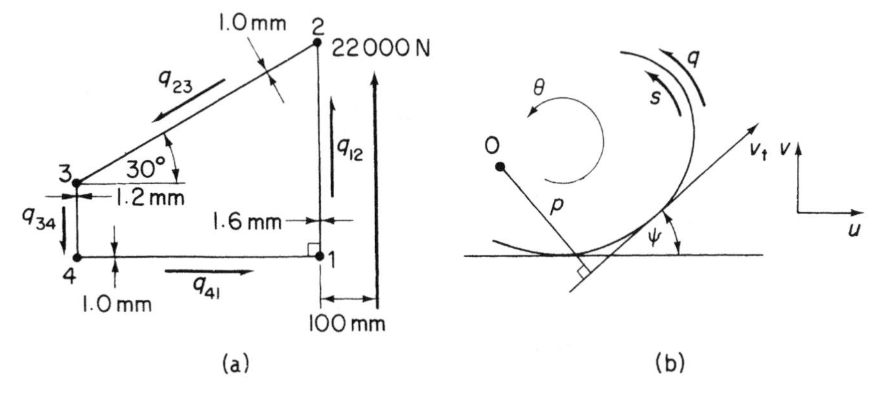

Fig. 26.2 (a) Beam cross-section at built-in end; (b) notation and sign convention.

$\theta' = \mathrm{d}\theta/\mathrm{d}z$, $u' = \mathrm{d}u/\mathrm{d}z$ and $v' = \mathrm{d}v/\mathrm{d}z$, we obtain from Eq. (26.1)

$$q_{12} = 1.6Gv' \tag{i}$$

$$q_{23} = 1.0G(375 \times 0.886\theta' - 0.886u' - 0.5v') \tag{ii}$$

$$q_{34} = 1.2G(500 \times 0.866\theta' - v') \tag{iii}$$

$$q_{41} = 1.0Gu' \tag{iv}$$

For horizontal equilibrium

$$500 \times 0.886q_{41} - 500 \times 0.886q_{23} = 0$$

giving

$$q_{41} = q_{23} \tag{v}$$

For vertical equilibrium

$$375q_{12} - 125q_{34} - 250q_{23} = 22\,000 \tag{vi}$$

For moment equilibrium about point 1

$$500 \times 375 \times 0.886q_{23} + 125 \times 500 \times 0.886q_{34} = 22\,000 \times 100$$

or

$$3q_{23} + q_{34} = 40.6 \tag{vii}$$

Substituting for q_{12}, etc. from Eqs (i), (ii), (iii) and (iv) into Eqs (v), (vi) and (vii), and solving for θ', u' and v', gives $\theta' = 0.122/G$, $u' = 9.71/G$, $v' = 42.9/G$. The values of θ', u' and v' are now inserted in Eqs (i), (ii), (iii) and (iv), giving $q_{12} = 68.5$ N/mm, $q_{23} = 9.8$ N/mm, $q_{34} = 11.9$ N/mm, $q_{41} = 9.8$ N/mm from which

$$\tau_{12} = 42.8\,\mathrm{N/mm^2} \quad \tau_{23} = \tau_{41} = 9.8\,\mathrm{N/mm^2} \quad \tau_{34} = 9.9\,\mathrm{N/mm^2}$$

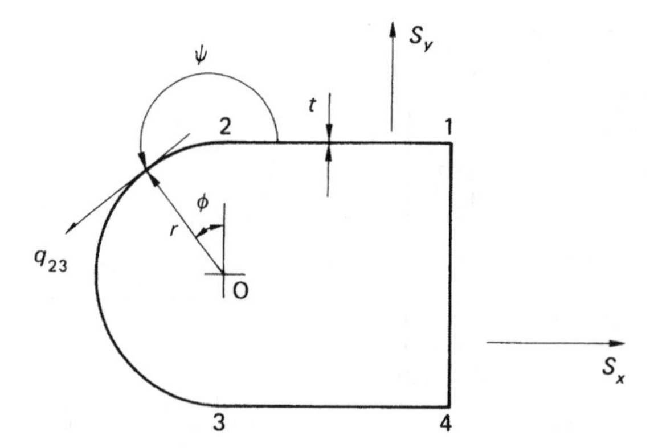

Fig. 26.3 Built-in end of a beam section having a curved wall.

We note in Example 26.1 that there is a discontinuity of shear flow at each of the corners of the beam. This implies the existence of axial loads at the corners which would, in practice, be resisted by booms, if stress concentrations are to be avoided. We see also that in a beam having straight walls the shear flows are constant along each wall so that, from Eq. (17.2), the direct stress gradient $\partial\sigma_z/\partial z = 0$ in the walls at the built-in end although not necessarily in the booms. Finally, the centre of twist of the beam section at the built-in end may be found using Eq. (17.11), i.e.

$$x_R = -\frac{v'}{\theta'} \quad y_R = \frac{u'}{\theta'}$$

which, from the results of Example 26.1, give $x_R = -351.6$ mm, $y_R = 79.6$ mm. Thus, the centre of twist is 351.6 mm to the left of and 79.6 mm above corner 1 of the section and will not, as we noted in Section 26.1, coincide with the shear centre of the section.

The method of analysis of beam sections having curved walls is similar to that of Example 26.1 except that in the curved walls the shear flow will not be constant since both p and ψ in Eq. (26.1) will generally vary. Consider the beam section shown in Fig. 26.3 in which the curved wall 23 is semicircular and of radius r. In the wall 23, $p = r$ and $\psi = 180 + \phi$, so that Eq. (26.1) gives

$$q_{23} = Gt(r\theta' - u' \cos\phi - v' \sin\phi)$$

The resultants of q_{23} are then

$$\text{Horizontally}: \int_0^{\pi} q_{23} \cos\phi r \, \mathrm{d}\phi$$

$$\text{Vertically}: \int_0^{\pi} q_{23} \sin\phi r \, \mathrm{d}\phi$$

$$\text{Moment (about 0)}: \int_0^{\pi} q_{23} r^2 \, \mathrm{d}\phi$$

The shear flows in the remaining walls are constant and the solution proceeds as before.

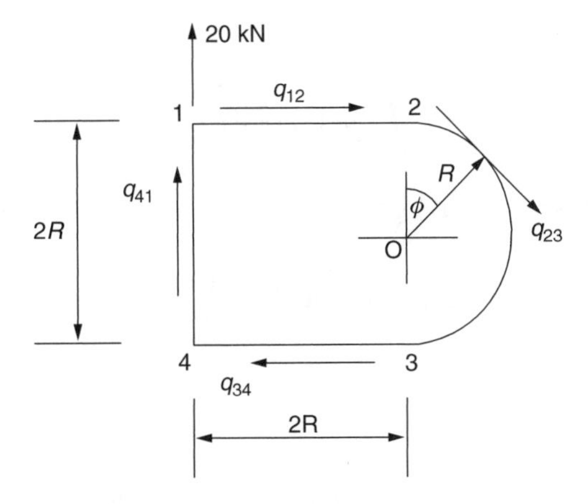

Fig. 26.4 Beam section of Example 26.2.

Example 26.2

Determine the shear flow distribution at the built-in end of a beam whose cross-section is shown in Fig. 26.4. All walls have the same thickness t and shear modulus G; $R = 200$ mm.

In general at a built-in end (see Eq (26.1))

$$q = Gt\left(p\frac{d\theta}{dz} + \frac{du}{dz}\cos\psi + \frac{dv}{dz}\sin\psi\right)$$

Therefore, taking O as the origin and writing $\theta' = d\theta/dz$, $u' = du/dz$ and $v' = dv/dz$

$$q_{41} = Gt(-2R\theta' + v') \tag{i}$$

$$q_{12} = Gt(-R\theta' + u') \tag{ii}$$

$$q_{34} = Gt(-R\theta' - u') \tag{iii}$$

$$q_{23} = Gt(-R\theta' + u'\cos\phi - v'\sin\phi) \tag{iv}$$

From symmetry

$$q_{12} = q_{34}$$

i.e.

$$Gt(-R\theta' + u') = Gt(-R\theta' - u')$$

Therefore

$$u' = 0$$

Resolving vertically

$$q_{41}2R - \int_0^{\pi} q_{23}\sin\phi\, R\, d\phi = 20 \times 10^3$$

i.e.

$$q_{41} - \frac{1}{2}\int_0^{\pi} q_{23} \sin\phi \, d\phi = \frac{10\,000}{R}$$

Substituting from Eqs (i) and (iv) gives

$$-R\theta' + 1.79v' = \frac{10\,000}{GtR} \tag{v}$$

Now taking moments about O

$$q_{41}\,2R\,2R + q_{12}\,2R\,R + q_{34}\,2R\,R + \int_0^{\pi} q_{23}\,R^2 d\phi = 20\,000 \times 2R$$

which gives

$$2q_{41} + q_{12} + q_{34} + \frac{1}{2}\int_0^{\pi} q_{23}\, d\phi = \frac{20\,000}{R}$$

Substituting from Eqs (i), (ii), (iii) and (iv)

$$2Gt(-2R\theta' + v') - 2GtR\theta' + \frac{Gt}{2}\int_0^{\pi} (-R\theta' - v'\sin\phi)\, d\phi = \frac{20\,000}{R}$$

from which

$$R\theta' - 0.13v' = -\frac{2641.7}{GtR} \tag{vi}$$

Solving Eqs (v) and (vi)

$$v' = \frac{4432.7}{GtR}, \quad R\theta' = -\frac{2065.4}{GtR}$$

Therefore

$$q_{41} = Gt\left(\frac{2 \times 2065.4}{200Gt} + \frac{4432.7}{200Gt}\right) = 42.8\,\text{N/mm}$$

Similarly

$$q_{12} = q_{34} = 10.3\,\text{N/mm}$$

Finally

$$q_{23} = 10.3 - 22.2\sin\phi\,\text{N/mm}$$

26.3 Thin-walled rectangular section beam subjected to torsion

In Example 18.2 we determined the warping distribution in a thin-walled rectangular section beam which was not subjected to structural constraint. This free warping distribution (w_0) was found to be linear around a cross-section and uniform along the length of the beam having values at the corners of

$$w_0 = \pm \frac{T}{8abG}\left(\frac{b}{t_b} - \frac{a}{t_a}\right)$$

The effect of structural constraint, such as building one end of the beam in, is to reduce this free warping to zero at the built-in section so that direct stresses are induced which subsequently modify the shear stresses predicted by elementary torsion theory. These direct stresses must be self-equilibrating since the applied load is a pure torque.

The analysis of a rectangular section beam built-in at one end and subjected to a pure torque at the other is simplified if the section is idealized into one comprising four corner booms which are assumed to carry all the direct stresses together with shear–stress-only carrying walls. The assumption on which the idealization is based is that the direct stress distribution at any cross-section is directly proportional to the warping which has been suppressed. Therefore, the distribution of direct stress is linear around any cross-section and has values equal in magnitude but opposite in sign at opposite corners of a wall. This applies at all cross-sections since the free warping will be suppressed to some extent along the complete length of the beam. In Fig. 26.5(b) all the booms will have the same cross-sectional area from anti-symmetry and, from Eq. (20.1) or (20.2)

$$B = \frac{at_a}{6}(2-1) + \frac{bt_b}{6}(2-1) = \frac{1}{6}(at_a + bt_b)$$

To the boom area B will be added existing concentrations of area such as connecting angle sections at the corners. The contributions of stringers may be included by allowing for their direct stress carrying capacity by increasing the actual wall thickness by an amount equal to the total stringer area on one wall before idealizing the section.

We have seen in Chapter 20 that the effect of structural idealization is to reduce the shear flow in the walls of a beam to a constant value between adjacent booms.

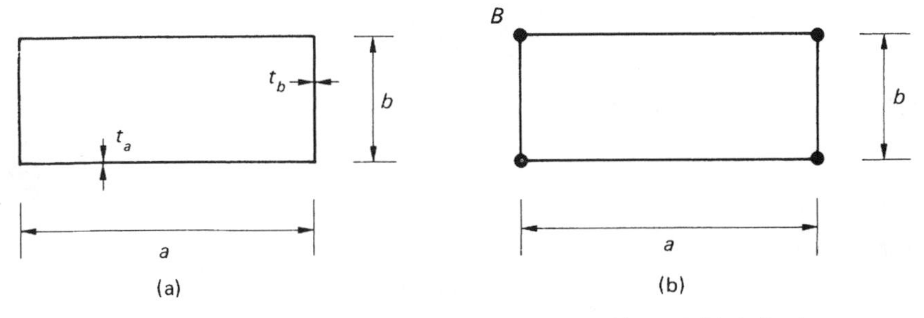

Fig. 26.5 Idealization of a rectangular section beam subjected to torsion: (a) actual; (b) idealized.

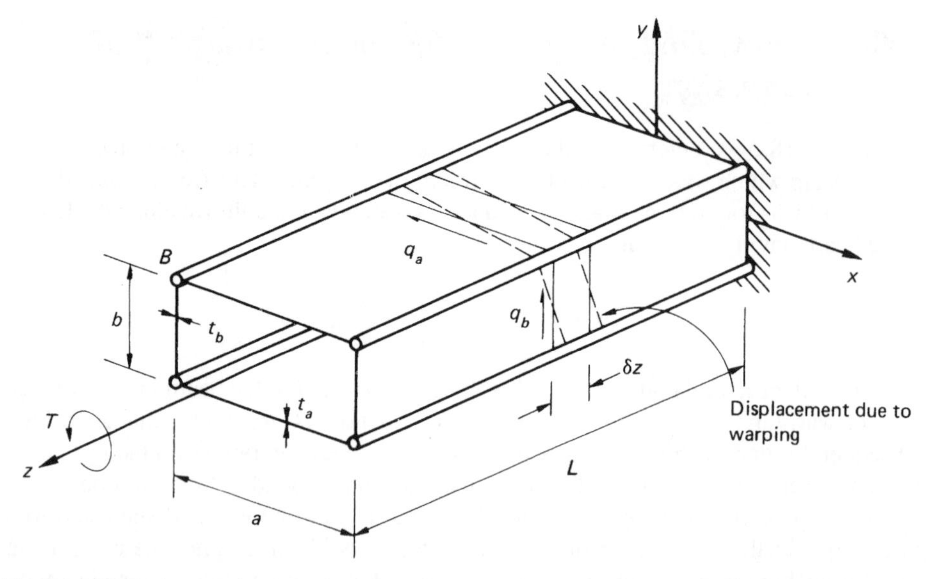

Fig. 26.6 Idealized rectangular section beam built-in at one end and subjected to a torque at the other.

In Fig. 26.6 suppose that the shear flows in the covers and webs at any section are q_a and q_b, respectively; from antisymmetry the shear flows in both covers will be q_a and in both webs q_b. The resultant of these shear flows is equivalent to the applied torque so that

$$T = \oint qp \, \mathrm{d}s = 2q_a a \frac{b}{2} + 2q_b b \frac{a}{2}$$

or

$$T = ab(q_a + q_b) \tag{26.4}$$

We now use Eq. (17.19), i.e.

$$q = Gt \left(\frac{\partial w}{\partial s} + \frac{\partial v}{\partial z} \right)$$

to determine q_a and q_b. Since the beam cross-section is doubly symmetrical the axis of twist passes through the centre of symmetry at any section so that, from Eq. (17.8)

$$\frac{\partial v_{\mathrm{t}}}{\partial z} = p_{\mathrm{R}} \frac{\mathrm{d}\theta}{\mathrm{d}z} \tag{26.5}$$

Therefore for the covers of the beam

$$\frac{\partial v_{\mathrm{t}}}{\partial z} = \frac{b}{2} \frac{\mathrm{d}\theta}{\mathrm{d}z} \tag{26.6}$$

and for the webs

$$\frac{\partial v_{\mathrm{t}}}{\partial z} = \frac{a}{2} \frac{\mathrm{d}\theta}{\mathrm{d}z} \tag{26.7}$$

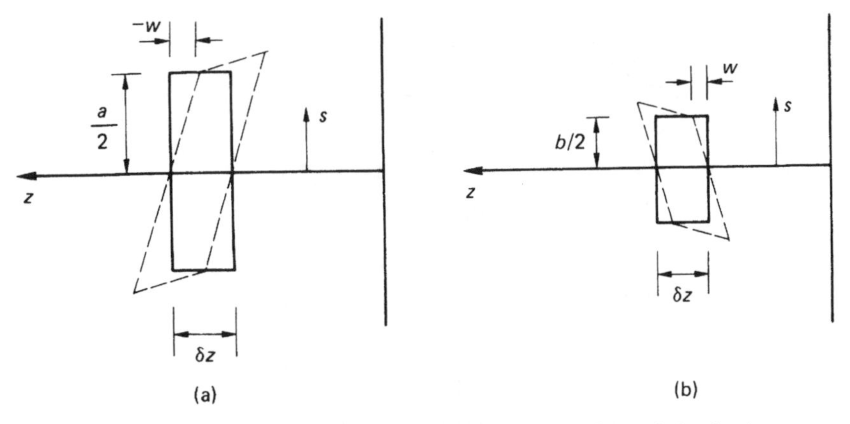

Fig. 26.7 Shear distortion of (a) an element of the top cover; (b) an element of the right hand web.

The elements of length δz of the covers and webs of the beam will warp into the shapes shown in Fig. 26.6 if T is positive (anticlockwise) and $b/t_b > a/t_a$. Clearly there must be compatibility of displacement at adjacent edges of the elements. From Fig. 26.7(a)

$$\frac{\partial w}{\partial s} = \frac{-w}{a/2} \tag{26.8}$$

and from Fig. 26.7(b)

$$\frac{\partial w}{\partial s} = \frac{w}{b/2} \tag{26.9}$$

Substituting for $\partial w/\partial s$ and $\partial v_t/\partial z$ in Eq. (17.19) separately for the covers and webs, we obtain

$$q_a = Gt_a\left(\frac{-2w}{a} + \frac{b}{2}\frac{\mathrm{d}\theta}{\mathrm{d}z}\right) \qquad q_b = Gt_b\left(\frac{2w}{b} + \frac{a}{2}\frac{\mathrm{d}\theta}{\mathrm{d}z}\right) \tag{26.10}$$

Now substituting for q_a and q_b in Eq. (26.4) we have

$$T = abG\left[t_a\left(\frac{-2w}{a} + \frac{b}{2}\frac{\mathrm{d}\theta}{\mathrm{d}z}\right) + t_b\left(\frac{2w}{b} + \frac{a}{2}\frac{\mathrm{d}\theta}{\mathrm{d}z}\right)\right]$$

Rearranging

$$\frac{\mathrm{d}\theta}{\mathrm{d}z} = \frac{4w(bt_a - at_b)}{ab(bt_a + at_b)} + \frac{2T}{abG(bt_a + at_b)} \tag{26.11}$$

If we now substitute for $\mathrm{d}\theta/\mathrm{d}z$ from Eq. (26.11) into Eqs (26.10) we have

$$q_a = \frac{-4wGt_bt_a}{bt_a + at_b} + \frac{Tt_a}{a(bt_a + at_b)} \qquad q_b = \frac{4wGt_bt_a}{bt_a + at_b} + \frac{Tt_b}{b(bt_a + at_b)} \tag{26.12}$$

Equations (26.11) and (26.12) give the rate of twist and the shear flows (and hence shear stresses) in the beam in terms of the warping w and the applied torque T. Their derivation is based on the compatibility of displacement which exists at the cover/boom/web

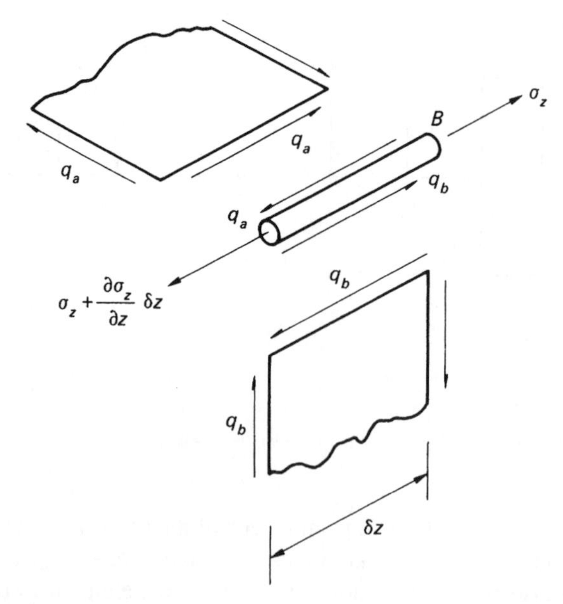

Fig. 26.8 Equilibrium of boom element.

junctions. We shall now use the further condition of equilibrium between the shears in the covers and webs and the direct load in the booms to obtain expressions for the warping displacement and the distributions of boom stress and load. Thus, for the equilibrium of an element of the top right-hand boom shown in Fig. 26.8

$$\left(\sigma_z + \frac{\partial \sigma_z}{\partial z}\delta z\right)B - \sigma_z B + q_a \delta z - q_b \delta z = 0$$

i.e.

$$B\frac{\partial \sigma_z}{\partial z} + q_a - q_b = 0 \tag{26.13}$$

Now

$$\sigma_z = E\frac{\partial w}{\partial z} \quad \text{(see Chapter 1)}$$

Substituting for σ_z in Eq. (26.13) we obtain

$$BE\frac{\partial^2 w}{\partial z^2} + q_a - q_b = 0 \tag{26.14}$$

Replacing q_a and q_b from Eqs (26.12) gives

$$BE\frac{\partial^2 w}{\partial z^2} - \frac{8Gt_b t_a}{bt_a + at_b}w = -\frac{T}{ab}\frac{(bt_a - at_b)}{(bt_a + at_b)}$$

or

$$\frac{\partial^2 w}{\partial z^2} - \mu^2 w = -\frac{T}{abBE}\frac{(bt_a - at_b)}{(bt_a + at_b)} \tag{26.15}$$

where

$$\mu^2 = \frac{8Gt_bt_a}{BE(bt_a + at_b)}$$

The differential equation (26.15) is of standard form and its solution is

$$w = C\cosh\mu z + D\sinh\mu z + \frac{T}{8abG}\left(\frac{b}{t_b} - \frac{a}{t_a}\right) \tag{26.16}$$

in which the last term is seen to be the free warping displacement w_0 of the top right-hand corner boom. The constants C and D in Eq. (26.16) are found from the boundary conditions of the beam. In this particular case the warping $w = 0$ at the built-in end and the direct strain $\partial w/\partial z = 0$ at the free end where there is no direct load. From the first of these

$$C = -\frac{T}{8abG}\left(\frac{b}{t_b} - \frac{a}{t_a}\right) = -w_0$$

and from the second

$$D = w_0 \tanh\mu L$$

Then

$$w = w_0(1 - \cosh\mu z + \tanh\mu L \sinh\mu z) \tag{26.17}$$

or rearranging

$$w = w_0\left[1 - \frac{\cosh\mu(L - z)}{\cosh\mu L}\right] \tag{26.18}$$

The variation of direct stress in the boom is obtained from $\sigma_z = E\partial w/\partial z$ and Eq. (26.18), i.e.

$$\sigma_z = \mu E w_0 \frac{\sinh\mu(L - z)}{\cosh\mu L} \tag{26.19}$$

and the variation of boom load P is then

$$P = B\sigma_z = B\mu E w_0 \frac{\sinh\mu(L - z)}{\cosh\mu L} \tag{26.20}$$

Substituting for w in Eqs (26.12) and rearranging, we obtain the shear stress distribution in the covers and webs. Thus

$$\tau_a = \frac{q_a}{t_a} = \frac{T}{2abt_a}\left[1 + \frac{(bt_a - at_b)}{(bt_a + at_b)}\frac{\cosh\mu(L - z)}{\cosh\mu L}\right] \tag{26.21}$$

$$\tau_b = \frac{q_b}{t_b} = \frac{T}{2abt_b}\left[1 - \frac{(bt_a - at_b)}{(bt_a + at_b)}\frac{\cosh\mu(L - z)}{\cosh\mu L}\right] \tag{26.22}$$

Inspection of Eqs (26.21) and (26.22) shows that the shear stress distributions each comprise two parts. The first terms, $T/2abt_a$ and $T/2abt_b$, are the shear stresses predicted by elementary theory (see Section 18.1), while the hyperbolic second terms

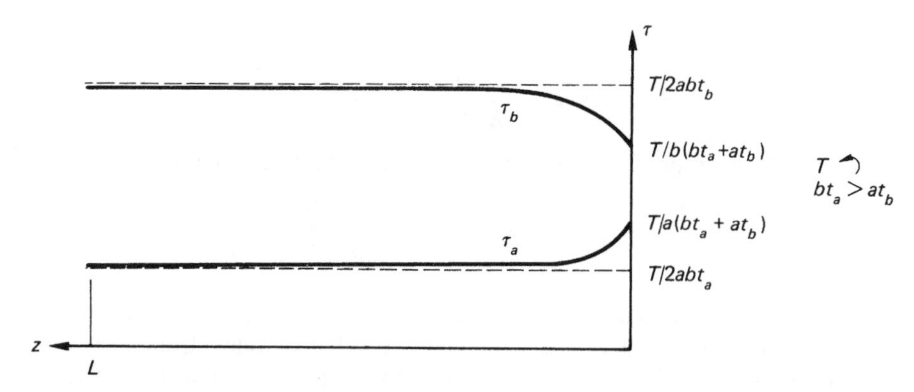

Fig. 26.9 Shear stress distributions along the beam of Fig. 11.5.

represent the effects of the warping restraint. Clearly, for an anticlockwise torque and $bt_a > at_b$, the effect of this constraint is to increase the shear stress in the covers over that predicted by elementary theory and decrease the shear stress in the webs. It may also be noted that for bt_a to be greater than at_b for the beam of Fig. 26.6, in which $a > b$, then t_a must be appreciably greater than t_b so that $T/2abt_a < T/2abt_b$. Also at the built-in end ($z = 0$), Eqs (26.21) and (26.22) reduce to $\tau_a = T/a(bt_a + at_b)$ and $\tau_b = T/b(bt_a + at_b)$ so that even though τ_b is reduced by the axial constraint and τ_a increased, τ_b is still greater than τ_a. It should also be noted that these values of τ_a and τ_b at the built-in end may be obtained using the method of Section 26.2 and that these are the values of shear stress irrespective of whether the section has been idealized or not. In other words, the presence of intermediate stringers and/or direct stress carrying walls does not affect the shear flows at the built-in end since the direct stress gradient at this section is zero (see Section 26.2 and Eq. (17.2)) except in the corner booms. Finally, when both z and L become large, i.e. at the free end of a long, slender beam

$$\tau_a \to \frac{T}{2abt_a} \quad \text{and} \quad \tau_b \to \frac{T}{2abt_b}$$

The above situation is shown in Fig. 26.9.

In the particular case when $bt_a = at_b$ we see that the second terms on the right-hand side of Eqs (26.21) and (26.22) disappear and no constraint effects are present; the direct stress of Eqs (26.19) is also zero since $w_0 = 0$ (see Example 18.2).

The rate of twist is obtained by substituting for w from Eq. (26.18) in Eq. (26.11). Thus

$$\frac{\mathrm{d}\theta}{\mathrm{d}z} = \frac{T}{2a^2b^2G}\left(\frac{b}{t_b} + \frac{a}{t_a}\right)\left[1 - \left(\frac{bt_a - at_b}{bt_a + at_b}\right)^2 \frac{\cosh \mu(L - z)}{\cosh \mu L}\right] \tag{26.23}$$

in which we see that again the expression on the right-hand side comprises the rate of twist given by elementary theory, $T(b/t_b + a/t_a)/2a^2b^2G$ (see Section 18.1), together with a correction due to the warping restraint. Clearly the rate of twist is always reduced by the constraint since $(bt_a - at_b)^2$ is always positive. Integration of Eq. (26.23) gives the distribution of angle of twist along the length of the beam, the boundary condition in this case being $\theta = 0$ at $z = 0$.

Example 26.3

A uniform four boom box of span 5 m is 500 mm wide by 20 mm deep and has four corner booms each of cross-sectional area 800 mm^2, its wall thickness is 1.0 mm. If the box is subjected to a uniformly distributed torque loading of 20 Nm/mm along its length and it is supported at each end such that complete freedom of warping exists at the end cross-sections calculate the angle of twist at the mid-span section. Take $G = 20\,000$ N/mm^2 and $G/E = 0.36$.

The reactive torques at each support are $= 20 \times 5000/2 = 50\,000$ Nm

Taking the origin for z at the mid-span of the beam the torque at any section is given by

$$T(z) = 20(2500 - z) - 50\,000 = -20z\ \text{Nm}$$

Substituting in Eq. (26.16) we obtain

$$w = C \cosh \mu z + D \sinh \mu z - \frac{20z \times 10^3(b - a)}{8abGt}$$

The boundary conditions are:

$w = 0$ when $z = 0$ from symmetry and $\partial w/\partial z = 0$ when $z = L$ ($L = 2500$ mm)

From the first of these $C = 0$ while from the second

$$D = \frac{20 \times 10^3(b - a)}{8\mu abGt \cosh \mu L}$$

Therefore

$$w = \frac{20(b - a) \times 10^3}{8abGt}\left(\frac{\sinh \mu z}{\mu \cosh \mu L} - z\right) \qquad \text{(i)}$$

Further

$$\mu^2 = \frac{8Gt}{AE(b + a)} = \frac{8 \times 0.36 \times 1.0}{800(200 + 500)} = 5.14 \times 10^{-6}$$

so that Eq. (i) becomes

$$w = -3.75 \times 10^{-4}(3.04 \sinh \mu z - z) \qquad \text{(ii)}$$

Substituting for w, etc. in Eq. (26.11)

$$\frac{d\theta}{dz} = 10^{-8}(1.95 \sinh \mu z - 3.49z)$$

Hence

$$\theta = 10^{-8}\left(\frac{1.95}{\mu} \cosh \mu z - 1.75z^2 + F\right) \qquad \text{(iii)}$$

When $z = L$ (2500 mm) $\theta = 0$. Then, from Eq. (iii)

$$F = 10.8 \times 10^6$$

so that

$$\theta = 10^{-8}(859 \cosh \mu z - 1.75z^2 + 10.8 \times 10^6) \quad \text{(iv)}$$

At mid-span where $z = 0$, from Eq. (iv)

$$\theta = 0.108 \text{ rad} \quad \text{or} \quad \theta = 6.2^\circ$$

26.4 Shear lag

A problem closely related to the restrained torsion of rectangular section beams is that generally known as *shear lag*. We have seen in Chapter 18 that torsion induces shear stresses in the walls of beams and these cause shear strains which produce warping of the cross-section. When this warping is restrained, direct stresses are set up which modify the shear stresses. In a similar manner the shear strains in the thin walls of beams subjected to shear loads cause cross-sections to distort or warp so that the basic assumption of elementary bending theory of plane sections remaining plane is no longer valid. The direct and shear stress distributions predicted by elementary theory therefore become significantly inaccurate. Further modifications arise when any form of structural constraint prevents the free displacement of the cross-sections of a beam. Generally, shear lag becomes a problem in wide, relatively shallow, thin-walled beams such as wings in which the shear distortion of the thin upper and lower surface skins causes redistribution of stress in the stringers and spar caps while the thicker and shallower spar webs experience little effect.

Consider the box beam shown in Fig. 26.10. Elementary bending theory predicts that the direct stress at any section AA would be uniform across the width of the covers so that the stringers and web flanges would all be subjected to the same stress. However, the shear strains at the section cause the distortion shown so that the intermediate stringers carry lower stresses than the web flanges. Since the resultant of the direct stresses must

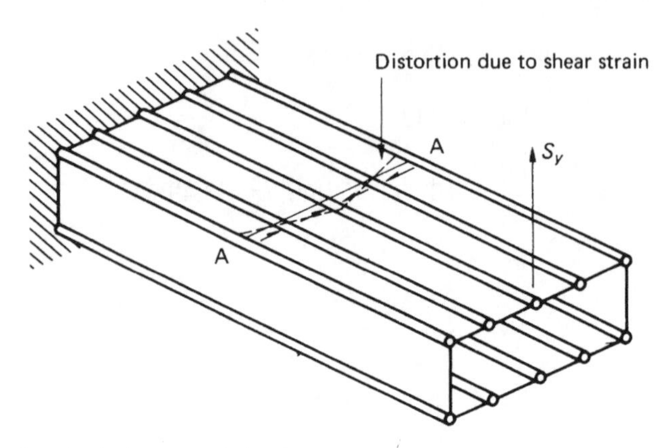

Fig. 26.10 Shear distortion in the covers of a box beam.

be equivalent to the applied bending moment this means that the direct stresses in the web flanges must be greater than those predicted by elementary bending theory. Our investigation of the shear lag problem will be restricted to idealized six- and eight-boom doubly symmetrical rectangular section beams subjected to shear loads acting in the plane of symmetry and in which the axis of twist, the flexural axis and the zero warping axis coincide; the shear loads therefore produce no twist and hence no warping due to twist. In the analysis we shall assume that the cross-sections of beams remain undistorted in their own plane.

Figure 26.11 shows an idealized six-boom beam built-in at one end and carrying a shear load at the other; the corner booms have a cross-sectional area B while the central booms have a cross-sectional area A. At any section the vertical shear load is shared equally by the two webs. Also, since the beam has been idealized, the shear flow at any section will be constant between the booms so that, for a web, the situation is that shown in the free body diagram of Fig. 26.12, in addition, the corner booms are subjected to equal and opposite loads P_B. The complementary shear flows $S_y/2h$ are applied to the corner booms as shown so that the top cover, say, is subjected to loads as shown in Fig. 26.13. We assume that suitable edge members are present at the free end of the cover to equilibrate the shear flows; we also assume that strains in the transverse direction are negligible.

It is advantageous to adopt a methodical approach in the analysis. Thus, use may be made of the symmetry of the cover so that only one edge boom, one panel and the central boom need to be considered as long as the symmetry is allowed for in the assumed directions of the panel shear flows q, as shown in Fig. 26.13. Further, the origin for z may be taken to be at either the free or built-in end. A marginally simpler solution is obtained if the origin is taken to be at the free end, in which case the solution represents that for an infinitely long panel. Considering the equilibrium of an element of an edge boom (Fig. 26.14), in which we assume that the boom load is positive (tension)

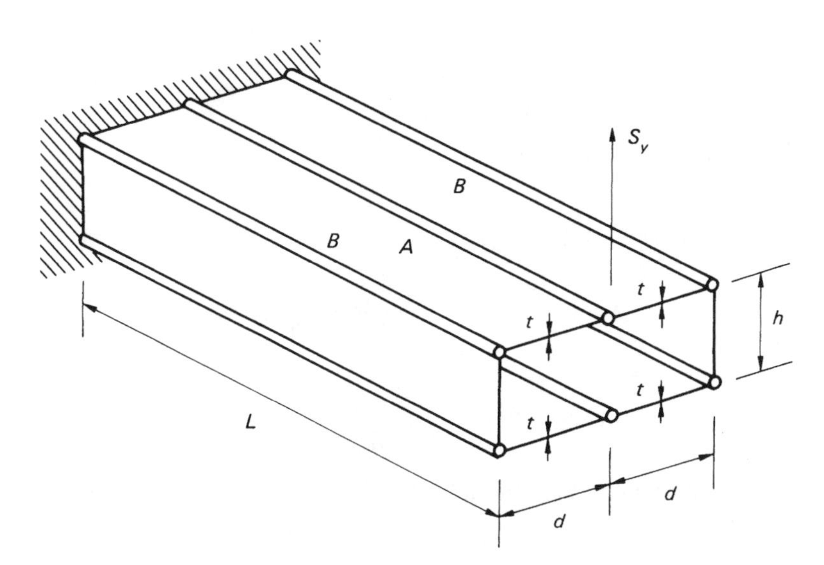

Fig. 26.11 Six-boom beam subjected to a shear load.

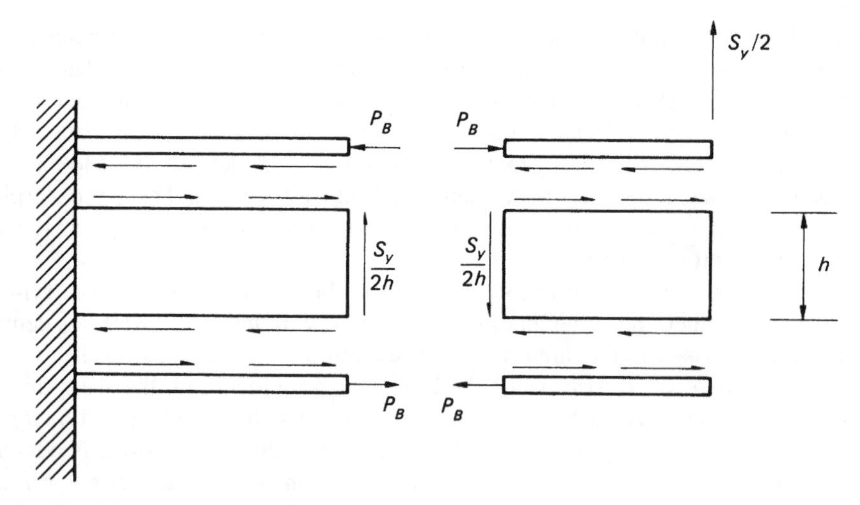

Fig. 26.12 Loads on webs and corner booms of the beam of Fig. 26.11.

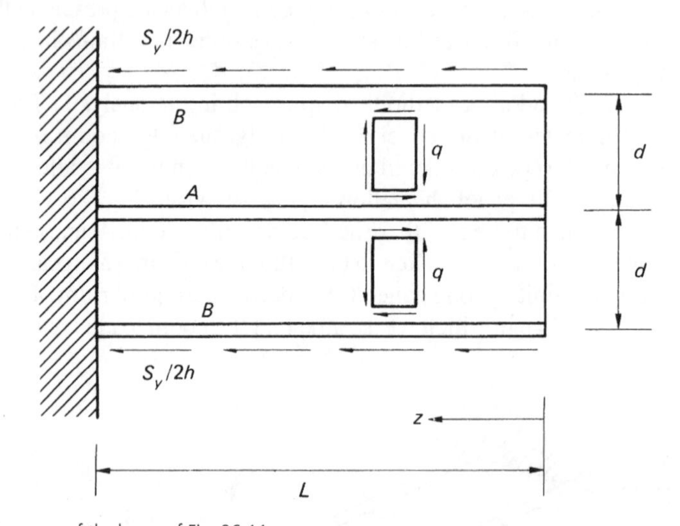

Fig. 26.13 Top cover of the beam of Fig. 26.11.

and increases with increasing z, we have

$$P_B + \frac{\partial P_B}{\partial z}\delta z - P_B - q\delta z + \frac{S_y}{2h}\delta z = 0$$

or

$$\frac{\partial P_B}{\partial z} - q + \frac{S_y}{2h} = 0 \tag{26.24}$$

Similarly, for an element of the central boom (Fig. 26.15)

$$\frac{\partial P_A}{\partial z} + 2q = 0 \tag{26.25}$$

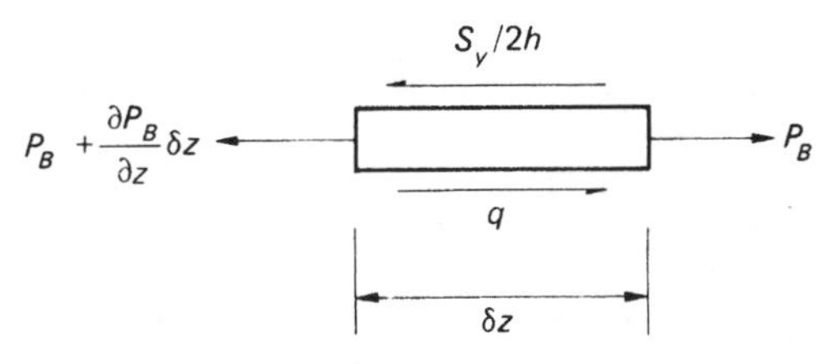

Fig. 26.14 Equilibrium of boom element.

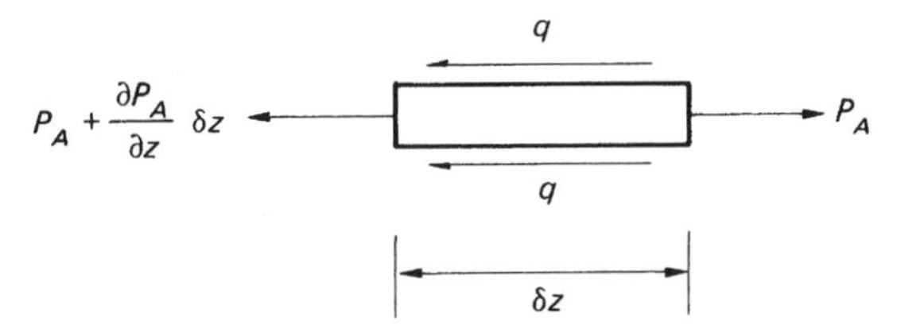

Fig. 26.15 Equilibrium of element of central boom.

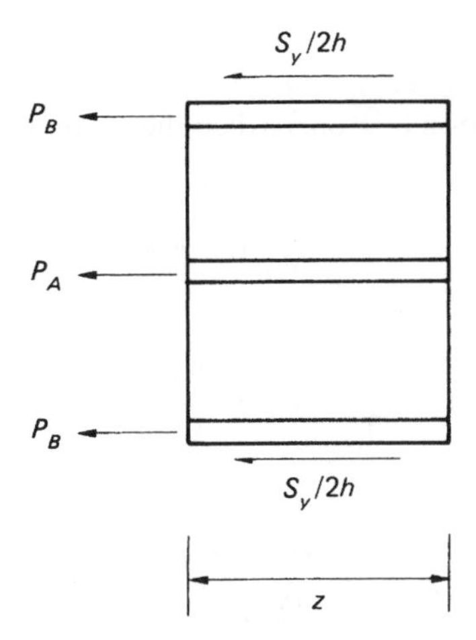

Fig. 26.16 Equilibrium of a length z of cover.

Now considering the overall equilibrium of a length z of the cover (Fig. 26.16), we have

$$2P_B + P_A + \frac{S_y}{h}z = 0 \tag{26.26}$$

We now consider the compatibility condition which exists in the displacements of elements of the booms and adjacent elements of the panels. Figure 26.17(a) shows the displacements of the cover and an element of a panel and the adjacent elements of the boom. Note that the element of the panel is distorted in a manner which agrees with the assumed directions of the shear flows in Fig. 26.13 and that the shear strain increases with z. From Fig. 26.17(b)

$$(1 + \varepsilon_B)\delta z = (1 + \varepsilon_A)\delta z + d\frac{\partial \gamma}{\partial z}\partial z$$

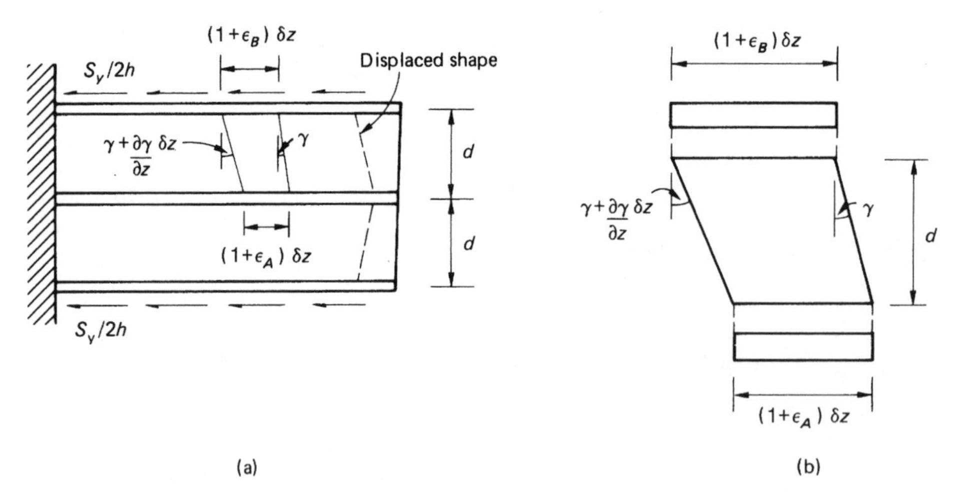

Fig. 26.17 Compatibility condition.

in which ε_B and ε_A are the direct strains in the elements of boom. Then, rearranging and noting that γ is a function of z only when the section is completely idealized, we have

$$\frac{\mathrm{d}\gamma}{\mathrm{d}z} = \frac{1}{d}(\varepsilon_B - \varepsilon_A) \tag{26.27}$$

Now

$$\varepsilon_B = \frac{P_B}{BE} \quad \varepsilon_A = \frac{P_A}{AE} \quad \gamma = \frac{q}{Gt}$$

so that Eq. (26.27) becomes

$$\frac{\mathrm{d}q}{\mathrm{d}z} = \frac{Gt}{dE}\left(\frac{P_B}{B} - \frac{P_A}{A}\right) \tag{26.28}$$

We now select the unknown to be determined initially. Generally, it is simpler mathematically to determine either of the boom load distributions, P_B or P_A, rather than the shear flow q. Thus, choosing P_A, say, as the unknown, we substitute in Eq. (26.28) for q from Eq. (11.25) and for P_B from Eq. (26.26). Hence

$$-\frac{1}{2}\frac{\partial^2 P_A}{\partial z^2} = \frac{Gt}{dE}\left(-\frac{P_A}{2B} - \frac{S_y z}{2Bh} - \frac{P_A}{A}\right)$$

Rearranging, we obtain

$$\frac{\partial^2 P_A}{\partial z^2} - \frac{Gt(2B + A)}{dEAB}P_A = \frac{GtS_y z}{dEBh}$$

or

$$\frac{\partial^2 P_A}{\partial z^2} - \lambda^2 P_A = \frac{GtS_y z}{dEBh} \tag{26.29}$$

in which $\lambda^2 = Gt(2B+A)/dEAB$. The solution of Eq. (26.29) is of standard form and is

$$P_A = C\cosh\lambda z + D\sinh\lambda z - \frac{S_y A}{h(2B+A)}z$$

The constants C and D are determined from the boundary conditions of the cover of the beam namely, $P_A = 0$ when $z = 0$ and $\gamma = q/Gt = -(\partial P_A/\partial z)/2Gt = 0$ when $z = L$ (see Eq. (26.25)). From the first of these $C = 0$ and from the second

$$D = \frac{S_y A}{\lambda h(2B+A)\cosh\lambda L}$$

Thus

$$P_A = -\frac{S_y A}{h(2B+A)}\left(z - \frac{\sinh\lambda z}{\lambda\cosh\lambda L}\right) \tag{26.30}$$

The direct stress distribution $\sigma_A(= P_A/A)$ follows, i.e.

$$\sigma_A = -\frac{S_y}{h(2B+A)}\left(z - \frac{\sinh\lambda z}{\lambda\cosh\lambda L}\right) \tag{26.31}$$

The distribution of load in the edge booms is obtained by substituting for P_A from Eq. (26.30) in Eq. (26.26), thus

$$P_B = -\frac{S_y B}{h(2B+A)}\left(z + \frac{A}{2B\lambda}\frac{\sinh\lambda z}{\cosh\lambda L}\right) \tag{26.32}$$

whence

$$\sigma_B = -\frac{S_y}{h(2B+A)}\left(z + \frac{A}{2B\lambda}\frac{\sinh\lambda z}{\cosh\lambda L}\right) \tag{26.33}$$

Finally, from either pairs of Eqs (26.25) and (26.30) or (26.24) and (26.32)

$$q = \frac{S_y A}{2h(2B+A)}\left(1 - \frac{\cosh\lambda z}{\cosh\lambda L}\right) \tag{26.34}$$

so that the shear stress distribution $\tau(=q/t)$ is

$$\tau = \frac{S_y A}{2ht(2B+A)}\left(1 - \frac{\cosh\lambda z}{\cosh\lambda L}\right) \tag{26.35}$$

Elementary theory gives

$$\sigma_A = \sigma_B = -\frac{S_y z}{h(2B+A)}$$

and

$$q = \frac{S_y A}{2h(2B+A)}$$

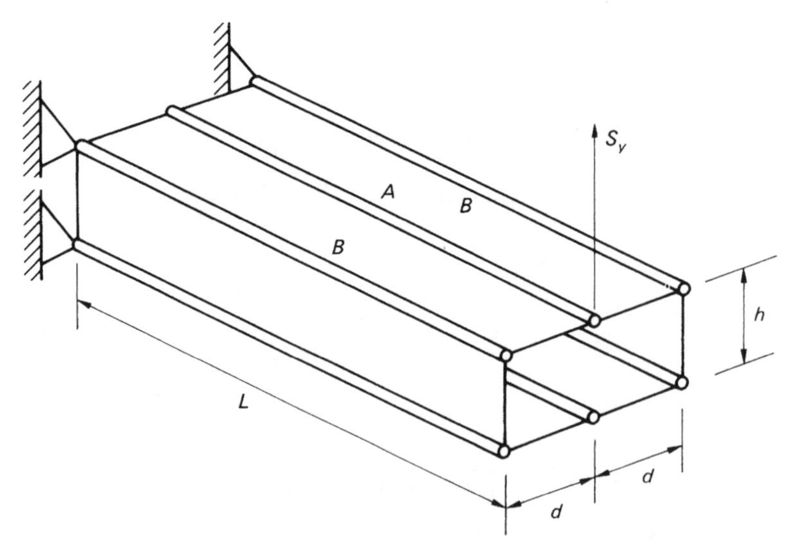

Fig. 26.18 Rectangular section beam supported at corner booms only.

so that, as in the case of the torsion of a four boom rectangular section beam, the solution comprises terms corresponding to elementary theory together with terms representing the effects of shear lag and structural constraint.

Many wing structures are spliced only at the spars so that the intermediate stringers are not subjected to bending stresses at the splice. The situation for a six boom rectangular section beam is then as shown in Fig. 26.18. The analysis is carried out in an identical manner to that in the previous case except that the boundary conditions for the central stringer are $P_A = 0$ when $z = 0$ and $z = L$. The solution is

$$P_A = -\frac{S_y A}{h(2B + A)}\left(z - L\frac{\sinh \lambda z}{\sinh \lambda L}\right) \tag{26.36}$$

$$P_B = -\frac{S_y B}{h(2B + A)}\left(z + \frac{AL}{2B}\frac{\sinh \lambda z}{\sinh \lambda L}\right) \tag{26.37}$$

$$q = \frac{S_y A}{2h(2B + A)}\left(1 - \lambda L\frac{\cosh \lambda z}{\sinh \lambda L}\right) \tag{26.38}$$

where $\lambda^2 = Gt(2B + A)/dEAB$. Examination of Eq. (26.38) shows that q changes sign when $\cosh \lambda z = (\sinh \lambda L)/\lambda L$, the solution of which gives a value of z less than L, i.e. q changes sign at some point along the length of the beam. The displaced shape of the top cover is therefore as shown in Fig. 26.19. Clearly, the final length of the central stringer is greater than in the previous case and appreciably greater than the final length of the spar flanges. The shear lag effect is therefore greater than before. In some instances this may be beneficial since a larger portion of the applied bending moment is resisted by the heavier section spar flanges. These are also restrained against buckling in two directions by the webs and covers while the lighter section stringers are restrained in one direction only. The beam is therefore able to withstand higher bending moments than those calculated from elementary theory.

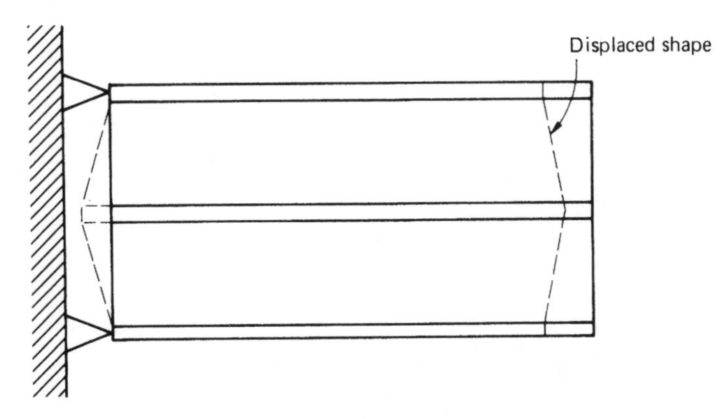

Fig. 26.19 Displaced shape of top cover of box team of Fig. 26.18.

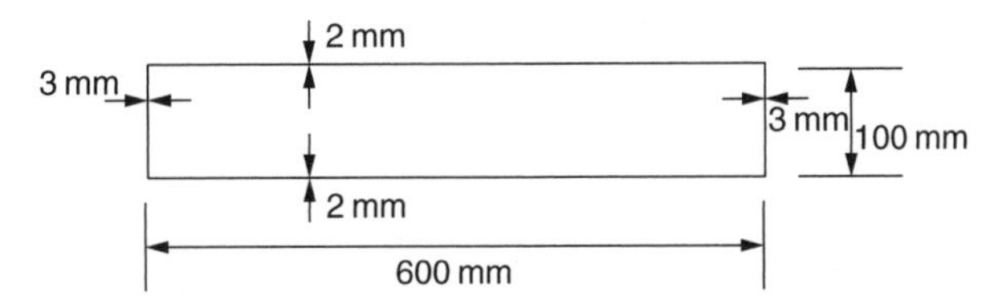

Fig. 26.20 Beam section of Example 26.4.

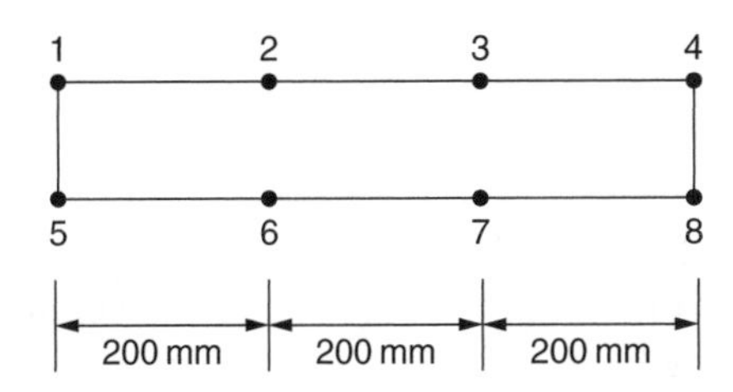

Fig. 26.21 Idealized beam section of Example 26.4.

Example 26.4

A shallow box section beam whose cross-section is shown in Fig. 26.20 is simply supported over a span of 2 m and carries a vertically downward load of 20 kN at mid-span. Idealise the section into one suitable for shear lag analysis, comprising eight booms, and hence determine the distribution of direct stress along the top right-hand corner of the beam. Take $G/E = 0.36$.

The idealized section is shown in Fig. 26.21.
Using either Eqs (20.1) or (20.2)

$$B_1 = B_4 = B_8 = B_5 = \frac{100 \times 3}{6}(2 - 1) + \frac{200 \times 2}{6}(2 + 1) = 250\,\text{mm}^2$$

$$B_2 = B_3 = B_6 = B_7 = \frac{200 \times 2}{6}(2 + 1) \times 2 = 400\,\text{mm}^2$$

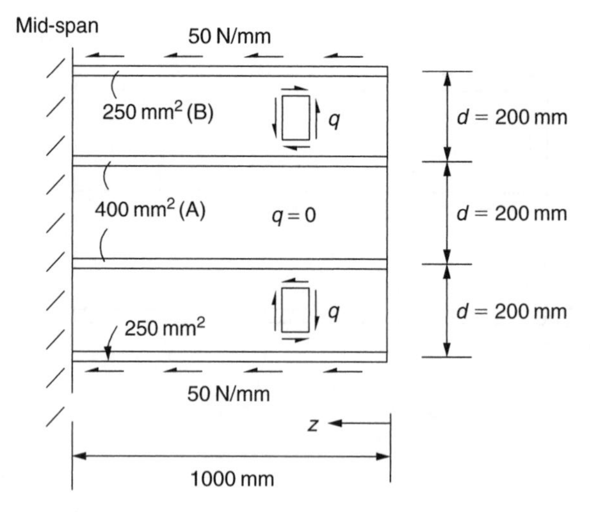

Fig. 26.22 Shear flows acting on top cover of idealized beam section of Example 26.4.

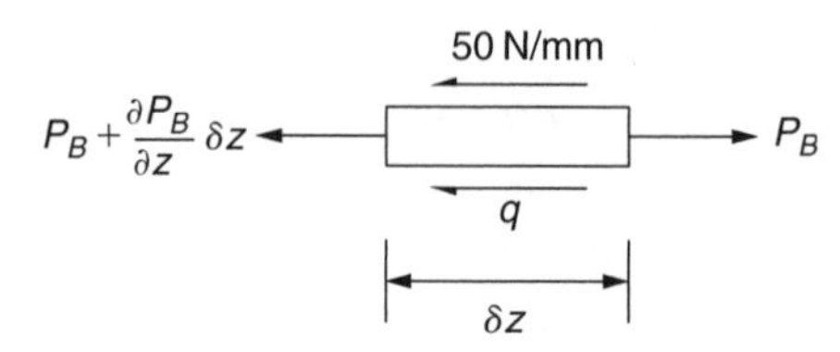

Fig. 26.23 Element of boom B.

The support reactions of 10 kN produce loads of 5 kN on each vertical web. These, in turn, produce shear flows of 50 N/mm along each corner boom as shown in Fig. 26.22 for the top cover of the beam.

Considering the equilibrium of elements of the booms we have, for the top boom, Fig. 26.23

$$P_B + \frac{\partial P_B}{\partial z}\delta z - P_B + q\delta z + 50\delta z = 0$$

which gives

$$\frac{\partial P_B}{\partial z} = -q - 50 \tag{i}$$

Similarly for an element of boom A

$$\frac{\partial P_A}{\partial z} = q \tag{ii}$$

Overall equilibrium of a length z of the panel gives

$$2P_B + 2P_A + 2 \times 50z = 0$$

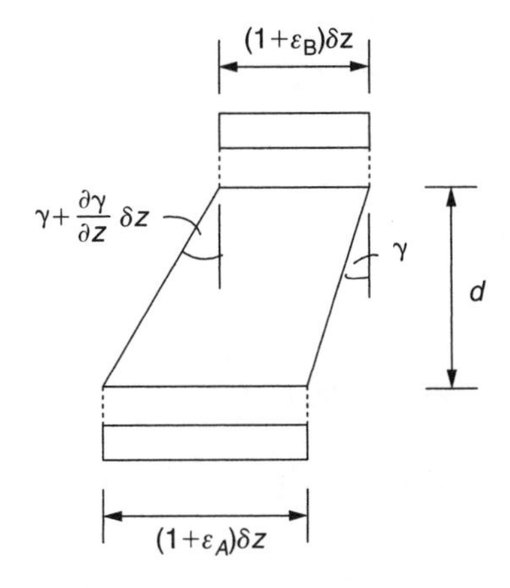

Fig. 26.24 Compatibility condition for top cover of beam of Example 26.4.

i.e

$$P_B + P_A + 50z = 0 \tag{iii}$$

The compatibility of displacement between elements of boom and adjacent panel, Fig. 26.24 gives

$$\frac{\partial \gamma}{\partial z} = \frac{1}{d}(\varepsilon_A - \varepsilon_B) \tag{iv}$$

But

$$\varepsilon_A = P_A/E_A \quad \varepsilon_B = P_B/E_B \quad \gamma = q/Gt$$

Substituting in Eq. (iv) we obtain

$$\frac{\partial q}{\partial z} = \frac{Gt}{dE}\left(\frac{P_A}{A} - \frac{P_B}{B}\right) \tag{v}$$

From Eq. (iii)

$$P_A = -P_B - 50z$$

From Eq. (i)

$$\frac{\partial q}{\partial z} = -\frac{\partial^2 P_B}{\partial z^2}$$

Substituting in Eq. (v)

$$\frac{\partial^2 P_B}{\partial z^2} - \mu^2 P_B = \frac{50Gt}{dEA}z \tag{vi}$$

in which

$$\mu^2 = \frac{Gt}{dE}\left(\frac{A+B}{AB}\right)$$

The solution of Eq. (vi) is

$$P_B = C \cosh \mu z + D \sinh \mu z - \frac{50B}{A+B} z$$

The boundary conditions are;

$$\text{when } z = 0, \ P_B = 0$$

$$\text{and when } z = 100\,\text{mm} \quad \frac{\partial P_B}{\partial z} = -50 \text{ (from Eq (i) since } q = 0 \text{ at } z = 1000\,\text{mm)}$$

From the first of these $C = 0$ while from the second

$$D = \frac{-50A}{(A+B)\mu \cosh 1000\mu}$$

Therefore

$$\sigma_B = \frac{P_B}{B} = \frac{-50A}{B(A+B)\mu \cosh 1000\mu} = \sinh \mu z - \frac{50}{A+B} z$$

Substituting the boom areas, etc. gives

$$\sigma_B = -0.4 \sinh \mu z - 0.08z$$

In certain situations beams, or parts of beams, carry loads which cause in-plane bending of the covers. An example is shown in Fig. 26.25 where the loads P cause bending in addition to axial effects. Shear lag modifies the stresses predicted by elementary theory in a similar manner to the previous cases. From symmetry we can consider either the

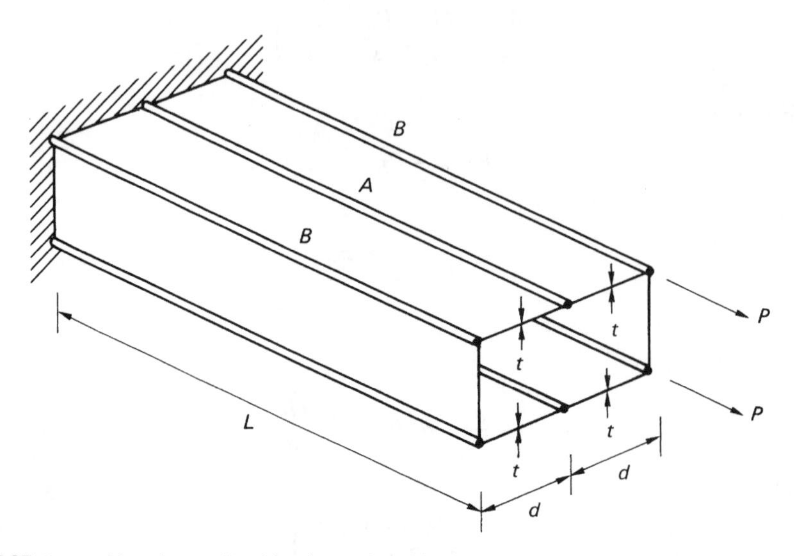

Fig. 26.25 Beam subjected to combined bending and axial load.

top or bottom cover in isolation as shown in Fig. 26.26(a). In this case the load P causes bending as well as extension of the cover so that at any section z the beam has a slope $\partial v/\partial z$ (Fig. 26.26(b)). We shall again assume that transverse strains are negligible and that the booms carry all the direct load.

Initially, as before, we choose directions for the shear flows in the top and bottom panels. Any directions may be chosen since the question of symmetry does not arise. The equilibrium of an element δz of each boom is first considered giving

$$\frac{\partial P_{B1}}{\partial z} = -q_1 \quad \frac{\partial P_A}{\partial z} = q_1 - q_2 \quad \frac{\partial P_{B2}}{\partial z} = q_2 \tag{26.39}$$

where P_{B1} is the load in boom 1 and P_{B2} is the load in boom 2. Longitudinal and moment equilibrium about boom 2 of a length z of the cover give, respectively

$$P_{B1} + P_{B2} + P_A = P \quad P_{B1}2d + P_A d = P2d \tag{26.40}$$

The compatibility condition now includes the effect of bending in addition to extension, as shown in Fig. 26.27. Note that the panel is distorted in a manner which agrees with the assumed direction of shear flow and that γ_1 and $\partial v/\partial z$ increase with z. Thus

$$(1 + \varepsilon_A)\delta z = (1 + \varepsilon_{B1})\delta z + d\left(\frac{\mathrm{d}\gamma_1}{\mathrm{d}z} + \frac{\mathrm{d}^2 v}{\mathrm{d}z^2}\right)\delta z$$

where γ_1 and v are functions of z only. Thus

$$\frac{\mathrm{d}\gamma_1}{\mathrm{d}z} = \frac{1}{d}(\varepsilon_A - \varepsilon_{B1}) - \frac{\mathrm{d}^2 v}{\mathrm{d}z^2} \tag{26.41}$$

Similarly, for an element of the lower panel

$$\frac{\mathrm{d}\gamma_2}{\mathrm{d}z} = \frac{1}{d}(\varepsilon_{B2} - \varepsilon_A) - \frac{\mathrm{d}^2 v}{\mathrm{d}z^2} \tag{26.42}$$

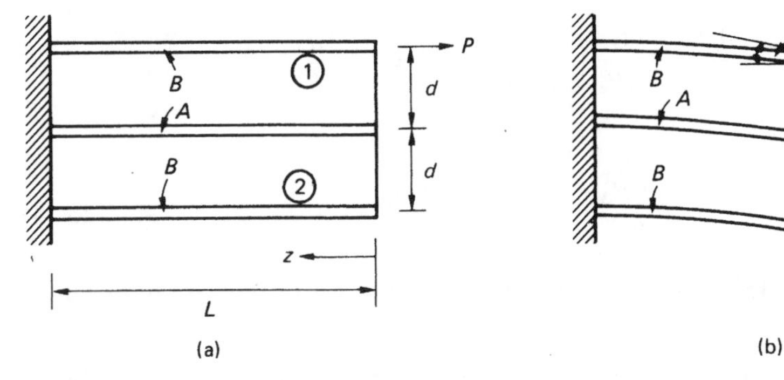

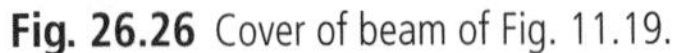

Fig. 26.26 Cover of beam of Fig. 11.19.

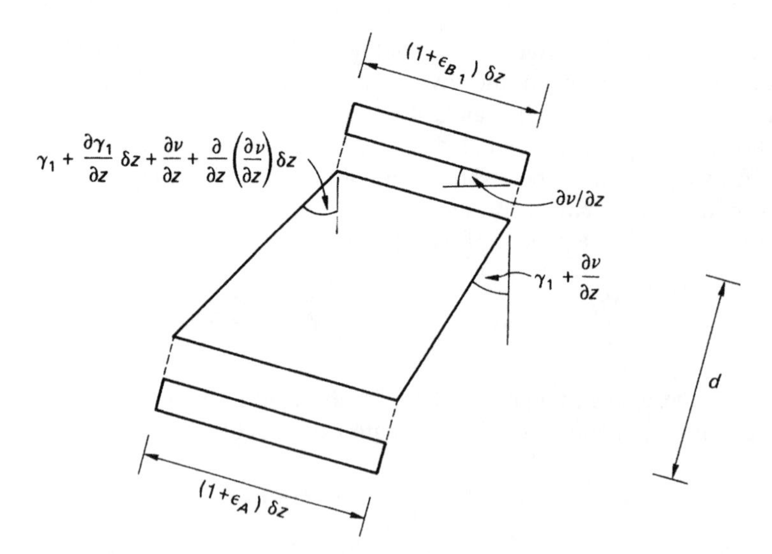

Fig. 26.27 Compatibility condition for combined bending and axial load.

Subtraction of Eq. (26.42) from Eq. (26.41) eliminates $\mathrm{d}^2v/\mathrm{d}z^2$, i.e.

$$\frac{\mathrm{d}\gamma_1}{\mathrm{d}z} - \frac{\mathrm{d}\gamma_2}{\mathrm{d}z} = \frac{1}{d}(2\varepsilon_A - \varepsilon_{B1} - \varepsilon_{B2})$$

or, as before

$$\frac{\mathrm{d}q_1}{\mathrm{d}z} - \frac{\mathrm{d}q_2}{\mathrm{d}z} = \frac{Gt}{dE}\left(\frac{2P_A}{A} - \frac{P_{B1}}{B} - \frac{P_{B2}}{B}\right) \tag{26.43}$$

In this particular problem the simplest method of solution is to choose P_A as the unknown since, from Eqs (26.39)

$$\frac{\mathrm{d}q_1}{\mathrm{d}z} - \frac{\mathrm{d}q_2}{\mathrm{d}z} = \frac{\partial^2 P_A}{\partial z^2}$$

Also substituting for P_{B1} and P_{B2} from Eq. (26.40), we obtain

$$\frac{\partial^2 P_A}{\partial z^2} - \frac{Gt}{dE}\left(\frac{2B+A}{AB}\right)P_A = -\frac{PGt}{dEB}$$

or

$$\frac{\partial^2 P_A}{\partial z^2} - \lambda^2 P_A = -\frac{PGt}{dEB} \tag{26.44}$$

where $\lambda^2 = Gt(2B+A)/dEAB$. The solution of Eq. (26.44) is of standard form and is

$$P_A = C\cosh\lambda z + D\sinh\lambda z + \frac{PA}{2B+A} \tag{26.45}$$

The boundary conditions are $P_A = 0$ when $z = 0$ and $q_1 = q_2 = 0 = \partial P_A/\partial z$ at the built-in end (no shear loads are applied). Hence

$$P_A = \frac{PA}{2B + A}(1 - \cosh \lambda z + \tanh \lambda L \sinh \lambda z)$$

or, rearranging

$$P_A = \frac{PA}{2B + A}\left[1 - \frac{\cosh \lambda(L - z)}{\cosh \lambda L}\right] \quad (26.46)$$

Hence

$$\sigma_A = \frac{P}{2B + A}\left[1 - \frac{\cosh \lambda(L - z)}{\cosh \lambda L}\right] \quad (26.47)$$

Substituting for P_A in the second of Eqs (26.40), we have

$$P_{B1} = \frac{PA}{2(2B + A)}\left[\frac{4B + A}{A} + \frac{\cosh \lambda(L - z)}{\cosh \lambda L}\right] \quad (26.48)$$

whence

$$\sigma_{B1} = \frac{PA}{2B(2B + A)}\left[\frac{4B + A}{A} + \frac{\cosh \lambda(L - z)}{\cosh \lambda L}\right] \quad (26.49)$$

Also from Eqs (26.40)

$$P_{B2} = -\frac{P_A}{2}$$

so that

$$P_{B2} = \frac{-PA}{2(2B + A)}\left[1 - \frac{\cosh \lambda(L - z)}{\cosh \lambda L}\right] \quad (26.50)$$

and

$$\sigma_{B2} = \frac{-PA}{2B(2B + A)}\left[1 - \frac{\cosh \lambda(L - z)}{\cosh \lambda L}\right] \quad (26.51)$$

Finally, the shear flow distributions are obtained from Eqs (16.39), thus

$$q_1 = \frac{-\partial P_{B1}}{\partial z} = \frac{PA\lambda}{2(2B + A)} \frac{\sinh \lambda(L - z)}{\cosh \lambda L} \quad (26.52)$$

$$q_2 = \frac{\partial P_{B2}}{\partial z} = \frac{-PA\lambda}{2(2B + A)} \frac{\sinh \lambda(L - z)}{\cosh \lambda L} \quad (26.53)$$

Again we see that each expression for direct stress, Eqs (26.47), (26.49) and (26.51), comprises a term which gives the solution from elementary theory together with a correction for the shear lag effect. The shear flows q_1 and q_2 are self-equilibrating, as can be seen from Eqs (26.52) and (26.53), and are entirely produced by the shear lag effect (q_1 and q_2 must be self-equilibrating since no shear loads are applied).

Example 26.5

The unsymmetrical panel shown in Fig. 26.28 comprises three direct stress carrying booms and two shear stress carrying panels. If the panel supports a load P at its free end and is pinned to supports at the ends of its outer booms determine the distribution of direct load in the central boom. Determine also the load in the central boom when $A = B = C$ and shear lag effects are absent.

As before we consider the equilibrium of elements of the booms, say A and B. This gives

$$\frac{\partial P_A}{\partial z} = -q_1 \tag{i}$$

and

$$\frac{\partial P_B}{\partial z} = q_1 - q_2 \tag{ii}$$

For overall equilibrium of a length z of the panel

$$P_A + P_B + P_C = P \tag{iii}$$

and taking moments about boom C

$$2P_A + P_B = P \tag{iv}$$

The compatibility condition is shown in Fig. 26.29 and gives

$$\frac{\partial \gamma_1}{\partial z} = \frac{1}{d}(\varepsilon_A - \varepsilon_A) - \frac{\partial^2 v}{\partial z^2} \tag{v}$$

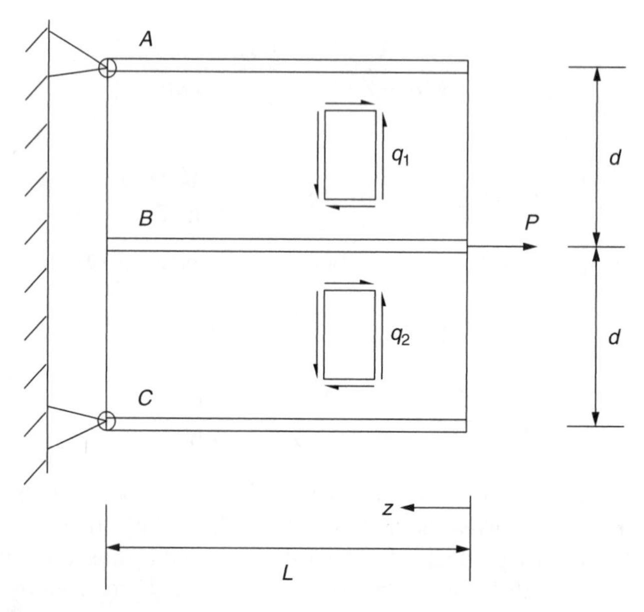

Fig. 26.28 Panel of Example 26.5.

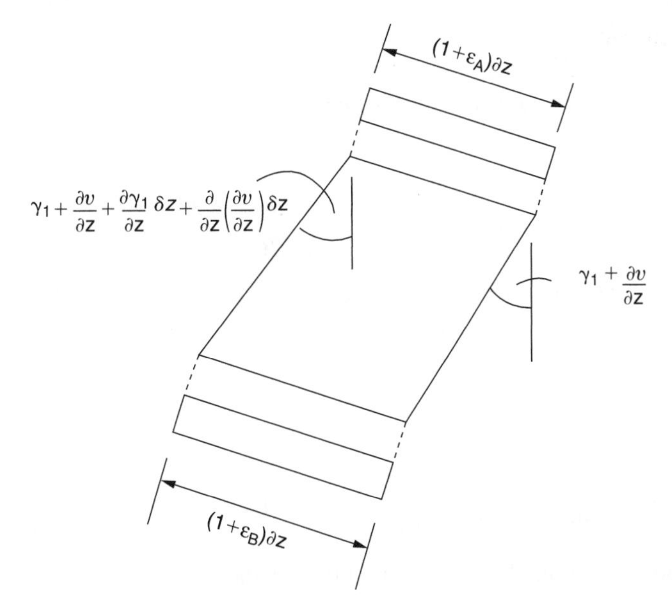

Fig. 26.29 Compatibility condition for the panel of Example 26.5.

Similarly, for elements of the booms B and C

$$\frac{\partial \gamma_2}{\partial z} = \frac{1}{d}(\varepsilon_C - \varepsilon_B) - \frac{\partial^2 v}{\partial z^2} \tag{vi}$$

Subtracting Eq. (vi) from (v) gives

$$\frac{\partial \gamma_1}{\partial z} - \frac{\partial \gamma_2}{\partial z} = \frac{1}{d}(2\varepsilon_B - \varepsilon_A - \varepsilon_C) \tag{vii}$$

Also

$$\gamma_1 = \frac{q_1}{Gt} \quad \gamma_2 = \frac{q_2}{Gt} \quad \varepsilon_A = \frac{P_A}{AE} \quad \varepsilon_B = \frac{P_B}{BE} \quad \text{and} \quad \varepsilon_C = \frac{P_C}{CE}$$

Substituting these expressions in Eq. (vii) gives

$$\frac{\partial q_1}{\partial z} - \frac{\partial q_2}{\partial z} = \frac{Gt}{dE}\left(\frac{2P_B}{B} - \frac{P_A}{A} - \frac{P_C}{C}\right) \tag{viii}$$

From Eqs (iv) and (iii)

$$P_A = \frac{1}{2}(P - P_B), \quad P_C = \frac{1}{2}(P - P_B)$$

Substituting in Eq. (viii), using Eq. (ii) and rearranging we have

$$\frac{\partial^2 P_B}{\partial z^2} - \frac{Gt}{dE}\left(\frac{4AC + BC + AB}{2ABC}\right)P_B = -\frac{GtP}{2dE}\left(\frac{A + C}{AC}\right)$$

the solution of which is

$$P_B = D\cosh\mu z + F\sinh\mu z + \frac{B(A+C)P}{(4AC+BC+AB)}$$

where

$$\mu^2 = \frac{Gt}{dE}\left(\frac{4AC+BC+AB}{2ABC}\right)$$

The boundary conditions are: when $z=0$, $P_B=P$ and when $z=L$, $P_B=0$. From the first of these

$$D = \frac{4AC}{4AC+BC+AB}P$$

while from the second

$$F = -\frac{P}{\sinh\mu L}\left[\frac{4AC}{4AC+BC+AB}\cosh\mu L + \frac{B(A+C)}{4AC+BC+AB}\right]$$

The expression for the load in the central boom is then

$$P_B = \frac{P}{4AC+BC+AB}\left[4AC\cosh\mu z - \left(\frac{4AC\cosh\mu L + AB + BC}{\sinh\mu L}\right) \times \sinh\mu z + B(A+C)\right]$$

If there is no shear lag the hyperbolic terms disappear and when $A=B=C$

$$P_B = P/3$$

Reference

1 Argyris, J. H. and Dunne, P. C., The general theory of cylindrical and conical tubes under torsion and bending loads, *J. Roy. Aero. Soc.*, Parts I–IV, February 1947; Part V, September and November 1947; Part VI, May and June 1949.

Problems

P.26.1 A thin-walled beam with the singly symmetrical cross-section shown in Fig. P.26.1, is built-in at one end where the shear force $S_y = 111\,250$ N is applied through the web 25. Assuming the cross-section remains undistorted by the loading, determine the shear flow and the position of the centre of twist at the built-in end. The shear modulus G is the same for all walls.

Ans: $q_{12} = q_{56} = 46.6$ N/mm, $q_{52} = 180.8$ N/mm,
$q_{32} = q_{54} = 1.4$ N/mm, $q_{43} = 74.6$ N/mm,
$x_R = -630.1$ mm, $y_R = 0$ (relative to mid-point of 52).

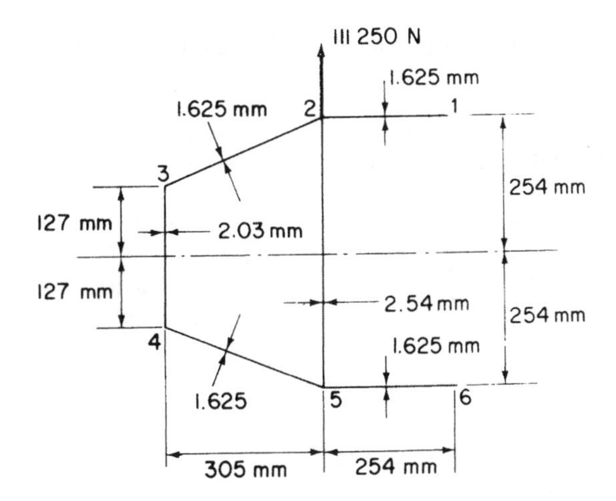

Fig. P.26.1

P.26.2 A thin-walled two-cell beam with the singly symmetrical cross-section shown in Fig. P.26.2 is built-in at one end where the torque is 11 000 Nm. Assuming the cross-section remains undistorted by the loading, determine the distribution of shear flow and the position of the centre of twist at the built-in end. The shear modulus G is the same for all walls.

Ans: $q_{12} = q_{45} = 44.1\,\text{N/mm}$, $q_{23} = q_{34} = 42.9\,\text{N/mm}$,
$q_{51} = 80.2\,\text{N/mm}$, $q_{24} = 37.4\,\text{N/mm}$,
$x_R = -79.5\,\text{mm}$, $y_R = 0$ (referred to mid-point of web 24).

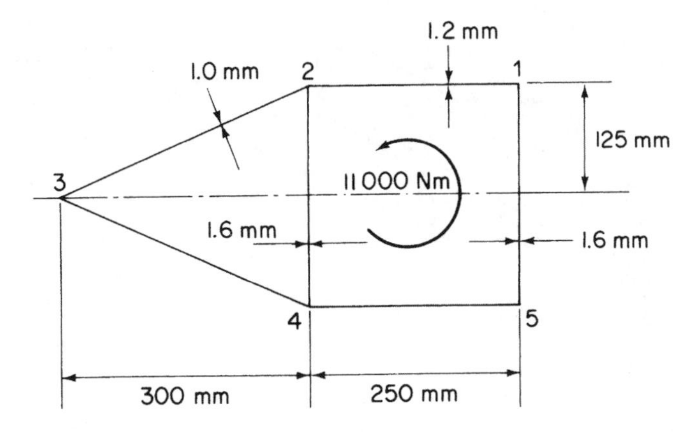

Fig. P.26.2

P.26.3 A singly symmetrical, thin-walled, closed section beam is built-in at one end where a shear load of 10 000 N is applied as shown in Fig. P.26.3. Calculate the resulting shear flow distribution at the built-in end if the cross-section of the beam remains undistorted by the loading and the shear modulus G and wall thickness t are each constant throughout the section.

Ans: $q_{12} = 3992.9/R$ N/mm, $q_{23} = 711.3/R$ N/mm,
$q_{31} = (1502.4 - 1894.7\cos\phi - 2102.1\sin\phi)/R$ N/mm.

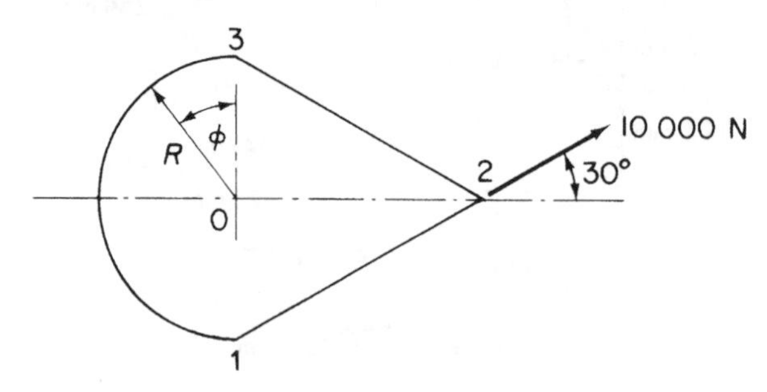

Fig. P.26.3

P.26.4 A uniform, four-boom beam, built-in at one end, has the rectangular cross-section shown in Fig. P.26.4. The walls are assumed to be effective only in shear, the thickness and shear modulus being the same for all walls while the booms, which are of equal area, carry only direct stresses. Assuming that the cross-section remains undistorted by the loading, calculate the twist at the free end due to a uniformly distributed torque loading $T = 20$ N m/mm along its entire length. Take $G = 20\,000$ N/mm^2 and $G/E = 0.36$.

Ans: 5.9° anticlockwise.

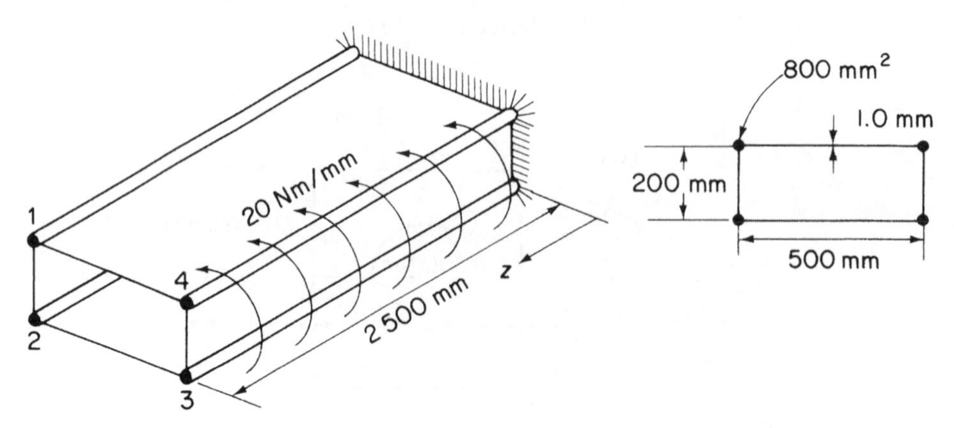

Fig. P.26.4

P.26.5 Figure P.26.5 shows the doubly symmetrical idealized cross-section of a uniform box beam of length l. Each of the four corner booms has area B and Young's modulus E, and they constitute the entire direct stress carrying area. The thin walls all have the same shear modulus G. The beam transmits a torque T from one end to the other, and at each end warping is completely suppressed. Between the ends, the shape of the cross-section is maintained without further restriction of warping.

Obtain an expression for the distribution of the end load along the length of one of the corner booms. Assuming $bt_1 > at_2$, indicate graphically the relation between torque direction and tension and compression in the boom end loads.

$$\textit{Ans.} \quad P = \frac{\mu BET}{8abGt_1t_2}(bt_1 - at_2)\left[-\sinh\mu z + \frac{(\cosh\mu l - 1)}{\sinh\mu l}\cosh\mu z\right]$$

where

$$\mu^2 = 8Gt_1t_2/BE(at_2 + bt_1).$$

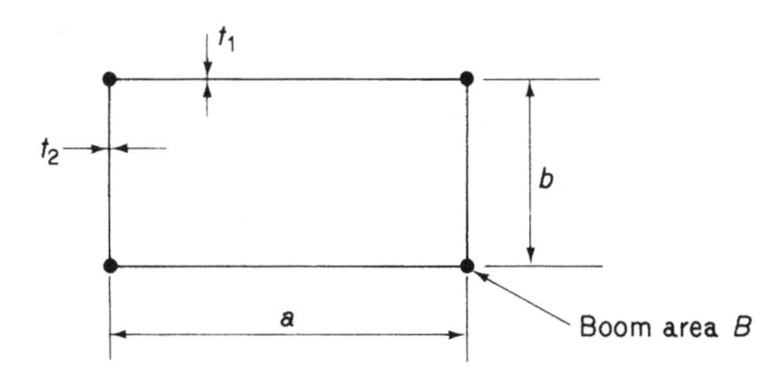

Fig. P.26.5

P.26.6 The idealized cross-section of a beam is shown in Fig. P.26.6. The beam is of length L and is attached to a flexible support at one end which only partially prevents warping of the cross-section; at its free end the beam carries a concentrated torque T.

Assuming that the warping at the built-in end is directly proportional to the free warping, ie $w = kw_o$, derive an expression for the distribution of direct stress along the top right-hand corner boom. State the conditions corresponding to the values $k = 0$ and $k = 1$.

$$\textit{Ans.} \quad \sigma = -\mu Ew_0(k-1)\frac{\sinh\mu(L-z)}{\cosh\mu L}, \quad \mu^2 = \frac{8Gt_bt_a}{BE(bt_a + at_b)}$$

$$\text{when } k = 0, \quad \sigma = \mu Ew_0\frac{\sinh\mu(L-z)}{\cosh\mu L} \quad \text{(i.e a rigid foundation)}$$

$$\text{when } k = 1, \quad \sigma = 0 \text{ (i.e free warping)}$$

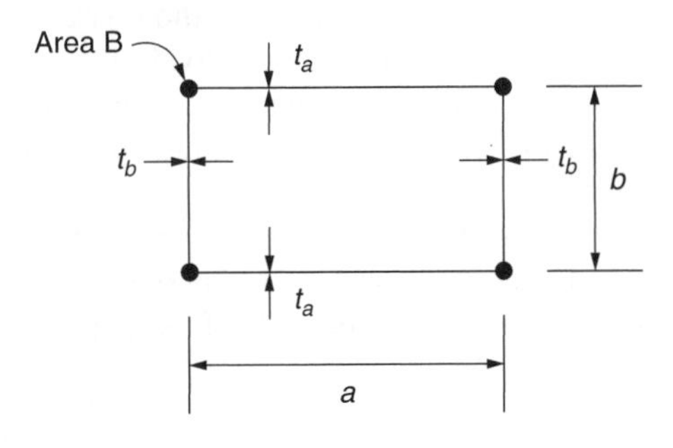

Fig. P.26.6

P.26.7 In the panel shown in Fig. P.26.7 the area, A_s, of the central stringer is to be designed so that the stress in it is 80% of the constant stress, σ_e, in the edge members, each of area B.

Assuming that the sheet, which is of constant thickness, t, carries only shear stress and that transverse strains are prevented, derive expressions for A_s and B in terms of the applied loads and the appropriate elastic moduli, E for the longitudinal members and G for the sheet.

Evaluate these expressions in the case where $P = 450\,000$ N; $P_s = 145\,000$ N; $S = 350$ N/mm; $\sigma_e = 275$ N/mm^2; $l = 1250$ mm; $b = 250$ mm; $t = 2.5$ mm and $G = 0.38E$. Find the fraction of the total tension at the abutment which is carried by the stringer.

Ans. $A_s = \dfrac{Gt}{2Eb}\left(lz - \dfrac{z^2}{2}\right) + \dfrac{1.25P_s}{\sigma_e}$,

$$B = \frac{0.1Gt}{Eb}z^2 + \frac{1}{\sigma_c}\left[\left(S - \frac{0.2Gt\sigma_e l}{bE}\right)z + P\right], \quad 0.25.$$

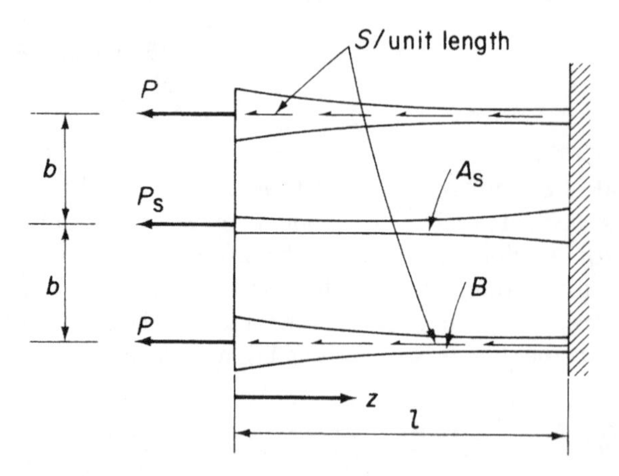

Fig. P.26.7

P.26.8 A symmetrical panel has the form shown in Fig. P.26.8. The longerons are of constant area, B_1 for the edge members and B_2 for the central member, and the sheet is of uniform thickness t. The panel is assembled without stress.

Obtain an expression for the distribution of end load in the central longeron if it is then raised to a temperature T (constant along its length) above the edge members. Also give the longitudinal displacement, at one end of the panel, of the central longeron relative to the edge members.

Assume that end loads are carried only by the longerons, that the sheet carries only shear, and that transverse members are provided to prevent transverse straining and to ensure shear effectiveness of the sheet at the ends of the panel.

Ans. $$P_2 = E\alpha T\left(\cosh\mu z - \tanh\frac{\mu l}{2}\sinh\mu z - 1\right) \Big/ \left(\frac{1}{2B_1} + \frac{1}{B_2}\right)$$

$$\text{Disp.} = \frac{\alpha T}{\mu} \tanh \mu \frac{l}{2}$$

where

$$u^2 = \frac{2Gt}{dE} \left(\frac{1}{2B_1} + \frac{1}{B_2} \right).$$

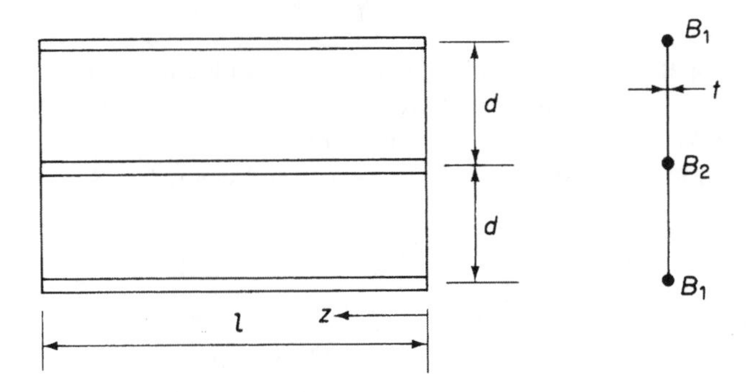

Fig. P.26.8

P.26.9 The flat panel shown in Fig. P.26.9 comprises a sheet of uniform thickness t, a central stringer of constant area A and edge members of varying area. The panel is supported on pinned supports and is subjected to externally applied shear flows S_1 and S_2, together with end loads $P_{1,0}$ and $P_{2,0}$ as shown. The areas of the edge members vary such that the direct stresses σ_1 and σ_2 in the edge members are constant.

Assuming that transverse strains are prevented, that the sheet transmits shear stress only and that each part has suitable end members to take the complementary shear stresses, derive expressions for the variation of direct stress σ_3 in the stringer and for the variation of shear flow in the upper panel in terms of the dimensions given and the elastic moduli E and G for the material.

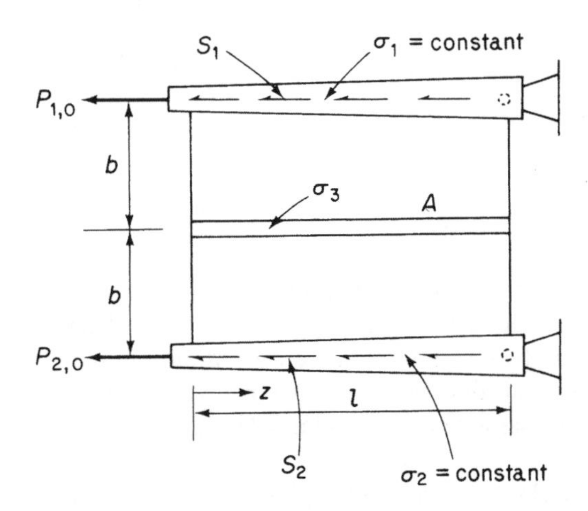

Fig. P.26.9

$$Ans. \quad \sigma_3 = \left(\frac{\sigma_1 + \sigma_2}{2}\right)\left[1 - \cosh \mu z - \frac{\sinh \mu z}{\sinh \mu l}(1 - \cosh \mu l)\right]$$

$$q_1 = A\left(\frac{\sigma_1 + \sigma_2}{4}\right)\mu\left[\sinh \mu z + \frac{\cosh \mu z}{\sinh \mu l}(1 - \cosh \mu l)\right]$$

where

$$\mu^2 = 2Gt/bAE$$

P.26.10 The panel shown in Fig. P.26.10 has been idealized into a combination of direct stress carrying booms and shear stress carrying plates; the boom areas are shown and the plate thickness is t. Derive expressions for the distribution of direct load in each boom and state how the load distributions are affected when $A = B$.

$$Ans. \quad P_1 = \frac{6P}{2A + B}\left[-\left(\frac{B + 8A}{6}\right) - \left(\frac{B - A}{3}\right)\frac{\cosh \mu(L - z)}{\cosh \mu L}\right]$$

$$P_2 = \frac{6P}{2A + B}\left[-B + \frac{2}{3}(B - A)\frac{\cosh \mu(L - z)}{\cosh \mu L}\right]$$

$$P_3 = \frac{6P}{2A + B}\left[-\left(\frac{4A - B}{6}\right) - \left(\frac{B - A}{3}\right)\frac{\cosh \mu(L - z)}{\cosh \mu L}\right]$$

When $A = B, P_1 = -3P, P_2 = -2P, P_3 = -P$, i.e no shear lag.

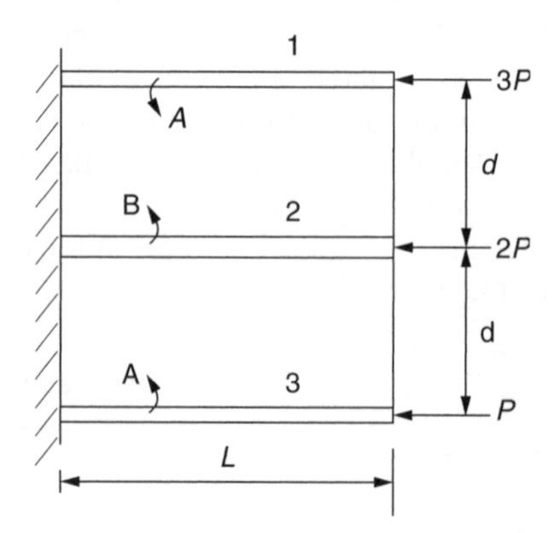

Fig. P.26.10

P.26.11 A uniform cantilever of length l has the doubly symmetrical cross-section shown in Fig. P.26.11. The section shape remains undistorted in its own plane after loading. Direct stresses on the cross-section are carried only in the concentrated longeron areas shown, and the wall thickness dimensions given relate only to shearing effects. All longerons have the same Young's modulus E and all walls the same effective shear modulus G.

The root of the cantilever is built-in, warping being completely suppressed there, and a shearing force S is applied at the tip in the position indicated.

Derive an expression for the resultant end load in a corner longeron. Also calculate the resultant deflection of the tip, including the effects of both direct and shear strains.

Ans. $P = -\dfrac{S}{8h}\left(\dfrac{\sinh \mu z}{\mu \cosh \mu l} + 3z\right)$

where $\mu^2 = 4Gt/3dBE$ (top right hand) (origin for z at free end)

$$\text{Def.} = \frac{Sl}{12h}\left(\frac{11}{4Gt} + \frac{l^2}{EBh}\right).$$

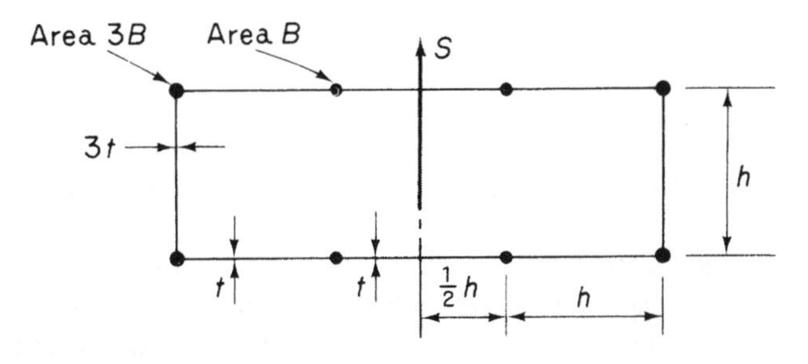

Fig. P.26.11

P.26.12 The idealized cantilever beam shown in Fig. P.26.12 carries a uniformly distributed load of intensity w. Assuming that all direct stresses are carried by the booms while the panels are effective only in shear determine the distribution of direct stress in the central boom in the top cover. Young's modulus for the booms is E and the shear modulus of the walls is G.

Ans. $P_A = -\dfrac{wA}{h(2B + A)}\left[\dfrac{\cosh \mu z}{\mu^2} + \left(\dfrac{\mu L - \sinh \mu L}{\mu^2 \cosh \mu L}\right)\sinh \mu z - \dfrac{1}{\mu^2} - \dfrac{z^2}{2}\right]$

where

$$\mu^2 = \frac{Gt(2B + A)}{dEAB}$$

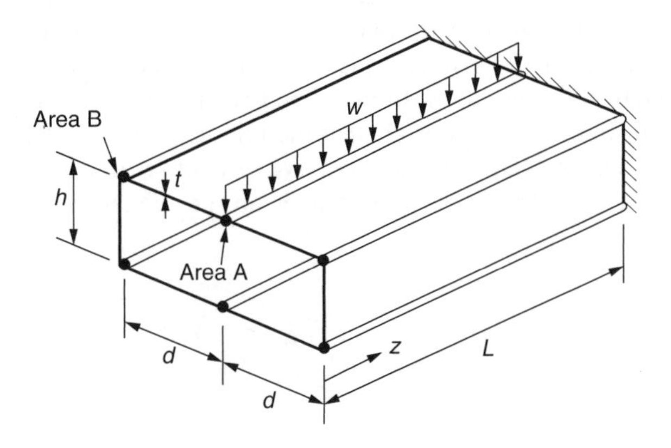

Fig. P.26.12

27

Open section beams

Instances of open section beams occurring in isolation are infrequent in aircraft structures. The majority of wing structures do, however, contain cut-outs for undercarriages, inspection panels and the like, so that at these sections the wing is virtually an open section beam. We saw in Chapter 23 that one method of analysis for such cases is to regard the applied torque as being resisted by the differential bending of the front and rear spars in the cut-out bay. An alternative approach is to consider the cut-out bay as an open section beam built-in at each end and subjected to a torque. We shall now investigate the method of analysis of such beams.

27.1 I-section beam subjected to torsion

If such a beam is axially unconstrained and loaded by a pure torque T the rate of twist is constant along the beam and is given by

$$T = GJ\frac{\mathrm{d}\theta}{\mathrm{d}z} \quad \text{(from Eq. (18.12))}$$

We also showed in Section 18.2 that the shear stress varies linearly across the thickness of the beam wall and is zero at the middle plane (Fig. 27.1). It follows that although the beam and the middle plane warp (we are concerned here with primary warping), there is no shear distortion of the middle plane. The mechanics of this warping are more easily understood by reference to the thin-walled I-section beam of Fig. 27.2(a). A plan view of the beam (Fig. 27.2(b)) reveals that the middle plane of each flange remains rectangular, although twisted, after torsion. We now observe the effect of applying a restraint to one end of the beam. The flanges are no longer free to warp and will

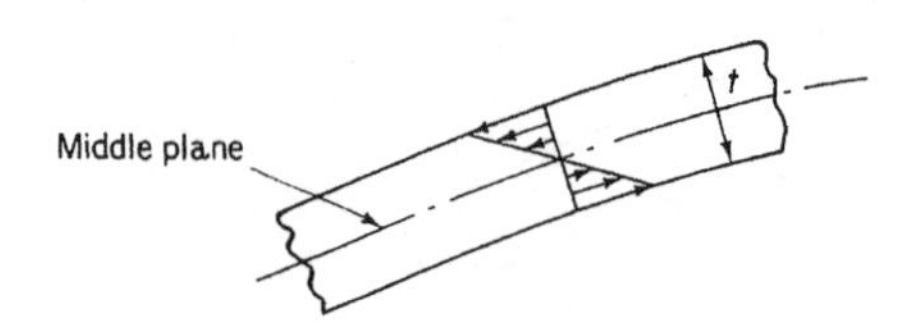

Fig. 27.1 Shear stress distribution across the wall of an open section beam subjected to torsion.

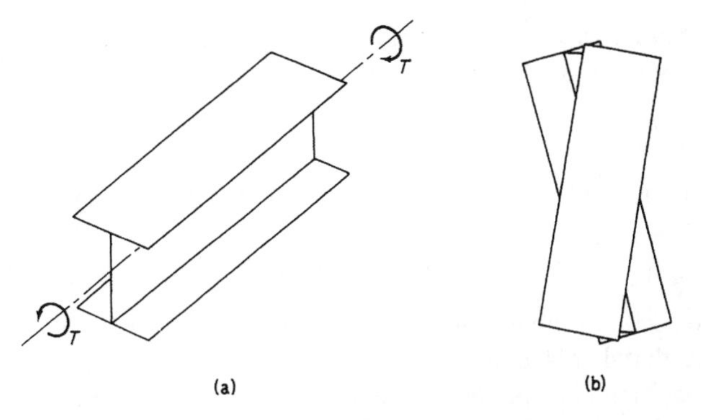

Fig. 27.2 (a) Torsion of I-section beam; (b) plan view of beam showing undistorted shape of flanges.

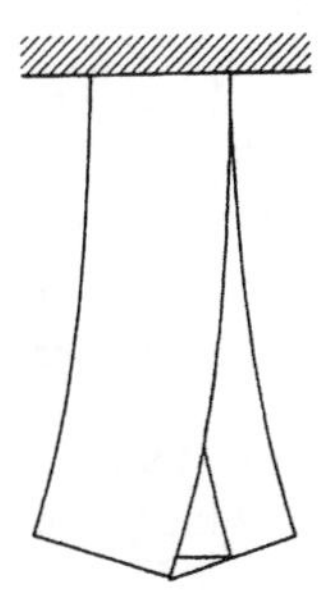

Fig. 27.3 Bending effect of axial constraint on flanges of I-section beam subjected to torsion.

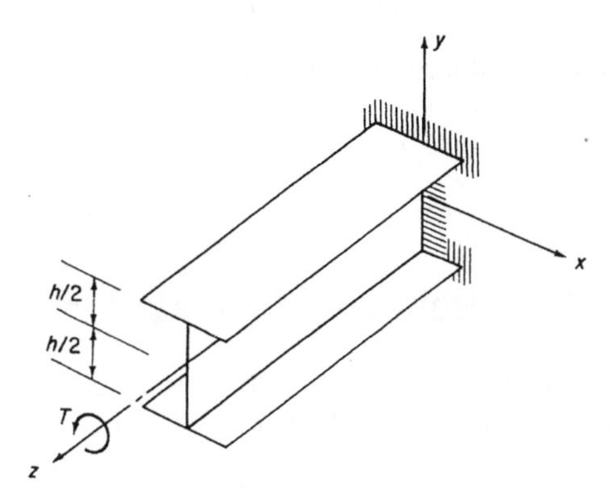

Fig. 27.4 Torsion of I-section beam fully built-in at one end.

bend in their own planes into the shape shown in plan in Fig. 27.3. Obviously the beam still twists along its length but the rate of twist is no longer constant and the resistance to torsion is provided by the St. Venant shear stresses (unrestrained warping) plus the resistance of the flanges to bending. The total torque may therefore be written $T = T_J + T_\Gamma$, where $T_J = GJ\,\mathrm{d}\theta/\mathrm{d}z$ from the unconstrained torsion of open sections but in which $\mathrm{d}\theta/\mathrm{d}z$ is not constant, and T_Γ is obtained from a consideration of the bending of the flanges. It will be instructive to derive an expression for T_Γ for the I-section beam of Fig. 27.4 before we turn our attention to the case of a beam of arbitrary section.

Suppose that at any section z the angle of twist of the I-beam is θ. Then the lateral displacement u of the lower flange is

$$u = \theta \frac{h}{2}$$

and the bending moment M_F in the plane of the flange is given by

$$M_F = -EI_F \frac{d^2u}{dz^2}$$

where I_F is the second moment of area of the *flange* cross-section about the y axis. It is assumed here that displacements produced by shear are negligible so that the lateral deflection of the flange is completely due to the self-equilibrating direct stress system σ_Γ set up by the bending of the flange. We shall not, however, assume that the shear stresses in the flange are negligible. The shear S_F in the flange is then

$$S_F = \frac{dM_F}{dz} = -EI_F \frac{d^3u}{dz^3}$$

or substituting for u in terms of θ and h

$$S_F = -EI_F \frac{h}{2} \frac{d^3\theta}{dz^3}$$

Similarly, there is a shear force in the top flange of the same magnitude but opposite in direction. Together they form a couple which represents the second part T_Γ of the total torque, thus

$$T_\Gamma = S_F h = -EI_F \frac{h^2}{2} \frac{d^3\theta}{dz^3}$$

and the expression for the total torque may be written

$$T = GJ \frac{d\theta}{dz} - EI_F \frac{h^2}{2} \frac{d^3\theta}{dz^3}$$

27.2 Torsion of an arbitrary section beam

The insight into the physical aspects of the problem gained in the above will be found helpful in the development of the general theory for the arbitrary section beam shown in Fig. 27.5.

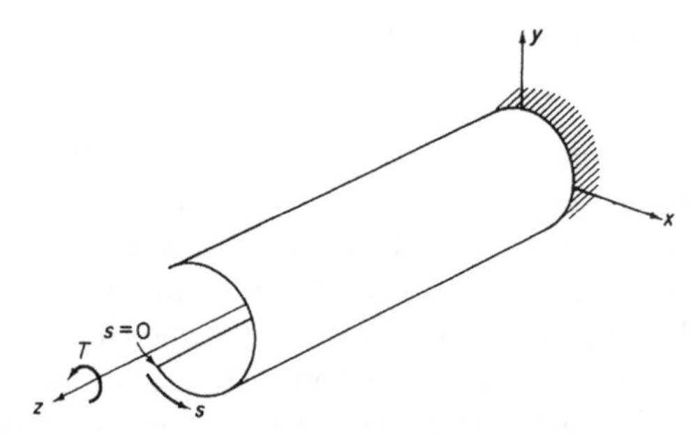

Fig. 27.5 Torsion of an open section beam fully built-in at one end.

The theory, originally developed by Wagner and Kappus, is most generally known as the Wagner torsion bending theory. It assumes that the beam is long compared with its cross-sectional dimensions, that the cross-section remains undistorted by the loading and that the shear strain γ_{zs} of the middle plane of the beam is negligible although the stresses producing the shear strain are not. From similar assumptions is derived, in Section 18.2.1, an expression for the primary warping w of the beam, viz.

$$w = -2A_R \frac{d\theta}{dz} \qquad \text{(Eq. (18.19))}$$

In the presence of axial constraint, $d\theta/dz$ is no longer constant so that the longitudinal strain $\partial w/\partial z$ is not zero and direct (also shear) stresses are induced. Then

$$\sigma_\Gamma = E\frac{\partial w}{\partial z} = -2A_R E \frac{d^2\theta}{dz^2} \qquad (27.1)$$

The σ_Γ stress system must be self-equilibrating since the applied load is a pure torque. Therefore, at any section the resultant end load is zero and

$$\int_c \sigma_\Gamma t\, ds = 0 \quad \left(\int_c \text{denotes integration around the beam section}\right)$$

or, from Eq. (27.1) and observing that $d^2\theta/dz^2$ is a function of z only

$$\int_c 2A_R t\, ds = 0 \qquad (27.2)$$

The limits of integration of Eq. (27.2) present some difficulty in that A_R is zero when w is zero at an unknown value of s. Let

$$2A_R = 2A_{R,0} - 2A'_R$$

where $A_{R,0}$ is the area swept out from $s=0$ and A'_R is the value of $A_{R,0}$ at $w=0$ (see Fig. 27.6). Then in Eq. (27.2)

$$\int_c 2A_{R,0} t\, ds - 2A'_R \int_c t\, ds = 0$$

and

$$2A'_R = \frac{\int_c 2A_{R,0} t\, ds}{\int_c t\, ds}$$

giving

$$2A_R = 2A_{R,0} - \frac{\int_c 2A_{R,0} t\, ds}{\int_c t\, ds} \qquad (27.3)$$

The axial constraint shear flow system, q_Γ, is in equilibrium with the self-equilibrating direct stress system. Thus, from Eq. (17.2)

$$\frac{\partial q_\Gamma}{\partial s} + t\frac{\partial \sigma_\Gamma}{\partial z} = 0$$

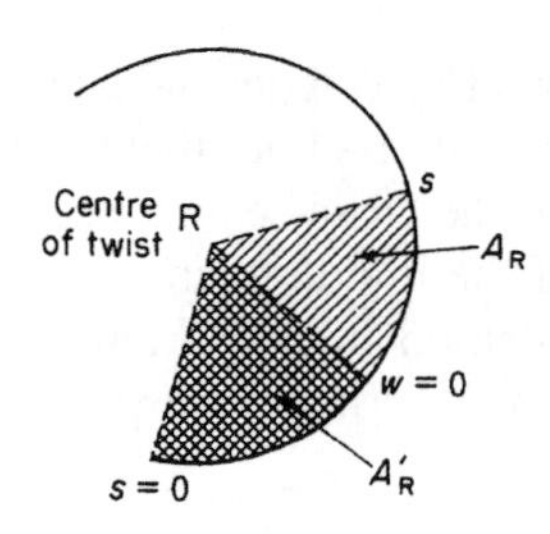

Fig. 27.6 Computation of swept area A_R.

Hence

$$\frac{\partial q_\Gamma}{\partial s} = -t\frac{\partial \sigma_\Gamma}{\partial z}$$

Substituting for σ_Γ from Eq. (27.1) and noting that $q_\Gamma = 0$ when $s = 0$, we have

$$q_\Gamma = \int_0^s 2A_R E t \frac{d^3\theta}{dz^3} ds$$

or

$$q_\Gamma = E\frac{d^3\theta}{dz^3}\int_0^s 2A_R t \, ds \tag{27.4}$$

Now

$$T_\Gamma = \int_c p_R q_\Gamma \, ds$$

or, from Eq. (27.4)

$$T_\Gamma = E\frac{d^3\theta}{dz^3}\int_c p_R \left(\int_0^s 2A_R t \, ds\right) ds$$

The integral in this equation is evaluated by substituting $p_R = (d/ds)(2A_R)$ and integrating by parts. Thus

$$\int_c \frac{d}{ds}(2A_R)\left(\int_0^s 2A_R t \, ds\right) ds = \left[2A_R \int_0^s 2A_R t \, ds\right]_c - \int_c 4A_R^2 t \, ds$$

At each open edge of the beam q_Γ, and therefore $\int_0^s 2A_R t \, ds$, is zero so that the integral reduces to $-\int_c 4A_R^2 t \, ds$, giving

$$T_\Gamma = -E\Gamma_R \frac{d^3\theta}{dz^3} \tag{27.5}$$

where $\Gamma_R = \int_c 4A_R^2 t \, ds$, the *torsion-bending constant*, and is purely a function of the geometry of the cross-section. The total torque T, which is the sum of the St. Venant torque and the Wagner torsion bending torque, is then written

$$T = GJ\frac{d\theta}{dz} - E\Gamma_R \frac{d^3\theta}{dz^3} \tag{27.6}$$

(*Note*: Compare Eq. (27.6) with the expression derived for the I-section beam.)

In the expression for Γ_R the thickness t is actually the direct stress carrying thickness t_D of the beam wall so that Γ_R, for a beam with n booms, may be generally written

$$\Gamma_R = \int_c 4A_R^2 t_D \, ds + \sum_{r=1}^{n} (2A_{R,r})^2 B_r$$

where B_r is the cross-sectional area of the rth boom. The calculation of Γ_R enables the second order differential equation in $d\theta/dz$ (Eq. (27.6)) to be solved. The constraint shear flows, q_Γ, follow from Eqs (27.4) and (27.3) and the longitudinal constraint stresses from Eq. (27.1). However, before illustrating the complete method of solution with examples we shall examine the calculation of Γ_R.

So far we have referred the swept area A_R, and hence Γ_R, to the centre of twist of the beam without locating its position. This may be accomplished as follows. At any section of the beam the resultant of the q_Γ shear flows is a pure torque (as is the resultant of the St. Venant shear stresses) so that in Fig. 27.7

$$\int_c q_\Gamma \sin\psi \, ds = S_y = 0$$

Therefore, from Eq. (27.4)

$$E\frac{d^3\theta}{dz^3}\int_c \left(\int_0^s 2A_R t \, ds\right) \sin\psi \, ds = 0$$

Now

$$\sin\psi = \frac{dy}{ds} \qquad \frac{d}{ds}(2A_R) = p_R$$

and the above expression may be integrated by parts, thus

$$\int_c \frac{dy}{ds}\left(\int_0^s 2A_R t \, ds\right) ds = \left[y\int_0^s 2A_R t \, ds\right]_c - \int_c y 2A_R t \, ds = 0$$

The first term on the right-hand side vanishes as $\int_0^s 2A_R t \, ds$ is zero at each open edge of the beam, leaving

$$\int_c y 2A_R t \, ds = 0$$

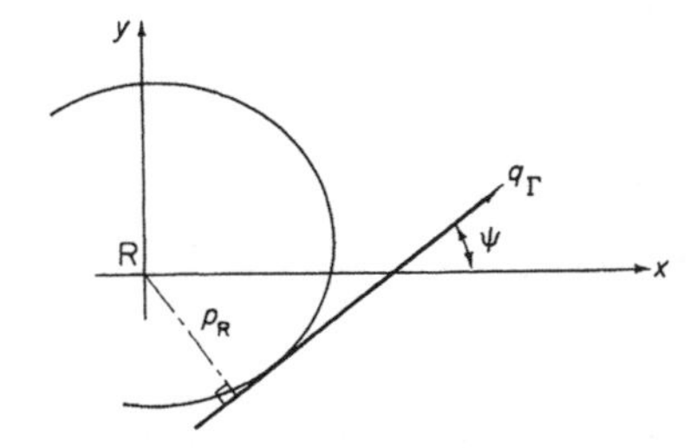

Fig. 27.7 Determination of the position of the centre of twist.

Again integrating by parts

$$\int_c y2A_{\mathrm{R}}t\,\mathrm{d}s = \left[2A_{\mathrm{R}}\int_0^s yt\,\mathrm{d}s\right]_c - \int_c p_{\mathrm{R}}\left(\int_0^s yt\,\mathrm{d}s\right)\mathrm{d}s = 0$$

The integral in the first term on the right-hand side of the above equation may be recognized, from Chapter 17, as being directly proportional to the shear flow produced in a singly symmetrical open section beam supporting a shear load S_y. Its value is therefore zero at each open edge of the beam. Hence

$$\int_c p_{\mathrm{R}}\left(\int_0^s yt\,\mathrm{d}s\right)\mathrm{d}s = 0 \tag{27.7}$$

Similarly, for the horizontal component S_x to be zero

$$\int_c p_{\mathrm{R}}\left(\int_0^s xt\,\mathrm{d}s\right)\mathrm{d}s = 0 \tag{27.8}$$

Equations (27.7) and (27.8) hold if the centre of twist coincides with the shear centre of the cross-section. To summarize, the centre of twist of a section of an open section beam carrying a pure torque is the shear centre of the section.

We are now in a position to calculate Γ_{R}. This may be done by evaluating $\int_c 4A_{\mathrm{R}}^2 t\,\mathrm{d}s$ in which $2A_{\mathrm{R}}$ is given by Eq. (27.3). In general, the calculation may be lengthy unless the section has flat sides in which case a convenient analogy shortens the work considerably. For the flat-sided section in Fig. 27.8(a) we first plot the area $2A_{\mathrm{R},0}$ swept out from the point 1 where we choose $s=0$ (Fig. 27.8(b)). The swept area $A_{\mathrm{R},0}$ increases linearly from zero at 1 to $(1/2)p_{12}d_{12}$ at 2 and so on. Note that movement along side 23 produces no increment of $2A_{\mathrm{R},0}$ as $p_{23}=0$. Further, we adopt a sign convention for p such that p is positive if movement in the positive s direction of the foot of p along the tangent causes anticlockwise rotation about R. The increment of $2A_{\mathrm{R},0}$ from side 34 is therefore negative.

In the derivation of Eq. (27.3) we showed that

$$2A'_{\mathrm{R}} = \frac{\int_c 2A_{\mathrm{R},0}t\,\mathrm{d}s}{\int_c t\,\mathrm{d}s}$$

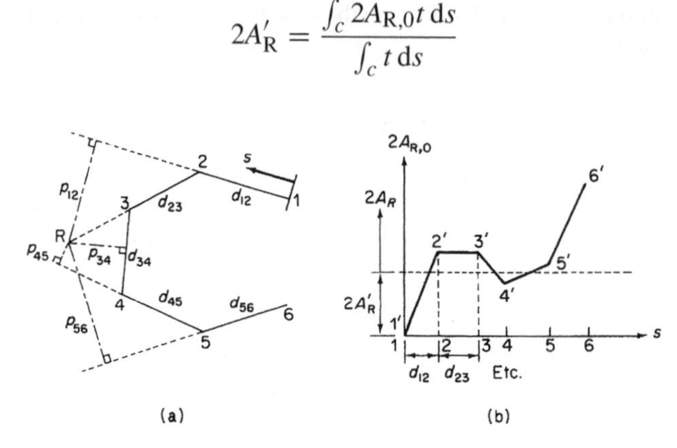

Fig. 27.8 Computation of torsion bending constant Γ_{R}: (a) dimensions of flat-sided open section beam; (b) variation of $2A_{\mathrm{R},0}$ around beam section.

Suppose now that the line $1'2'3'\ldots6'$ is a wire of varying density such that the weight of each element $\delta s'$ is $t\delta s$. Thus the weight of length $1'2'$ is td_{12}, etc. The y coordinate of the centre of gravity of the 'wire' is then

$$\bar{y} = \frac{\int yt\,\mathrm{d}s}{\int t\,\mathrm{d}s}$$

Comparing this expression with the previous one for $2A'_{\mathrm{R}}$, y and $\bar{y}$ are clearly analogous to $2A_{\mathrm{R},0}$ and $2A'_{\mathrm{R}}$, respectively. Further

$$\Gamma_{\mathrm{R}} = \int_c (2A_{\mathrm{R}})^2 t\,\mathrm{d}s = \int_c (2A_{\mathrm{R},0} - 2A'_{\mathrm{R}})^2 t\,\mathrm{d}s$$

Expanding and substituting

$$2A'_{\mathrm{R}} \int_c t\,\mathrm{d}s \quad \text{for} \quad \int_c 2A_{R,0} t\,\mathrm{d}s$$

gives

$$\Gamma_{\mathrm{R}} = \int_c (2A_{\mathrm{R},0})^2 t\,\mathrm{d}s - (2A'_{\mathrm{R}})^2 \int_c t\,\mathrm{d}s \tag{27.9}$$

Therefore, in Eq. (27.9), Γ_{R} is analogous to the moment of inertia of the 'wire' about an axis through its centre of gravity parallel to the s axis.

Example 27.1

An open section beam of length L has the section shown in Fig. 27.9. The beam is firmly built-in at one end and carries a pure torque T. Derive expressions for the direct stress and shear flow distributions produced by the axial constraint (the σ_Γ and q_Γ systems) and the rate of twist of the beam.

The beam is loaded by a pure torque so that the axis of twist passes through the shear centre S(R) of each section. We shall take the origin for s at the point 1 and initially plot $2A_{\mathrm{R},0}$ against s to determine Γ_{R} (see Fig. 27.10). The position of the centre of gravity, $(2A'_{\mathrm{R}})$, of the wire $1'2'3'4'$ is found by taking moments about the s axis. Then

$$t(2d+h)2A'_{\mathrm{R}} = td\left(\frac{hd}{4}\right) + th\left(\frac{hd}{2}\right) + td\left(\frac{hd}{4}\right)$$

from which

$$2A'_{\mathrm{R}} = \frac{hd(h+d)}{2(h+2d)} \tag{i}$$

Γ_{R} follows from the moment of inertia of the 'wire' about an axis through its centre of gravity. Hence

$$\Gamma_{\mathrm{R}} = 2td\frac{1}{3}\left(\frac{hd}{2}\right)^2 + th\left(\frac{hd}{2}\right)^2 - \left[\frac{hd(h+d)}{2(h+2d)}\right]^2 t(h+2d)$$

which simplifies to

$$\Gamma_{\mathrm{R}} = \frac{t\,d^3h^2}{12}\left(\frac{2h+d}{h+2d}\right) \tag{ii}$$

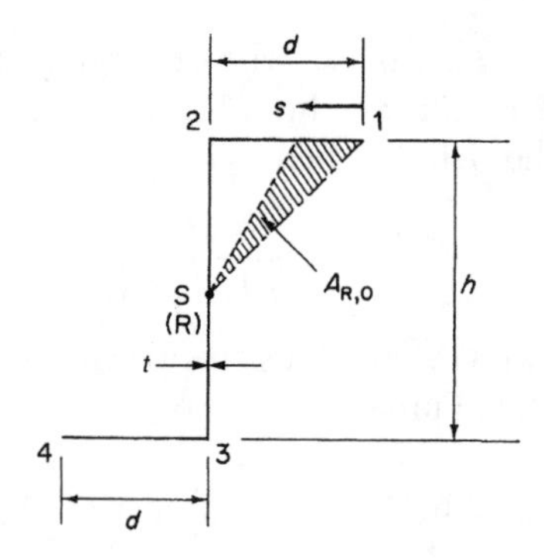

Fig. 27.9 Section of axially constrained open section beam under torsion.

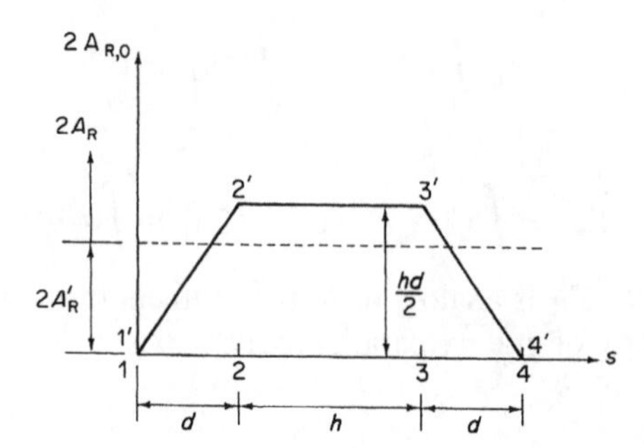

Fig. 27.10 Calculation of Γ_R for the section of Example 27.1.

Equation (27.6), i.e.

$$T = GJ\frac{d\theta}{dz} - E\Gamma_R\frac{d^3\theta}{dz^3}$$

may now be solved for $d\theta/dz$. Rearranging and writing $\mu^2 = GJ/E\Gamma_R$ we have

$$\frac{d^3\theta}{dz^3} - \mu^2\frac{d\theta}{dz} = -\mu^2\frac{T}{GJ} \tag{iii}$$

The solution of Eq. (iii) is of standard form, i.e.

$$\frac{d\theta}{dz} = \frac{T}{GJ} + A\cosh\mu z + B\sinh\mu z$$

The constants A and B are found from the boundary conditions:

(1) At the built-in end the warping $w = 0$ and since $w = -2A_R d\theta/dz$ then $d\theta/dz = 0$ at the built-in end.

(2) At the free end $\sigma_\Gamma = 0$, as there is no constraint and no externally applied direct load. Therefore, from Eq. (27.1), $d^2\theta/dz^2 = 0$ at the free end.

From (1)

$$A = -T/GJ$$

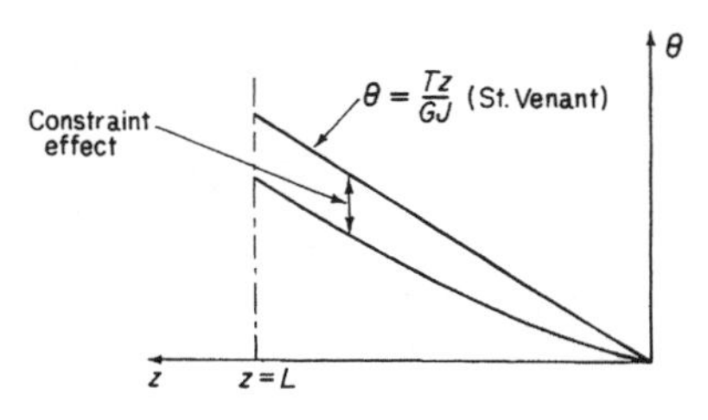

Fig. 27.11 Stiffening effect of axial constraint.

From (2)

$$B = (T/GJ)\tanh \mu L$$

so that

$$\frac{\mathrm{d}\theta}{\mathrm{d}z} = \frac{T}{GJ}(1 - \cosh \mu z + \tanh \mu L \sinh \mu z)$$

or

$$\frac{\mathrm{d}\theta}{\mathrm{d}z} = \frac{T}{GJ}\left[1 - \frac{\cosh \mu(L-z)}{\cosh \mu L}\right] \tag{iv}$$

The first term in Eq. (iv) is seen to be the rate of twist derived from the St. Venant torsion theory. The hyperbolic second term is therefore the modification introduced by the axial constraint. Equation (iv) may be integrated to find the distribution of angle of twist θ, the appropriate boundary condition being $\theta = 0$ at the built-in end, i.e.

$$\theta = \frac{T}{GJ}\left[z + \frac{\sinh \mu(L-z)}{\mu \cosh \mu L} - \frac{\sinh \mu L}{\mu \cosh \mu L}\right] \tag{v}$$

and the angle of twist, $\theta_{\mathrm{F,E}}$, at the free end of the beam is

$$\theta_{\mathrm{F,E}} = \frac{TL}{GJ}\left(1 - \frac{\tanh \mu L}{\mu L}\right) \tag{vi}$$

Plotting θ against z (Fig. 27.11) illustrates the stiffening effect of axial constraint on the beam.

The decrease in the effect of axial constraint towards the free end of the beam is shown by an examination of the variation of the St. Venant (T_J) and Wagner (T_Γ) torques along the beam. From Eq. (iv)

$$T_J = GJ\frac{\mathrm{d}\theta}{\mathrm{d}z} = T\left[1 - \frac{\cosh \mu(L-z)}{\cosh \mu L}\right] \tag{vii}$$

and

$$T_\Gamma = -E\Gamma_{\mathrm{R}}\frac{\mathrm{d}^3\theta}{\mathrm{d}z^3} = T\frac{\cosh \mu(L-z)}{\cosh \mu L} \tag{viii}$$

T_J and T_Γ are now plotted against z as fractions of the total torque T (Fig. 27.12). At the built-in end the entire torque is carried by the Wagner stresses, but although the

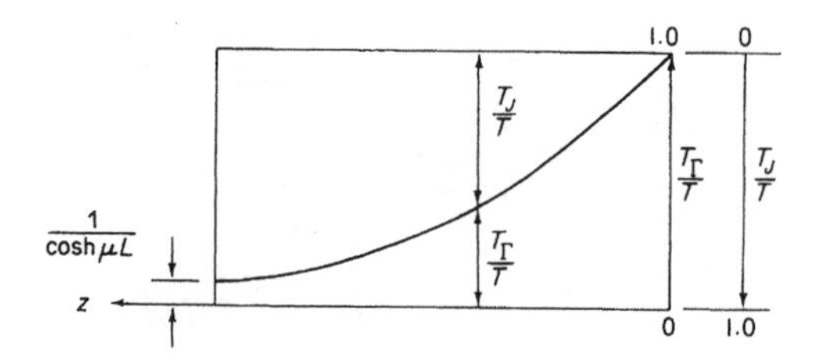

Fig. 27.12 Distribution of St. Venant and torsion-bending torques along the length of the open section beam shown in Fig. 27.9.

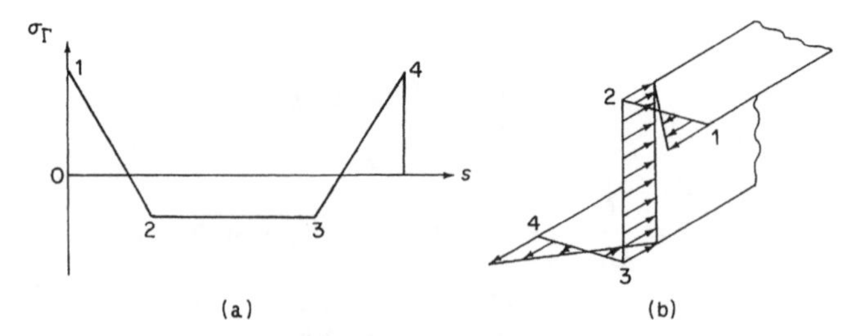

Fig. 27.13 Distribution of axial constraint direct stress around the section.

constraint effect diminishes towards the free end it does not disappear entirely. This is due to the fact that the axial constraint shear flow, q_Γ, does not vanish at $z = L$, for at this section (and all other sections) $\mathrm{d}^3\theta/\mathrm{d}z^3$ is not zero.

Equations (iii)–(viii) are, of course, valid for open section beams of any cross-section. Their application in a particular case is governed by the value of the torsion bending constant Γ_R and the St. Venant torsion constant $J[=(h + 2d)t^3/3$ for this example]. With this in mind we can proceed, as required by the example, to derive the direct stress and shear flow distributions. The former is obtained from Eqs (27.1) and (iv), i.e.

$$\sigma_\Gamma = -2A_\mathrm{R}E\frac{T}{GJ}\mu\frac{\sinh \mu(L - z)}{\cosh \mu L}$$

or writing $\mu^2 = GJ/E\Gamma_\mathrm{R}$ and rearranging

$$\sigma_\Gamma = -\sqrt{\frac{E}{GJ\Gamma_\mathrm{R}}}T2A_\mathrm{R}\frac{\sinh \mu(L - z)}{\cosh \mu L} \tag{ix}$$

In Eq. (ix) E, G, J and Γ_R are constants for a particular beam, T is the applied torque, A_R is a function of s and the hyperbolic term is a function of z. It follows that at a given section of the beam the direct stress is proportional to $-2A_\mathrm{R}$, and for the beam of this example the direct stress distribution has, from Fig. 27.10, the form shown in Figs 27.13(a) and (b). In addition, the value of σ_Γ at a particular value of s varies along the beam in the manner shown in Fig. 27.14.

Finally, the axial constraint shear flow, q_Γ, is obtained from Eq. (27.4), namely

$$q_\Gamma = E\frac{\mathrm{d}^3\theta}{\mathrm{d}z^3}\int_0^s 2A_\mathrm{R}t\,\mathrm{d}s$$

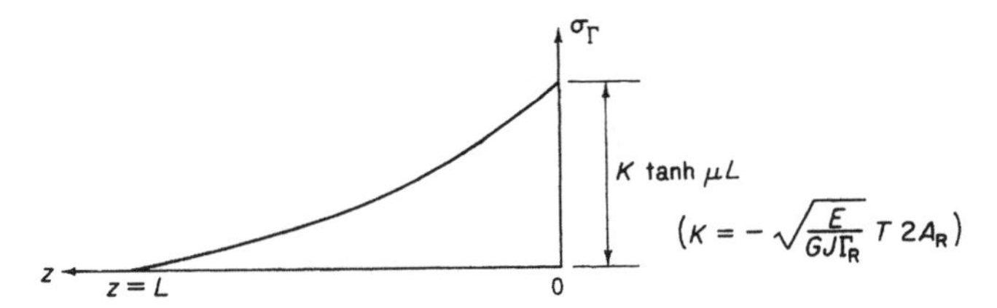

Fig. 27.14 Spanwise distribution of axial constraint direct stress.

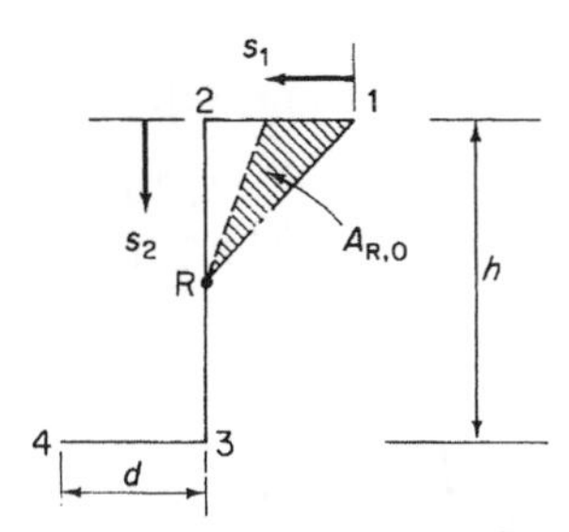

Fig. 27.15 Calculation of axial constraint shear flows.

At any section z, q_Γ is proportional to $\int_0^s 2A_R t\,ds$ and is computed as follows. Referring to Fig. 27.15, $2A_R = 2A_{R,0} - 2A'_R$ so that in flange 12

$$2A_R = \frac{hs_1}{2} - \frac{hd}{2}\left(\frac{h+d}{h+2d}\right)$$

Hence

$$\int_0^s 2A_R t\,ds = t\left[\frac{hs_1^2}{4} - \frac{hd}{2}\left(\frac{h+d}{h+2d}\right)s_1\right]$$

so that

$$q_{\Gamma,1} = 0 \quad \text{and} \quad q_{\Gamma,2} = -E\frac{d^3\theta}{dz^3}\frac{h^2d^2t}{4(h+2d)}$$

Similarly

$$q_{\Gamma,23} = E\frac{d^3\theta}{dz^3}\left[\frac{hd^2t}{2(h+2d)}s_2 - \frac{h^2d^2t}{4(h+2d)}\right]$$

whence

$$q_{\Gamma,2} = -E\frac{d^3\theta}{dz^3}\frac{h^2d^2t}{4(h+2d)} \qquad q_{\Gamma,3} = E\frac{d^3\theta}{dz^3}\frac{h^2d^2t}{4(h+2d)}$$

Note that in the above $d^3\theta/dz^3$ is negative (Eq. (viii)). Also at the mid-point of the web where $s_2 = h/2$, $q_\Gamma = 0$. The distribution on the lower flange follows from antisymmetry and the distribution of q_Γ around the section is of the form shown in Fig. 27.16.

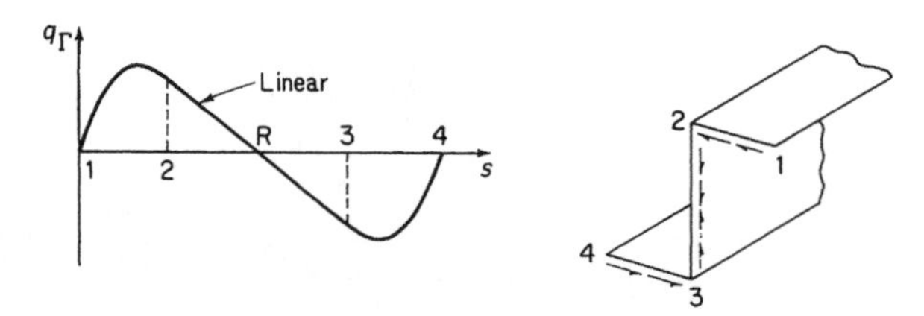

Fig. 27.16 Distribution of axial constraint shear flows.

The spanwise variation of q_Γ has the same form as the variation of T_Γ since

$$T_\Gamma = -E\Gamma_R \frac{d^3\theta}{dz^3}$$

giving

$$q_\Gamma = -\frac{T_\Gamma}{\Gamma_R}\int_0^s 2A_R t\,ds \quad \text{from Eq. (27.4)}$$

Hence for a given value of s, ($\int_0^s 2A_R t\,ds$), q_Γ is proportional to T_Γ (see Fig. 27.12).

27.3 Distributed torque loading

We now consider the more general case of a beam carrying a distributed torque loading. In Fig. 27.17 an element of a beam is subjected to a distributed torque of intensity $T_i(z)$, i.e. a torque per unit length. At the section z the torque comprises the St. Venant torque T_J plus the torque due to axial constraint T_Γ. At the section $z + \delta z$ the torque increases to $T + \delta T (= T_J + \delta T_J + T_\Gamma + \delta T_\Gamma)$ so that for equilibrium of the beam element

$$T_J + \delta T_J + T_\Gamma + \delta T_\Gamma + T_i(z)\delta z - T_J - T_\Gamma = 0$$

or

$$-T_i(z)\delta z = \delta T_J + \delta T_\Gamma = \delta T$$

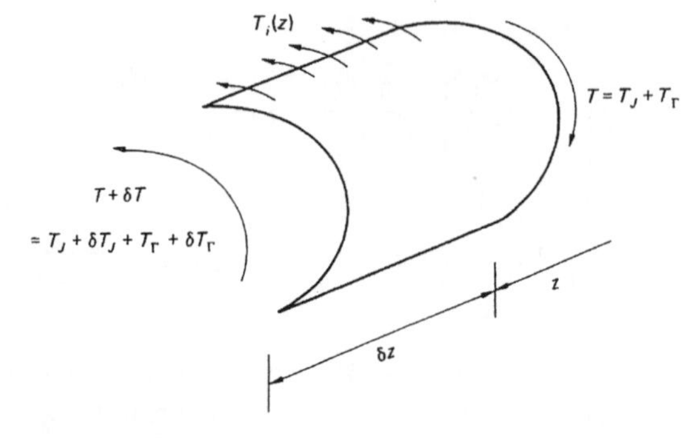

Fig. 27.17 Beam carrying a distributed torque loading.

Hence

$$\frac{\mathrm{d}T}{\mathrm{d}z} = -T_i(z) = \frac{\mathrm{d}T_J}{\mathrm{d}z} + \frac{\mathrm{d}T_\Gamma}{\mathrm{d}z} \tag{27.10}$$

Now

$$T_J = GJ\frac{\mathrm{d}\theta}{\mathrm{d}z} \quad \text{(Eq. (18.12))}$$

and

$$T_\Gamma = -E\Gamma\frac{\mathrm{d}^3\theta}{\mathrm{d}z^3} \quad \text{(Eq. (27.5))}$$

so that Eq. (27.10) becomes

$$E\Gamma\frac{\mathrm{d}^4\theta}{\mathrm{d}z^4} - GJ\frac{\mathrm{d}^2\theta}{\mathrm{d}z^2} = T_i(z) \tag{27.11}$$

The solution of Eq. (27.11) is again of standard form in which the constants of integration are found from the boundary conditions of the particular beam under consideration. For example, for a cantilever beam of length L in which the origin for z is at the built-in end and which is subjected to a uniform torque loading, the boundary conditions are:

when $z = L$, $\mathrm{d}^2\theta/\mathrm{d}z^2 = 0$ (from Eq. (27.1))
when $z = 0$, $\mathrm{d}\theta/\mathrm{d}z = 0$ (since the warping is zero at the built-in end, see Eq. (18.19))
when $z = L$, $\mathrm{d}^3\theta/\mathrm{d}z^3 = 0$ (since $T_\Gamma = T_J = T = 0$ at the free end, see Eq. (27.5))
when $z = 0$, $\theta = 0$ (there is no rotation at the built-in end).

27.4 Extension of the theory to allow for general systems of loading

So far we have been concerned with open section beams subjected to torsion in which, due to constraint effects, axial stresses are induced. Since pure torsion can generate axial stresses it is logical to suppose that certain distributions of axial stress applied as external loads will cause twisting. The problem is to determine that component of an applied direct stress system which causes twisting.

Figure 27.18 shows the profile of a thin-walled open section beam subjected to a general system of loads which produce longitudinal, transverse and rotational displacements of its cross-section. In the analysis we assume that the cross-section of the beam is undistorted by the loading and that displacements corresponding to the shear strains are negligible. In Fig. 27.18 the tangential displacement v_t is given by Eq. (17.7), i.e.

$$v_\mathrm{t} = p_\mathrm{R}\theta + u\cos\psi + v\sin\psi \tag{27.12}$$

Also, since shear strains are assumed to be negligible, Eq. (17.6) becomes

$$\gamma = \frac{\partial w}{\partial s} + \frac{\partial v_\mathrm{t}}{\partial z} = 0 \tag{27.13}$$

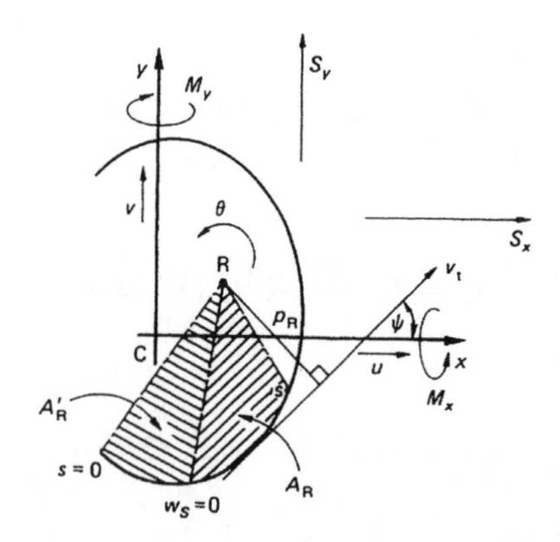

Fig. 27.18 Cross-section of an open section beam subjected to a general system of loads.

Substituting for v_t in Eq. (27.13) from (27.12) and integrating from the origin for s to any point s around the cross-section, we have

$$w_s - w_0 = -\frac{d\theta}{dz}2A_{R,0} - \frac{du}{dz}(x - x_0) - \frac{dv}{dz}(y - y_0) \tag{27.14}$$

where $2A_{R,0} = \int_0^s p_R\, ds$. The direct stress at any point in the wall of the beam is given by

$$\sigma_z = E\frac{\partial w_s}{\partial z}$$

Therefore, from Eq. (27.14)

$$\sigma_z = E\left[\frac{\partial w_0}{\partial z} - \frac{d^2\theta}{dz^2}2A_{R,0} - \frac{d^2u}{dz^2}(x - x_0) - \frac{d^2v}{dz^2}(y - y_0)\right] \tag{27.15}$$

Now $A_{R,0} = A'_R + A_R$ (Fig. 27.18) so that Eq. (27.15) may be rewritten

$$\sigma_z = f_1(z) - E\frac{d^2\theta}{dz^2}2A_R - E\frac{d^2u}{dz^2}x - E\frac{d^2v}{dz^2}y \tag{27.16}$$

in which

$$f_1(z) = E\left(\frac{\partial w_0}{\partial z} - \frac{d^2\theta}{dz^2}2A'_R + \frac{d^2u}{dz^2}x_0 + \frac{d^2v}{dz^2}y_0\right)$$

The axial load P on the section is given by

$$P = \int_c \sigma_z t\, ds = f_1(z)\int_c t\, ds - E\frac{d^2\theta}{dz^2}\int_c 2A_R t\, ds - E\frac{d^2u}{dz^2}\int_c tx\, ds - E\frac{d^2v}{dz^2}\int_c ty\, ds$$

where $\int_c$ denotes integration taken completely around the section. From Eq. (27.2) we see that $\int_c 2A_R t\, ds = 0$. Also, if the origin of axes coincides with the centroid of the section $\int_c tx\, ds = \int_c ty\, ds = 0$ and $\int ty\, ds = 0$ so that

$$P = \int_c \sigma_z t\, ds = f_1(z)A \tag{27.17}$$

in which A is the cross-sectional area of the material in the wall of the beam.

The component of bending moment, M_x, about the x axis is given by

$$M_x = \int_c \sigma_z t y \, \mathrm{d}s$$

Substituting for σ_z from Eq. (27.16) we have

$$M_x = f_1(z) \int_c ty \, \mathrm{d}s - E\frac{\mathrm{d}^2\theta}{\mathrm{d}z^2} \int_c 2A_\mathrm{R} ty \, \mathrm{d}s - E\frac{\mathrm{d}^2 u}{\mathrm{d}z^2} \int_c txy \, \mathrm{d}s - E\frac{\mathrm{d}^2 v}{\mathrm{d}z^2} \int_c ty^2 \, \mathrm{d}s$$

We have seen in the derivation of Eqs (27.7) and (27.8) that $\int_c 2A_\mathrm{R} ty \, \mathrm{d}s = 0$. Also since

$$\int_c ty \, \mathrm{d}s = 0 \quad \int_c txy \, \mathrm{d}s = I_{xy} \quad \int_c ty^2 \, \mathrm{d}s = I_{xx}$$

$$M_x = -E\frac{\mathrm{d}^2 u}{\mathrm{d}z^2} I_{xy} - E\frac{\mathrm{d}^2 v}{\mathrm{d}z^2} I_{xx} \tag{27.18}$$

Similarly

$$M_y = \int_c \sigma_z tx \, \mathrm{d}s = -E\frac{\mathrm{d}^2 u}{\mathrm{d}z^2} I_{yy} - E\frac{\mathrm{d}^2 v}{\mathrm{d}z^2} I_{xy} \tag{27.19}$$

Equations (27.18) and (27.19) are identical to Eqs (16.31) so that from Eqs (16.29)

$$E\frac{\mathrm{d}^2 u}{\mathrm{d}z^2} = \frac{M_x I_{xy} - M_y I_{xx}}{I_{xx} I_{yy} - I_{xy}^2} \quad E\frac{\mathrm{d}^2 v}{\mathrm{d}z^2} = \frac{-M_x I_{yy} + M_y I_{xy}}{I_{xx} I_{yy} - I_{xy}^2} \tag{27.20}$$

The first differential, $\mathrm{d}^2\theta/\mathrm{d}z^2$, of the rate of twist in Eq. (27.16) may be isolated by multiplying throughout by $2A_\mathrm{R} t$ and integrating around the section. Thus

$$\int_c \sigma_z 2A_\mathrm{R} t \, \mathrm{d}s = f_1(z) \int_c 2A_\mathrm{R} t \, \mathrm{d}s - E\frac{\mathrm{d}^2\theta}{\mathrm{d}z^2} \int_c (2A_\mathrm{R})^2 t \, \mathrm{d}s - E\frac{\mathrm{d}^2 u}{\mathrm{d}z^2} \int_c 2A_\mathrm{R} tx \, \mathrm{d}s$$
$$- E\frac{\mathrm{d}^2 v}{\mathrm{d}z^2} \int_c 2A_\mathrm{R} ty \, \mathrm{d}s$$

As before

$$\int_c 2A_\mathrm{R} t \, \mathrm{d}s = 0 \quad \int_c 2A_\mathrm{R} tx \, \mathrm{d}s = \int_c 2A_\mathrm{R} ty \, \mathrm{d}s = 0$$

and

$$\int_c (2A_\mathrm{R})^2 t \, \mathrm{d}s = \Gamma_\mathrm{R}$$

so that

$$\int_c \sigma_z 2A_\mathrm{R} t \, \mathrm{d}s = -E\Gamma_\mathrm{R}\frac{\mathrm{d}^2\theta}{\mathrm{d}z^2}$$

or

$$\frac{d^2\theta}{dz^2} = -\int_c \frac{\sigma_z 2A_R t\, ds}{E\Gamma_R} \tag{27.21}$$

Substituting in Eq. (27.16) from Eqs (27.17), (27.20) and (27.21), we obtain

$$\sigma_z = \frac{P}{A} + \left(\frac{M_y I_{xx} - M_x I_{xy}}{I_{xx}I_{yy} - I_{xy}^2}\right)x + \left(\frac{M_x I_{yy} - M_y I_{xy}}{I_{xx}I_{yy} - I_{xy}^2}\right)y + \frac{2A_R \int_c \sigma_z 2A_R t\, ds}{\Gamma_R} \tag{27.22}$$

The second two terms on the right-hand side of Eq. (27.22) give the direct stress due to bending as predicted by elementary beam theory; note that the above approach provides an alternative method of derivation of Eq. (16.18).

Comparing the last term on the right-hand side of Eq. (27.22) with Eq. (27.1), we see that

$$\frac{2A_R \int_c \sigma_z 2A_R t\, ds}{\Gamma_R} = \sigma_\Gamma$$

It follows therefore that the external application of a direct stress system σ_z induces a self-equilibrating direct stress system σ_Γ. Also, the first differential of the rate of twist ($d^2\theta/dz^2$) is related to the applied σ_z stress system through the term $\int_c \sigma_z 2A_R t\, ds$. Therefore, if $\int_c \sigma_z 2A_R t\, ds$ is interpreted in terms of the applied loads at a particular section then a boundary condition exists (for $d^2\theta/dz^2$) which determines one of the constants in the solution of either Eq. (27.6) or (27.11).

27.5 Moment couple (bimoment)

The units of $\int_c \sigma_z 2A_R t\, ds$ are *force* × (*distance*)^2 or *moment* × *distance*. A simple physical representation of this expression would thus consist of two equal and opposite moments applied in parallel planes some distance apart. This combination has been termed a *moment couple*[1] or a *bimoment*[2] and is given the symbol M_Γ or B_ω. Equation (27.22) is then written

$$\sigma_z = \frac{P}{A} + \left(\frac{M_y I_{xx} - M_x I_{xy}}{I_{xx}I_{yy} - I_{xy}^2}\right)x + \left(\frac{M_x I_{yy} - M_y I_{xy}}{I_{xx}I_{yy} - I_{xy}^2}\right)y + \frac{M_\Gamma 2A_R}{\Gamma_R} \tag{27.23}$$

As a simple example of the determination of M_Γ consider the open section beam shown in Fig. 27.19 which is subjected to a series of concentrated loads $P_1, P_2, \ldots, P_k, \ldots, P_n$ parallel to its longitudinal axis. The term $\sigma_z t\, ds$ in $\int_c \sigma_z 2A_R t ds$ may be regarded as a concentrated load acting at a point in the wall of the beam. Thus, $\int_c \sigma_z 2A_R t\, ds$ becomes $\sum_{k=1}^{n} P_k 2A_{Rk}$ and hence

$$M_\Gamma = \sum_{k=1}^{n} P_R 2A_{Rk} \tag{27.24}$$

M_Γ is determined for a range of other loading systems in Ref. [2].

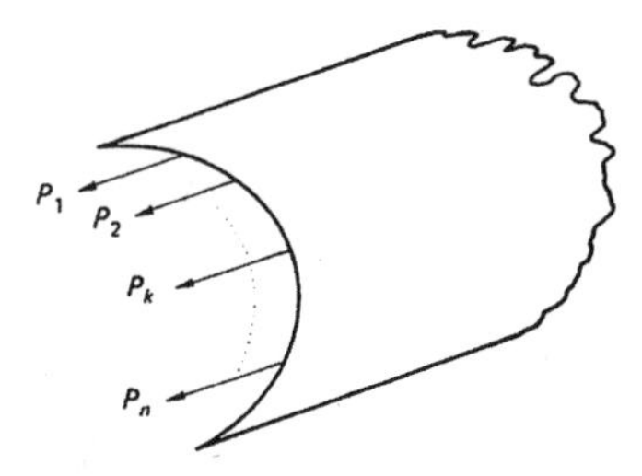

Fig. 27.19 Open section beam subjected to concentrated loads parallel to its longitudinal axis.

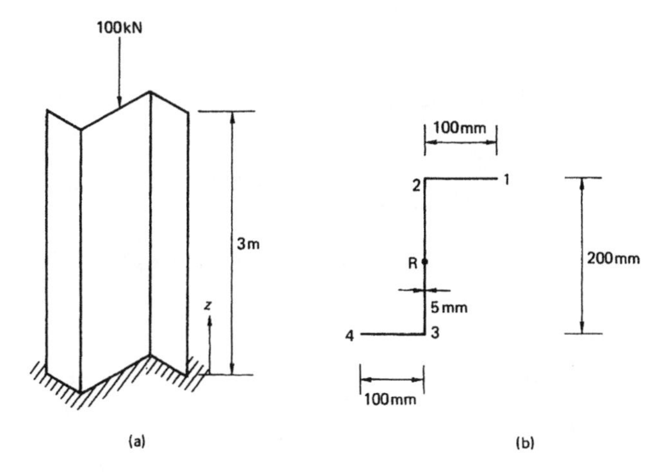

Fig. 27.20 Column of Example 27.2.

Example 27.2

The column shown in Fig. 27.20(a) carries a vertical load of 100 kN. Calculate the angle of twist at the top of the column and the distribution of direct stress at its base. $E = 200\,000$ N/mm^2 and $G/E = 0.36$.

The centre of twist R of the column cross-section coincides with its shear centre at the mid-point of the web 23. The distribution of $2A_\mathrm{R}$ is obtained by the method detailed in Example 27.1 and is shown in Fig. 27.21. The torsion bending constant Γ_R is given by Eq. (ii) of Example 27.1 and has the value 2.08×10^{10} mm^6. The St. Venant torsion constant $J = \Sigma st^3/3 = 0.17 \times 10^5$ mm^4 so that $\sqrt{GJ/E\Gamma_\mathrm{R}}$ ($=\mu$ in Eq. (iii) of Example 27.1) $= 0.54 \times 10^{-3}$. Since no torque is applied to the column the solution of Eq. (iii) in Example 27.1 is

$$\frac{\mathrm{d}\theta}{\mathrm{d}z} = C\cosh \mu z + D \sinh \mu z \qquad \text{(i)}$$

At the base of the column warping of the cross-section is suppressed so that, from Eq. (18.19), $\mathrm{d}\theta/\mathrm{d}z = 0$ when $z = 0$. Substituting in Eq. (i) gives $C = 0$. The moment couple at the top of the column is obtained from Eq. (27.24) and is

$$M_\Gamma = P2A_\mathrm{R} = -100 \times 2.5 \times 10^3 = -25 \times 10^5 \text{ kN mm}^2$$

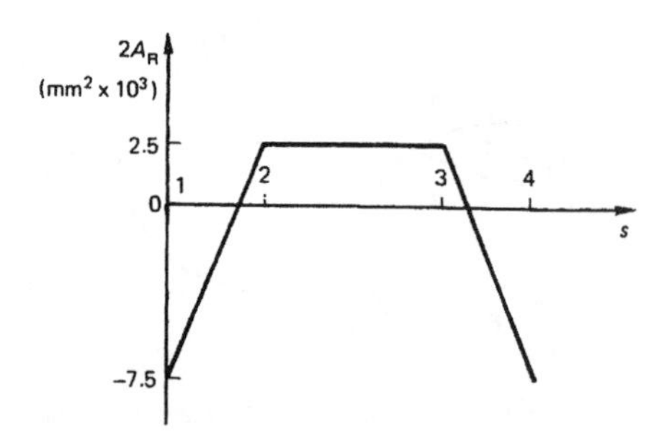

Fig. 27.21 Distribution of area $2A_R$ in the column of Example 27.2.

Therefore, from Eq. (27.21) and noting that $\int_c \sigma_z 2A_R t\,ds = M_\Gamma$, we have

$$\frac{d^2\theta}{dz^2} = \frac{2.5 \times 10^5 \times 10^3}{200\,000 \times 2.08 \times 10^{10}} = 0.06 \times 10^{-6}/\text{mm}^2$$

at $z = 3000$ mm. Substitution in the differential of Eq. (i) gives $D = 0.04 \times 10^{-3}$ so that Eq. (i) becomes

$$\frac{d\theta}{dz} = 0.04 \times 10^{-3} \sinh 0.54 \times 10^{-3}\, z \qquad \text{(ii)}$$

Integration of Eq. (ii) gives

$$\theta = 0.08 \cosh 0.54 \times 10^{-3}\, z + F$$

At the built-in end ($z = 0$) $\theta = 0$ so that $F = -0.08$. Hence

$$\theta = 0.08(\cosh 0.54 \times 10^{-3} z - 1) \qquad \text{(iii)}$$

At the top of the column ($z = 3000$ mm) the angle of twist is then

$$\theta(\text{top}) = 0.08 \cosh 0.54 \times 10^{-3} \times 3000 = 0.21\ \text{rad}(12.01°)$$

The axial load is applied through the centroid of the cross-section so that no bending occurs and Eq. (27.23) reduces to

$$\sigma_z = \frac{P}{A} + \frac{M_\Gamma 2A_R}{\Gamma_R} \qquad \text{(iv)}$$

At the base of the column

$$(M_\Gamma)_{z=0} = -E\Gamma_R \left(\frac{d^2\theta}{dz^2}\right)_{z=0} \qquad \text{(see Eq. (27.21))}$$

Therefore, from Eq. (ii)

$$(M_\Gamma)_{z=0} = -200\,000 \times 2.08 \times 10^{10} \times 0.02 \times 10^{-6} = -83.2 \times 10^6\ \text{N mm}^2$$

The direct stress distribution at the base of the column is then, from Eq. (iv)

$$\sigma_z = -\frac{100 \times 10^3}{400 \times 5} - \frac{83.2 \times 10^6}{2.08 \times 10^{10}} 2A_R$$

or

$$\sigma_z = -50 - 4.0 \times 10^{-3}\, 2A_R$$

The direct stress distribution is therefore linear around the base of the column (see Fig. 27.21) with

$$\sigma_{z_1} = \sigma_{z_4} = 20.0\,\text{N/mm}^2$$

$$\sigma_{z_2} = \sigma_{z_3} = -68.0\,\text{N/mm}^2$$

27.5.1 Shear flow due to M_Γ

The self-equilibrating shear flow distribution, q_Γ, produced by axial constraint is given by

$$\frac{\partial q_\Gamma}{\partial s} = -t\frac{\partial \sigma_\Gamma}{\partial z} \qquad \text{(see derivation of Eq. (27.4))}$$

From the last term on the right-hand side of Eqs (27.23)

$$\frac{\partial \sigma_\Gamma}{\partial z} = \frac{\partial M_\Gamma}{\partial z}\frac{2A_R}{\Gamma_R}$$

From Eq. (27.21)

$$M_\Gamma = -E\Gamma_R\frac{\mathrm{d}^2\theta}{\mathrm{d}z^2}$$

so that

$$\frac{\partial M_\Gamma}{\partial z} = -E\Gamma_R\frac{\mathrm{d}^2\theta}{\mathrm{d}z^3} = T_\Gamma \qquad \text{(see Eq. (27.5))}$$

Hence

$$\frac{\partial q_\Gamma}{\partial s} = -T_\Gamma\frac{2A_R t}{\Gamma_R}$$

and

$$q_\Gamma = -\frac{T_\Gamma}{\Gamma_R}\int_0^s 2A_R t\,\mathrm{d}s \qquad \text{(as before)}$$

References

1 Megson, T. H. G., Extension of the Wagner torsion bending theory to allow for general systems of loading, *The Aeronautical Quarterly*, Vol. XXVI, August 1975.

2 Vlasov, V. Z., Thin-walled elastic beams, *Israel Program for Scientific Translations*, Jerusalem, 1961.

Problems

P.27.1 An axially symmetric beam has the thin-walled cross-section shown in Fig. P.27.1. If the thickness t is constant throughout and making the usual assumptions for a thin-walled cross-section, show that the torsion bending constant Γ_R calculated about the shear centre S is

$$\Gamma_R = \frac{13}{12} d^5 t$$

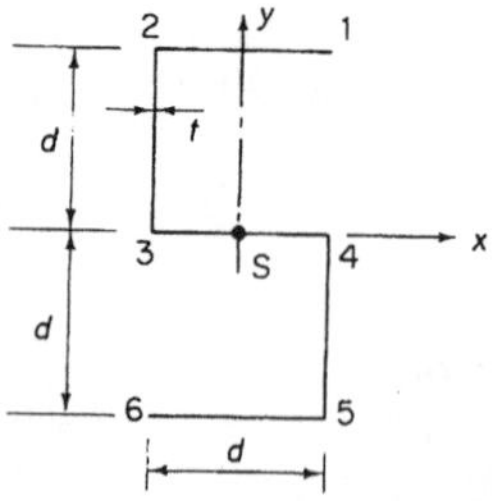

Fig. P.27.1

P.27.2 A uniform beam has the point-symmetric cross-section shown in Fig. P.27.2. Making the usual assumptions for a thin-walled cross-section, show that the torsion-bending constant Γ calculated about the shear centre S is $\Gamma = \frac{8}{3} a^5 t \sin^2 2\alpha$. The thickness t is constant throughout.

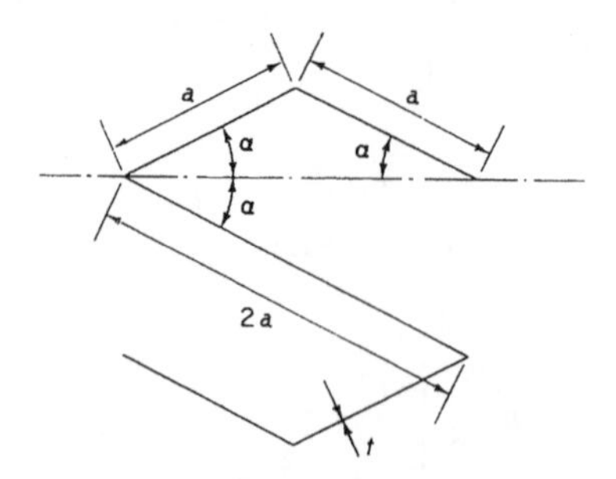

Fig. P.27.2

P.27.3 The thin-walled section shown in Fig. P.27.3 consists of two semicircular arcs of constant thickness t. Show that the torsion bending constant about the shear centre S is

$$\Gamma = \pi^2 r^5 t \left(\frac{\pi}{3} - \frac{3}{\pi} \right)$$

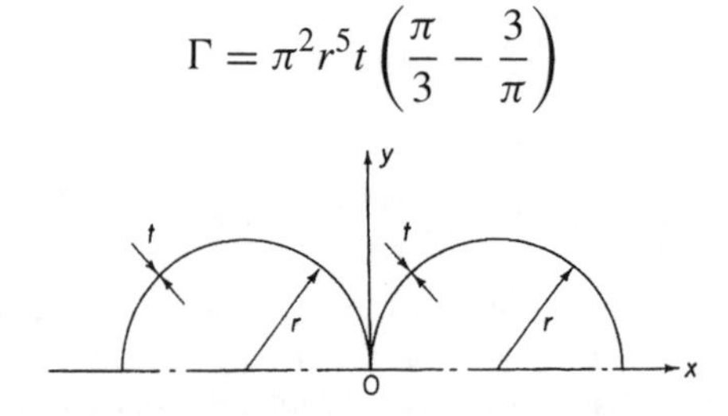

Fig. P.27.3

P.27.4 A thin-walled, I-section beam, of constant wall thickness t, is mounted as a cantilever with its web horizontal. At the tip, a downward force is applied in the plane of one of the flanges, as shown in Fig. P.27.4. Assuming the necessary results of the elementary theory of bending, the St. Venant theory of torsion and the Wagner torsion-bending theory, determine the distribution of direct stress over the cross-section at the supported end.

Take

$$E/G = 2.6 \quad P = 200\,\text{N}$$
$$h = 75\,\text{mm} \quad d = 37.5\,\text{mm}$$
$$t = 2.5\,\text{mm} \quad l = 375\,\text{mm}$$

Ans. $-\sigma_1 = \sigma_3 = 108.9\,\text{N/mm}^2$, $\sigma_6 = -\sigma_5 = 18.9\,\text{N/mm}^2$, $\sigma_2 = \sigma_4 = \sigma_{24} = 0$.

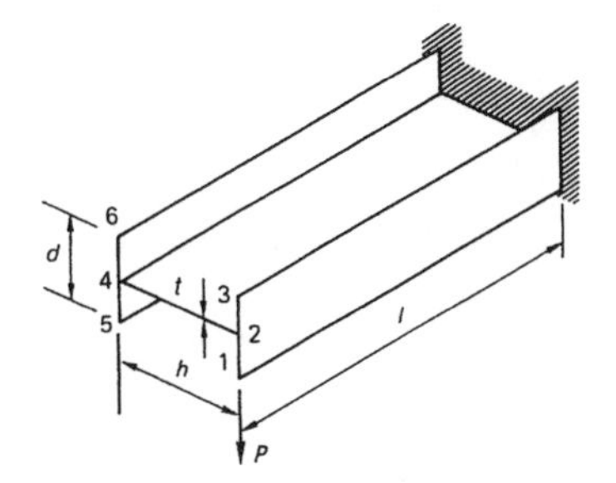

Fig. P.27.4

P.27.5 An open section beam of length $2l$, whose ends are free to warp, consists of two uniform portions of equal length l, as shown in Fig. P.27.5. The cross-sections of the two halves are identical except that the thickness in one half is t and in the other $2t$. If the St. Venant torsion constant and the torsion-bending constant for the portion of thickness t are J and Γ, respectively, show that when the beam is loaded by a constant torque T the relative twist between the free ends is given by

$$\theta = \frac{Tl}{8GJ}\left[9 - \frac{49\sinh 2\mu l}{2\mu l(10\cosh^2 \mu l - 1)}\right]$$

where

$$\mu^2 = GJ/E\Gamma \text{ and } G = \text{shear modulus (constant throughout)}$$

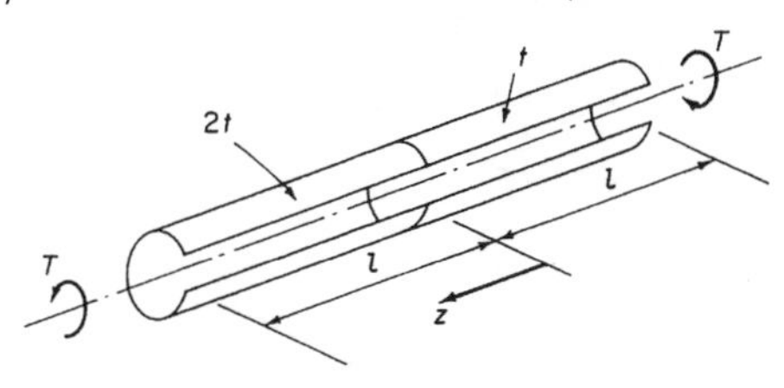

Fig. P.27.5

P.27.6 A thin-walled cantilever beam of length L has the cross-section shown in Fig. P.27.6 and carries a load P positioned as shown at its free end. Determine the torsion bending constant for the beam section and derive an expression for the angle of twist θ_T at the free end of the beam. Calculate the value of this angle for $P = 100$ N, $a = 30$ mm, $L = 1000$ mm, $t = 2.0$ mm, $E = 70\,000$ N/mm^2 and $G = 25\,000$ N/mm^2

Ans. $\Gamma = 1.25a^5t \quad \theta_T = 6.93°$.

$$\theta_\Gamma = \frac{TL}{GJ}\left[1 - \frac{\tanh \mu L}{\mu L}\right]$$

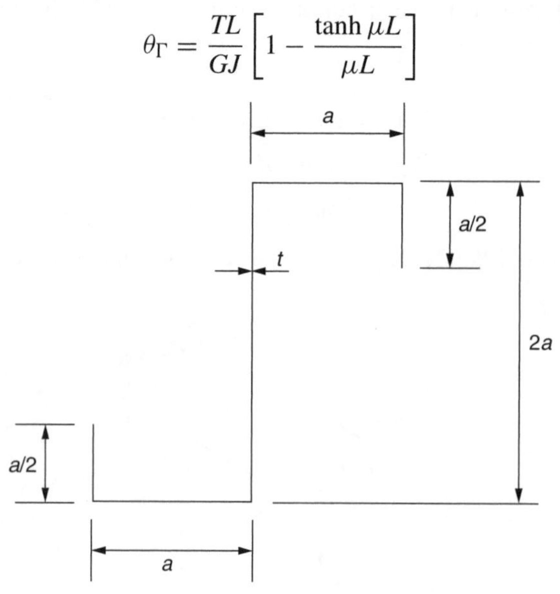

Fig. P.27.6

P.27.7 Determine the torsion bending constant for the thin-walled beam shown in Fig. P.27.7 and also derive an expression for the angle of twist at its free end.

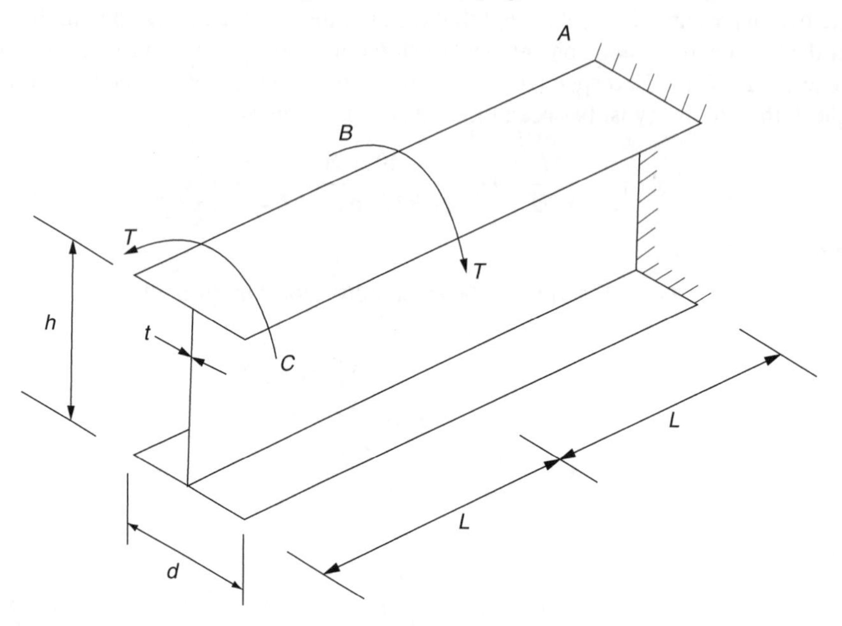

Fig. P.27.7

$$Ans. \quad \Gamma = th^2d^3/24 \quad \theta_T = \frac{T}{GJ}\left(L - \frac{\sinh \mu L}{\mu L \cosh 2\mu L}\right)$$

P.27.8 A thin-walled cantilever beam of length L has the cross-section shown in Fig. P.27.8 and carries an anticlockwise torque T at its free end. Determine the torsion bending constant for the beam section and derive an expression for the rate of twist along the length of the beam.

In a practical case the beam supports a shear load of 150 N at its free end applied vertically upwards in the plane of the web. If $L = 500$ mm, $a = 20$ mm, $t = 1.0$ mm and $G/E = 0.3$ calculate the value of direct stress at the point 2 including both axial constraint and elementary bending stresses.

$$Ans. \quad \Gamma = 7a^5t/24 \quad \frac{\mathrm{d}\theta}{\mathrm{d}z} = \frac{T}{GJ}\left(1 - \frac{\cosh \mu(L-z)}{\cosh \mu L}\right)$$

125.7 N/mm^2 (compression).

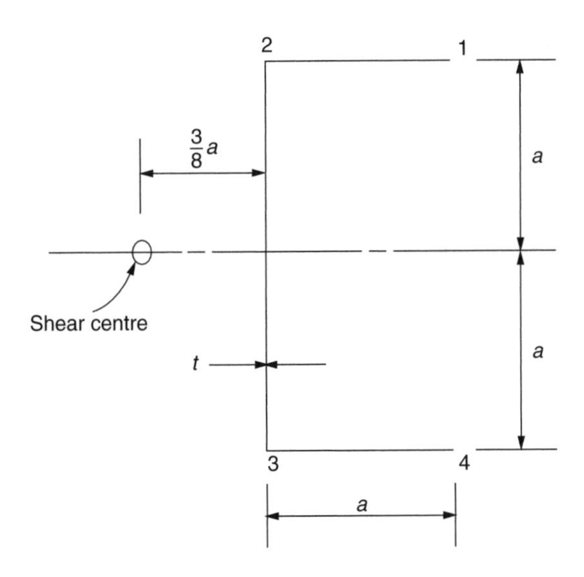

Fig. P.27.8

P.27.9 Calculate the direct stress distribution (including both axial constraint and elementary bending stresses) at the built-in end of the cantilever beam shown in

Fig. P.27.9 for the case when $w = 0.5$ N/mm, $L = 1500$ mm, $h = 200$ mm, $d = 50$ mm, $t = 5$ mm and $E/G = 3.0$.

Ans. $\sigma_1 = -\sigma_3 = 197.5\ \text{N/mm}^2 \quad \sigma_2 = \sigma_5 = 0 \quad \sigma_4 = -\sigma_6 = -72.5\ \text{N/mm}^2$.

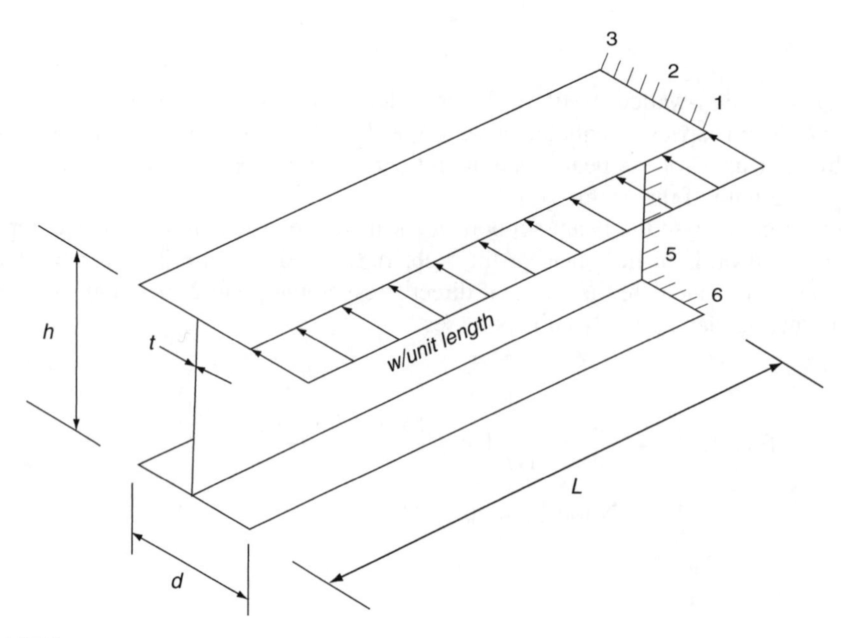

Fig. P.27.9

Section B6 Introduction to Aeroelasticity

28

Wing problems

Aircraft structures, being extremely flexible, are prone to distortion under load. When these loads are caused by aerodynamic forces, which themselves depend on the geometry of the structure and the orientation of the various structural components to the surrounding airflow, then structural distortion results in changes in aerodynamic load, leading to further distortion and so on. The interaction of aerodynamic and elastic forces is known as *aeroelasticity.*

28.1 Types of problem

Two distinct types of aeroelastic problem occur. One involves the interaction of aerodynamic and elastic forces of the type described above. Such interactions may exhibit divergent tendencies in a too flexible structure, leading to failure, or, in an adequately stiff structure, converge until a condition of stable equilibrium is reached. In this type of problem *static or steady state* systems of aerodynamic and elastic forces produce such aeroelastic phenomena as *divergence* and *control reversal*. The second class of problem involves the inertia of the structure as well as aerodynamic and elastic forces. Dynamic loading systems, of which gusts are of primary importance, induce oscillations of structural components. If the natural or resonant frequency of the component is in the region of the frequency of the applied loads then the amplitude of the oscillations may diverge, causing failure. Also, as we observed in Chapter 15, the presence of fluctuating loads is a fatigue hazard. For obvious reasons we refer to these problems as *dynamic*. Included in this group are flutter, buffeting and dynamic response.

The various aeroelastic problems may be conveniently summarized in the form of a 'tree' as follows:

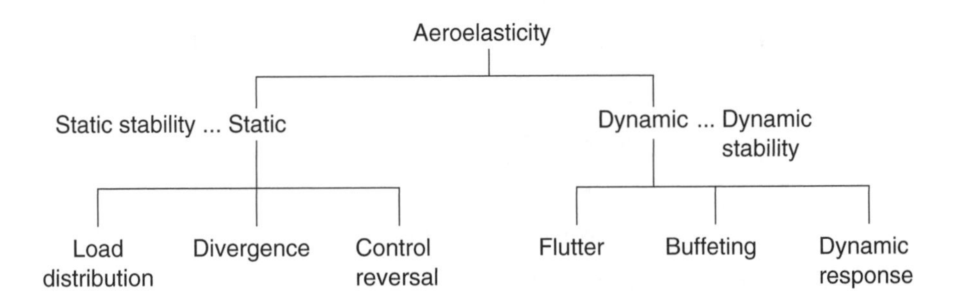

In this chapter we shall concentrate on the purely structural aspects of aeroelasticity; its effect on aircraft static and dynamic stability is treated in books devoted primarily to aircraft stability and control.[1,2]

28.2 Load distribution and divergence

Redistribution of aerodynamic loads and divergence are closely related aeroelastic phenomena; we shall therefore consider them simultaneously. It is essential in the design of structural components that the aerodynamic load distribution on the component is known. Wing distortion, for example, may produce significant changes in lift distribution from that calculated on the assumption of a rigid wing, especially in instances of high wing loadings such as those experienced in manoeuvres and gusts. To estimate actual lift distributions the aerodynamicist requires to know the incidence of the wing at all stations along its span. Obviously this is affected by any twisting of the wing which may be present.

Let us consider the case of a simple straight wing with the centre of twist behind the aerodynamic centre (see Fig. 28.1). The moment of the lift vector about the centre of twist causes an increase in wing incidence which produces a further increase in lift, leading to another increase in incidence and so on. At speeds below a critical value, called the *divergence speed*, the increments in lift converge to a condition of stable equilibrium in which the torsional moment of the aerodynamic forces about the centre of twist is balanced by the torsional rigidity of the wing. The calculation of lift distribution then proceeds from a knowledge of the distribution of twist along the wing. For a straight wing the redistribution of lift usually causes an outward spanwise movement of the centre of pressure, resulting in greater bending moments at the wing root. In the case of a swept wing a reduction in streamwise incidence of the outboard sections due to bending deflections causes a movement of the centre of pressure towards the wing root.

All aerodynamic surfaces of the aircraft suffer similar load redistribution due to distortion.

28.2.1 Wing torsional divergence (two-dimensional case)

The most common divergence problem is the torsional divergence of a wing. It is useful, initially, to consider the case of a wing of area S without ailerons and in a

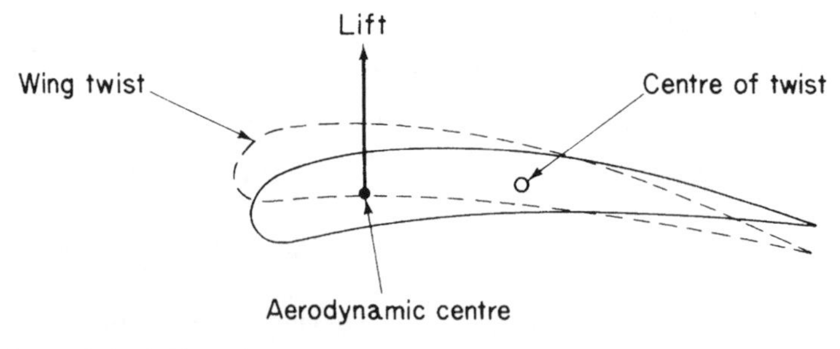

Fig. 28.1 Increase of wing incidence due to wing twist.

two-dimensional flow, as shown in Fig. 28.2. The torsional stiffness of the wing, which we shall represent by a spring of stiffness, K, resists the moment of the lift vector, L, and the wing pitching moment, M_0, acting at the aerodynamic centre of the wing section. For moment equilibrium of the wing section about the aerodynamic centre we have

$$M_0 + Lec = K\theta \tag{28.1}$$

where ec is the distance of the aerodynamic centre forward of the flexural centre expressed in terms of the wing chord, c, and θ is the elastic twist of the wing. From aerodynamic theory

$$M_0 = \frac{1}{2}\rho V^2 ScC_{\mathrm{M,0}} \quad L = \frac{1}{2}\rho V^2 SC_{\mathrm{L}}$$

Substituting in Eq. (28.1) yields

$$\frac{1}{2}\rho V^2 S(cC_{\mathrm{M,0}} + ecC_{\mathrm{L}}) = K\theta$$

or, since

$$C_{\mathrm{L}} = C_{\mathrm{L,0}} + \frac{\partial C_{\mathrm{L}}}{\partial \alpha}(\alpha + \theta)$$

in which α is the initial wing incidence or, in other words, the incidence corresponding to given flight conditions assuming that the wing is rigid and $C_{\mathrm{L,0}}$ is the wing lift coefficient at zero incidence, then

$$\frac{1}{2}\rho V^2 S\left[cC_{\mathrm{M,0}} + ec_{\mathrm{L,0}} + ec\frac{\partial C_{\mathrm{L}}}{\partial \alpha}(\alpha + \theta)\right] = K\theta$$

where $\partial C_{\mathrm{L}}/\partial \alpha$ is the wing lift curve slope. Rearranging gives

$$\theta\left(K - \frac{1}{2}\rho V^2 Sec\frac{\partial C_{\mathrm{L}}}{\partial \alpha}\right) = \frac{1}{2}\rho V^2 Sc\left(C_{\mathrm{M,0}} + eC_{\mathrm{L,0}} + e\frac{\partial C_{\mathrm{L}}}{\partial \alpha}\alpha\right)$$

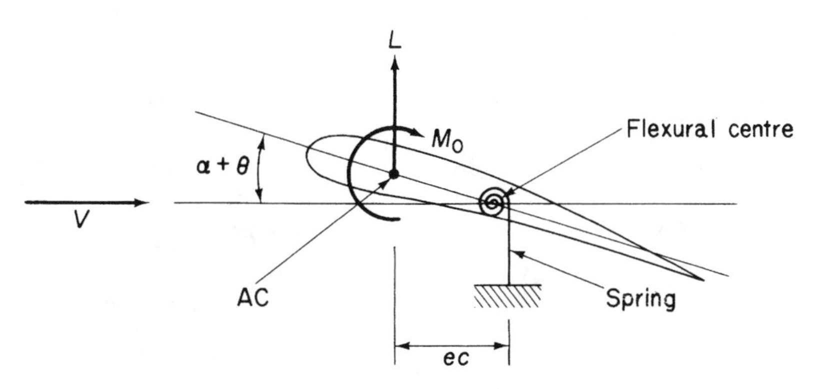

Fig. 28.2 Determination of wing divergence speed (two-dimensional case).

or

$$\theta = \frac{\frac{1}{2}\rho V^2 Sc[C_{M,0} + eC_{L,0} + e(\partial C_L/\partial\alpha)\alpha]}{K - \frac{1}{2}\rho V^2 Sec(\partial C_L/\partial\alpha)} \tag{28.2}$$

Equation (28.2) shows that divergence occurs (i.e. θ becomes infinite) when

$$K = \frac{1}{2}\rho V^2 Sec\frac{\partial C_L}{\partial\alpha}$$

The divergence speed V_d is then

$$V_d = \sqrt{\frac{2K}{\rho Sec(\partial C_L/\partial\alpha)}} \tag{28.3}$$

We see from Eq. (28.3) that V_d may be increased either by stiffening the wing (increasing K) or by reducing the distance ec between the aerodynamic and flexural centres. The former approach involves weight and cost penalties so that designers usually prefer to design a wing structure with the flexural centre as far forward as possible. If the aerodynamic centre coincides with or is aft of the flexural centre then the wing is stable at all speeds.

28.2.2 Wing torsional divergence (finite wing)

We shall consider the simple case of a straight wing having its flexural axis nearly perpendicular to the aircraft's plane of symmetry (Fig. 28.3(a)). We shall also assume that wing cross-sections remain undistorted under the loading. Applying strip theory

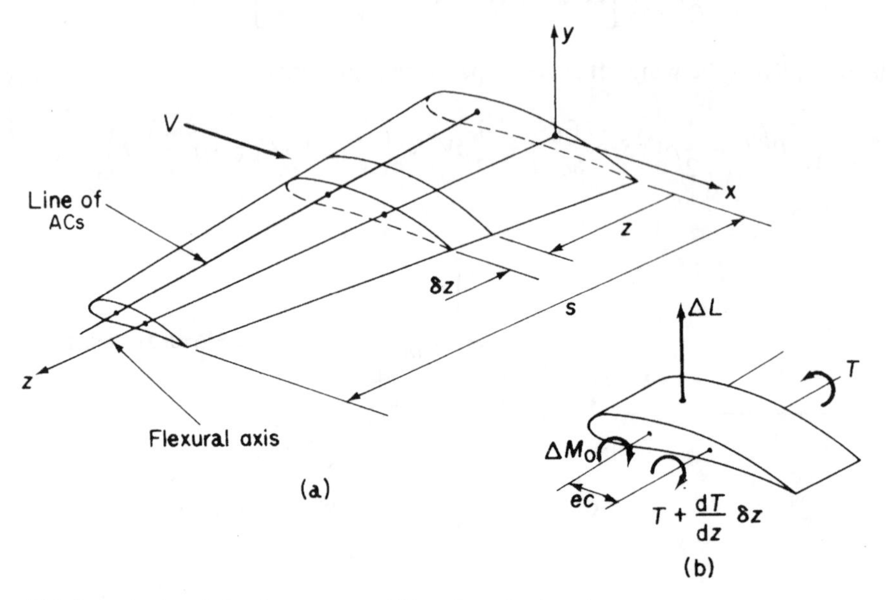

Fig. 28.3 Determination of wing divergence speed (three-dimensional case).

in the usual manner, i.e. we regard a small element of chord c and spanwise width δz as acting independently of the remainder of the wing and consider its equilibrium, we have from Fig. 28.3(b), neglecting wing weight

$$\left(T + \frac{\mathrm{d}T}{\mathrm{d}z}\delta z\right) - T + \Delta Lec + \Delta M_0 = 0 \tag{28.4}$$

where T is the applied torque at any spanwise section z and ΔL and ΔM_0 are the lift and pitching moment on the elemental strip acting at its aerodynamic centre, respectively. As δz approaches zero, Eq. (28.4) becomes

$$\frac{\mathrm{d}T}{\mathrm{d}z} + ec\frac{\mathrm{d}L}{\mathrm{d}z} + \frac{\mathrm{d}M_0}{\mathrm{d}z} = 0 \tag{28.5}$$

In Eq. (28.4)

$$\Delta L = \frac{1}{2}\rho V^2 c\delta z\frac{\partial c_1}{\partial \alpha}(\alpha + \theta)$$

where $\partial c_1/\partial\alpha$ is the local two-dimensional lift curve slope and

$$\Delta M_0 = \frac{1}{2}\rho V^2 c^2 \delta z c_{\mathrm{m},0}$$

in which $c_{\mathrm{m},0}$ is the local pitching moment coefficient about the aerodynamic centre. Also from torsion theory (see Chapter 3) $T = GJ\,\mathrm{d}\theta/\mathrm{d}z$. Substituting for L, M_0 and T in Eq. (28.5) gives

$$\frac{\mathrm{d}^2\theta}{\mathrm{d}z^2} + \frac{\frac{1}{2}\rho V^2 ec^2(\partial c_1/\partial\alpha)\theta}{GJ} = \frac{-\frac{1}{2}\rho V^2 ec^2(\partial c_1/\partial\alpha)\alpha}{GJ} - \frac{\frac{1}{2}\rho V^2 c^2 c_{\mathrm{m},0}}{GJ} \tag{28.6}$$

Equation (28.6) is a second-order differential equation in θ having a solution of the standard form

$$\theta = A\sin\lambda z + B\cos\lambda z - \left[\frac{c_{\mathrm{m},0}}{e(\partial c_1/\partial\alpha)} + \alpha\right] \tag{28.7}$$

where

$$\lambda^2 = \frac{\frac{1}{2}\rho V^2 ec^2(\partial c_1/\partial\alpha)}{GJ}$$

and A and B are unknown constants that are obtained from the boundary conditions; namely, $\theta = 0$ when $z = 0$ at the wing root and $\mathrm{d}\theta/\mathrm{d}z = 0$ at $z = s$ since the torque is zero at the wing tip. From the first of these

$$B = \left[\frac{c_{\mathrm{m},0}}{e(\partial c_1/\partial\alpha)} + \alpha\right]$$

and from the second

$$A = \left[\frac{c_{\mathrm{m},0}}{e(\partial c_1/\partial\alpha)} + \alpha\right]\tan\lambda s$$

Hence

$$\theta = \left[\frac{c_{m,0}}{e(\partial c_l/\partial \alpha)} + \alpha\right](\tan \lambda s \sin \lambda z + \cos \lambda z - 1) \tag{28.8}$$

or rearranging

$$\theta = \left[\frac{c_{m,0}}{e(\partial c_l/\partial \alpha)} + \alpha\right]\left[\frac{\cos \lambda(s-z)}{\cos \lambda s} - 1\right] \tag{28.9}$$

Therefore, at divergence when the elastic twist, θ, becomes infinite

$$\cos \lambda s = 0$$

so that

$$\lambda s = (2n+1)\frac{\pi}{2} \quad \text{for} \quad n = 0, 1, 2, \ldots, \infty \tag{28.10}$$

The smallest value corresponding to the divergence speed V_d occurs when $n = 0$, thus

$$\lambda s = \pi/2$$

or

$$\lambda^2 = \pi^2/4s^2$$

from which

$$V_d = \sqrt{\frac{\pi^2 GJ}{2\rho e c^2 s^2 (\partial c_l/\partial \alpha)}} \tag{28.11}$$

Mathematical solutions of the type given in Eq. (28.10) rarely apply with any accuracy to actual wing or tail surfaces. However, they do give an indication of the order of the divergence speed, V_d. In fact, when the two-dimensional lift-curve slope, $\partial c_l/\partial \alpha$, is used they lead to conservative estimates of V_d. It has been shown that when $\partial c_l/\partial \alpha$ is replaced by the three-dimensional lift-curve slope of the finite wing, values of V_d become very close to those determined from more sophisticated aerodynamic and aeroelastic theory.

The lift distribution on a straight wing, accounting for the elastic twist, is found by introducing a relationship between incidence and lift distribution from aerodynamic theory. In the case of simple strip theory the local wing lift coefficient, c_l, is given by

$$c_l = \frac{\partial c_l}{\partial \alpha}(\alpha + \theta)$$

in which the distribution of elastic twist θ is known from Eq. (28.9).

28.2.3 Swept wing divergence

In the calculation of divergence speeds of straight wings the flexural axis was taken to be nearly perpendicular to the aircraft's plane of symmetry. Bending of such wings has no influence on divergence, this being entirely dependent on the twisting of the

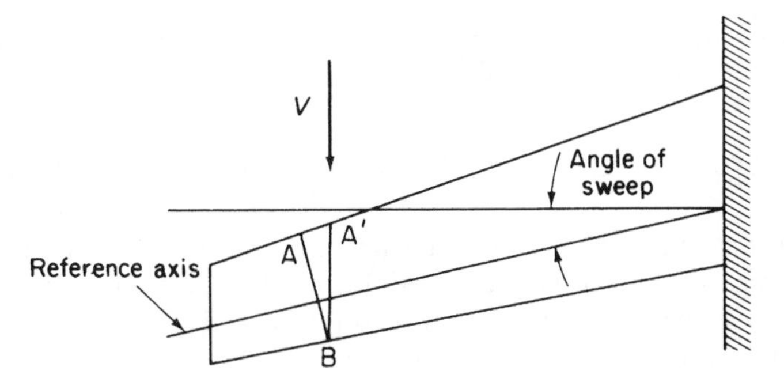

Fig. 28.4 Effect of wing sweep on wing divergence speed.

wing about its flexural axis. This is no longer the case for a swept wing where the spanwise axes are inclined to the aircraft's plane of symmetry. Let us consider the swept wing of Fig. 28.4. The wing lift distribution causes the wing to bend in an upward direction. Points A and B on a line perpendicular to the reference axis will deflect by approximately the same amount, but this will be greater than the deflection of A′ which means that bending reduces the streamwise incidence of the wing. The corresponding negative increment of lift opposes the elastic twist, thereby reducing the possibility of wing divergence. In fact, the divergence speed of swept wings is so high that it poses no problems for the designer. Diederich and Budiansky in 1948 showed that wings with moderate or large sweepback cannot diverge. The opposite of course is true for swept-forward wings where bending deflections have a destabilizing effect and divergence speeds are extremely low. The determination of lift distributions and divergence speeds for swept-forward wings is presented in Ref. [3].

28.3 Control effectiveness and reversal

The flexibility of the major aerodynamic surfaces (wings, vertical and horizontal tails) adversely affects the effectiveness of the corresponding control surfaces (ailerons, rudder and elevators). For example, the downward deflection of an aileron causes a nose-down twisting of the wing which consequently reduces the aileron incidence. Thus, the wing twist tends to reduce the increase in lift produced by the aileron deflection, and thereby the rolling moment to a value less than that for a rigid wing. The aerodynamic twisting moment on the wing due to aileron deflection increases as the square of the speed but the elastic restoring moment is constant since it depends on the torsional stiffness of the wing structure. Therefore, ailerons become markedly less effective as the speed increases until, at a particular speed, the *aileron reversal speed*, aileron deflection does not produce any rolling moment at all. At higher speeds reversed aileron movements are necessary in that a positive increment of wing lift requires an upward aileron deflection and vice versa.

Similar, less critical, problems arise in the loss of effectiveness and reversal of the rudder and elevator controls. They are complicated by the additional deformations of the fuselage and tailplane–fuselage attachment points, which may be as important as the deformations of the tailplane itself. We shall concentrate in this section on the problem of aileron effectiveness and reversal.

28.3.1 Aileron effectiveness and reversal (two-dimensional case)

We shall illustrate the problem by investigating, as in Section 28.1, the case of a wing-aileron combination in a two-dimensional flow. In Fig. 28.5 an aileron deflection ξ produces *changes* ΔL and ΔM_0 in the wing lift, L, and wing pitching moment, M_0; these in turn cause an elastic twist, θ, of the wing. Thus

$$\Delta L = \left(\frac{\partial C_L}{\partial \alpha}\theta + \frac{\partial C_L}{\partial \xi}\xi\right)\frac{1}{2}\rho V^2 S \tag{28.12}$$

where $\partial C_L/\partial\alpha$ has been previously defined and $\partial C_L/\partial\xi$ is the rate of change of lift coefficient with aileron angle. Also

$$\Delta M_0 = \frac{\partial C_{M,0}}{\partial \xi}\xi\frac{1}{2}\rho V^2 Sc \tag{28.13}$$

in which $\partial C_{M,0}/\partial\xi$ is the rate of change of wing pitching moment coefficient with aileron deflection. The moment produced by these increments in lift and pitching moment is equilibrated by an increment of torque ΔT about the flexural axis. Hence

$$\Delta T = K\theta = \frac{1}{2}\rho V^2 Sc\left[\left(\frac{\partial C_L}{\partial \alpha}\theta + \frac{\partial C_L}{\partial \xi}\xi\right)e + \frac{\partial C_{M,0}}{\partial \xi}\xi\right] \tag{28.14}$$

Isolating θ from Eq. (28.14) gives

$$\theta = \frac{\frac{1}{2}\rho V^2 Sc[(\partial C_L/\partial\xi)e + \partial C_{M,0}/\partial\xi]\xi}{K - \frac{1}{2}\rho V^2 Sce(\partial C_L/\partial\alpha)} \tag{28.15}$$

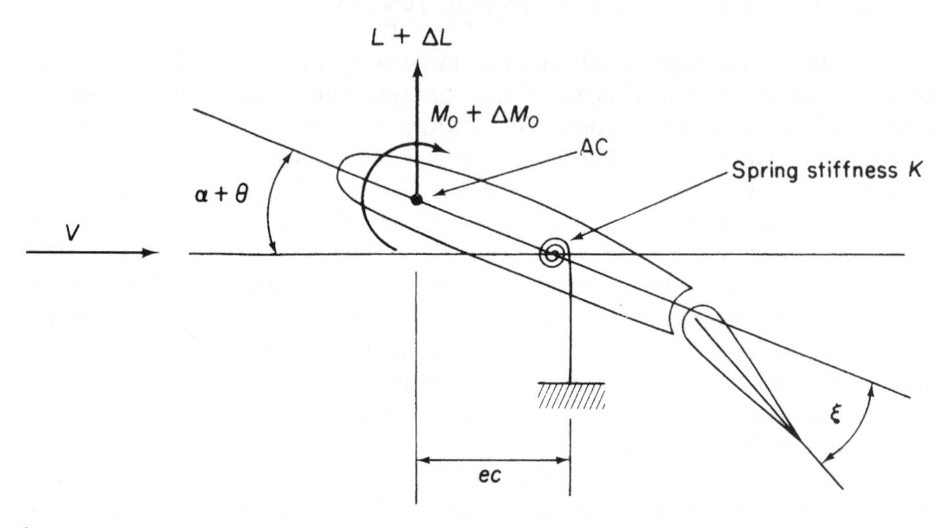

Fig. 28.5 Aileron effectiveness and reversal speed (two-dimensional case).

Substituting for θ in Eq. (28.12) we have

$$\Delta L = \frac{1}{2}\rho V^2 S\left[\frac{\frac{1}{2}\rho V^2 Sc\{(\partial C_L/\partial\xi)e + \partial C_{M,0}/\partial\xi\}}{K - \frac{1}{2}\rho V^2 Sce(\partial C_L/\partial\alpha)}\frac{\partial C_L}{\partial\alpha} + \frac{\partial C_L}{\partial\xi}\right]\xi$$

which simplifies to

$$\Delta L = \frac{1}{2}\rho V^2 S\left[\frac{[\frac{1}{2}\rho V^2 Sc(\partial C_{M,0}/\partial\xi)(\partial C_L/\partial\alpha) + K(\partial C_L/\partial\xi)]}{K - \frac{1}{2}\rho V^2 Sce(\partial C_L/\partial\alpha)}\right]\xi \tag{28.16}$$

The increment of wing lift is therefore a linear function of aileron deflection and becomes zero, i.e. aileron reversal occurs, when

$$\frac{1}{2}\rho V^2 Sc\frac{\partial C_{M,0}}{\partial\xi}\frac{\partial C_L}{\partial\alpha} + K\frac{\partial C_L}{\partial\xi} = 0 \tag{28.17}$$

Hence the aileron reversal speed, V_r, is, from Eq. (28.17)

$$V_r = \sqrt{\frac{-K(\partial C_L/\partial\xi)}{\frac{1}{2}\rho SC(\partial C_{M,0}/\partial\xi)(\partial C_L/\partial\alpha)}} \tag{28.18}$$

We may define aileron effectiveness at speeds below the reversal speed in terms of the lift ΔL_R produced by an aileron deflection on a rigid wing. Thus

$$\text{Aileron effectiveness} = \Delta L/\Delta L_R \tag{28.19}$$

where

$$\Delta L_R = \frac{\partial C_L}{\partial\xi}\xi\frac{1}{2}\rho V^2 S \tag{28.20}$$

Hence, substituting in Eq. (28.19) for ΔL from Eq. (28.16) and ΔL_R from Eq. (28.20), we have

$$\text{Aileron effectiveness} = \frac{\frac{1}{2}\rho V^2 Sc(\partial C_{M,0}/\partial\xi)(\partial C_L/\partial\alpha) + K(\partial C_L/\partial\xi)}{[K - \frac{1}{2}\rho V^2 Sce(\partial C_L/\partial\alpha)]\partial C_L/\partial\xi} \tag{28.21}$$

Equation (28.21) may be expressed in terms of the wing divergence speed V_d and aileron reversal speed V_r, using Eqs (28.3) and (28.18), respectively; hence

$$\text{Aileron effectiveness} = \frac{1 - V^2/V_r^2}{1 - V^2/V_d^2} \tag{28.22}$$

We see that when $V_d = V_r$, which occurs when $\partial C_L/\partial\xi = -(\partial C_{M,0}/\partial\xi)/e$, then the aileron is completely effective at all speeds. Such a situation arises because the nose-down wing twist caused by aileron deflection is cancelled by the nose-up twist produced by the increase in wing lift.

Although the analysis described above is based on a two-dimensional case, it is sometimes used in practice to give approximate answers for finite wings. The method is to apply the theory to a representative wing cross-section at an arbitrary spanwise station and use the local wing section properties in the formulae.

28.3.2 Aileron effectiveness and reversal (finite wing)

We shall again apply strip theory to investigate the aeroelastic effects of aileron deflection on a finite wing. In Fig. 28.6(a) the deflection of the aileron through an angle ξ produces a rolling velocity p rad/s, having the sense shown. The wing incidence at any section z is thus reduced due to p by an amount pz/V. The downward aileron deflection shown here coincides with an upward deflection on the opposite wing, thereby contributing to the rolling velocity p. The incidence of the opposite wing is therefore increased by this direction of roll. Since we are concerned with aileron effects we consider the antisymmetric lift and pitching moment produced by aileron deflection. Thus, in Fig. 28.6(b), the forces and moments are *changes* from the level flight condition.

The lift ΔL on the strip shown in Fig. 28.6(b) is given by

$$\Delta L = \frac{1}{2}\rho V^2 c\delta z\left[\frac{\partial c_1}{\partial \alpha}\left(\theta - \frac{pz}{V}\right) + \frac{\partial c_1}{\partial \xi}f_a(z)\xi\right] \tag{28.23}$$

where $\partial c_1/\partial\alpha$ has been previously defined and $\partial c_1/\partial\xi$ is the rate of change of local wing lift coefficient with aileron angle. The function $f_a(z)$ represents aileron forces and moments along the span; for $0 \le z \le s_1, f_a(z) = 0$ and for $s_1 \le z \le s, f_a(z) = 1$. The pitching moment ΔM_0 on the elemental strip is given by

$$\Delta M_0 = \frac{1}{2}\rho V^2 c^2 \delta z \frac{\partial c_{m,0}}{\partial \xi} f_a(z)\xi \tag{28.24}$$

in which $\partial c_{m,0}/\partial\xi$ is the rate of change of local pitching moment coefficient with aileron angle.

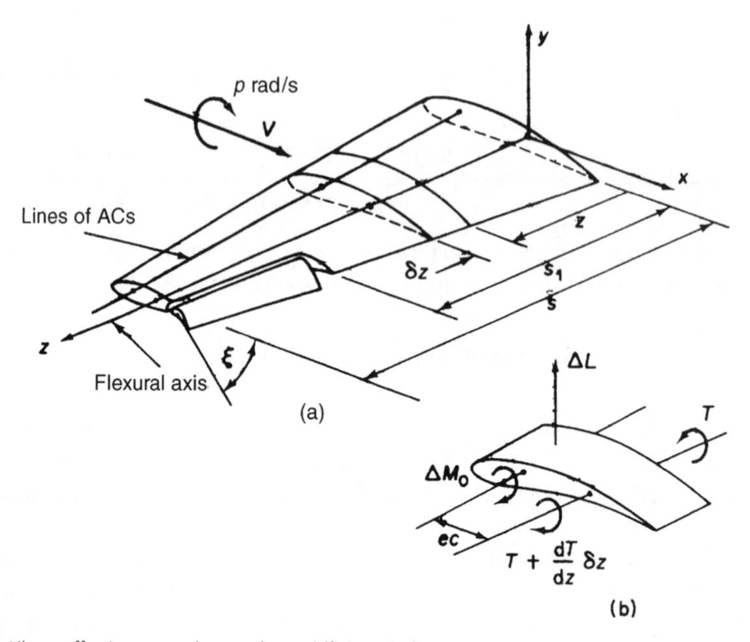

Fig. 28.6 Aileron effectiveness and reversal speed (finite wing).

Considering the moment equilibrium of the elemental strip of Fig. 28.6(b) we obtain, neglecting wing weight

$$\frac{\mathrm{d}T}{\mathrm{d}z}\delta z + \Delta Lec + \Delta M_0 = 0 \tag{28.25}$$

or substituting for ΔL and ΔM_0 from Eqs (28.23) and (28.24)

$$\frac{\mathrm{d}T}{\mathrm{d}z} + \frac{1}{2}\rho V^2 ec^2\left[\frac{\partial c_1}{\partial \alpha}\left(\theta - \frac{pz}{V}\right) + \frac{\partial c_1}{\partial \xi}f_{\mathrm{a}}(z)\xi\right] + \frac{1}{2}\rho V^2 c^2 \frac{\partial c_{\mathrm{m},0}}{\partial \xi}f_{\mathrm{a}}(z)\xi = 0 \tag{28.26}$$

Substituting for T in Eq. (28.26) from torsion theory ($T = GJ\,\mathrm{d}\theta/\mathrm{d}z$) and rearranging we have

$$\frac{\mathrm{d}^2\theta}{\mathrm{d}z^2} + \frac{\frac{1}{2}\rho V^2 ec^2(\partial c_1/\partial\alpha)}{GJ}\theta = \frac{\frac{1}{2}\rho V^2 c^2}{GJ}\left[e\frac{\partial c_1}{\partial \alpha}\frac{pz}{V} - e\frac{\partial c_1}{\partial \xi}f_{\mathrm{a}}(z)\xi - \frac{\partial c_{\mathrm{m},0}}{\partial \xi}f_{\mathrm{a}}(z)\xi\right] \tag{28.27}$$

Writing

$$\frac{\frac{1}{2}\rho V^2 ec^2(\partial c_1/\partial\alpha)}{GJ} = \lambda^2$$

we obtain

$$\frac{\mathrm{d}^2\theta}{\mathrm{d}z^2} + \lambda^2\theta = \lambda^2\frac{pz}{V} - \frac{\lambda^2}{\partial c_1/\partial\alpha}\left(\frac{\partial c_1}{\partial \xi} + \frac{1}{e}\frac{\partial c_{\mathrm{m},0}}{\partial \xi}\right)f_{\mathrm{a}}(z)\xi \tag{28.28}$$

It may be shown that the solution of Eq. (28.28), satisfying the boundary conditions

$$\theta = 0 \quad \text{at } z = 0 \quad \text{and} \quad d\theta/dz = 0 \quad \text{at } z = s$$

is

$$\theta = \frac{p}{V}\left(z - \frac{\sin\lambda z}{\lambda\cos\lambda s}\right) - \frac{1}{\partial c_1/\partial\alpha}\left(\frac{\partial c_1}{\partial \xi} + \frac{1}{e}\frac{\partial c_{\mathrm{m},0}}{\partial \xi}\right) \times \left[f_{\mathrm{a}}(z)\{1 - \cos\lambda(z - s_1)\} - \frac{\sin\lambda(s - s_1)}{\cos\lambda s}\sin\lambda z\right]\xi \tag{28.29}$$

where

$$\cos\lambda(z - s_1) = 0 \quad \text{when} \quad z < s_1$$

The spanwise variation of total local wing lift coefficient is given by strip theory as

$$c_1 = \frac{\partial c_1}{\partial\alpha}\left(\alpha + \theta - \frac{pz}{V}\right) + \frac{\partial c_1}{\partial\xi}f_{\mathrm{a}}(z)\xi \tag{28.30}$$

where θ is known from Eq. (28.29) and α is the steady flight wing incidence.

The aileron effectiveness is often measured in terms of the wing-tip helix angle (ps/V) per unit aileron displacement during a steady roll. In this condition the rolling moments due to a given aileron deflection, ξ, wing twist and aerodynamic damping are in equilibrium so that from Fig. 28.6(a) and Eq. (28.23) and noting that ailerons on opposite wings both contribute to the rolling, we have

$$2\int_0^s \frac{1}{2}\rho V^2 c\left[\frac{\partial c_l}{\partial \alpha}\left(\theta - \frac{pz}{V}\right) + \frac{\partial c_l}{\partial \xi} f_a(z)\xi\right] z\,dz = 0 \tag{28.31}$$

from which

$$\int_0^s \frac{\partial c_l}{\partial \alpha}\left(\theta - \frac{pz}{V}\right) z\,dz = -\xi\int_0^s \frac{\partial c_l}{\partial \xi} f_a(z) z\,dz \tag{28.32}$$

Substituting for θ from Eq. (28.29) into Eq. (28.32) gives

$$\int_0^s \frac{\partial c_l}{\partial \alpha}\left\{\frac{ps}{V}\frac{\sin\lambda z}{\lambda s\cos\lambda s} - \frac{1}{\partial c_l/\partial \alpha}\left(\frac{\partial c_l}{\partial \xi} + \frac{1}{e}\frac{\partial c_{m,0}}{\partial \xi}\right)\right.$$
$$\left.\times\left[f_a(z)\{1 - \cos\lambda(z - s_1)\} - \frac{\sin\lambda(s - s_1)}{\cos\lambda s}\sin\lambda z\right]\xi\right\} z\,dz$$
$$= -\xi\int_0^s \frac{\partial c_l}{\partial \xi} f_a(z) z\,dz$$

Hence

$$\xi\int_0^s\left\{\left(\frac{\partial c_l}{\partial \xi} + \frac{1}{e}\frac{\partial c_{m,0}}{\partial \xi}\right)\left[f_a(z)\{1 - \cos\lambda(z - s_1)\} - \frac{\sin\lambda(s - s_1)}{\cos\lambda s}\sin\lambda z\right]\right.$$
$$\left. - \frac{\partial c_l}{\partial \xi} f_a(z)\right\} z\,dz = \frac{ps}{V}\int_0^s \frac{\partial c_l}{\partial \alpha}\frac{\sin\lambda z}{\lambda s\cos\lambda s} z\,dz \tag{28.33}$$

Therefore, aileron effectiveness $(ps/V)/\xi$ is given by

$$\frac{(ps/V)}{\xi} = \frac{\displaystyle\int_0^s\left[-\frac{\partial c_l}{\partial \varepsilon} f_a(z) + \left(\frac{\partial c_l}{\partial \xi} + \frac{1}{e}\frac{\partial c_{m,0}}{\partial \xi}\right)\times\left[f_a(z)\{1 - \cos\lambda(z - s_1)\} - \frac{\sin\lambda(s - s_1)}{\cos\lambda s}\sin\lambda z\right]\right] z\,dz}{\int_0^s \frac{\partial c_l}{\partial \alpha}\frac{\sin\lambda z}{\lambda s\cos\lambda s} z\,dz}$$

Integration of the right-hand side of the above equation gives

$$\frac{(ps/V)}{\xi} = \frac{\left(\dfrac{\cos\lambda s_1}{\cos\lambda s} - 1\right)\dfrac{1}{\partial c_l/\partial \alpha}\dfrac{\partial c_l}{\partial \xi} + \left(\dfrac{\cos\lambda s_1}{\cos\lambda s} - 1 - \lambda^2\dfrac{s^2 - s_1^2}{2}\right)\dfrac{1}{e(\partial c_l/\partial \alpha)}\dfrac{\partial c_{m,0}}{\partial \xi}}{\left(\dfrac{\tan\lambda s}{\lambda s} - 1\right)} \tag{28.34}$$

The aileron reversal speed occurs when the aileron effectiveness is zero. Thus, equating the numerator of Eq. (28.34) to zero, we obtain the transcendental equation

$$\left(\frac{\partial c_l}{\partial \xi} + \frac{1}{e}\frac{\partial c_{m,0}}{\partial \xi}\right)(\cos\lambda s - \cos\lambda s_1) + \left(\lambda^2\frac{s^2 - s_1^2}{2}\frac{1}{e}\frac{\partial c_{m,0}}{\partial \xi}\right)\cos\lambda s = 0 \tag{28.35}$$

Alternative methods of obtaining divergence and control reversal speeds employ matrix or energy procedures. Details of such treatments may be found in Ref. [3].

28.4 Introduction to 'flutter'

We have previously defined flutter as the dynamic instability of an elastic body in an airstream. It is found most frequently in aircraft structures subjected to large aerodynamic loads such as wings, tail units and control surfaces. Flutter occurs at a critical or flutter speed V_f which in turn is defined as the lowest airspeed at which a given structure will oscillate with sustained simple harmonic motion. Flight at speeds below and above the flutter speed represents conditions of stable and unstable (that is divergent) structural oscillation, respectively.

Generally, an elastic system having just one degree of freedom cannot be unstable unless some peculiar mechanical characteristic exists such as a negative spring force or a negative damping force. However, it is possible for systems with two or more degrees of freedom to be unstable without possessing unusual characteristics. The forces associated with each individual degree of freedom can interact, causing divergent oscillations for certain phase differences. The flutter of a wing in which the flexural and torsional modes are coupled is an important example of this type of instability. Some indication of the physical nature of *wing-bending–torsion-flutter* may be had from an examination of aerodynamic and inertia forces during a combined bending and torsional oscillation in which the individual motions are 90° out of phase. In a pure bending or pure torsional oscillation the aerodynamic forces produced by the effective wing incidence oppose the motion; the geometric incidence in pure bending remains constant and therefore does not affect the aerodynamic damping force, while in pure torsion the geometric incidence produces aerodynamic forces which oppose the motion during one-half of the cycle but assist it during the other half so that the overall effect is nil. Thus, pure bending or pure torsional oscillations are quickly damped out. This is not the case in the combined oscillation when the maximum twist occurs at zero bending and vice versa; i.e. a 90° phase difference.

Consider the wing shown in Fig. 28.7 in various stages of a bending–torsion oscillation. At the position of zero bending the twisting of the wing causes a positive geometric incidence and therefore an aerodynamic force in the same direction as the motion of the wing. A similar but reversed situation exists as the wing moves in a downward direction; the negative geometric incidence due to wing twist causes a downward aerodynamic force. It follows that, although the effective wing incidence produces aerodynamic forces which oppose the motion at all stages, the aerodynamic forces associated with the geometric incidence have a destabilizing effect. At a certain speed – the flutter speed V_f – this destabilization action becomes greater than the stabilizing forces and the oscillations diverge. In practical cases the bending and torsional oscillations would not be as much as 90° out of phase; however, the same basic principles apply.

The type of flutter described above, in which two distinctly different types of oscillating motion interact such that the resultant motion is divergent, is known as *classical flutter*. Other types of flutter, *non-classical flutter*, may involve only one type of motion.

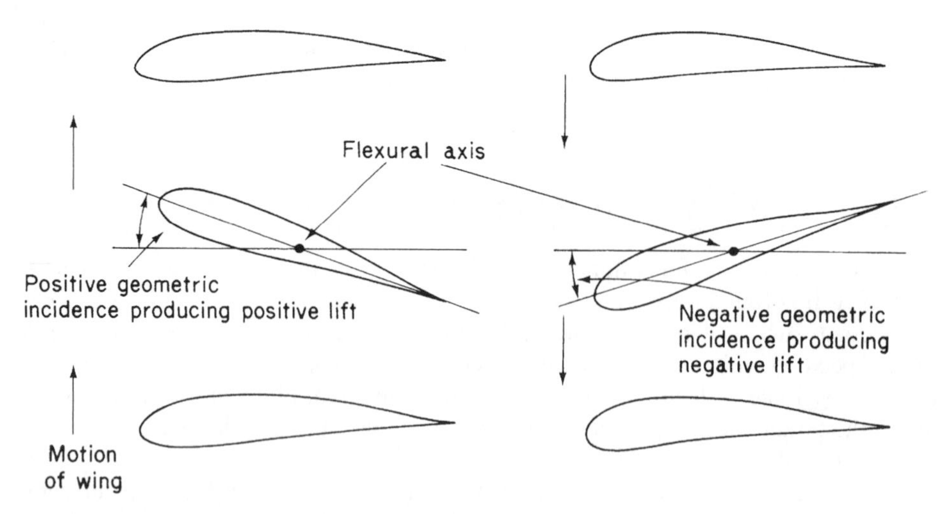

Fig. 28.7 Coupling of bending and torsional oscillations and destabilizing effect of geometric incidence.

For example, *stalling flutter* of a wing occurs at a high incidence where, for particular positions of the spanwise axis of twist, self-excited twisting oscillations occur which, above a critical speed, diverge.

Another non-classical form of flutter, *aileron buzz,* occurs at high subsonic speeds and is associated with the shock wave on the wing forward of the aileron. If the aileron oscillates downwards the flow over the upper surface of the wing accelerates, intensifying the shock and resulting in a reduction in pressure in the boundary layer behind the shock. The aileron, therefore, tends to be sucked back to its neutral position. When the aileron rises the shock intensity reduces and the pressure in the boundary layer increases, tending to push the aileron back to its neutral position. At low frequencies these pressure changes are approximately 180° out of phase with the aileron deflection and therefore become aerodynamic damping forces. At higher frequencies a component of pressure appears in phase with the aileron velocity which excites the oscillation. If this is greater than all other damping actions on the aileron a high frequency oscillation results in which only one type of motion, rotation of the aileron about its hinge, is present, i.e. aileron buzz. Aileron buzz may be prevented by employing control jacks of sufficient stiffness to ensure that the natural frequency of aileron rotation is high.

Buffeting is produced most commonly in a tailplane by eddies caused by poor airflow in the wing wake striking the tailplane at a frequency equal to its natural frequency; a resonant oscillation having one degree of freedom could then occur. The problem may be alleviated by proper positioning of the tailplane and clean aerodynamic design.

28.4.1 Coupling

We have seen that the classical flutter of an aircraft wing involves the interaction of flexural and torsional motions. Separately neither motion will cause flutter but together, at critical values of amplitude and phase angle, the forces produced by one motion excite

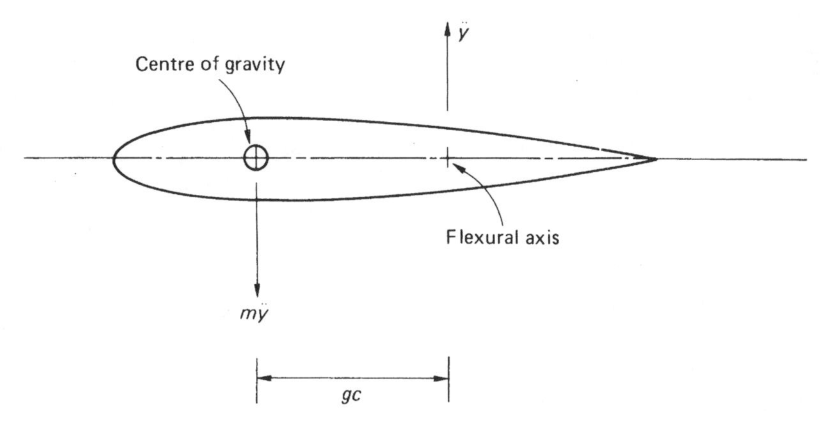

Fig. 28.8 Inertial coupling of a wing.

the other; the two types of motion are then said to be *coupled*. Various forms of coupling occur: inertial, aerodynamic and elastic.

The cross-section of a small length of wing is shown in Fig. 28.8. Its centre of gravity is a distance gc ahead of its flexural axis, c is the wing section chord and the mass of the small length of wing is m. If the length of wing is subjected to an upward acceleration $\ddot{y}$ an accompanying inertia force $m\ddot{y}$ acts at its centre of gravity in a downward direction, thereby producing a nose-down torque about the flexural axis of $m\ddot{y}gc$, causing the wing to twist. The vertical motion therefore induces a twisting motion by virtue of the inertia forces present, i.e. *inertial coupling*. Conversely, an angular acceleration $\ddot{\alpha}$ about the flexural axis causes a linear acceleration of $gc\ddot{\alpha}$ at the centre of gravity with a corresponding inertia force of $mgc\ddot{\alpha}$. Thus, angular acceleration generates a force producing translation, again inertial coupling. Note that the inertia torque due to unit linear acceleration (mgc) is equal to the inertia force due to unit angular acceleration (mgc); the inertial coupling therefore possesses symmetry.

Aerodynamic coupling is associated with changes of lift produced by wing rotation or translation. A change of wing incidence, i.e. a rotation of the wing, induces a change of lift which causes translation while a translation of velocity $\dot{y}$, say, results in an effective change in incidence, thereby yielding a lift which causes rotation. These aerodynamic forces, which oscillate in a flutter condition, act through a centre analogous to the aerodynamic centre of a wing in steady motion; this centre is known as the *centre of independence*.

Consider now the wing section shown in Fig. 28.9 and suppose that the wing stiffness is represented by a spring of stiffness k positioned at its flexural axis. Suppose also that the displacement of the wing is defined by the vertical deflection y of an arbitrary point O (Fig. 28.9(a)) and a rotation α about O (Fig. 28.9(b)). In Fig. 28.9(a) the vertical displacement produces a spring force which causes a clockwise torque (kyd) on the wing section about O, resulting in an increase in wing incidence α. In Fig. 28.9(b) the clockwise rotation α about O results in a spring force $kd\alpha$ acting in an upward direction on the wing section, thereby producing translations in the positive y direction. Thus, translation and rotation are coupled by virtue of the elastic stiffness of the wing, hence *elastic coupling*. We note that, as in the case of inertial coupling, elastic coupling possesses symmetry since the moment due to unit displacement (kd) is equal to the

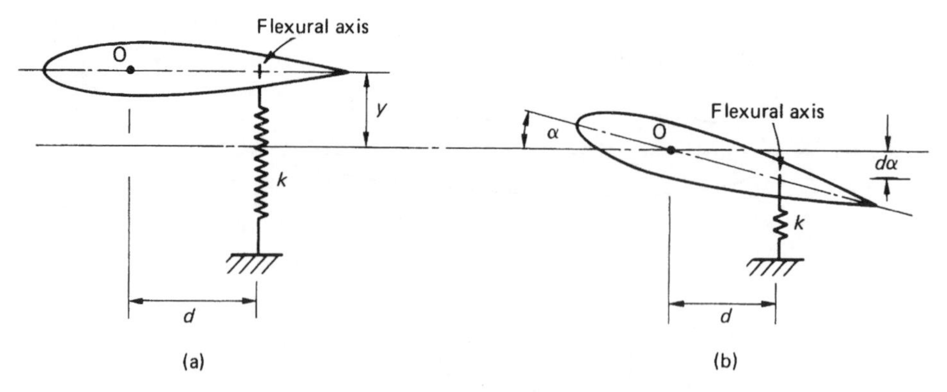

Fig. 28.9 Elastic coupling of a wing.

force produced by the unit rotation (kd). Also, if the arbitrarily chosen point O is made to coincide with the flexural axis, $d = 0$ and the coupling disappears.

From the above it can be seen that flutter will be prevented by uncoupling the two constituent motions. Thus, inertial coupling is prevented if the centre of gravity coincides with the flexural axis, while aerodynamic coupling is eliminated when the centre of independence coincides with the flexural axis. This, in fact, would also eliminate elastic coupling since O in Fig. 28.9 would generally be the centre of independence. Unfortunately, in practical situations, the centre of independence is usually forward of the flexural axis, while the centre of gravity is behind it giving conditions which promote flutter.

28.4.2 Determination of critical flutter speed

Consider a wing section of chord c oscillating harmonically in an airflow of velocity V and density ρ and having instantaneous displacements, velocities and accelerations of, rotationally, α, $\dot{\alpha}$, $\ddot{\alpha}$, and, translationally, y, $\dot{y}$, $\ddot{y}$. The oscillation causes a reduction in lift from the steady state lift[4] so that, in effect, the lift due to the oscillation acts downwards. The downward lift corresponding to α, $\dot{\alpha}$ and $\ddot{\alpha}$ is, respectively

$$l_\alpha \rho c V^2 \alpha = L_\alpha \alpha$$

$$l_{\dot{\alpha}} \rho c^2 V \dot{\alpha} = L_{\dot{\alpha}} \dot{\alpha}$$

$$l_{\ddot{\alpha}} \rho c^3 \ddot{\alpha} = L_{\ddot{\alpha}} \ddot{\alpha}$$

in which l_α, $l_{\dot{\alpha}}$, $l_{\ddot{\alpha}}$ are non-dimensional coefficients analogous to the lift-curve slopes in steady motion. Similarly, downward forces due to the translation of the wing section occur and are

$$l_y \rho c V^2 y/c = L_y y$$

$$l_{\dot{y}} \rho c^2 V \dot{y}/c = L_{\dot{y}} \dot{y}$$

$$l_{\ddot{y}} \rho c^3 \ddot{y}/c = L_{\ddot{y}} \ddot{y}$$

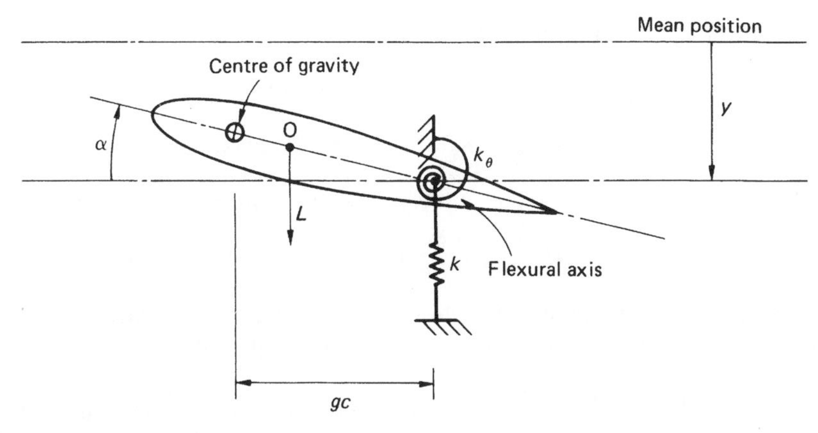

Fig. 28.10 Flutter of a wing section.

Thus, the total aerodynamic lift on the wing section due to the oscillating motion is given by

$$L = L_y y + L_{\dot{y}}\dot{y} + L_{\ddot{y}}\ddot{y} + L_\alpha \alpha + L_{\dot{\alpha}}\dot{\alpha} + L_{\ddot{\alpha}}\ddot{\alpha} \tag{28.36}$$

We have previously seen that rotational and translational displacements produce moments about any chosen centre. Thus, the total nose-up moment on the wing section is

$$M = M_y y + M_{\dot{y}}\dot{y} + M_{\ddot{y}}\ddot{y} + M_\alpha \alpha + M_{\dot{\alpha}}\dot{\alpha} + M_{\ddot{\alpha}}\ddot{\alpha} \tag{28.37}$$

where

$$M_y y = l_y \rho c^2 V^2 y/c$$
$$M_{\dot{y}}\dot{y} = l_{\dot{y}} \rho c^3 V \dot{y}/c$$
$$M_{\ddot{y}}\ddot{y} = l_{\ddot{y}} \rho c^4 \ddot{y}/c$$
$$M_\alpha \alpha = m_\alpha \rho c^2 V^2 \alpha/c$$
$$M_{\dot{\alpha}}\dot{\alpha} = m_{\dot{\alpha}} \rho c^3 V \dot{\alpha}/c$$
$$M_{\ddot{\alpha}}\ddot{\alpha} = m_{\ddot{\alpha}} \rho c^4 \ddot{\alpha}/c$$

in which m_α, etc. are analogous to the steady motion local pitching moment coefficients.

Now consider the wing section shown in Fig. 28.10. The wing section is oscillating about a mean position and its flexural and torsional stiffnesses are represented by springs of stiffness k and k_θ, respectively. Suppose that its instantaneous displacement from the mean position is y, which is now taken as positive downwards. In addition to the aerodynamic lift and moment forces of Eqs (28.36) and (28.37) the wing section experiences inertial and elastic forces and moments. Thus, if the mass of the wing section is m and I_O is its moment of inertia about O, instantaneous equations of vertical force and moment equilibrium may be written as follows. For vertical force equilibrium

$$L - m\ddot{y} + mgc\ddot{\alpha} - ky = 0 \tag{28.38}$$

and for moment equilibrium about O

$$M - I_{\mathrm{O}}\ddot{\alpha} + mgc\ddot{y} - k_{\theta}\alpha = 0 \tag{28.39}$$

Substituting for L and M from Eqs (28.36) and (28.37) we obtain

$$(m - L_{\ddot{y}})\ddot{y} - L_{\dot{y}}\dot{y} + (k - L_y)y - (mgc + L_{\ddot{\alpha}})\ddot{\alpha} - L_{\dot{\alpha}}\dot{\alpha} - L_{\alpha}\alpha = 0 \tag{28.40}$$

$$(-mgc - M_{\ddot{y}})\ddot{y} - M_{\dot{y}}\dot{y} - M_y y + (I_{\mathrm{O}} - M_{\ddot{\alpha}})\ddot{\alpha} - M_{\dot{\alpha}}\dot{\alpha} + (k_{\theta} - M_{\alpha})\alpha = 0 \tag{28.41}$$

The terms involving y in the force equation and α in the moment equation are known as *direct* terms, while those containing α in the force equation and y in the moment equation are known as *coupling* terms.

The critical flutter speed V_{f} is contained in Eqs (28.40) and (28.41) within the terms L_y, $L_{\dot{y}}$, L_{α}, $L_{\dot{\alpha}}$, M_y, $M_{\dot{y}}$, M_{α} and $M_{\dot{\alpha}}$. Its value corresponds to the condition that these equations represent simple harmonic motion. Above this critical value the equations represent divergent oscillatory motion, while at lower speeds they represent damped oscillatory motion. For simple harmonic motion

$$y = y_0\,\mathrm{e}^{\mathrm{i}\omega t} \qquad \alpha = \alpha_0\,\mathrm{e}^{\mathrm{i}\omega t}$$

Substituting in Eqs (28.40) and (28.41) and rewriting in matrix form we obtain

$$\begin{bmatrix} -\omega^2(m - L_{\ddot{y}}) - \mathrm{i}\omega L_{\dot{y}} + k - L_y & \omega^2(mgc + L_{\ddot{\alpha}}) - \mathrm{i}\omega L_{\dot{\alpha}} - L_{\alpha} \\ \omega^2(mgc + M_{\ddot{y}}) - \mathrm{i}\omega M_{\dot{y}} + M_y & -\omega^2(I_{\mathrm{O}} - M_{\ddot{\alpha}}) - \mathrm{i}\omega M_{\dot{\alpha}} + k_{\theta} - M_{\alpha} \end{bmatrix} \begin{Bmatrix} y_0 \\ \alpha_0 \end{Bmatrix} = 0 \tag{28.42}$$

The solution of Eq. (28.42) is most readily obtained by computer[4] for which several methods are available. One method represents the motion of the system at a general speed V by

$$y = y_0\,\mathrm{e}^{(\delta + \mathrm{i}\omega)t} \qquad \alpha = \alpha_0^{(\delta + \mathrm{i}\omega)t}$$

in which $\delta + \mathrm{i}\omega$ is one of the complex roots of the determinant of Eq. (28.42). For any speed V the imaginary part ω gives the frequency of the oscillating system while δ represents the exponential growth rate. At low speeds the oscillation decays (δ is negative) and at high speeds it diverges (δ is positive). Zero growth rate corresponds to the critical flutter speed V_{f}, which may therefore be obtained by calculating δ for a range of speeds and determining the value of V_{f} for $\delta = 0$.

28.4.3 Prevention of flutter

We have previously seen that flutter can be prevented by eliminating inertial, aerodynamic and elastic coupling by arranging for the centre of gravity, the centre of independence and the flexural axis of the wing section to coincide. The means by which this may be achieved are indicated in the coupling terms in Eqs (28.40) and (28.41).

In Eq. (28.41) the inertial coupling term is $mgc + M_{\ddot{y}}$ in which $M_{\ddot{y}}$ is usually very much smaller than mgc. Thus, inertial coupling may be virtually eliminated by adjusting the position of the centre of gravity of the wing section through *mass balancing* so that it coincides with the flexural axis, i.e. $gc = 0$. The aerodynamic coupling term $M_{\dot{y}}\dot{y}$ vanishes, as we have seen, when the centre of independence coincides with the flexural axis. Further, the terms $M_y y$ and $L_{\dot{\alpha}}\dot{\alpha}$ are very small and may be neglected so that Eqs (28.40) and (28.41) now reduce to

$$(m - L_{\ddot{y}})\ddot{y} - L_{\dot{y}}\dot{y} + (k - L_y)y - L_\alpha \alpha = 0 \tag{28.43}$$

and

$$(I_{\mathrm{O}} - M_{\ddot{\alpha}})\ddot{\alpha} - M_{\dot{\alpha}}\dot{\alpha} + (k_\theta - M_\alpha)\alpha = 0 \tag{28.44}$$

The remaining coupling term $L_\alpha \alpha$ cannot be eliminated since the vertical force required to maintain flight is produced by wing incidence.

Equation (28.44) governs the torsional motion of the wing section and contains no coupling terms so that, since all the coefficients are positive at speeds below the wing section torsional divergence speed, any torsional oscillation produced, say, by a gust will decay. Also, from Eq. (28.43), it would appear that a vertical oscillation could be maintained by the incidence term $L_\alpha \alpha$. However, rotational oscillations, as we have seen from Eq. (28.44), decay so that the lift force $L_\alpha \alpha$ is a decaying force and cannot maintain any vertical oscillation.

In practice it is not always possible to prevent flutter by eliminating coupling terms. However, increasing structural stiffness, although carrying the penalty of increased weight, can raise the value of V_{f} above the operating speed range. Further, arranging for the centre of gravity of the wing section to be as close as possible to and forward of the flexural axis is beneficial. Thus, wing mounted jet engines are housed in pods well ahead of the flexural axis of the wing.

28.4.4 Experimental determination of flutter speed

The previous analysis has been concerned with the flutter of a simple two degrees of freedom model. In practice the structure of an aircraft can oscillate in many different ways. For example, a wing has fundamental bending and torsional modes of oscillation on which secondary or overtone modes of oscillation are superimposed. Also it is possible for fuselage bending oscillations to produce changes in wing camber thereby affecting wing lift and for control surfaces oscillating about their hinges to produce aerodynamic forces on the main surfaces.

The equations of motion for an actual aircraft are therefore complex with a number N, say, of different motions being represented (N can be as high as 12). There are, therefore, N equations of motion which are aerodynamically coupled. At a given speed, solution of these N equations yields N different values of $\delta + \mathrm{i}\omega$ corresponding to the N modes of oscillation. Again, as in the simple two degrees of freedom case, the critical flutter speed for each mode may be found by calculating δ for a range of speeds and determining the value of speed at which $\delta = 0$.

A similar approach is used experimentally on actual aircraft. The aircraft is flown at a given steady speed and caused to oscillate either by exploding a small detonator on the wing or control surface or by a sudden control jerk. The resulting oscillations are recorded and analysed to determine the decay rate. The procedure is repeated

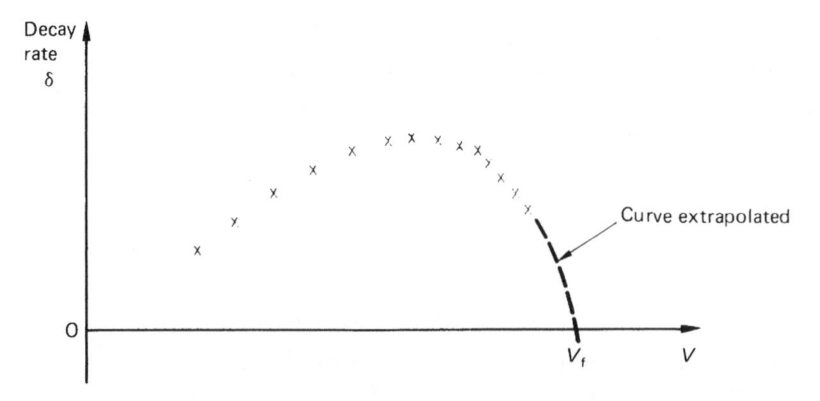

Fig. 28.11 Experimental determination of flutter speed.

at increasing speeds with smaller increments being used at higher speeds. The measured decay rates are plotted against speed, producing a curve such as that shown in Fig. 28.11. This curve is then extrapolated to the zero decay point which corresponds to V_f. Clearly this approach requires as accurate as possible a preliminary estimation of flutter speed since induced oscillations above the flutter speed diverge leading to possibly catastrophic results.

Other experimental work involves wind tunnel tests on flutter models, the results being used to check theoretical calculations.[3]

28.4.5 Control surface flutter

If a control surface oscillates about its hinge, oscillating forces are induced on the main surface. For example, if a wing oscillates in bending at the same time as the aileron oscillates about its hinge, flutter can occur provided there is a phase difference between the two motions. In similar ways elevator and rudder flutter can occur as the fuselage oscillates in bending. Other forms of control surface flutter involve more than two different types of motion. Included in this category are wing bending/aileron rotation/tab rotation and elevator rotation/fuselage bending/rigid body pitching and translation of the complete aircraft.

It can be shown[4] that control surface flutter can be prevented by eliminating the inertial coupling between the control rotation and the motion of the main surface. This may be achieved by mass balancing the control surface whereby weights are attached to the control surface forward of the hinge line.

All newly designed aircraft are subjected early in the life of a prototype to a *ground resonance test* to determine actual normal modes and frequencies. The primary objectives of such tests are to check the accuracy of the calculated normal modes on which the flutter predictions are based and to show up any unanticipated peculiarities in the vibrational behaviour of the aircraft. Usually the aircraft rests on some low frequency support system or even on its deflated tyres. Electrodynamic exciters are mounted in pairs on the wings and tail with accelerometers as the measuring devices. The test procedure is generally first to discover the resonant frequencies by recording amplitude and phase of a selected number of accelerometers over a given frequency range. Having obtained the resonant frequencies the aircraft is then excited at each of these frequencies in turn and all accelerometer records taken simultaneously.

References

1 Babister, A. W., *Aircraft Stability and Control*, Pergamon Press, London, 1961.
2 Duncan, W. J., *The Principles of the Control and Stability of Aircraft*, Cambridge University Press, Cambridge, 1959.
3 Bisplinghoff, R. L., Ashley, H. and Halfman, R. L., *Aeroelasticity*, Addison-Wesley Publishing Co. Inc., Cambridge, MA, 1955.
4 Dowell, E. H. *et al.*, *A Modern Course in Aeroelasticity*, Sijthoff and Noordhoff, Alphen aan den Rijn, The Netherlands, 1978.

Problems

P.28.1 An initially untwisted rectangular wing of semi-span s and chord c has its flexural axis normal to the plane of symmetry, and is of constant cross-section with torsional rigidity GJ. The aerodynamic centre is ec ahead of the flexural axis, the lift-coefficient slope is a and the pitching moment coefficient at zero lift is $C_{m,0}$. At speed V in air of density ρ the wing-root incidence from zero lift is α_0.

Using simple strip theory, i.e. ignoring downwash effects, show that the incidence at a section distant y from the plane of symmetry is given by

$$\alpha_0 + \theta = \left(\frac{C_{m,0}}{ea} + \alpha_0\right)\frac{\cos\lambda(s-y)}{\cos\lambda s} - \frac{C_{m,0}}{ea}$$

where

$$\lambda^2 = \frac{ea\frac{1}{2}\rho V^2 c^2}{GJ}$$

Hence, assuming $C_{m,0}$ to be negative, find the condition giving the speed at which the lift would be reduced to zero.

Ans. $V_d = \sqrt{\dfrac{\pi^2 GJ}{2\rho ec^2 s^2 a}}$.

P.28.2 The rectangular wing shown in Fig. P.28.2 has a constant torsional rigidity GJ and an aileron of constant chord. The aerodynamic centre of the wing is at a constant distance ec ahead of the flexural axis while the additional lift due to operation of the aileron acts along a line a distance hc aft of the flexural axis; the local, two-dimensional lift-curve slopes are a_1 for the wing and a_2 for aileron deflection. Using strip theory and considering only the lift due to the change of incidence arising from aileron movement, show that the aileron reversal speed is given by

$$\tan\lambda ks\int_0^{ks} y\sin\lambda y\,\mathrm{d}y - \tan\lambda s\int_0^{s} y\sin\lambda y\,\mathrm{d}y - \int_{ks}^{s} y\cos\lambda y\,\mathrm{d}y = \frac{(e+h)}{2h\cos\lambda ks}[(ks)^2 - s^2]$$

where

$$\lambda^2 = \frac{1}{2}\rho V^2 a_1\, ec^2/GJ$$

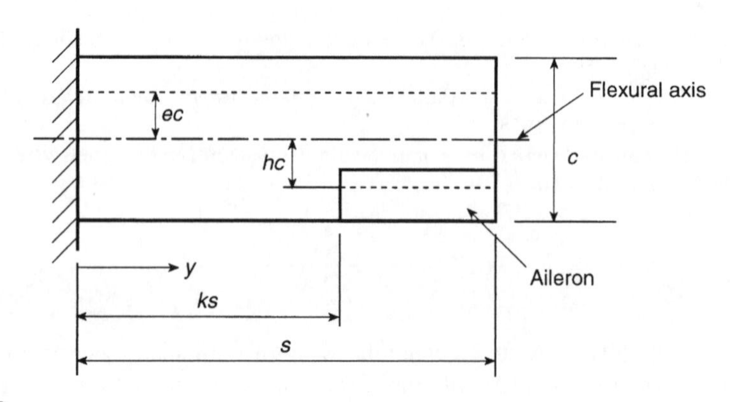

Fig. P.28.2

Appendix: Design of a rear fuselage

Figure A.1 shows the elevation of a two seater trainer/semi-aerobatic aircraft. It is required to carry out the detailed structural design of the portion of the rear fuselage between the sections AA and BB.

A.1 Specification

The required flight envelope for this particular aircraft is shown in Fig. A.2 (refer also to Fig. 13.10) where

$$n_1 = 6.28, \quad V_D \text{ (design diving speed)} = 183.8\,\text{m/s}$$

Also

$$V_C = 0.8V_D = 147.0\,\text{m/s}$$
$$n_2 = 0.75n_1 = 4.71$$
$$n_3 = 0.5n_1 = 3.14$$

Note also that airworthiness requirements specify that since $n_1 > 3$ the point D_2 lies on the $n = 0$ axis.

Further requirements are that:

(i) at any point in the flight envelope an additional pitching acceleration given by

$$\left(20 + \frac{475}{W}\right)\frac{n}{V}\ \text{rad/s}^2 \tag{A.1}$$

be applied where W is the total weight of the aircraft in kN and V is the velocity of the aircraft in m/s.

(ii) for asymmetric flight an angle of yaw given by

$$\psi = 0.7n_1 + \frac{457.2}{V_D}\text{degrees} \tag{A.2}$$

must be allowed for; the angle of yaw increases the overall pitching moment coefficient of the aircraft by −0.0015/degree of yaw.

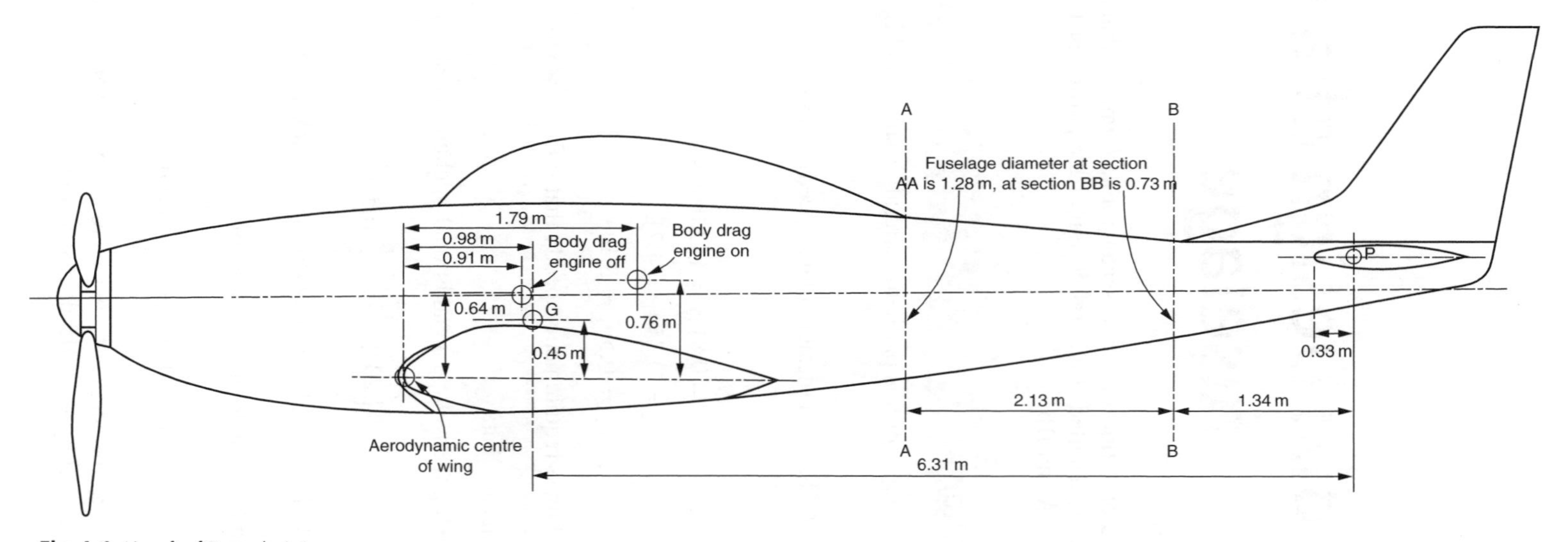

Fig. A.1 Aircraft of Example A.1.

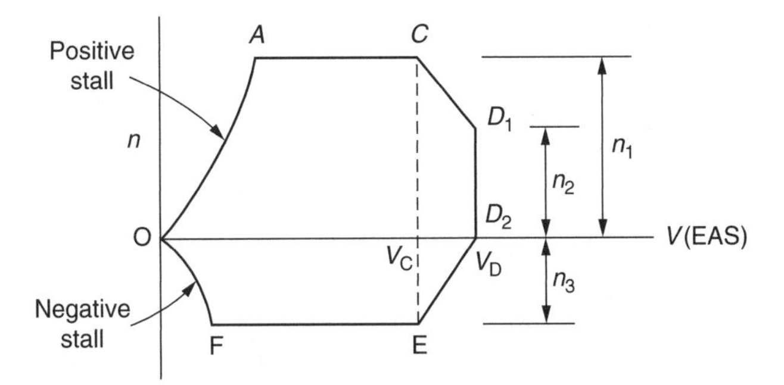

Fig. A.2 Flight envelope.

A.2 Data

Preliminary design work has produced the following data.

Aircraft

Fully loaded weight = 37.43 kN.

Moment of inertia of fully loaded aircraft about the centre of gravity (G in Fig. A.1)

$$= 22\,235\ \text{kg m}^2$$

Position of G and the body drag centres, engine on and off, are shown in Fig. A.1.

The body drag coefficients are

$$C_{D,B}\ (\text{engine on}) = 0.01583$$
$$C_{D,B}\ (\text{engine off}) = 0.0576$$

The engine has a maximum horse power of 905 and the propeller efficiency is 90%.

Wing

The wing has a span of 14.07 m and gross area of 29.64 m^2. Its aerodynamic mean chord, $c = 2.82$ m and the variations of lift and drag coefficients with incidence are shown in Fig. A.3.

Also, the pitching moment coefficient is given by

$$C_M = -0.238 C_L$$

and, due to a rigger's incidence of $-1.5°$, there is an additional pitching moment coefficient equal to -0.036.

Tailplane

The tailplane has a span of 6.55 m and a gross area of 8.59 m^2; the position of the aerodynamic centre, P, of the tailplane is shown in Fig. A.1.

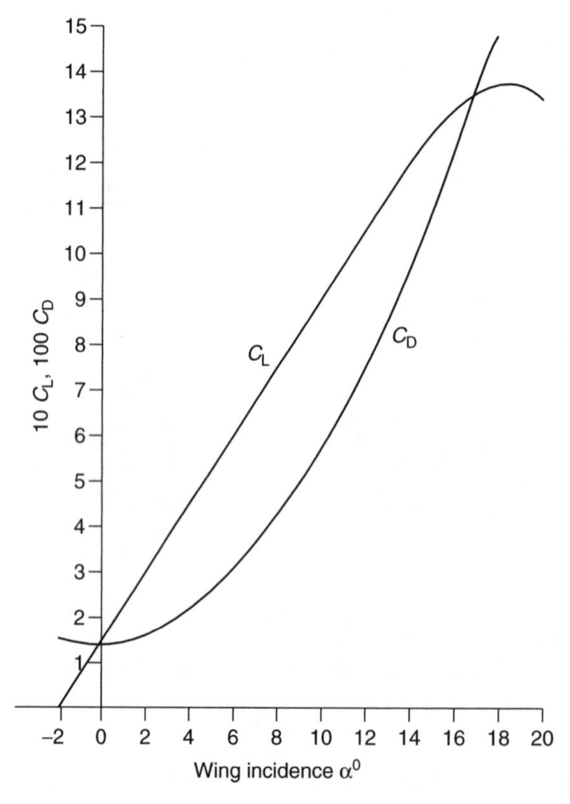

Fig. A.3 Wing characteristics.

Due to the asymmetry of the slipstream induced by yaw asymmetric loads are induced on the tailplane. These loads, upwards on one side and downwards on the other, result in a torque given by

$$\frac{0.00125}{\sqrt{1-M^2}}\rho V^2 S_t b_t \psi \tag{A.3}$$

where M is the mach number, S_t the tailplane area and b_t the tailplane span.

Fin

The fin has a height of 1.65 m, an area of 1.80 m^2 and an aspect ratio of 1.5. Also, it may be shown that the lift-curve slope, a_1, of the fin is given by

$$a_1 = \frac{5.5A}{A+2} \tag{A.4}$$

in which A is the aspect ratio of a wing which is equivalent to two fins.

In yawed flight the incidence of the fin to the air flow is ψ so that a fin load equal to $\frac{1}{2}\rho V^2 S_F a_1 \psi$ is generated where V is the aircraft speed and S_F the fin area.

The position of the centre of pressure of the fin depends upon the geometry of the pressure distribution. Calculations show that the centre of pressure is 1.13 m above the axis of the rear fuselage and a distance of 3.7 m aft of the section AA.

A.3 Initial calculations

Flight envelope

The positive stall curve in the flight envelope of Fig. A.2 is found from basic aerodynamic wing theory and is given by

$$C_{\mathrm{L,max}} = \frac{nW}{\frac{1}{2}\rho V_{\mathrm{s}}^2 S}$$

where V_{s} is the stalling speed and S the wing area. Then

$$V_{\mathrm{s}} = \left(\frac{2nW}{\rho S C_{\mathrm{L,max}}}\right)^{1/2} \tag{A.5}$$

Substituting the values given in the preceding data and taking ρ, the air density at sea level, as 1.226 kg/m^3

$$V_{\mathrm{s}} = \left(\frac{2 \times 37.43 \times 10^3}{1.226 \times 29.64 \times 1.38}\right)^{1/2} (n)^{1/2}$$

i.e.

$$V_{\mathrm{s}} = 38.6(n)^{1/2} \tag{A.6}$$

The positive stall curve is found by assigning a series of values to n and then calculating the corresponding stalling speeds. For $n = n_1 = 6.28$

$$V_{\mathrm{s}} = 38.6(6.28)^{1/2} = 96.7\ \mathrm{m/s} \quad \text{(A on flight envelope)}$$

Fin lift-curve slope

From Eq. (A.4)

$$a_1 = \frac{5.5 \times 3.0}{3.0 + 2.0} = 3.3$$

Speed of sound

At sea level at a temperature of 15°C the speed of sound is 340.8 m/s.

A.4 Balancing out calculations

The tailplane and fin loads corresponding to the various critical points in the flight envelope will now be calculated so that, subsequently, values of shear force, bending moment and torque acting on the rear fuselage may be determined. The cases to be investigated are:

Case A (point A on the flight envelope, engine on)
Case A′ (point A on the flight envelope, engine off)
Case C (point C on the flight envelope, engine off)
Case D_1 (point D_1 on the flight envelope, engine off)
Case D_2 (point D_2 on the flight envelope, engine off)

Case A

From the flight envelope $n = 6.28$, $V = 96.7$ m/s and from Fig. A.3 the wing incidence α corresponding to $C_{L,max} = 1.38$ is 18°. The forces acting on the aircraft and their lines of action are shown in Fig. A.4, the dimensions may be scaled from an actual drawing (the simplest approach) or calculated.

Since 1 hp = 746 W = 746 mN/s, the thrust T of the engine is given by

$$T = \frac{\eta \times hp \times 746}{V}$$

i.e.

$$T = \frac{0.9 \times 905 \times 746}{96.7}$$

so that

$$T = 6284\,\text{N}$$

Also

$$nW = 6.28 \times 37.43 \times 10^3 = 235\,060\,\text{N}$$

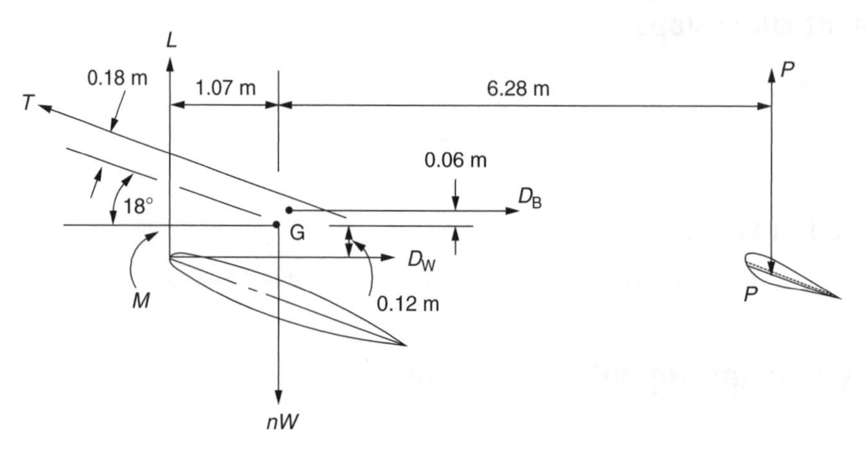

Fig. A.4 Balancing out calculations.

A first approximation for the wing lift, L, is obtained by neglecting the tailplane lift, P, i.e.

$$L = nW - T\sin(18^\circ - 1.5^\circ)$$

which gives

$$L = 235\,060 - 6284\sin 16.5^\circ = 233\,275\,\text{N}$$

From Fig. A.3 the wing drag coefficient, $C_{D,W}$, is 0.149 so that the wing drag, which is given by

$$D_W = C_{D,W}\frac{1}{2}\rho V^2 S$$

is

$$D_W = 0.149 \times 1.226 \times 96.7^2 \times 29.64/2 = 25\,315\,\text{N}$$

The body drag coefficient is 0.01583 so that

$$D_B = 0.01583 \times 1.226 \times 96.7^2 \times 29.64/2 = 2690\,\text{N}$$

The angle of yaw is given by Eq. (A.2), i.e.

$$\psi = 0.7 \times 6.28 + \frac{457.2}{183.8} = 6.9^\circ$$

The total pitching moment coefficient is then

$$C_M = -0.238 \times 1.38 - 0.036 - 0.0015 \times 6.9 = -0.375$$

so that

$$M = C_M\frac{1}{2}\rho V^2 Sc = -0.375 \times 1.226 \times 96.7^2 \times 29.64 \times 2.82/2 = -179\,669\,\text{N m}$$

The additional pitching moment acceleration is, from Eq. (A.1)

$$\left(20 + \frac{475}{37.43}\right)\frac{6.28}{96.7} = 2.12\,\text{rad/s}^2$$

Then, taking moments about G (refer to Fig A.4)

$$1.07L - 0.18T + 0.06D_B - 0.12D_W - 6.28P - 179\,669 = 22\,235 \times 2.12$$

i.e.

$$1.07L - 0.18 \times 6284 + 0.06 \times 2690 - 0.12 \times 25\,315 - 6.28P - 179\,669 = 22\,235 \times 2.12$$

which simplifies to

$$5.78P = L - 215\,715$$

First approximation, $L = 233\,275$ N gives $P = 2991$ N
Second approximation, $L = 233\,275 - 2991 = 230\,284$ N gives $P = 2482$ N
Third approximation, $L = 233\,275 - 2482 = 230\,793$ N gives $P = 2569$ N
Fourth approximation, $L = 233\,275 - 2569 = 230\,706$ N gives $P = 2554$ N
Fifth approximation, $L = 233\,275 - 2554 = 230\,721$ N gives $P = 2556$ N

Therefore the tail load $P = 2556$ N.

The torque produced by the asymmetric loading on the tailplane is given by Eq. (A.3), i.e.

$$\text{Tailplane torque} = \frac{0.00125}{\sqrt{1 - (96.7/340.8)^2}} \times 1.226 \times 96.7^2 \times 8.59 \times 6.55 \times 6.9 = 5802\text{ N m}$$

The load on the fin caused by the yawed flight is given by $\frac{1}{2}\rho V^2 S\psi a_1$, i.e.

$$\text{Fin load} = 1.226 \times 96.7^2 \times 1.8(6.9 \times \pi/180) \times 3.3/2 = 4100\text{ N}$$

The torque produced on the fuselage by this fin load is $4100 \times 1.13 = 4633$ N m. The total torque on the rear fuselage is therefore given by

$$\text{Total torque (real fuselage)} = 5802 + 4633 = 10\,435\text{ N m}$$

The tail and fin loads and the rear fuselage torque corresponding to the remaining flight envelope cases are calculated in an identical manner and are listed in Table A.1.

Table A.1

Case	Tail load (N) (+ ↑)	Fin load (N) (+ →)	Fuselage torque (N m) (+ ↓)
A	2556	4100	10435
A′	2292	4100	10435
C	596	9501	24957
D_1	−4997	12460	34031
D_2	−9412	5340	14635

A.5 Fuselage loads

The dimensions of the portion of the rear fuselage to be designed are given in Fig. A.1.

Fuselage section

The construction of structural components was discussed in Chapter 12 where it was seen that fuselages generally comprise arrangements of stringers, frames and skin. For this particular aircraft the fuselage is unpressurized so that the frames will not support significant loads. However they will be required to maintain the fuselage shape but may therefore be nominal in size, suitable frame sections will be suggested later. The combination of stringers and skin will resist the shear forces, bending moments and torques produced by self-weight and aerodynamic loads. For this purpose a circular

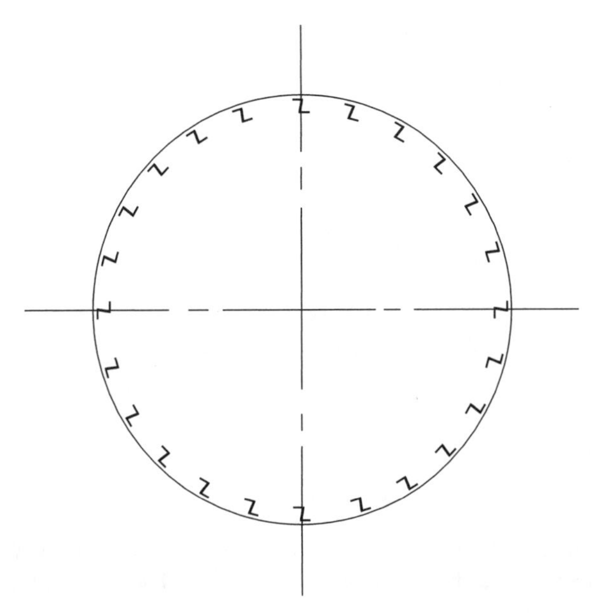

Fig. A.5 Stringer arrangement in rear fuselage.

cross-section will meet the design requirements of the aircraft and be simple to fabricate and design.

Figure A.5 shows a possible section. Twenty-four stringers arranged symmetrically, each having the same cross-sectional area, would be spaced at approximately 168 mm at the section AA and at 96 mm at the section BB.

Material

An aluminium alloy will be used for both stringers and skin and has the following properties:

$$0.1\% \text{ Proof stress} = 186\,\text{N/mm}^2$$

$$\text{Shear strength} = 117\,\text{N/mm}^2$$

Self-weight

In a conventional single-engined aircraft of the type shown in Fig. A.1 it is usual to assume that the fuselage weight is from 4.8% to 8.0% of the total weight and that the weight of the tailplane/fin assembly is from 1.2% to 2.5% of the total weight. It will be further assumed in this case that half of the fuselage weight is aft of the section AA and that the weight distribution varies directly as the skin surface area. Therefore, taking average values

$$\text{Weight of rear fuselage} = \frac{37.43 \times 10^3 \times 6.4}{2 \times 100} = 1198\,\text{N}$$

$$\text{Weight of tailplane/fin} = \frac{37.43 \times 10^3 \times 1.8}{100} = 674\,\text{N}$$

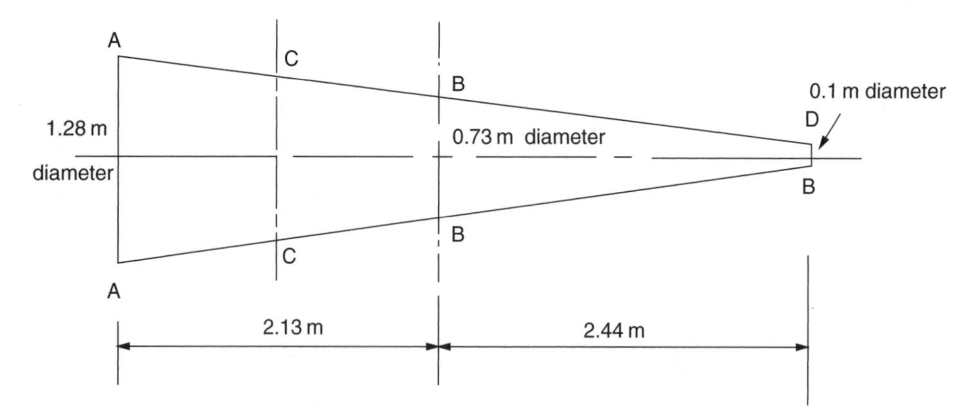

Fig. A.6 Rear fuselage sections.

For ease of calculation the rear fuselage is assumed to taper uniformly as shown in Fig. A.6; CC is a section midway between AA and BB. The total skin area is given by

$$\text{Skin area} = \pi(1.28 + 0.1) \times 4.57/2 = 9.91\,\text{m}^2$$

At the section AA the weight/m of fuselage $= 1198 \times \pi \times 1.28/9.91 = 486.1$ N/m
At the section CC the weight/m $= 1198 \times \pi \times 1.01/9.91 = 383.6$ N/m
At the section BB the weight/m $= 1198 \times \pi \times 0.73/9.91 = 277.2$ N/m
At the section DD the weight/m $= 1198 \times \pi \times 0.1/9.91 = 38.0$ N/m

Also the centre of gravity of the tailplane/fin assembly has been estimated to be 4.06 m from the section AA on a line parallel to the fuselage centre line.

Shear forces and bending moments due to self-weight

At the section AA

$$\text{SF} = (1198 + 674)n = 1872n\,\text{N} \tag{A.7}$$

$$\begin{aligned}\text{BM} &= [(38.0 \times 4.57^2/2) + (448.1 \times 4.57^2/2 \times 3) + 674 \times 4.06]n\cos\alpha \\ &= 4693n\cos\alpha\,\text{N m}\end{aligned} \tag{A.8}$$

where n is the normal acceleration coefficient and α the wing incidence.

At the section CC

$$\text{SF} = [1872 - (486.1 + 383.6) \times (2.13/2 \times 2)]n = 1409n\,\text{N} \tag{A.9}$$

$$\begin{aligned}\text{BM} &= [(38.0 \times 3.51^2/2) + (345.6 \times 3.51^2/2 \times 3) + 674 \times 2.99]n\cos\alpha \\ &= 2959n\cos\alpha\,\text{N m}\end{aligned} \tag{A.10}$$

At the section BB

$$SF = [1872 - (486.1 + 277.2) \times (2.13/2)]n = 10059n\,\text{N} \tag{A.11}$$

$$\begin{aligned} BM &= [(38.0 \times 2.44^2/2) + (239.2 \times 2.44^2/2 \times 3) + 674 \times 1.93]n\cos\alpha \\ &= 1651n\cos\alpha\,\text{N m} \end{aligned} \tag{A.12}$$

Total shear forces, bending moments and torques

The values of shear force, bending moment and torque at the sections AA, BB and CC will now be calculated for the flight envelope cases listed in Section A.4.

Case A ($n = 6.28$, $\alpha = 18°$)

Section AA
The shear force due to the self-weight and tail load is, from Eq. (A.7) and Table A.1
SF (S_y) = $1872 \times 6.28 - 2556 = 9200$ N (acting vertically downwards)
The shear force due to the fin load is, from Table A.1
SF (S_x) = 4100 N (acting horizontally to the right)
The bending moment due to the self-weight and tail load is, from Eq. (A.8) and Table A.1 (see also Fig. A.1)
BM (M_x) = $4693 \times 6.28 \cos 18° - 2556 \times 3.47 = 19160$ N m
The bending moment due to the fin load is, from Table A.1
BM (M_y) = $4100 \times 3.7 = 15170$ N m
The torque due to asymmetric flight and the fin load is, from Table A.1
$T = 10435$ N m

The values of shear force, bending moment and torque at the section AA due to the remaining flight envelope cases are calculated in an identical manner. The complete procedure is then repeated for the sections CC and BB. The results are listed in Table A.2 with the positive directions and senses of the forces, moments and torques shown in Fig. A.7; these are as specified in Section 16.2.1 for an internal section when viewed

Table A.2

Section	Case	S_x (N)	S_y (N)	M_x (N m)	M_y (N m)	T (N m)
AA	A	4100	9200	19160	15170	10435
	A′	4100	9434	19938	15170	10435
	C	9501	11125	27534	28958	24957
	D_1	12460	13350	39470	37978	34031
	D_2	5340	9412	32688	16276	14635
CC	A	4100	6675	10308	8301	10435
	A′	4100	7120	11393	8301	10435
	C	9501	8811	16276	18813	24957
	D_1	12460	11837	26490	24686	34031
	D_2	5340	9412	22651	10580	14635
BB	A	4100	4673	4747	3824	10435
	A′	4100	5118	5358	3824	10435
	C	9501	6809	8003	8687	24957
	D_1	12460	10547	13347	11393	34031
	D_2	5340	9412	12614	4883	14635

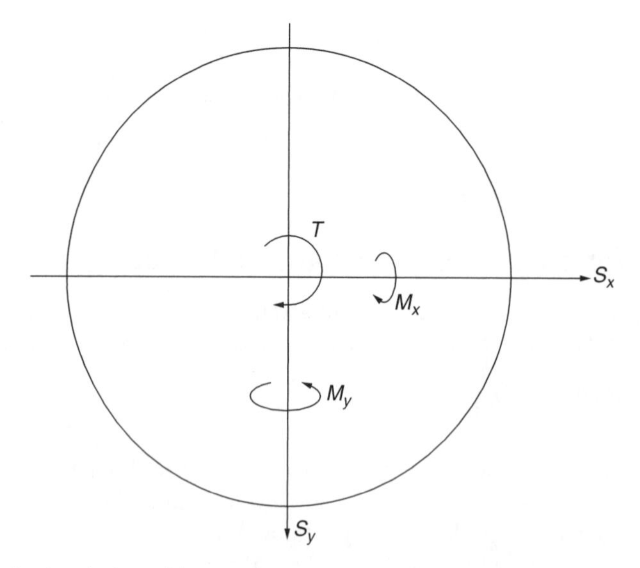

Fig. A.7 Positive directions for internal forces and moments.

in the direction Oz (see Fig. 16.9) except for torsion where it is assumed that a positive fin load produces a positive torque.

A.6 Fuselage design calculations

Two approaches to the actual design are possible. *Elastic design* uses *allowable* or *working* stresses which are obtained from, say, the 0.1% proof stress by incorporating a factor of safety, usually 1.5; these stresses are then combined with the actual loads to produce skin and stringer sizes. Alternatively, *ultimate load design* is based on the actual loads multiplied by an ultimate load factor (see Section 13.1) which then produces failure loads, the stresses involved are therefore the ultimate stresses. For linear systems the methods produce identical results so that, in this case, since the 0.1% proof stress is given, elastic design will be used. The working, or allowable, stresses are then

$$\text{Direct} = 186/1.5 = 124\,\text{N/mm}^2$$
$$\text{Shear} = 117/1.5 = 78\,\text{N/mm}^2$$

The proposed fuselage section is circular as previously shown in Fig. A.5. The design process is required to produce suitable stringer sections and a skin thickness. Suppose that each stringer (or boom) has a cross-sectional area B mm^2 and that the skin thickness is t mm. The idealized fuselage section (see Section 20.2) is shown in Fig A.8.

Stringer sections

The direct stress in each stringer produced by bending moments M_x and M_y is given by Eq. (16.20), i.e.

$$\sigma_z = \frac{M_x}{I_{xx}}y + \frac{M_y}{I_{yy}}x \tag{A.13}$$

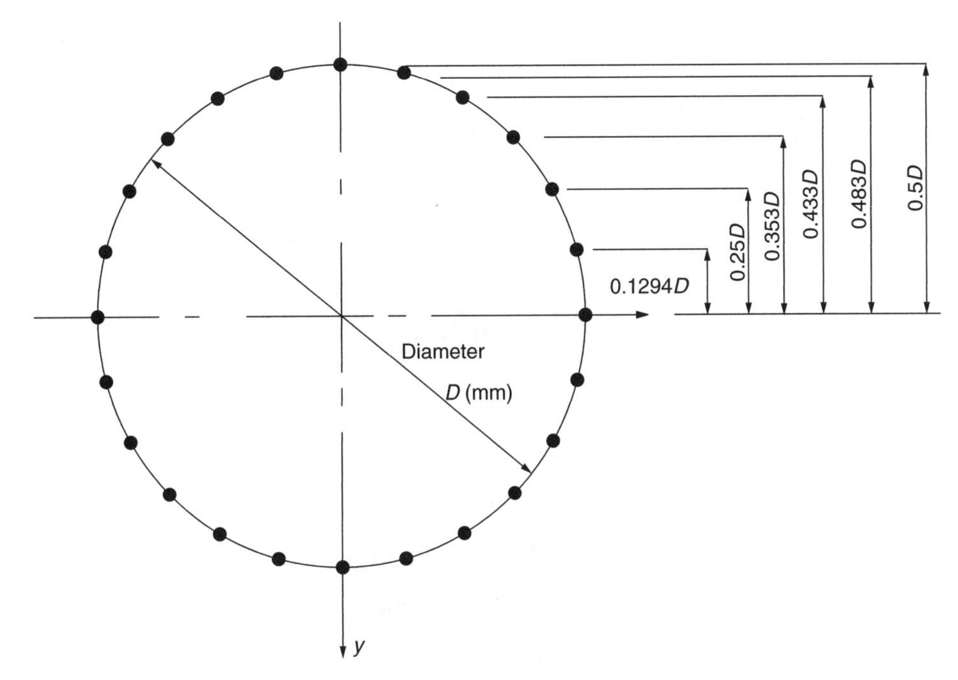

Fig. A.8 Idealized fuselage cross-section.

where

$$I_{xx} = I_{yy} = 4BD^2(0.1294^2 + 0.25^2 + 0.353^2 + 0.433^2 + 0.483^2 + 0.5^2/2)$$

i.e.

$$I_{xx} = I_{yy} = 3.0BD^2 \text{ mm}^4$$

A positive value of M_x will cause tensile stresses in stringers 2 to 12 (there will be zero stress in stringers 1 and 13) and compressive stresses in stringers 14 to 24. A positive value of M_y will produce tensile stresses in stringers 8 to 18 and compressive stresses in stringers 6 to 20 (zero stress in stringers 7 and 19). Therefore M_x and M_y both produce tensile stresses in stringers 7 to 13 and compressive stresses in stringers 19 to 1; in the remaining stringers the stresses due to M_x and M_y are of opposite sign.

Inspection of Table A.2 shows that M_x and M_y reach their greatest values at each fuselage section in Case D_1.

Section AA (diameter $D = 1.28$ m)

$$M_x = 39\,470 \text{ N m} \quad M_y = 37\,978 \text{ N m}$$

Equation (A.13) becomes

$$\sigma = \frac{39\,470 \times 10^3}{3.0B \times 1.28^2 \times 10^6} y + \frac{37\,978 \times 10^3}{3.0B \times 1.28^2 \times 10^6} x$$

i.e.

$$\sigma = (8.03y + 7.73x)/B$$

At stringer 7, $x = 0$, $y = 0.64$ m
Then

$$\sigma_7 = 8.03 \times 0.64 \times 10^3/B$$

i.e.

$$\sigma_7 = 5139/B \, \text{N/mm}^2$$

Similarly

$$\sigma_8 = 6245/B \, \text{N/mm}^2$$
$$\sigma_9 = 6924/B \, \text{N/mm}^2$$
$$\sigma_{10} = 7121/B \, \text{N/mm}^2$$
$$\sigma_{11} = 6854/B \, \text{N/mm}^2$$
$$\sigma_{12} = 6109/B \, \text{N/mm}^2$$
$$\sigma_{13} = 4947/B \, \text{N/mm}^2$$

Section CC (diameter $D = 1.01$ m)

$$M_x = 26\,490 \, \text{N m} \quad M_y = 24\,686 \, \text{N m}$$

Equation (A.13) becomes

$$\sigma = (8.66y + 8.07x)/B$$

Then

$$\sigma_7 = 4373/B \, \text{N/mm}^2$$
$$\sigma_8 = 5279/B \, \text{N/mm}^2$$
$$\sigma_9 = 5825/B \, \text{N/mm}^2$$
$$\sigma_{10} = 5965/B \, \text{N/mm}^2$$
$$\sigma_{11} = 5716/B \, \text{N/mm}^2$$
$$\sigma_{12} = 5069/B \, \text{N/mm}^2$$
$$\sigma_{13} = 4075/B \, \text{N/mm}^2$$

Section BB (diameter $D = 0.73$ m)

$$M_x = 13\,347 \, \text{N m} \quad M_y = 11\,393 \, \text{N m}$$

Equation (A.13) becomes

$$\sigma = (8.35y + 7.13x)/B$$

Then

$$\sigma_7 = 3048/B\,\text{N/mm}^2$$
$$\sigma_8 = 3618/B\,\text{N/mm}^2$$
$$\sigma_9 = 3941/B\,\text{N/mm}^2$$
$$\sigma_{10} = 3989/B\,\text{N/mm}^2$$
$$\sigma_{11} = 3834/B\,\text{N/mm}^2$$
$$\sigma_{12} = 3303/B\,\text{N/mm}^2$$
$$\sigma_{13} = 2602/B\,\text{N/mm}^2$$

From the above it can be seen that the maximum direct stress at each fuselage section occurs in stringer 10. Also the stress in stringer 10 (and all other stringers) is lower at section CC than at section AA and lower at section BB than at section CC. Therefore if fuselage frames are positioned at each of these sections lighter stringers may be used between CC and BB than between AA and CC. An additional frame will be positioned midway between AA and CC and between CC and BB, and will be slotted to allow the stringers to pass through. The arrangement is shown diagrammatically in Fig. A.9 and in detail in Fig. A.13.

The allowable direct stress in a stringer is 124 N/mm^2. The maximum direct stress in stringer 10 at the section AA is 7121/B N/mm^2. The required stringer area of cross-section is then given by

$$7121/B = 124$$

i.e.

$$B = 57.4\,\text{mm}^2$$

The Z-section stringer shown in Fig. A.10 has a cross-sectional area = 58.1 mm^2 and will therefore be satisfactory

The maximum direct stress in stringer 10 at the section CC is 5965/B N/mm^2.

The required stringer area is then given by

$$5965/B = 124$$

i.e.

$$B = 48.1\,\text{mm}^2$$

The cross-section shown in Fig. A.11 has a cross-sectional area of 51.9 mm^2 and is therefore satisfactory.

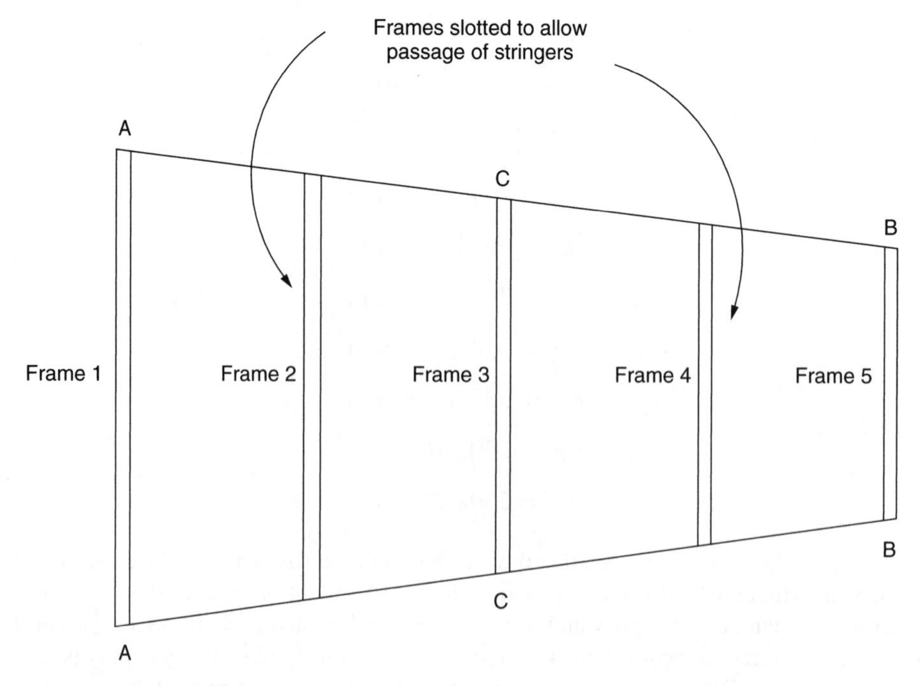

Fig. A.9 Arrangement of fuselage frames.

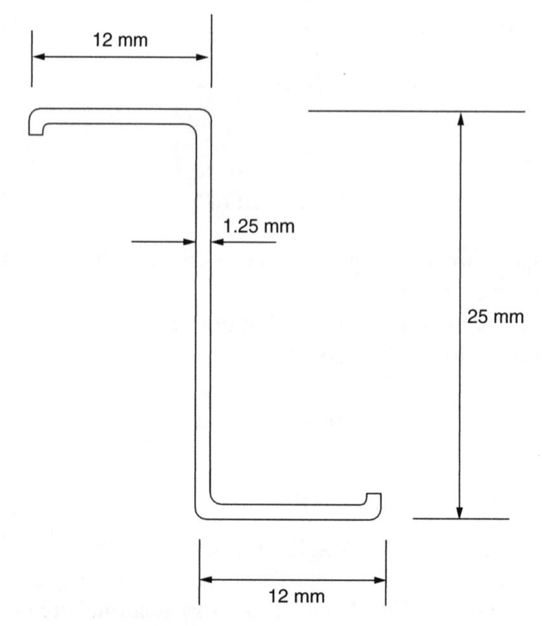

Fig. A.10 Stringer section, AA to CC (Type A).

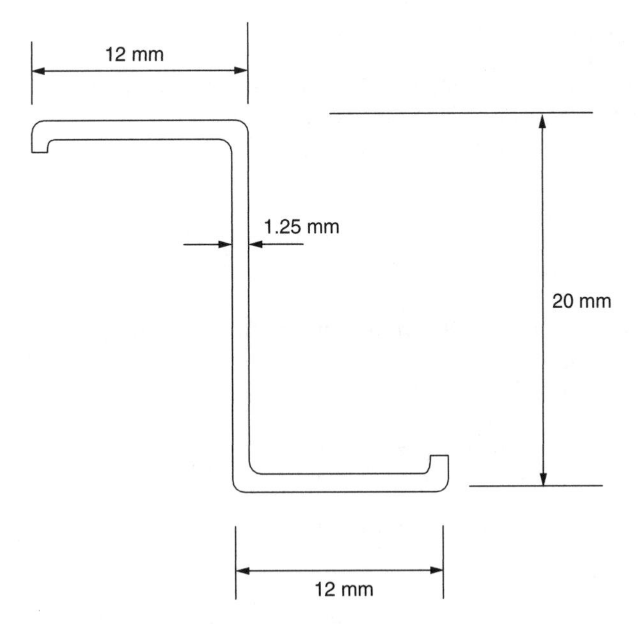

Fig. A.11 Stringer section, CC to BB (Type B).

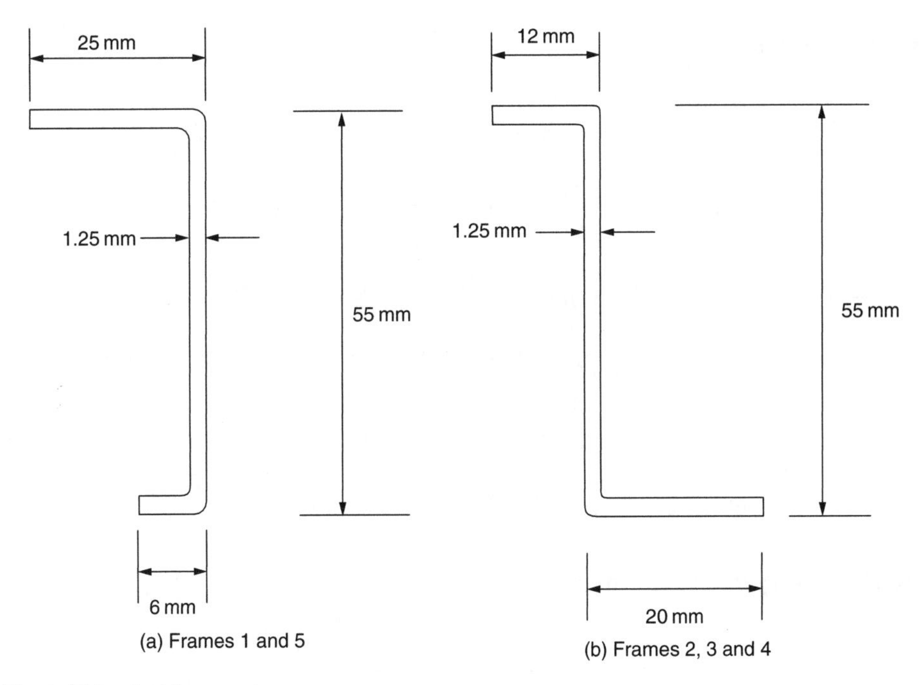

Fig. A.12 Fuselage frame sections.

Although the fuselage frames are non-load bearing the frames at AA, CC and BB must be of sufficient size to allow the ends of the stringers to be connected to them via brackets while intermediate frames must be of sufficient size to allow slots to be cut so that the stringers can pass through them. The frame sections to be used are shown in Fig. A.12.

Skin thickness

The fuselage cross-section is subjected to shear loads S_x and S_y along two perpendicular axes of symmetry. Equation (20.5) gives the change in shear flow as a boom, i.e. a stringer, is crossed and, due to symmetry, reduces to

$$q_2 - q_1 = -\frac{S_x}{I_{yy}} B_r x_r - \frac{S_y}{I_{xx}} B_r y_r \tag{A.14}$$

Then, since, B_r (=B) is constant round the fuselage section and $I_{xx} = I_{yy} = 3.0BD^2$ Eq. (A.14) reduces to

$$q_2 - q_1 = -\frac{S_x}{3.0D^2} x_r - \frac{S_y}{3.0D^2} y_r$$

Consider the action of S_y (or S_x) only. Then

$$q_2 - q_1 = -\frac{S_y}{3.0D^2} y_r \tag{A.15}$$

Referring now to Fig. A.8

$$q_{23} = q_{12} + \frac{S_y}{3.0D^2} \times 0.1294D = q_{12} + 0.043S_y/D$$

$$q_{34} = q_{23} + \frac{S_y}{3.0D^2} \times 0.25D = q_{12} + 0.126S_y/D$$

$$q_{45} = q_{34} + \frac{S_y}{3.0D^2} \times 0.353D = q_{12} + 0.244S_y/D$$

$$q_{56} = q_{45} + \frac{S_y}{3.0D^2} \times 0.433D = q_{12} + 0.388S_y/D$$

$$q_{67} = q_{56} + \frac{S_y}{3.0D^2} \times 0.483D = q_{12} + 0.549S_y/D$$

$$q_{78} = q_{67} + \frac{S_y}{3.0D^2} \times 0.5D = q_{12} + 0.716S_y/D$$

From symmetry $q_{78} = -q_{67}$ so that

$$q_{12} + \frac{0.716S_y}{D} = -q_{12} - \frac{0.549S_y}{D}$$

giving

$$q_{12} = -\frac{0.633S_y}{D}$$

Then

$$q_{23} = (-0.633 + 0.043)S_y/D = -0.59S_y/D$$

Similarly

$$q_{34} = -0.507S_y/D$$
$$q_{45} = -0.389S_y/D$$
$$q_{56} = -0.245S_y/D$$
$$q_{67} = -0.084S_y/D$$

Consider now the action of S_x only. Equation (A.14) becomes

$$q_2 - q_1 = -\frac{S_x}{3.0D^2}x_r$$

Again referring to Fig. A.8

$$q_{65} = q_{76} - \frac{S_x}{3.0D^2} \times 0.1294D = q_{76} - 0.043\,S_x/D$$
$$q_{54} = q_{65} - \frac{S_x}{3.0D^2} \times 0.25D = q_{76} - 0.126\,S_x/D$$
$$q_{43} = q_{54} - \frac{S_x}{3.0D^2} \times 0.353D = q_{76} - 0.244\,S_x/D$$
$$q_{32} = q_{43} - \frac{S_x}{3.0D^2} \times 0.433D = q_{76} - 0.388\,S_x/D$$
$$q_{21} = q_{32} - \frac{S_x}{3.0D^2} \times 0.483D = q_{76} - 0.549\,S_x/D$$
$$q_{1\,24} = q_{21} - \frac{S_x}{3.0D^2} \times 0.5D = q_{76} - 0.716\,S_x/D$$

But $q_{21} = -q_{1\,24}$ from symmetry so that

$$q_{76} - 0.549S_x/D = -q_{76} + 0.716S_x/D$$

i.e.

$$q_{76} = \frac{0.633S_x}{D}$$

Then

$$q_{65} = 0.59S_x/D$$
$$q_{54} = 0.507S_x/D$$
$$q_{43} = 0.389S_x/D$$
$$q_{32} = 0.245S_x/D$$
$$q_{21} = 0.084S_x/D$$

Note that the shear flows due to S_x and S_y in skin panels 76 to 21 inclusive are in the same direction. An identical situation arises in panels 19 18 to 14 13 but in the remaining panels the shear flows are opposed.

The shear flow produced by the applied torque is given by Eq. (18.1), i.e.

$$q = \frac{T}{2A}$$

where A is the area enclosed by the fuselage skin. Then

$$q = \frac{T}{2(\pi D^2/4)}$$

or

$$q = 0.637T/D^2 \qquad \text{(A.16)}$$

It can be seen from Table A.2 that all the applied torques are positive, i.e. clockwise. The shear flow is then in the same sense in skin panels 76 to 21 as the shear flows due to S_x and S_y; these panels are therefore subjected to the greatest shear stresses.

The total shear flow in each of the panels 76 to 21 is then

$$q_{76} = 0.084S_y/D + 0.633S_x/D + 0.637T/D^2$$

$$q_{65} = 0.245S_y/D + 0.590S_x/D + 0.637T/D^2$$

$$q_{54} = 0.389S_y/D + 0.507S_x/D + 0.637T/D^2$$

$$q_{43} = 0.507S_y/D + 0.389S_x/D + 0.637T/D^2$$

$$q_{32} = 0.590S_y/D + 0.245S_x/D + 0.637T/D^2$$

$$q_{21} = 0.633S_y/D + 0.084S_x/D + 0.637T/D^2 \qquad \text{(A.17)}$$

From Table A.2 the maximum values of S_y, S_x and T at each section are produced by Case D_1 in the flight envelope.

Section AA (diameter $D = 1.28$ m)

$$S_x = 12\,460\,\text{N} \quad S_y = 13\,350\,\text{N} \quad T = 34\,031\,\text{N m}$$

Then, from Eqs (A.17)

$$q_{76} = 0.084 \times 13\,350/(1.28 \times 10^3) + 0.633 \times 12\,460/(1.28 \times 10^3) + 0.637 \times 34\,031 \times 10^3/(1.28 \times 10^3)^2$$

i.e.

$$q_{76} = 20.3\,\text{N/mm}$$

Similarly

$$q_{65} = 21.5\,\text{N/mm}$$
$$q_{54} = 22.2\,\text{N/mm}$$
$$q_{43} = 22.3\,\text{N/mm}$$
$$q_{32} = 21.8\,\text{N/mm}$$
$$q_{21} = 20.7\,\text{N/mm}$$

Section CC (diameter $D = 1.01$ m)

$$S_x = 12\,460\,\text{N} \quad S_y = 11\,837\,\text{N} \quad T = 34\,031\,\text{N m}$$

Then, from Eqs (A.17)

$$q_{76} = \frac{0.084 \times 11\,837}{1.01 \times 10^3} + \frac{0.633 \times 12\,460}{1.01 \times 10^3} + \frac{0.637 \times 34\,031 \times 10^3}{(1.01 \times 10^3)^2}$$

i.e.

$$q_{76} = 30.0\,\text{N/mm}$$

Similarly

$$q_{65} = 31.4\,\text{N/mm}$$
$$q_{54} = 33.1\,\text{N/mm}$$
$$q_{43} = 32.0\,\text{N/mm}$$
$$q_{32} = 31.2\,\text{N/mm}$$
$$q_{21} = 29.7\,\text{N/mm}$$

Section BB (diameter $D = 0.73$ m)

$$S_x = 12\,460\,\text{N} \quad S_y = 10\,547\,\text{N} \quad T = 34\,031\,\text{Nm}$$

From Eqs (A.17)

$$q_{76} = \frac{0.084 \times 10\,547}{0.73 \times 10^3} + \frac{0.633 \times 12\,460}{0.73 \times 10^3} + \frac{0.637 \times 34\,031 \times 10^3}{(0.73 \times 10^3)^2}$$

i.e.

$$q_{76} = 52.7\,\text{N/mm}$$

Similarly

$$q_{65} = 54.3\,\text{N/mm}$$
$$q_{54} = 55.0\,\text{N/mm}$$
$$q_{43} = 54.6\,\text{N/mm}$$
$$q_{32} = 53.4\,\text{N/mm}$$
$$q_{21} = 51.3\,\text{N/mm}$$

The skin will be of constant thickness so that the maximum shear stress in the skin will occur in the panel in which the shear flow is a maximum. This, from the above, is 55.0 N/mm in panel 54 at section BB. From Section A.5 the maximum allowable shear stress is 78 N/mm^2, therefore

$$\frac{55.0}{t} = 78$$

which gives

$$t = 0.71\,\text{mm}$$

A skin thickness of, say, 0.75 mm would not meet the requirements of a minimum thickness for rivet diameters equal to or greater than 2.5 mm (the probable rivet diameter but determined later). A skin thickness of 1.0 mm will therefore be used.

Rivet size

Skin/stringer rivets

The *change* in end load over a unit length of stringer can be found using the method of Section 20.3.4. This *change* in end load is then the shear force on the stringer/skin connection, i.e. the rivets. Using this approach, the bending moment due to S_x at a section 1 mm (say) from the section in which S_x is applied is $S_x \times 1$ N mm. The direct stress in the rth stringer produced by the bending moment is given by the second of Eqs (16.21), i.e.

$$\sigma_z = \frac{S_x \times 1}{I_{yy}}$$

The end load in the stringer is then

$$P_r = \sigma_z B_r = \frac{S_x B_r x_r}{I_{yy}}$$

Similarly, due to S_y

$$P_r = \frac{S_y B_r y_r}{I_{xx}}$$

Since $I_{xx} = I_{yy} = 3.0\text{BD}^2$ the total change in end load over the 1 mm length of stringer is given by

$$\text{Total change in end load } P_r = \frac{S_x}{3.0D^2} x_r + \frac{S_y}{3.0D^2} y_r \qquad \text{(A.18)}$$

Clearly the change in end load will be a maximum when S_x and S_y have the same sign and x_r and y_r have the same sign; this occurs in stringers 7 to 13 and 19 to 1. In the former case the change in end load is tensile while in the latter it is compressive. Further, the maximum values of S_x and S_y at sections AA, CC and BB all occur for Case D_1 (see Table A.2); these cases will now be investigated.

Section AA (diameter $D = 1.28$ m)

$$S_x = 12\,460\,\text{N} \quad S_y = 13\,350\,\text{N}$$

Stringer 7

$$P_7 = \frac{12\,460}{3.0(1.28 \times 10^3)^2}(0) + \frac{13\,350}{3.0(1.28 \times 10^3)^2}(0.5D)$$

i.e.

$$P_7 = 1.74\,\text{N/mm}$$

Similarly

$$P_8 = 2.10\,\text{N/mm}$$
$$P_9 = 2.32\,\text{N/mm}$$
$$P_{10} = 2.37\,\text{N/mm}$$
$$P_{11} = 2.27\,\text{N/mm}$$
$$P_{12} = 2.02\,\text{N/mm}$$
$$P_{13} = 1.62\,\text{N/mm}$$

Section CC (Diameter $D = 1.01$ m)

$$S_x = 12\,460\,\text{N} \quad S_y = 11\,837\,\text{N}$$

$$P_7 = 1.95\,\text{N/mm}$$
$$P_8 = 2.42\,\text{N/mm}$$
$$P_9 = 2.72\,\text{N/mm}$$
$$P_{10} = 2.83\,\text{N/mm}$$
$$P_{11} = 2.76\,\text{N/mm}$$
$$P_{12} = 2.49\,\text{N/mm}$$
$$P_{13} = 2.06\,\text{N/mm}$$

Section BB (diameter $D = 0.73$ m)

$$S_x = 12\,460\,\text{N} \quad S_y = 10\,547\,\text{N}$$

$$P_7 = 2.41\,\text{N/mm}$$
$$P_8 = 3.06\,\text{N/mm}$$
$$P_9 = 3.51\,\text{N/mm}$$
$$P_{10} = 3.71\,\text{N/mm}$$
$$P_{11} = 3.67\,\text{N/mm}$$
$$P_{12} = 3.37\,\text{N/mm}$$
$$P_{13} = 2.84\,\text{N/mm}$$

From the above it can be seen that the maximum load on the rivets occurs at section BB in stringer 10 and is 3.71 N/mm. Assuming 2.5 mm diameter countersunk rivets which have, in a skin thickness of 1.0 mm, an allowable load in shear of 668 N the number of rivets/m given by

$$n = \frac{3.71 \times 10^3}{668} = 5.6 \quad \text{say 6 rivets/m}$$

However this would give a rivet pitch of approximately 167 mm which is not sufficient to ensure a rigid structure. Therefore 2.5 mm diameter rivets will be used at a pitch of 25 mm.

Frame/stringer rivets

The maximum stringer load at the section AA is 7121 N and this is resisted by the rivets connecting the skin to the frame over a length equal to the stringer spacing of 167.6 mm. Therefore the number of 2.5 mm diameter rivets required is $7121/668 = 10.7$, say 11. This gives a rivet pitch of $167.6/11 \simeq 15$ mm.

At the section BB the maximum stringer load is 3989 N so that the number of rivets required is $3989/668 \simeq 6$. This gives a rivet pitch of $0.73 \times 10^3 \times \pi/(24 \times 6) \simeq 16$ mm.

At the section CC the maximum stringer load is 5965 N so that the number of rivets required is $5965/668 \simeq 9$. The required rivet pitch is then $1.01 \times 10^3 \times \pi/(24 \times 9) \simeq 14$ mm.

Therefore for all frames a rivet pitch of 12.5 mm will be used.

The layout of a quarter of the rear fuselage is shown in Fig. A.13 with the detail design shown in Figs A.14(a)–(e).

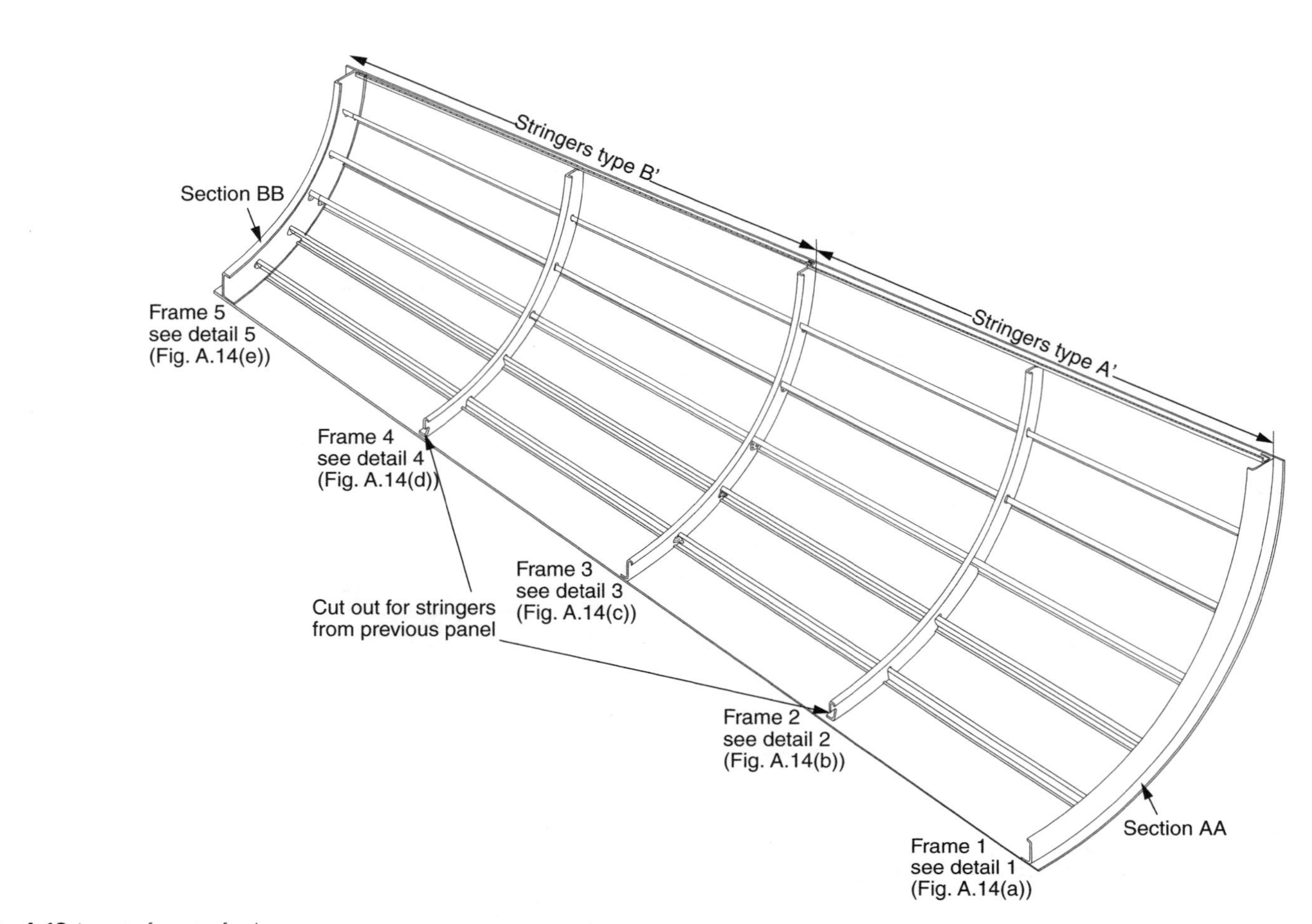

Fig. A.13 Layout of quarter fuselage.

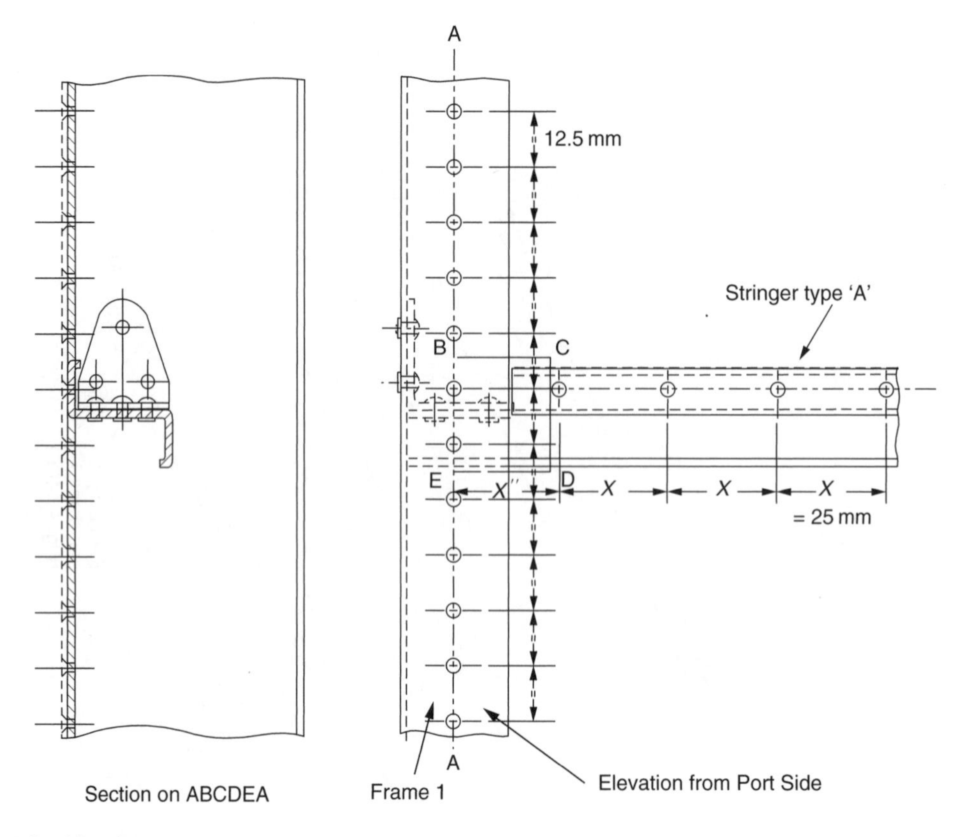

For Details of Bracket
see Fig. A.14 (c)
Skins from Previous
Sections Overlap Where
Necessary.
All Rivets 2.5 mm Countersunk
Except for Bracket.

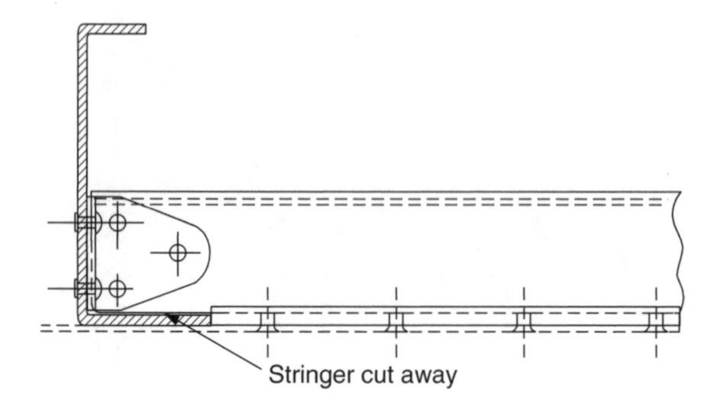

Fig. A.14(a) Detail, Frame 1.

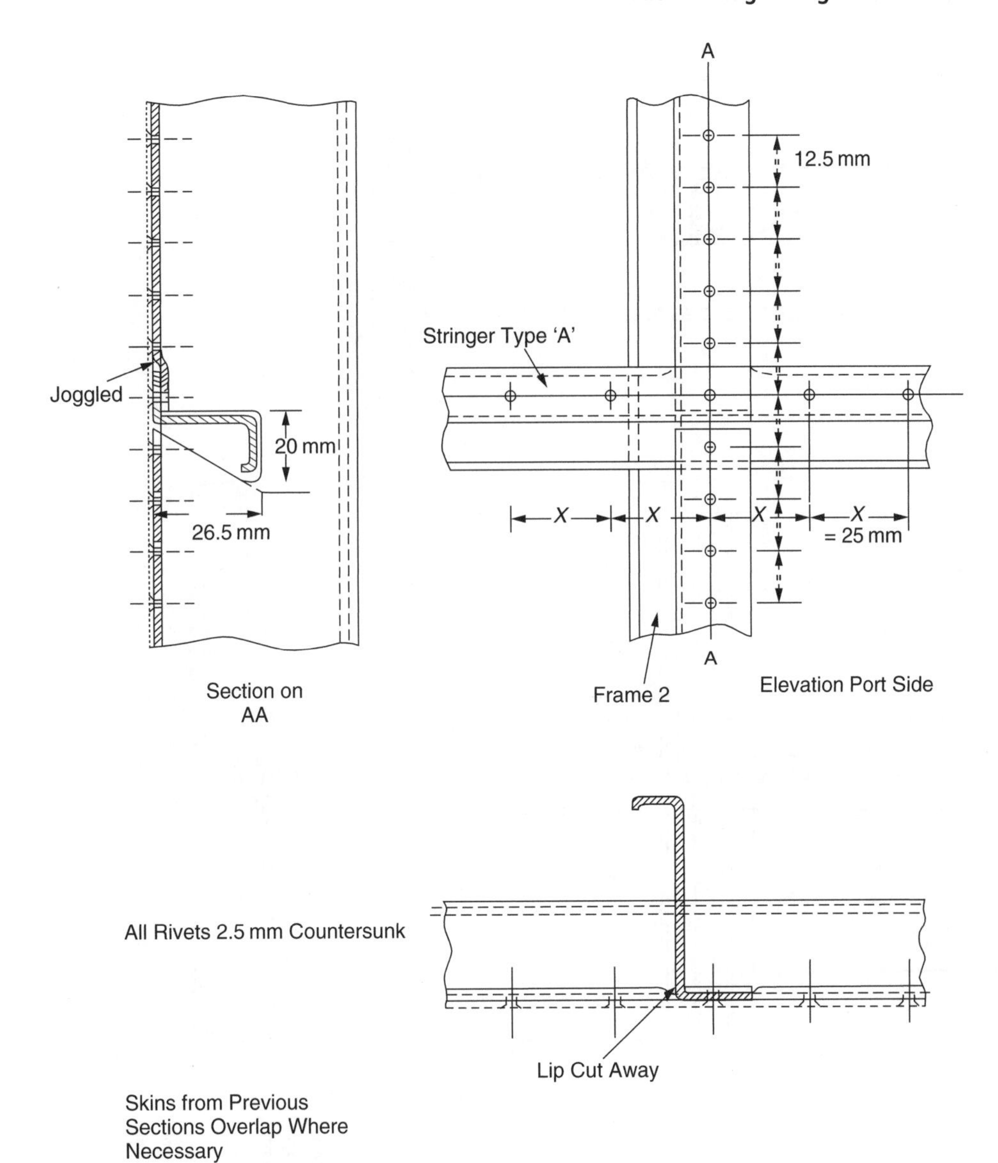

Fig. A.14(b) Detail 2, Frame 2.

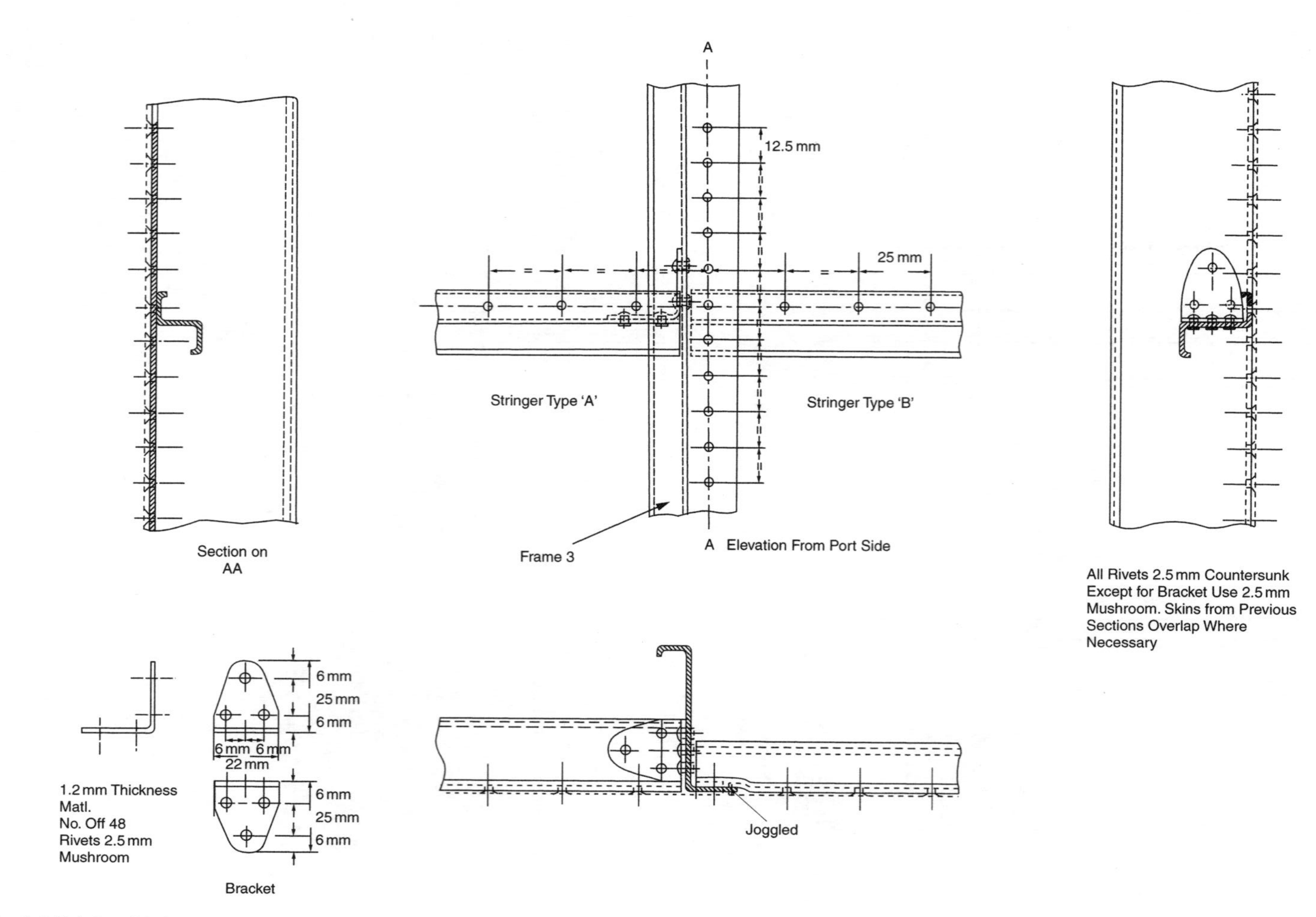

Fig. A.14(c) Detail 3, Frame 3.

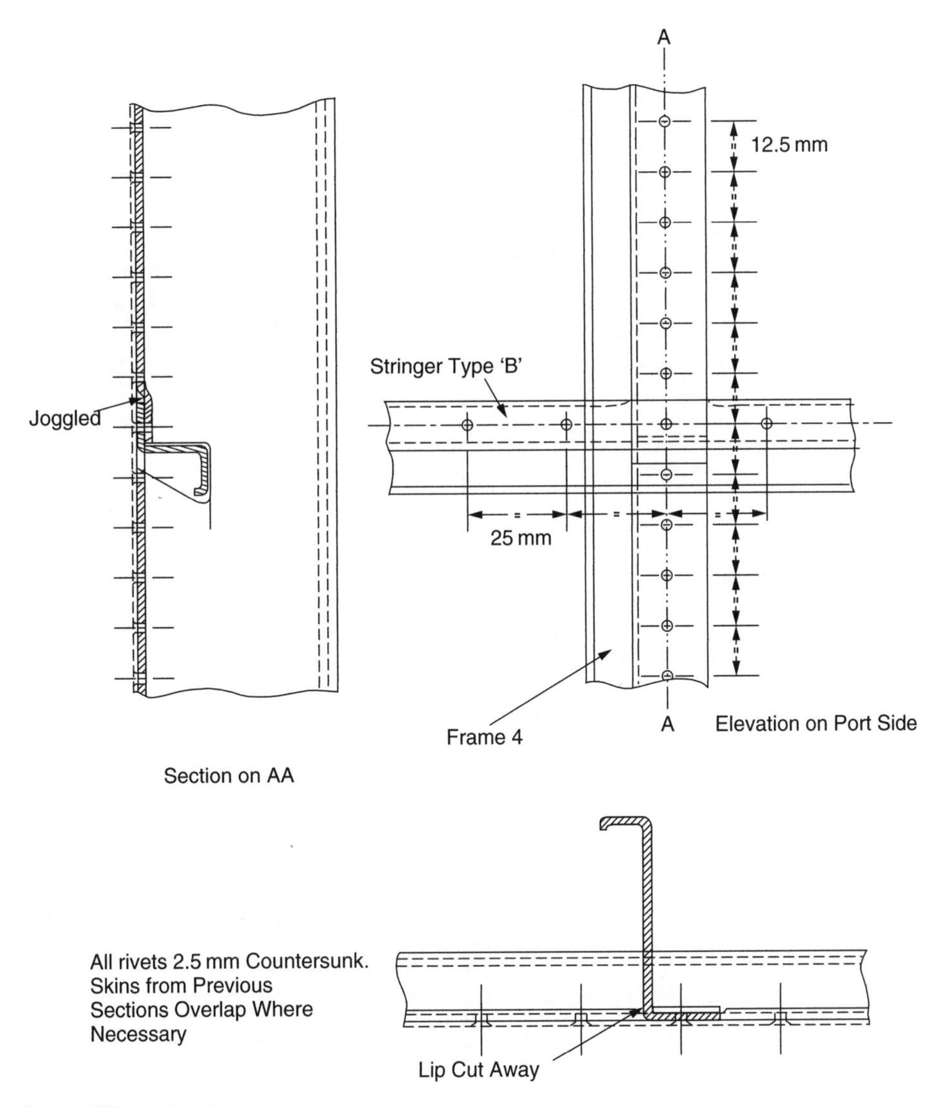

Fig. A.14(d) Detail 4, Frame 4.

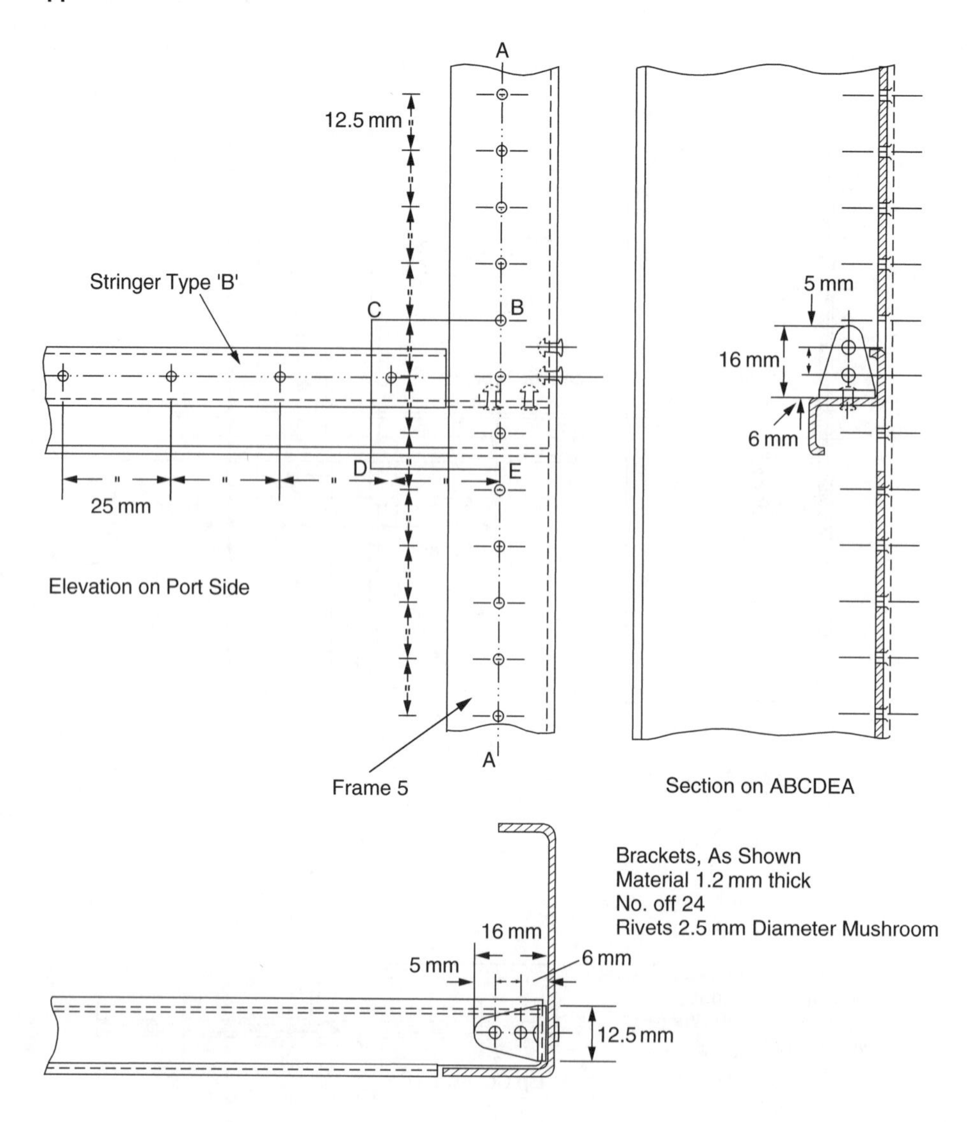

Fig. A.14(e) Detail 5, Frame 5.

Index